THE YEASTS

Volume 6

Second Edition

THE YEASTS

Edited by

ALAN E. WHEALS
ANTHONY H. ROSE

School of Biology and Biochemistry,
University of Bath, Claverton Down, Bath, UK

J. STUART HARRISON

Ashley House, Upper Frog Street, Tenby,
Dyfed, UK

Volume 6

Second Edition

Yeast Genetics

ACADEMIC PRESS
Harcourt Brace & Company, Publishers

London San Diego New York
Boston Sydney Tokyo Toronto

ACADEMIC PRESS LIMITED
24/28 Oval Road
London NW1 7DX

United States Edition published by
ACADEMIC PRESS INC.
San Diego, CA 92101

Copyright © 1995 by
ACADEMIC PRESS LIMITED

A CIP record for this book is available from the British Library
ISBN 0-12-596416-1

Typeset by Colset Pty Ltd, Singapore
and printed in Great Britain by T. J. Press Ltd, Padstow, Cornwall

Obituary: Professor Anthony H. Rose

Anthony (Tony) H. Rose studied at the University of Birmingham where, in 1954, he obtained a PhD degree in biochemistry with a thesis entitled "Growth of a yeast and a lactic-acid bacterium in mixed culture". This was followed in 1969 by a DSc Microbiology on "Studies in Microbial Physiology". During 1967 to 1974 he was a member of the Council of the Society for General Microbiology, and became the International Secretary of the Society and Secretary-General of the Federation of European Microbiological Societies, a post which he held from 1974 to 1980. From 1968 he was Professor of microbiology at the School of Biological Sciences, University of Bath, and in 1973 was appointed Head of the School of Biological Sciences.

He was an indefatigable contributor of scientific publications, particularly as relating to yeast, with which he became more closely concerned over the years, to the extent that of eighty-seven research papers in which he was involved, seventy-five have "yeast" (or a named species) in the title. Tony was also the inspiration for, and co-editor of, "The Yeasts", a three-volume edition of which was published in 1969–1971. This was followed in 1987 to 1995 by six volumes of an enlarged second edition; the present volume of which completes the series.

Tony's enthusiasm for his chosen scientific speciality put him in personal contact with many of the leading workers in this field; in consequence of which he travelled widely, visiting laboratories in a number of countries.

His far-seeing vision of future developments in biology led him to become the leading protagonist for the adoption of *Saccharomyces cerevisiae* as the archetypal eukaryotic organism in microbial biology on account of its many well-studied and consistent properties and ready availability in the form of baker's yeast.

Finally, I must record my personal indebtedness to Tony for his keen interest and selfless co-operation during the many years of our rewarding association throughout the editing of this treatise.

J. Stuart Harrison

Contributors

C. Bennett, Laboratory of Molecular Genetics, National Institute of Environmental Health Sciences, Box 12233, Research Triangle Park, NC 27709, USA

K. A. Bostian, Department of Microbiology and Molecular Genetics, Merck Research Laboratories, Rahway, NJ 07065, USA

C. R. Contopoulou, Yeast Genetic Stock Center, MCB/Biophysics and Cell Physiology, 102 Donner Laboratory, Berkeley, California 94720, USA

B. S. Cox, Department of Plant Sciences, Oxford University, South Parks Road, Oxford OX1 3RA, UK

R. Dölz, Biocomputing, Biozentrum, 70, Klingelbergstrasse, CH-4056, Basel, Switzerland

C. Gjermansen, Department of Yeast Genetics, Carlsberg Laboratory, Gamle Carlsberg Vej 10, DK-2500 Copenhagen Valby, Denmark

S. Holmberg, Institute of Genetics, University of Copenhagen, Øster Farimagsgade 2A, 1353 Copenhagen K, Denmark

D. B. Kaback, Department of Microbiology and Molecular Genetics, University of Medicine and Dentistry, New Jersey Medical School, 185 South Orange Avenue, Newark, NJ 07103, USA

M. C. Kielland-Brandt, Department of Yeast Genetics, Carlsberg Laboratory, Gamle Carlsberg Vej 10, DK-2500 Copenhagen Valby, Denmark

J. S. King, Yeast Genetic Stock Center, MCB/Biophysics and Cell Physiology, 102 Donner Laboratory, Berkeley, California 94720, USA

J. Kohli, Institute of General Microbiology, University of Bern, Baltzer-Str. 4, CH-3012 Bern, Switzerland

J. Lazowska, Centre de Génétique Moléculaire, Laboratoire propre du CNRS associé à l'Université Pierre et Marie Curie, F-91190, Gif-sur-Yvette, France

H. Lehrach, Genome Analysis Laboratory, Imperial Cancer Research Fund, PO Box 123, 44 Lincoln's Inn Fields, London WCZA 3PX, UK

P. Linder, Dept. of Microbiology, Biozentrum, 70, Klingelbergstrasse, CH-4056, Basel, Switzerland

E. Maier, Genome Analysis Laboratory, Imperial Cancer Research Fund, PO Box 123, 44 Lincoln's Inn Fields, London WCZA 3PX, UK

R. K. Mortimer, Yeast Genetic Stock Center, MCB/Biophysics and Cell Physiology, 102 Donner Laboratory, Berkeley, California 94720, USA

M.-O. Mossé, Centre de Génétique Moléculaire, Laboratoire propre du CNRS associé à l'Université Pierre et Marie Curie, F-91190, Gif-sur-Yvette, France

P. Munz, Institute of General Microbiology, University of Bern, Baltzer-Str. 4, CH-3012 Bern, Switzerland

T. Nilsson-Tillgren, Institute of Genetics, University of Copenhagen, Øster Farimagsgade 2A, 1353 Copenhagen K, Denmark

S. A. Parent, Department of Microbiology and Molecular Genetics, Merck Research Laboratories, Rahway, NJ 07065, USA

M. B. Pedersen, Carlsberg Research Laboratory, Gamle Carlsberg Vej 10, DK 2500 Copenhagen Valby, Denmark

E. Perkins, Laboratory of Molecular Genetics, National Institute of Environmental Health Sciences, Box 12233, Research Triangle Park, NC 27709, USA

G. Porter, Laboratory of Molecular Genetics, National Institute of Environmental Health Sciences, Box 12233, Research Triangle Park, NC 27709, USA

R. T. M. Poulter, Department of Biochemistry, University of Otago, Dunedin, New Zealand

S. D. Priebe, Laboratory of Molecular Genetics, National Institute of Environmental Health Sciences, Box 12233, Research Triangle Park, NC 27709, USA

M. A. Resnick, Laboratory of Molecular Genetics, National Institute of Environmental Health Sciences, Box 12233, Research Triangle Park, NC 27709, USA

M. D. Rose, Department of Molecular Biology, Lewis Thomas Laboratory, Princeton University, Princeton, NJ 08544, USA

F. Sherman, Departments of Biochemistry and Biophysics, University of Rochester School of Medicine and Dentistry, Rochester, NY 14642, USA

P. P. Slonimski, Centre de Génétique Moléculaire, Laboratoire propre du CNRS associé à l'Université Pierre et Marie Curie, F-91190, Gif-sur-Yvette, France

G. F. Sprague, Jr., Institute of Molecular Biology and Department of Biology, University of Oregon, Eugene, OR 97403, USA

P. Sudbery, Department of Molecular Biology and Biotechnology, University of Sheffield, Sheffield S10 2TN, UK

A. E. Wheals, University of Bath, School of Biology and Biochemistry, Bath BA2 7AY, UK

R. B. Wickner, Section on Genetics of Simple Eukaryotes, LBP, National Institutes of Diabetes, Digestive and Kidney Diseases, NIH Bethesda, MD 20892, USA

Preface

The days when a single author was able to produce a series of volumes which could satisfactorily cover the whole gamut of a general science such as chemistry, physics or biology are long past. More recently these continuously expanding disciplines, and others such as mathematics and chemical engineering, have become essential basics for the complete study of a single class of living organisms.

We have endeavoured in this multivolume treatise to provide as wide a coverage as possible of the different areas of investigation that comprise the science of zymology, the study of yeasts. Six volumes, involving 65 chapters, have resulted. Prominent workers in each relevant discipline present an overall account of their chosen field with particular attention to the results of the most recent advances. Inevitably there will be gaps when dealing with this very broad field: for such we apologise and trust that any deficiencies may not seriously detract from the value of the present work.

The first requirement has been the recognition and definition of the members of the group, that is taxonomy and classification, closely followed by such essentials as biological functions related to the external environment, and intracellular mechanisms. Finally the nature and expression of genetic material has completed the series. Advances since the publication of the first edition of *The Yeasts* in 1969–1971 have been quite phenomenal. The many branches of knowledge discussed in detail are too numerous to mention here, but each volume has been planned to cover a group of related subjects, and so form a useful book in its own right.

We gratefully acknowledge the debt we owe to the contributors to these volumes and others concerned for their efforts, often under difficult circumstances. This comment is made more poignant to us by the untimely death of our co-editor, Anthony Rose, during the preparation of this final volume. We hope that this treatise will provide a useful contemporary view of the present state of the art and a fitting memorial to him as a member of the yeast community who furthered its aims for over 30 years. With the ever burgeoning literature it is doubtful whether others will have sufficient breadth to embark on such a task again.

Alan Wheals
J. Stuart Harrison

Contents

6 Genetics of Brewing Yeasts
Morten C. Kielland-Brandt, Torsten Nilsson-Tillgren, Claes Gjermansen, Steen Holmberg and Mogens B. Pedersen

Contents of Volume 1

Contents of Volume 2

Contents of Volume 3

34

Contents of Volume 4

Contents of Volume 5

Abbreviations

PD	parental ditype (p. 29)
TT	tetratype (p. 29)
NPD	non-parental type (p. 29)
MBC	methyl-benzyimidazole carbamate
YGSC	Yeast Genetics Stock Center
PFGE	pulsed-field gel electrophoresis
OFAGE	orthogonal field alternation gel electrophoresis
FIGE	field inversion gels
CHEF	clamped hexagonal electric fields
ars	autonomously replicating sequence
cen	centromeres
tel	telomeres
5-FOA	5-fluoro-orotic acid
YIp	yeast-integrating plasmid
PCR	polymerase chain reaction
PEG	polyethylene glycol
YAC	yeast artificial chromosome
CRE	cyclic AMP response element
SIR	silent information regulator
DHFR	dihydrofolate reductase
YRp	yeast-replicating plasmids
YCp	yeast-centromeric plasmids
YEp	yeast-episomal plasmids
YLp	yeast linear plasmid
YPp	yeast-promoter plasmid
YXp	yeast expression plasmid
YHp	yeast hybrid plasmid
YMp	yeast mutagenesis plasmid
MCR	multiple cloning site regions
TEL	telomeres
λYES	lambda phage cDNA expression vectors
CaMV	cauliflower mosaic virus
ORFs	open-reading frames
LTR	long terminal repeat
EMS	ethylmethane sulphonate

NTG	N-ethyl-N'-nitro-N-nitrosoguanidine
OMD	orotidine monophosphate decarboxylase
HIV	human immunodeficiency virus
OFAGE	orthogonal field agarose gel electrophoresis
FIGE	field inversion gel electrophoresis
CHEF	clamped homogeneous field gel electrophoresis
TAFE	transverse alternating field electrophoresis
RFLP	restriction fragment length polymorphism
IRE	internal replication entrances
PMS	post-meiotic segregation
UAS	upstream activation sequence
SRF	serum response factor

1 Introduction

Alan E. Wheals

University of Bath, School of Biology and Biochemistry, Bath BA2 7AY, UK

This volume, devoted solely to genetics, completes the second edition of *The Yeasts*. When the first edition of this series was being written, genetics warranted a single chapter, genetic modification did not exist, no genes had been isolated and the number of chromosomes in the yeast *Saccharomyces cerevisiae* was still unclear (Mortimer and Hawthorne, 1969). Now, many thousands of genes have been isolated and much of the entire genome of *Sacch. cerevisiae* has been sequenced. This remarkable transformation has come about for many reasons. Although *Saccharomyces* yeasts have been in the forefront of microbiological and biochemical research for over a century, on the genetic stage, yeasts have been mere spear carriers for a long time. After the pioneering work of Winge and Lindegren with *Sacch. cerevisiae* and that of Leupold for *Schizosaccharomyces pombe*, these two yeasts in particular have gradually overtaken the other eukaryotic "genetic" micro-organisms, *Neurospora crassa* and *Aspergillus nidulans*. The reasons are complex and well described in a recent history (Hall and Linder, 1993). Certainly ease of safe culture, economic importance, a well-understood biochemistry, ability to grow haploids and diploids with similar morphologies, and simplicity of mating have all played their part. The Cold Spring Harbor Yeast Courses have also been crucial in developing two generations of yeast geneticists. Some of the most able scientists have used yeasts to investigate a variety of fundamental cell biological problems (Guthrie and Fink, 1991) and consequently yeasts have become the *Escherichia coli* of the eukaryote world (Watson *et al.*, 1987).

When the genomes of *Sacch. cerevisiae* and *Schiz. pombe* have been completely sequenced, projects that should be completed before the end of

1

the decade, it will not signal the end of yeast genetics. On the contrary, the isolation of so many gene sequences whose functions we do not understand (Oliver *et al.*, 1992; Kaback, Chapter 5, this volume) will require even more sophisticated genetic analysis. Deletions and disruptions, random and site-directed mutagenesis, overexpression and two-hybrid systems, position effects and gene dosage, gel-retardation studies and *in situ* hybridization, and heterologous cloning and hybrid genes will all be needed for elucidation of gene function.

Sacch. cerevisiae and *Schiz. pombe* have a pre-eminent position as organisms for the genetic study of biological problems. One 1993 database, which includes 1495 articles with both the words "yeast" and "gene" as keywords, has 387 of those articles with the specific name *Sacch. cerevisiae* in the title. *Schiz. pombe* follows a long way behind with only a tenth of those citations. However, this volume is not restricted to just two yeasts. A number of other yeasts are now being studied in ever greater genetic detail from the pathogen *Candida albicans* to the methylotrophs such as *Yarrowia* and *Pichia*. One of the key features of this volume is that, except for organism-specific chapters, e.g. the one on *Candida albicans*, the articles have been designed to cover yeasts in general. This has proved demanding for authors but we are sure that readers will see the comparative benefits. The aim of this volume has been to cover the essential genetic techniques that can be used with yeasts, to look at the special genetic features of organisms of industrial or medical importance, and to cover in more detail some specific aspects of yeast genetics.

One of the advantages of working with yeasts is that there can be a synergistic relationship between the techniques of classical and molecular genetics. Chapters 2, 3 and 4 have emphasized the practicalities of genetic approaches. This has been achieved not by writing laboratory guides but by explaining in very clear terms the underlying strategies for analysing problems using genetics. Classical approaches are first described by Brian Cox when he describes "Genetic analysis in *Sacch. cerevisiae* and *Schiz. pombe*" (p. 7). His approach has been to put the work in a historical perspective before embarking on a lucid account of tetrad analysis, a technique that may still deter the newcomer. The range of alternative lifestyles is then described and how these have been turned to our advantage. Finally, an extremely practical section is devoted to genetic mapping strategies, which shows that, even with current cloning technologies, good mapping approaches may be a quicker or even the only method to locate a gene. Even when all *Sacch. cerevisiae* genes have been sequenced there will still be a need to assign phenotypic novelty to a locus. As the need for the genetic exploitation of other species grows, so it will be possible to see the transfer of this technology to them.

The range of molecular techniques available is indicated in the following two articles. First, Mark Rose (p. 69), in an article entitled "Modern and post-modern genetics in *Saccharomyces cerevisiae*", gives a comprehensive overview of modern molecular yeast genetic technology. Since the article concentrates on strategies rather than on species-specific technical details, it should be of great value to all yeast workers. One of the points he makes is that this technology is so powerful there is a demand to apply it to other yeasts, and other eukaryotes, and this demand is slowly being met. The heart of the methodology revolves around the range of plasmid vectors that can be used for isolating genes, manipulating them and reintroducing them back into hosts as either independently replicating vectors or to yield stably integrated genes. This aspect is described by Stephen Parent and Keith Bostian (p. 121) in an article entitled "Recombinant DNA technology: yeast vectors". The molecular details of the vectors will continually change but the approach of the authors is one that emphasizes generic features that will remain constant even after changes to constructs, genetic markers and restriction sites. Nevertheless, an exhaustive set of tables showing widely used vectors is included.

One of the benefits of both classical and molecular genetics approaches is revealed in Chapter 5 when David Kaback discusses our current understanding of "Yeast genome structure" (p. 179). There is now sufficient information to piece together a comprehensive overview of the components of chromosomes, the distribution and nature of individual genomes and some clues as to chromosome evolution. Once again *Schiz. pombe* and *Sacch. cerevisiae* are to the fore, but *Candida albicans*, *Cryptococcus neoformans* and *Kluyveromyces* spp. also feature.

After this general introduction, three chapters are devoted to the genetics of yeasts of commercial, industrial and medical importance. In Chapter 6 Morton Kielland-Brandt and his colleagues (p. 223) concentrate on the "Genetics of brewing yeasts" and the variety of *Sacch. cerevisiae* that is known as *Sacch. carlsbergensis*. Although selection of suitable strains has been going on for centuries, breeding of strains with desired characteristics is relatively recent. The problem that has thwarted breeders is that it is easy to alter the genetic composition and brewing characteristics of particular features of the yeast by mutation or recombination, but these changes have invariably led to undrinkable products, since the chemistry of flavour is very complex and subtle. The new genetic approaches have used more directed methods, which tend to involve either single-gene changes or single-chromosome changes. Although technically successful, the brewing industry is cautious in allowing the products of genetically modified yeasts on to the market. The genetic improvement of baking yeasts has preceded that of brewing yeasts and the first product of genetic modification has already been approved. Chapter 7, the second of these three articles, is an exploration by Peter Sudbery (p. 255) of the genetic importance of yeasts other than *Sacch. cerevisiae* and *Schiz. pombe*. Most work has been done on *Candida utilis*, *Candida tropicalis*, *Yarrowia lipolytica*, *Kluyveromyces lactis* and *Pachysolen tannophilus*. The industrial importance of these yeasts, some of which do not have perfect sexual stages, will ensure that genetic exploitation of them will be intensified using approaches that have proved successful in *Sacch. cerevisiae* and *Schiz. pombe*. The last of these three chapters by Russell Poulter deals with the human pathogen *Candida albicans* (p. 285). It is only recently that the genetic nature of *C. albicans* has been discovered and this has led to a rapid increase in the effort devoted to the genetic analysis of this yeast. The realistic hope is that such approaches will lead to a better understanding of pathogenesis and perhaps to novel strategies of treatment.

The last three chapters deal in more detail with aspects of the genetic systems of yeasts. First, Reed Wickner has produced a guide to "Non-Mendelian genetic elements in *Sacch. cerevisiae*: RNA viruses, 2 μm DNA, ψ, [URE3], 20S RNA and other wonders of nature" (p. 309). The title indicates the range of often poorly characterized genetic elements that occur in just one yeast species and similar elements are likely to be widespread in other yeasts. Indeed, the plasmids of *Kluyveromyces lactis* are just one example that is described in this article. Since recombination and DNA repair are fundamental to an understanding of both mutagenesis and genetic exchange, the chapter by Michael Resnick and his colleagues (p. 357) is devoted to the subject. Entitled "Double-strand breaks and recombinational repair: the role of processing, signalling and DNA homology", the paper presents a unifying hypothesis underpinning an understanding of all of these molecular events, and their relevance to both cellular events and evolutionary processes. Finally, with the mating system of yeasts being a prerequisite for combining whole genomes, George Sprague describes "Mating and mating-type interconversion in *Sacch. cerevisiae* and *Schiz. pombe* yeast" (p. 411). Understanding of these systems has not only provided means of regulating cytogamy and karyogamy, but has led to ways of utilizing the "homothallic" genes to create homozygous diploids.

To complete the volume, we have included an extensive reference section that includes, for both *Sacch. cerevisiae* and *Schiz. pombe*, sections on rules of genetic nomenclature, physical and genetic maps and lists of genes. The decision was taken to include gene data, even though the lists will rapidly become outdated. Although updated lists are now available on remote databases, this is not always immediately accessible and is not readily portable. This volume will provide a handy laboratory sourcebook for information on the ever-growing collection of genes, which have been isolated and characterized in some detail.

Some 25 years ago, when prokaryotic genetics was to the fore, the concept of Project K was advanced by Francis Crick whereby funding should be made available world-wide for a "complete" analysis of *Escherichia coli* K12, the strain of choice of the bacterium of choice for all molecular genetic studies. It would involve a comprehensive programme of genetics, biochemistry and physiology. In the spirit of this, my co-editor shortly afterwards suggested the complementary Project Y (for yeast) in which a similar strategy would be applied to *Sacch. cerevisiae* (Rose and Harrison, 1971) but unsurprisingly neither was seriously promulgated. Now, however, the independent research results of thousands of scientists have resulted in both organisms, by virtue of their intrinsic advantages, being placed in such a position whereby something akin to Projects K and Y will happen *de facto* and serendipitously rather than by design. The *Sacch. cerevisiae* genome sequencing project will be completed within a few years together with the yeast *Schiz. pombe*, the bacterium *Bacillus subtilis*, the nematode *Caenorhabditis elegans* and the plant *Arabidopsis thaliana*. By proving attractive to "small scientists", really big science is being accomplished. *Sacch. cerevisiae* and *Schiz. pombe* in particular have provided unique contributions to the development of our understanding of both yeast microbiology in particular and eukaryote biology in general (Alberts *et al.*, 1994).

References

Alberts, B., Bray, D., Lewis, J., Raff, M., Roberts, K. and Watson, J.D. (1994). "The Molecular Biology of the Cell". Garland Publishing, New York.

Guthrie, C. and Fink, G.R. (1991). "Guide to Yeast Genetics and Molecular Biology". Academic Press, New York.

Hall, M.N. and Linder, P. (1993). "The Early History Of Yeast Genetics". Cold Spring Harbor Press, Cold Spring Harbor, NY.

Mortimer, R.K. and Hawthorne, D.C. (1969). *In* "The Yeasts" (A.H. Rose and S.E. Harrison, eds), vol. 1, pp. 386–460. Academic Press, London.

Oliver, S.G. and 147 other authors. (1992). *Nature* **357**, 38.

Rose, A.H. and Harrison, S.E. (1971). *In* "The Yeasts" (A.H. Rose and S.E. Harrison, eds), vol. 2, pp. 1–2. Academic Press, London.

Watson, J.D., Hopkins, N.H., Roberts, J.W., Steitz, J.A. and Weiner, A.M. (1987). "Molecular Biology of the Gene", 3rd edn. Benjamin/Cummings, Menlo Park, CA.

2 Genetic Analysis in *Saccharomyces cerevisiae* and *Schizosaccharomyces pombe*

Brian S. Cox

Department of Plant Sciences, Oxford University, South Parks Road, Oxford, OX1 3RA, UK.

The Yeasts Vol. 6, 2nd edition
ISBN 0-12-596416-1

8   B.S. Cox

I. INTRODUCTION

A. Genetics with budding and fission yeasts

Genetics is a system of experimentation that allows the exploration of biological functions *in vivo*. Function is analysed by making new combinations of genes and observing the phenotypic consequences. By the same means, through recombination experiments, the structure and organization of the determinants of function, the genetic material, may be analysed and related to function. Both aspects of genetics depend upon bringing novel combinations of genes together and, in classical genetic experimentation, this has been done by exploiting the sexual cycle of alternating generations separated by cell fusion and meiosis, as well as by mutagenesis.

Simple eukaryotes, particularly fungi, are ideally suited to this, as was recognized by Dodge and his student Lindegren, whom he introduced to *Neurospora*. They grow on simple media and rapidly generate hundreds of millions of unicellular progeny among which can be sought products of rare genetic events such as mutation and recombination. This article is an account of the genetic systems of two eukaryote micro-organisms that in recent years have proven to be ideal vehicles for genetic research. *Saccharomyces cerevisiae* and *Schizosaccharomyces pombe* have come to be known, respectively, as the budding and the fission yeast.

The genetics of *Sacch. cerevisiae* was the first to be explored by Ørvind Winge, working at the Carlsberg Laboratories in Denmark. In due course, Carl Lindegren at Carbondale, Illinois, also turned his attention to this species, abandoning *Neurospora*.

Winge, with his colleague, Catherine Roberts, explored the genetic systems of a variety of yeasts, including those involved in lager brewing and was the first to demonstrate, by micromanipulation of single cells and spores, and by microscopic observation, the alternation of haploid and diploid generations in *Saccharomyces* (Winge, 1935). His task was hampered by the fact that the brewing yeasts he worked with were homothallic, in the sense that the haploid products of meiosis diploidized during growth and so were able autonomously to complete successive rounds of sexual reproduction. Lindegren's strain of yeast was heterothallic: haploid cultures derived from ascospores remained haploid until brought into contact with another haploid culture of the appropriate type, when cell fusion occurred leading to a diploid phase (Lindegren and Lindegren, 1943). The ability to mate was controlled by two alleles at a single locus, which Lindegren called "a" and "α". In due course, Winge showed that the homothallism in his strains was secondary, imposed on the underlying heterothallism by the presence of a single dominant gene, D (Winge and Roberts, 1949). Gene D is now called HO. The heterothallism is controlled by the same pair of alleles, **a** and α, as described by Lindegren.

The distant repercussions of these three papers are still felt today, nearly half a century later. Demonstration of the alternation of haploid and diploid generations, and of its underlying heterothallic control has been the practical means of making the genetic system of *Sacch. cerevisiae* so readily accessible to yeast geneticists. In analysing how that heterothallism works, the hierarchical determination of cell type and function by transcription factors has been revealed (Herskowitz, 1989). The HO gene has proved to be the pivot of two fundamental developmental paradigms. It controls the determination of cell type by DNA re-arrangement (Herskowitz, 1988) and is itself controlled by a process that makes a distinction between a mother and a daughter cell; the very act of cellular differentiation (Nasmyth *et al.*, 1987).

At a symposium held at the Carlsberg laboratories in 1950, which brought together the handful of people by then working on the genetics of yeasts, results of research with a novel yeast that divided its cells by fission were presented by Urs Leupold (Leupold, 1950). Cells of *Schiz. pombe* were of four varieties: two homothallic, one heterothallic " − " and the fourth heterothallic " + ". Only one of the two homothallic strains, namely h^{90}, appears to have survived.

Alternation of generations, homothallism and the control of ploidy in these yeasts are illustrated in Fig. 1. The crucial event in these cycles for genetic analysis is retention of the four products of each meiosis in tetrads of ascospores. Both Winge and Lindegren invented ways of recovering all four spores by micromanipulation and germinating them to form cultures. The ways in which they used their invention illustrate three of the aspects of the power of tetrad analysis.

Winge's interest was in genetic control of fermentation of sugars. In *Sacch. cerevisiae* this is complex, involving unlinked redundant copies of dominant genes. Counting the number of loci involved would have been almost impossible by statistical analysis of random progeny. However, a very limited tetrad analysis could establish immediately whether the genetic control was simple or complex, and could furnish an estimate of the degree of complexity (i.e. how many loci were involved), which would provide a basis for designing the next step in the analysis (Winge and Roberts, 1950). The reason, of course, is that two alleles at a single locus segregate 2:2 in nearly every tetrad and a handful of tetrads is sufficient to demonstrate this unequivocally (Fig. 2). This is still the method of choice for establishing, for example, whether novel phenotypes, particularly pleiotropic ones, are under single-locus control (Fig. 2) and, more recently, whether any genes integrated after transformation with DNA is at its native locus.

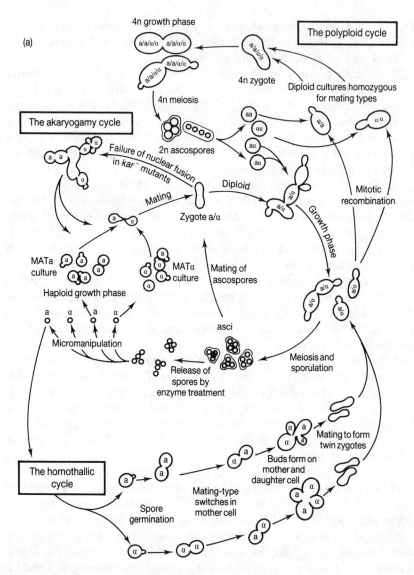

Fig. 1. Life Cycles. (a) The life cycle of *Saccharomyces cerevisiae*. The cycle in the centre of the diagram is that of most laboratory yeasts, which spend most of their time as haploids. Wild and industrial yeasts, if they ever sporulate, would generally by-pass the haploid growth phase by forming zygotes between germinating spores (Winge, 1935). Many are also homothallic (Winge and Roberts, 1949), so the natural condition for such yeasts is diploidy. Entry into the polyploid cycle is inevitable because of mitotic recombination and some industrial processes seem to favour proliferation of such variants. (b) The life cycles of *Schizosaccharomyces pombe*. On the left is the cycle found in homothallic strains (h^{90}); on the right is the behaviour of the heterothallic mutants h^+ and h^-. "P" and "M" in cells represent the active mating-type alleles, plus or minus, respectively. Transitions from one part of the cycle to another are shown as being determined by transfers between medium containing a low concentration of nitrogenous nutrients (low N) and nutritionally rich media (high N).

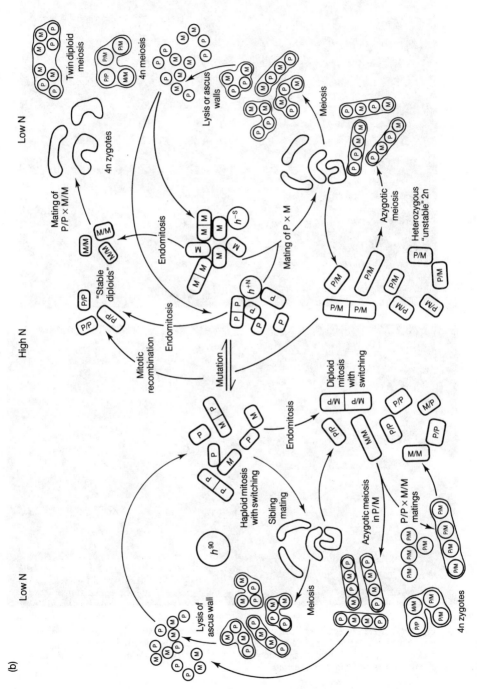

Fig. 1b.

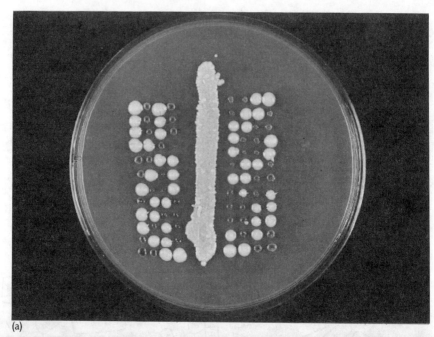

(a)

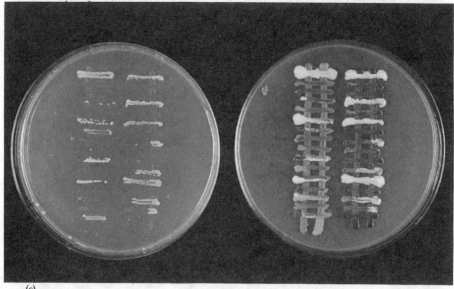

(c)

Fig. 2. (a) A 2:2 segregation of colour in tetrads of *Saccharomyces cerevisiae*. (c) Also shown is an alternative method of conducting complementation tests, namely cross-streaking. Right, master plate; left, replica; streaks down from left to right—*ade1a*, *ade1α*, *ade2a*, *ade2α*; streaks across, individual segregants.

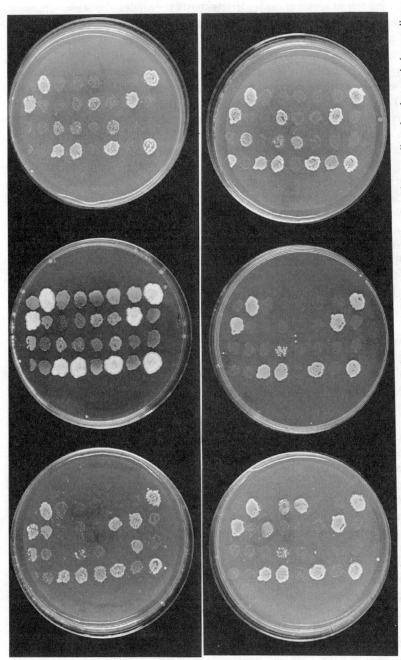

Fig. 2. (b) Complementation tests. Eight tetrads segregating both *ade1* and *ade2* are shown. They have been replica-plated to omission medium and to lawns of *ade1* or *ade2* cells of either mating type. The lawns were on omission medium with enough complete growth medium to allow mating. Top row: centre, master plate on complete medium; left, replica to *ade2*a; right, replica to *ade1*a; bottom row: centre, replica to minimal medium; left, replica to *ade2*α; right, replica to *ade1*α.

The 2:2 tetrad segregation is also the basis for analysis of the mechanism of recombination, since it turns out that this involves a process of gene conversion whereby two alleles segregate other than 2:2. This has been investigated in several tetrad-forming species of fungus. In *Schiz. pombe*, analysis of these processes reached great levels of elegance and quantification with the work of Leupold and his associates (Hawthorne and Leupold, 1974) and in *Sacch. cerevisiae* with that of Fogel, Mortimer and their associates (Fogel *et al.*, 1981). It was Lindegren who first observed non-2:2 segregations in *Sacch. cerevisiae* (Lindegren, 1949).

Lindegren, however, used tetrads principally to map genes and gene mapping is the common perception of what has been the major contribution of tetrad analysis to yeast genetics. His pioneering work in genetic analysis of chromosome organization (establishing linkage groups) and behaviour in *Neurospora* meant that he was ideally placed to develop mapping methods in *Sacch. cerevisiae* (Lindegren, 1945). His first map, published in 1949, had 11 linkage groups associated with centromeres (Lindegren, 1949). It prompted commitment to the yeast map, over the next 40 years, by Hawthorne, Mortimer and their later associates (Hawthorne, 1955; Mortimer and Hawthorne, 1969, 1973, 1975; Mortimer and Schild 1980, 1985; Mortimer *et al.*, 1989). Their most recent version describes the locations of 769 genetic loci. In *Schiz. pombe*, the latest published map is by Munz *et al.* (1989) in which 162 genes have been mapped, and a further 77 assigned to chromosomes but not to loci.

In *Schiz. pombe*, a most elegant example of the power of tetrad analysis is to be found in Leupold's first paper dealing with its genetics. From his crosses between two homothallic and two heterothallic strains, he demonstrated that homothallism and heterothallism were controlled by allelemorphs at the same locus, that second-division meiotic spindles formed at opposite ends of the ascus and did not overlap, and that the locus was at a distance from the centromere. How all of this may be derived from statistics of tetrad segregations is explained in the later section on centromere linkage.

The fourth contribution of tetrad analysis to genetics was in establishing and analysing non-Mendelian inheritance. This began in yeast with Boris Ephrussi's great discovery of the "petite" mutation (Ephrussi *et al.*, 1949; Ephrussi, 1953) and led to the sophisticated analysis of mitochondrial genomes. Just as a chromosomal gene betrays its location by a 2:2 segregation in tetrads, so do non-chromosomal genes by failing to do this. In the case of the *petite* mutation, Ephrussi found that, when crossed with the normal *grande* type of yeast, *petites* failed to segregate in the tetrads; instead they yielded 4 *grande*:0 *petite*. Several other non-Mendelian determinants have been discovered now in *Sacch. cerevisiae*, some as a result of tetrad analysis, some of them still poorly characterized. They include killer (Somers and Bevan, 1969), psi (Cox, 1965), URE3 (Aigle and Lacroute, 1975) and the 2 μm circle (Livingston, 1977). For many of these phenotypes, there are chromosomal genes involved in their heredity as well as the extra-chromosomal determinant itself, and analysing this relationship has depended heavily on tetrads. The clearly distinct segregation patterns of chromosomal and non-chromosomal genes makes discrimination very clear (Fig. 3).

The Mendelian pattern of 2:2 inheritance in tetrads has in more recent times been applied to the demonstration of centromere activity in plasmids and artificial chromosomes both in *Sacch. cerevisiae* (Newlon, 1988, 1989) and in *Schiz. pombe* (Niwa *et al.*, 1986, 1989).

B. Working with *Saccharomyces cerevisiae*

In *Sacch. cerevisiae*, alternation of generations is controllable by very simple manipulations. Propagation and storage of diploid cells is achievable as long as they are maintained on growth media containing sufficient nitrogenous nutrients, and the switch into meiosis and formation of haploid spores is induced by transfer of vigorously growing cultures to a starvation medium. Both environmental and genetic controls are involved; heterozygosity at the mating-type (*MAT*) locus is a prerequisite for meiosis and sporulation (Kassir *et al.*, 1988).

Similarly, haploid cultures can be maintained as such indefinitely, provided they are *ho⁻* and therefore heterothallic. Return to the diploid state is brought about by mixing cells of opposite mating type, which then fuse. This makes assay of diploidy-dependent functions very simple and the most useful of these is the complementation test. When two recessive mutations having the same phenotype are in different genes, the diploid, which is doubly heterozygous, has the dominant phenotype: the mutations complement one another. Sorting arrays of mutants of identical phenotype, for example, temperature-sensitive or radiation-sensitive variants, into complementation groups each of which corresponds to a single gene is very easy in budding yeast. Similarly, when more than one gene controlling a phenotype is segregating in tetrads, segregation of each or all of them can be followed by simple tests for complementation. The easiest of these is the patch test, originally devised by Woods and Bevan to analyse complex patterns of allelic complementation among mutations in the ADE2 gene (Woods and Bevan, 1966). It is done by replica-plating patches of the strains to be tested on to a lawn of cells of opposite mating type. Once

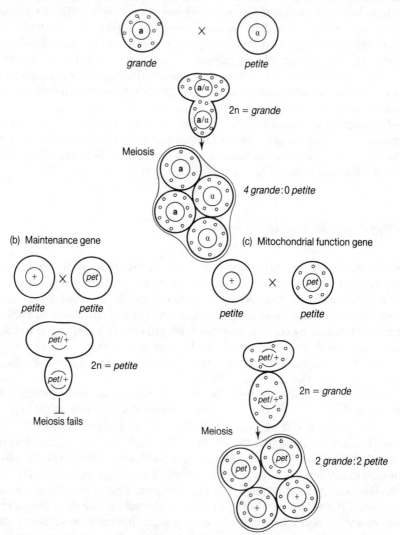

(a) Mitochondrial DNA mutation or loss

grande   ×   petite

2n = grande

Meiosis

4 grande : 0 petite

(b) Maintenance gene

petite   ×   petite

2n = petite

Meiosis fails

(c) Mitochondrial function gene

petite   ×   petite

2n = grande

Meiosis

2 grande : 2 petite

Fig. 3. Inheritance of a cytoplasmic determinant such as mitochondrial DNA in *Saccharomyces cerevisiae*, and interactions between the cytoplasmic determinant and nuclear genes. In formation of the zygote, the cytoplasm of both parents becomes mixed and, since there is no means of segregating its components at meiosis, all four spores inherit similar cytoplasmic genes. The *grande* phenotype segregates 4:0 in tetrads (after Ephrussi, 1953). (a) The *petite* phenotype is caused by loss or mutation of mitochondrial DNA. (b) The *petite* phenotype is due to a nuclear gene affecting inheritance of mitochondrial DNA. (c) The *petite* phenotype is due to a mutation affecting mitochondrial function. Mating the *pet* mutant of either (b) of (c) examples with a *grande* strain gives an identical result, i.e. a 2 *grande*: 2 *petite* segregation.

diploids have formed, the whole plate can be replicated again for assays of phenotype. In Woods and Bevan's experiments, further replication was not necessary, since *ade2* mutations accumulate a red pigment, and mutant and wild-type phenotypes can be determined by their colour. In Fig. 2, the patch test is illustrated, as well as an alternative method, the cross-streak.

The test is so simple that diploid phenotypes much more difficult to assay than merely the ability to grow can be tested. For example, complementation of genes controlling mutation or ultraviolet-induced mutagenesis can be assayed by the patch test. So can phenotypes that depend on recombination, taking advantage of the mitotic recombination that occurs spontaneously at a high rate in yeast. Using the same procedures, prototrophs arising from mitotic recombination in diploid patches can be used as an allelism test when non-complementing alleles are segregating in crosses. Diploid cells containing non-identical alleles may recombine and the recombinant cells grow into papillae so that such patches are covered in papillae. When the alleles are identical, no papillae form. This is what made it possible for Fogel and Mortimer and their associates to analyse segregation of arrays of heteroallelic mutations in tens of thousands of tetrads, establishing the principles of meiotic fine-structure recombination (Fogel *et al.*, 1979). Similarly, allelism of mutants controlling the ability to complete recombination and meiosis can be checked by first replicating diploid patches on to sporulation medium and then on to a medium selecting prototrophs formed by recombination.

C. Working with *Schizosaccharomyces pombe*

Exploiting the alternation of generations in *Schiz. pombe* is similar both in principle and in practice, but differences in the control of events in the fission-yeast cell cycle make details of procedures different. The strains commonly used, all derived from a single culture by Leupold, are of three types: a homothallic strain, h^{90}, and two heterothallic types, h^- and h^+. The latter are heterothallic as a result of mutations at the mating-type loci (see below), which make mating-type switching either rare in the case of h^+ or ineffective in h^-. For this reason, in Fig. 1 the life cycle of fission yeast has been shown as two separate cycles, one appropriate to the non-mutant h^{90} types and the other for the heterothallic mutants.

In the homothallic strains, mating-type switching occurs constantly in a growing haploid culture (Egel and Gutz, 1981; Gutz 1990; see Section II later), so that the population of cells is a mixture of "+" and "−" types. Mating is a rare (but not unknown) event in conditions promoting active growth, but is induced and occurs throughout the population when cells are transferred to a medium containing low levels of nitrogenous nutrients (Leupold, 1950). It is followed immediately by meiosis and sporulation. The starvation conditions that terminate colony growth are sufficient to induce significant numbers of mating and sporulation events, and this provides a means by which homothallic strains can be distinguished on plates from heterothallic ones: spores of *Schiz. pombe* contain starch so that homothallic colonies go black when exposed to iodine vapour because of the large numbers of asci they contain.

Mating strains of *Schiz. pombe* for genetic analysis requires different strategies depending on the purpose. If the objective is analysis of genetic recombination following meiosis, it is necessary to ensure that only hybrid zygotes sporulate. Mating in homothallic strains, when it occurs, characteristically does so between sister cells, an indication that one of them switched mating type at the preceding mitotic division (see Fig. 7c). Nevertheless, homothallic strains can be mated, either with another homothallic strain or with either of the heterothallic mutants. Matings are carried out by mixing strains on a medium containing low levels of nitrogenous nutrients, which induces the process. However, such mixtures contain a preponderance of self-matings and, if either a hybrid diploid culture or asci containing tetrads from hybrid zygotes are required from it, selection for hybrid diploid cells must be applied. This is usually done by making the two parents complementary for auxotrophic markers and selecting prototrophs. Provided this is done before zygotes go through meiosis, hybrid diploid rather than recombinant haploid prototrophs can be obtained. Convenient auxotrophs for selection of diploids are complementing allelic mutants of the *ADE6* locus (Leupold and Gutz, 1964). This is the homologue of the *ADE2* locus of *Sacch. cerevisiae* and, in both yeasts, mutants accumulate a red pigment. Complementing diploids are white and easily distinguishable from haploid parents.

Diploid vegetative growth is promoted by transfer from low-nitrogen back to a high-nitrogen medium. Diploid colonies, however, contain many dead cells and this is exploited as a means of identifying them. Media for growing *Schiz. pombe* are usually made containing a red dye (magdala red). Diploid colonies go a deep-red colour on this while haploids remain pink. Diploids formed and identified in this way from crosses involving homothallic parents can be induced to go through what is referred to as azygotic meiosis and to form tetrads of spores in linear asci by transfer to low-nitrogen medium.

Diploid cultures are usually needed for two purposes, namely assays of complementation and for genetic analysis by recombination or by haploidization (see below). Both zygotic and azygotic asci autolyse when left for a length of time on non-growth medium and tetrads of spores can be separated mechanically with ease. This property has made random-spore analysis particularly easy with fission yeast. Diploid cells that have failed to sporulate and could confuse the analysis can be selectively eliminated by treatment with 30% ethanol.

Although all essential genetic procedures can be carried out using homothallic strains of *Schiz. pombe*, most workers find that it is more convenient to carry out these procedures with heterothallic mutants. For tetrad analysis, for example, mixing h^- and h^+ strains on sporulation medium leads in due course to formation of 100% hybrid asci without the need for prototrophic selection.

Complementation tests for allelism in *Schiz. pombe* demand that diploids are selected before being tested for the appropriate phenotype irrespective of whether homothallic or heterothallic strains are involved. This is because sporulation is part of the developmental pathway initiated by mating and unless it is pre-empted by selecting for vegetatively growing diploid cells, meiosis will generate recombinants with the same wild-type phenotype as complementing markers. An alternative stratagem for making non-sporulating diploids is to use a mutant of the *mat2* (P) cassette, *mat2-B102* (Egel, 1973). This has the mating type of h^+ and mates normally with h^- (or h^{90}) cultures, but the mutation inactivates the P_i gene product of the *mat2* (P = plus) cassette. This product is needed for meiosis so that all zygotes revert to vegetative growth on transfer back to a rich medium. If one of the parents in complementation tests is *mat2-B102*, complementation tests can be done, as they can with *Sacch. cerevisiae*, by patch-matings or cross-streaks on low-nitrogen medium, followed by replica-plating to the appropriate test medium.

There are some important procedures in genetic analysis that depend on so-called "stable" diploids. These are diploid cultures that do not go through meiosis and form asci. There are two commonly used sources of such diploids. First, they may be formed spontaneously by mitotic recombination in diploids initially heterozygous at the mating-type locus. The transcripts of all four reading-frames of the two mating-type cassettes are essential for meiosis in *Schiz. pombe* (Kelly *et al.*, 1988; see later), so homozygosis of a non-switching mating-type locus absolutely prevents it. Naturally, using the *mat2-B102* mutant as one of the parents, will also ensure the formation of "stable" diploids.

"Stable" hybrid diploids can be identified in populations of sporulating hybrid diploids by the iodine test on colonies; they do not go black. They are especially useful for two kinds of genetic analysis in *Schiz. pombe*, namely haploidization and mitotic recombination, each of which plays an important part in locating genes on one of the three linkage groups of this yeast.

Diploids also arise in haploid cultures, apparently by endomitosis. They are, of course, homozygous at all loci, except that, when they occur in homothallic strains, mating-type switching occurs and, in due course, they become a source of triploid and tetraploid zygotes. Cultures of haploids can be screened for the presence of diploid variants by plating on red-dye medium.

In recent years, gene cloning and the ability to transform yeast with naked DNA has extended its flexibility to the point at which almost any genetic demand can be made of yeast cells and any question which is relevant to their biology asked, by genetic manipulation. Two particular properties of the yeast system seem to be crucial in providing this flexibility. The first is the ability readily to clone genes by complementation of functional loss. The second property is unique to *Sacch. cerevisiae*. It is that all recombination in this species is limited to exchange between homologous sequences. Integration of transforming DNA in other eukaryotes, including *Schiz. pombe* is as often as not at apparently random, certainly nonhomologous, sites. In other fungi, it is often accompanied by seemingly meaningless DNA re-arrangements. In *Sacch. cerevisiae*, DNA sequences can be faithfully targeted to their homologous sites in the genome. This property has been extensively exploited in genome manipulation. It can be used unequivocally to confirm the identity of a cloned gene by homology, in addition to function. It can also be used to change the sequence of genes in a targeted way, integrating versions with point mutations, deletions or additions so as to analyse their function *in vivo*; to recover mutant forms of genes for further analysis or to construct new sequences and linkage groups so that complex genetic processes like recombination can be dissected at a fine-structure level. In fission yeast, although non-homologous recombination events are common, homologous recombination occurs sufficiently frequently for all of the manipulations that are available in budding yeast to be undertaken successfully (Moreno *et al.*, 1991). A comprehensive account of these methodologies can be found in the volume edited by Guthrie and Fink (1991).

This review describes classical methods used for analysis of the structure and organization of genetic material, down to the level of the gene. Fine-structure mapping, by which I mean structure below this level, by classic means is more or less a thing of the past and will only be touched upon.

II. TETRAD ANALYSIS

Over 90% of the genetic map of *Sacch. cerevisiae* published most recently (see Appendix 2) was gained by tetrad analysis. To recover the four spores in a single tetrad, the ascus wall is digested enzymically, a procedure first used by Johnston and Mortimer (1959). The spores themselves are impervious to the enzyme and stick together in their groups of four after the digest has been spread on an agar plate. There they are separated by micromanipulation (Sherman and Hicks, 1991).

Two of the spores in every tetrad will carry the *MATa* and two the *MATα* allele of the mating-type locus. If another gene, at a different locus, is also heterozygous $+/-$, then it too will segregate its alleles 2:2, but independently of the mating-type alleles. The result when several tetrads are analysed is just three classes of tetrad:

I	II	III
a+	a−	a+
a+	a−	a−
α−	α+	α+
α−	α+	α−
"ditype"		"tetratype"

It is easy to show that, if the a/α and the $+/-$ loci are on different chromosomes, and either is distant from the centromere of its chromosome, then these types in a large sample of tetrads occur in the proportions 1:1:4, and this is true of any pair of segregating genes (see, e.g. Table III).

Departures from these free-segregation ratios are the basis of mapping yeast genes. There are two reasons for other ratios, namely (i) two loci are linked on the same chromosome, or (ii) both loci are linked to a centromere not necessarily on the same chromosome.

A. Linkage on the same chromosome

When two genes being considered are on the same chromosome, all three types of tetrad may appear, but each depends on particular configurations of chiasmata. These are illustrated in Fig. 4. The interpretation involves a convention for describing the two kinds of ditype tetrad in relation to the input configuration of the alleles: Parental ditype (PD) means there are two types of spore in the tetrad, representing the parental configuration of markers. Tetratype (TT) means there are four different spores, two parental and two recombinant and NPD, for non-parental ditype, indicates two types of spore, but both recombinant.

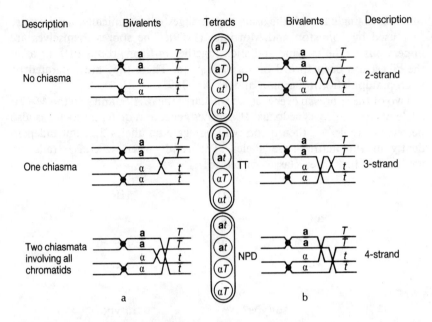

Fig. 4. Segregation in meiosis of two linked loci. The markers chosen for illustration are the *MAT* locus (*a*/α) and *THR4* (*T/t*) on chromosome III. (a) This illustrates the minimum requirements in chiasma distribution, which achieve each of the three types of tetrad. (b) Consequences of various arrangements of double chiasmata in intervals. Between (a) and (b), the column labelled tetrads shows the results of bivalent configurations illustrated to the right and left, and the abbreviated description of each type of tetrad segregation. PD indicates parental ditype; TT, tetratype and NPD non-parental ditype.

Fig. 4(a) shows that the only way of arriving at a TT or NPD tetrad is if one or more chiasmata occur in the interval between markers. Any factor, such as physical proximity, which limits the frequency of formation of chiasmata, lowers the prevalence of these types relative to PDs. Fig. 4(b) shows that, with higher orders of chiasma formation, depending on their distribution among chromatids, any of the three types of tetrad can be formed. Type (b) (iii) doubles, which involve all four chromatids of a bivalent, are 1/4 of the total number of double cross-over tetrads. This is the only kind of event that produces NPD tetrads. Consequently, the frequency of NPDs is very sensitive to any decrease in chiasma frequency. Parental ditype tetrads, on the other hand, can be formed not only from 1/4 of the bivalents with two chiasmata in the interval ((b)(i)) but from all of those with none ((a)(i)), which of course are enriched when recombination is decreased. A significant departure from a 1PD:1NPD ratio in

Table I. The inference of linkage from tetrad data

Numbers of ditype tetrads	Significance level		cM	Tetrads
	5%	1%		
5	5 : 0	–	20	10
10	9 : 1	10 : 0	40	30
50	32 : 18	34 : 16	60	150

The ratio of parental ditypes (PD) to non-parental ditypes (NPD) that give a significant indication of linkage. The fourth column (cM) shows the map distance in centimorgans that would generally give the ratios of numbers shown in the second column. The fifth column shows approximately how many tetrads one would have to dissect in order to achieve these levels of significance at the distances shown.

tetrads is generally the first indication that two loci are linked, since it is detectable in only a handful of tetrads (Table I).

B. Conversion of tetrad data into gene maps

Various systems have been used to convert tetrad data, in which the numbers of PD, NPD and TT types are counted for any interval defined by a pair of marked loci, into gene maps. Gene mapping is based on the proposition that the frequency of cross-overs occurring in a particular length of chromosome, bounded or marked by two genes, is proportional to that length. This frequency is estimated from recombination data, but the procedure is not straightforward. This is because it is actually impossible to tell from the appearance of any segregant chromosome exactly how many cross-overs occurred in the marked interval at the preceding meiosis. All one can state with certainty is that, if it is recombinant, at least one cross-over must have occurred. This does not give one sufficient information from which to calculate x, the frequency of crossing over. Consider, for example, tetrad data.

All three tetrad types, PD, NPD or TT, may be generated by a variety of cross-over events in an interval. Any tetrad with a cross-over rank of two or greater (i.e. with two or more cross-overs in the interval being considered) may generate any one of the three types. Thus a PD tetrad may represent $0, 2, 3, \ldots n$ cross-overs, NDP $2, 3, \ldots n$ and TT $1, 2, 3, \ldots n$ cross-overs. It is clearly impossible to count the precise number of cross-overs that have occurred in an interval directly in any set of tetrads and thus to arrive at an estimate of their frequency.

One can, however, make simplifying assumptions. For example, if an interval is small enough that the probability of crossing over is low (say 20%), then it is reasonable to assume that double cross-overs will be rare (in this case 4%) and higher ranks rarer still (equal to or less than 1%). If ranks higher than two are then assumed to make a negligible contribution to the tetrads observed, map distance can be estimated using the numbers of NPD tetrads observed to estimate the numbers of double cross-overs. Making this assumption, Perkins (1949) derived the expression:

$$\text{Map distance (cM)} = 50 \times (TT + 6NPD)/(PD + NPD + TT)$$

As may be seen from the examples given in Table II, this works very well for short intervals. The method does, however, have shortcomings. The first derives from its assumption, namely that tetrads ranking three or more

Table II(a). A selection of tetrad data taken from Mortimer *et al.* (1989). It illustrates the results of applying different methods of calculating linkage to them. Various chromosomes are illustrated, and the data are grouped to show how interference affects the results. Ma and Mortimer (1983) plotted distances calculated by Perkins' method with those using Snow's data to show that, at small intervals, the two methods give identical estimates but, as distances increase, Perkins' method underestimates. The two are also more concordant when interference is high ($k \leq 0.2$). X is the map distance in cM according to Perkins (Per), Papazian (Pap) or Snow (S). k is the interference according to Papazian (Pap) or Snow (S). See the text for these methods of calculation. P indicates parental ditypes, N non-parental ditypes and T tetratypes

Interval	Chromosome	P	N	T	X(Per)	X(Pap)	X(S)	k(Pap)	k(S)
(i) $k \leqslant 0.6$									
cdc15-flo1	I	52	3	116	39.2	62.5	39.6	0.12	0.2
cdc15-pho11	I	94	8	226	41.8	66.9	42.5	0.17	0.2
MEL1-gal1	II	33	16	113	64.5	112.7	82.2	0.67	0.6
lys2-tyr1	II	75	8	153	42.6	63.0	44.2	0.26	0.3
tyr1-his7	II	48	14	175	54.6	97.1	59.3	0.40	0.4
MAT-ros1	III	33	14	146	59.6	115.9	66.9	0.49	0.4
MAT-his4	III	97	7	174	38.8	56.4	58.3	0.41	0.5
trp1-rad55	IV	31	7	108	51.4	90.3	54.3	0.33	0.3
trp4-ade8	IV	43	3	72	38.1	54.1	39.1	0.26	0.3
trp4-crl16	IV	75	1	36	18.8	20.7	19.0	0.53	0.5
ura3-crl4	V	60	3	54	30.8	36.0	32.1	0.62	0.6
crl1-aar2	VI	61	2	85	32.8	46.0	33.0	0.17	0.2
trp5-leu1	VII	1036	3	330	12.7	14.1	12.7	0.25	0.3
trp5-leu1		2216	13	1086	17.6	20.4	17.6	0.22	0.2
trp5-ade6	VII	43	4	67	39.9	53.6	41.9	0.41	0.5
cdc11-hom6	X	71	14	200	49.8	80.5	53.1	0.33	0.4
(ii) $k \geqslant 0.8$									
hom6-dal5	X	235	1	49	9.64	9.86	9.7	0.83	0.8
rec1-rad54	VII	16	1	13	31.7	34.7	34.2	0.95	1.0
ser2-pet54	VII	157	1	37	11.0	11.2	11.1	1.0	1.0
ade6-cdc62	VII	45	1	20	19.7	20.3	20.4	1.0	1.0
hap2-lys5	VII	14	4	22	57.5	69.3	86.3	1.5	1.4
rad4-rem1	V	117	5	35	20.7	16.9	26.5	4.3	4.1

Table II(b). Some tetrad data from *Schizosaccharomyces pombe* (Munz *et al.*, 1989)

Interval	Chromosome	P	N	T	X(Per)	X(Pap)	X(S)	k(Pap)	k(S)
ade2-ade4	I	62	36	192	70.3	121.0	120.6	0.84	0.87
ade3-lys-2	I	67	24	143	54.1	70.6	70.7	0.91	0.92
ade3-ura1	I	69	37	200	69.0	112.8	112.9	0.88	0.91
ade2-leu2	I	244	9	114	27.7	32.0	32.1	0.57	0.60
ade4-ura2	I	184	141	585	78.6	152.6	152.6	1.25	1.41

cross-overs make a negligible contribution to the sample. It follows that, as the interval gets larger, and ranks of three or greater start to make a significant contribution, so the expression tends to underestimate map distance, that is the real frequency of crossing over. The second is that it presents a statistical problem. The map distance derived from using this expression depends heavily on the numbers of NPD tetrads in the sample (for estimating double cross-overs). These numbers are usually small for linked markers and subject to large fluctuations due to sampling error. Both the numbers of NPDs and their associated statistical errors are multiplied by six in Perkins' expression.

Less arbitrary methods of estimating cross-over frequency make use of the Poisson equation. This states that the probability of observing n crossovers in an interval where the frequency of cross-over overall is x, is:

$$p_{(n)} = \frac{x^n \cdot e^{-x}}{n!}$$

which for $n = 0$ reduces to:

$$p_{(0)} = e^{-x}$$

Quite a simple method was suggested by Papazian (1952) and makes use of this equation. To estimate $p_{(0)}$ from tetrad data, one counts the frequency of PD tetrads, since these are the only ones formed from a nil cross-over meiosis. One must allow for those PDs formed from higher order cross-over events, and this can be estimated from the number of NPDs observed. Thus:

$$p_{(0)} = \frac{PD - NPD}{PD + NPD + TT}$$

and

$$x \text{ (the frequency of crossing over)} = -\ln p_{(0)}$$

The map distances this formula generates are shown in column 7

"*X*(Pap)" of Table II. It is clear that most of the time they are greatly in excess of those calculated by the Perkins method. The reason is that, in most parts of the yeast genome, crossing over does not occur at random. Instead, the probability of a cross-over occurring is strongly influenced by whether one has already occurred in the neighbourhood. This influence is called "interference" (*k*) and may be positive, when the occurrence of a cross-over lowers the probability of another occurring in its vicinity, or negative. Over most of the yeast genome, interference is positive, decreasing the number of events compared to those expected on a random basis. As a result, most of the estimates produced by Papazian's method are over-estimates of the true amount of crossing over. However, in Table II, data have been included from regions where interference is negligible or negative (*k* equal to or greater than 0.8). Comparing the $p_{(0)}$ method with the values shown in the column headed "*X*(S)" shows that, indeed, the method gives the expected matches or underestimates, depending on whether interference is absent (*k* = 1) or negative (*k* > 1) Table II(a)(ii) and Table II(b).

In *Schiz. pombe*, it turns out that interference is indeed negligible [Snow, 1979; Table II(b)] and, provided one accepts the statistical implications, this would be a very quick and suitable method for mapping genes.

Clearly what is needed is a method of calculating cross-over frequency, which makes no simplifying assumptions about the numbers of multiple cross-overs, and also calculates interference and takes it into account. Such a method, devised by Snow (1979), was used to calculate the values in the "*X*(S)" column of Table II. Like the data, they are taken from Mortimer *et al.* (1989).

The principle of this method is as follows. First, the numbers of each type of tetrad, PD, NPD or TT in each cross-over rank are found. This is done using expressions derived by Mather (1935, 1957) and applied to tetrads by Barratt *et al.* (1954). The Poisson distribution is then used to relate the cross-over frequency, *x*, to the probability of getting each rank *r*, of cross-over events in an interval (*r* = 0, 1, 2, 3,... *n*).

Mather (1935, 1957) showed that the number of equational segregations of bivalents was related to the cross-over rank, *r*, by the function:

$$p(E) = 2/3[1 - (-1/2)^r]$$

In tetrads, equational segregations (*E*) yield tetratypes (TT). Since, when there are two or more cross-overs (*r* = 2), PD and NPD are equal and together equal to $1 - p(\text{TT})$, three expressions relating each type of tetrad to cross-over rank can be derived, remembering that all tetrads of rank 0 yield only PDs and all of rank 1 yield TTs. The Poisson expression relates each of these to *x*, the frequency of crossing over, by summing over all of the relevant ranks. Finally, differentiating these expressions allows *x* to be solved for given values of PD, NPD and TT, using the maximum-likelihood method. It is an iterative procedure and requires a computer, for which a program is available (Snow, 1979).

As described so far, this method gives values for x that are identical to those obtained by using Papazian's estimate from only the $p_{(0)}$ term of the Poisson expression. See, for example, Table II(b). However, the value of Snow's approach, apart from the fact that it uses all of the data, gives it all equal weight and makes no simplifying assumptions, is that allowance can be made for interference in a way originally suggested by Barratt *et al.* (1954). If each term of the Poisson distribution in the expressions is multiplied by k^{r-1}, they can then be differentiated with respect either to x or k, setting the other to 0. Furthermore, standard errors can be computed for each set of data for both x and k; this is a dubious procedure with other methods.

Once again, there are simpler methods for obtaining good approximations to these estimates of x and k. Firstly, Ma and Mortimer (1983) found an empirical expression that, for data from *Sacch. cerevisiae*, converts map distances found by Perkins' simple formula to close approximations to X(S). The expression is:

$$X_{(S)} = \frac{80.7x_{(Per)} - 0.883x_{(Per)}^{2}}{83.3 - x_{(Per)}}$$

This approximation works best when k falls around the common values for the budding-yeast genome of between 0.2 and 0.6.

Interference can be estimated with quite good results from the suggestion, again by Papazian (1952), that k is given by the ratio between the NPDs observed and those expected. When there is interference, the observed numbers fall short of those expected, and the ratio is less than 1. The NPDs expected can be calculated as already indicated from the numbers of tetratypes. Thus:

$$N(\exp) = A \times 1/2[1 - T - (1 - 3T/2)^{2/3}]$$

where A, N and T are the numbers of all tetrads, NPD tetrads and TT tetrads, respectively. Values of k calculated in this way are shown in column 9 alongside those from Mortimer *et al.* (1989) (column 10).

How do the map distances arrived at compare to physical lengths of DNA? Pulsed-field gels have made it possible to measure the lengths of the 16 small chromosomes of *Sacch. cerevisiae*. Summing over all intergene map distances gives a total "length" for the yeast genome of 4295 cM. The amount of DNA present in the chromosomes is 12.5 Mb, so that the ratio is 0.34 cM for each kb (Olson, 1992). This ratio is maintained for individual chromosomes whose size has been measured on pulse-field gels, except that

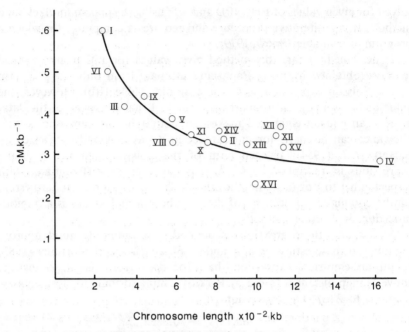

Fig. 5. Genetic length (cM) per kilobase as a function of physical length for each of the 16 yeast chromosomes. The data plotted are those in Mortimer *et al.* (1989).

shorter chromosomes have more recombination (Fig. 5). At finer levels of resolution, there is considerable variation. For example, the map distance between *CEN3* and *LEU2* is 8.1 cM and the length of DNA separating them is 22 kb, the ratio of $0.37\,\text{cM kb}^{-1}$ being close to standard. In the neighbouring region, however, a length of DNA between the centromere and *PGK1* of 17 kb has a map length of 2 cM equal to $0.12\,\text{cM kb}^{-1}$ (Clark and Carbon, 1980). Even within the *CEN3–LEU2* interval, localized regions of high and low recombination have been measured (Symington and Petes, 1988). Discrepancies like these will be of use and interest to those studying the mechanism of recombination and organization of chromatin in the meiotic nucleus, since they identify hotspots of recombination and regions where the process seems to be inhibited. Around such genetic variation, experiments can be designed that will elucidate this mysterious process.

The genome of *Schiz. pombe* has been almost as intensively analysed by meiotic recombination. The latest map published (Munz *et al.*, 1989) is very extensive and includes 153 genes. The three chromosomes visible cytologically (Robinow, 1977) and by pulse-field electrophoresis (Smith *et al.*, 1987), with sizes of 5.7, 4.6 and 3.5 Mb, respectively, have a total genetic length of 2271 cM (0.165 cM kb^{-1}). Chromosomes I (987 cM = 0.17 cM kb^{-1}) and II (762 cM = 0.16 cM kb^{-1}) have the genome average rate of recombination. Chromosome III apparently has less (523 cM = 0.15 cM kb^{-1}) but the meiotic map is not complete, a fragment having been associated with this chromosome by mitotic analysis. If it is assumed that the recombination rate is indeed average for this chromosome, the final genetic length for this chromosome will turn out to be 577 cM.

Both of these yeasts have a genetic length that falls within the normal range of genetically analysed organisms, although they have the smallest physical genomes. However, if the amount of recombination in each meiotic chromosome (i.e. numbers of cross-overs or chiasmata for each bivalent) is calculated, it seems to be unusually high, especially in *Schiz. pombe*, which has been calculated at 12.5 cross-overs per bivalent. The numbers usually observed over a wide range of genome sizes and karyotypes are between 1.0 and 3.0 chiasmata per bivalent. This, together with the lack of interference and of synaptonemal complexes (Olson *et al.*, 1978) and the observation that, unlike the situation in *Sacch. cerevisiae* or *Ascobolus immersus*, the frequency of gene conversion per gene shows a random distribution (Gygax and Thuriaux, 1984) suggest that fission yeast is likely to have unusual and interesting features of meiosis.

C. Centromere linkage

Fungi like *Neurospora* or *Sordaria* form linear asci and ordering of the products of meiosis is regularly achieved in them by the positions of the meiotic spindles (Fig. 6). Lindegren (1933) pointed out that this allowed equational gene segregations to be distinguished from reductional ones. Equational divisions (often called second division segregations, since two alleles are not separated to different nuclei until the second meiotic division) depend on one or more cross-overs occurring in the interval. This is illustrated in Fig. 6. This allows the map distance between the centromere and the gene to be calculated from Mather's (1935) expression (already stated). All cross-over configurations, which would give rise to tetratype tetrads if one were scoring two gene markers, give equational segregations; all others (the equivalent of PDs and NPDs) result in reductional segregations.

Linear asci are a characteristic of *Schiz. pombe*, and Leupold (1950) in his early work went to the trouble of dissecting them while retaining the

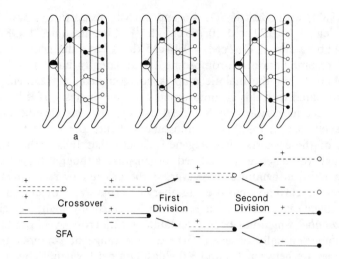

Fig. 6. Tracings of Lindegren's (1933) diagrams of linear asci in *Neurospora crassa* and, below, his explanation for the reductional (a) or equational (b, c) patterns. SFA indicates spindle fibre attachment.

order of the spores. Among his data are some from crosses of h^{90} with each of the heterothallic isolates. These tetrads demonstrated that the homothallic and two heterothallic forms were determined by alleles at the same locus, since every tetrad segregated 2hom:2het strains, with the heterothallic segregants being "+" or "−" depending on the heterothallic parent. The ordered tetrads yielded about 30% of (hom, hom, het, het) and 70% of ((hom, het) (hom, het)) types. This in itself indicates nothing about spindle orientation in the developing asci but, included in his analysis, were crosses of the two heterothallic strains with a second homothallic strain, h^{40}, in which all the ordered tetrads segregated 2hom:2het and in the order (hom, hom, het, het). This immediately established that the hom^{40} type was also determined by an allele at the hom/het locus (i.e. the *MAT* locus) but also that: (i) second-division spindles invariably formed at opposite ends of the ascus; and (ii) that the segregating alleles were closely centromere linked. Since, in crosses with h^{90}, these two statements could not both be true, the simple explanation is that the usual position of the *MAT* locus is at some distance from the centromere, but that the h^{40} mutation for some reason, possibly because of an inversion, suppressed crossing over in the *CEN–MAT* interval.

Linear asci do occur occasionally in *Sacch. cerevisiae* and, in 1955, Hawthorne carried out a similar analysis (Hawthorne, 1955) dissecting a large number from a diploid, which had various genes segregating. He found that one gene, *trp1*, showed a consistent order of segregation of (+ − + −), the ends, right or left of a yeast linear tetrad, not being distinguishable. He interpreted this as having resulted from 100% reductional segregations of this gene, indicating that it was very closely linked to its centromere and therefore that the alternating patterns of alleles must have been due to a spindle conformation in the sporulating cells whereby the spindles of the second meiotic division overlap. Hawthorne confirmed the overlapping spindles by cytological observation but, although many centromere-linked genes have been shown by molecular methods to be indeed very close to a centromere (q.v. *LEU2* and *PGK1*), ironically *TRP1* is not yet one of them. Hawthorne's experiment required considerable skills with a micromanipulator and a lot of patience. Unlike the asci of *Schiz. pombe*, those of *Sacch. cerevisiae* do not autolyse and break apart easily. Nor could they be digested enzymically, but had to be broken with the dissecting needle because Hawthorne wished to retain the spore order. The method results in a lot of spore inviability, so collecting complete sets of tetrads took time.

Few strains of *Sacch. cerevisiae* in current laboratory use form linear asci, although, when zygotes sporulate, linear asci may be formed with overlapping spindles (Thomas and Botstein, 1987). Not even in *Schiz. pombe* do workers these days attempt to retain the order of spores. Nevertheless, tetrad data can be used for centromere mapping, making use of gene markers. In Table III, two gene markers on different chromosomes are shown segregating in linear tetrads. The kinds of tetrad (PD, NPD or TT) that result from various possible combinations of equational and reductional segregations of the two chromosomes in each tetrad are shown. It is clear that, only when there is an equational segregation of one or other marker, is there a TT tetrad formed. PD or NPD tetrads may be derived from any kind of event. It follows that, if there is a reduction in cross-overs in *both* centromere intervals, the frequency of tetratypes relative to ditypes declines. The expected relative frequencies from frequent cross-overs in either interval are 2TT:1 (PD + NPD) (cf. Leupold's data for h^{90} × het) so that ratios of less than two indicate that both markers are linked to their centromeres.

This property can be exploited in two ways. In 1949, Lindegren suggested the use of three markers, considered pairwise and solving three simultaneous equations for recombination in each of the gene–centromere intervals: x, y and z. The expression for each pair of markers is:

$$f(T_{x,y}) = x + y - 3xy/2$$

Since then, markers like *TRP1* have been found, which are so close to

Table III. A table showing the segregation of two markers in linear tetrads

		1st		2nd			
		a	+	a	+	a	+
		a	+	+	a	+	a
		+	a	a	+	+	a
		+	a	+	a	a	+
1st	b b + +	P	N				
	+ + b b	N	P				
2nd	b + b +			P	N		
	+ b + b			N	P		
	b + + b					P	N
	+ b b +					N	P

Notional linear arrays are shown for reductional (first division) and equational (second-division) segregations and the tetrad types that result when the arrays of "a" and "b" combine. P indicates parental ditype, N indicates non-parental ditype, unlabelled squares indicate tetratype tetrads.

a centromere that they can be used as indicators of spindle configurations in meiosis. If, in Table III, the alleles "Bb" were such a marker, the only tetrads that would be formed would be those in the unshaded part and the number of tetratypes is a direct count of the equational divisions of the "Aa" gene. Snow's iteration can be applied to this marker as already indicated. However, interference cannot be calculated, since that requires information about the precise numbers of $(0, 1, 2, \ldots n)$ cross-over events, and that is not extractable from these data. If Snow's programme is not available, map distance in centimorgans is the frequency of second-division segregations multiplied by 50.

D. Homologous integrations, reversions and the numbers of mutations

In classical yeast genetics, the questions asked of a marker are: "is it linked to any other gene or to a centromere?" and, if so, "what is the map distance?". As already described, linkage of two markers is indicated by a departure from the expected ratio of 1PD:1NPD and of a marker to its centromere by an excess of ditype segregations with a *CEN*-linked marker.

There is another common use of tetrads, which in effect is the inverse of the first of these questions. Situations arise in which it is asked whether a particular phenotype is determined by one or more than one locus. For example, one of the most important mutant phenotypes sought by research workers is the temperature-sensitive (*ts*) lethal. Temperature-sensitive mutations have defined a few hundreds of genes in the most significant cellular processes, namely those controlling the cell cycle, macromolecular synthesis and processing, secretion and transport. Many mutagenesis protocols generate multiple mutations (Lindegren *et al.*, 1965) and it is clearly important to know whether the phenotype of a mutant is due to one or to more than one mutation. To do this, the mutant is crossed with wild type and tetrads are dissected. A 2:2 segregation defines the singularity. This procedure is also used when characterizing reverse mutation. Again a reversion from mutant to wild type may be due either to a mutation at the original mutant locus or to one in a suppressor gene elsewhere. The events are distinguished normally by back-crossing the revertant to a true wild type and analysing for segregation of the mutant phenotype. The problem is, how can one be sure it really is a single mutation (or a reversion at the same site)? It could equally be that two mutations have occurred but are quite closely linked.

Similarly, when a novel gene is cloned, the identity of the cloned DNA with the gene being sought is demonstrated by linkage of the integrated clone with the original locus. The cloned sequence is transformed into a strain, which is mutant at that locus. The integration is targeted so that it occurs at a homologous site (Orr-Weaver *et al.*, 1981; Rothstein, 1991). The transformants obtained are crossed with a wild-type strain and tetrads dissected. Linkage is defined again by a uniform PD segregation, in this case 4WT:0mutant. If the mutant phenotype segregates, the cloned sequence must be derived from elsewhere in the genome. There is an alternative procedure, which achieves the same ends. In this, a disrupted non-functional version of the clone is integrated in transforming a wild-type strain. This is then crossed with a mutant strain, and linkage of the disrupted allele and the mutant allele assessed by failure of the wild-type allele to segregate. Although this method involves a few more DNA manipulations, it has the added advantage that the identity of the cloned DNA and the targeted gene can be established from the mutant phenotype of the diploid, without resorting to tetrad analysis.

In every case, singularity or linkage is indicated by the absence of NPD or TT tetrads. A single one of either kind means that there are two determinants segregating. What one needs is a means of estimating the maximum distance apart two markers might be, given that there are indeed no NPD or TT types but only PDs. Thus, if one isolates from an integrative transformation eight tetrads with no mutant segregating, what can one say about the homology of the cloned sequence with the wild-type gene? What is the greatest map distance one is likely to be from the other?

From the Poisson formula, the probability of getting a PD (no crossover) tetrad in an interval with a mean-exchange frequency of x, $p_{(PD)}$ is e^{-x}. The probability of getting N such tetrads is therefore:

$$p_{(N)} = (e^{-x})^N$$

This can be written:

$$-\ln p_{(N)} = x \cdot N$$

There are two approaches to deciding the number of PD tetrads that will suggest that the two markers map together. One may set a value for $p_{(N)}$ that is acceptable. For example, one may decide that a 5% or smaller probability that the two are not the same is acceptable, or one may want to be more stringent and ask for 1%. When the chosen value is substituted for $p_{(N)}$, it is easy to work out the number of tetrads needed to put the genes within Y cM of one another (Y cM $= 50x$). Alternatively, one may already have 10 or 20 such tetrads, in which case one can find the maximum distance they might be apart at any given level of probability. Table IV shows the relationship between cM and N for different choices of $p_{(N)}$. Although the numbers of tetrads required to put sequences remotely in the same neighbourhood might seem depressing for many who have used this method of establishing "identity", some perspective may be gained from the consideration that 40 cM represents about 1% of the yeast genome and 3.4 cM is 10 kb (Mortimer et al., 1989). In Schiz. pombe, 10 kb is 1.6 cM and 1% of the genome is about 23 cM (Munz et al., 1989).

III. ALTERNATIVE LIFESTYLES

In addition to depicting the standard alternation of haploid and diploid generations that occurs in Sacch. cerevisiae, Fig. 1 also illustrates other

Table IV. Numbers of PD tetrads needed to achieve the probability shown that two markers are within x cM of each other

	$p_{(N)}$		
cM	0.1	0.05	0.01
1.0	115	150	230
2.0	57.5	75	115
3.4	34	44	68
10.0	11.5	15	23
(for $N = 10$,	cM = 11.5	15	23)

behavioural patterns found in cultures of budding yeast. These are shown as the homothallic cycle, the akaryogamy cycle and the polyploid cycle.

Schiz. pombe is, perhaps, the mirror image of this. In this yeast species, homothallism is the default condition with alternation of diploid with haploid generations, as it is observed in budding yeast, a rather unnatural alternative. By the same token perhaps, akaryogamy, which in budding yeast is an aberration brought about by mutation, is the norm in fission yeast; or at least nuclear fusion is pre-empted by meiosis. In any case, each alternative offers special facilities for genetic analysis.

A. The homothallic cycle

The best known of these cycles in budding yeast, that promoted by the *HO* gene, has been referred to already. The remarkable observation is that, in cells growing from spores carrying the *HO* gene, the mating type is changed in an ordered pattern so that a *MATα* cell may become *MATa* or a *MATa* cell, *MATα*. The change only occurs in the mother cell of a pair and occurs with about 60% efficiency. The consequence, if the cells are undisturbed, is sib mating (Fig. 7(a) and (b)). Once mated, the diploid cells formed stop switching, since the *HO* gene is repressed in a/α cells. If switched cells are prevented from mating, they may switch back again, as may their daughters after they themselves have budded. The switch occurs by substitution of the active allele at the *MAT* locus by DNA carrying the gene for the opposite mating type. The DNAs for the substitutions are located towards the ends of chromosome III, *MATα* at the left (*HMLα*) in most strains and *MATa* (*HMRa*) at the right. Both of these sequences, called cassettes, are unexpressed. The event is a substitution, not an exchange, since mutations at the expressed *MAT* locus are lost in the process and not recovered in subsequent switches. About 70% of substitutions use the opposite mating-type cassette. It is not only sequence which determines this; position may be involved, since *HMLα MATα-5 HMRα* cells switch to *MATα* (Rine *et al.*, 1981).

The whole process is the subject of intense interest for reasons given by Nasmyth *et al.* (1987) and as already outlined. It has been used as a model system for: investigating mechanisms of transcriptional regulation, since repression of the silent cassettes is essential to the function of a cell as a haploid (Klar *et al.*, 1979); investigating the structure and action of transcription factors coded by each *MAT* gene, and the way they control the pathways leading to diploid and haploid cell types; and as a paradigm for understanding the basis of cell differentiation (Herskowitz, 1989).

(a)

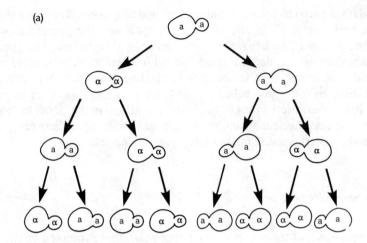

RULE; mother cells switch; daughters do not

(b)

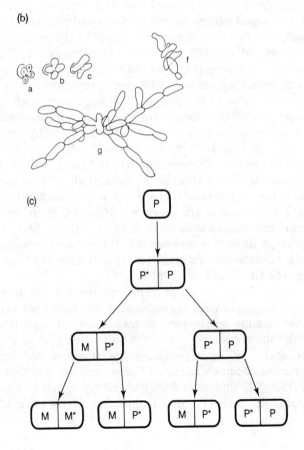

(c)

(d)

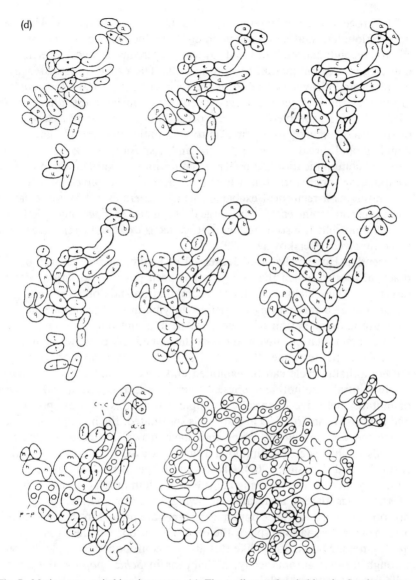

Fig. 7. Mating-type switching in yeasts. (a) The pedigree of switching in *Saccharomyces cerevisiae*. The rule is that mother cells switch, daughters do not. Indeed, Herskowitz (1989) draws this in such a way as to show that daughter cells behave like the stem-cell lines found in the early development of animals. (b) A tracing of one of Winge's (1935) drawings. It shows a single spore germinating to form first, four haploid cells, "a", which then mate to form twin zygotes, "b" and "c", and finally, "f" and "g", a diploid microcolony. (c) The pedigree of switching in *Schizosaccharomyces pombe*. An asterisk indicates an allele that will switch in one of the daughter cells after the next mitosis. P and M are the two mating-types, h^+ and h^-. Siblings then mate as is shown in (d), Leupold's original drawings of sibling matings in h^{90}. As time goes on, first zygotes are formed, then asci and, finally spores are released (Leupold, 1950).

The process is also of interest for the study of recombination. Although observationally, mating-type switching falls into the category of site-specific recombination, it proceeds by using mechanisms involved in normal homologous recombination in mitotic cells. The event is instigated by a double-strand cleavage of DNA at the recipient active *MAT* locus by the product of the *HO* gene, a target-specific endonuclease (Strathern *et al.*, 1982). However, completion of the process requires activity of at least three of the genes required for normal mitotic homologous recombination and double-strand break repair (Malone and Esposito, 1980; J.C. Game, personal communication). It differs from normal recombination in that re-combination, once started, nearly always resolves as a conversion and only very rarely as a reciprocal exchange, which generates either a circular or a linear chromosome each with a large deletion and each leading to lethality in a haploid cell. It also differs in that choice is exercised over the cassette to be involved (Herskowitz, 1989).

Recently, Herskowitz and Jensen (1991) summarized several of the practical uses to which the *HO* gene has been put in association with its target site. These include studies of recombination and double-strand break repair, use as a rare-cutting restriction enzyme in studies of chromosome structure and generation of isogenic **a**, α, **a**/α and polyploid cell lines.

These manipulations involved use of the cloned *HO* gene, which may for convenience be put under the control of the *GAL10* promoter. However, the homothallic cycle can be exploited and genetic analysis easily carried out in naturally homothallic lines. An example is in isolation and analysis of mutants defective in specifically diploid functions such as sporulation and meiosis. If an *HO/HO* **a**/α strain is sporulated, the haploid spores can be treated with mutagens. When they have grown into colonies, most of the cells will have become diploid and can be assayed for ability to sporulate (Esposito and Esposito, 1969). Crosses can be performed with homothallic strains by mixing a sporulating culture of the homothallic strain with cells of the other parent and, after allowing time for mating, performing prototroph selection. Since both mating types are present in the homothallic asci, neither the mating type of the other parent nor whether it is HO^+ or ho^- matters. Mass matings like this are not confounded by diploids going through meiosis automatically, as happens in *Schiz. pombe*; the diploid cells formed remain diploid.

The fission-yeast equivalent of the *HO* gene has not yet been isolated. No doubt it exists, because switching in *Schiz. pombe* is also initiated by a double-strand break at the *MAT* locus and, presumably, this is caused by an endonuclease. If it exists, it is not regulated in the same way as is the *HO* gene of *Sacch. cerevisiae*, for the breaks are found in diploids heterozygous at *MAT* as well as in haploids. These breaks are not limited to mother cells, nor to the G1 phase of the cell cycle, nor to homothallic strains, since the basis of the heterothallism is different in the two yeasts (Beach, 1983; Beach and Klar, 1984). Indeed, the pedigree of switching in *Schiz. pombe* is different (Fig. 7(d)) and seems to be determined not by intracellular distribution of a regulatory factor, but by DNA imprinting (Klar, 1989).

In *Schiz. pombe* heterothallism constitutes the alternative lifestyle and its basis is mutation, or more commonly re-arrangement at the *MAT* locus that either prevents or decreases frequency of mating-type switching. A number of variants of the *MAT* locus have been described by Beach and Klar (1984). In Fig. 8, arrangement of information at the normal, h^{90}, locus and two commonly used heterothallic variants is illustrated. Switching, in the sense of cassette substitution, occurs in both variants but does not normally change mating type. In the h^- form, a deletion has removed the P information, so that it can neither switch to P nor mutate back to homothallic. It is stable and, for that reason, has been referred to in the literature as h^{-s}. The h^+ version has a duplication, which has the effect of making not the left-most cassette but the next proximal cassette (*mat3:1*) the target for switching. The active left-most cassette is a P (plus) and is more or less

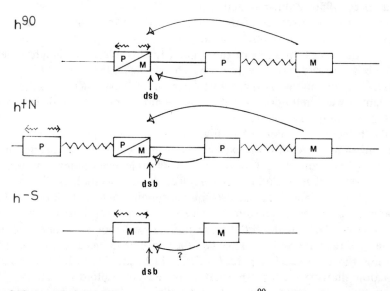

Fig. 8. Structure of the mating-type region in wild-type (h^{90}) *Schizosaccharomyces pombe* and two heterothallic mutants found by Leupold (1950), namely h^{+N} and h^{-s} (after Beach and Klar, 1984). The left-most cassette is the expressed one, indicated by transcription symbols above the box (Kelly *et al.*, 1988). Boxes labelled P/M switch from P to M and back as a result of transpositions indicated by arrows. Arrows labelled dsb indicate where double-strand breaks that initiate these transpositions are found (Klar and Miglio, 1986).

immune from switching. Nevertheless, switching does sometimes occur and there are occasional M (minus) sectors in h^+ colonies. These are not stable like h^{-s}, but behave like h^{+N} (the name given to this allele of *MAT*).

B. Cytoduction

Another important cycle for genetic analysis and manipulation (and so for exploring many fundamental aspects of cell function) is the akaryogamy cycle, promoting cytoduction. It is a phenomenon limited so far to *Sacch. cerevisiae* and is abnormal in that it is brought about as a result of mutation. Three loci, *kar1–kar3* have been identified (Conde and Fink, 1976; Polaina and Conde, 1982) in which, following normal cell fusion between *MATa* and *MATα* cells, nuclear fusion fails in up to 80% of subsequent cell divisions. The result is segregation of haploid cells of either mating type from the zygote and of heterokaryons (Benitez *et al.*, 1984), which in turn segregate haploid cells. Cytoduction also occurs spontaneously at a low frequency, about one mating in 10^3 (Wright and Lederberg, 1956; Zakharov and Yarovoy, 1977).

There are two features of this process that are of value to geneticists. First, while it is true that nuclei of the parent cells segregate from each other in subsequent cell divisions of the zygote, the cytoplasm of daughter cells is a mix of the two parent cytoplasms, a heteroplasmon (Jinks, 1963, 1964). Use of the *kar1* mutation to generate heteroplasmic cytoductants is the most straightforward and unequivocal way of demonstrating that a determinant has an extranuclear location.

The principle is illustrated in Fig. 9 and some examples of cytoplasmic transmission of well-known extrachromosomal determinants are given in Table V. Of course, not only are cytoplasmic determinants mixed in cytoduction but so are other macromolecules and metabolites. One use of this has been to induce meiosis in sporulation-deficient diploids. For example, in some studies it is necessary to make diploids with the constitution *MATα/MATα0*. These diploids will not sporulate but will mate as if they are *MATα*. They can therefore be induced to sporulate by mating them with *MATa kar1* cells, which provide wild-type *MATa* function allowing them to sporulate, but prevent triploid formation. The same can be achieved, obviously, with diploids homozygous for mating type. Klar (1980), who first described these events, showed that the haploid (*kar1*) nucleus also engaged in a meiosis-like sequence, forming two spores. The procedure included recombination in the haploid nucleus when it was disomic for chromosome III (*MATa/MATa, kar1*). However, chromo-

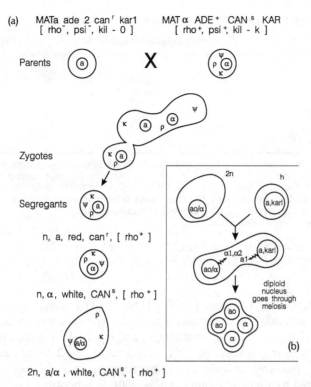

Fig. 9. Illustration of cytoduction. (a) Cytoduction of cytoplasmic particles, namely mitochondria (ρ), killer (k) and psi (Ψ). Three types of cell segregate from the heterokaryotic zygote, a diploid and the two haploid types. All three may carry the cytoplasmic determinants contributed by the α parent. Use of appropriate recessive markers (colour or canavanine resistance) allows the cytoductant haploid (*MATa, kar1, can*[R], [rho+]) to be selected or screened from the resulting population of mixed white, canavanine-sensitive and red, canavanine-resistant, petite cells. (b) Induction of sporulation in a non-sporulating diploid, mutant at the *MATa* locus. Klar (1980) found that zygotes may form six-spored asci, two of the spores containing the *kar1* nucleus (see the text). However, four-spored asci may also be formed from diploid cells segregating from the zygote.

somes in the haploid nucleus did not engage in recombination with those in the diploid nucleus. These observations are the equivalent of the twin meioses described by Gutz (1967) in *Schiz. pombe*. Another use is for studying effects of induction in one cell on the nucleus of another. For example, mating ultraviolet-irradiated cells would allow one to study ultraviolet-inducible responses in nuclei of the opposite mating type, which had been treated differently when either parent is *kar1*.

It is worth commenting on the observation that two determinants, mitDNA (*rho*) and the 2 μm circle, show less than 100% transmission in

Table V. Frequency of transmission of various determinants during cytoduction

Determinant	Percentage transmission	Copy number
rho	50^a	10–50
psi	100^a	?
kil-k	100^a	10^4
2 µm circle	25^b	70
ADE1 (chr I)	10^{-1c}	1
HIS7 (chr II)	6×10^{-3c}	1
MAT, HIS4 (chr. III)	10^{-4d}	1

In each example, cytoductants were selected using an independent cytoplasmic marker. For example, to measure cytoduction of [rho], [psi] and [kil-k], the recipient haploid was MATα, kar1, ade2.1, SUQ5, can^r, [rho−], [psi−], [kil-0]. The donor was MATa, ade2.1, CAN1, [rho+], [psi+], [kil-k]. Canavanine-resistant cytoductants were selected either as [rho+] or [psi+], and the numbers of cytoductants carrying unselected markers scored.
^{a} B.S. Cox (unpublished observation).
^{b} Livingston (1977).
^{c} Sigurdson et al. (1981).
^{d} Nilsson-Tillgren et al. (1980).

cytoduction studies (Livingston, 1977; Table V). The reasons are different. Deficiency in mitDNA transmission is probably explained by poor cytoplasmic mixing and compartmentalization of mitDNA in organelles with a low copy number (see Stevens, 1981). The reason why 2 µm circles are not efficiently cytoduced in spite of a high copy number (70) was elegantly demonstrated by Kielland-Brandt et al. (1980). It is because the 2 µm circle is also compartmentalized but in the nucleus. What is interesting about these data, of course, is the apparent leakage of 2 µm circles from one nucleus to another; they are not absolutely confined to the input nucleus. The same workers and others went on to ask whether chromosomes could also leak from one nucleus to another and indeed they can (Nillson-Tillgren et al., 1980; Dutcher, 1981; Sigurdson et al., 1981).

C. The polyploid cycle

The third cycle illustrated involves formation of polyploids. A rare event in diploid mitoses is a reciprocal crossing over (q.v.). The effect of this is, on 50% of occasions, to make all loci distal to the cross-over homozygous (Fig. 10). Mitotic crossing over seldom occurs in more than one chromosome at a time. When, in Sacch. cerevisiae, it occurs in chromosome III between the centromere and MAT, the reciprocal products segregate as MATa/MATa and MATα/MATα, mother and daughter cells. These may then mate to form a tetraploid. All diploid cultures contain some tetraploid cells.

Tetraploid strains of Sacch. cerevisiae are functionally diploid; they fail

a. Reciprocal

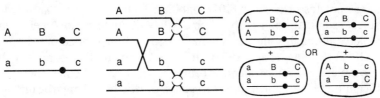

b. Non—reciprocal

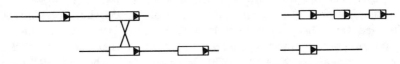

c. Unequal

d. Intrachromosomal

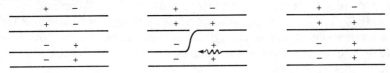

Fig. 10. Various modes of mitotic recombination. (a) and (b) are modes distinguished by Roman (1956) occurring between homologous chromosomes. Non-reciprocal recombination was observed when allelic recombinants were recovered from heteroallelic diploids. The non-reciprocality refers to the 3:1 segregation of one of the non-mutant sites in the interacting pair of alleles. No particular mechanism of recombination is implied by the diagram. The equally diagrammatic structure shown in (a) as a cross-over between homologous chromosomes at the four-strand stage may be resolved as either of the two pairs of progeny cells shown, depending on the orientation of centromeres at mitosis. (c) and (d) represent two kinds of ectopic recombination, which may occur between repeated sequences anywhere in the genome. Those shown are on the same chromosome and are direct repeats. Inverted repeats can generate inversions (intrachromosomal), or dicentric bridges and acentric fragments. Ectopic recombination between repeats on non-homologous chromosomes generates reciprocal translocations, or dicentrics and fragments depending on the orientation of repeats with respect to their centromeres.

to mate but can sporulate. Although such a strain is an autopolyploid, products of meiosis can show 100% viability. To the best of my knowledge, no critical examination of chromosome behaviour in tetraploid meiosis has been made, although some genetic analyses have been carried out (Pomper *et al.*, 1954; Roman *et al.*, 1955). It is not known, therefore, whether chromosomes pair as bivalents or multivalents, or whether there is a significant number of aneuploid segregants, or whether all are true diploids. Yeast can tolerate a high degree of aneuploidy without losing viability (Parry and Cox, 1970). Analysis suggests, nevertheless, that segregation can be very regular with few unequal disjunctions. Thus in Roman *et al.* (1955), segregation of mating type was 81 tetrads with four non-maters (a/α, a/α, a/α, a/α; 16 with 2a:2α (a/a, a/a, α/α, α/α) and 71 with two non-maters: 1a:1α (a/α, a/α, a/a, α/α). Aneuploid segregations, which would yield such segregations as ($a/a/\alpha$, α, a/α, a/α), i.e. three non-maters: one mater or ($\alpha/a/a$, α/α, a/a, α): three maters: one non-mater, were found in only three of these 168 tetrads.

The polyploid cycle adds a degree of flexibility to yeast genetic analysis, which is underexploited. One of the first uses of polyploids was in radiobiology, where survival and mutagenesis of yeast of different ploidies were measured, perhaps with a view to relating the results through target theory to organization of genetic material (Mortimer, 1958; Laskowski, 1960). This led to the significant observation that the lethal lesions induced by ionizing radiation were dominant. Polyploidy was also used by Wickner for gene mapping (Wickner, 1979; Wickner and Liebowitz, 1979). The method is based on segregation of aneuploids ($n + 1$, $n + 2$ ...) in meiosis in triploids. Aneuploidy was detected very easily in these segregants by irregular segregation of recessive markers from trisomes of constitution $+/+/-$. Most tetrads from such a trisomic constitution segregate $4+ : 0-$ or $3+ : 1-$, with only a minority segregating 2:2 (Cox and Bevan, 1962). A handful of tetrads is all that is necessary to find the irregularity.

There are other uses to which polyploidy can be put. It is ideal for manipulating gene dosage in a controlled fashion. The standard life cycle allows three classes of genotype: $+/+$, $+/-$ and $-/-$, while tetraploids allow intermediate ratios of alleles to be assayed for their phenotypic effects. The polyploid cycle can also be used for rescuing diploids that fail to sporulate so that they can be genetically analysed. For example, a *petite* diploid will not sporulate; a diploid homozygous for *rad50/rad50* sporulates but the spores are inviable. Such strains may be given a genetic future by mating them with diploids homozygous for mating type when the tetraploid formed will sporulate.

It sometimes happens that stock cultures of haploid strains diploidize, especially when stored in cryoprotective solutions like glycerol. Mating them with standard haploids gives non-fertile triploids. The genes they carry can be recovered through the tetraploid cycle (or by cytoduction-induced meiosis, q.v.). By the same token, many strains of yeast in industrial use for brewing, baking or wine-making are diploid, triploid or in some intermediate aneuploid state and not very fertile after meiosis. Strain improvement can be hindered by these factors. At least some of the difficulties could be circumvented by raising them to a higher ploidy and, at the same time, producing hybrids, by mating with diploids homozygous for mating type.

Many phenomena are better analysed in diploids than in haploids, for example, recombination and sporulation, various forms of DNA repair, cell-cycle morphologies, positioning of buds in cell division and processes depending on interaction of mating-type loci. For research in these areas, it is probably more convenient to build stock diploid strains of appropriate genotypes, since diploids and the polyploid cycle offer a controlled approach. It is also the case that diploid yeasts, which are the native form of the organism, are more viable and vigorous, show less variation and suffer less from clumping than haploids, all of which would tend to make genetic experiments easier.

Herskowitz and Jensen (1991) suggested the use of controlled *HO*-induced mating-type switching for generating polyploids. The system proposed has the disadvantage that cultures grown in this way are a mixture of haploid, diploid and polyploid cells. It may be preferable to enter the polyploid cycle in a controlled way as follows. A diploid homozygous for an auxotrophic marker (e.g. *ade1/ade1*) is plated to form colonies after a low dose of ultraviolet radiation (say 50 J m^{-2}) to induce mitotic crossing over. Colonies are then replica-plated on to a lawn of haploids of either mating type, which have a complementing auxotrophy (e.g. *ade2*). Diploids which have become homozygous for *MAT* form complementing triploids, which can be identified either by colour or by replica-plating to omission medium. If lawns of both **a** and α cells are used, it is often possible to identify the reciprocal, *MATa/MATa* and *MATα/MATα* products of a cross-over event as opposite halves of the same colony. These can then be recovered from the master plate.

In *Schiz. pombe*, polyploid cells arise as a consequence of endomitosis, which gives a diploid cell, followed by mating. In h^{90}, this happens spontaneously following mating-type switching and, in heterothallic strains, diploid cells will mate just like haploids of the same mating type. Meiosis in tetraploids may occur either in a tetraploid nucleus or, if conditions are chosen that prevent karyogamy, twin meioses will yield asci with eight haploid spores (Gutz, 1967).

D. Aneuploidy

Meiosis in triploids is expected to generate many aneuploid segregants and in both yeasts is highly infertile (Parry and Cox, 1970; Niwa and Yanagida, 1985). In the latter study of *Schiz. pombe*, although the majority of asci formed (96%) contained four spores, only 23% of the spores germinated to form colonies. About 12% of the asci contained four viable spores and nearly all of these segregated two diploid and two haploid colonies. All the triploids used in these crosses were heterozygous for complementing alleles of the *ade6* locus, namely *ade6-210/ade6-216/ADE6⁺*. Most of the remaining survivors of these triploid meioses were either diploid, haploid or unstable for the *ade6* markers, segregating red sectors. Niwa and Yanagida (1985) showed, by tetrad analysis and cytologically, that these segregants were aneuploid *n* + 1 for chromosome III. They failed to find evidence of aneuploidy for the other two chromsomes. Presumably the genetic imbalance caused by aneuploidy of chromosomes I and II affects viability. Niwa *et al.* (1986, 1989) constructed deletion derivatives of chromosome III, which could be sustained as aneuploid variants, and used them in studies of chromosome segregation in meiosis and mitosis.

By contrast, budding yeast can tolerate a great deal of aneuploidy. All, 16 possible $2n - 1$ strains are viable; this is the basis of the chromosome-loss method of assigning mutations to chromosomes (Falco and Botstein, 1983, q.v.). When a triploid is sporulated and the spores dissected, viability is found to be about 25% (Parry and Cox, 1970). The probability of obtaining an euploid spore from a triploid meiosis, assuming regular formation of trivalents and random segregation of chromosomes, is about $3 \cdot 10^{-5}$. If viability depended on euploidy, triploids would be essentially sterile. In fact, in this study in which markers for 14 of the 16 chromosomes were involved, aneuploid ($n + 1$ or more) segregants were found for all except chromosome X (II and VI not being represented). The majority of the segregants were multiply aneuploid, with up to five disomic chromosomes. Naturally occurring aneuploids ($n + 1$) are commonly found in yeast cultures and have been exploited for gene mapping (Mortimer and Hawthorne, 1973).

From what I have already written in this article, it is clear that making aneuploids at will in *Sacch. cerevisiae* is now quite feasible. Falco and Botstein (1983) have made it possible to manufacture a $2n - 1$ set and any $n + 1$ or $2n + 1$ type could be created by the cytoduction method (Nillson-Tillgren *et al.*, 1980). Apart from allocating mutations to chromosomes (Mortimer and Hawthorne, 1973; Wickner, 1979; Falco and Botstein, 1983), aneuploids can be useful in other ways for genetic analysis.

Strains that are aneuploid $n + 1$ have been used to isolate mutants defective in recombination (Rodarte *et al.*, 1968). An $n + 1$ aneuploid in which the duplicated chromosome carried two different alleles of *arg4* was the treated strain. Recombination in colonies of this strain could be detected by replica-plating them to arginine-less omission medium, where normally every colony is found to generate many ARG^+ papillae. Clones that fail to do this are easily identified. The same strain can be used to quantify allelic recombination after various treatments, ultraviolet irradiation for example, and even to score for meiotic recombination when it is made doubly aneuploid with heterozygosity at the *MAT* locus. The *CUP1* gene on the same chromosome (VIII) can be used to detect and measure the occurrence of aneuploidy, either spontaneous or induced. The resistance of yeast to copper ions is very sensitive to the copy number of the *CUP1* gene, and the difference between one and two copies in n and $n + 1_{(VIII)}$ strains can be detected on appropriate media (Fogel *et al.*, 1983). This assay uniquely detects chromosome gain. Much more commonly, aneuploids are used to assay stability of chromosomes by systems that detect chromosome loss. Either basically diploid ($2n$ or $2n + 1$) or basically haploid ($n + 1$) strains may be used in conjunction with recessive selectable or morphological markers such as *can1*, *ade1* or *ade2* (Shero *et al.*, 1991). Such studies have led to identification of chemicals or mutant genes that promote chromosome instability. Methyl benzylimidazole carbamate (MBC) is one such compound (Wood, 1987). Mutations of *rad52* (Malone and Esposito, 1980), *rad51* and *rad54* (J.C. Game, personal communication), *ch11* (Liras *et al.*, 1978) and the *ts* mutants *cdc6* and *cdc14* (Kawasaki, 1979) also promote chromosome instability in diploids homozygous for them. Both MBC and the *rad52* and *cdc6* mutations are the basis of methods devised to locate genes to chromosomes that have been used in various laboratories, before development of the *FLP*-induced loss method (Mortimer and Contopoulou, 1987). In recent times, genetic control of chromosome stability has been more conveniently studied using artificial chromosomes, usually one carrying a colour marker (Shero *et al.*, 1991; Newlon, 1988, 1989). They are generally already fairly unstable compared with natural chromosomes, and mutations increasing or decreasing stability can be screened using them.

Aneuploidy of short chromosomes is quite commonly observed in laboratory strains. Aneuploidy of chromosome III in particular is quite conspicuous because it leads to formation of non-mater a/α haploids in tetrads. It has been used to maintain the circular form of chromosome III, which is formed by rare reciprocal exchanges between *MAT* and *HMLα* during switching of *MATa* to *MATα*. If this circle were the only representative of chromosome III in a cell, it would not survive because of the loss of parts distal to *MAT*. Circular chromosome III has been useful

48   B.S. Cox

in a number of studies. It has, for example, been used to monitor and correlate physical events such as strand breaks with recombination during meiosis (Game *et al.*, 1990b), since it runs in a unique position on pulse-field gels. It has also made it possible to purify this large segment of DNA from the rest of the genome, a useful feature for sequencing studies and for studies of structure and organization (Strathern *et al.*, 1979; Nillson-Tillgren *et al.*, 1981).

Aneuploids that are $2n - 1$ may be used to expose a particular chromosome to mutation. In a recent paper, for example, Harris and Pringle (1991) used such a diploid lacking one copy of chromosome I to attempt to saturate the essential loci on it with *ts* mutations. Obviously, using a haploid would have meant having to screen out the hundreds of *ts* mutations that occur on other chromosomes before they could address the problem.

E. Mitotic recombination

Recombination occurs at a low rate during vegetative growth of yeast and many other organisms. Its frequency may be increased by chemical or physical mutagens. It may be exploited for mapping genes. Indeed, in some organisms such as *Aspergillus nidulans*, it is the method of choice or even, in imperfect fungi, of necessity (Pontecorvo *et al.*, 1953; Pontecorvo, 1956).

It has had two uses in yeast. First of all, in gene mapping, it has been the means of connecting linkage groups in situations where conventional tetrad analysis has proved inadequate because of high recombination rates in certain intervals in meiosis (Mortimer *et al.*, 1989). Mitotic crossing over is rare, seldom affects more than one chromosome at a time and then seldom more than once in that chromosome. This means that when a centromere-proximal marker becomes homozygous as a result of mitotic crossing over, the large segment of the chromosome distal to it will also be homozygous. Any recessive marker linked to it will be exposed and detectable (Fig. 10(a)). This method of connecting linkage groups has been extensively exploited by Mortimer and his colleagues. In the latest map of the genome of *Sacch. cerevisiae*, there are still three chromosome arms whose integrity is determined by mitotic analysis. Of historical interest is the fact that it can be used for both chromosome mapping and fine-structure mapping. For mapping linked genes, co-incident homozygosis is scored for all of the markers involved. Distal genes will show 100% coincident homozygosis with proximal genes. The order and relative distance apart of proximal genes can be determined by the frequency with which they are coincidentally homozygous with distal markers. Mitotic recombination has been equally extensively used in mapping in *Schiz. pombe* where a combination of fewer chromosomes and longer linkage groups has made it a particularly useful method (Kohli *et al.*, 1977; Gygax and Thuriaux, 1984).

In 1958, Roman and Jacob found that, when prototrophs formed by mitotic allelic recombination between *ade2* mutations of *Sacch. cerevisiae* were analysed, the events were non-reciprocal; the doubly mutant reciprocal recombinant allele was never found (Fig. 10(b)). They also reported that outside markers failed to recombine in association with these events. Since then, Fabre (1978) has shown that much of the allelic recombination in mitosis occurs at the G1 stage in this yeast, and outside marker recombination at this stage would be difficult to demonstrate. Although Esposito (1978) and Esposito and Wagstaff (1981) have since demonstrated that the events are related, as they are in meiosis, it is worth retaining the verbal distinction between allelic recombination (frequently called gene conversion) and intergenic recombination (sometimes called crossing over), since the mechanisms by which they occur are not yet understood and there are respects in which the two phenomena behave differently. Not all of the differences can be put down to artefacts arising from the very different ways in which allelic recombination and crossing over are detected.

A great deal of fine-structure mapping has been carried out by mitotic recombination, which is just as effective as meiosis in generating recombinants in sufficient numbers to score, particularly when mutagenic treatments are used. Manney and Mortimer (1964) showed that, in *Sacch. cerevisiae*, the relationship between allelic recombination frequency and X-ray dose was linear and that different allele combinations gave truly additive maps. The findings suggest that X-ray-induced recombination frequency was directly related to physical distance apart, unlike meiotic allelic recombination, which shows map expansion and negative interference. Similarly, mitotic recombination, either spontaneous or induced by methyl methane sulphonate, has been used for fine-structure mapping of genes in *Schiz. pombe* (Hofer *et al.*, 1979). As with *Sacch. cerevisiae*, both marker effects and map expansion disappear when mitotic recombination is induced by mutagens.

The most important use of mitotic recombination, however, has been as a means of analysing the process of recombination itself and its role in the life of the cell. All of the systems described in Fig. 10 have been exploited. There is no question that recombination is the principal method of repair of DNA damaged by X-rays and other mutagens that cause strand breaks (Resnick, 1976; Brunborg *et al.*, 1980). It may also be an important pathway of repair of ultraviolet-induced lesions, although mutants defective in recombination are not very sensitive to ultraviolet irradiation unless other repair pathways are blocked. Nevertheless, both reciprocal and allelic recombination are highly inducible by ultraviolet irradiation (Parry and Cox, 1968) and both are stimulated much more if conventional ultraviolet-repair pathways such as excision repair are blocked (Hunnable and Cox, 1971). Allelic recombination after ultraviolet irradiation is further stimulated by dark holding, a treatment that improves survival but decreases intergenic recombination.

Mitotic recombination is again becoming an interesting metabolic process, since a great deal of modern genetic manipulation in yeasts and other organisms depends on it. *Sacch. cerevisiae* seems to be unique in that, when transformed with plasmids or DNA fragments that have to be integrated to succeed, the integration always takes place by homologous recombination and therefore occurs at a site in the genome homologous to some of the DNA on the plasmid. Integration can be targeted (Orr-Weaver and Szostack, 1985). This gives *Sacch. cerevisiae* an enviable advantage compared with other organisms. There are two aspects of this that prompt interest. One is that it makes *Sacch. cerevisiae* a unique test bed for studies of mitotic recombination metabolism, considerably aided by the ability to make sophisticated constructs in which DNA transactions can be monitored. The other aspect is to know what it is that is missing in *Sacch. cerevisiae* that makes non-homologous recombination so common in other organisms. Since *Schiz. pombe* evidently has what is missing and most plasmids used for genetic manipulations in each organism work in the other, the opportunity clearly exists for an informative collaboration between the two species.

F. Ectopic recombination

More recently, genetic manipulation has been used to construct arrangements of genes in which more penetrating questions about recombination in either mitosis or meiosis, or both, can be asked. Very many of these systems depend on ectopic recombination, namely that occurring between repeated sequences in non-homologous positions in the genome (Fig. 10(c) and (d)). Ectopic recombination occurs between native repeats, for example, within tandemly repeated clusters such as the rDNA or *CUP1* loci of *Sacch. cerevisiae* (Petes, 1980; Jackson and Fink, 1981; Szostak and Wu, 1980; Fogel *et al.*, 1984), between Ty elements (Chaleff and Fink, 1980; Liebman *et al.*, 1981; Roeder and Fink, 1983) and between the subtelomeric Y' repeats of telomeres (Louis and Haber, 1990), as well as between repeats introduced artificially. In *Schiz. pombe*, ectopic recombination has been observed between dispersed members of a family of serine-accepting tRNAs (Munz and Leupold, 1981).

Governed by the orientation of recombining elements, ectopic recombination may generate viable chromosome re-arrangements such as inversions, deletions, duplications and translocations as well as lethal dicentric and acentric fragments (Fogel *et al.*, 1984; Szankasi *et al.*, 1986; Louis and Haber, 1990). Ectopic recombination assays exploit these consequences, usually by including a marker whose loss can be monitored between two direct repeats (Petes and Hill, 1988).

IV. MAPPING STRATEGIES

A. *Saccharomyces cerevisiae*

The genetic map of *Sacch. cerevisiae* most recently published by Mortimer *et al.* (1989) describes the location of 769 genetic loci, including 16 centromeres on 16 linkage groups. Summing all the intervals suggests that the total genetic length of the genome of *Sacch. cerevisiae* is about 4300 cM (about the same as the human genetic length) and this represents a ratio of $0.34\,\text{cM}\,\text{kb}^{-1}$. Finding the location of a gene occupying less than 5 kb of DNA is a "needle in a haystack" proposition, but some helpful short cuts have been described. There are two kinds of strategy, depending on whether one has a mutant phenotype or a cloned fragment of DNA to map.

1. Mapping mutations

A number of systems have been described for locating a mutation to a chromosome quickly. Only those that are simple to use will be described.

The most straightforward system, constructed at the Yeast Genetics Stock Center (YGSC), Berkeley, principally by Rebecca Contopoulou, consists of a set of 18 strains, duplicates in each mating type of nine multiply marked strains, which between them carry their markers spaced evenly throughout the yeast genome, all more or less within 50 cM of one another (Mortimer and Contopoulou, 1987) (Fig. 11). Linkage of a mutation that lies within 34 cM (100 kb) of one of these markers should be established with as few as 12 tetrads from the relevant cross. These would be enough to give a significant departure from the 1PD:1NPD ratio expected of non-linkage about nine times out of ten, particularly in regions of high interference ($k = 0.3$), which is most of the genome. Greater

52   B.S. Cox

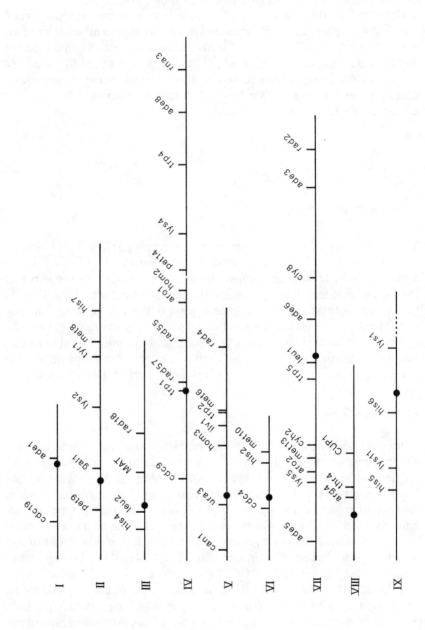

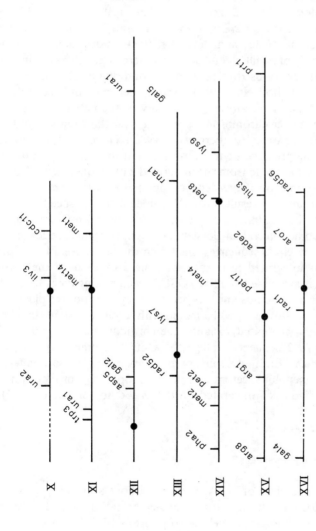

Fig. 11. Markers used to construct mapping strains at the Yeast Genetics Stock Center and by Ben-Ichiru Ono (Mortimer and Contopoulou, 1987). The data used to construct this diagram were taken from Mortimer *et al.* (1989). See Appendix II for details of gene symbols.

certainty can be achieved by further tetrad analysis of the appropriate diploid as can a more precise map position. A different set of strains, all *MATα* and with the markers spread approximately 30 cM apart, has been constructed by Dr Ben-Ichiro Ono and is also available from the YGSC.

A second system depends on chromosome loss. Loss of a chromosome from a diploid strain is not usually lethal, so that $2n - 1$ aneuploids survive and grow well. This fact can be exploited for locating mutations to particular chromosomes. There are two general methods of inducing chromosome loss, namely, genetic or chemical (Liras *et al.*, 1978; Kawasaki, 1979; Mortimer *et al.*, 1981; Falco and Botstein, 1983; Wood, 1987). Of these, one (Falco and Botstein, 1983) is outstandingly simple to use and is the only one that will be described here. The others are ingenious but much more laborious (Mortimer and Contopoulou, 1987).

The Falco and Botstein (1983) system depends on the fact that chromosomes which have an integrated copy of the native yeast plasmid, the 2 μm circle, suffer chromosome breakage near the 2 μm site and often chromosome loss as well (Homberg *et al.*, 1982). These events can be made to depend on the presence of the 2 μm plasmid in cells at the same time by excluding the *FLP* gene from the integrated 2 μm sequences (Falco *et al.*, 1982). Chromosome breakage depends upon the FLP function of 2 μm circles, and presumably results from FLP-mediated site-specific recombination between the *FLP* target sequences (*FRT*) on a 2 μm circle and that in the chromosome. Falco and Botstein (1983) have made a series of [cir-0] strains with 2 μm circle sequences, including the target for FLP-mediated recombination, integrated near the centromere of each of the chromosomes. The set is available from the YGSC in each mating type, 32 strains in all. They suffer no chromosome loss because they have no free plasmid and no FLP function. However, when mated with a [cir-+] strain, the diploid segregates $2n - 1$ aneuploid variants, the chromosome lost being the one with the 2 μm *FRT* sequences integrated at its centromere. If the [cir-+] parent was *mutX*, the mutX phenotype is expressed in such aneuploid colonies or as aneuploid sectors of colonies on plating out that particular diploid which had *FRT* integrated on the *mutX* homologue (Fig. 12).

2. Mapping cloned DNA

(a) Pulsed-field gel electrophoresis

Mapping a cloned piece of DNA, with or without a known gene on it, is most easily done physically, using molecular techniques. The simplest methods employ pulsed-field gel electrophoresis (PFGE) systems such as orthogonal field alternation gel electrophoresis (OFAGE; Carle and Olson,

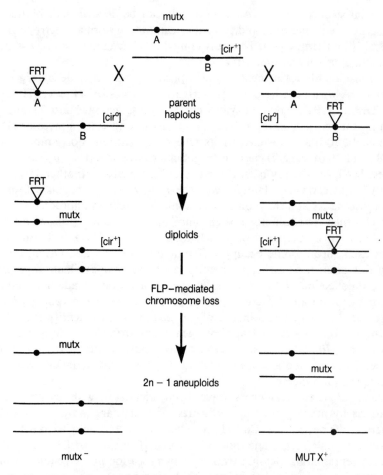

Fig. 12. The *FLP*-mediated chromosome-loss method for mapping mutations to chromosomes. Two chromosomes are shown, A and B. In one strain, a portion of the 2 μm circle, including the *FLP* target sequence *FRT*, has been integrated at the centromere of A, in the other at the centromere of B. Both of these strains are [cir-0] and therefore flp⁻. The consequences of mating a [cir+] strain with a mutation *mutx* on chromosome A to each of these is shown. One strain segregates *mutx*⁻ sectors, the other does not.

1984, 1985), field inversion gels (FIGE; Schwarz and Cantor, 1984), clamped hexagonal electric fields (CHEF; Chu *et al.*, 1986) or "waltzer" gels (Southern *et al.*, 1987). These systems separate the 16 chromosomes of *Sacch. cerevisiae* into between 13 and 16 bands, depending on the strain (Mortimer *et al.*, 1991). The strain dependence is due to extensive chromosomal polymorphisms and translocations amongst laboratory and industrial strains. The DNA in these gels can be depurinated, blotted to nitrocellulose membranes and probed as can that in normal electrophoresis analysis (Southern, 1975). Thus any cloned DNA fragment can be assigned immediately to its chromosome.

A considerable refinement of this approach can be made by digesting yeast DNA with the rare-cutting restriction enzymes, *Not*I and *Sfi*I (Link and Olson, 1991). These enzymes cut the yeast genome into 77 and 52 identifiable fragments, respectively, allowing one to place a piece of DNA within about 100 kb (34 cM) of its true map position. Forty-nine of the *Not*I- and 36 of the *Sfi*I-derived fragments can be identified by the use of genes that have already been cloned as unique probes. Another 15 can be identified as carrying telomere sequences. Only seven fragments, four on chromosome XI and three on chromosome XV, remain unordered relative to each other. The technique of chromosome transplanting allows a single membrane to be prepared on which the *Not*I- or *Sfi*I-derived fragments from each chromosome occupy different tracks (Link and Olson, 1991).

In due course, a further refinement could become available through a random collection of 15 kb clones of the yeast genome made in a lambda vector by Olson and his colleagues as part of a physical mapping project (Olson *et al.*, 1986). The principle of this exercise was to identify each clone by the set of restriction fragments generated from it by digestion with *Eco*RI and *Hin*dIII, together. Sets were computer matched for fragments of identical size, and a series of overlapping fragments (contigs) produced in arrays called merges.

Many of these fragments have been identified with chromosomal locations by probing with cloned genes, and they are available from the American Type Culture Collection. Eventually, it should be possible to produce a single-membrane filter with about 1000 spots of DNA bound to it, each representing a contig from a known location in the genome, which can then be probed with a cloned DNA fragment.

(b) Targeted chromosome breakage and random chromosome breakage

Meanwhile, two methods have been described that can locate a cloned DNA to within about 10 kb of its location on a chromosome, the accuracy depending on the resolving power of a pulsed-field gel. The method developed by Vollrath *et al.* (1988) causes chromosome breakage at a point to which the cloned DNA is targeted by homologous recombination. Measurement of the sizes of the fragments produced locates the target gene. Fragmentation is produced by cloning the DNA to be mapped into the plasmids as shown in Fig. 13, left. The consequences of integration of the gene into its homologous sequence on the chromosome are shown in

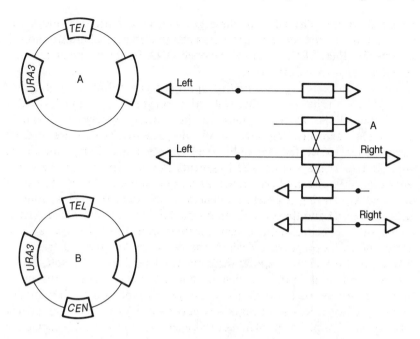

Fig. 13. The left diagram is illustrating gene mapping by targeted integration. Two plasmids are used to carry the cloned DNA (open box). Each has a selectable yeast marker (*URA3*) and a Y′ sequence from a yeast telomere: one (B) also has a centromere. A recipient strain is transformed with each of these constructs. Right, "A" integrates to form a shortened version of the chromosome carrying the homologous sequence of DNA (Left) and B with its centromere can form a short complete chromosome with the distal arm of the same chromosome (Right). The size of these short chromosomes can be determined on a pulse-field gel.

Fig. 13(Right). Either the new *CEN*-GENE-*TEL* fragment or the reciprocal CHROMOSOME-GENE-*TEL* fragment can be measured by PFGE to give the distance of the clone from one end of the chromosome.

The random-breakage method developed by Game *et al.* (1990a) depends on the fact that, if a chromosome is broken exactly once but at random sites, the population of molecules produced will show a discontinuous size distribution if it is probed with a sequence from any one particular site. This is explained in Fig. 14. This shows that, although the total population has a random size distribution, the subpopulation of molecules identified by the probe does not. Two subdistributions are produced, one with a minimum size equivalent to the distance of the probe from the left-hand end of the chromosome, the other with a minimum size equivalent to its distance from the right-hand end. Thus, all breaks in LA will produce detectable molecules larger than AR and all breaks in AR will give

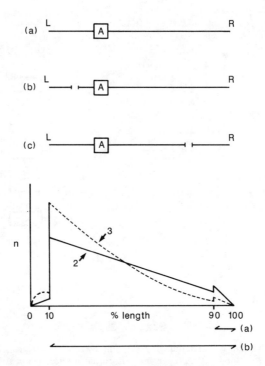

Fig. 14. Random-break mapping. Top: (a) A chromosome with a gene at position A. L and R are the left- and right-hand ends, respectively. After electrophoresis, if probed with A, it would be found to be full size. (b) The chromosome with a break in LA. The probe identifies a smaller fragment than in (a). All breaks in LA produce fragments which are equal to or larger than AR. (c) A break in AR again gives a smaller fragment than (a) but the smallest fragment produced by a break in AR will be of size LA. Bottom: Distribution of numbers of fragments (*n*) of various sizes (abscissa) identified by probe A from all chromosomes with exactly two or exactly three breaks. A is assumed to be 10% of the distance along the chromosome. The discontinuities in the distributions are at the same sizes. (a) represents all of those fragments with breaks in LA and (b) includes all those with breaks in AR. When there is more than one break in each molecule, some breaks span A, so some molecules smaller than LA are produced. Adapted from Game *et al.* (1990a).

detectable molecules larger than LA. The effect of this is to produce two discontinuities in the size distribution of probed molecules, one at the size represented by LA and the other at AR, with a smear of larger probed molecules above these points and nothing below. Achieving exactly one break in each chromosome is not possible but it can be shown, indeed it is intuitively obvious, that discontinuities would occur at the same positions if the chromosomes were broken randomly in exactly two, exactly three or exactly *n* places (Fig. 14). Therefore, a treatment such as X-irradiation, which can cause a few random double-strand breaks, will generate

2. Genetic Analysis in Sacch. cerevisiae and Schiz. pombe 59

discontinuities. An illustration of the discontinuities observed with a *HIS3* fragment probe (chromosome XV) probing extracts of yeast DNA treated with various doses of X-rays and analysed by PFGE is shown in Fig. 15.

B. Schizosaccharomyces pombe

1. Mapping mutations

Mapping a new mutation in *Schiz. pombe* is probably simpler than it is in budding yeast. First, with only three chromosomes, it involves less effort to identify the linkage group to which it belongs. This may be done meiotically, by mitotic recombination or by haploidization. Chromosome loss occurs spontaneously in diploid vegetative cultures, and loss of one chromosome is followed rapidly by loss of the remaining two disomics, to

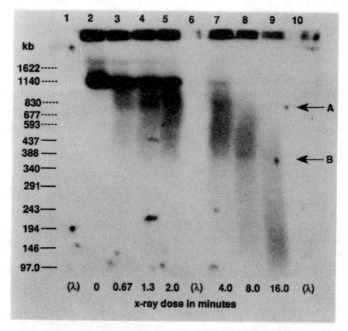

Fig. 15. Autoradiograph showing the discontinuities that develop when yeast DNA irradiated with various doses of X-rays is analysed on a pulse-field gel and probed with *HIS3*. *HIS3* is 34% of the distance from one end of chromosome XV. Discontinuities develop at 770 and 380 kb, representing 67% and 33% of the 1150 kb chromosome. Reproduced from Game *et al.* (1990b).

yield a haploid cell. Chromosomes are lost at random so, from a diploid that is heterozygous for a marker on each chromosome, A/a, B/b, C/c, any of the eight possible haploid segregants can arise. An unknown mutation will always cosegregate with the marker on its chromosome. The method has been in use for years with species of *Aspergillus* (Pontecorvo, 1956). As it is, in that fungus, the frequency of haploidization is enhanced by treatment with p- or m-fluorophenylalanine (Gutz, 1966; Flores da Cunha, 1970). Diploids for this kind of analysis must be stable, i.e. incapable of going through meiosis. It is essential to start with heterothallic parents, but then either of the strategies for obtaining stable diploids may be used, namely screening for homozygosis at the *mat* locus or exploiting the *mat2-B102* allele.

The position of a mutation relative to other markers can also be established using mitotic crossing over. The *mat* locus is a hotspot for mitotic crossing over (Kohli *et al.*, 1977), because of the double-strand breaks that are constantly induced at that site as part of the mating-type switching regimen. Such cross-over products are homozygous for mating type and, if the locus is not h^{90}, are stable diploids identifiable by the iodine test. A coincidence of homozygosity of an unknown mutation and a stable diploid sector or colony is an indication that the gene is distal to *mat* on chromosome IIR. Linkage to other chromosome arms may also be determined by mitotic crossing over. This yields homozygotes at about 0.2%, which is sufficient for establishing coincident events after replica-plating colonies. It is possible to enhance these frequencies by treatment with ultraviolet irradiation, X-rays or other DNA-damaging agents as it is in *Sacch. cerevisiae*. However, it is probably best to avoid such treatments, since multiple events may occur at higher than random expectation (cf. Gygax and Thuriaux, 1984; Lehman and Munz, 1987). Once again, if mitotic crossing over is to be used, it is necessary to select hybrid diploids from heterothallic matings and preferable if one of the parents is *mat2-B102*.

Thirdly, a good estimate of a mutation's linkage relationship can be gained by random-spore analysis, crossing the mutant with a well-marked strain, allowing zygotic asci to form and, after autolysis and alcohol treatment, plating out to score segregant phenotypes. This precludes use of the *mat2-B102* mutation but obviously requires the mating to be a heterothallic one.

Although, in theory, random-spore analysis should be sufficient to establish the position of genes very precisely, in practice it is necessary and difficult to allow for differential viability of allele combinations and the final mapping depends upon tetrads. The choice of crosses to make in order to minimize the labour involved would be made on the basis of the results of one or a combination of the preliminary analyses already described. There seems to be no equivalent in *Schiz. pombe* of the mapping strains developed by Contopoulou or by Ono for *Sacch. cerevisiae.*

Most workers seem to use a combination of these methods. For example, Gygax and Thuriaux (1984) used the following procedure. (i) A strain carrying the marker to be mapped was crossed with one carrying the centromere-linked marker *lys1*. About 30 tetrads were dissected and this provided preliminary evidence about whether the marker was linked to a centromere, to *lys1* or to *mat*. It also allowed an h^-, *lys1*, *mut* strain to be chosen for (ii) crossing to a *mat2-B102 ade6-704* strain. The diploids were treated with *m*-fluorophenylalanine, and haploid colonies identified and screened for mating type, adenine requirement (identifiable by the red pigment accumulated by *ade6* mutants) or lysine requirement and the mut phenotype. The genes *lys1*, *mat* and *ade6* are on chromosomes I, II and III, respectively, and cosegregation of *mut* with any of these locates it. The final map position was determined by tetrad analysis after crosses with appropriately marked strains.

2. Mapping cloned DNA

Recently, a map of the fragments produced by *Not*I digestion of DNA from *Schiz. pombe* was published (Smith and Cantor, 1987; Fan *et al.*, 1988). The map has 14 sites giving rise to 17 fragments, with no sites on chromosome III. Miyake *et al.* (1991) gave examples of how this might be used to map cloned genes (in their case, of the *ras* supergene family). The genes were cloned on a plasmid carrying the *LEU2* gene of *Sacch. cerevisiae* and transformants, which had stable integrations of the plasmid screened. Homologous integrants were identified by Southern blots of DNA extracts that were cut with an enzyme that did not cut within the transforming plasmid, followed by probing with the relevant plasmid DNA. Homologous integration is signalled by a single band of higher molecular weight than is found on untransformed DNA cut with the same enzyme. The original plasmid also contained a single *Not*I site, so that the location of the gene and its position on any *Not*I fragment relative to the ends of the fragment could be found by probing *Not*I digests of transformed and integrant strains. Finally, the precise location of the integrated sequence could be found by tetrad analysis, scoring for linkage of the budding yeast *LEU2* gene. Although it has not been tried, there is no doubt that the random-breakage method developed for *Sacch. cerevisiae* could also be used for mapping cloned DNA sequences of *Schiz. pombe*.

V. CONCLUSIONS

A casual glance at the literature makes it obvious that the majority of research enterprises involving the genetics, cell biology and molecular biology of yeast now make use almost exclusively of the facilities for genetic analysis offered by the ability to clone genes and transform yeast with plasmids which can also be grown in *Escherichia coli*. It would seem that there is little need for classical techniques for gene manipulation that are described in this article. All one needs is a transformable yeast strain with appropriate selectable markers and the rest follows. Indeed many, perhaps most, yeast geneticists nowadays do not dissect tetrads, do not indeed have microscopes, and may never have looked at living yeast cells, because so much can be done with cultures on plates or in flasks.

I believe this judgement is premature. It is worth recalling what it is that has made microbial biochemical genetics such a productive experimental system since Tatum and Beadle (1945) showed what could be done. Before they started work with *Neurospora crassa*, use was made of whatever mutations were available, such as flower colour or eye-colour mutants, for example. The question asked was what was different about the organism. Beadle and Tatum's innovation was to reverse the order in which questions were asked. They now asked: "what kind of mutations would one expect to get if genes controlled this or that biochemical reaction?". This is the question that has been asked ever since of bacteriophages, bacteria, fungi, flies and worms. Yeast cells do very interesting things, interesting that is to people who work on more complex organisms. Quite apart from proceeding through an ordered cell cycle, replicating their DNA in an organized way, and dividing nuclei, synthesizing, processing, transporting, compartmentalizing and secreting macromolecules and the other cell traffic common to all organisms, they engage in alternative developmental pathways, respond to environmental signals, communicate, differentiate between mother and daughter cells if necessary, make decisions about where to place their daughter buds or septa, and grow to a defined size and shape. Each and any of these systems is susceptible to genetic analysis by the Beadle and Tatum paradigm. It does not look as though the list of cell structures, organizations and processes that interest people and that occur in yeast is yet exhausted and it is likely that workers will want to isolate mutants conforming to particular phenotypic parameters in yeast for some time. At least the earliest stages of describing and classifying mutant sets will be done using the basic facilities which these genetically versatile organisms provide.

In addition to the facilities it provides for genetic manipulation, the life cycle itself and its component parts, namely mating, nuclear fusion, meiosis, recombination and chromosome transmission will for some time continue to attract the attention of modern molecular genetic analysis. Tetrad analysis, complementation and the other manoeuvres described in this review will be among the assays involved in analysing these processes. Finally, it is worth quoting from Olson *et al.* (1986). Their comments about physical mapping could apply as well to genetic mapping, particularly should the two depend on each other for correlations and idiosyncracies:

"... a strong case can be made for the value of constructing physical maps of the genomes of intensively studied organisms. We expect the main value of these maps to lie in facilitating the organization of molecular genetic information. Just as conventional cartography provides an indispensible framework for organizing data in fields as diverse as demography and geophysics, it is reasonable to suppose that 'DNA cartography' will prove equally useful in organizing the vast quantities of molecular genetic data that may be expected to accumulate in the coming decades. Furthermore, the principal by-product of these projects—global clone collections that are cross-indexed to the physical maps—could be expected to improve the efficiency of subsequent structural and functional studies of local regions."

VI. ACKNOWLEDGEMENT

I would like to dedicate this article to my former supervisor, Professor Alan Bevan of Queen Mary and Westfield College, University of London, on the occasion of his retirement. Thanks for the light.

References

Aigle, M. and Lacroute, F. (1975). *Molecular and General Genetics* **136**, 327.
Barratt, R.W., Newmeyer, D., Perkins, D.D. and Garnjobst, L. (1954). *Advances in Genetics* **6**, 1.
Beach, D. (1983). *Nature* **305**, 682.
Beach, D. and Klar, A.J.S. (1984). *EMBO Journal* **3**, 603.
Benitez, T., del Castillo, L., Aguilera, A. and Conde, J. (1984). *Current Genetics* **8**, 345.
Brunborg, G., Resnick, M.A. and Williamson, D.H. (1980). *Radiation Research* **82**, 547.
Carle, G.F. and Olson, M.V. (1984). *Nucleic Acids Research* **12**, 5647.
Carle, G.F. and Olson, M.V. (1985). *Proceedings of the National Academy of Sciences, USA* **82**, 3756.
Chaleff, D.T. and Fink, G.R. (1980). *Cell* **21**, 227.
Chu, G., Vollrath, D. and Davis, R.W. (1986). *Science* **234**, 1582.
Clark, L. and Carbon, J. (1980). *Nature* **287**, 504.
Conde, J. and Fink, G.R. (1976). *Proceedings of the National Academy of Sciences, USA* **73**, 3651.
Cox, B.S. (1965). *Heredity* **20**, 505.
Cox, B.S. and Bevan, E.A. (1962). *New Phytologist* **61**, 342.
Dutcher, S.K. (1981). *Molecular and Cellular Biology* **1**, 245.

64   B.S. Cox

Egel, R. (1973). *Molecular and General Genetics* **121**, 277.
Egel, R. and Gutz, H. (1981). *Current Genetics* **3**, 5.
Ephrussi, B. (1953). "Nucleocytoplasmic Relations in Microorganisms". Clarendon Press, Oxford.
Ephrussi, B., Hottinguer, H. and Chimenes, A.M. (1949). *Annales de l'Institut Pasteur* **76**, 351.
Esposito, M.S. (1978). *Proceedings of the National Academy of Sciences, USA* **75**, 4436.
Esposito, M.S. and Esposito, R.E. (1969). *Genetics* **61**, 79.
Esposito, M.S. and Wagstaff, J.E. (1981). *In* "The Molecular Biology of the Yeast *Saccharomyces*. I. Life Cycle and Inheritance" (J.N. Strathern, E.W. Jones and J.R. Broach, eds), pp. 341–370. Cold Spring Harbor Laboratories, Cold Spring Harbor.
Fabre, F. (1978). *Nature* **272**, 795.
Falco, S.C. and Botstein, D. (1983). *Genetics* **105**, 857.
Falco, S.C., Li, Y., Broach, J.R. and Botstein, D. (1982). *Cell* **29**, 573.
Fan, J.-B., Chikashige, Y., Smith, C.L., Niwa, O., Yanagida, M. and Cantor, C.R. (1988). *Nucleic Acids Research* **17**, 2801.
Flores da Cunha, M. (1970). *Genetical Research* **16**, 127.
Fogel, S., Mortimer, R.K., Lusnak, K. and Tavares, F. (1979). *Cold Spring Harbor Symposium on Quantitative Biology* **43**, 1325.
Fogel, S., Mortimer, R.K. and Lusnak, K. (1981). *In* "The Molecular Biology of the Yeast *Saccharomyces*. I. Life Cycle and Inheritance" (J.N. Strathern, E.W. Jones and J.R. Broach, eds), pp. 289–339. Cold Spring Harbor Laboratories, Cold Spring Harbor.
Fogel, S., Welch, J.W., Cathala, G. and Karin, M. (1983). *Current Genetics* **7**, 347.
Fogel, S., Welch, J.W. and Louis, E.J. (1984). *Cold Spring Harbor Symposium on Quantitative Biology* **49**, 55.
Game, J.C., Bell, M., King, J.S. and Mortimer, R.K. (1990a). *Nucleic Acids Research* **18**, 4453.
Game, J.C., Sitney, K.C., Cook, V.E. and Mortimer, R.K. (1990b). *Genetics* **123**, 695.
Guthrie, C. and Fink, G.R. (1991). "Guide to Yeast Genetics and Molecular Biology". Methods in Enzymology, Vol. 194. Academic Press, San Diego.
Gutz, H. (1966). *Journal of Bacteriology* **92**, 1567.
Gutz, H. (1967). *Science* **158**, 796.
Gutz, H. (1990). *Seminars in Developmental Biology* **1**, 169.
Gygax, A. and Thuriaux, P. (1984). *Current Genetics* **8**, 85.
Harris, S.D. and Pringle, J.R. (1991). *Genetics* **127**, 279.
Hawthorne, D.C. (1955). *Genetics* **40**, 511.
Hawthorne, D.C. and Leupold, U. (1974). *Current Topics in Microbiology and Immunology* **64**, 1.
Herskowitz, I. (1988). *Microbiological Reviews* **52**, 536.
Herskowitz, I. (1989). *Nature* **342**, 749.
Herskowitz, I. and Jensen, R.E. (1991). *Methods in Enzymology* **194**, 132.
Hofer, F., Hollenstein, H., Janner, F., Minet, M., Thuriaux, P. and Leupold, U. (1979). *Current Genetics* **1**, 45.
Holmberg, S., Nilsson-Tillgren, T., Kielland-Brandt, M.C. and Petersen, J.G.L. (1982). *Carlsberg Research Communications* **47**, 355.
Hunnable, E.G. and Cox, B.S. (1971). *Mutation Research* **13**, 297.
Jackson, J.A. and Fink, G.R. (1981). *Nature* **292**, 306.
Jinks, J.L. (1963). *In* "Methodology in Basic Genetics" (W.J. Burdette, ed.). Holden-Day Inc., San Francisco.
Jinks, J.L. (1964). "Extrachromosomal Inheritance". Prentice-Hall, Englewood Cliffs, NJ.
Johnston, J.R. and Mortimer, R.K. (1959). *Journal of Bacteriology* **78**, 292.
Kassir, Y., Granot, D. and Simchen, G. (1988). *Cell* **52**, 853.
Kawasaki, G. (1979). "Karyotic instability and carbon source effect in cell-cycle mutants of *Saccharomyces cerevisiae*." PhD Thesis: University of Washington, Seattle.
Kelly, M., Burke, J., Smith, M. and Klar, A.J.S. (1988). *EMBO Journal* **7**, 1537.

Kielland-Brandt, M.C., Wilken, B., Holmberg, S., Petersen, J.G.L. and Nilsson-Tillgren, T. (1980). *Carlsberg Research Communications* **45**, 119.
Klar, A.J.S. (1980). *Genetics* **94**, 597.
Klar, A.J.S. (1989). *In* "Mobile DNA". (D.E. Berg and M.M. Howe, eds). American Society of Microbiology, Washington, DC.
Klar, A.J.S. and Miglio, L.M. (1986). *Cell* **46**, 725.
Klar, A.J.S., Fogel, S. and McLeod, K. (1979). *Genetics* **93**, 37.
Kohli, J., Hottinger, H., Munz, P., Strauss, A. and Thuriaux, P. (1977). *Genetics* **87**, 471.
Laskowski, W. (1960). *Zeitschrift für Naturforschung* **156**, 495.
Lehman, E. and Munz, P. (1987). *Current Genetics* **11**, 419.
Leupold, U. (1950). *Comptes Rendus des Travailles de la Laboratoire Carlsberg, Serie Physiologique* **24**, 381.
Leupold, U. and Gutz, H. (1964). *Proceedings of the XI International Congress of Genetics* **2**, 31.
Liebman, S., Shalit, P. and Picologlou, S. (1981). *Cell* **26**, 401.
Lindegren, C.C. (1933). *Bulletin of the Torrey Botanical Club* **60**, 133.
Lindegren, C.C. (1945). *Bacteriological Reviews* **11**, 111.
Lindegren, C.C. (1949). "The Yeast Cell: Its Genetics and Cytology". Educational Publishers Inc., St Louis.
Lindegren, C.C. and Lindegren, G. (1943). *Proceedings of the National Academy of Sciences, USA* **29**, 306.
Lindegren, G., Hwang, Y.L., Oshima, Y. and Lindegren, C.C. (1965). *Canadian Journal of Genetics and Cytology* **7**, 491.
Link, A.J. and Olson, M.V. (1991). *Genetics* **127**, 681.
Liras, P., McCusker, J., Masciolo, S. and Haber, J.E. (1978). *Genetics* **88**, 651.
Livingston, D.M. (1977). *Genetics* **83**, 73.
Louis, E.J. and Haber, J.E. (1990). *Genetics* **124**, 547.
Ma, C. and Mortimer, R.K. (1983). *Molecular and Cellular Biology* **3**, 1886.
Malone, R.E. and Esposito, R.E. (1980). *Proceedings of the National Academy of Sciences, USA* **77**, 503.
Manney, T.R. and Mortimer, R.K. (1964). *Science* **143**, 581.
Mather, K. (1935). *Journal of Genetics* **30**, 53.
Mather, K. (1957). "The Measurement of Linkage in Heredity". Methuen, London.
Miyake, S., Tanaka, A. and Yamamoto, M. (1991). *Current Genetics* **20**, 277.
Moreno, S., Klar, A.J.S. and Nurse, P. (1991). *Methods in Enzymology* **194**, 795.
Mortimer, R.K. (1958). *Radiation Research* **9**, 312.
Mortimer, R.K. and Contopoulou, C.R. (1987). "Yeast Genetic Stock Center Catalogue". University of California Berkeley, Berkeley.
Mortimer, R.K., Contopoulou, C.R. and Schild, D. (1981). *Proceedings of the National Academy of Sciences, USA* **78**, 5778.
Mortimer, R.K., Game, J.C., Bell, M. and Contopoulou, C.R. (1991). *Methods in Pulsed Field Gels* **1**, 1.
Mortimer, R.K. and Hawthorne, D.C. (1969). *In* "The Yeasts" (A.H. Rose and J.S. Harrison, eds), Vol. 1, pp. 385–460. Academic Press, London.
Mortimer, R.K. and Hawthorne, D.C. (1973). *Genetics* **74**, 33.
Mortimer, R.K. and Hawthorne, D.C. (1975). *In* "Methods in Cell Biology" (D.M. Prescott, ed.), Vol. 11, pp. 221–233. Academic Press, New York.
Mortimer, R.K. and Schild, D. (1980). *Microbiological Reviews* **44**, 519.
Mortimer, R.K. and Schild, D. (1985). *Microbiological Reviews* **49**, 181.
Mortimer, R.K., Schild, D., Contopoulou, C.R. and Kans, J.A. (1989). *Yeast* **5**, 321.
Munz, P. and Leupold, U. (1981). *In* "Molecular Genetics in Yeast" (D. von Wettstein, J. Friis, M. Kielland-Brandt and A. Stenderup, eds). Munksgaard, Copenhagen.
Munz, P., Wolf, K., Kohli, J. and Leupold, U. (1989). *In* "Molecular Biology of the Fission Yeast" (A. Nasim, P. Young and B.F. Johnson, eds), pp. 1–30. Academic Press, San Diego.

Nasmyth, K.A., Seldon, K. and Ammerer, G. (1987). *Cell* **49**, 549.

Newlon, C. (1988). *Microbiological Reviews* **52**, 568.

Newlon, C.S. (1989). *In* "The Yeasts" (A.H. Rose and J.S. Harrison, eds), pp. 57–116. Academic Press, London.

Nilsson-Tillgren, T., Petersen, J.G., Holmberg, S. and Kielland-Brandt, M.C. (1980). *Carlsberg Research Communications* **45**, 113.

Nilsson-Tillgren, T., Gjermansen, C., Kielland-Brandt, M.C., Petersen, J.G.L. and Holmberg, S. (1981). *Carlsberg Research Communications* **46**, 65.

Niwa, O., Matsumoto, T., Chikashiga, Y. and Yanagida, M. (1989). *EMBO Journal* **8**, 3045.

Niwa, O., Matsumoto, T. and Yanagida, M. (1986). *Molecular and General Genetics* **203**, 397.

Niwa, O. and Yanagida, M. (1985). *Current Genetics* **9**, 463.

Olson, M.V. (1992) *In* "Molecular and Cellular Biology of the Yeast *Saccharomyces* (J.R. Broach, J.R. Pringle and E.W. Jones, eds), Vol. 1, pp. 1–49. Cold Spring Harbor Press, Cold Spring Harbor.

Olson, M.V., Datchik, J.E., Graham, M.Y., Brodeur, G.M., Helms, C., Frank, M., MacCollin, M. and Scheinman, R. (1986). *Proceedings of the National Academy of Sciences, USA* **83**, 7826.

Olson, M.W., Eden, V., Egel-Mitani, M. and Egel, R. (1978). *Hereditas* **89**, 189.

Orr-Weaver, T.L. and Szostak, J.W. (1985). *Microbiological Reviews* **49**, 33.

Orr-Weaver, T.L., Szostak, J.W. and Rothstein, R.L. (1981). *Proceedings of the National Academy of Sciences, USA* **78**, 6354.

Papazian, H.P. (1952). *Genetics* **37**, 175.

Parry, E.M. and Cox, B.S. (1970). *Genetical Research* **16**, 333.

Parry, J.M. and Cox, B.S. (1968). *Genetical Research* **12**, 187.

Perkins, D.D. (1949). *Genetics* **34**, 607.

Petes, T.D. (1980). *Cell* **19**, 756.

Petes, T.D. and Hill, C.W. (1988). *Annual Reviews of Genetics* **22**, 147.

Polaina, J. and Conde, J. (1982). *Molecular and General Genetics* **186**, 253.

Pomper, S., Daniels, K.M. and McKee, D.W. (1954). *Genetics* **39**, 343.

Pontecorvo, G. (1956). *Annual Review of Microbiology* **10**, 393.

Pontecorvo, G., Roper, J.A., Hemmons, L.M., MacDonald, K.D. and Bufton, A.W.J. (1953). *Advances in Genetics* **5**, 141.

Resnick, M.A. (1976). *Journal of Theoretical Biology* **59**, 97.

Rine, J.D., Jensen, R., Hagen, D., Blair, L. and Herskowitz, I. (1981). *Cold Spring Harbor Symposium on Quantitative Biology* **45**, 95.

Robinow, C.F. (1977). *Genetics* **87**, 491.

Rodarte, U., Fogel, S. and Mortimer, R.K. (1968). *Genetics* **60**, 216.

Roeder, G.S. and Fink, G.R. (1983). *In* "Mobile Genetic Elements" (J.A. Shapiro, ed.), pp. 299–328. Academic Press, New York.

Roman, H. (1956). *Cold Spring Harbor Symposium on Quantitative Biology* **21**, 175.

Roman, H. and Jacob, F. (1958). *Cold Spring Harbor Symposium on Quantitative Biology* **23**, 155.

Roman, H., Phillips, M.M. and Sands, S.M. (1955). *Genetics* **40**, 546.

Rothstein, R. (1991). *Methods in Enzymology* **194**, 281.

Schwarz, D.C. and Cantor, C.R. (1984). *Cell* **37**, 67.

Sherman, F. and Hicks, J. (1991). *Methods in Enzymology* **194**, 21.

Shero, J.H., Koval, M., Spencer, F., Palmer, R.E., Hieter, P. and Koshland, D. (1991). *Methods in Enzymology* **194**, 749.

Sigurdson, D.C., Gaarder, M.E. and Livingston, D.M. (1981). *Molecular and General Genetics* **183**, 59.

Smith, C.L. and Cantor, C.R. (1987). *Methods in Enzymology* **155**, 449.

Smith, C.L., Matsumoto, T., Niwa, O., Kleo, S., Fan, J.-B., Yanagida, M. and Cantor, C.R. (1987). *Nucleic Acids Research* **15**, 4485.

Snow, R. (1979). *Genetics* **92**, 231.

Somers, J.M. and Bevan, E.A. (1969). *Genetical Research* **13**, 71.

Southern, E.M. (1975). *Journal of Molecular Biology* **98**, 503.

Southern, E.M., Anand, R., Brown, W.R.A. and Fletcher, D.S. (1987). *Nucleic Acids Research* 15, 5925.

Stevens, B. (1981). *In* "Molecular Biology of the Yeast *Saccharomyces*. I. Life Cycle and Inheritance" (J.N. Strathern, E.W. Jones and J.R. Broach, eds), pp. 471–504. Cold Spring Harbor Laboratories, Cold Spring Harbor.

Strathern, J.N., Klar, A.J.S., Hicks, J.B., Abraham, J.A., Ivy, J.M., Nasmyth, K.A. and McGill, C. (1982). *Cell* 31, 183.

Strathern, J.N., Newlon, C.S., Herskowitz, I. and Hicks, J.B. (1979). *Cell* 18, 309.

Symington, L.S. and Petes, T.D. (1988). *Molecular and Cellular Biology* 8, 595.

Szankasi, P., Gysler, C., Zehnter, U., Leupold, U., Kohli, J. and Munz, P. (1986). *Molecular and General Genetics* 202, 394.

Szostak, J.W. and Wu, R, (1980). *Nature* 284, 426.

Tatum, E.L. and Beadle, G.W (1945). *Annals of the Missouri Botanic Garden* 32, 125.

Thomas, J.H. and Botstein, D. (1987). *Genetics* 115, 229.

Vollrath, D., Davis, R.W., Connolly, C. and Hieter, P. (1988). *Proceedings of the National Academy of Sciences, USA* 85, 6027.

Wickner, R.B. (1979). *Genetics* 92, 803.

Wickner, R.B. and Liebowitz, M.J. (1979). *Journal of Bacteriology* 140, 154.

Winge, Ø. (1935). *Comptes Rendus des Travailles de la Laboratoire Carlsberg, Serie Physiologique* 21, 77.

Winge, Ø. and Roberts, C. (1949). *Comptes Rendus des Travailles de la Laboratoire Carlsberg, Serie Physiologique* 24, 341.

Winge, Ø. and Roberts, C. (1950). *Comptes Rendus des Travailles de la Laboratoire Carlsberg, Serie Physiologique* 25, 35.

Wood, J.S. (1987). *Molecular and Cellular Biology* 2, 1080.

Woods, R.A. and Bevan, E.A. (1966). *Heredity* 21, 121.

Wright, R.E. and Lederberg, J. (1956). *Proceedings of the National Academy of Sciences, Washington* 43, 919.

Zakharov, I.A. and Yarovoy, B.P. (1977). *Molecular and Cellular Biochemistry* 14, 15.

3 Modern and Post-modern Genetics in *Saccharomyces cerevisiae*

Mark D. Rose

Department of Molecular Biology, Lewis Thomas Laboratory, Princeton University, Princeton, New Jersey 08544, USA

The Yeasts Vol. 6, 2nd edition
ISBN 0-12-596416-1

I. INTRODUCTION

The yeast *Saccharomyces cerevisiae* has the most facile genetics of any
eukaryotic organism. The biological features that contribute to its ease
of use are a small genome, simplicity of culture conditions, stable haploid
and diploid states, and rapid growth. The genetic features that contribute
to its utility are: several efficient methods of transformation, a high rate
of homologous recombination that allows the targeting of plasmids to
specific loci, and a stable diploid state that greatly simplifies tests of
complementation. Furthermore, a long history of modern yeast genetics has
produced a large number of useful tools for the molecular geneticist. Many
of the genes have been cloned, mapped and sequenced, including several
genes whose function can be selected either for or against. All of the
elements known to be required for the stable propagation of yeast
chromosomes have been molecularly cloned. These are the chromosomal
origins of DNA replication, ars (autonomously replicating sequence),
centromeres (cen) and telomeres (tel). An endogenous stable, high-copy
plasmid called the 2 µm circle has been extensively characterized. All of
these pieces in various combinations have allowed the construction of a
dizzying array of plasmids having a large variety of specific uses. Finally,
a variety of schemes for regulated gene expression, including strongly
regulated promoters, have been described, which allow the modern
practitioner to express their favourite genes at will. Taken together, the
variety and proliferation of useful tools for yeast genetics has made the
organism an indispensable workshop for the characterization of basic
biological processes in eukaryotes.

   This review will attempt to describe the myriad modern ways of studying
Sacch. cerevisiae. It will stress strategies and approaches rather than
describe protocols, many of which are carefully and clearly described either

in the primary literature or in a number of recent useful methods books
(Rose *et al.*, 1990; Guthrie and Fink, 1991). It is hoped that this will serve
both as a guideline for the novice in *Sacch. cerevisiae* and as inspiration
for the expert trying to develop molecular genetics in other related
organisms.

Given the extreme breadth of the topic, this review cannot pretend to be
exhaustive nor would much useful purpose be served by an encyclopaedic
catalogue of examples from the recent literature. Instead, each approach
will be illustrated with a few particularly apt examples from the field.
Apologies are therefore due in advance to those who have utilized these
methods or helped develop them, but have not been cited.

This review is organized into five broad sections. First will be a general
discussion of methods for identifying the gene of interest. Second will be
approaches to studying gene function by constructing variant forms of the
gene or protein. Third, because all of these methods require efficient
methods for transformation, there will be a brief discussion about the
different methods that have been developed. The fourth section will focus
on the structural concerns of mapping mutations within a gene and
mapping a gene within the genome. Finally, different strategies for finding
other genes whose products may interact or functionally overlap with the
first genes will be discussed.

II. FINDING THE GENE

A. The benefits of a small genome

One of the great technical virtues of *Sacch. cerevisiae* is the ease by which
genes can be cloned using the suppression of mutant phenotypes conferred
by previously identified genetic variants. This facility is dependent upon
efficient transformation, the existence of numerous vectors (see Parent and
Bostian, Chapter 4, this volume), the small size of the genome and the
availability of a number of excellent plasmid libraries containing yeast
genomic sequences. The yeast genome contains approximately 14 000 kb
(Olson, 1991), about 3–4 times the *Escherichia coli* genome. Several
projects are currently underway to sequence entire yeast chromosomes and
the entire sequence of chromosome III has been reported (Oliver *et al.*,
1992). The 315 kb of chromosome III contains some 182 open reading
frames larger than 300 bp, corresponding well with transcript mapping data
that indicated the presence of at least 160 transcripts from this chromosome

(Yoshikawa and Isono, 1990). The average size of a yeast gene including its associated regulatory sequences on chromosome III is approximately 1.7 kb. This small size accrues in part from the rarity of introns (which when present are small, averaging about 300 bp in size) and from the small size of intergenic regions, averaging only 300 bp. Extrapolating to the entire genome there should be a total of between 7500 and 8000 genes (allowing for between 100 and 200 copies of the ribosomal DNA).

The relatively small size of the genome and the small size of its genes have profound implications for thinking about the numbers of mutants and clones that must be examined in any thorough screen of the genome. For the initial isolation of mutations, it is tempting to speculate on the expected frequencies of mutations having observable phenotypes. However, a number of factors complicate the analysis. First, such calculations assume an equal target size. However, this is a poor assumption, owing to unknown differences in gene size, susceptibility to mutations that lead to loss of function, and the viability and recovery of such mutants. Second, although many genes that are members of gene families in higher cells are uniquely represented in yeast, this is not universally true. Currently, many examples of repeated or functionally redundant genes have been reported. As both sequencing and more sophisticated genetic screens are completed, it is certain that redundancy will be found more frequently. Recessive mutations in such genes may be impossible to uncover by classic methods and dominant mutations are likely to be rare. Third, it remains unclear as to how many genes are required for normal yeast growth. Estimates of the fraction of yeast genes in which gene disruptions lead to easily discernible phenotypes suggest that 60–70% are "dispensable" for growth (Goebl and Petes, 1986). Probably many of the dispensable genes are functionally redundant with other genes in the genome. It is also likely that many of the genes are required for growth regimes that are quite different from those that are generally encountered in the laboratory. A recent example is the recognition that, under certain poor nutritional conditions, yeast cells can undergo a switch to a "foraging" mode of growth forming pseudohyphal colonies (Gimeno et al., 1992). For these reasons it can be quite difficult to predict the frequency at which simple loss of function mutations can give rise to a relevant phenotype. Nevertheless, the frequency of loss of function mutations for a given gene in a population of heavily mutagenized cells can be fairly high. This can greatly facilitate mutant isolation if a relatively difficult assay must be performed. For example, only 150 temperature-sensitive mutants needed to be examined before a top2 allele was isolated and only 250 were examined to identify an allele of top1 (DiNardo et al., 1984; Thrash et al., 1984).

Aside from these concerns, once a mutation is in hand, the small size of the genome implies that the number of plasmids that must be examined to find a gene can be quite small. Most libraries of the yeast genomic DNA sequences use inserts of at least 10–15 kb. Thus the insert size is generally much larger than the average gene size and one can usually safely ignore the problem of ensuring that the entire structural gene is contained within the insert. Thus the expected frequency for a functional gene to be present in most libraries is between 1/1000 and 1/1500. Given a sufficiently large library, transformation with greater than five genome equivalents should ensure isolation of the gene of interest with a probability of greater than 99%. Of course, other considerations can distort this number, and these include underrepresentation in the library due to toxicity in *E. coli* of the desired gene or a close neighbour, and inability to transform yeast due to the presence of neighbouring chromosomal structural sequences. Therefore, it is common practice to use even larger libraries to overcome selection artefacts in *E. coli* or allow selection of the rare insert that has not included the deleterious sequence.

B. Cloning by "complementation"

Methods of screening transformants for complementation vary according to the specific phenotype being scored. In some cases direct selection can be applied. However, in cases where the original mutation reverts, this can lead to a high frequency of false positives as any transformant can potentially give rise to revertant colonies. An alternative method utilizing the indirect selection of replica-plating is preferred. In this case, transformant colonies containing truly complementing plasmids can be distinguished by the homogeneity of the colony phenotype. In contrast, reversion events that arise in the transformant colony give rise only to sparse "papillae" on the replica-plate. However, by this method, the desired transformants cannot be distinguished from pre-existing revertants that become transformed by irrelevant plasmids. Again the small size of the genome is an advantage as the frequency of revertants in the population should be several orders of magnitude rarer than the authentic complementing plasmids.

Nevertheless, the possibility of reversion (or other modifying host mutations) indicates that a direct test must be employed to assure that complementation is conferred by the plasmid. This can be done in either of two ways. In the first, purified transformants are cultured under non-selective conditions to allow the plasmid to segregate. After plating the culture (again under non-selective conditions) for individual colonies, the colonies are replica-plated to determine whether loss of the plasmid (scored by the vector's selectable marker) correlates with loss of the complementing activity. Where the plasmid can be selected against (e.g. 5-fluoro-orotic acid (5-FOA) selection against *URA3*; Boeke *et al.*, 1984), plasmid-free colonies are selected directly and then tested for loss of complementation. Potential pitfalls can arise in either case if multiple plasmids are present in the transformant. In the first case, complementing activity can be lost without loss of the vector marker. In the second case, the existence of the second plasmid is masked and this can lead to confusion in later steps. Ultimately, a synthetic approach is more reliable. The plasmid is first recovered in *E. coli* and then retransformed back into the mutant strain. These secondary transformants are then tested for complementing activity. Although less ambiguous, the retransformation test can be fairly time consuming. Thus, the cosegregation test is often completed first and then confirmed by re-transformation.

Once candidate plasmids have been identified and recovered in *E. coli*, restriction mapping will determine whether they contain DNA from one or more genetic loci. At this point, a test must be applied to prove that the DNA is derived from the locus of the mutation. Although the authentic gene is one of the expected classes of plasmid, cases of genetic redundancy, overlapping function and suppression can potentially complicate the analysis. The initial isolation of the gene relied on a test of gene function, therefore an independent test should rely on a test of position or structure. Position can be readily determined by use of homologous recombination to determine whether the complementing DNA is derived from a region linked to the mutation. In practice, the cloned DNA is subcloned into a YIp (yeast-integrating plasmid) vector. If a unique restriction site is found in the insert, cleavage at this site will stimulate homologous recombination and target plasmid integration to the site corresponding to the insert DNA (see Fig. 1). While not strictly necessary, cleavage is preferred because it favours the desired integration event and prevents integration at the site of the plasmid marker. In addition, it will allow integration to occur even if an ars element is present in the insert DNA.

Integration at the site of the insert can be confirmed by Southern blot hybridization. Cleavage with an enzyme that does not cut in the integrating plasmid should show a larger band in the genome and loss of the corresponding wild-type band. Cleavage in the vector (but not in the insert) will yield two new bands, neither of which should correspond to the wild-type genomic band. Cleavage in the insert will yield both a plasmid-sized band and the wild-type band. This last digest may be useful to confirm that the vector is intact but otherwise yields no useful information about where the plasmid has integrated.

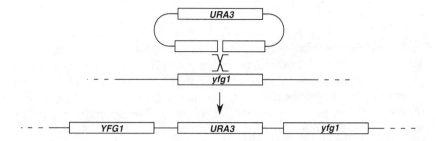

Fig. 1. Plasmid integration. The plasmid is cleaved with a restriction enzyme within the yeast DNA insert. The DNA ends catalyse a recombination event between the homologous DNA sequences. Integration results in a direct repeat of the yeast sequences. Relevant yeast genes are indicated by the thick lines; plasmid sequences and surrounding chromosomal sequences are indicated by thin lines. *YFG1*, wild-type copy of your favourite gene; *yfg1*, mutant copy.

Once an integrant is obtained, its location is mapped by meiotic crosses. In this case, mapping *must* be performed relative to the *wild-type* allele of the mutant locus. If the mutant was transformed to obtain the integrant, the transformant should be crossed to a wild-type strain. If the wild type was transformed, then the cross should be performed with a mutant strain. In the first case, if the plasmid has integrated at the site of the mutation, usually the integrant will be wild type, and the cross to wild type will result in a 4:0 segregation of wild type to mutant. If the plasmid had integrated elsewhere, then the cross will give rise to the appearance of mutant spores as a consequence of the random segregation of non-homologous chromosomes. Had the cross been performed against a mutant strain then a 2:2 segregation would have been observed owing to the segregation of the single wild-type locus introduced by the plasmid. It is important to realize that the 2:2 segregation pattern would occur *regardless* of the site of integration of the plasmid. Note, however, that this latter cross is an important control for the first cross as it demonstrates that the plasmid integrated at a single locus. The alternative (and preferable) experiment is to transform a wild-type strain and cross to a mutant. Whereas in the initial experiment, the two copies of wild-type information are in "repulsion", in the latter cross they are "coupled". Therefore, a 2:2 segregation of the wild phenotype indicates that the gene and plasmid are linked, whereas an excess of wild-type spores would indicate that they are unlinked. Note also that the plasmid marker can be followed and showed to be in linkage with the wild-type information.

C. No mutations in hand

Unfortunately, it is often the case that no appropriate mutation is currently available to allow the identification of the gene by complementation. Luckily, in addition to isolation of the relevant mutations *de novo*, there are at least two strategies that can be applied. One relies on the physical structure of the gene or protein of interest. The second utilizes the alterations in phenotype that may arise from the overexpression of a particular protein.

D. Physical methods of gene identification

Physical methods of gene identification can be based upon either the DNA sequence or the protein sequence. DNA-based methods have been described in a number of current manuals. In the recent past, a DNA sequence would be used directly as a hybridization probe to identify the gene in a library of genomic clones. Currently, it is more likely that some variant of the polymerase chain reaction (PCR) would be utilized to amplify a segment of the yeast genome. The DNA sequence of the PCR-amplified product is usually determined to ensure that the sequence matches in regions that were not part of the primers. The PCR product is then be used as a hybridization probe to obtain the entire gene. The initial DNA sequence might be from a gene identified in a different organism or be derived from the reverse translation of a known protein sequence. In either case, the small genome of yeast allows fairly degenerate PCR primers to be used; in some cases, specific amplification is seen using a mixture of more than 100 000 different oligonucleotides (Roof *et al.*, 1992).

Several approaches directed specifically towards protein sequences have been utilized. The most frequently used method is the lambda gt11 cloning system in which yeast genomic (or cDNA) sequences are expressed during a lytic phage infection of *E. coli* as proteins fused to β-galactosidase (Snyder *et al.*, 1987; Young and Davis, 1991). Typically proteins expressed upon induction with IPTG are detected using an antibody directed against either the authentic or a cross-reacting protein. In some cases, the technique can be extended by using other proteins that might bind to the protein of interest. One instance where this approach has been used to some success has been in the identification of yeast calmodulin binding proteins (Pausch *et al.*, 1991). In either case, the interaction must not rely on modifications that occur only in the yeast. Because multiple proteins can be translated from the same mRNA in *E. coli*, it is fairly common for the yeast proteins to be expressed as independent translation products and not as fusion proteins. This can be quickly established by a Western blot against proteins induced by IPTG in a lysogen. This information will help in determining whether the intact gene is present, and whether the DNA sequence of the gene should begin at the insert junction.

The use of methods that are based upon the recognition of specific protein structures necessitates a careful consideration of controls to demonstrate that the gene actually encodes the target protein of interest. This can be a serious problem in the case where there is little or no protein sequence data available. In many cases, particularly in cloning genes from higher eukaryotes, the ability of the protein expressed in *E. coli* to elicit an immune response against the original protein is often taken as a demonstration of identity. In yeast, the situation is somewhat more amenable to analysis as the protein can be easily overexpressed in yeast (e.g. on the 2 μm plasmid) or deleted entirely (assuming that it is not essential). In either case, the quantity of the corresponding protein band should be seen to be altered in predicted ways.

E. Functional methods

In principle, any alteration of phenotype can be used as a method for screening a plasmid library for the gene of interest. It is frequently the case that overexpression of a protein may lead to an easily recognizable and predictable phenotype (reviewed in Rine, 1991). For example, transformants of a high copy number plasmid library were screened for resistance to drugs whose target proteins had been previously identified (Rine *et al.*, 1983; Basson *et al.*, 1986). In these cases, drug-induced inhibition of the target proteins made them rate limiting for growth. Overexpression of the desired proteins resulted in increased growth.

In some cases, overexpression leads to a novel phenotype (reviewed in Rine, 1991), possibly by causing an imbalance in the stoichiometry of essential structural proteins. For example, overexpression of histones H2A and H2B, or H3 and H4 lead to a substantial decrease in chromosome stability (Meeks-Wagner and Hartwell, 1986). In contrast, balanced overexpression of all four histones did not effect chromosome stability, thereby demonstrating that overexpression *per se* was not the cause of the problem. This phenomenon was then exploited in a screen for other genes

(*MIF1* and *MIF2*), whose overexpression leads to decreased chromosome stability (Meeks-Wagner *et al.*, 1986).

In many cases, overexpression of a protein can suppress a mutation in a different gene (Hinnebusch and Fink, 1983; MacKay, 1983; Natsoulis *et al.*, 1986; Bender and Pringle, 1989; Hoyt *et al.*, 1992; Vallen *et al.*, 1994; to cite only a few pertinent examples). In principle, a fairly large number of different physical phenomena might lead to suppression. Although it is not clear that all of these have been identified, some useful mechanisms of suppression include: (i) overexpression of an interacting protein overcoming a mutationally weakened binding constant, by the law of mass action; (ii) increased flux of a rate-limiting metabolite compensating for the decreased activity of a mutant enzyme; (iii) increased substrate concentration increasing the stability of a mutant protein; (iv) increased concentration of a protein of partially overlapping function compensating for loss of the optimal protein; (v) increased modifying enzyme causing increased activity; and (vi) increased activity of a downstream enzyme by-passing the requirement for an earlier step.

In principle, there are also mechanisms of high copy suppression that are much more indirect and therefore less useful to the investigator. These include: (i) increased ionic or osmotic strength directly suppressing the defective protein; (ii) increased ribosomal misreading producing wild-type protein; and (iii) decreased growth rate compensating for a rate-limiting mutant enzyme. While the likelihood of encountering such suppressors may be low, this author has personal experience with both the second and third cases. They should be considered to be more than simply formal possibilities. Therefore, the investigator is required to establish their relevance by independent means.

Given multiple possible modes of suppression, how is one to know whether a gene obtained by high copy number suppression is one that will be useful for further progress? Ultimately, this is a question of taste and how much one is willing to be distracted by observations that are intriguing but not central to one's research effort. Two criteria that are often applied are: (i) that the sequence of the gene is explanatory; or (ii) that disruption of the gene produces a phenotype that is related to the initial phenotype. In either case, the usefulness of the suppressor can only be assessed by the analysis of the function of the suppressor gene. If, as is often the case, the disruption has no phenotype, then gene redundancy should be ruled out before the gene is abandoned.

There are several technical limitations of the high copy number cloning strategy. In some cases overexpression of a particular gene is toxic and the transformant effectively inviable (Rose and Fink, 1987). In other cases, the relevant gene may not be appreciably overexpressed on 2 μm plasmids. Of course, it is also true that overexpression of any particular gene may not have any phenotypic effect. It may not be possible to arrive at conditions where the wild-type protein is rate limiting for growth. Otherwise the high copy number technique is limited only by the imagination of the investigator in terms of defining appropriate phenotypes to screen.

To overcome some of the problems associated with the high copy number plasmid approach, two different groups have developed methods that utilize a strong inducible promoter from the *GAL1* gene. In one case a library of plasmids was constructed containing yeast cDNAs fused to P_{GAL1} (Liu *et al.*, 1992). In the second case, a library of sheared yeast genomic fragments were fused to P_{GAL1} (Ramer *et al.*, 1992). Both libraries are extremely large, in the case of the cDNA library there are a total of 1.1×10^7 independent plasmids, some 50 times larger than the predicted number of different mRNAs in yeast. In the case of the genomic library, there are 5×10^7 different plasmids, or more than enough to fuse P_{GAL1} to every nucleotide in yeast. The major differences between the two libraries concern the abundance of sequences in the plasmids. The cDNA library should mirror the mRNA content in the cell under the specific conditions of growth used by the authors. The genomic library should have all sequences equally represented. Therefore, rarely expressed or unusually regulated transcripts may be found more readily in the genomic library.

F. Brute-force approaches

The primary utility of high copy number approaches lies in the extreme ease of application. A secondary benefit is the fact that genes are effectively cloned by the same manipulation that originally identified them. The fact that only a few thousand transformants need be examined to cover the genome completely makes a number of fairly vague or laborious approaches to gene cloning practical. For example, one extremely powerful but laborious method is the microscopic examination of single cells, an approach that has been extremely successful for the identification of new nuclear-division mutants (Winey *et al.*, 1991). Direct biochemical assays for the overexpressed protein are also possible. If the activity can be easily measured *in vitro*, then this could be applied as a means for screening a population of transformants, assuming that the assay is simple enough and the investigator is sufficiently motivated. If an antibody for the protein exists, then overexpression in yeast can be assessed directly (Lyons and Nelson, 1984) and this may obviate the problem inherent with an epitope that includes an essential eukaryotic modification.

III. TRANSFORMATION OF YEAST

A. Sphaeroplast transformation

The first reproducible reports of yeast transformation were based upon earlier methods for the formation and fusion of yeast sphaeroplasts (Hinnen *et al.*, 1978). In essence, the action of cell-wall hydrolytic enzymes is used to remove portions of cell wall in the presence of osmotic stabilizers. Either a snail-gut extract, Glusulase, or a preparation of the secreted enzymes from *Arthrobacter luteus*, Zymolyase, are used, sometimes after a reducing agent has removed disulphide cross-links in the cell wall. The sphaeroplasts are stabilized with 1 M sorbitol, which is not utilized as a carbon source by *Sacch. cerevisiae*. DNA is then added to the sphaeroplasts, often along with non-specific "carrier" DNA. The DNA and sphaeroplasts are then co-precipitated using a mixture of polyethylene glycol (PEG) and Ca^{2+}. After resuspending the cells in sorbitol solution, the sphaeroplasts are mixed with molten 2% agar and then poured on the surface of a selective plate containing sorbitol. This protocol is particularly laborious and not every strain can be transformed readily by this method. Nevertheless, for some strains it allows a very high frequency of transformation, approximately $1-5 \times 10^4$ transformants/μg DNA. Furthermore the method can be carefully optimized to allow for extremely high rates of transformation (Burgers and Percival, 1987). These optimized methods are of particular importance for transformation with the very large fragments of DNA that are used in yeast artificial chromosome (YAC) cloning.

Aside from the labour involved, problems with the sphaeroplast transformation method include high strain variation in transformation efficiency, sensitivity to residual detergent, recovery of the transformants embedded in an agar matrix and a reported high frequency of diploidization (Harashima *et al.*, 1984). The first problem can be solved by repeated back-crosses to a strain showing high transformation frequencies. The second problem can be avoided by scrupulous care with reagents and glassware. One place that detergents can be inadvertently introduced is through filter sterilization units, which can contain small amounts of wetting agents. The third problem can be partially solved by plating the sphaeroplasts on the surface of plates containing osmotic protectant but the transformation frequency is reduced. Although the severity of the problem of diploids being formed at high frequency by sphaeroplast fusion has been questioned (Rose *et al.*, 1986), the optimized protocol was developed to avoid it (Burgers and Percival, 1987).

B. Chemical transformation

Because of the difficulties inherent in the previous protocol, most investigators now use lithium salts to prepare cells for transformation (Ito *et al.*, 1983). Cells are treated with lithium acetate, which appears to permeabilize the cell wall. DNA is added and the cells are co-precipitated with PEG. After a brief heat shock, cells are washed free of the PEG and lithium acetate, and spread on ordinary selective plates.

While not as efficient as the sphaeroplast transformation method (approximately 1 × 10³ transformants/μg DNA), several refinements have led to considerably improved transformation frequencies. These include the use of single-stranded carrier DNA (Schiestl and Gietz, 1989) or organic solvents, such as ethanol or dimethyl sulphoxide (Soni *et al.*, 1993), either of which can increase the efficiency as much as 10–50-fold.

At the same time it is often useful to scale down the protocol to a rapid method that allows for much less manipulation and higher sample throughput, albeit at the cost of reduced yields. To this end a modified method using overnight cultures or cells scraped off agar plates has been very successful (Elble, 1992). One occasion where a faster protocol is useful is in the rescreening of mutants in which the selection criteria partly relies on the ability of specific genes to suppress the phenotype. A further utility of the lithium acetate protocol is that the competent cells can be frozen and thawed with only slight (10-fold) loss of transformation efficiency for routine transformation into standard strains.

C. Electrical

The recent adaptation of methods for electroporation of DNA into yeast (Hashimoto *et al.*, 1985; Becker and Guarente, 1991) has greatly improved the efficiency and convenience of transformation. Fresh overnight cultures are washed free of salt, suspended in an osmotic protectant such as sorbitol, DNA is added and the cells are pulsed in an electroporation device. After the electrical pulse, cells are spread on the surface of selective plates. Transformation frequencies are in the range of 1–5 × 10⁵ transformants/μg DNA, if care is taken to have cells in the right growth phase, cells are kept chilled during the electrical shock and selective plates containing an osmotic protectant (sorbitol) are used. While electroporation is the most efficient method of transformation and the procedure is simple, the requirement for specialized equipment and the cost of the cuvettes used for the electrical shock can be prohibitive for routine use.

D. Bolistic

Cellular organelles cannot be transformed by the methods that work for nuclear transformation. However, mitochondria can be transformed by a "bolistic" technique that utilizes high-speed tungsten particles to puncture the cell (Fox *et al.*, 1988; Johnston *et al.*, 1988). The tungsten particles are adsorbed with the DNA and then fired at the cells using either gunpowder charges or helium gas as the propellent. Devices for bolistic transformation are commercially available.

A second type of bolistic transformation involves the use of small glass beads of the sort typically used in the preparation of yeast extracts. Agitation of the yeast with the glass beads apparently causes the transient rupturing of cells and, at this time, DNA may be taken up by the cells (Costanza and Fox, 1988). While not an efficient method of transformation, agitation with glass beads has the distinct advantage that it is extremely rapid and can be performed on small samples of saturated cell cultures. Thus a large number of different strains may be transformed at the same time.

E. Trans-kingdom sexuality

Certain endogenous prokaryotic plasmids have extraordinarily broad host ranges as they encode the ability to be transferred between and maintained in very distantly related prokaryotic species. They can also mobilize other resident plasmids into the new host cell. The truly astonishing finding by Heinemann and Sprague (1989) was that one such broad host-range plasmid could also mediate the transfer of plasmids from *E. coli* to *Sacch. cerevisiae*! The great utility of this method is that its extreme technical simplicity allows rapid and simultaneous transfer of plasmids into a wide variety of different host strains. For example, one can envisage mutant screens that employ tranformation as one of the criteria for selection. In addition, one can rapidly test a series of mutant plasmid constructs in a large number of mutant backgrounds. The caveats to the protocol are that it is relatively inefficient, does not work for all yeast shuttle plasmids and requires some facility with *E. coli* genetics to set up the appropriate donor strains. Furthermore, since the broad host-range plasmids are transmissible between different prokaryotic species, experiments using this technique must utilize a higher level of biological or physical containment. Transfer to *Schizosaccharomyces pombe* has also been demonstrated (Sikorski *et al.*, 1990), so this technique should be generally useful for other yeasts.

IV. MUTAGENIZING THE GENE

Because of the emphasis on phenotype, genetic analysis is primarily the study of function *in vivo*, utilizing genetic variation to generate functional variation. Classically, geneticists were limited to the study of quasi-random mutant alleles that could be generated either spontaneously or by mutagens. With the advent of recombinant DNA techniques, the range of variation has been greatly extended and the precision of the genetic alterations has been brought to the level of the nucleotide. The advantages are three-fold. First, unconscious bias in the acquisition of an initial set of mutations can lead to an extremely misleading set of phenotypes. The ability to construct additional mutations can either reveal new functions for the gene or strengthen the interpretation of the original mutant phenotype. Second, knowledge of the precise alterations (e.g. large deletion or higher level expression) allows for greater accuracy in interpretation. Finally, sophisticated genetic alterations that would have been unthinkable earlier are now used routinely for the analysis of protein function. The next few sections will address some of the variety of ways that genes have been altered *in vitro* and re-introduced into yeast.

A. Gene disruption

One of the fundamental advantages of modern yeast genetics is the ease with which recessive mutations can be introduced into the chromosome in place of the wild-type allele. Both in a functional sense and a technical sense the most useful mutation is a "null" allele in which all gene structure and function has been ablated. For this one mutation alone the interpretation of phenotype is free of the ambiguity of partial function. Furthermore, since the null allele is recessive to all other mutations, it is the starting point for the identification of additional conditional and partial loss of function mutations. Several different methods have been described to produce null mutations and introduce them into the genome. Each has certain advantages.

B. The two-step gene replacement

The first method described for gene disruption in yeast (Scherer and Davis, 1979) was based upon the prokaryotic paradigm of gene replacement by specialized transducing phage. As shown diagrammatically in Fig. 2, homologous recombination allows the integration of a YIp plasmid containing the gene of interest, along with a selectable plasmid marker. The recombination event results in a tandem repeat of the gene of interest, separated by the plasmid sequences. Ordinarily the introduced mutation will be present in one of the two copies; however, gene conversion associated with the cross-over event will occasionally result in both copies containing either the mutant or wild-type allele. Therefore, it is important to isolate a few independent transformants. The next step relies on a second homologous cross-over event in the repeated DNA to "loop-out" the plasmid. The plasmid, which is devoid of an origin of replication, is then lost by segregation. The second cross-over can occur in either of two topographically distinct regions defined by the locations of the disruption mutation and the integrating cross-over event. A second cross-over on the same side of the mutation as the first cross-over (I) regenerates the wild-type allele on the chromosome and the mutation is lost with the plasmid. A cross-over on the opposite side of the mutation (II) leaves the mutation on the chromosome and the wild-type allele is lost with the plasmid. Typically, the site of the initial cross-over event is experimentally determined by a double-strand break introduced by the investigator at a convenient restriction site. The position of the second cross-over is

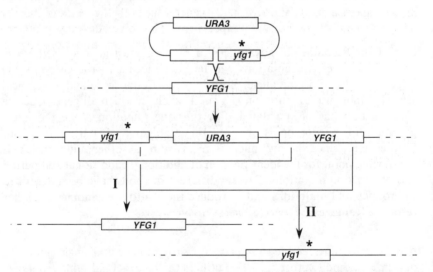

Fig. 2. Two-step gene replacement. A mutant form of *yfg1* is introduced on a YIp vector and integration occurs by homologous recombination. The plasmid can "loop-out" by homologous recombination, leaving either a mutant or wild-type copy of *YFG1* on the chromosome, depending upon the site of the cross-over.

probably random so that both types of loop-outs are obtained. In principle, the relative frequencies of the two types of loop-outs are influenced by the relative extent of homology on either side of the mutation. Therefore, it is preferable to linearize the plasmid by cleavage in the smaller region of homology to one side of the mutation.

Three other considerations affect the design of the experiment. First, the loop-out is performed in a diploid cell in case the mutation results in lethality or slow growth. The phenotype is then scored after sporulation. Second, a plasmid marker is used that can be selected either for or against. *URA3* can be selected for in a *ura3* auxotroph or against by using 5-FOA (Boeke *et al.*, 1984). Alternatively, *LYS2* can be selected for or against using a *lys2* auxotroph and α-amino-adipic acid (Chattoo *et al.*, 1979). However, if either of these markers cannot be used, the frequency of recombination between the direct repeats is often high enough (1/1000–1/10 000) to allow the loop-out events to be found by replica-plating. Finally, to establish that the resulting mutant phenotype is the result of the specific mutation, it is useful to incorporate a second scorable genetic marker at the site of the mutation. The resulting mutant phenotype can then be shown to arise from a mutation that is tightly linked to the integrated marker.

Although the method is somewhat more laborious than some described later, there are some technical advantages to the two-step method. First, two different isogenic strains (i.e. the two different loop-outs) can be derived from a single transformant, generating both the mutant and the wild-type control in the same experiment. This precludes any problems associated with the inadvertent generation of irrelevant mutations during transformation. Second, since the loop-out event is detected by a phenotype that is independent of the introduced mutation, a wide variety of mutations can be integrated including simple deletions and point mutations. There is no requirement to leave a selectable marker integrated at the site. Thus, subtle mutations can be integrated without leaving a large disruption of the normal chromosome structure and precious selectable markers are not "used up" in the construction.

C. The one-step gene replacement – "omega" transformation

It is sometimes preferable to replace or disrupt a gene in a single step. This one-step method (Rothstein, 1983) utilizes the fact that a double cross-over event (or gene conversion) can result in the complete replacement of the wild-type gene by the mutant form. To force the desired recombination event, the plasmid is cleaved twice, on both sides of the mutation. Both ends of the linear DNA will recombine with the chromosome, resulting in the integration of the mutation and the consequent loss of the wild-type DNA. The mutation is selected by the use of a marker integrated within the yeast sequences. A hypothetical DNA intermediate for the integration of the marker DNA looks much like an omega (Fig. 3).

Because the double cross-over event occurs at a considerably lower frequency than the single cross-over, some care must be taken to ensure that both sites are cleaved to completion. Typically, the transformants are checked to determine whether plasmid sequences have integrated, the signal of a single cross-over event. Obviously, transformation to disrupt an essential gene must be accomplished in a diploid strain. Since transformation of diploids sometimes generates recessive lethal mutations (possibly due to chromosome loss), it is especially important to check that recessive lethal mutations are linked to the integrated marker gene. One disadvantage to this method is that the mutant strain is separated from its isogenic parent by a potentially mutagenic transformation event. A second disadvantage is that each gene disruption uses up another marker gene. This can greatly complicate the construction of multiply mutant strains. A third disadvantage is the requirement for convenient restriction sites on either side of the mutation.

D. Recovery of useful genetic markers – disruption by excisable cassettes

Alani and Kleckner (1987) devised a clever method to reduce one of the problems associated with the one-step gene replacement whereby useful

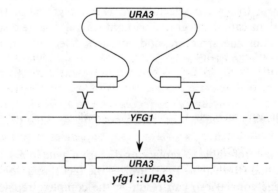

yfg1 :: URA3

Fig. 3. One-step gene replacement. A selectable marker is introduced into *YFG1* by *in vitro* recombinant DNA methods. Cleavage of the plasmid in yeast sequences on both sides of the gene produces a linear fragment that integrates by recombination at the two ends of the fragment. Integration is concomitant with replacement of the chromosomal sequences.

markers are used up at each round of gene disruption. In their method, a cassette was created in which *URA3* was embedded within DNA derived from the bacterial *hisG* gene. Importantly, the fragment from the *hisG* gene was duplicated in a direct fashion on either side of *URA3*. The *hisG–URA3–hisG* cassette was then used to produce a gene replacement by the one-step method, selecting for the *URA3* gene. Homologous recombination between the *hisG* repeats would then result in excision of the integrated *URA3* gene leaving a single copy of the *hisG* fragment on the chromosome at the site of the disruption. Thus the *ura3* auxotrophy was regenerated and the gene disruption was marked by DNA sequences that could be readily detected by Southern blot hybridization or PCR techniques. The great advantage of this method is that many more rounds of gene replacement using this extraordinarily useful marker could be applied. Because *ura3⁻* mutations can be directly selected in many organisms using 5-FOA, this is a generally useful approach. Thus only a single genetic marker is needed to disrupt multiple genes, a distinct benefit for working with any organism with less well-developed genetics.

Recently, this approach has been taken one step further by the construction of a series of plasmids that contain cassettes of selectable markers flanked by cyclic AMP response element (CRE) sites (Sauer, 1994). These sites are recognized by the very efficient LOX recombinase encoded by the bacterial phage P1. After the gene is disrupted in a standard fashion, the marker can be excised by regulated expression in yeast of the LOX recombinase. As a result, the auxotrophic marker is rapidly and efficiently restored while leaving the gene disrupted by a small fragment that is easily detected by PCR methods.

E. One-step gene disruption

An alternative method for one-step gene disruption has the advantage of rapidity (Shortle *et al.*, 1982a). A restriction fragment wholly internal to the structural gene of interest is placed on a YIp vector. Cleavage at a unique site internal to the structural gene forces the plasmid to integrate within the structural gene via a single cross-over event (Fig. 4). Like the two-step gene disruption, a tandem repeat of the yeast sequences is formed. However, because the cloned DNA fragment is wholly internal to the structural gene, both copies of the gene are truncated, one at the 5′ end and one at the 3′ end. The plasmid marker then serves as a marker for the site of the integration.

This method has the advantage of providing a quick answer and, because it relies on a single cross-over event, the desired transformant can be

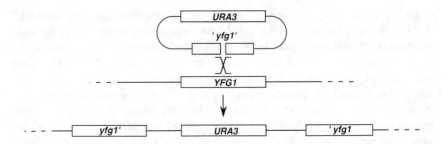

Fig. 4. One-step gene disruption—the xapper. The fragment of *YFG1* is truncated at both ends and integration leads to two singly truncated forms of the gene. *'yfg1*, N-terminal deletion; *yfg1'*, C-terminal deletion.

obtained efficiently. However, there are several disadvantages. First, the disrupted gene may not actually be a "null" allele. Truncated proteins may be produced, which can have confusing effects on phenotype. Second, unlike the previous two methods, the disruption can revert by recombination, a relatively frequent event. Third, only a single class of mutation can be produced, limiting its utility.

F. Gamma-transformation

Like the one-step gene disruption, gamma-transformation (Sikorski and Hieter, 1989) uses a specially constructed plasmid, which leads to the disruption of the gene upon integration. Because gamma-transformation can use the whole structural gene and any convenient flanking sequences, only one internal site must be identified. To construct the plasmid, two yeast DNA fragments are cloned into a YIp vector in reversed order while maintaining the same relative orientation. Thus the DNA region encoding the carboxy-terminal portion of the protein is placed 5' to the region encoding the amino-terminal part. The site originally internal to the gene is where the plasmid is fused and a novel junction is formed where sequences originally 3' are fused to sequences from the 5' end of the gene. When the plasmid is cleaved at the novel junction, integration can occur by a crossing-over event directed by the two free ends (Fig. 5). Thus the resulting integration event looks much like the one-step gene replacement but the entire plasmid has become integrated into the gene.

Although the construction of the plasmid appears to be somewhat convoluted, like the one-step gene disruption, the plasmid provides a rapid

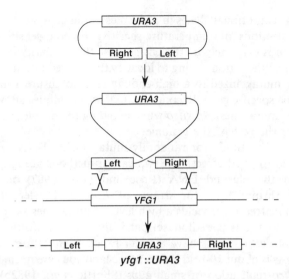

Fig. 5. One-step gene disruption—gamma-transformation. Two fragments from *YFG1* are placed on a YIp vector in inverted order. Cleavage between the yeast fragments causes recombination at two sites, leading to integration of the intact vector plasmid.

means for disrupting a gene in a number of different strains. However, unlike the one-step gene disruption, the gamma-transformant cannot revert. Moreover, if flanking DNA sequences are used, large portions or all of the structural gene may be deleted, providing confidence in the construction of the null allele.

G. Isolation of conditional alleles

Null alleles are most useful for establishing whether a gene is essential for a given process. However, there are several occasions when a conditional allele is required for further analysis. First, it is impossible to infer the function of a gene from an inviable spore. Second, even for viable null mutants, physiological or genetic adaptation by the mutant strain can lead to misleading phenotypes. This is particularly evident for slow-growing mutants where faster growing variants quickly outgrow the parent mutant strain. Finally, more sophisticated genetic studies (e.g. protein–protein interactions, structure/function analysis, and execution point analysis) require point mutations.

Classically, conditional alleles in a given essential gene were identified by screening collections of temperature-sensitive or cold-sensitive mutants. This approach is extremely inefficient because the similarity in phenotypes requires complementation testing to identify the desired mutations and cells could not be mutagenized to a high enough level to ensure a high yield of mutations in a specific gene. Recombinant DNA techniques allow extremely high levels of mutagenesis *in vitro* while avoiding the problem of randomly mutagenizing the rest of the genome.

A number of methods for randomly mutagenizing DNA *in vitro* have been exploited to good effect. The two easiest methods are hydroxylamine treatment of the plasmid DNA (Rose and Fink, 1987) or passage of the plasmid through mutator strains of *E. coli* (Fowler *et al.*, 1974). Hydroxylamine treatment yields a high level of mutagenesis (up to 5% loss of function mutations per kilobase) but is limited to transition mutations of C to T. Mutator strains have a wider range of mutations but the rate of mutagenesis is about 10-fold lower. More laborious are misincorporation during *in vitro* replication of small gaps (Shortle *et al.*, 1982b), and PCR using manganese (Leung *et al.*, 1989) or dITP to increase the error rate (Spee *et al.*, 1993). Both methods are highly efficient and capable of generating a wide spectrum of point mutations. Mutagenic PCR has the additional feature that the level of mutagenesis can be controlled by the concentration of manganese and so can be optimized for the DNA fragment.

A final method that is being used more frequently is the systematic replacement of small clusters of charged amino acids with alanines (Wertman *et al.*, 1992). The rationale is that clusters of charged residues are most likely to reside on the protein surface. Replacement with alanines should increase the stability of those protein conformations in which these residues are buried. Although some of the mutant proteins are inactive and others are unaffected, a surprising number are temperature sensitive. Presumably the transition to the buried conformation is favoured at the higher temperatures. The alanine-scanning technique has been particularly effective for cytoskeletal proteins such as actin (Wertman *et al.*, 1992) and β-tubulin (Reijo *et al.*, 1994). These proteins are otherwise quite difficult to mutate to temperature sensitivity, probably because of extensive stabilizing interactions in the polymer. The disadvantage of this method is that it is both labour intensive and expensive.

Three general methods have been described for identification of conditional mutations among mutagenized plasmids. Each has been designed to solve a similar set of technical problems. First, if the mutagenized gene is essential for viability, then the host cell will have a wild-type copy of the gene. Since the wild-type gene will usually be dominant to the conditional alleles, some method for eliminating the wild-type gene must be present. Second, it has generally been observed that a sizeable fraction of colonies appearing after transformation are slow growing and appear to be temperature sensitive. The background of these cells may be as high as several per cent of the transformant colonies and so may be an order of magnitude more frequent than the desired mutations. Therefore, the method must incorporate a rapid means for distinguishing plasmid-based mutations from the background. Ultimately, all methods must demonstrate that the mutation is due to an alteration in the coding region of the gene of interest.

H. Hemi-xapping

The first described method for identifying conditional mutations in an essential gene is a variation on the scheme for one-step gene disruption (Shortle *et al.*, 1984). If the yeast DNA fragment on the YIp plasmid is not completely internal to the structural gene, then integration of the plasmid will yield one intact copy of the gene and one truncated copy (Fig. 6). Restriction sites are judiciously chosen such that a major portion of the intact copy of the gene will be derived from the integrating plasmid. If the plasmid was mutagenized first, then the transformants will display the phenotype of the introduced gene. Because of the presence of repeated

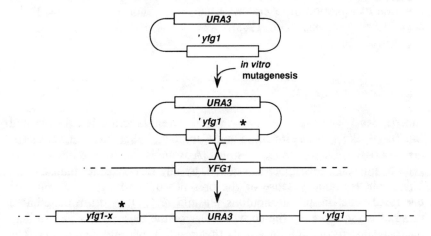

Fig. 6. Screening for mutations in essential genes—the hemi-xapper. A partially truncated form of *yfg1* is mutagenized *in vitro*. Integration leads to truncation of the unmutagenized chromosomal copy of the gene and transfer of the mutagenized sequences to the intact copy.

DNA sequences, the plasmid can loop-out by homologous recombination. Depending upon the cross-over site and the mutation site, plasmid excision will result in a high frequency of revertants to the wild phenotype. This provides a useful means for distinguishing plasmid-derived mutations from background mutations. The *URA3* gene is utilized as the plasmid marker and the plasmid excision events are selected by growth on 5-FOA media. If the mutation was within the DNA brought into the cell by the plasmid, then the excision strains should be greatly enriched for revertants. If the mutation was elsewhere, then there should be no enrichment for revertants among the excision strains. Once identified, a stable integrant of the mutation can be obtained as an excision strain that retained the mutant phenotype.

There are two major advantages of this method. First, the phenotype is scored on cells that carry a single copy of the mutant gene. This avoids gene dosage effects that arise in other mutant screens to be described below. Second, the integration of the mutant gene on to the chromosome is an inherent part of the method. The equivalent step in other methods requires additional plasmid construction followed by gene replacement. There are, however, some potential disadvantages that are related to the frequency of mutant isolation. Several steps along the protocol may reduce mutant recovery. Since recombination is required to integrate the mutagenized plasmid, this step is inherently less efficient than a simple transformation. Gene conversion extending from the cross-over site may lead to loss of the mutation by repair. Only part of the gene is examined in any one experiment. Nevertheless, this method has been extremely successful for the isolation of mutation in several essential genes (Shortle *et al.*, 1984; Holm *et al.*, 1985; Huffaker *et al.*, 1988).

J. The plasmid shuffle

Independently developed in at least two different laboratories, the plasmid shuffle (Boeke *et al.*, 1986; Budd and Campbell, 1987) makes clever use of genes carried on autonomous plasmids. A strain is constructed in which the gene of interest is disrupted or deleted, by any marker other than *URA3*. The strain is viable because of the presence of a wild-type copy of the disrupted gene on an autonomous plasmid (Fig. 7). A centromere-based plasmid is used to keep the copy number of the plasmid low and to help select against recombination events that result in plasmid integration. The covering plasmid is marked with a gene that can be selected either for or against, usually *URA3*. The strain is constructed by transforming the

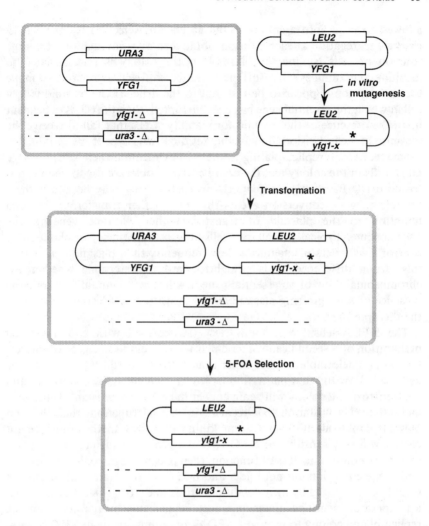

Fig. 7. Screening for mutations in essential genes—the plasmid shuffle. The chromosomal copy of *YFG1* is deleted and cell viability is maintained by an autonomous plasmid (YCp or YEp) containing an intact copy of the gene. The cell is transformed with a mutagenized plasmid carrying a second copy of the gene on a plasmid bearing a different selectable marker. Loss of the original *URA3*-based plasmid by selection on 5-FOA uncovers recessive mutations in *YFG1*.

covering plasmid into a diploid that is heterozygous for the disruption and then sporulating.

The basis of the plasmid shuffle is that the autonomous plasmid is not particularly stable; even a centromere-based plasmid may be lost at a rate of 1–2% per generation. If the plasmid were in a wild-type strain and allowed to grow into a colony in the absence of selection for the *URA3* marker, segregation at each division would lead to accumulation of a large fraction of cells lacking the plasmid. If the colony is replica-plated to medium containing 5-FOA, sufficient numbers of cells grow so as to make the entire colony appear to be resistant to the drug. If the plasmid is in a cell in which the essential gene has been disrupted, then the cell is dependent upon the presence of the plasmid for viability. Cells that fail to receive the plasmid at cell division die as the residual wild-type gene product is exhausted. Upon replica-plating to the 5-FOA medium there are no viable Ura⁻ cells in the colony and the entire colony would appear to be sensitive to the drug. In practice, a few cells in each colony may be Ura⁻; these arise from gene conversion events that have either transferred the *ura3* mutation to the plasmid or transferred the wild-type gene to the chromosome. Typically, the Ura⁻ cells arise at a frequency of about 10^{-6}, four or five orders of magnitude less frequently than plasmid loss. Given this strong difference, it is straightforward to determine whether the chromosomal copy of an essential gene is wild type or mutant, despite the presence of a suppressing plasmid and even without a selectable marker in the disrupted gene.

The 5-FOA-sensitive strain is now transformed with a mutagenized preparation of a second centromeric plasmid containing the gene of interest but using a selectable marker other than *URA3*. A wild-type copy of the gene will allow the first plasmid to be lost during non-selective growth and the transformant colony will again appear to be 5-FOA resistant. If the gene had suffered a mutation that resulted in loss of function, then the first plasmid could not be lost without killing the cell and the transformant colony will stay sensitive to 5-FOA. If the gene had suffered a mutation that led to conditional loss of function, then plasmid loss would be tolerated under one condition but not under another. Thus the mutations of interest would be recognized by their conditional resistance to 5-FOA. For example, a temperature-sensitive mutation would be detected by the ability of the replica plated colony to grow on 5-FOA-containing media at 23°C but not at 37°C. To distinguish the temperature-sensitive background mutations, the same colonies are replica-plated to media selecting for the original plasmid at both temperatures. Host mutations will be temperature sensitive on either medium. The desired mutations will be temperature resistant because the original plasmid containing the wild-type gene is still present.

There are several distinct advantages to the plasmid shuffle. First, recombination is not required at any step, therefore it is simple to prepare the transforming DNA and the transformation is efficient. Second, the entire gene is a potential target for mutagenesis. Third, it is easy to recover the mutant version of the gene by transformation of crude preparations of yeast DNA into *E. coli*. This allows rapid verification that the mutation is indeed contained within the plasmid, and also facilitates subsequent mapping and sequencing experiments. In one study, the plasmid shuffle successfully identified several mutations in the α-tubulin gene whereas few were found by the hemi-xapping method (Schatz *et al.*, 1988).

There are also several disadvantages to the plasmid shuffle that may mitigate its use under certain circumstances. Most of these stem from the fact that the mutation is carried on an autonomous centromeric plasmid. Such plasmids can vary in copy number from one to three or more copies per cell. Accordingly, different cells in the population can have very different phenotypes due to differing gene dosages. Indeed, additional copies of some mutant alleles can produce a phenotype that is indistinguishable from wild type. Therefore, a bias against subtle mutations is present in the plasmid shuffle. In addition, most kinds of further genetic analysis require that the mutation be integrated on to the chromosome so that a YIp plasmid must be constructed for a gene replacement. This can be tedious for large numbers of mutations. Finally, the cost of 5-FOA makes large mutant hunts prohibitively expensive.

A variation on the plasmid shuffle has been described that overcomes the requirement for 5-FOA (Bender and Pringle, 1991). This scheme makes use of mutations, *ade1* or *ade2*, that confer a red colour to the yeast. When either gene is mutant an intermediate in adenine biosynthesis accumulates and is then converted to a red pigment. The scheme is essentially identical to the plasmid shuffle with the exception that plasmid loss is detected by the appearance of red sectors in the colony due to plasmid segregation. For example, suppose the first plasmid contains the wild-type *ADE2* gene along with the gene of interest. Plasmid loss uncovers a recessive lethal mutation, so no red sectors appear in the colony which is uniformly white. Introduction of the second plasmid allows the first plasmid to segregate and frequent red sectors are observed. Mutations in the gene of interest again result in the death of the segregants and these are detected as uniform white colonies. This scheme saves the expense of the 5-FOA and because it works on primary transformants saves some manipulation of the colonies. One disadvantage to this scheme is that other frequent events can cause the cells to remain white, such as the loss of mitochondrial function. Therefore, there may be relatively higher numbers of false positives to be examined.

K. Regulated expression

An independent way to study gene function is to vary the concentration of the wild-type protein in the cell. There are two general ways in which regulated expression experiments are useful. First, as discussed above, overexpression frequently leads to novel phenotypes that can be informative as to the function of the protein (reviewed in Rine, 1991). Second, gene expression can be shut off using a well-regulated promoter like that associated with the *GAL1* gene (Deshaies *et al.*, 1988; Han and Grunstein, 1988; Vogel *et al.*, 1990). Like overexpression, shut-off experiments have several uses. The simplest of these is to determine the phenotype of cells in which the protein has been depleted by continued growth after synthesis of the protein has been shut off. One potential hazard with the shut-off experiment is that several generations may be required to deplete the wild-type protein. During the last generations, cells may adapt to lowered protein levels in ways that can confuse the analysis of the primary defect. Therefore, it is important to demonstrate that depletion of the wild-type protein confers the same phenotype as loss-of-function mutations. In addition, concordance between the two phenotypes can help verify that the mutant phenotype is not the result of an unusual mutant allele. Finally, a strong regulated promoter can allow a pulse-chase experiment whereby a protein is synthesized for a short period and then its synthesis is shut off to allow its post-translational fate to be assessed.

Often the simplest way to achieve overexpression is by simply placing the gene on a high copy number vector derived from the endogenous 2 μm plasmid. Many genes have been reported for which transformation on a high copy number plasmid results in novel phenotypes. However, some genes are not expressed well on high copy plasmids or are toxic when overexpressed. In addition, variability in the copy number of the plasmid can lead to non-uniformity in the phenotype. Therefore, the best experiment utilizes a strong inducible promoter, which can be integrated into the chromosome. The most useful promoter is P_{GAL1}, which can be induced rapidly over 1000-fold by the addition of galactose to cells growing on a non-fermentable carbon source. At the same time, P_{GAL1} is also controlled by catabolite repression; the gene can be quickly and effectively shut off by addition of glucose to a culture growing on galactose. Methods for constructing and using promoter fusions for overexpression in yeast are described in Schneider and Guarente (1991) and Schena *et al.* (1991).

An alternative way to achieve the regulated expression of a gene is by use of the regulated "silencing" that causes repression of the non-transcribed "silent" copies of the mating-type loci. The repressing effect extends into the chromosomal regions flanking the silent loci such that neighbouring genes introduced into these regions are also silenced (Schnell and Rine, 1986). There are several mutations in genes called *SIR* (for silent information regulator) that destroy the repression, and thereby allow the expression of the silent loci and surrounding sequences. A temperature-sensitive mutation in *SIR4* allows conditional expression of the region. To use the silencing effect, the gene of interest is integrated into or adjacent to a silent locus, and the temperature-sensitive *sir4* mutation is used to regulate expression in a temperature-dependent manner (van Zyl *et al.*, 1989). Either the gene's own promoter or a stronger one may be used to express the gene. One caveat to this method is that any given gene may not be silenced to the same extent as the mating-type loci.

One general problem with shut-off experiments is the very long time that some proteins require to be depleted from the cell. The number of generations required before the protein is effectively depleted is dependent upon the level of overexpression relative to the normal level of the protein, the stability of the protein and the fraction of protein activity that is required for growth. A notable exception to the general observation that multiple generations are required are the cyclins, which are rapidly depleted because they are specifically degraded in each cell cycle (Cross, 1990). Following this lead, a strategy was devised to cause any protein to be turned over rapidly by making use of the N-end rule for protein turnover (Park *et al.*, 1992). To utilize this system a hybrid is made between the gene of interest and an amino-terminal ubiquitin gene. Normally, the ubiquitin protein is rapidly cleaved from the hybrid protein at a specific site leaving a defined amino-terminal residue. According to the N-end rule the identity of the N-terminal residue will determine the rate of turnover of the protein. In practice, the level of the protein would be determined by a competition between synthesis and turnover. As before, the hybrid would be expressed from P_{GAL1} and then synthesis would be shut off by glucose addition. Judicious use of the N-end rule would then allow rapid degradation and protein depletion. In principle, degradation can occur with half-lives on the order of a minute or less. With such rapid kinetics, experiments identical to those performed with temperature-sensitive mutations can be performed. Indeed, by providing an independent means of inactivating proteins, the determination of the order of functions in a pathway is greatly facilitated.

A variation on the N-end rule degradation scheme has recently been described in which a temperature-sensitive dihydrofolate reductase (DHFR) is included between the protein of interest and the N-terminal ubiquitin domain to form a trihybrid protein (Dohmen *et al.*, 1994). As before, the ubiquitin domain is cleaved rapidly from the protein. Because of the N-end rule, the degradation rate of the entire hybrid protein is governed by the turnover of the DHFR domain. The DHFR mutation was selected to be stable at 25°C but rapidly degraded at 37°C. Thus the DHFR domain serves as a portable signal for the temperature-dependent degradation of any protein to which it is fused. This method should find utility for any protein for which it is difficult to isolate a temperature-sensitive mutation and where the hybrid protein is functional.

V. HYBRID GENES

One large class of mutations that have found considerable use are gene fusions. So many different types of fusions have been constructed that it is difficult to generalize about their uses. However, in the broadest sense, they all serve to couple an easily assayed activity to a region of the gene or protein that contains a functional determinant of interest. The extent of the gene that is incorporated into the hybrid can vary from almost none to all. At one extreme are the familiar gene fusions designed to study promoter activity. These may contain only a handful of nucleotides from the upstream non-coding sequences fused to the structural gene of a readily measured enzyme such as β-galactosidase. At the opposite extreme are fusions designed to study the localization of the native protein. These hybrids may contain all of the protein fused to a small peptide that serves as an epitope for a monoclonal antibody. In the middle of this range are hybrids designed to identify a variety of functional domains and localization sequences. Some examples of each will be discussed below.

A. Regulation — cis-acting sequences

A very large body of literature exists that describes the use of gene fusion technology to identify and analyse DNA sequences that control transcription. In general, the following strategy is followed. The E. coli lacZ gene is fused close to the 5′ end of the structural gene of interest. Expression of β-galactosidase is then shown to be regulated in a manner appropriate to the original gene. This demonstrates that all of the relevant regulatory elements lie upstream of the coding sequences. A series of 3′ and 5′ nested deletions are then made in the upstream non-coding region. The eventual loss of expression and/or regulation delineates the outer boundaries of the regions required for expression. Smaller deletions and replacements are then used to determine which regions are actually required for expression or regulation. The analysis is complicated by the frequent observation of multiple and repeated regulatory elements whose action may be redundant or synergistic. Thus the final proof that the sites are those that are both necessary and sufficient requires the reconstruction of the functional regulatory elements upstream of a promoterless gene, e.g. one that has already had its upstream region deleted. In this manner, a very large number of regulatory elements have been delineated in yeast and other organisms.

B. Localization

Often one of the most useful pieces of information about a protein of unknown function is its localization within the cell. Often a large number of hypotheses can be excluded by the location of the protein. For example, localization within a membranous organelle immediately rules out any model of the protein acting within the cytoplasm. One of the major issues in the genetic analysis is whether a protein acts directly or indirectly to exert its effects. A strong argument for a direct action can be made if the protein actually localizes to the presumed site of action.

The best localization experiments use antibodies that recognize the authentic protein expressed at its normal level within the cell (reviewed in Pringle *et al.*, 1991). However, there are often technical limitations that interfere with this ideal. The most severe problems are associated with proteins that are not expressed at high enough level to produce a detectable signal by immunofluorescence. In some cases antibody "sandwiching" techniques can boost the signal into a detectable range (Page and Snyder, 1992). In other cases overexpression is required to detect the protein. This last method has the severe problem that overexpressed proteins may mislocalize or otherwise obscure the authentic signal. Therefore any localization experiment based upon overexpressed protein must include an independent means for assessing the legitimacy of the results.

A second approach to localization is to fuse small peptide epitopes to the protein. The hybrid proteins can often be shown to be fully active *in vivo* and in such cases it is likely that their localization patterns reflect the localization of the authentic proteins (Davis and Fink, 1990; Roof *et al.*, 1992). One advantage of peptide epitopes is that they are recognized by commercially available monoclonal antibodies. Thus the proteins can be detected by antibodies of high affinity and very low cross-reactivity to other yeast proteins. Two epitopes that are often used are segments from the *myc* oncogene and the influenza haemagglutinin. Since the antibodies are bivalent, affinity can be further enhanced by incorporation of multiple copies of the epitope. Very good results have been obtained from using three repeats of the haemagglutinin epitope (Roof *et al.*, 1992). However, there are some limitations to the epitope tagging approach. First, regardless of the affinity of the antibody for binding, ultimately the fluorescent signal is limited by the number of antibody molecules that can bind per protein and this in turn is limited by the number of distinct epitopes. Therefore, polyclonal antibodies should inherently produce a brighter immuno-fluorescent signal. Second, because only a single epitope is recognized, antibody recognition can be sensitive to both protein context and cell fixation. For example, the haemagglutinin epitope can be very sensitive to formaldehyde fixation (Davis and Fink, 1990).

An alternative to the epitope-tagging method is to fuse the protein to another whole protein, such as β-galactosidase (Hall *et al.*, 1984; Vallen *et al.*, 1992). One advantage to this approach is that both monoclonal and polyclonal antibodies are commercially available. A second is that β-galactosidase has been intensively studied and its characteristics are well known. An obvious disadvantage is that fusion to the large β-galactosidase molecule may seriously disrupt function. Nevertheless, some β-galactosidase hybrids retain normal function and so these hybrids probably localize normally. For one particular use, the dissection of proteins into functional domains, there are also some important benefits that accrue from the use of β-galactosidase. First, the β-galactosidase hybrids are ordinarily quite stable in yeast. Second, because β-galactosidase forms a tetramer in the cell, small functional domains present in the hybrids will be displayed in an essentially oligomeric manner. The oligomerization deters problems of reduced binding due to the loss of a dimerization domain from the protein of interest. It may even serve to enhance binding to polymers. Clearly there is substantial opportunity for artefact when using the hybrids to localize proteins and much care must be taken to ensure that the detected localization pattern is biologically relevant.

C. Localization and targeting sequences

A number of different localization or targeting sequences have been discovered in proteins that determine their location within the cell. The most familiar targeting sequence is the hydrophobic "signal" sequence found at the N-terminus of secreted proteins (von Heijne, 1990). Also present at the N-terminus is the mitochondrial targeting sequence comprised of an excess of basic residues, often arrayed to form an amphipathic α-helix (reviewed in Pon and Schatz, 1991). Small basic regions can serve as nuclear targeting signals (reviewed in Silver, 1991). In each of these examples, proteins are directed to the appropriate compartment by the display of specific determinants that exist as connected linear elements in the primary protein sequence. This property has allowed the signal sequences to be readily dissected from their native proteins and then to be fused to other marker proteins, where they retain their targeting activity and direct the hybrid protein to the specific compartment.

Most studies have sought to determine a specific targeting signal within a specific protein, in much the same manner as regulatory sequences have been delineated. One very original study sought to characterize the class of sequences required for import into the secretory system (Kaiser *et al.*, 1987). In this approach, random DNA fragments were fused on to a truncated form of the *SUC2* gene, which expressed a form of invertase lacking the N-terminal signal sequence. A remarkably large number of sequences would suffice for some level of secretion and these all shared hydrophobicity and the absence of charged residues as a common determinant.

While many targeting sequences are located at the N-terminus, others are found at the C-terminus. Two examples of these are the endoplasmic reticulum retention signal (Munro and Pelham, 1987) and regions directing association with various membranes (e.g. *ras*). In principle, each of these can be defined by appropriate gene fusions.

D. Membrane topology

One use for hybrid proteins is mapping of the topology of integral membrane proteins. The strategy is to fuse the marker protein at various positions within the membrane protein. The marker protein domain will end up on either the cytoplasmic or ER luminal face depending upon the number and orientation of membrane spanning domains in the N-terminal portion of the membrane protein. Two different proteins have been used: invertase (Feldheim *et al.*, 1992) and the C-terminal domain of the tripartite product of the *HIS4* gene, histidinol dehydrogenase (Sengstag *et al.*, 1990). Invertase is used because it has up to 12 potential glycosylation sites, so its appearance in the ER lumen is marked by a very large increase in molecular weight, which is sensitive to drugs, such as tunicamycin, that block glycosylation. In the case of histidinol dehydrogenase, the enzyme must act in the cytoplasm to convert histidinol to histidine. Thus the location of the protein is signalled by the ability of a histidine auxotroph to grow on histidinol. In principle, β-galactosidase could also be used as glycosylation inactivates the protein (M. Rose, unpublished observation). As elegant as this genetic approach is, it is not without its caveats. In the case of HMG-CoA reductase, although most of the His4C hybrids gave results consistent with expectation from protein sequence, some were confusing and inconsistent (Sengstag *et al.*, 1990).

E. Domain mapping

One of the most powerful concepts in protein structure is that many large and complex proteins are modular. They are essentially constructed from a number of separate structural domains joined together as "natural" gene fusions. Often the protein domains can fold into a functional structure independent of other domains. This phenomenon has been exploited by the use of artificial gene fusions to map and identify functional domains in complex proteins. There are many examples of the use of gene fusions for domain mapping; a few examples will serve to illustrate.

One of the most elegant examples of domain mapping is the detailed dissection of DNA binding regulatory proteins such as that done for Gal4 protein by a number of different laboratories (reviewed in Ptashne, 1988). Gal4 protein contains separate domains for DNA binding, transcriptional activation and binding to a negative regulatory protein encoded by *GAL80*. The basic principle upon which these experiments were based was that equivalent functional domains from different proteins are essentially interchangeable. Once one functional domain was identified, it could be replaced by the equivalent domain from another well-characterized protein and then the next functional domain could be mapped. Thus the DNA binding domain from a bacterial DNA binding protein, lexA, could replace that of Gal4p as long as the appropriate DNA binding site was supplied. Likewise transcriptional activator domains from viral proteins (or random DNA encoding acidic stretches) could replace the activation domain of Gal4p. At the extreme a completely artificial protein composed of domains cobbled together from separate sources could be shown to activate transcription in yeast functionally (Ma and Ptashne, 1987). It is rare indeed, that the chemist's criterion of synthesis as the final proof of structure has been utilized in any biological system of such complexity.

VI. GENETIC MAPPING – IN AND OUT OF THE GENE

Ultimately, all genetic arguments are founded upon the idea that phenotypic differences are based upon structural changes in the gene. Thus, there are many conclusions that rest upon the identity and positions of specific alleles. While complementation tests provide a rough functional test for determining whether different mutations are likely to be allelic, there are numerous examples of misleading results from the standard complementation test. Ultimately allelism must be established by a test that establishes allele position. Another inherent weakness to the complementation test as a practical means of testing allelism is that one can only perform the test between mutations having a similar mutant phenotype. Unfortunately, the full spectrum of phenotypes may not have been identified or reported for many mutations, some genes might only be known by rather specific mutant alleles and some combinations of mutations may not be testable (e.g. cold sensitivity and temperature sensitivity). Thus there can be no guarantee that all of the appropriate genes will have been tested for allelism. It is often more efficient to map a given mutation or gene positionally because this allows one to rule out all previously mapped genes as being allelic.

A. Chromosomal mapping

Classic mapping techniques in yeast are hindered by the high rate of recombination and relatively large number of chromosomes. Consequently, unless a gene happened to be centromere linked, establishing linkage to known genes was often a hit-or-miss affair. Although several genetic techniques have been devised to effectively reduce recombination, these have been supplanted by a simple physical method. Using the special geometry and alternating electrical fields of clamped hexagonal electric fields (CHEF) gels, all of the chromosomes of *Sacch. cerevisiae* can be physically separated on a single gel (Carle and Olson, 1985; Olson, 1991). The separated chromosomes are then blotted to a filter. The filter can then be probed by DNA hybridization techniques using DNA from the gene of interest to identify the chromosome of origin. Thus, if a mutation is recessive, the fastest way of mapping it is to clone the gene and then to use it as a hybridization probe.

In some rare cases, a gene cannot be cloned easily. In addition, the cloning of dominant mutations requires that a new library of genomic plasmid clones be generated for each gene to be cloned. Therefore, an additional method of identifying the chromosome is required. Perhaps the fastest method uses a variation of the 2 μm mapping method first described by Falco and Botstein (1983). This method is based upon the observation that an integrated copy of the 2 μm plasmid destabilizes a chromosome into which it has been integrated. Destabilization arises from FLP-recombinase catalysed recombination occurring between specific inverted repeat sequences present on the plasmid. Cross-overs occurring between the inverted repeats on sister chromatids lead to the formation of a dicentric chromosome and an acentric fragment. Consequently, all of the DNA that is centromere-distal to the integrated plasmid is lost. The remaining sequences are lost with a lower frequency depending upon when and where the dicentric chromosome breaks and how the break is "healed". If the gene had been cloned, then it would be used to integrate a plasmid containing 2 μm sequences. The strain was then mated to tester strains containing a number of recessive mutations on a variety of chromosomes. The 2 μm catalysed chromosome-loss events were detected by loss of the plasmid marker gene. The presumptive $2n - 1$ aneuploids were then tested to determine which recessive mutations were uncovered. The frequency with which the recessive marker was uncovered provided both chromosomal and positional information. The current incarnation of the 2 μm mapping method utilizes a set of 16 strains; in each strain one chromosome contains an integrated copy of a 2 μm plasmid (Wakem and Sherman, 1990). Thus any given recessive mutant would be crossed to the complete set to produce 16 diploids. Aneuploids resulting from chromosome loss would be collected and tested for whether the recessive mutation had been uncovered, thereby identifying the chromosome of interest. Dominant mutations are more difficult to map by this method, but this can be accomplished by crossing the dominant mutation into each 2 μm integrant strain and then mating the resulting strains to the recessive wild-type haploid.

Once the chromosome has been identified, the next step is to identify the gene's position on the chromosome by genetic and physical methods. The ease of this step has been revolutionized by the development of complete ordered set of chromosome fragments cloned into lambda phage (Riles *et al.*, 1993). A few filters can be obtained from the ATCC that contain essentially the entire yeast genome divided into about 1000 phages. The cloned gene is hybridized to the filter and the hybridizing spot identified. Given a DNA clone and a set of filters, a gene can be mapped within a few days. After identifying the physical location, the mutation of interest can be mapped against neighbouring genes by standard genetic techniques.

B. Fine-structure mapping

Once a series of mutations have been identified that appear to be allelic, the next level of analysis is to determine their exact location within the gene. While any given project may have its own rationale, some reasons for fine-structure mapping include: (i) the clustering of mutant alleles can provide important information about protein structure and function; (ii) specific mutations may provide important evidence about enzyme mechanism; (iii) given the possibility of multiple mutations, fine-structure mapping provides proof that an observed phenotypic change is due to a specific mutation; and (iv) the DNA sequence of the entire mutagenized gene need not be determined if the region of the gene containing the relevant mutation has been functionally determined.

Several different schemes have been developed to allow fine-structure mapping in yeast. The most efficient methods use recombinant DNA techniques coupled to the high level of recombination in yeast *in vivo*. All methods begin with the construction of a set of nested or overlapping deletions of the cloned gene. For these purposes, deletions are essential because they cannot revert and they provide unambiguous physical reference points. The deletions can be constructed using convenient restriction cleavage sites or as a by-product of DNA sequencing. The deletion constructs are then transformed into yeast strains containing point mutations to be mapped. Recombination events between the point mutations, and the deletions are selected and classified as to whether wild-type recombinants appear. A priori, the appearance of wild-type recombinants indicates that the point mutation is not within the region covered by the deletion. The failure to recover wild-type recombinants indicates that the mutation is close to or covered by the deletion. The specific strategy for generating and selecting the recombinants varies but the discovery that double-strand breaks stimulate recombination has greatly facilitated mapping (Orr-Weaver *et al.*, 1981). If the novel junction at the site of the deletion is cleaved with a restriction enzyme, then successful transformation will require a recombination event to heal the gap (Orr-Weaver *et al.*, 1983). This is true whether the vector is of the integrating type or has sequences that allow autonomous replication. Selecting transformants by virtue of the plasmid marker allows one to collect a large number of recombination events in the gene of interest. Whether any of the transformants have been restored to the wild phenotype is then determined. If the point mutation is not covered by the deletion then as many as 50% of the transformants may be wild type. Mutations close to a deletion end-point may give only a few per cent wild-type transformants suggesting that the gap may be enlarged during transformation or recombination. By these means mutations can be mapped to a specific deletion interval, even if the phenotype is not one that can be selected for directly (e.g. altered levels of β-galactosidase activity). One additional advantage is that by avoiding direct selection for the wild type, reversion events are minimal.

So far it has been assumed that the point mutations are present on the chromosome. However, this need not be the case. As long as the entire gene is deleted on the chromosome to prevent it from recombining, then point mutations may be on autonomous plasmids or even (if the gene is not essential) transformed into yeast along with the gapped plasmid (Kunes et al., 1987).

Finally, one special case of mapping should be mentioned. In many instances interesting mutant alleles will be present on the chromosome and need to be recovered for other purposes. Because recombinational repair of the gapped plasmid occurs with high fidelity, plasmids can be repaired with mutant sequences subsequently recovered in E. coli (Orr-Weaver et al., 1983). Indeed, this principle can be extended to allow recombination in yeast to construct new plasmid combinations, as long as a double-strand break is introduced in appropriate regions of homologous sequences (Ma et al., 1987).

VII. INTERACTION GENETICS – FINDING NEW GENES

The ultimate aims of the genetic approach are to define all of the components in a pathway and then to order the components by analysis of their interactions. The simple isolation of mutations that block function is a first step in the analysis, but such procedures are fundamentally limited by the assumptions that are built into the selection scheme. Therefore additional methods are required to identify other components. Usually, the next step is to search for mutations that modify, either positively or negatively, the phenotypes of the original mutations. The putative interacting genes must then be shown to be directly involved in the process under study. Secondarily, the pattern of genetic interactions may reveal important clues as to the order of the pathway and any physical interactions between the components.

A. Suppressors

In the classic method of identifying genetic interactions, suppressor analysis, reversion of the phenotype conferred by the original mutation is simply selected for. Earlier, suppression by genes present on high copy number plasmids was discussed. In this section suppressors that arise by mutation will be discussed. There are many possible mechanisms of suppression, which act at different stages of gene expression and protein function. Given the current emphasis on protein function, the most interesting suppressors are those that act in the same pathway or where proteins directly interact.

How are the different suppressors distinguished? First, it should be appreciated that the initial mutation will ultimately govern which set of suppressors are obtained. For example, nonsense mutations will be likely to yield mostly informational suppressors, tRNAs with anti-codons complementary to the termination codon. Therefore, if suppressors in interacting proteins are desired, it is best to start with mutations that will produce an intact protein. Often temperature-sensitive alleles are used but even with these mutations care must be taken. Some temperature-sensitive alleles result from chain-termination mutations. Some temperature-sensitive mutations produce inactive proteins only when the protein is assembled at the high temperature; the pre-existing protein remains functional. Suppressors of such a mutation, if they can be obtained, are likely to affect interactions with the nascent protein and not the mature protein. Therefore, it is best to start with a mutant protein that is temperature-sensitive for function or stability. In that case, alteration of an interacting protein may stabilize the mutant protein.

After candidate suppressors are obtained they are crossed to wild-type strains to distinguish intragenic revertants from extragenic suppressors. The intragenic revertants will yield only wild-type progeny from such a cross. Unlinked extragenic suppressors will yield 1/4 progeny bearing the original phenotype, assuming that the suppressor has no phenotype on its own. If the suppressor has a mutant phenotype then an additional mutant class will appear. The ratios will depend upon whether the suppressor's mutant phenotype is suppressed by the original mutation. In either case, it is the reappearance of the original mutation that is diagnostic for the extragenic suppressors.

One criterion that is determined early in the analysis is whether the suppressor mutation is dominant or recessive. Most intragenic revertants will be dominant as they correspond to a gain of function. Extragenic suppressors may be either recessive or dominant depending upon the mechanism of suppression. Either class may arise from mutations that indirectly affect the physiology of the cell or the regulation of the gene. However, most suppressor mutations in interacting proteins are expected to be dominant, because in those cases suppression is thought to arise from an alteration in the physical interaction between the two proteins. In contrast, most suppressors that act by blocking a pathway are expected to be recessive, since simple loss of function should be sufficient. Clearly, it is difficult to predict which suppressors will be the most informative and therefore both dominant and recessive classes are chosen for further study.

In some cases the dominant suppressors may be quite rare and lost among a great excess of recessive suppressors. In this case dominant suppressors may be selected for directly in a diploid strain that is homozygous for the original mutation. After sporulation, the progeny are examined to identify those that are true dominant mutations and those that are recessive, but which became homozygous as a result of mitotic recombination. It is convenient to integrate a genetic marker next to one of the copies of the original mutation, as this will allow the intragenic revertants to be easily distinguished by their linkage to the marker.

In principle, mutations that affect steps in the same pathway or protein components of the same complex should result in similar phenotypes. To facilitate the analysis, suppressor mutations are chosen that have a secondary phenotype separate from suppression. For example, suppressors of a temperature-sensitive mutation might be screened for those that confer a cold-sensitive defect. In addition to allowing a quick determination of the loss-of-function phenotype, this procedure also provides a recessive phenotype for the cloning of the suppressor gene. However, the requirement of a secondary phenotype is quite stringent. Some interacting proteins may not easily yield conditional phenotypes. In some examples, the secondary phenotype was only apparent after the suppressor was crossed into a wild-type background (Adams and Botstein, 1989; Vallen et al., 1994) If necessary, the suppressor gene can be cloned by making a plasmid library from the suppressor strain.

Two important criteria for assessing the relevance of a suppressor are gene and allele specificity. Ideally, the class of suppression that arises from direct physical interaction between two mutant proteins should be both gene and allele specific. That is, the suppressor should act only on mutations in the original gene and only on a very restricted subset of mutant alleles. The latter constraint arises from the assumption that suppression is the manifestation of a precise stereospecific interaction between the two proteins; therefore only certain combinations should be effective. In contrast, some suppressors that act indirectly (e.g. nonsense suppressors or modifiers of the intracellular pH) should be capable of suppressing mutations in other genes. Other suppressors that act indirectly but which have a relevant function (e.g. by acting downstream in a pathway) should be capable of suppressing multiple alleles in the same gene and may suppress other mutations in the same pathway. In practice, these criteria are not often tested rigorously. Often a sufficient number of different mutant alleles do not exist to allow a credible test to be performed. However, one allele that can and should be tested is the null allele, as suppression of the null would rule out suppression by direct interaction. In contrast, gene specificity may be difficult to assess as it is not always clear which genes/alleles should be tested. There is always the possibility that the next gene would reveal a lack of specificity. Furthermore, examples are known in which suppressor alleles in proteins that do interact are neither strictly allele or gene specific. There are also examples of proteins that do not physically interact but which nevertheless appear to give allele-specific patterns of suppression. Therefore, although specificity is an important method of characterizing suppressors, it is not a rigorous test of whether proteins interact physically. Instead, suppression should be utilized as a means of identifying strong candidates for proteins that might interact or act in the same pathway. Demonstrations of physical interaction require direct tests *in vitro*.

B. Synthetic lethals

The flip side of suppression is called "synthetic lethality", the observation that certain pairs of mutations in different genes result in a phenotype that is more severe than either mutation alone (reviewed in Huffaker *et al.*, 1987). For example, two different mutations might cause temperature-sensitive growth but the double mutant is inviable at any temperature. In the past, such synergism was dismissed as resulting from the non-specific interaction between two different "sick" mutations. However, in one study of mutations in the secretory pathway (Kaiser and Schekman, 1990), it was demonstrated that double mutant combinations are not generally more severe than single mutants. A strong correlation was observed between the appearance of synthetic–lethal interactions and whether the two mutations affected the same step in the secretory pathway. Double mutants constructed with mutations acting at different steps were not more severe than the most defective single mutant; mutations affecting the same step had a high likelihood of producing synthetic lethality. The interpretation of this phenomenon is that synthetic lethality is likely to arise when proteins act together in a complex. Partial loss of function for either protein might not be severe enough to inactivate the complex at the permissive temperature. However, the combination of both mutations would so compromise the complex that function would be lost even at the permissive temperature. In confirmation of this hypothesis it was demonstrated that two of the genes showing synthetic lethality, *SEC17* and *SEC18*, encode proteins that interact *in vitro*.

The phenomenon of synthetic lethality (or more properly "synthetic phenotypes", since under some conditions the double mutant may be viable) has been observed in several systems where proteins are known to interact directly. Therefore, the observation is often cited as evidence for direct physical interaction. However, there are also several clear examples in which synthetic lethality arises between mutations in genes that almost certainly do not interact directly. For example, synthetic lethality arises for proteins that are functionally redundant, such that neither mutation is inviable, but both together are inviable. For example, this is observed for members of the kinesin gene family (Hoyt *et al.*, 1992; Roof *et al.*, 1992). A different type of synthetic lethality may arise for physiological reasons, as when a mutant remains viable because of the indirect activity of a second gene. For example, arginine auxotrophs are viable because a specific permease, encoded by the *CAN1* gene, allows the uptake of arginine. Thus, a *can1 arg4* double mutant is inviable on synthetic media even when arginine is supplied. Although the connection between these two genes is obvious, in the more usual situation one has very little information about the nature of two interacting genes. Therefore, synthetic lethality should not be taken as a priori evidence for interaction. However, like suppression, the observation of synthetic lethality presents a strong case to justify further investigation.

Because synthetic lethality relies on loss-of-function mutations, rather than gain-of-function mutations, it is likely that it is inherently easier to detect interactions by this method than by suppression. Theoretically, for multiprotein complexes, any mutation that destabilizes one component should lead to an overall destabilization of the complex. Thus many different combinations of double mutants should result in synthetic phenotypes. Based upon empirical observations, synthetic lethal interactions are, in fact, relatively insensitive to allele specificity.

Three different methods have been described to allow the isolation of mutations that confer a synthetic lethal phenotype. In two of the schemes, a plasmid marker is used as a reporter to signal the presence of a synthetic lethal mutation (Bender and Pringle, 1991; Roof *et al.*, 1992). A mutation in the first gene is present on the chromosome and an intact copy of the same gene is present on an autonomous plasmid (Fig. 8). *URA3* and *ADE3* are both used as plasmid markers. During the course of growth on non-selective media, plasmid mis-segregation at each division results in some cells in the colony that have lost the plasmid. For the *URA3* gene, these are detected by replica-plating to 5-FOA-containing media; frequent plasmid loss yields a sufficient number of Ura$^-$ cells to make the colony appear to be wholly 5-FOA resistant. For the *ADE3* marker, the host strain is both *ade2* and *ade3*. The *ade3* mutation prevents the accumulation of the pigment that turns *ade2* mutants red. Loss of the *ADE3* plasmid is observed as sectors of red cells in otherwise white colonies. In either case, the mutagenized colonies are grown under conditions where plasmid loss should be tolerated, either because the gene is not essential or because the cells are incubated at the permissive temperature for the original conditional mutation. Colonies that fail to lose the plasmid are detected either by

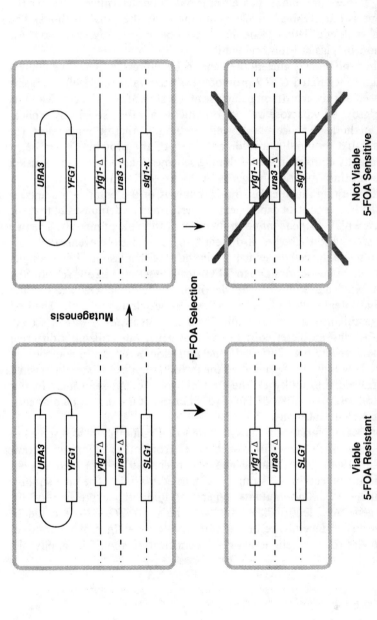

Fig. 8. Synthetic-lethal mutant screen. A chromosomal mutation in *YFG1* is covered by an intact copy of the gene. For convenience it is assumed here that *YFG1* is not essential for viability and that a deletion is used; however, a mild mutation in an essential gene could be used. In the parental strain, loss of the covering plasmid is tolerated and Ura⁻ segregants grow on 5-FOA. In a strain bearing a mutation in a gene that is redundant or which interacts with *YFG1*, loss of the plasmid is not tolerated and Ura⁻ segregants are inviable.

sensitivity to 5-FOA or by the absence of red sectors. The putative mutants are then sorted into complementation groups and determined whether the phenotype is due to single nuclear mutations by the usual methods. The putative synthetic lethal gene can then be cloned by screening for restoration of plasmid mis-segregation.

The third method engineers the gene to be expressed under the control of the tightly regulated *GAL1* promoter (Mitsuzawa *et al.*, 1989). The gene is expressed during growth on galactose and is shut off on glucose. Mutants that can grow on galactose but not on glucose are then sought by replica-plating methods. The caveat to this method is that it may select for mutations that required extraordinary levels of the protein. Otherwise, it is conceptually equivalent to the cloning schemes that look for suppression of a pre-existing mutation by genes on high copy number plasmids.

Before investing much effort on the analysis of the putative mutants, it is essential that a control experiment be performed to demonstrate that the mutants are really dependent upon function of the gene of interest. The two plasmid segregation schemes also identify other undesired events that only appear to make the cell dependent upon the specific plasmid. For example, mutants that are supersensitive to 5-FOA can be selected in one scheme and upstream *ade*⁻ mutations can be identified in the other. Gene conversion events that integrate the plasmid marker can be selected in both. The best way to distinguish the desired mutations is to transform with a second plasmid bearing an intact copy of the first gene, but utilizing a different selectable marker, e.g. *LEU2*. If the mutation has truly made the cell dependent upon the first gene, then the introduction of the second plasmid should restore the ability of the first plasmid to mis-segregate. In our experience, less than 10% of putative mutations turn out to be authentic synthetic lethal mutations.

An interesting variant on the synthetic lethal mutant hunt is a screen to identify yeast homologues of cloned genes from other organisms (Kranz and Holm, 1990). The cloned gene is engineered to be expressed in yeast and then introduced into a strain on a marked plasmid. The host strain is then mutagenized and mutants dependent upon the expression of the foreign gene are identified. In this way the yeast *TOP2* gene could be identified using dependence on the *Drosophila* gene. In principle, any of the strategies for synthetic lethal gene identification should be applicable.

C. Intragenic complementation and unlinked non-complementation

In a standard complementation test, two mutants bearing recessive mutations are crossed together to produce a diploid. If the mutations are in different genes then the diploid will necessarily be in the heterozygous state for both genes. Since the mutations are both recessive, the resulting diploid should be phenotypically a wild type. If the mutations are both alleles of the same gene, then the diploid is effectively homozygous for loss-of-function and should display the mutant phenotype. Two exceptions have been observed for this simple scheme. Perhaps the easiest to understand is the phenomenon of intragenic complementation for which there are two straightforward interpretations. In the first case, the protein has more than one independent functional domain and any single allele may knock out one or more of the domains. As long as the domains can act independently as, for example, if they produce diffusible intermediates or bind to different proteins, then mis-sense mutations in the different domains would complement each other, since functional copies of each domain are present in the diploid. The gene would be recognized as being a single entity by mutations such as deletions or nonsense mutations that would fail to complement alleles in either domain and by the genetic linkage of the various alleles. In this example, intragenic complementation is expected to be observed frequently among a large set of point mutations within the gene.

The second case of intragenic complementation is akin to the phenomenon of suppression. In this instance it is assumed that the protein is a homo-dimer or homo-oligomer. Any mutation alone might reduce the activity or stability of the protein complex. However, in certain instances when two different mutant proteins are assembled into the dimer or oligomer, the combination fortuitously restores more normal activity or stability to the complex. One way to envisage the suppression is by a type of complementarity on the protein face that is responsible for protein dimerization. One mutation might introduce a negative charge at a site that was formerly neutral, a suppressing mutation at the site of interaction might introduce a positive charge and the resulting ionic interaction would stabilize the complex. However it might arise, only rare mutant combinations are expected to exhibit this class of intragenic complementation.

The second exception to the complementation test is the phenomenon of "unlinked non-complementation" (Stearns and Botstein, 1988). In practice, certain recessive mutations in two different genes (known to be different because they are unlinked by recombination) fail to complement in the doubly heterozygous diploid. It is as if the wild-type proteins are not functional in the doubly heterozygous diploid even though they are clearly functional in each singly heterozygous diploid. Two explanations have been advanced to account for this surprising phenomenon; one is based upon protein stoichiometry and the second is based upon a "poison" protein model.

In either explanation, the proteins are assumed to be subunits in a hetero-dimeric protein but higher order oligomers should also exhibit the effect.

It is also proposed that the alleles exhibiting unlinked non-complementation produce proteins that are stable enough to fold and become incorporated into the dimer. The stoichiometry model makes a careful consideration of the amount of functional dimer that might be present in the different diploids. In either singly heterozygous diploid, 50% of the dimer would be inactive due to the incorporation of the mutant protein. This level of activity must be sufficient to produce a wild phenotype because the mutation is recessive. In the doubly heterozygous state, each mutation would independently inactivate 50% of the dimers. The active dimer would thereby be reduced to only 25% of normal. It is then assumed that this level of activity is below a critical threshold for the cell to be wild type. This effect could be significantly enhanced if the two mutant proteins cannot be co-assembled. In that case, a greater fraction of the mutant proteins would be co-assembled with wild-type proteins, thereby reducing the fraction of active wild-type dimers.

The first explanation assumes that the doubly mutant dimer is effectively inert, simply acting as a sink for mutant proteins. In contrast, the poison protein model proposes that the doubly mutant dimer has a novel dominant negative activity, which is unlike that of either singly mutant dimer. Although it is difficult to specify a precise mechanism for the activity of the poison protein, given all of the phenomena of allele-specific suppression and synthetic lethality, it is not too difficult to imagine that the properties of the doubly mutant dimer might be different form the singly mutant. To distinguish between these two models, a careful test of the effects of gene dosage is carried out. Essentially, the effects of the poison protein should be relatively resistant to effects of increased wild-type gene dosage. The stoichiometry model predicts that small increases in the wild-type gene dosage should restore a wild phenotype.

Regardless of the mechanism, unlinked non-complementation provides another means to analyse gene interaction and to identify new interacting genes. As the basis of a genetic screen, a large number of mutagenized colonies are replica-mated to a specific mutant strain for which unlinked non-complementing mutants are sought. Diploids that exhibit a failure in complementation of the original mutation are then identified. In parallel, mutagenized colonies are mated to a wild-type strain to identify steriles and dominant negative mutants. Non-complementing diploids are sporulated to look for mutations that are not linked to the original mutations. A screen of this type was carried out looking for non-complementers of mutations in the gene for β-tubulin; alleles of α-tubulin were identified.

D. Two-hybrid screens

Perhaps the most original and powerful new method for identifying interacting proteins is called the "two-hybrid" screen. The screen is based upon several earlier observations about the nature of transcriptional activation. First, transcriptional activators tend to be modular in their design and function; separate protein domains mediate DNA binding and activation. Second, transcriptional activation requires localization of the activation domain to the proximity of the promoter region. Third, activation domains need not be covalently linked to DNA binding domains to activate transcription. Therefore, close association with a separate DNA binding domain via protein–protein interactions is sufficient to localize the activation domain and stimulate transcription. These facts together are the basis for the two-hybrid screen. In practice, three components are required: a DNA binding domain is fused to one protein (termed the "bait"), an activation domain is fused to another protein that potentially interacts with the first protein (the "prey"), and a reporter gene whose transcriptional activation signals the interaction.

Two different two-hybrid systems have been developed for use in yeast (Fields and Song, 1989; Gyuris *et al.*, 1993). One utilizes functional domains from the Gal4p transcription activation protein (Fields and Song, 1989). The "bait" plasmid contains a fragment encoding the Gal4p DNA binding domain with a convenient restriction site engineered at the 3' end of the coding sequence. A separate "prey" plasmid expresses the activation domain of Gal4p, which also contains a useful restriction site at the 3' end. Both plasmids contain strong constitutive promoters to drive expression of the hybrid genes. To run the screen, a gene encoding the "bait" protein is fused in frame with the Gal4 DNA binding domain. A gene encoding a putative interacting protein is fused in frame with the Gal4 activation domain. In this system, the reporter plasmid contains the upstream activating sequences from the *GAL1–GAL10* region to promote transcription of the *E. coli lacZ* gene. Association of the two hybrid proteins via their interacting domains then leads to transcription of the *lacZ* reporter gene. Reporter gene expression is detected by the blue colour of a colony on medium containing the chromogenic substrate Xgal.

In addition to testing whether any two specific proteins interact, one can also screen for unknown genes that potentially produce interacting proteins. To this end, libraries have been constructed that contain yeast genomic DNA (or various other organisms genomic or cDNAs) fused to the activation domain. One particularly useful set of libraries contain *Sau*3A partial digests of genomic yeast DNA cloned into a *Bam*HI site in each of three different plasmids. In each, the *Bam*HI site is in a different reading frame with respect to the Gal4p activation domain. This ensures that any given *Sau*3A site within an open reading frame will make a productive fusion in one of the three plasmid libraries.

The second independent system avoids Gal4p and instead uses the DNA binding domain from *E. coli lexA* repressor protein and the *lexA* operator sequence (Gyuris *et al.*, 1993). In this case the activator domain is a fragment of bacterial DNA that expresses an acidic peptide, which acts as a transcriptional activator in yeast when fused to a DNA binding domain. Built into the activator construct are a nuclear localization signal to direct the protein into the nucleus and an epitope tag to determine whether the hybrid protein is expressed. The reporter construct can be either the *E. coli lacZ* gene or the yeast *LEU2* gene downstream of *lexA* operators. The virtue of the *LEU2* fusion is that it provides an independent selectable assay for transcription. Therefore, one can select for transformants that are Leu$^+$ and then screen for those that are also expressing *lacZ*. The combination helps reduce the number of false positives obtained in the selection.

Regardless of the system, several controls are necessary to demonstrate that the interaction is specific. First, it is critical to demonstrate that neither hybrid protein can activate on its own. This might be the case if either interacting protein is a transcriptional activator or the "bait" contains a region that is particularly acidic. Second, it must be shown that the interaction is actually dependent upon the two test domains. This level of specificity is demonstrated by, first, showing that the unfused domains do not activate when expressed in the presence of the other hybrid protein and, second, that other extraneous hybrids do not activate.

There are several distinct advantages to analysing protein–protein interactions by two hybrid screens. First, essentially wild-type protein domains are utilized so that the screen is not dependent upon potentially misleading mutant forms of the proteins. Second, the screen takes place in yeast, *in vivo*. In principle, therefore, it is capable of detecting weak or unstable interactions that would be difficult to detect by *in vitro* methods. Third, interactions that are regulated or which require protein modifications can be detected. For example, cell-cycle-specific interactions ought to be detected as cell-cycle-specific reporter gene expression.

A number of artefacts may also appear in the two hybrid screen; the most pernicious is the binding of proteins that do not normally interact *in vivo*. One potential source of artefact comes from having both proteins expressed at very high levels, thereby favouring interactions with very weak affinities. This may be avoided by using weaker promoters to express the hybrids. A second possible source of error is that many proteins contain related structural motifs, which can interact in a promiscuous fashion. For example, any filamentous coiled-coil protein may interact with any other member of this class. A second problem may arise because the proteins must localize to the nucleus in spite of other possible localization signals. In this case the screen may fail to identify interacting proteins because one of the proteins is not in the nucleus. A third potential problem arises from the fact that some overexpressed proteins are quite toxic to yeast. This may obviate the ability to construct the desired DNA binding hybrid or the identification of the desired interacting protein. There are two routes around this impasse. One way is to dissect the protein, because sometimes it is possible to separate the interacting domain from the domain causing toxicity. The second approach is to use a regulated promoter to express the hybrids. In this case, after induction of the hybrids, activation may be detected by expression of β-galactosidase in the dying cells.

Given the potential for artefact, the investigator must establish independent criteria for the interaction between two proteins. For this purpose, any of the many methods, both genetic and biochemical, for establishing interaction may be used. Many of these have been described elsewhere in this review. However, as for any type of interaction, the most compelling case for a relevant association would include evidence that the two proteins are required for similar functions in the cell.

VIII. CONCLUSION

Modern yeast genetics has developed because of the fortuitous confluence of a rich history of classical genetic methods and the advent of modern recombinant DNA techniques. Much of the progress over the last decade has been based upon the use of recombinant DNA methods to construct *in vitro* counterparts of the powerful *in vivo* methods developed earlier by bacterial geneticists. However, most if not all of the bacterial methods have now been replicated. We are entering a post-modern era in which the new methods arise, *ab initio*, within the context of current yeast genetics. For example, the two-hybrid screen emanated from the abundance of detailed knowledge garnered about the behaviour of yeast transcriptional activators. Similarly, YAC cloning techniques arose from the unique contribution that yeast molecular genetics has made to the understanding of chromosome structure. Nevertheless, because *in vitro* techniques have been largely developed to be independent of the specific biology of yeast, many can be, and are, easily adapted to research in other organisms. Undoubtedly the future will see the creation of similar rich systems of molecular genetic analysis in all the organisms to which scientists have turned their attention.

References

Adams, A.E., and Botstein, D. (1989). *Genetics* **121**, 675.
Alani, E., Cao, L. and Kleckner, N. (1987). *Genetics* **116**, 541.
Basson, M.E., Thorsness, M. and Rine, J. (1986). *Proceedings of the National Academy of Sciences, USA* **83**, 5563.
Becker, D.M. and Guarente, L. (1991). *Methods in Enzymology* **194**, 182.
Bender, A. and Pringle, J.R. (1989). *Proceedings of the National Academy of Sciences, USA* **86**, 9976.
Bender, A. and Pringle, J.R. (1991). *Cellular Biology* **11**, 1295.
Boeke, J.D., Lacroute, F. and Fink, G.R. (1984). *Molecular and General Genetics* **181**, 288.
Boeke, J.D., Trueheart, J., Natsoulis, G. and Fink, G.R. (1986). *Methods in Enzymology* **154**, 164.
Budd, M. and Campbell, J.L. (1987). *Proceedings of the National Academy of Sciences, USA* **84**, 2838.
Burgers, P.M.J. and Percival, K.J. (1987). *Analytical Biochemistry* **163**, 391.
Carle, G. and Olson, M.V. (1985). *Proceedings of the National Academy of Sciences, USA* **82**, 3756.
Chattoo, B.B., Sherman, F., Azubalis, D.A., Fjellstedt, T.A., Mehnert, D. and Ogur, M. (1979). *Genetics* **93**, 51.
Costanza, M.C. and Fox., T.D. (1988). *Genetics* **120**, 667.
Cross, F.R. (1990). *Molecular and Cellular Biology* **10**, 6482.
Davis, L. and Fink G.R. (1990). *Cell* **61**, 965.
Deshaies, R.J., Koch, B.D., Werner-Washburne, M., Craig, E.A. and Schekman, R. (1988) *Nature* **332**, 800.
DiNardo, S., Voelkil, K. and Sternglanz, R. (1984). *Proceedings of the National Academy of Sciences, USA* **81**, 2616.
Dohmen, R.J., Wu, P. and Varshavsky, A. (1994). *Science* **263**, 1273.
Elble, R. (1992). *Biotechniques* **13**, 18.
Falco, S.C. and Botstein, D. (1983). *Genetics* **105**, 857.
Feldheim, D., Rothblatt, D. and Schekman, R. (1992). *Molecular and Cellular Biology* **12**, 3288.
Fields, S. and Song, O. (1989). *Nature* **340**, 245.
Fowler, R., Degnen, G. and Cox, E. (1974). *Molecular and General Genetics* **133**, 179.
Fox, T.D., Sanford, J.C. and McMullin, T.W. (1988). *Proceedings of the National Academy of Sciences, USA* **85**, 7288.
Gimeno, C.J., Ljundahl, P.O., Styles, C.A. and Fink, G.R. (1992). *Cell* **68**, 1077.
Goebl, M.G. and Petes, T.D. (1986). *Cell* **46**, 983.
Guthrie, C. and Fink, G.R. (1991). *Methods in Enzymology* **194**, 1.
Gyuris, J., Golemis, E., Chertkov, H. and Brent, R. (1993). *Cell* **75**, 791.
Hall, M.N., Hereford, L. and Herskowitz, I. (1984). *Cell* **36**, 1057.
Han, M. and Grunstein, M. (1988). *Cell* **55**, 1137.
Harashima, S., Takagi, A. and Oshima, Y. (1984). *Molecular and Cellular Biology* **4**, 771.
Hashimoto, H.H., Morikawa, Y., Yamada, Y. and Kimura, A. (1985). *Applied Microbiology and Biotechnology* **21**, 336.
Heinemann, J.A. and Sprague, G.F., Jr (1989). *Nature* **340**, 205.
Hinnebusch, A.G. and Fink, G.R. (1983). *Proceedings of the National Academy of Sciences, USA* **80**, 5374.

Hinnen, A., Hicks, J.B. and Fink, G.R. (1978). *Proceedings of the National Academy of Sciences, USA* **75**, 1929.

Holm, C., Goto, T., Wang, J.C. and Botstein, D. (1985). *Cell* **41**, 553.

Hoyt, M.A., He, L., Loo, K.K. and Saunders, W.S. (1992). *Journal of Cell Biology* **118**, 109.

Huffaker, T., Hoyt, M.A. and Botstein, D. (1987). *Annual Review of Genetics* **21**, 259.

Huffaker, T.C., Thomas, J.H. and Botstein, D. (1988). *Journal of Cell Biology* **106**, 1997.

Ito, H., Fukada, Y., Murata, K. and Kimura, A. (1983). *Journal of Bacteriology* **153**, 163.

Johnston, S.A., Anziano, P., Shark, K., Sanford, J.C. and Butow, R.A. (1988). *Science* **240**, 1538.

Kaiser, C.A., Preuss, D., Grisafi, P. and Botstein, D. (1987). *Science* **235**, 312.

Kaiser, C.A. and Schekman, R. (1990). *Cell* **61**, 723.

Kranz, J.E. and Holm, C. (1990). *Proceedings of the National Academy of Sciences, USA* **87**, 6629.

Kunes, S., Ma, H., Overbye, K., Fox, M.S. and Botstein, D. (1987). *Genetics* **115**, 73.

Leung, D.W., Chen, E. and Goeddel, D.V. (1989). *Technique* **1**, 11.

Liu, H., Krizek, J. and Bretscher, A. (1992). *Genetics* **132**, 665.

Lyons, S. and Nelson, N. (1984). *Proceedings of the National Academy of Sciences, USA* **81**, 7426.

Ma, H. and Ptashne, M. (1987). *Cell* **51**, 113.

Ma, H., Kunes, S., Schatz, P. and Botstein, D. (1987). *Gene* **58**, 201.

MacKay, V.L. (1983). *Methods in Enzymology* **101**, 325.

Meeks-Wagner, D.W. and Hartwell, L. (1986). *Cell* **44**, 43

Meeks-Wagner, D.W., Wood, J.S., Garvik, B. and Hartwell, L. (1986). *Cell* **44**, 53.

Mitsuzawa, H., Uno, I., Oshima, T. and Ishikawa, T. (1989). *Genetics* **123**, 739.

Munro, S. and Pelham, H.R.B. (1987). *Cell* **48**, 899.

Natsoulis, G., Hilger, F. and Fink, G.R. (1986). *Cell* **46**, 235.

Oliver, S.G. *et al.* (1992). *Nature* **357**, 38.

Olson, M.V. (1991). *In* "The Molecular and Cellular Biology of the Yeast *Saccharomyces*" (J. Broach, J. Pringle and E. Jones, eds.), Vol. 1, pp. 1–40. Cold Spring Harbor Press, Cold Spring Harbor.

Orr-Weaver, T.L., Szostak, J.W. and Rothstein, R.J. (1981). *Proceedings of the National Academy of Sciences, USA* **78**, 6354.

Orr-Weaver, T.L., Szostak, J.W. and Rothstein, R.J. (1983). *Methods in Enzymology* **101**, 228.

Page, B. and Snyder, M. (1992). *Genes and Development* **6**, 1414.

Park, E.C., Finlay, D. and Szostak, J.W. (1992). *Proceedings of the National Academy of Sciences, USA* **89**, 1249.

Pausch, M.H., Kaim, D., Kunisawa, R., Admon, A. and Thorner, J. (1991). *EMBO Journal* **10**, 1511.

Pon, L. and Schatz, G. (1991). *In* "The Molecular and Cellular Biology of the Yeast *Saccharomyces*" (J. Broach, J. Pringle and E. Jones, eds.), Vol. 1, pp. 333–406. Cold Spring Harbor Press, Cold Spring Harbor.

Pringle, J.R., Adams, A.E.M., Drubin, D.G. and Haarer, B.K. (1991). *Methods in Enzymology* **194**, 565.

Ptashne, M. (1988). *Nature* **335**, 683.

Ramer, S.W., Elledge, S.J. and Davis, R.W. (1992). *Proceedings of the National Academy of Sciences, USA* **89**, 11589.

Reijo, R.A., Cooper, E.M., Beagle, G.J. and Huffaker, T.C. (1994). *Molecular Biology of the Cell* **5**, 29.

Riles, L., Dutchik, J.E., Baktha, A., McCauley, B.K., Thayer, E.C., Leckie, M.P., Braden, V.V., Depke, J.E. and Olson, M.V. (1993). *Genetics* **134**, 81.

Rine, J. (1991). *Methods in Enzymology* **194**, 239.

Rine, J., Hansen, W., Hardeman, E. and Davis, R.W. (1983). *Proceedings of the National Academy of Sciences, USA* **80**, 6750.

Roof, D., Meluh, P.B. and Rose, M.D. (1992). *Journal of Cell Biology* **118**, 95.

Rose, M.D. and Fink, G.R. (1987). *Cell* **48**, 1047.

Rose, M.D., Price, B.R. and Fink, G.R. (1986). *Molecular and Cellular Biology* **6**, 3490.

Rose, M.D., Winston, F. and Hieter, P. (1990). "Methods in Yeast Genetics, A Laboratory Course Manual". Cold Spring Harbor Laboratory, Cold Spring Harbor.

Rothstein, R.J. (1983). *Methods in Enzymology* **101**, 202.

Sauer, B., (1994). *Biotechniques* **16**, 1086.

Schatz, P.J., Solomon, F. and Botstein, D. (1988). *Genetics* **120**, 681.

Schena, M., Picard, D. and Yamamoto, K.R. (1991). *Methods in Enzymology* **194**, 389.

Scherer, S. and Davis, R.W. (1979). *Proceedings of the National Academy of Sciences, USA* **76**, 4951.

Schiestl, R.H. and Gietz, R.D. (1989). *Current Genetics* **16**, 339.

Schneider, J.C. and Guarente, L. (1991). *Methods in Enzymology* **194**, 373.

Schnell, R. and Rine, J. (1986). *Molecular and Cellular Biology* **6**, 494.

Sengstag, C., Stirling, C., Schekman, R. and Rine, J. (1990). *Molecular and Cellular Biology* **10**, 672.

Shortle, D., Haber, J.E. and Botstein, D. (1982a). *Science* **217**, 371.

Shortle, D., Grisafi, P., Benkovic, S. J. and Botstein, D. (1982b). *Proceedings of the National Academy of Sciences, USA* **79**, 1588.

Shortle, D., Novick, P., and Botstein, D. (1984). *Proceedings of the National Academy of Sciences, USA* **81**, 4889.

Sikorski, R.S. and Hieter, P. (1989). *Genetics* **122**, 19.

Sikorski, R.S., Michaud, W. Levin, H.L., Boeke, J.D. and Hieter, P. (1990). *Nature* **345**, 581.

Silver, P.A. (1991). *Cell* **64**, 489.

Snyder, M., Elledge, S., Sweetser, D., Young, R.A. and Davis, R.W. (1987). *Methods in Enzymology* **154**, 107.

Soni, R., Carmichael, J.P. and Murray, J.A. (1993). *Current Genetics* **24**, 455.

Spee, J.H., de Vos, W.M. and Kuipers, O.P. (1993). *Nucleic Acids Research* **21**, 777.

Stearns, T. and Botstein, D. (1988). *Genetics* **119**, 249.

Thrash, C. Voekel, K., DiNardo, S. and Sternglanz, R. (1984). *Journal of Biological Chemistry* **259**, 1375.

Vallen, E.A., Ho, W., Winey, M. and Rose, M.D. (1994). *Genetics* **137**, 407.

Vallen, E.A., Roberts, T., Van Zee, K. and Rose, M.D. (1992). *Cell* **69**, 505.

van Zyl, W.H., Wills, N. and Broach, J.R. (1989). *Genetics* **123**, 55.

Vogel, J.P., Misra, L.M. and Rose, M.D. (1990). *Journal of Cell Biology* **110**, 1885.

von Heijne, G. (1990). *Journal of Membrane Biology* **115**, 195.

Wakem, L.P. and Sherman, F. (1990). *Genetics* **125**, 333.

Wertman, K.F., Drubin, D.G. and Botstein, D. (1992). *Genetics* **132**, 337.

Winey, M., Goetsch, L., Baum, P. and Byers, B. (1991). *Journal of Cell Biology* **114**, 745.

Yoshikawa, A. and Isono, K. (1990). *Yeast* **6**, 383.

Young, R.A. and Davis, R.W. (1991). *Methods in Enzymology* **194**.

4 Recombinant DNA Technology: Yeast Vectors

Stephen A. Parent and Keith A. Bostian

Department of Microbiology and Molecular Genetics, Merck Research Laboratories, Rahway, NJ 07065, USA

I. INTRODUCTION

Advances in recombinant DNA technologies have accelerated our understanding of many biological problems ranging from mechanisms of gene regulation to aspects of cell structure, reproduction and development. The impact of these technologies is no better seen than in the simple eukaryote *Saccharomyces cerevisiae*. Despite its unicellular nature, it has served as a

The Yeasts Vol. 6, 2nd edition
ISBN 0-12-596416-1

popular experimental model because it shares most fundamental characteristics of cell biology with multicellular organisms and it offers several advantages over its counterparts. Laboratory strains of *Sacch. cerevisiae* can be grown vegetatively as haploid α or **a** cells or as an α/**a** diploid cell. The short generation times of these cell types allow their easy cultivation in large quantities on defined medium. Diploid cells readily undergo meiosis, resulting in production of a tetrad ascus containing four haploid spores, all of which can be recovered and propagated. These features have been exploited by geneticists and biochemists for decades.

The advent of a DNA transformation system in yeast (Beggs, 1978; Hinnen *et al.*, 1978), and early efforts to refine and elaborate these techniques (Scherer and Davis, 1979; Struhl *et al.*, 1979; Orr-Weaver *et al.*, 1981, 1983; Shortle *et al.*, 1982; Rothstein, 1983; Struhl, 1983), along with the development of a wide range of yeast-vector systems (for detailed reviews, see Parent *et al.*, 1985; West, 1988; Rose and Broach, 1990), has resulted in yeast (*Sacch. cerevisiae*) becoming a useful and popular choice for experimentalists interested in understanding basic aspects of cellular and molecular phenomena. Well-established and advanced yeast genetic and molecular techniques make it possible to isolate and characterize relevant mutations affecting virtually all yeast cellular pathways of interest, many of which are conserved evolutionarily. Once mutants are available on a given pathway, the genes defined by the mutations can be readily isolated. For a mutant exhibiting a recessive phenotype, this is typically achieved by introducing a genomic DNA library from a wild-type strain into the mutant by DNA transformation, and either selecting or screening for clones that have complemented the mutant phenotype. The transforming DNA is then isolated, evaluated in *Escherichia coli* and retransformed into yeast for verification. Cloning a gene defined by a dominant mutation requires more work but the overall strategy is similar. A specific genomic DNA library from the mutant is constructed in a shuttle vector, and wild-type transformants exhibiting the dominant mutant phenotype are identified by selection or screening techniques. Yeast genes of interest can also be isolated by reverse genetics, if sequence information about an RNA or protein product is known or if antibodies against the protein are available (Young and Davis, 1984; Snyder *et al.*, 1987). However, this approach is seldom used unless the former strategy is too difficult or impossible to apply. Once isolated, the structure and cellular regulation of the gene can be studied immediately. The cloned gene can also be used to express, isolate and characterize the gene product either in yeast or other organisms. Expression systems can be especially useful when the product is rare or limiting in the cell.

Molecular genetic structure–function studies provide new insights into the roles of genetic sequences and can be rapidly performed on cloned

DNA. Yeast systems provide powerful approaches to this end. DNA subjected to *in vitro* and/or *in vivo* mutational analyses can be rapidly examined in yeast to determine the phenotypic consequences of the mutations. DNA introduced into yeast can be propagated extrachromosomally or integrated into and replicated with endogenous yeast chromosomes (Beggs, 1978; Hinnen *et al.*, 1978). Chromosomal integration typically proceeds via homologous recombination in yeast (Hinnen *et al.*, 1978) and this recombination event can be directed to specific site(s) of interest (Orr-Weaver *et al.*, 1981). The high degree of homologous recombination taking place during yeast transformation has allowed researchers to manipulate the genome of this organism with a great degree of precision. Since *in vitro* manipulated exogenous yeast DNA can be re-introduced into its native genomic site, researchers can study mutant sequences in their native environment and avoid complicating issues of abnormal chromosome location.

In this chapter, we describe yeast vector systems and components generating the variety of techniques already summarized above. Basic aspects of the plasmid vectors are described, along with general notes about their components, construction and use. Tables describing relevant information about the vectors are included to help assist readers in the use and design of their own vector systems. Examples of vector types are also included, although many more vectors are also available. For more detailed and comprehensive listings of available plasmid vectors, we refer the reader to several recent reviews (Parent *et al.*, 1985; Rose and Broach, 1990).

II. ASPECTS OF YEAST VECTOR SYSTEMS

A. Vector types

Several general types of yeast vectors exist, categorized initially by their mode of replication (Botstein *et al.*, 1979; Struhl *et al.*, 1979). Yeast-integrating plasmid (YIp) vectors are unable to propagate autonomously because they lack sequences that serve as yeast origins of DNA replication (*ori*). To be maintained in the cell, these plasmids integrate into chromosomal DNA, usually in single copy and are replicated with chromosomal DNA (Hinnen *et al.*, 1978). Yeast-replicating plasmid (YRp) vectors contain an *ori* or autonomously replicating sequence (*ARS*) and are maintained extrachromosomally (Struhl *et al.*, 1979). The *ARS* elements promoting this replication are usually provided by yeast chromosomal sequences but can also originate from other sources (Stinchcomb *et al.*, 1980). These plasmids are generally propagated at high copy number. However, they are unstable during mitosis and meiosis, and the plasmid copy number typically varies widely within individual cells of a population. The instability, high copy number and copy number variation result largely from asymmetrical plasmid segregation at division (Murray and Szostak, 1983a). For these reasons, YRp plasmids are now rarely used. Yeast centromeric plasmid (YCp) vectors contain a centromeric sequence in addition to an *ori*. The vectors stably segregate and are maintained at low copy number autonomously. At division, stable segregation of the newly replicated YCp plasmid results in two progeny cells typically receiving one or two plasmids. Yeast episomal plasmid (YEp) vectors are derived from the endogenous yeast 2 μm plasmid and are maintained extrachromosomally (for a review, see Rose and Broach, 1990). The native 2 μm plasmid is a stable, double-stranded DNA, present at 50–100 copies in each cell. Yeast episomal plasmid vectors contain the 2 μm sequences required for stable propagation at high copy number. These sequences typically include the 2 μm *ori* and a *cis*-acting sequence termed *REP3* or *STB*, which is responsible, in conjunction with other plasmid sequences, for stable plasmid segregation. Despite the 2 μm sequences, these vectors segregate irregularly, leading to some instability and clonal plasmid copy number variation within individual cells of a population. They are generally, however, more stable than YRp vectors.

All of the vectors already described above are circular in nature. Linear artificial chromosomes have also been constructed in yeast and are called yeast linear plasmid (YLp) vectors. In addition to the *ARS* elements required for their autonomous replication, these artificial chromosomal plasmids contain telomeric sequences, which are responsible for maintaining the plasmid in its linear state (Szostak and Blackburn, 1982). These vectors provide ideal molecules for analysing chromosomal and cellular components necessary for stable chromosome maintenance. Murray and Szostak (1983b) demonstrated that very large artificial linear chromosomes (containing yeast genes, *ARS* and *CEN* elements) could be engineered and maintained very stably in yeast. Their findings indicate that larger linear artificial chromosomes are more stable than their smaller linear or circular counterparts, which led to the development of a yeast artificial chromosome (YAC) vector system for creation of yeast libraries containing very large DNA inserts (Burke *et al.*, 1987).

Several years ago, we reviewed the extensive literature on yeast vector systems and compiled a comprehensive documentation of plasmid vectors and details about their origins (Parent *et al.*, 1985). This provided investigators with a resource to trace the genealogies and structures of plasmids of interest. In the review, we extended the classification of plasmids, to assist readers in locating vectors. Yeast promoter plasmid (YPp) vectors contain easily assayable or scoreable reporter sequences to which promoter-containing transcriptional and/or translational signals can be fused. They have been used extensively to investigate regulatory sequences involved in controlling gene expression. Yeast expression plasmid (YXp) vectors contain transcriptional promoter sequences, and typically a transcriptional terminator to which homologous or heterologous gene sequences can be fused for expression in yeast. Yeast hybrid plasmid (YHp) vectors are complex, typically consisting of hybrid gene sequences providing model systems for studying aspects of gene expression in yeast.

For the purpose of the present review, we continue to use this extended nomenclature to assist readers in identifying vector systems for their individual purposes. Since publication of our earlier review, a new class of yeast mutagenesis plasmid (YMp) vectors has been developed. They may be used to mutagenize rapidly and evaluate cloned DNA sequences *in vitro* and/or *in vivo*, and are useful in structure–function analyses of cloned genes. They will be described in a section devoted to vectors designed for mutagenesis.

B. Vector components

Most yeast plasmids are shuttle vectors, which contain sequences allowing them to be propagated in *E. coli* or other bacterial hosts (Table I). This allows for convenient amplification and subsequent *in vitro* manipulation. The vectors usually contain a bacterial-derived backbone, which allows for maintenance and propagation of the plasmid in *E. coli*, the most common one originating from pBR322 (Bolivar *et al.*, 1977). This plasmid and its derivatives contain an origin of replication promoting high copy-number maintenance and two selectable antibiotic markers, the β-lactamase gene (*bla*), conferring resistance to ampicillin, and the tetracycline-resistance gene (*tet*), encoding resistance to tetracycline. Plasmid pBR322 also contains a number of unique restriction enzyme sites for cloning purposes. Plasmid pJRD158 is a derivative of pBR322, which contains 28 unique cloning sites and has been used to construct two yeast–*E. coli* shuttle vectors, pMH158 and pJO158, which contain 21 and 23 unique restriction sites, respectively (Heusterspreute *et al.*, 1985). More recently, smaller yeast–*E. coli* shuttle vectors derived from pUC plasmids have been developed (see Tables I and XII). These pUC derivatives typically contain the pBR322 origin of replication and the *bla* gene, as well as the *lacZ* α fragment, with in-frame polylinker or multiple cloning site regions (MCR) (Yanisch-Perron *et al.*, 1985).

Table I. Bacterial/phage components of yeast shuttle vectors

Plasmid/phage	Origin of replication	Size (kb)	Features	Bacterial hosts	Source
pBR322	ColE1	4.36	Amp^r, Tet^r	Escherichia coli	Bolivar et al. (1977)
pJRD158	ColE1	3.9	Amp^r, Tet^r	Escherichia coli	Davison et al. (1984, 1987)
pSC101	pSC101	—	Amp^r	Escherichia coli	Rose and Broach (1991), Rose et al. (1989), Armstrong et al. (1984)
pC194	pC194	2.9	Cam^r	Escherichia coli, Staphylococcus aureus, Bacillus subtilis	Horinouchi and Weisblum (1982a,b), Goursot et al. (1982), Polumienko et al. (1986)
pEMBL9	ColE1, F1 ori	4.0	Amp^r, lacZ α	Escherichia coli	Baldari and Cesareni (1985)
pUC19	ColE1	2.68	Amp^r, lacZ α	Escherichia coli	Yanisch-Perron et al. (1985)
pBluescript	ColE1, F1 ori	2.95	Amp^r, lacZ α	Escherichia coli	Stratagene Cloning Systems
pHC79	ColE1	6.43	Amp^r, Tet^r, λ cos	Escherichia coli	Hohn and Hinnen (1982)
pTR262	ColE1	4.6	λ cI, λ P$_R$-Tet^r	Escherichia coli	Roberts et al. (1980), Wright et al. (1986a)
pDPT51	oriT-R	—	Amp^r, Tp^r, tra-R, mob-R oriT-R, mob-C	Broad-host range in Gram-negative and Gram-positive species	Heinemann and Sprague (1989)
pFLEU2	oriT-F	—	mob-F oriT-F, LEU2, tra-F		Heinemann and Sprague (1989)
M13mp18	M13 ori	7.25	lacZ α	Escherichia coli	Yanisch-Perron et al. (1985)

Amp^r, ampicillin resistance gene; Tet^r, tetracycline resistance gene; Cam^r, chloramphenicol resistance gene; Tp^r, trimethoprim resistance gene.

The *lacZ* α fragment in these vectors allows use of the well-defined α complementation system developed by Messing and his colleagues (Yanisch-Perron *et al.*, 1985) to screen for recombinant clones containing inserts within the polylinker region. They also have the advantage of usually being smaller than pBR322 derivatives and achieving higher copy numbers in *E. coli*. Yeast vectors containing polylinker regions other than those derived from the pUC series also exist, such as the more extensive MCR of the pBLUESCRIPT vectors (Stratagene Cloning Systems, La Jolla, CA, USA). Several additional features have been incorporated in these vectors. They contain bacteriophage T3 and T7 promoters flanking the polylinker regions, which allow efficient *in vitro* synthesis of strand-specific RNA transcripts from DNA inserted into the polylinker region. Transcripts generated from these vectors *in vitro* can be used for a variety of purposes, including synthesis of radiolabelled probes for Southern and Northern hybridization analyses. Another useful feature incorporated into several of these vectors is addition of appropriately positioned restriction-enzyme sites, which allow generation of unidirectional nested deletions using ExoIII and mung-bean nuclease (Henikoff, 1984). Several yeast–*E. coli* shuttle vectors also contain the intergenic regions of the single-stranded filamentous phages M13 or f1 from *E. coli*. These sequences encompass the phage origins of replication, and enable efficient synthesis and packaging of single-stranded plasmid DNA into phage particles upon co-infection of *E. coli* transformants with helper phage. Yeast–*E. coli* shuttle vectors with many of these features are discussed later and listed in Table XII at the end of this chapter (p. 169).

Other bacterial sequences have been developed in yeast plasmids to facilitate vector propagation in bacterial hosts, including the broad host-range plasmid pC194, single-stranded M13 phage and the cosmid pHC79 (Table I). In addition, Rose and his colleagues have developed a YCp yeast–*E. coli* shuttle vector, which is a recombinant between YCp50 (Rose *et al.*, 1987) and a derivative of the low copy-number vector pSC101 from *E. coli* (see Table I). This plasmid is useful for propagating genes which are toxic to *E. coli* when they are present at high copy number. Several groups have also constructed yeast–*E. coli* shuttle vectors which enable positive selection for recombinant clones in *E. coli* (Gent *et al.*, 1985; Wright *et al.*, 1986a; Sengstag and Hinnen, 1987). The vectors typically contain the pBR322 *ori*, the *bla* gene, the λ cI gene and a *tet* selectable marker fused to a λ cI repressible promoter (i.e. λ P_R or λ P_L). Upon cloning into a unique restriction-enzyme site within cI, *tet* is expressed, allowing direct antibiotic selection for recombinant plasmids.

Typically, yeast–*E. coli* shuttle plasmids are propagated in *E. coli*, purified and transformed into yeast by one of several routinely used procedures. Original transformation procedures entailed removing the yeast cell wall with digestive enzymes, incubating DNA with osmotically stabilized sphaeroplasts in the presence of calcium chloride and polyethylene glycol, and plating cells in osmotically supported medium (Beggs, 1978; Hinnen et al., 1978). However, plasmid transfer from E. coli to yeast can also be achieved directly by fusing protoplasts of the bacterium with yeast sphaeroplasts, obviating the need to prepare DNA from E. coli (Broach et al., 1979). Several transformation procedures that do not require sphaeroplast preparation are available and include the LiAc protocol (Ito et al., 1983), the particle bombardment method (Johnston et al., 1988) and electroporation (Hashimoto et al., 1985; Karube et al., 1985; Uno et al., 1988; Becker and Guarente, 1991). These protocols offer advantages over sphaeroplast methods. However, in each instance, plasmid DNA is usually prepared from E. coli. Heinemann and Sprague (1989) demonstrated plasmid transfer directly from viable E. coli to yeast so long as the bacterial sequences of the shuttle vector are capable of undergoing conjugative transfer. Bacterial conjugative systems encoded by pDPT51 or F factor (Table I) are capable of mediating E. coli-to-yeast transfer of vectors, which contain ori-T from R751, and ColE1 or F plasmids, respectively. The conjugation process is dependent upon transfer (tra) functions of the mobile plasmids, as well as mob systems, which are compatible with the ori-T of the vector to be transferred. This procedure obviates vector DNA preparation from E. coli.

In addition to the backbone sequences that permit efficient and rapid propagation in E. coli, yeast vectors all contain selectable markers, which allow plasmid-bearing yeast transformants to be selected. A variety of selectable markers are available in yeast. The earliest and most commonly used encode the yeast genes HIS3, LEU2, TRP1 and URA3, which complement specific auxotrophies. Information regarding these markers is included in Table II. Each marker also complements an E. coli auxotrophic mutant defective in the respective bacterial gene.

The utility of a selectable marker depends on the nature and stability of the chromosomal mutation being complemented. The mutations are typically recessive and non-reverting. For example, the ura3-52 mutation is the result of a Ty retrotransposon insertion within the coding sequences of the URA3 gene (Rose and Winston, 1984). It is extremely stable with a reversion frequency of less than 10^{-9} (Botstein et al., 1979). The his3, leu2, trp1 and ura3 alleles listed in Table II are also stable. Another important attribute of a marker are the restriction sites associated with it. Gietz and Sugino (1988) constructed LEU2, TRP1 and URA3 alleles, which lack all of the 6-bp restriction-enzyme sites that occur in the pUC19 polylinker (see footnotes to Table II). Wild-type alleles for each of these genes have also been inserted into various polylinker regions to form convenient

Table II. Properties of commonly used yeast selectable markers

Selectable marker	Complementable yeast mutations	Complementable bacterial mutations	Routinely used DNA fragments	Mini-map of common restriction enzyme sites[a]	Reference
HIS3	his3-Δ1, his3-Δ200	hisB463	1.7 kb BamHI	AccI (1185, 1211), AsuII (423, 873), AvaI (1327, 1760), AvaII (14, 546, 656), AvaIII (1248), BalI (678), BamHI (1,1765), BclI (1019), BglII (866, 926, 1667), BstBI (423, 873), BstXI, (860), ClaI (1698), DraII (788), EagI (1675), HindIII (775, 962), KpnI (1093), NheI (975), NsiI (1250), PstI (1154), XhoI (1327), XmaIII (1675)	Struhl (1985), Botstein et al. (1979), Struhl et al. (1979), Sikorski and Hieter (1989)
LEU2	leu2-3,112, leu2-Δ1	leuB6	2.2 kb XhoI/SalI	AccI (1840, 2214), AseI (1971), AsuII (280, 768), AvaI (1), AvaII (537, 881, 1209), BstBI (280, 768), BstXI (1361), ClaI (798), DraII (537), EcoNI (915, 1605), EcoRI (1283), EcoRV (1396), HincII (241, 521, 1543, 2215), HpaI (241), KpnI (897), SalI (2213), SspI (36, 527, 1975), Tth111I (2210), XhoI (1)	Andreadis et al. (1984), Botstein et al. (1979), Sikorski and Hieter (1989)
LYS2	lys2-201	—	4.85 kb XbaI-HindIII	Mini-map of 4.98 kb XbaI-HindIII fragment: AccI (1595, 4065), AseI (3208), AsuII (843, 2011), AvaI (3160), AvaII (694, 869, 1527, 2310, 2531), BamHI (3542), BclI (937, 1490, 2393, 2518), BglII (683), BspMII (4220), BstBI (843, 2011), BstXI (3128), Bsu36I (1579), DraII (694, 3797), EcoRV (277, 3923), EcoRI (56), EspI (900, 1674, 3578), HincII	Barnes and Thorner (1986), Fleig et al. (1986)

Table II. Continued

Selectable marker	Complementable yeast mutations	Complementable bacterial mutations	Routinely used DNA fragments	Mini-map of common restriction enzyme sites [a]	Reference
				(2111), HindIII (4851, 4975), HpaI (2111), KpnI (1441, 1981), NcoI (1834, 4825), NheI (656, 2676), NruI (1114), PmaCI (1323), PvuI (2123), PvuII (4631), SacII (3037), SnaBI (3830, 4488), SpeI (1364, 2102), SspI (560, 885, 3168, 3401, 4045, 4601), StuI (3027), XbaI (1), XhoI (3160)	
TRP1	trp1-Δ1, trp1-Δ63, trp1-Δ901	trpC9830 trpC1117	1.45 kb EcoRI	AccI (68), AseI (113), AvaII (133, 1381, 1408), BglII (852), BstXI (400), Bsu36I (344), EcoRV (387), HincII (953), HindIII (615), NheI (1037), PmaCI (62), PstI (827), StuI (829), XbaI (186)	Sikorski and Hieter (1989), Tschumper and Carbon (1980), Stearns et al. (1990)
URA3	ura3–52	pyrF::Tn5	1.1 kb HindIII	AccI (552, 1057), ApaI (605), AsuII (492), AvaI (1106), AvaII (380), AvaIII (1052), BstBI (492), DraII (380), EcoRV (415), HincII (733, 869), NcoI (432), NsiI (1054), PstI (209), SmaI (1108), StuI (663)	Rose et al. (1984), Rose and Winston (1984)

[a] Restriction-enzyme sites (positions in parentheses) present in the indicated fragment. The orientation of genes is 5′ to 3′ with respect to their transcription. Sites were identified in sequences within GenBank and EMBL databases by the MAP program (UW GCC Sequence Analysis Software Package; Devereux, 1989). Enzymes searched include AatII, AccI, ApaI, AseI, AsuII, AvaI, AvaII, AvaIII, BalI, BamHI, BclI, BglII, BspMII, BstBI, BstXI, Bsu36I, ClaI, DraII, EagI, EcoNI, EcoRI, EcoRV, EspI, HincII, HindIII, HpaI, KpnI, NcoI, NheI, NotI, NruI, NsiI, PmaCI, PstI, PvuI, PvuII, SacII, SalI, SnaBI, SmaI, SpeI, SphI, SspI, StuI, Tth111I, XbaI, XmaIII, XhoI. Complete nucleotide sequence of fragments are included in referenced papers. Gietz and Sugino (1988) have constructed LEU2, TRP1 and URA3 alleles, which lack the following restriction-enzyme sites: LEU2 (EcoRI and KpnI), TRP1 (HindIII, PstI, and XbaI), URA3 (PstI).

marker cassettes (Table III), providing additional flexibility in the design of new vectors.

Yeast *URA3* and *LYS2* genes have an additional advantage as markers because positive and negative selections are available for each. Positive selection is carried out by auxotrophic complementation, while negative selection relies on the ability to overcome a block in cell growth by an inhibitor when cells become deficient in expression of the gene. The negative-selection agents for the *URA3* and *LYS2* genes are 5-fluoro-orotic acid and α-aminoadipate, respectively (Chattoo *et al.*, 1979; Boeke *et al.*, 1984, 1987). The yeast *URA3* gene has been the most widely used of the two systems, due in large part to its smaller size and the existence of sites for many common restriction enzymes in the *LYS2* gene.

Several dominant drug-resistance genes have also been developed as yeast plasmid selectable markers (Table IV). Although not widely used, these markers are useful in areas where nutritional markers cannot be employed. One of their disadvantages is that selection for the marker requires high concentrations of drug to inhibit growth of non-transformed cells, and these concentrations can lower the recovery frequency of viable transformants.

In general, yeast plasmids are constructed by standard *in vitro* recombinant DNA manipulations and isolated in *E. coli*. However, circular plasmid vectors can be constructed in *Sacch. cerevisiae* by swapping vector regions by homologous recombination (Ma *et al.*, 1987; Stearns *et al.*, 1990). In this procedure, circular plasmids are generated by transforming yeast with a linearized plasmid and a second DNA fragment containing sufficient and appropriate homology to serve as a substrate for recombinational repair. Using this procedure and the pBR322 backbone of commonly used yeast vectors as the homologous regions, a series of YCp, YEp and YRp vectors with various combinations of yeast selectable markers (i.e. *LEU2*, *HIS3*, *LYS2*, *URA3* and *TRP1*) has been constructed (Ma *et al.*, 1987). This procedure can introduce new markers into plasmids, transfer markers between plasmids, replace markers of a plasmid, as well as replace resident alleles of a plasmid marker with new alleles. The other components of yeast vectors, including *ARS*, *CEN*, and *TEL* elements as well as reporter genes, promoter sequences and secretion signals, are discussed in individual sections of this chapter, describing the utilities of each of the basic yeast vector systems.

III. YIp VECTORS

Yeast integrating plasmid vectors do not replicate autonomously and, as such, transform yeast at low frequency by integration into the genome of

Table III. Portable cassettes containing yeast selectable markers

Vector component	Plasmid	Unique restriction sites flanking marker[a]	Remarks	Reference
pUC18 HIS3 fragment	pJJ215	**EcoRI, SacI, SmaI,** NsiI, BssHII, XhoI, EagI, ClaI, XbaI, SalI, SphI, NarI	1.7 kb BamHI HIS3 fragment inserted into BamHI site of pUC18 such that orientation of HIS3 is in the same direction as bla. Plasmid pJJ217 is similar but contains HIS3 in opposite orientation.	Jones and Prakash (1990)
pUC18 LEU2 fragment	pJJ250	**SacI, SmaI, BamHI, XbaI,** SalI, PstI, SphI, HindIII, NarI	2 kb Klenow treated HpaI-SalI LEU2 fragment from YEp13 (Broach et al., 1979) inserted into the XbaI site (Klenow treated) of pUC18 such that orientation of LEU2 is the same as bla. Plasmid pJJ252 is similar but contains LEU2 in the opposite orientation.	Jones and Prakash (1990)
pUC9 LYS2 fragment	pDP6	**HindIII, PstI, SalI XbaI,** HindIII, PstI, SalI XbaI, SmaI, SstI, EcoRI	pUC9 derivative containing a LYS2 cassette flanked by several unique restriction sites. The cassette contains in the 5' to 3' orientation, a HindIII-XbaI MCR from M13mp10, a XbaI-HindIII LYS2 fragment, followed by the complete M13mp10 HindIII-EcoRI MCR.	Fleig et al. (1986)
pUC18 TRP1 fragment	pJJ248	**SacI, KpnI, SmaI, BamHI,** StuI, BglII, SalI, SphI, NarI	0.93 kb EcoRI-RsaI TRP1 fragment from YRp17 (Stinchcomb et al., 1982) inserted into the XbaI site (Klenow treated) of pUC18 such that orientation of TRP1 is the same as bla. Plasmid pJJ246 is similar but contains TRP1 in the opposite orientation.	Jones and Prakash (1990)
pUC18 URA3 fragment	pJJ242	**EcoRI, SacI, KpnI, BamHI, XbaI, SalI, SphI,** NsiI, NarI	1.1 kb HindIII URA3 fragment from plasmid YEp24 (Botstein et al., 1979) inserted into HindIII sites of pUC18 such that orientation of URA3 is the same as bla. Plasmid pJJ244 is similar but contains URA3 in the opposite orientation.	Jones and Prakash (1990)

Table III. Continued

Vector component	Plasmid	Unique restriction sites flanking marker[a]	Remarks	Reference
pUC18 URA3 fragment	pJJ236	**EcoRI, SacI, KpnI, SmaI, BamHI, XbaI**, NsiI, SphI, NarI	1.1 kb HindIII-SmaI (Klenow treated) URA3 fragment from plasmid YEp24 (Botstein et al., 1979) inserted into SalI (Klenow treated) site of pUC18 such that orientation of URA3 is the same as bla. Plasmid pJJ238 is similar but contains URA3 in the opposite orientation.	Jones and Prakash (personal communication)
pUC8 URA3 fragment	pMF21	**EcoRI, SmaI**, SmaI, EcoRI	1.1 kb HindIII URA3 fragment inserted into the HindIII site of pUC8, followed by SmaI digestion to remove sequences between SmaI site at 3′ end of URA3 and pUC SmaI site, and subsequent cleavage by HindIII and insertion of HindIII-SmaI and SmaI-EcoRI adapter-linkers.	Fagan and Scott (1985)

[a] Restriction-enzyme sites in bold and normal type are present at the 5′ and 3′ end of the selectable marker, respectively. In addition to the cassettes listed, HIS3, LEU2, LYS2 and URA3 YIp plasmids with a variety of flanking polylinkers have been described (Stearns et al., 1990).

Table IV. Dominant drug-resistance selectable markers

Dominant selectable marker	Drug	Original[a] source	Yeast plasmid	Features[b]	Reference
Cam^r (Cat)	Chloramphenicol	Tn9	pCH100	2 μm ori, *TRP1*, *SalI-ADC1p-CAT-CYC1t-HindIII*	Hadfield et al. (1986)
			pCH115	*HIS3*, *SalI-ADC1p-CAT-CYC1t-SalI*	
Ble^r	Phleomycin	Tn5	pUT308	2 μm ori, *ARS1*, *URA3*, *CYC1pΔ-Ble-CYC1t*	Gatignol et al. (1987)
Hyg^r	Hygromycin B	pJR225	pLG89	2 μm ori, *ARS1*, *URA3*, *CYC1-Hph-CYC1t*	Gritz and Davies (1983)
Km^r	G418	Tn903	YEp13C	2 μm ori, *LEU2*, Km^r	Webster and Dickson (1983)
Rdhf^r	Methotrexate and sulphanilamide	R388	pADA1	2 μm ori, *LEU2*, *BamHI-ADC1p-Rdhfr-ADC1t-BamHI*	Miyajima et al. (1984)
DFR1	Methotrexate and sulphanilamide	Yeast	pPL241	2 μm ori, *LEU2*, *DFR1*	Lagosky et al. (1987)
TK	Sulphanilamide and amethopterin	HSV	pJM81	2 μm ori, *HIS3*, *LEU2*, HSV-TK	McNeil and Friesen (1981)
			pAYE56	2 μm ori, *LEU2*, HSV-TK	Zealey et al. (1988)

[a] Original plasmid source of the marker are indicated and cited in references.
[b] Vector components necessary for yeast function are indicated. Suffices *p* and *t* refer to promoter and terminator sequences of the described gene, respectively.

transformed cells by homologous recombination (Hinnen *et al.*, 1978). The number of integration sites depends upon the number of complementary genomic sequences that are present in the vector. Plasmids with two yeast genes have the potential to integrate at the genomic loci of either gene, while vectors containing repetitive DNA sequences, such as Ty elements, may integrate into several sites within the genome. Integration of circular plasmid DNA by homologous recombination leads to a copy of the backbone vector sequence flanked by two direct copies of the homologous yeast DNA sequence. The site of integration can be targeted by restriction digestion at a site within the gene of interest (Orr-Weaver *et al.*, 1981). Linearizing the plasmid DNA increases the transformation frequency dramatically.

YIp vectors offer several advantages and we refer the reader to a recent review by Stearns *et al.* (1990), which describes several utilities in greater detail. The integrative vectors are typically present in a single copy in each genome. However, multiple integration events can occur and this may be a useful experimental feature. Strains constructed with YIp plasmids must be checked by restriction digestion and Southern analysis to confirm that the desired integration events have taken place. YIp transformants are also generally extremely stable, even in the absence of selective pressure and transformants can be grown for many generations under non-selective growth conditions without appreciable loss of vector sequences (Hinnen *et al.*, 1978). When plasmid loss does occur, it is typically by a homologous recombination reaction between tandemly repeated homologous DNA sequences, leading to a "looping or popping out" of the vector sequence and one copy of the duplicated sequence. The *in vivo* recombination event can occur anywhere throughout the homologous region, and occurs at a frequency of approximately 1×10^{-3} to 1×10^{-4} (Stearns *et al.*, 1990). This process has led to development of a "transplacement" technique whereby wild-type sequences can be replaced by *in vitro*-generated mutant sequences (Scherer and Davis, 1979). Integration of mutant sequences, followed by excision of the vector, leads to a transplacement event, which exchanges mutant and wild-type sequences. Since the excision process can also remove mutant sequences, the progeny resulting from such an eviction procedure must be examined phenotypically and, if possible, structurally. Use of *URA3* or *LYS2* genes in this process can readily facilitate the transplacement process because of the forward and reverse selections available for each marker. Since the vector does not self-replicate, it is rapidly lost from the population.

One specific but wide utility of these vectors is to map the chromosomal loci of genes whose wild-type or mutant phenotypes are unknown or difficult to score. As an example, the gene of interest can be cloned into YIp5 (Struhl *et al.*, 1979), containing the *URA3* gene, and the resulting recombinant plasmid used to transform cells by targeted recombination at the locus of the gene under study. The resulting transformants are mated with appropriate *ura3* mapping strains, and the segregation of the plasmid-linked selectable marker (i.e. *URA3*) followed through crosses by tetrad analysis. YIp plasmids can also be used to clone mutations created *in vivo*. A copy of the wild-type gene is integrated into the mutant allele, and genomic DNA from the transformant is prepared and digested with an appropriate restriction enzyme. After re-circularization of the DNA by *in vitro* ligation, it is transformed into *E. coli* (Winston *et al.*, 1983). It is also possible, through appropriate engineering of a target gene residing on a YIp vector, to isolate recessive target gene mutations in one step (for details, see Stearns *et al.*, 1990). Likewise, null mutations can be isolated readily in one step by homologous integration of a YIp plasmid containing an internal fragment of the gene. This process creates two truncated gene fragments, and assumes that neither fragment retains full or partial activity. Most YIp plasmids contain the standard yeast selectable markers, namely *HIS3*, *LEU2* and *URA3*, described in Table II.

IV. YEp VECTORS

Yeast episomal plasmid vectors replicate autonomously via a segment of the endogenous yeast 2 μm plasmid on the vector, which serves as an origin of replication. Properties of the 2 μm *ori* promoting the high copy-number extrachromosomal existence of this plasmid confer high-frequency transformation properties on YEp vectors (Beggs, 1978; Gerbaud *et al.*, 1979; Hicks *et al.*, 1978; Struhl *et al.*, 1979). We refer readers to several recent, comprehensive reviews on native 2 μm plasmid and 2 μm circle-based vectors (Murray, 1987; Armstrong *et al.*, 1988; Rose and Broach, 1990). These authors describe aspects of 2 μm plasmid replication and segregation, which are relevant to understanding copy-number control and stability of YEp plasmids.

This small double-stranded circular plasmid exists within the yeast nucleus, there being 50–100 copies in each cell. It consists of two unique repeat sequences separated by two precise 599 bp repeats, which are in inverted orientation with respect to one another (Hartley and Donelson, 1980). The *ori* spans one of the inverted repeats and the larger of the two unique regions. Four large open reading frames on the plasmid are transcribed into poly(A)+ RNAs. Two of these, *REP1* and *REP2*, encode products involved in promoting equi-partitioning of plasmid between daughter cells at division (Jayaram *et al.*, 1983; Kikuchi, 1983; Cashmore *et al.*, 1986). An additional plasmid sequence element, called *REP3* or *STB*, is required in *cis* to the *ori* to mediate action of the *trans*-acting *REP1* and *REP2* products. The precise molecular mechanisms by which the *REP* partitioning system operates is unknown at this time.

Stable plasmid inheritance is also influenced by copy-number amplification, which is mediated by a novel mechanism first proposed by Futcher (1986). The product of the plasmid-encoded *FLP* gene catalyses a recombination event between the inverted repeat units of the plasmid, resulting in an interconversion of the two unique regions with respect to each other. This results in a shift from bidirectional, theta to rolling-circle replication. Plasmid replication normally occurs in a bidirectional manner through the theta intermediate. Under conditions of low plasmid-copy number, *FLP*-mediated recombination within the theta intermediate gives rise to rolling-circle replication intermediates and multiple plasmid copies from a single parent. Experimental evidence for this model has been obtained (Volkert and Broach, 1986; Reynolds *et al.*, 1987). Expression of the *FLP* gene is also co-ordinately repressed by action of the *REP1* and *REP2* gene products, providing a plausible mechanism for copy-number control (Murray *et al.*, 1987; Reynolds *et al.*, 1987; Som *et al.*, 1988). At lower copy number, diminished dosage of the *REP* gene products results in derepression of the *FLP* gene and copy-number amplification. When the copy number rises, the concentration of *REP* products increases leading to repression of *FLP* activity and inhibition of copy-number amplification. Two micrometre copy-number control is also fine tuned by an antirepressor activity of the D open reading frame, and autoregulation of *REP1* expression by products of the *REP1* and *REP2* genes.

YEp vectors contain either a full copy of the 2 µm plasmid (i.e. pCV20 and pCV21) or a region of the 2 µm circle, which encompasses the *ori* and *REP3* region (i.e. YEp24 and YEp13) (Rose and Broach, 1990). This region is usually derived from a 2.2 kb *EcoRI* fragment or a 2.1 kb *HindIII* fragment of the B form of the native plasmid (Table V). Since products of 2 µm *REP1* and *REP2* are also required for stable segregation, plasmids containing only the *ori*-*REP3* region are routinely propagated in cir[+] hosts containing native 2 µm plasmid.

YEp plasmids are relatively stable but are lost at frequencies of around $1/10^2$ cells in each generation, which is at least two orders of magnitude higher than native 2 µm plasmid. Consequently, growth for several generations in the absence of selection will lead to a decline in the fraction of plasmid-containing cells within a population and a decrease in average plasmid-copy number. Even under conditions of selective growth, the fraction of plasmid-bearing cells in a transformant population may range from 60% to 95%. Several factors, including simple insertion of sequences into

Table V. ARS and CEN components of vectors

Vector component	Routinely used DNA fragment	Plasmid	Remarks[a]	Reference
ARS1	1.45 kb EcoRI, 0.85 kb EcoRI-HindIII	YRp7	The 1.45 kb EcoRI fragment encompasses the adjacent TRP1 gene. See Table II for a mini-map of restriction sites.	Struhl et al. (1979)
ARS2	0.6 kb XhoI	pGT22	A mini-map of 1.5 kb fragment encompassing tRNA GLN, delta sequence and ARS2. AccI (724, 1394), AsuII (235, 290), AvaI (749, 1376), AvaII (224, 276, 342), BspMII (304), BstBI (235,290), ClaI (47), DraII (276), EcoRV (1042), HindIII (1), SpeI (491, 1019), SspI (799, 1097, 1269, 1411), XhoI (749, 1376).	Tschumper and Carbon (1982)
ARS3	1.35 kb EcoRI	pBR322-sup11	This fragment also contains the SUP11-1 ochre-suppressing tRNA and inefficient ARS3.	St John et al. (1981)
ARSH4	0.37 kb MFS	pAB9	Contains a 374 bp Sau3A Histone H4-associated ARS fragment inserted into the BamHI site of pUC8.	Bouton and Smith (1986)
2 μm ori	2.2 kb EcoRI, 2.2 kb HindIII, 1.5 kb Sau3A	YEp24, pMJ5	Each fragment contains the 2 μm ori cis-acting REP3 region. Mini-map of 2.2 kb EcoRI B fragment: AccI (278, 2206, 2215), AvaI (1978), AvaII (1390), AvaII (33, 1294, 2179), AvaIII (338, 1398), BalI (266), BclI (304), BstBI (1978), DraII (205, 2179), EcoRI (1,2241), EspI (2125), HincII (219, 1686), HindIII (105), HpaI (1686), NruI (2092), NsiI (340, 1400), PstI (2000), PvuI (1922), SnaBI (1044), SpeI (319, 2218), SspI (808), StuI (2162), XbaI (703).	Botstein et al. (1979); Broach (1983); Hartley and Donelson (1980)
CEN3	2 kb BamHI-HindIII, 1.1 kb BamHI-ClaI, 0.6 kb Sau3A-BamHI	pYe(CEN3)11	The 2 kb BamHI-HindIII fragment contains 0.4 kb of bacterial pBR313 sequence. The smaller 0.6 kb fragment is deficient in CEN function. The mini-map of the 627 bp BamHI-Sau3A fragment is: AvaI (619), HincII (477), NheI (36), SmaI (621), SspI (153).	Clarke and Carbon (1980)

Table V. Continued

Vector component	Routinely used DNA fragment	Plasmid	Remarks[a]	Reference
CEN4	3.6 kb BamHI-EcoRI 1.6 kb XhoI 1.75 kb PvuII-EcoRI 0.85 kb PvuII-HpaI	YCp19	Mitotically stable YRp17 derivative (ARS1, TRP1, URA3) containing a 3.6 kb BamHI-EcoRI CEN4 fragment. The mini-map of the 2095 bp fragment is: AsuII (37), AvaI (300, 1792), BalI (2022), BclI (413, 460), BglII (314), BstBI (37), BstXI (1746), EcoRI (2090), HincII (429, 445, 1191), HpaI (1191), KpnI (1321) NheI (1668), PvuII (339), SpeI (384), SspI (630), StuI (1402, 1709), Tth111I (776), XhoI (301, 1792).	Stinchcomb et al. (1982) Mann and Davis (1986)
CEN5	1.5 kb BamHI	pYe(MSS1)3	Contains a 1.5 kb BamHI CEN5 fragment cloned into ARS1, LEU2, URA3 vector YRp114.	Maine et al. (1984)
CEN6	0.12 kb MFS	pUC19-CEN6:32	Contains a 125 bp MboII/MboII CEN6 fragment cloned into the HincII site of pUC19. The mini-map of a larger 800 bp fragment is: AseI (86), KpnI (122), SnaBI (539), SspI (30), PmaCI (268).	Sikorski and Hieter (1989)
CEN11	1.6 kb SalI 0.9 kb Sau3A	pYe(CEN11)12	YRp7′ derivative containing a 1.6 kb SalI CEN11 fragment. The smaller CEN11 fragment appears somewhat defective in supporting meiotic segregation or copy-number control. The mini-map of the 858 bp fragment is: AseI (256), AvaI (133), AvaIII (576), BamHI (1), HincII (36,448), KpnI (786), NruI (693), NsiI (578), PmaCI (852), SpeI (91), SspI (73), XbaI (651).	Fitzgerald-Hayes et al. (1982a) Fitzgerald-Hayes et al. (1982b)
ARS1-CEN4	2.6 kb SpeI-NdeI 1.7 kb HindIII	YCp50 pTC1	YCp50 (ARS1, CEN4, URA3) has been described previously. pTC1 is a pBR322 derivative containing a Klenow-treated 1.45 kb.	Elledge and Davis (1988) Johnston and Davis (1984)

Continued

Table V. Continued

Vector component	Routinely used DNA fragment	Plasmid	Remarks[a]	Reference
			EcoRI *TRP1 ARS1* fragment inserted into the *Eco*RI site and an 850 bp *Pvu*II-*Hpa*I *CEN4* fragment in the *Cla*I site.	Gietz and Sugino (1988)
	*Nar*I-*Hind*III	pUN10	pUN10 is an *ARS1, CEN4, TRP1* vector (Table XI), which contains an *ARS1-CEN4* cassette.	
ARSH4-CEN6.32	0.5 kb MFS	pRSS84	pUC derivative containing an engineered *ARSH4/CEN6* cassette with flanking restriction sites as follows: *Hind*III, *Pst*I, *Sal*I, *ARSH4/CEN6, Pst*I, *Sph*I, *Hind*III.	Sikorski and Hieter (1989)

[a] Restriction-enzyme sites (positions in parentheses) present in the fragments indicated are the same as those in Table II. Sites were identified in sequences within GenBank and EMBL databases by the MAP program (UW GCG Sequence Analysis Software Package; Devereux, 1989). Complete nucleotide sequences of DNA fragments are included in the referenced papers.

the plasmid, the nature of the sequences inserted, as well as vector size can influence vector stability. These factors are discussed in detail by Rose and Broach (1990). Plasmid copy number can also influence stability. Higher copy number YEp vectors tend to be more stable than those with lower copy numbers.

The copy number of most YEp plasmids in cir⁺ hosts averages 10–40 in each cell. However, they are not symmetrically distributed among all cells in the transformant population. Consequently, the copy number of the plasmid is not uniform throughout the population of cells, and the plasmids tend to assume a distribution of copy numbers within individual cells of the population. The average of this distribution is the population copy number in each cell. Typically, the copy-number fluctuations of these plasmids has little or no significant impact on their use, although in some instances these distributions may be significant. For example, selections or screens involving plasmid-borne markers may enrich for transformant cells, which contain plasmids at either very low or very high copy number.

YEp plasmids have proven very useful in several areas, including general cloning experiments and gene-expression studies. Yeast genomic DNA libraries in YEp13 and YEp24 have been used to isolate many yeast genes by complementation (Broach *et al.*, 1979; Carlson and Botstein, 1982). Two problems related to the high copy-number of plasmids in these libraries, however, need to be considered. First, genes whose overexpression is deleterious to the cell may be absent from or underrepresented in the libraries, and YCp libraries should be considered to circumvent this problem. Secondly, complementation of a mutation of interest can occur by overexpression of heterologous genes, as well as by transformation with the desired wild-type gene. Although these clones may represent useful suppressors of a given mutation, they complicate the original cloning effort. Another common use of these vectors is for overexpression of a gene for a variety of purposes (for discussion, see Rine, 1991). Several factors that may limit the level of YEp-based overexpression include plasmid distribution and copy number within the population, plasmid stability, as well as the availability of cellular factors, which may regulate expression of the gene.

Several systems have been developed for ultra high copy-number propagation of YEp plasmids *in vivo*. The most common method relies on use of the partially defective *leu2-d* allele, originating from the YEp vector pJDB219 (Beggs, 1978; Erhart and Hollenberg, 1983). Vector pJDB219 and its derivatives are present at copy numbers ranging from 200 to 300 in each cell. The vector contains a randomly sheared DNA fragment spanning the *LEU2* gene inserted into the *Pst*I site of the D region of the 2 μm plasmid by homopolymer (dA/dT) tailing. The allele contains 29 bp 5′ to the translational initiation codon of *LEU2*, in the opposite orientation to that of open-reading frame D. Expression of *leu2-d* is several orders of magnitude lower than the chromosomally encoded gene and several lines of evidence suggest that this defective expression of *leu2-d* is responsible for the high copy-number properties of the plasmid (for a review, see Rose and Broach, 1990). Strains containing the 2 μm plasmid are the desired host for *leu2-d* vectors containing only the plasmid *ori* and *REP3* loci. However, the high copy number of these plasmids can cure the cell of endogenous 2 μm plasmid, leading to instability of the vector. Strains lacking the 2 μm plasmid are the preferred host for *leu2-d* vectors containing the entire 2 μm plasmid. Recombination events between the native 2 μm circles and the vector in a cir$^+$ host can result in a *leu2-d* recombinant lacking the bacterial vector. *Leu2-d* vectors, such as pJDB219, persist at very high copy numbers for many generations after removal of selective pressure, making them useful in large-scale cultures where plasmid selection is not always feasible.

A second system for high copy-number amplification is based upon *FLP*-controlled amplification of the plasmid. Several groups have demonstrated that like the native 2 μm plasmid (Murray *et al.*, 1987; Reynolds *et al.*, 1987; Som *et al.*, 1988), YEp vectors derived from the 2 μm plasmid show increased steady-state levels when expression of *FLP* is increased (for a detailed review, see Rose and Broach, 1990). Two amplifiable vectors constructed by P. Hieter and his colleagues (P. Hieter, personal communication; for details, see Rose and Broach, 1990) contain the entire 2 μm circle with shuttle-vector sequences (pHSS6; Seifert *et al.*, 1986) and the yeast *TRP1* or *URA3* markers, inserted within the *FLP* region. The vectors also contain polylinker regions for insertion of cloned genes. Galactose induction of a chromosomal *GAL-FLP* gene in the host leads to overamplification of the plasmid or, alternatively, the plasmid may be propagated in strains containing a constitutively expressed *FLP* gene.

V. YRp VECTORS

Yeast replicating plasmid vectors are maintained as extrachromosomal elements, due to DNA sequences (*ARS*) on the vector that serve as an *ori* and promote autonomous replication. Typically these sequences are derived from the yeast genome (Kingsman *et al.*, 1979; Stinchcomb *et al.*, 1979; Struhl *et al.*, 1979). However, they have also been isolated from other organisms (Stinchcomb *et al.* 1980). YRp plasmids transform yeast at high frequencies (i.e. 500–2000 transformants per microgram of DNA) and exist at high copy numbers in a population of cells, but, because they are unstable during mitosis and meiosis, copy number varies widely within the population. The instability, high copy number and copy-number variation result in large part from asymmetrical plasmid segregation at division (Murray and Szostak, 1983a). Mitotic losses of YRp plasmids can be significant (i.e. 30% of cells in each generation). Although the high transformation frequencies of these plasmids make them popular vectors for general cloning purposes, their unstable and asymmetrical segregation properties make them undesirable for other purposes such as analyses of gene regulatory mechanisms. They have provided useful systems for isolating genomic elements such as centromeric DNA (CEN), which are important for stable chromosomal segregation and maintenance *in vivo* (see YCp vectors below). Various *ARS* sequences that have been incorporated into a number of yeast vectors are listed in Table V.

VI . YCp VECTORS

Yeast centromere plasmid vectors are autonomously replicating vectors containing centromere (CEN) sequences conferring mitotic and meiotic stability. These vectors are typically present at very low copy number (approximately 1–3 per cell). Clarke and Carbon (1980) isolated the first yeast centromeric sequence (CEN3) by its ability to stabilize the defective inheritance properties of YRp plasmids. Since discovery of the intrinsic stability properties of CEN3, several other yeast centromeres have been isolated and characterized (Fitzgerald-Hayes *et al.*, 1982a,b; Stinchcomb *et al.*, 1982; Maine *et al.*, 1984; Hieter *et al.*, 1985; Panzeri *et al.*, 1985; Mann and Davis, 1986). Molecular genetic analysis of CEN components required for proper CEN function have identified the presence of three conserved domains, designated I, II and III. All of these sequences appear to be important for mitotic stabilization by CEN (Carbon, 1984). Despite their increased stability relative to YRp vectors, YCp vectors are lost at frequencies (approximately once in 10^2 divisions), which are much higher than native yeast chromosomes. Hartwell *et al.* (1982) estimated that chromosome V is lost at a rate of once in each 10^5 divisions. The intrinsic stability and low copy number of YCp autonomously replicating plasmids make them popular choices as general subcloning vectors, as well as vectors for construction of yeast genomic DNA libraries (Rose *et al.*, 1987), thereby avoiding a variety of problems associated with use of stable multicopy YEp plasmids (see the earlier discussion). Two visual colony-sectoring assays utilizing mutants of *Sacch. cerevisiae* defective in purine biosynthesis have been developed with YCp vectors to monitor changes in chromosome ploidy during mitosis (Koshland and Hieter, 1987). Both systems provide useful means for monitoring chromosome transmission quantitatively and qualitatively. The systems have also been developed to analyse the mutational state of essential genes (Elledge and Davis, 1988), and identify mutations in yeast genes that are functionally equivalent to genes from other organisms (Kranz and Holm, 1990). The most commonly used *CEN* sequences are listed in Table V.

VII. SPECIALIZED YEAST VECTORS

A. YLp VECTORS

Yeast linear plasmid vectors contain homologous and heterologous DNA sequences that function as telomeres (TEL) *in vivo* (Szostak and Blackburn, 1982; Dani and Zakian, 1983; Shampay *et al.*, 1984). These vectors have been used as model substrates for examining telomere structure and function, and some of the plasmids have been used to construct artificial chromosomes containing, in addition to TEL sequences, all known essential chromosome components, including genes, origins of replication and centromeres (Murray and Szostak, 1983b). The stabilities of these artificial chromosomes are generally several orders of magnitude lower than natural chromosomes, possibly due to their small size.

The vectors have also been useful in cloning. Olson and his colleagues have developed yeast host/vector systems, which allow cloning of large genes or clusters of genes in yeast (Burke *et al.*, 1987). The YAC (Yeast Artificial Chromosome) vectors contain a variety of cloning sites and selectable markers (Table VI). A YLp host/vector system has also been developed as a tool in a mapping procedure for determining the physical location of a DNA sequence on a natural or artificial yeast chromosome (Vollrath *et al.*, 1988). The principle of this chromosome-fragmentation method relies on physically splitting the chromosome containing the target sequence at a specific site by transformation with small linear DNA molecules containing the sequence of interest at one end and a TEL at the other. The end of the linear molecule containing the cloned sequence recombines with its homologous target sequence leading to formation of chromosome fragments containing all sequences that are distal or proximal to the mapping site, depending upon the orientation of the target sequences in the small linear transforming DNA. The position of the cloned sequence is determined relative to the chromosome ends by sizing the generated chromosomal fragments by CHEF gel electrophoresis. Table VI contains information about several useful YLp vectors.

Table VI. Linear plasmids used as vectors in yeasts

Telomere	Plasmid	Restriction sites[a]	Features[b]	Reference
Tetrahymena	pYAC2	SmaI	TRP1, ARS1, CEN4, SUP4 (SmaI), URA3, TEL-HIS3-TEL	Burke et al. (1987)
thermophila	pYAC3	SnaBI	TRP1, ARS1, CEN4, SUP4 (SnaBI), URA3, TEL-HIS3-TEL	Burke et al. (1987)
rDNA	pYAC4	EcoRI	TRP1, ARS1, CEN4, SUP4 (EcoRI), URA3, TEL-HIS3-TEL	Burke et al. (1987)
	pYAC5[c]	NotI	TRP1, ARS1, CEN4, SUP4 (NotI), URA3, TEL-HIS3-TEL	Burke et al. (1987)
	pYAC-RC	MCR	TRP1, ARS1, CEN4, SUP4 (NotI, SacII, SalI, MluI, ClaI, SnaBI), URA3, TEL-HIS3-TEL	Marchuk and Collins (1988)
	pYAC4-NEO	MCR	TRP1, ARS1, CEN4, SUP4 (EcoRI), Neo[r], URA3, TEL-HIS3-TEL	Cooke and Cross (1988)
Y'	YCF3	MCR	Y' TEL-BglII, NotI, EcoRI, HindIII, BamHI, URA3, SUP11	Vollrath et al. (1988)
Y'	YCF4	MCR	Y' TEL-BglII, NotI, EcoRI, HindIII, BamHI, CEN4, URA3, SUP11	Vollrath et al. (1988)

[a] Restriction-enzyme sites, which can be used for cloning purposes. MCR indicates multiple cloning region.
[b] Features include vector components, which allow for maintenance of the plasmid in yeast, as well as cloning sites.
[c] pYAC5 is now being distributed as pYAC55 to distinguish it from several pYAC5s with incorrect structures, which were inadvertently distributed (M. Olson, personal communication).

B. YPp VECTORS

Yeast promoter plasmid vectors contain reporter genes to which promoter sequences containing transcriptional and/or translational signals can be fused. The reporter gene typically encodes an easily assayable protein, the most widely used being the β-galactosidase gene *lacZ* of *E. coli* (Table VII; Guarente and Ptashne, 1981; Guarente *et al.*, 1982; Casadaban *et al.*, 1983). It has been well studied and characterized by molecular-genetic dissection, and several useful chromogenic and fluorogenic enzyme substrates are available. Another useful and unique feature of β-galactosidase is the dispensable nature of its amino-terminus. The system also requires no special host strain, since *Sacch. cerevisiae* does not contain an endogenous β-galactosidase. However, despite its widespread use, some limitations of this reporter do exist. The principal disadvantage is that the enzyme is a large tetramer, which exists in the cytosol and many of the standard substrates are impermeable to the cell. Detection of transformant colonies on solid medium with the substrate X-gal sometimes requires several days of prolonged incubation before the standard blue-coloured enzymic product develops sufficiently. A significant fraction of detected enzyme results from lysed or dying cells within the colony. To avoid some of these problems, total cellular protein extracts can be rapidly prepared and assayed, but this is not feasible if a large number of clones need to be surveyed.

The yeast *PHO5* gene encoding the major secreted repressible acid phosphatase has also been developed as a reporter gene (Table VII; Sidhu and Bollon, 1987; Hwang *et al.*, 1988; Zvonok *et al.*, 1988), and this system offers several advantages and disadvantages over the use of *lacZ*. The principal advantage of *PHO5* is that the enzyme is secreted into the periplasmic space, where it is more readily accessible to several chromogenic and fluorogenic substrates. In contrast, β-galactosidase translocates through plasma membranes inefficiently, even when it is fused to protein secretory signals (Emr *et al.*, 1983; Kaiser and Botstein, 1990), and has limited utility when analysing genes whose protein products traverse cell membranes. The principal disadvantage of the reporter is that *Sacch. cerevisiae* contains several genes encoding acid phosphatases (Bostian *et al.*, 1980; Kramer and Andersen, 1980; Rogers *et al.*, 1982). The *PHO5* gene encodes the major isozyme accounting for approximately 80% of cellular enzyme. Two other genes, *PHO10* and *PHO11*, encode minor repressible isozymes. The *PHO3* gene encodes an acid phosphatase isozyme, which is expressed at low levels and induced in phosphate-rich medium (Tait-Kamradt *et al.*, 1986). Plasmids containing *PHO5* reporter genes are used in *pho5 pho3* host strains, which lack the products of the respective genes (Hwang *et al.*, 1988; Zvonok *et al.*, 1988). Like acid phosphatase, the secreted form of invertase (the product

Table VII. Reporter and fusion genes in yeast vectors

Reporter gene	Plasmid	Restriction sites[a]	Features[b]	Reference
LacZ	pLG670-Z	XhoI	2 μm ori, URA3, XhoI-CYC1-LacZ	Guarente and Ptashne (1981)
	pMC1585	BamHI	2 μm ori, URA3, BamHI-LacZ	Casadaban et al. (1983)
	pMC1587	SmaI, BamHI	2 μm ori, LEU2, SmaI, BamHI-LacZ	Casadaban et al. (1983)
	pMC1790	EcoRI, SmaI, BamHI	ARS1, TRP1, EcoRI, SmaI, BamHI-LacZ	Casadaban et al. (1983)
	pMC2010	EcoRI, SmaI, BamHI	2 μm ori, TRP1, EcoRI, SmaI, BamHI-LacZ	Casadaban et al. (1983)
	YEp350 series	MFS	2 μm ori, URA3, MCR-LacZ	Myers et al. (1986)
	YIp350 series	MFS	URA3, MCR-LacZ	Myers et al. (1986)
	YEp360 series	MFS	2 μm ori, LEU2, MCR-LacZ	Myers et al. (1986)
	YIp360 series	MFS	LEU2, MCR-LacZ	Myers et al. (1986)
PHO5	pNZ8	MFS	EcoRI, SacI, KpnI, SmaI, BamHI-PHO5-PstI, SphI, HindIII	Zvonok et al. (1988)
	pZHB81	MFS	2 μm ori, LEU2, SacI, SmaI, BamHI-PHO5	Zvonok et al. (1988)
	pZHB82	MFS	2 μm ori, LEU2, XhoI, BssHI, BalI, PvuII, XbaI, BglII, BclI, MluI, SacI, SmaI, BamHI-PHO5	Zvonok et al. (1988)
	pNZ8X	MFS	EcoRI, XhoI, EcoRI, SacI, SmaI, BamHI, BalI-PHO5-PstI, SphI, HindIII	Zvonok et al. (1988)
	YpMB1	MFS	2 μm ori, URA3, SacI, SmaI, BamHI-PHO5	G. Bajszár (personal communication)
	pVC701	EcoRI	ARS1, TRP1, CEN3, EcoRI-PHO5	Hwang et al. (1988)
MFα1-SUC2	pSEY210	MFS	2 μm ori, URA3, EcoRI-MFα1$_p$-MFα1$_L$-HindIII-SUC2-HindIII	Emr et al. (1983)
SUC2	pSEY303	MFS	2 μm ori, URA3, EcoRI, XmaI/SmaI, BamHI, SalI, PstI, HindIII-'SUC2	Emr et al. (1986)
	pRB576	SmaI	ARS1, CEN4, URA3, SmaI-suc2-450	Kaiser et al. (1987)
galK	YCpR2	XhoI (2) and polylinker	ARS1, TRP1, CEN3, XhoI-CYC1p-MCR-galK	Rymond et al. (1983)
CAT	pEP100	SalI	2 μm ori, LEU2, CAT	Altherr and Rodriguez (1988)
	pK15 and pKW4	BamHI	2 μm ori, TRP1, BamHI-CAT	Kwak et al. (1987)

Continued

Table VII. Continued

Reporter gene	Plasmid	Restriction sites[a]	Features[b]	Reference
URA3 cassettes	pNKY48	MFS	ClaI, XhoI, SalI, **XbaI**, **Bam**HI, **Hind**III-HIS4C::URA3-BamHI, HindIII, BamHI, BglII, NcoI, EcoRI	Alani and Kleckner (1987)
	pNKY70	MFS	ClaI, XhoI, SalI, **XbaI**, **Bam**HI, **EcoR**I-HIS4C::URA3-BamHI, HindIII, BamHI, BglII, NcoI, EcoRI	Alani and Kleckner (1987)

[a] Restriction-enzyme sites, which precede the reporter. MFS indicates multiple flanking sites.

[b] Features include vector components necessary for maintenance of the plasmid in yeast, as well as restriction sites preceding the reporter gene. MCR indicates multiple cloning region. See references for details of plasmid constructions and the relative positions of restriction-enzyme sites, with respect to the reporter. The YEp350 and YEp360 series of plasmids contain a MCR with unique restriction-enzyme sites including EcoRI, KpnI, SmaI, BamHI, XbaI, SalI, PstI, SphI and HindIII. The MCR region has been inserted in both orientations with respect to lacZ and is proximal to the eighth codon of the gene. In the series, all sites are phased to three different reading frames. In pZHB82, the 759 bp sequence of the restriction site bank between the XhoI and SacI sites is identical to that of pJRD158 (Table I), followed by pUC18 sequences between SacI and BamHI. SacI, SmaI and BamHI can be used to insert the 3′ end of the promoter and all other unique restriction enzyme sites upstream can be used to accept the 5′ end of the promoter. The MFα1-SUC2 fusion gene consists of the 5′ flanking and prepro-leader region of the α-factor gene (EcoRI-HindIII) fused to coding sequences for the SUC2 gene (HindIII fragment) beginning at codon 4: CAA GCT TTC. SUC2 sequences can be removed as a HindIII fragment. The URA3 protein-fusion cassettes contain the 91 amino-acid residue HIS4C gene fused to the URA3 gene. The flanking ClaI, XhoI, and SalI sites at the 5′ end of the cassette are followed by the sequence: TC TAG AGG ATC CAA GCT TTG HIS4C-URA3 (pNKY48). pNKY70 is a pNKY48 derivative containing a GGA ATT CCA GCT linker between the GCT and TTG codons of pNKY48.

of the *SUC2* gene) has been developed as a reporter, and it too is an attractive one because of its extracellular location (Emr *et al.*, 1983). Its product can also readily be assayed *in vitro* and *in vivo* (Zimmermann *et al.*, 1973; Goldstein and Lampen, 1975). The *SUC2* gene offers the additional advantage that the secreted form of invertase is necessary for sucrose utilization, and strains lacking it grow very poorly when sucrose is the sole carbon and energy source (Kaiser *et al.*, 1987). The yeast *URA3* gene has been developed as a reporter for analysis of promoter sequences and fusion proteins (Table VII; Alani and Kleckner, 1987) and offers several advantages over the systems already described. It is an attractive reporter because it can be selected for and against *in vivo* (Boeke *et al.*, 1984, 1987) and its product (orotidine-5'-phosphate decarboxylase) can be readily assayed *in vitro* (Rose *et al.*, 1984). Orotidine-5'-phosphate decarboxylase is a small cytosolic dimeric protein (25 kDa monomer), and, as such, may be more tractable than the larger β-galactosidase (116 kDa) and acid phosphatase (60 kDa) enzymes, which are active as tetramers.

Several other reporter systems have been used in yeast and include the *galK* gene in *E. coli* and the bacterial antibiotic resistance gene, *cat* (for a review, see Altherr and Rodriguez, 1988). The *galK* gene in *E. coli* encodes galactokinase, for which a sensitive enzymic radioassay is available. The *galK* system offers the advantages of positive and negative selection in appropriate yeast host strains (Rymond *et al.*, 1983). The yeast *GAL1* gene encodes the equivalent yeast enzyme, and forward and reverse selection for galactokinase activity can be achieved in a *gal1* host. Positive selection is achieved by selecting for galactokinase-proficient transformants in a *gal1* mutant. Negative selections are available in *gal7* and/or *gal10* host strains, which accumulate the toxic galactose metabolite galactose-1-phosphate. The Tn9-derived *cat* gene encodes chloramphenicol acetyltransferase, for which spectrophotometric and highly sensitive radioassays are available. Expression of chloramphenicol acetyltransferase in yeast also leads to a chloramphenicol-resistant phenotype (Cohen *et al.*, 1980; Kwak *et al.*, 1987). Useful yeast vectors containing reporter sequences described above are listed in Table VII.

C. YXp VECTORS

Homologous or heterologous gene expression in yeast offers several advantages over bacterial or mammalian and insect cell-culture expression, and a variety of yeast systems have been successfully developed. Yeast expression plasmid vectors typically contain transcriptional promoters and, in many instances, transcriptional terminator sequences to which sequences of interest may be fused. Many of the vectors also possess coding sequences, which result in the post-translational processing and secretion of protein products. For detailed descriptions of a variety of yeast expression systems, we refer the reader to several papers dedicated to gene-expression systems in yeast (see Emr, 1990, and accompanying papers in Volume 185 of *Methods in Enzymology*; Romanos *et al.*, 1992).

Most expression systems are based upon well-characterized and efficient promoters from a number of yeast genes. Several, including *ADC1*, *PGK*, *ENO1* and *GAPDH*, lead to high-level constitutive expression (Table VIII). These genes encode constitutive enzymes whose levels can amount to 1-5% of the total RNA and soluble protein in glycolytically growing cells. The major advantages of these promoter systems is that extremely high levels of expression can be readily attained. Plasmid pAB23BX offers the additional advantage that it is a cDNA cloning vector, which can be used to clone mammalian genes (Schild *et al.*, 1990). It has been used to clone human genes involved in *de novo* purine biosynthesis by functional complementation of mutations in *Sacch. cerevisiae*. Many other systems allow for controlled expression of a cloned gene, including promoters from the well-characterized *GAL* system, as well as those from the *ADH2*, *CUP1*, *PHO5* and *MFα1* genes (Table VIII). These systems rely either on induction of the promoter by the addition of an inducer (i.e. *GAL* and *CUP* systems) or the relief of strong transcriptional repression. The galactose-inducible *GAL1*, *GAL7* and *GAL10* promoters from *Sacch. cerevisiae* have been utilized in a number of YXp vectors, and are particularly useful for two reasons. These promoters are transcriptionally very silent under non-inducing conditions (growth medium lacking glucose and galactose) and upon galactose induction are rapidly turned on to very high levels. Expression of *GALp* from very high copy number plasmids, however, can be limiting due to the low cellular levels of the positive-acting *GAL4* protein (Baker *et al.*, 1987). To overcome this limitation, a *GAL10p-GAL4* expression cassette has been developed to express high levels of *GAL4* protein upon galactose induction (Schultz *et al.*, 1987; Mylin *et al.*, 1990). Used in conjunction with *GALp* expression vectors, the *GAL10p-GAL4* cassette can lead to higher expression from multiple, plasmid-borne *GAL* promoters than is attainable in the same strain lacking the *GAL10p-GAL4* cassette. Yeast cDNA expression libraries have also been constructed in a YCp vector containing the *GAL1* promoter (see pRS316-GAL1 in Table VIII; Liu *et al.*, 1992). Lambda phage cDNA expression vectors (λYES) containing the *Sacch. cerevisiae GAL1* and *E. coli* lac promoters have been developed (Elledge *et al.*, 1991). These vectors facilitate the construction and use of large cDNA libraries, and can easily be converted from phage to a smaller

Table VIII. Yeast promoter components

Promoter	Regulation	Promoter restriction fragment[a]	Plasmid source	Features[b]	Reference
ADH1	Constitutive	2 kb BamHI ~1.5 kb BamHI-EcoRI	pAAH5 pGBn	BamHI-ADH1p-HindIII-ADH1t-BamHI BamHI-ADH1p-EcoRI	Ammerer (1983) Hitzeman et al. (1981)
ENO1	Constitutive	EcoRI-SalI	pAC1	EcoRI-ENO1p-HindIII-ENO1t-SalI	Innis et al. (1985)
GAPDH	Constitutive	EcoRI-SalI — 1.3 kb BamHI or SalI	pGPD-1 pGPD-2 pPGAP1-3	EcoRI-GAPDHp-BamHI-EcoRI-TRP1-SalI EcoRI-GAPDHp-BamHI-PGKt 1: BamHI-GAPDHp-NcoI, EcoRI, SalI-GAPDHt-BamHI 2: BamHI-GAPDHp-NcoI, EcoRI, BglII-GAPDHt-BamHI 3: SalI-GAPDHp-NcoI, EcoRI, BglII-GAPDHt-SalI	Bitter and Egan (1984) Rosenberg et al. (1990)
		BamHI-BglII	pAB23BX	BamHI-GAPDHp-BglII, BstXI, XhoI, SacI, BstXI, BglII GAPDHt-BamHI	Schild et al. (1990)
		HindIII-XbaI	pG-1 pG-2 pG-3	HindIII-GAPDHp-BamHI, SalI-PGKt-XbaI HindIII-GAPDHp-BamHI, SalI-PGKt-XbaI HindIII-GAPDHp-polylinker/spacer-PGKt-XbaI	Schena et al. (1991) Schena et al. (1991) Schena et al. (1991)
PGK	Constitutive	1.87 kb HindIII 1.6 kb HindIII-EcoRI 0.97 kb XhoI-SalI	pMA91 YEpIPT yPGK	HindIII-PGKp-BglII-PGKt-HindIII HindIII-PGKp-EcoRI-HindIII XhoI-PGKp-EcoRI-HindIII, BamHI-PGKt-SalI	Kingsman et al. (1990) Hitzeman et al. (1986) Kang et al. (1990)
GAL1	Galactose inducible	EcoRI-XhoI — —	pCGE329 pRS316-GAL1 pRS series	EcoRI-GAL1p-XhoI KpnI-EcoRI-GAL1p-polylinker GAL1p-polylinker	Boeke et al. (1985) Liu et al. (1992) Phil Hieter (personal communication)
GAL10		0.5 kb Sau3A-SalI —	YEp51 pRS series	Sau3A-GAL10p-SalI GAL10p-polylinker	Broach et al. (1983) Phil Hieter (personal communication)
GAL7		1 kb EcoRI-BamHI	pBM756-8	EcoRI-GAL7p-BamHI	Mark Johnston (personal communication)

Continued

Table VIII. Continued

Promoter	Regulation	Promoter restriction fragment[a]	Plasmid source	Features[b]	Reference
GAL1/GAL10		0.85 kb BamHI-EcoRI 0.85 kb HindIII-EcoRI 0.85 kb with MFS	pBM272 pPH3	HindIII-BamHI-GAL1p/GAL10p-EcoRI HindIII, SphI, PstI, XbaI, BamHI-GAL1p/GAL10p-EcoRI	Mark Johnston (personal communication) Jagadeeswaran and Hass (1990)
CUP1	Copper inducible	HindIII-SacI 0.74 kb BamHI-HindIII	pYELC5 pCUP	HindIII-CUP1p-BamHI, PvuII, SalI, PstI, CUP1-SacI BamHI-CUP1p-NcoI-Asp718I-CYC1t-HindIII	Macreadie (1990) Kang et al. (1990)
ADH2	Glucose repressible	2.65 kb BamHI-XhoI	pADH2-2	BamHI-ADH2p-HindIII-ADH2t-XhoI	Price et al. (1990)
PHO5	Phosphate repressible	1.4 kb BamHI-BglII 0.52 kb BamHI-SalI 0.56 kb BamHI-AhaIII	p1A1R1-15 pYBDT-1 pAP20	BamHI-PHO5p-EcoRI-HindIII-TRP1-BglII BamHI-PHO5p-SalI BamHI-PHO5p-ATG TTT AAA	Parent et al. (1985) Lemontt et al. (1985) Andersen et al. (1983); Green et al. (1986)
MATα2 operator	Temperature regulation	BamHI —	pMTR series SXR series	BamHI-ADH2p-MATα2-BamHI BglII-TPI1p-MATα2	Sledziewski et al. (1990) Sledziewski et al. (1990)

[a] A convenient restriction fragment containing promoter cassette. MFS indicates multiple flanking sites.

[b] Features include details concerning restriction sites, which flank the promoter (p) cassette and, in cases where a terminator (t) is present, the cloning restriction site is between the promoter and terminator. The GAL1p pRS series includes pRS128 (TRP1), pRS129 (TRP1), pRS167 (TRP1 URA3) and pRS169 (URA3) and the polylinker SalI, EcoRI, SmaI and BamHI. The GAL10p pRS series includes pRS166 (TRP1 URA3) and pRS173 (URA3). The GAL10p pRS series contains SalI, EcoRI, SmaI and BamHI sites to construct protein fusions to an upstream ATG. Plasmids pBM756, pBM757, and pBM758 contain the 1 kb HinfI GAL7p fragment cloned into the EcoRI and BamHI sites of YCp50. Each vector contains a BamHI site for fusions to Gal7 protein in the following reading frames: pBM756, +1 beginning after amino acid 2; pBM757, +2 beginning after amino acid 14; pBM758, +1 beginning after amino acid 18. In pAP20, AhaIII site cleaves site italicized ATG TTT AAA leaving ATG TTT codons for fusions.

plasmid. Yeast cDNA libraries are useful for cloning genes by complementation and identifying genes whose overexpression confers specific phenotypes (for discussion, see Liu *et al.*, 1992).

The inducible metallothionein or *CUP1* promoter from yeast provides reasonably high expression levels rapidly induced by addition of copper ions to the growth medium (for a review of the utility of the *CUP1* promoter in expression systems, see Etcheverry, 1990). The choice of copper concentrations for induction depends upon the copper resistance of the host strain. Yeast strains amplify the *CUP1* locus in response to selective pressure of copper ions in their environment and, as the number of *CUP1* repeats in a strain increases, the concentration of copper ions required for induction also increases. Laboratory strains related to S228C or X2180 can generally tolerate 0.2–0.5 mM copper ions when grown in liquid medium (Etcheverry, 1990). The *CUP1* promoter has also been developed in a vector system for production of cytosolic ubiquitin fusion proteins (Butt *et al.*, 1988; McDonnell *et al.*, 1989). A host processing enzyme removes ubiquitin from the fusion protein and the system enhances protein expression by increasing protein solubility as well as protecting fusion proteins from proteolysis.

The yeast *ADH2* and *PHO5* promoters are tightly regulated promoters whose transcription is repressed by environmental concentrations of glucose and phosphate, respectively. Upon derepression, both promoters direct high levels of transcriptional expression. Each has been developed into yeast expression systems (Kramer *et al.*, 1984; Price *et al.*, 1990). Expression of the yeast alcohol dehydrogenase gene (*ADH2*) is tightly regulated by glucose repression; under repressing conditions, transcription of the gene is undetectable. Derepression occurs when cells are grown on nonfermentable carbon sources, and the primary advantage of this system is the ability to induce expression without changing growth medium, adding inducing compounds or changing temperature. Cultures can be grown to high densities in a glucose-containing medium and, when the glucose is metabolically depleted, derepression occurs. We refer the reader to a recent review on the use of the *ADH2* promoter for expression (Price *et al.*, 1990).

The *PHO5* promoter provides a similar means of transcriptional regulation. Expression of *PHO5* is very tightly negatively controlled by phosphate (Oshima, 1982; Bostian *et al.*, 1983). In the presence of high concentrations of phosphate, the promoter is tightly repressed but rapidly undergoes derepression to high levels upon phosphate starvation. Expression from the *PHO5* promoter can be readily achieved by growing cells to high densities in a rich medium with controlled concentrations of phosphate and, upon metabolic depletion of phosphate, derepression of the promoter occurs. Alternatively, derepression can be achieved by transferring cells from a medium containing a high concentration to one with a low concentration of phosphate. Regulation of *PHO5* occurs through several positive- and negative-acting regulatory genes, which provide a variety of means of promoter control. The yeast *PHO4* gene is a positive activator required for *PHO5* expression, while the *PHO80* gene negatively regulates *PHO5*. Kramer and his colleagues used *PHO5*p plasmids in *pho4^{ts} pho80* mutants to achieve derepression by temperature control (Kramer *et al.*, 1984). The *PHO5* promoter is repressed in *pho4^{ts} pho80* mutants grown at 35°C and derepressed when the temperature is lowered to 23°C. Overexpression of the positive-acting *PHO4* protein from multicopy plasmids or from *GAL10p* results in varying degrees of constitutive *PHO5* expression and higher levels of *PHO5* expression upon derepression (Parent *et al.*, 1987). These observations suggest that a *PHO5p-PHO4* expression system similar to the *GAL10p-GAL4* system described earlier might be feasible as a means of increasing expression from the *PHO5* promoter.

Temperature-regulated promoter systems, which can be simply induced by a temperature shift of the culture, have also been developed (Table VIII; for a review, see Sledziewski *et al.*, 1990). Moreover, intermediate levels of expression can be attained by maintaining growth at a temperature intermediate to those at which the promoter is either fully active or inactive. The most widely used temperature-regulated system is based upon the use of the *MATα2* operator from *Sacch. cerevisiae*. The *MATα2* repressor blocks expression of a-specific genes in α haploid cells by binding to a 31 bp operator region of their promoters. These genes are temperature-regulated when synthesis or activity of the repressor is temperature-dependent. This is most readily achieved in *MATa* cells carrying a temperature-sensitive mutation in any one of four *SIR* (silent information regulator) genes. The products of the *SIR* genes are required for repression of the *HMRa* and *HMLα* loci containing silent copies of *MATa* and *MATα* information, respectively. As an example, the SIR3 protein is inactive in *MATa sir3-8* mutants at 35°C, and the *MATα2* repressor is expressed, leading to repression of promoters containing the *MATα2* operator. At permissive temperatures (25°C), SIR3 protein is active, blocking transcription of the *HML* and *HMR* loci. This leads to repression of α2 protein synthesis and subsequent derepression of promoters containing the *MATα2* operator. The *MATα2* operators have also been introduced into a number of strong promoters to achieve temperature-regulated expression (Table VIII). Inducible expression systems based upon mammalian steroid receptors have been developed in yeast. Functional expression of the glucocorticoid and androgen receptors results in hormone-dependent induction of reporter plasmids containing multiple copies of the glucocorticoid and androgen response elements, respectively (Picard *et al.*, 1990; Purvis *et al.*, 1991; Schena *et al.*, 1991).

Systems in *Sacch. cerevisiae* that direct secretion of expressed proteins are also available and offer several advantages over cytoplasmic expression. For detailed discussions on protein secretion and processing events, we refer the reader to several recent reviews (Bitter *et al.*, 1987; Emr, 1990, and accompanying papers; Brake, 1990; Romanos *et al.*, 1992). Proteins that are toxic or unstable when produced in the cytoplasm can be expressed when secreted from the cell. Purification of the secreted protein can be achieved more readily because the extent of secretion of endogenous proteins from this organism is quite low. Proper disulphide-bond formation is also more likely to occur in proteins traversing the secretory pathway than in proteins expressed in the reducing environment of the cytoplasm and oxidized after cell disruption. Extensive biochemical and genetic evidence indicates that yeast protein-transport and secretory pathways share many common features with their counterparts in other eukaryotes. Several secretion–expression systems have been developed that utilize secretion signals from the α-factor pheromone precursor, invertase and acid phosphatase (Miyajima *et al.*, 1985; Smith *et al.*, 1985). Expression systems utilizing these signals are listed in Table IX.

The yeast Ty element has also been developed into two expression systems. In the first, yeast expression cassettes, as large as 2 kb, are inserted into plasmid-borne Ty elements fused to the *GAL1* promoter (Jacobs *et al.*, 1988). Upon galactose induction, the expression cassettes are transposed throughout the yeast genome and, once integrated, are stably segregated and maintained within the population in the absence of selective pressure. The insertions are made within a region of Ty1 near the 3′ LTR and require use of efficient promoters, which do not contain transcription terminators that block synthesis of the large Ty RNA molecule, which is required for transposition. In the second system, TyA fusion proteins are expressed from the *PGK* promoter on the 2 μm *LEU2* vector pMA5620 (Table IX; Adams *et al.*, 1987). TyA fusion proteins can produce hybrid Ty-VLPs, which are easily purified, and useful as reagents or to raise antisera.

D. YHp VECTORS

We previously described yeast hybrid plasmid vectors as those containing hybrid gene sequences and providing interesting model substrates for examining particular aspects of yeast-gene expression (i.e. RNA processing, nuclear or extracellular protein localization) (Parent *et al.*, 1985). In addition to the plasmids that we described in our earlier review, several new vectors falling into this category have been generated. Ma and Ptashne (1987) described a plasmid vector pMA424, which contains a region of the

Table IX. Signal sequences used for localizing proteins

Signal sequence	Plasmid	Features[a]	Reference
α-Factor	YEp70αT	MFα1p-α-Factor L-**EcoRI**-2 μmt	Hitzeman et al. (1986)
	pαC2	BamHI-MFα1p-α-Factor L-**HindIII**-**SalI**-MFα1t-BamHI	Bitter et al. (1987)
	pαC3	BamHI-MFα1p-α-Factor L-**HindIII**-**SalI**-MFα1t-BamHI	Bitter et al. (1987)
	pL-MFα1	GAL10p-α-Factor L-**BamHI**-**SpeI**-**XhoI**-**NotI**-MFα1t	Jacobson et al. (1989)
	pAB126	BamHI-GAPp-α-Factor L-**XbaI**-λ DNA-BamHI-λ DNA-**SalI**-MFα1t-BamHI	Brake (1990)
	pαADH2	EcoRI-ADH2p-α-Factor L-**KpnI**-polylinker	Price et al. (1990)
PHO5	p1A1	BamHI-PHO5p-PHO5L-**KpnI**-**SalI**	Tipper and Bostian (1984)
SUC2	pCGS681	TPI1p-SUC2-L-**NcoI**	Moir and Dumais (1987)
TyA	pMA5620	HindIII-PGKp-TyA-**BamHI**-PGKt	Adams et al. (1987)

[a]Features include details concerning restriction sites (in bold), which may be used to fuse amino-terminal signal sequences (L) of the vector to protein-coding sequences. Flanking promoter (p) and terminator (t) sequences are also indicated. Details about the reading frame of cloning sites can be found in the references. pL-MFα1 lacks the Lys-Arg KEX2 cleavage site at amino-acid position of prepro-leader.

amino-terminus of the GAL4 activator protein (amino-acid residues 1–147), which encompasses its DNA-binding domain immediately followed by a polylinker region from plasmid pEMBL9(+). Protein-coding sequences can be fused to the *GAL4* DNA-binding domain in this vector, and the hybrid proteins analysed for their effects upon transcription of reporter genes containing GAL4 DNA-binding sites. Plasmid vector pMA424 contains the pBR322 *ori*, and *bla* gene, as well as the yeast *HIS3* marker, 2 μm *ori*, and the *ADH1* promoter and terminator flanking GAL4(1–147). Similar vectors have been developed utilizing the DNA-binding region of the LexA protein of *E. coli* (Brent and Ptashne, 1985). For example, plasmid pSH2-1 is essentially identical to pMA424, except that the DNA-binding region of GAL4 (residues 1–147) has been replaced with the *lexA* DNA-binding domain (residues 1–87) (Hanes and Brent, 1989). Transcriptional properties of pSH2-1 LexA fusion proteins are monitored with reporter genes containing *lexA* operator sites (Brent and Ptashne, 1985).

Fields and his colleagues have developed a two-hybrid genetic system to study protein–protein interactions by taking advantage of the properties of the *Sacch. cerevisiae* transcriptional activator protein GAL4 (Fields and Song, 1989). The system consists of the GAL4 DNA-binding domain fused to a protein "X" and a GAL4 transcription activating region fused to a protein "Y". If X and Y interact to form a protein complex and reconstitute proximity of the GAL4 domains, transcription of genes regulated by GAL4 DNA-binding sites is activated in *gal4* deletion strains. The system has been developed to identify and clone genes whose products interact with a known protein of interest (Chien *et al.*, 1991). To do so, the known protein is fused to the DNA-binding domain of *GAL4* and expressed in a *gal4* deletion strain. Libraries consisting of the *GAL4* activation domain fused to protein sequences encoded by yeast genomic DNA fragments are screened for clones, which activate a *GAL4*-regulated promoter and are dependent upon the GAL4 DNA-binding domain fusion protein of interest.

The process of epitope tagging involves the fusion of a set of amino acids, the "epitope tag", to a protein of interest (for review, see Kolodziej and Young, 1991). Yeast proteins tagged with an epitope can be readily assayed for biological function *in vivo* by testing the ability of the tagged allele to complement a null mutation. The tagged proteins are detected with a specific monoclonal antibody, which recognizes the epitope. This procedure allows many properties of the tagged protein to be assayed, including its size, abundance, cellular location, post-translational modification and interactions with other proteins. Several advantages of monitoring a tagged protein with a well-characterized monoclonal antibody are discussed by Kolodziej and Young (1991). The antibodies, 12CA5 and 9E10, recognize epitopes derived from the influenza haemagglutinin protein and a portion of the human c-*myc* gene, respectively. Nucleotide sequences encoding the epitopes recognized by these antibodies can be introduced into genes of interest by site-directed mutagenesis or into yeast expression vectors with in-frame restriction enzyme sites for fusing to genes of interest.

Lastly, plasmids that contain the yeast *HO* gene can be used to change the mating type of a strain (Jensen *et al.*, 1983; Herskowitz and Jensen, 1991). For example, pGALHO contains a *GAL1p:HO* fusion gene and transformants containing it undergo a mating-type switch upon galactose induction (Rose and Broach, 1990; Herskowitz and Jensen, 1991).

E. Mutagenesis systems

Several systems have been developed recently for transposon insertional mutagenesis of yeast DNA sequences in *E. coli* or yeast. Of these, three have been developed for rapid mutagenesis of DNA cloned in *E. coli*, followed by re-introduction of the mutant DNA into yeast for evaluation (for a recent review, see Hoekstra *et al.*, 1991). Each facilitates rapid localization of a gene on large cloned DNA fragments (Table X). For details, we refer the reader to reports describing their development. Each system contains a transposon marked for selection in yeast, which transposes from the mutagenesis plasmid to the sequences of interest on a separate vector. DNA containing the transposon hops is then transformed into yeast and phenotypically evaluated. The high frequency of transpositions is sufficient to enable characterization of enough insertions to define the functional limits of the gene of interest. The mini-Tn3 system (Seifert *et al.*, 1986) and the Tn10-LUK system (Huisman *et al.*, 1987) have been developed for transposon mutagenesis of genes cloned on plasmid vectors and the mini-Tn10 system for genes contained on bacteriophage (Snyder *et al.*, 1986).

The mini-Tn3 system is marked with the ampicillin-resistance gene and the yeast DNA to be mapped is cloned into a target vector (pHSS4 or pHSS6) containing the kanamycin-resistance gene of *E. coli* (Km[r]) (Seifert *et al.*, 1986). After isolation and mapping of Tn3 inserts into the target DNA, linear yeast DNA containing the hops is isolated from the target vector by *Not*I digestion and transformed into yeast for phenotypic evaluation. The Tn10-LUK system (Huisman *et al.*, 1987) contains a promoter-less '*lacZ* reporter gene, the yeast *URA3* marker, and the kanamycin resistance gene (Km[r]) from *E. coli*. This system allows isolation and identification of mini-Tn10-LUK insertions into a yeast gene present on most standard multicopy yeast–*E. coli* shuttle plasmids. After isolation of the Tn10 insertions, the plasmids can be transformed into a *ura3* yeast host and phenotypically evaluated. Plasmid DNA can also be readily isolated for physical mapping of the inserts by restriction analysis. Insertions of the promoterless *'lacZ* reporter within a target gene can allow its orientation to be determined with a fair degree of confidence. In some instances, in-frame fusions of the *'lacZ* coding sequences with the target gene can also create stable LacZ fusion proteins. After insertional inactivation, Tn10-LUK null mutants can be excised from the vector and transformed into yeast to create chromosomal-gene disruptions *in vivo*. The mini-Tn10 "transplason" system (Snyder *et al.*, 1986) was developed to mutagenize phage clones, and has been used to identify rapidly and map antigenic coding sequences within λgt11 clones. Transposition events are selected in *E. coli* and typically evaluated by examining the effects on antigen expression by immunoscreening. Since the transposons contain yeast selectable markers, insertion mutations within a desired target gene can be introduced into the yeast genome by one-step gene transplacement.

The transposable element Ty has also been developed for transposon mutagenesis in yeast (Garfinkel *et al.*, 1988; Garfinkel and Strathern, 1991). High levels of Ty transposition can be induced by overexpression of active Ty elements fused to the *GAL1* promoter on multicopy plasmids (Boeke *et al.*, 1985). Garfinkel and his colleagues have modified two elements, Ty*H3* and Ty*917*, for use as transposon mutagens. Multicopy plasmids bearing *GAL1p*-Ty elements with unique restriction sites have been created, and can be tagged with yeast and bacterial selectable markers. Three such elements, which retain their ability to transpose and mutate target sequences *in vivo*, are listed in Table X. The marked pGTy system is used by transforming the marked Ty element into yeast, inducing high levels of transposition on galactose-containing medium, and selecting or screening for mutants with the desired phenotype(s). The plasmid-borne Ty element is segregated from the population and the induced mutants are analysed by tetrad or random-spore analyses, and by Southern analysis to confirm that the mutants originate from the marked Ty element. Since overexpression of the *GALp*-Ty elements also induces transposition of chromosomal Ty elements (Boeke *et al.*, 1985), the marked pGTy system can generate Ty-induced mutations by native chromosomal elements, as well as by the marked plasmid element. To lower the frequency of chromosomal Ty transpositions and circumvent this problem, the marked pGTy system can be used in a *spt3* mutant. The product of the *SPT3* gene is required for transposition of native elements but not that of *GAL1p*-Ty elements (Boeke *et al.*, 1986). Marked elements, which contain bacterial vector sequences (i.e. pGTy917πND and pGTy917πNI, see Table X) can be recovered from yeast and used to isolate the target sequences into which the element transposed.

Table X. Systems for the mutagenesis of yeast genes

System	Plasmid	Features[a]	Reference
mini-Tn3	m-Tn3(HIS3)	Tn3 tr, Amp[r], loxP site, *HIS3*, Tn3 tr	Seifert *et al.* (1986)
	m-Tn3(URA3)	Tn3 tr, Amp[r], loxP site, *URA3*, Tn3 tr	Seifert *et al.* (1986)
	m-Tn3(LEU2)	Tn3 tr, Amp[r], loxP site, *LEU2*, Tn3 tr	Seifert *et al.* (1986)
	m-Tn3(TRP1)	Tn3 tr, Amp[r], loxP site, *TRP1*, Tn3 tr	Seifert *et al.* (1986)
	m-Tn3(SUP11)	Tn3 tr, Amp[r], loxP site, *SUP11*, Tn3 tr	Seifert *et al.* (1986)
mini-Tn10	mTn10/URA3/tet[r]	Tn10 tr, *URA3*, tet[r], Tn10 tr	Snyder *et al.* (1986)
	mTn10/URA3/supF	Tn10 tr, *URA3*, supF, Tn10 tr	Snyder *et al.* (1986)
	mTn10/TRP1/kan[r]	Tn10 tr, *TRP1*, kan[r], Tn10 tr	Snyder *et al.* (1986)
mini-Tn10	Tn10-LUK	Tn10 tr, 'lacZ-*URA3*-kan[r], Tn10 tr	Huisman *et al.* (1987)
hisG::*URA3*::hisG	pNKY51	*Bgl*II-hisG::*URA3*::hisG-*Bam*HI	Alani *et al.* (1987)
Ty	pGTyH3*HIS3*	*URA3*, 2 μm-ori, *GAL1p*-TyH3-*HIS3*	Garfinkel *et al.* (1988)
	pGTy917πND	*URA3*, 2 μm-ori, *GAL1p*-Ty917-supF-neo[r]	Garfinkel *et al.* (1988)
	pGTy917πNI	*URA3*, 2 μm-ori, *GAL1p*-Ty917-neo[r]-supF	Garfinkel *et al.* (1988)
YRp14	YRp14/his3-Δ200	*URA3*, *SUP11*, *his3-Δ200*	Fasullo and Davis (1988)
transplacement	YRp14/trp1-Δ63	*URA3*, *SUP11*, *trp1-Δ63*	Sikorski and Hieter (1989)
	YRp14/trp1-Δ1	*URA3*, *SUP11*, *trp1-Δ1*	Sikorski and Hieter (1989)
YRp15 transplacement	YRp15/leu2-Δ1	*URA3*, *SUP11*, *leu2-Δ1*	Sikorski and Hieter (1989)

[a]Features include vector components, which allow selection in yeast. tr indicates Tn terminal repeat. Details about the use of each system can be found in references. Tn3-derived minitransposons containing a truncated *lacZ* gene are also available for making translational fusions between yeast genes and *E. coli* β-galactosidase (Hoekstra *et al.*, 1991). More specialized minitransposons are also available including m-Tn3(EIE), which inserts a small defined DNA segment consisting of two 38 bp ends of Tn3, a *Not*I site, and a *loxP* site (Hoekstra *et al.*, 1991).

Alani *et al.* (1987) have constructed a *Bgl*II-*Bam*HI *hisG::URA3::hisG* mutagenesis cassette, which can be used to create readily null mutants of a cloned target gene. The cassette is inserted within the coding region of the target gene, creating a *URA3* null allele, which is introduced into a *ura3* host by one-step gene disruption. The direct *hisG* repeats flanking the *URA3* insertion facilitate loss of the *URA3* gene from the null allele by homologous recombination between the direct repeats, leaving a null allele containing one copy of the *hisG* repeat. This allows re-use of the *URA3* marker in the same strain.

Sikorski and Hieter (1989) have constructed a series of transplacement vectors, which can be used to create non-reverting *his3-Δ*, *leu2-Δ* and *trp1-Δ* alleles in a yeast strain (see Section III on YIp vectors for a discussion of the process of transplacement). The vectors are derivatives of YRp14 and YRp15 (St John *et al.*, 1981), which contain the yeast *URA3* gene and *SUP11* ochre suppressor. The plasmids are digested within the *his3*, *leu2* or *trp1* sequence, and transformed into the appropriate host by selecting for uracil prototrophy. In suitable *ade2-101* diploid strains, pink transformants, carrying a single integrated copy of the *URA3/SUP11* plasmid, are isolated (Hieter *et al.*, 1985), sporulated and dissected. White haploid transformants are isolated, grown non-selectively in liquid YPD medium, and plated on solid YPD medium to isolate single colonies. Red segregants that have lost the *SUP11* gene are then tested for simultaneous loss of *URA3* and either the *HIS3*, *LEU2* or *TRP1* marker.

VIII. VECTOR SYSTEMS IN OTHER YEASTS

Of all yeasts, *Sacch. cerevisiae* has received most attention genetically and biochemically. Consequently, molecular genetic techniques are most refined for this organism. However, many of the technologies developed in *Sacch. cerevisiae* have been applied to other yeasts. Transformation and vector/ host systems have been reported in *Schiz. pombe* (Beach and Nurse, 1981), lactose-fermenting *Kluyveromyces lactis* (Das and Hollenberg, 1982; Sreekrishna *et al.*, 1984), various dimorphic *Candida* species (for reviews, see Kurtz *et al.*, 1990, and accompanying chapters), the industrially important *Yarrowia lipolytica* (Davidow *et al.*, 1985; Wing and Ogrydziak, 1985) and *Schwanniomyces occidentalis* (Klein and Favreau, 1988), methylotrophic strains *Pichia pastoris* (Cregg *et al.*, 1985; Tschopp *et al.*, 1987) and *Hansenula polymorpha* (Gleeson *et al.*, 1986; Tikhomirova *et al.*, 1986; Roggenkamp *et al.*, 1986), and the filamentous *Trichosporon cutaneum* (Glumoff *et al.*, 1989). The following section provides a brief description of vector systems, which are available in *Schiz. pombe* and *K. lactis*.

A. Systems for *Schizosaccharomyces pombe*

High-frequency DNA transformation of *Schiz. pombe* was first reported with a shuttle vector (pDB248), consisting of pBR322, a region of the 2 μm plasmid encompassing *ori*, and the *LEU2* gene from *Sacch. cerevisiae*, which complements *leul.32* mutations in *Schiz. pombe* (Beach and Nurse, 1981). Since that report, a number of useful plasmid vectors have been developed, allowing many of the powerful molecular genetic techniques developed in *Sacch. cerevisiae* to be used in *Schiz. pombe* (for a detailed discussion, see Moreno *et al.*, 1991). A gene bank is available for *Schiz. pombe* in a pDB248 derivative (Beach *et al.*, 1982). The *LEU2* plasmid YEp13 also transforms *Schiz. pombe* at high frequencies, and both vectors exhibit intermediate mitotic stability and low meiotic stability (Heyer *et al.*, 1986). Both 2 μm *LEU2* plasmids exist in low copy number (around 10 copies in each cell), and are unstable due to asymmetrical plasmid segregation at division. The *URA3* gene in *Sacch. cerevisiae* also works as a selectable marker for *Schiz. pombe* (Losson and Lacroute, 1983). Both markers are useful in high copy-number replicating plasmids because their DNA sequences have diverged sufficiently from their counterparts in *Schiz. pombe* and, as a result, homologous recombination between the markers and their genomic homologues is low. However, this genetic divergence results in inefficient production of enzyme when the genes are present at low copy number. Consequently, YIp vectors in *Sacch. cerevisiae* containing the *URA3* or *LEU2* genes transform *Schiz. pombe* to high copy number but at very low frequency (Heyer *et al.*, 1986).

Since the first reports of DNA transformation in *Schiz. pombe* and the ability to clone its genes by standard complementation strategies, several improved vectors have been developed, expanding the molecular genetic utility of this organism. Several vectors containing the *LEU2* gene and 2 μm *ori* from *Sacch. cerevisiae* have been developed, and many provide positive selections for recombinant clones, as described earlier (see the description of pCS19 in Section IIB; Wright *et al.* 1986a). The positive-selection vectors contain the bacterial *tet* gene under control of the λ P_r promoter, as well as the λ cI gene into which DNA inserts are cloned. Each of these multicopy vectors can be converted to integrative plasmids by removal of the 2 μm sequences and can be used to isolate *ARS* elements from *Schiz. pombe* (Maundrell *et al.*, 1985).

Despite the many apparent similarities between vectors for *Schiz. pombe* and *Sacch. cerevisiae*, there are notable differences. Typically, transformants in *Schiz. pombe* can be divided into low- and high-transformation frequency groups. However, the lowest transformation frequency observed is approximately 100 transformants for each microgram of DNA. Moreover, unlike YIp vectors in *Sacch. cerevisiae*, many recombinant plasmids in *Schiz. pombe* produce a fraction of mitotically unstable transformants (Wright *et al.*, 1986a,b). Non-*ARS* plasmid transformation in *Schiz. pombe* can result from homologous and non-homologous integration, and can lead to mitotic instability of the vector. Homologous and stable mitotic plasmid

Table XI. Vectors for *Schizosaccharomyces pombe* and *Kluyveromyces lactis*

Yeast/vector	Features[a]	Remarks	Reference
Schizosaccharomyces pombe			
pDB248	2 μm-ori, S.c. *LEU2*	Multicopy plasmid	Beach and Nurse (1981)
pDB248X	2 μm-ori, S.c. *LEU2*	Multicopy plasmid derived from pD248′	Durkacz et al. (1985)
pDB262	2 μm-ori, S.c. *LEU2*, λ cl, P_r–Tet^r	Multicopy positive selection vector	Wright et al. (1986a)
pWH4/pWH5	2 μm-ori, S.c. *LEU2*, λ cl, P_r–Tet^r	Multicopy positive selection vectors	Wright et al. (1986a)
pMAK262	S.c. *LEU2*, λ cl, P_r–Tet^r	Positive selection vector, which can be used to isolate ARS elements or integrate sequences into genome	Wright et al. (1986a)
pCG3	S.p. *ura4*	A pBR322 derivative containing 6.7 kb *ura4* fragment	Grimm et al. (1988)
pCG5	S.p. *ura4-D18*	A pBR322 derivative containing *ura4-D18* deletion allele	Grimm et al. (1988)
pFL20	S.c. *URA3*, S.p. *ARS STB*	Stable multicopy ARS plasmid	Losson and Lacroute (1983)
pWH102	S.c. *LEU2*, S.p. *ARS STB*	Stable Multicopy ARS plasmid	Heyer et al. (1986)
pIRT1	S.c. *URA3*, S.p. *ARS*, polylinker	Multicopy ARS plasmid with polylinker	Booher and Beach (1987); Moreno et al. (1991)
pIRT2	S.c. *LEU2*, S.p. *ARS*, M13 *ORI*, polylinker	Multicopy ARS plasmid with polylinker	Booher and Beach (1986); Moreno et al. (1991)
pIRT2U	S.p. *ura4*$^+$, S.p. *ARS*, M13 *ORI*, polylinker	Multicopy ARS plasmid with polylinker	Carr et al. (1989); Moreno et al. (1991)
pSTA12	S.p. *sup3-5*, polylinker	*sup3-5*/pUC12 derivative, which facilitates integration	Moreno et al. (1991)
pREP1	S.c. *LEU2*, S.p. *ARS*, M13 *ORI*; S.p. *nmt1*pr-polylinker-*nmt1*tr	Multicopy pUC119 derived expression vector using the thiamine repressible *nmt1*$^+$ promoter from *Schiz. pombe*	Maundrell (1990)
pCHY21	S.c. *URA3*, S.p. *ARS STB*, S.p. *fbp* pr	Multicopy expression vector using the *fbp* promoter from *Schiz. pombe*	Hoffman and Winston (1989)
pMB332	S.c. *URA3*, S.p. *ARS STB*, S.p. *adh* pr-polylinker-*act* tr	Multicopy expression vector using the constitutive *adh* promoter from *Schiz. pombe*	Bröker and Bäuml (1989)
pMB340	S.c. *URA3*, S.p. *ARS STB*, S.p. *adh* pr-*act* tr	Multicopy expression vector using the constitutive *adh* promoter from *Schiz. pombe* and an ATG translation initiation codon followed by *Bam*HI and *Bgl*II cloning sites	Bröker and Bäuml (1989)
pART1	S.c. *LEU2*, S.p. *ARS* M13 *ORI*, S.p. *adh* pr-polylinker	Multicopy expression vector using the constitutive *adh* promoter from *Schiz. pombe*	McLeod et al. (1987)
pEVP11	2 μm-ori, S.c. *LEU2*, S.p. *adh* pr-polylinker	Multicopy expression vector using the constitutive *adh* promoter from *Schiz. pombe*	Russell and Nurse (1986)

Continued

Table XI. Continued

Yeast/vector	Features[a]	Remarks	Reference
pBKO2	S.p. 54/1 pr-polylinker-54/1 tr	Expression vector containing the constitutive 54/1 promoter from Schiz. pombe followed by a polylinker and the 3′ end of the 54/1 gene	Kudla et al. (1988)
pBKphl	S.p. 54/1 pr:ble^r	Derivative of pBKO2 containing a 54/1 pr:ble^r gene, which confers resistance to bleomycin and phleomycin	Kudla et al. (1988)
pura4-1.1/ABD	S.p. ura4$^+$, S.p. ARS, CaMVp-Km^r-CaMV tr	Multicopy plasmid containing CaMVp-Km^r gene, which confers G418 resistance	Gmünder and Kohli (1989)
pura4-1.1/HP	S.p. ura4$^+$, S.p. ARS, CaMVp-Km^r-CaMV tr	Multicopy vector containing CaMVp-Km^r gene, which confers G418 resistance	Gmünder and Kohli (1989)
pMB-1	S.c. ARS1, TRP1, URA3, CEN4, TEL-HIS3-TEL, S.p. ura4, S.p. ARS	Modified YAC4 vector, which can be propagated in Sacch. cerevisiae and has been used to isolate CENs from Schiz. pombe	Hahnenberger et al. (1989)
pSp(cen3)-10C	S.c. ARS1, TRP1, URA3, CEN4, TEL S.p. ura4, S.p. ARS, S.p. cen3	pMB-1 derivative containing cen3 for Schiz. pombe	Hahnenberger et al. (1989)
pSp(cen1)-7L	S.c. ARS1, TRP1, URA3, CEN4, TEL S.p. ura4, S.p. ARS, S.p. cen1	pMB-1 derivative containing cen1 for Schiz. pombe	Hahnenberger et al. (1989)
Kluyveromyces lactis			
pKR1	K.l. ARS, Km^r	Multicopy ARS vector containing the dominant Km^r marker conferring G418 resistance	Sreekrishna et al. (1984)
pKR1B-YC19	K.l. ARS, Km^r, S.c. ARS1, CEN4, TRP1, URA3	Multicopy ARS vector containing dominant Km^r marker conferring G418 resistance, and the TRP1 and URA3 markers from Sacch. cerevisiae	Sreekrishna et al. (1984)
E1	pKD1 ori and REP, S.c. URA3	Mitotically stable pKD1 derivative containing ori and REP functions and the URA3 marker from Sacch. cerevisiae. Plasmid should be propagated in K. lactis cir° strain	Chen et al. (1988)

Table XI. Continued

Yeast/vector	Features[a]	Remarks	Reference
KEp6	pKD1 ori, S.c. URA3	A pKD1 derivative containing ori functions and the URA3 marker from Sacch. cerevisiae. Plasmid should be propagated in K. lactis cir[+] strain	Chen et al. (1988)
pKRF4	K.l. ARS, S.c. URA3	A multicopy ARS vector	Fabiani et al. (1990)
pYGU1	pGKL1 ori, S.c. LEU2, URA3	A multicopy broad host-range ARS vector	Gunge (1988)

[a] Features include vector components, which allow for selection in an appropriate yeast host. All are shuttle vectors which can be propagated in E. coli. Abbreviations include: S.c., Sacch. cerevisiae; S.p., Schiz. pombe; K.l., K. lactis; pr, promoter; tr, terminator. Details about the use of each system can be found in the references.

integration occurs in *Schiz. pombe*, but many integrative plasmids are maintained unstably as chromosomal integrants or re-arranged plasmids. Many of the unstable properties of these transformants result from homologous and non-homologous integration of transforming DNA into, and excision from, the genome. These observations suggest that unstable integration and plasmid re-arrangement result from the continuous excision of integrated plasmid, leading in some cases to aberrant excision events and removal of extra genomic sequences (Wright *et al.*, 1986b).

Procedures for efficient vector integration, gene disruption and gene replacement have been successfully developed in *Schiz. pombe*, despite the problems associated with illegitimate recombination between chromosomes and cloning vectors in this yeast (Grimm and Kohli, 1988; Grimm *et al.*, 1988). Homology-dependent recombination between plasmids and chromosomes in *Schiz. pombe* occurs at an estimated 30–40-fold higher frequency than homology-independent integration. This has been exploited in a cloning system enabling integrative transformation, gene disruption and gene replacement (Grimm *et al.*, 1988). The system utilizes the *ura4*⁺ marker in *Schiz. pombe*, which, like its counterpart (*URA3*) in *Sacch. cerevisiae*, provides both positive and negative selection (Table XI). The integrating plasmid, pCG5, can be used to integrate stably sequences at the *ura4* locus and contains chromosomal DNA sequences from *Schiz. pombe* flanking the *ura4* gene (see *ura4-D18* allele; Table XI). After insertion of foreign DNA between the flanking sequences of pCG5, the plasmid can be targeted to the *ura4*⁺ locus of a wild-type strain by selecting for 5-fluoro-orotic acid resistance. One-step gene disruption techniques pioneered in *Sacch. cerevisiae* (Rothstein, 1983) are possible in *Schiz. pombe* using these alleles. To construct a disruption, the *ura4*⁺ gene is inserted within the locus to be disrupted, and linear DNA is transformed into recipient *ura4*⁻ haploid or diploid hosts (Russell and Nurse, 1986; Grimm *et al.*, 1988). The *ura4-D18* system can also be used to introduce *in vitro* generated alleles of a gene back into the genome by targeting transforming DNA to a *ura4* disruption/deletion allele of the gene under study, and selecting for 5-fluoro-orotic acid-resistant transformants (Grimm *et al.*, 1988). The *ura4-D18* allele can also be used to construct *ura4* auxotrophic strains. Carr *et al.* (1989) have developed the *ade6-704/sup3-5* genetic system to facilitate targeted plasmid integration/*in vitro* excision methods and the cloning of mutant alleles from the *Schiz. pombe* genome. The system also enables the identification of transformant colonies in which a single copy of a *sup3-5*-bearing plasmid has been stably integrated into the genome of *Schiz. pombe ade6-704* cells.

Autonomously replicating plasmids are also available in *Schiz. pombe*. Autonomously replicating sequence elements with properties similar to their counterparts in *Sacch. cerevisiae* have been isolated from this organism (Sakaguchi and Yamamoto, 1982; Maundrell *et al.*, 1985; Heyer *et al.* 1986; Wright *et al.*, 1986b). Vectors containing these elements transform *Schiz. pombe* at high frequency and are mitotically unstable. One problem encountered when isolating *ARS* elements in this organism results from the mitotic instability of many plasmids with selectable markers. Vectors for *ARS* elements transform at high frequencies after long periods of incubation, and often exist as extrachromosomal re-arranged plasmids. Maundrell *et al.* (1988) argued that some of these vectors have acquired genomic *ARS* sequences from their hosts. A DNA fragment (*stb*) from *Schiz. pombe*, which improves the mitotic and meiotic stability of the *Schiz. pombe ars* vector pFL20 from this yeast, has been described (Losson and Lacroute, 1983; Heyer *et al.*, 1986). In conjunction with the *ars* element of pFL20, *stb* increases the frequency of symmetrical plasmid segregation through mitosis, and improves plasmid meiotic transmission. However, it does not behave like a centromere. Centromeres from *Schiz. pombe* confer stable mitotic and meiotic plasmid segregation, and have been isolated from a modified YAC vector containing an *ars* and the *ura4* gene from this yeast (Hahnenberger *et al.*, 1989). These mitotically stable artificial chromosomes can be recovered from *Schiz. pombe* in linear and circular states. They also segregate properly at meiosis. Centromeric (*CEN*) sequences isolated from *Sacch. cerevisiae* or *Schiz. pombe* are species specific and do not function in the heterologous host.

Vectors in *Schiz. pombe*, which allow for efficient constitutive or regulated expression of homologous or heterologous genes, have also been developed (Table XI). Expression from vectors containing the promoter of the glucose repressible *fkb* gene from this yeast can be regulated by glucose repression over a range greater than 100-fold by varying the carbon source in the growth medium (Hoffman and Winston, 1989). The one drawback of this system is that derepression also occurs when glucose-grown cells enter the stationary phase of growth. High-level constitutive expression vectors containing the alcohol dehydrogenase (*adh*) and 54/1 promoters in *Schiz. pombe* are available (Russell and Nurse, 1986; Kudla *et al.*, 1988; Bröker and Bäuml, 1989). The 54/1 promoter was isolated from a bank of DNA sequences fused to a promoter-less *lacZ* gene. It expresses β-galactosidase as approximately 5% of total cell protein (Kudla *et al.*, 1988). Transcription signals from SV40 have also been used for expression of heterologous sequences in *Schiz. pombe* (Lee and Nurse, 1987).

Plasmids containing dominant selectable markers have been constructed by fusing drug-resistance genes to efficient promoters of *Schiz. pombe* (Table XI). The *adh*::Km^r fusion genes, which confer a dominant G418 resistance in *Schiz. pombe*, have been developed (Gmünder and Kohli,

1989). These authors have also developed Km^r fusion genes with the 35S and 19S promoters from cauliflower mosaic virus (CaMV). In the Km^r fusion systems evaluated, the CaMV promoters are as efficient as the *adh* promoter in multicopy and integrative vectors. Vectors containing the *ble* gene from *E. coli* fused to the 54/1 promoter confer dominant resistance to bleomycin and phleomycin (Kudla *et al.*, 1988).

B. Systems in *Kluyveromyces lactis*

Transformation of *K. lactis* with DNA was first reported with vectors containing the *TRP1* gene from *Sacch. cerevisiae* and the Tn903 dominant selectable marker Km^r (Das and Hollenberg, 1982; Sreekrishna *et al.*, 1984; Table XI). High frequency YEp and YRp transformation vectors from *Sacch. cerevisiae* fail to transform *K. lactis*. However, *ARS* elements from *K. lactis* have been isolated functionally in *Sacch. cerevisiae* and *K. lactis*, and used to create high-frequency transforming plasmids (Das and Hollenberg, 1982; Sreekrishna *et al.*, 1984; Fabiani *et al.*, 1990). Integrative transformation with vectors lacking *ARS* sequences is also possible and integration can be targeted by cutting vector DNA in regions homologous to the genome.

The circular plasmid pKD1 in *K. drosophilarum* has been developed into a vector-transformation system for *K. lactis*. The principle of the system is similar to the 2 μm system in *Sacch. cerevisiae* (Chen *et al.*, 1988; Table XI), and the structural and functional organization of pKD1 is also similar to the 2 μm plasmid family. Two types of pKD1-derived vectors have been constructed. Type I vectors contain the entire pKD1 plasmid, transform host strains at high frequency, and are generally very stable in *K. lactis* hosts lacking endogenous pKD1. Native pKD1 lowers the transformation frequency and stability of Type I vectors, presumably by incompatibility resulting from replication competition. Type II vectors contain the pKD1 *ori* and, like their 2 μm counterparts, require resident pKD1 plasmids for stable replication. The selective markers in this system include the *URA3* and *TRP1* genes from *Sacch. cerevisiae*, and the Km^r gene. Chen *et al.* (1988) have also developed Km^r and *lacZ* gene fusion/reporter systems, as well as an expression vector utilizing the *PHO5* promoter from *Sacch. cerevisiae*. The *ADH1* and *GAL* promoters from *Sacch. cerevisiae* are also expressed in *K. lactis*, but not as strongly as in *Sacch. cerevisiae* (Chen *et al.*, 1988). The linear DNA plasmids, pGKL1 and pGKL2, from *K. lactis* have also been developed in circular and linear vector systems (for a review, see Gunge, 1988). The circular pGKL1-based vector pYGU1 exhibits a

Table XII. Shuttle vectors with lacZα-polylinker complementation systems

Polylinker cloning region[a]	Plasmid type	Yeast markers	Plasmid name	Other features[b]	Reference
pUC9	YIp	LEU2	pEMBLYi27	f1 ori	Baldari and Cesareni (1985)
	YIp	URA3	pEMBLYi22	f1 ori	Baldari and Cesareni (1985)
	YEp	URA3	pEMBLYe23	2 μm ori, f1 ori	Baldari and Cesareni (1985)
	YEp	LEU2	pEMBLYe30	2 μm ori, f1 ori	Baldari and Cesareni (1985)
	YRp	TRP1	pEMBLYr25	ARS1, f1 ori	Baldari and Cesareni (1985)
pUC18	YIp	LEU2	YIp351	—	Hill et al. (1986)
	YIp	URA3	YIp352	—	Hill et al. (1986)
	YEp	LEU2	YEp351	2 μm ori	Hill et al. (1986)
	YEp	LEU2	YEp-DE	2 μm ori	Khan et al. (1987)
	YEp	URA3	YEp352	2 μm ori	Hill et al. (1986)
	YCp	HIS3	pUN90	ARS1, CEN4	Elledge and Davis (1988)
	YCp	LEU2	pUN100	ARS1, CEN4	Elledge and Davis (1988)
	YCp	TRP1	pUN0, pUN1 and pUN10	ARS1, CEN4	Elledge and Davis (1988)
	YCp	TRP1	pUN30	ARS1, CEN4, M13 (+) ori	Elledge and Davis (1988)
	YCp	TRP1	pUN40	ARS1, CEN4, M13 (−) ori	Elledge and Davis (1988)
	YCp	URA3	pUN50	ARS1, CEN4	Elledge and Davis (1988)
	YCp	URA3	pUN70	ARS1, CEN4, M13 (+) ori	Elledge and Davis (1988)
	YCp	TRP1, SUP11	pUN80	ARS1, CEN4, M13 (−) ori	Elledge and Davis (1988)
	YCp	SUP11, URA3	pUN20	ARS1, CEN4	Elledge and Davis (1988)
			pUN60	ARS1, CEN4	Elledge and Davis (1988)
pUC19	YIp	LEU2	YIplac128		Gietz and Sugino (1988)
	YIp	TRP1	YIplac204		Gietz and Sugino (1988)
	YIp	URA3	YIplac211		Gietz and Sugino (1988)
	YIp	URA3	pEMBLYi32	f1 ori	Cesareni (1988)
	YEp	LEU2	YEplac181	2 μm ori	Gietz and Sugino (1988)
	YEp	TRP1	YEplac112	2 μm ori	Gietz and Sugino (1988)
	YEp	URA3	YEplac195	2 μm ori	Gietz and Sugino (1988)
	YCp	LEU2	YCplac111	ARS1, CEN4	Gietz and Sugino (1988)

Continued

Table XII. Continued

Polylinker cloning region[a]	Plasmid type	Yeast markers	Plasmid name	Other features[b]	Reference
	YCp	TRP1	YCplac22	ARS1, CEN4	Gietz and Sugino (1988)
	YCp	URA3	YCplac33	ARS1, CEN4	Gietz and Sugino (1988)
KS	YIp	HIS3	pRS303	f1 (+) ori, T3p and T7p	Sikorski and Hieter (1989)
	YIp	TRP1	pRS304	f1 (+) ori, T3p and T7p	Sikorski and Hieter (1989)
	YIp	LEU2	pRS305	f1 (+) ori, T3p and T7p	Sikorski and Hieter (1989)
	YIp	URA3	pRS306	f1 (+) ori, T3p and T7p	Sikorski and Hieter (1989)
	YCp	TRP1	pUN15	ARS1, CEN4, T7p and T3p	Elledge and Davis (1988)
	YCp	TRP1	pUN35	ARS1, CEN4, M13 (+) ori, T7p and T3p	Elledge and Davis (1988)
	YCp	TRP1	pUN45	ARS1, CEN4, M13 (-) ori, T7p and T3p	Elledge and Davis (1988)
	YCp	URA3	pUN55	ARS1, CEN4, T7p and T3p	Elledge and Davis (1988)
	YCp	URA3	pUN75	ARS1, CEN4, M13 (+) ori, T7p and T3p	Elledge and Davis (1988)
	YCp	URA3	pUN85	ARS1, CEN4, M13 (-) ori, T7p and T3p	Elledge and Davis (1988)
	YCp	HIS3	pUN95	ARS1, CEN4, T7p and T3p	Elledge and Davis (1988)
	YCp	LEU2	pUN105	ARS1, CEN4, T7p and T3p	Elledge and Davis (1988)
	YCp	TRP1, SUP11	pUN25	ARS1, CEN4, T7p and T3p	Elledge and Davis (1988)
	YCp	SUP11, URA3	pUN65	ARS1, CEN4, T3p and T7p	Elledge and Davis (1988)
	YCp	HIS3	pRS313	ARSH4, CEN6, f1 (+) ori, T3p and T7p	Sikorski and Hieter (1989)
	YCp	URA3	pRS316	ARSH4, CEN6, f1 (+) ori, T3p and T7p	Sikorski and Hieter (1989)
	YCp	LEU2	pRS315	ARSH4, CEN6, f1 (+) ori, T3p and T7p	Sikorski and Hieter (1989)
	YCp	TRP1	pRS314	ARSH4, CEN6, f1 (+) ori, T3p and T7p	Sikorski and Hieter (1989)

Table XII. Continued

Polylinker cloning region[a]	Plasmid type	Yeast markers	Plasmid name	Other features[b]	Reference
SK	YEp	*HIS3*	pRS423	2 μm ori, f1 (+) ori, T3p and T7p	Christianson *et al.* (1992)
	YEp	*LEU2*	pRS425	2 μm ori, f1 (+) ori, T3p and T7p	Christianson *et al.* (1992)
	YEp	*TRP1*	pRS424	2 μm ori, f1 (+) ori, T3p and T7p	Christianson *et al.* (1992)
	YEp	*URA3*	pRS426	2 μm ori, f1 (+) ori, T3p and T7p	Christianson *et al.* (1992)

[a]Polylinkers contain the following restriction-enzyme sites: pUC9, *HindIII, PstI, HincII/SalI/AccI, BamHI, SmaI/XmaI, EcoRI*; pUC18, *EcoRI, SacI, KpnI, SmaI/XmaI, BamHI, XbaI, HincII/SalI/AccI, PstI, SphI, HindIII*; pUC19 contains the same polylinker in opposite orientation; KS, *KpnI, ApaI/DraII, XhoI, SalI/HincII/AccI, ClaI, HindIII, EcoRV, EcoRI, PstI, SmaI, BamHI, SpeI, XbaI, NotI, EagI, BstXI/SacII, SacI* (Stratagene Cloning System, La Jolla, CA); SK contains the same polylinker in opposite orientation and the polylinker also has flanking *BssHII* sites. Plasmids in the pRS400 and pRS410 series are also available with SK+ polylinkers. The 400 series are YIp vectors and the 410 series are YCp vectors (Christianson *et al.*, 1992).
[b]Other features relevant to the plasmid are explained in the text. T3p and T7p are the bacteriophage T3 and T7 promoters, respectively, which flank the KS and SK polylinkers.

broad host range, and replicates in *K. lactis, Sacch. cerevisiae* and *Schiz. pombe.* Development of other native yeast plasmids into vector systems has also been successful. Several naturally occurring plasmids belonging to the 2 μm family have been isolated from *Zygosaccharomyces* species (Araki *et al.*, 1985; Toh-e and Utatsu, 1985; Utatsu *et al.*, 1986, 1987; Murray *et al.*, 1988; Araki and Oshima, 1989), and developed into vector systems for these yeasts.

IX. ACKNOWLEDGEMENTS

We are grateful to all of our colleagues who provided published and unpublished information on yeast vector systems. We also thank Ebony Daniels for assistance in preparing the manuscript and plasmid databases, members of our department for proofreading and commenting on the manuscript, and the staff of the Merck Information Services and Literature Resource Center for their support in providing published information.

References

Adams, S.E., Dawson, K.M., Gull, K., Kingsman, S.M. and Kingsman, A.J. (1987). *Nature* **329**, 68.
Alani, E. and Kleckner, N. (1987). *Genetics* **117**, 5.
Alani, E., Cao, L. and Kleckner, N. (1987). *Genetics* **116**, 541.
Altherr, M.R. and Rodriguez, R.L. (1988) *In* "Vectors: A Survey of Molecular Cloning Vectors and Their Uses" (R.L. Rodriguez and D.T. Denhardt, eds), pp. 405–417. Butterworths, Boston.
Ammerer, G. (1983). *Methods in Enzymology* **101**, 192.
Andersen, N., Thill, G.P. and Kramer, R.A. (1983). *Molecular and Cellular Biology* **3**, 562.
Andreadis, A., Hsu, Y.-P., Hermodson, M., Kohlaw, G. and Schimmel, P. (1984). *Journal of Biological Chemistry* **259**, 8059.
Araki, H. and Oshima, Y. (1989). *Journal of Molecular Biology* **207**, 757.
Araki, H., Jearnpipatkul, A., Tatsumi, H., Sakurai, T., Ushio, K., Muta, T. and Oshima, Y. (1985). *Journal of Molecular Biology* **182**, 191.
Armstrong, K.A., Acosta, R., Ledner, E., Machida, Y., Pancotto, M., McCormick, M., Ohtsubo, H. and Ohtsubo, E. (1984). *Journal of Molecular Biology* **175**, 331.
Armstrong, K.A., Som, T., Volkert, F.C. and Broach, J.R. (1988). *Cancer Cells/Eukaryotic DNA Replication* **6**, 213.
Baker, S.M., Johnston, S.A., Hopper, J.E. and Jaehning, J.A. (1987). *Molecular and General Genetics* **208**, 127.
Baldari, C. and Cesareni, G. (1985). *Gene* **35**, 27.
Barnes, D.A. and Thorner, J. (1986). *Molecular and Cellular Biology* **6**, 2828.
Beach, D. and Nurse, P. (1981). *Nature* **290**, 140.
Beach, D., Piper, M. and Nurse, P. (1982). *Molecular and General Genetics* **187**, 326.
Becker, D.M. and Guarente, L. (1991). *Methods in Enzymology* **194**, 182.
Beggs, J.D. (1978). *Nature* **275**, 102.

Bitter, G.A. and Egan, K.M. (1984). *Gene* **32**, 263.

Bitter, G.A., Egan, K.M., Koski, R.A., Jones, M.O., Elliott, S.G. and Giffin, J.C. (1987). *Methods in Enzymology* **153**, 516.

Boeke, J.D., LaCroute, F. and Fink, G.R. (1984). *Molecular and General Genetics* **197**, 345.

Boeke, J.D., Garfinkel, D.J., Styles, C.A. and Fink, G.R. (1985). *Cell* **40**, 491.

Boeke, J.D., Styles, C.A. and Fink, G.R. (1986). *Molecular and Cellular Biology* **6**, 3575.

Boeke, J.D., Trueheart, J., Natsoulis, G. and Fink, G.R. (1987). *Methods in Enzymology* **154**, 164.

Bolivar, F., Rodriguez, R.L., Greene, P.J., Betlach, M.C., Heyneker, H.L., Boyer, H.W., Crosa, J.H. and Falkow, S. (1977). *Gene* **2**, 95.

Booher, R. and Beach, D. (1986). *Molecular and Cellular Biology* **6**, 3523.

Booher, R. and Beach, D. (1987). *EMBO Journal* **6**, 3441.

Bostian, K.A., Lemire, J.M., Cannon, L.E. and Halvorson, H.O. (1980). *Proceedings of the National Academy of Sciences, USA* **77**, 4504.

Bostian, K.A., Lemire, J.M. and Halvorson, H.O. (1983). *Molecular and Cellular Biology* **3**, 839.

Botstein, D., Falco, S.C., Stewart, S.E., Brennan, M., Scherer, S., Stinchcomb, D.T., Struhl, K. and Davis, R.W. (1979). *Gene* **8**, 17.

Bouton, A.H. and Smith, M.M. (1986). *Molecular and Cellular Biology* **6**, 2354.

Brake, A.J. (1990). *Methods in Enzymology* **185**, 408.

Brent, R. and Ptashne, M. (1985). *Cell* **43**, 729.

Broach, J.R. (1983). *Methods in Enzymology* **101**, 307.

Broach, J.R., Strathern, J.N. and Hicks, J.B. (1979). *Gene* **8**, 121.

Broach, J.R., Li, Y.-Y., Wu, L.-C.C. and Jayaram, M. (1983). *In* "Experimental Manipulation of Gene Expression" (M. Inouye, ed.), pp. 83–117. Academic Press, New York.

Bröker, M. and Bäuml, O. (1989). *FEBS Letters* **248**, 105.

Burke, D.T., Carle, G.F. and Olson, M.V. (1987). *Science* **236**, 806.

Butt, T.R., Khan, M.I., Marsh, J., Ecker, D.J. and Crooke, S.T. (1988). *Journal of Biological Chemistry* **263**, 16364.

Carbon, J. (1984). *Cell* **37**, 351.

Carlson, M. and Botstein, D. (1982). *Cell* **28**, 145.

Carr, A.M., MacNeill, S.A., Hayles, J. and Nurse, P. (1989). *Molecular and General Genetics* **218**, 41.

Casadaban, M.J., Martinez-Arias, A., Shapira, S.K. and Chou, J. (1983). *Methods in Enzymology* **100**, 293.

Cashmore, A.M., Albury, M.S., Hadfield, C. and Meacock, P.A. (1986). *Molecular and General Genetics* **203**, 154.

Cesareni, G. (1988). *In* "Vectors: A Survey of Molecular Cloning Vectors and Their Uses" (R.L. Rodriguez and D.T. Denhardt, eds), pp. 103–111. Butterworths, Boston.

Chattoo, B.B., Sherman, F., Azubalis, D.A., Fjellstedt, T.A., Mehnert, D. and Ogur, M. (1979). *Genetics* **93**, 51.

Chen, X.J., Wésolowski-Louvel, M., Tanguy-Rougeau, C., Bianchi, M.M., Fabiani, L., Saliola, M., Falcone, C., Frontali, L. and Fukuhara, H. (1988). *Journal of Basic Microbiology* **28**, 211.

Chien, C.-T., Bartel, P.L., Sternglanz, R. and Fields, S. (1991). *Proceedings of the National Academy of Sciences, USA* **88**, 9578.

Christianson, T.W., Sikorski, R.S., Dante, M., Shero, J.H. and Hieter, P. (1992). *Gene* **110**, 119.

Clarke, L. and Carbon, J. (1980). *Nature* **287**, 504.

Cohen, J.D., Eccleshall, T.R., Needleman, R.B., Federoff, H., Buchferer, B.A. and Marmur, J. (1980). *Proceedings of the National Academy of Sciences, USA* **77**, 1078.

Cooke, H. and Cross, S. (1988). *Nucleic Acids Research* **16**, 11817.

Cregg, J.M., Barringer, K.J., Hessler, A.Y. and Madden, K.R. (1985). *Molecular and Cellular Biology* **5**, 3376.

Dani, G.M. and Zakian, V.A. (1983). *Proceedings of the National Academy of Sciences, USA* **80**, 3406.

Das, S. and Hollenberg, C.P. (1982). *Current Genetics* **6**, 123.

Davidow, L.S., Apostolakos, D., O'Donnell, M.M., Proctor, A.R., Ogrydziak, D.M., Wing, R.A., Stasko, I. and DeZeeuw, J.R. (1985). *Current Genetics* **10**, 39.

Davison, J., Heusterspreute, M., Merchez, M. and Brunel, F. (1984). *Gene* **28**, 311.

Davison, J., Heusterspreute, M. and Brunel, F. (1987). *Methods in Enzymology* **153**, 34.

Devereux, J. (1989). "The GCG Sequence Analysis Software Package, Version 6.0." Genetics Computer Group, University of Wisconsin Biotechnology Centre, 1710 University Ave., Madison, WI 53705, USA.

Durkacz, B., Beach, D., Hayles, J. and Nurse, P. (1985). *Molecular and General Genetics* **201**, 543.

Elledge, S.J. and Davis, R.W. (1988). *Gene* **70**, 303.

Elledge, S.J., Mulligan, J.T., Ramer, S.W., Spottswood, M. and Davis, R.W. (1991). *Proceedings of the National Academy of Sciences, USA* **88**, 1731.

Emr, S.D. (1990). *Methods in Enzymology* **185**, 231.

Emr, S.D., Schekman, R., Flessel, M.C. and Thorner, J. (1983). *Proceedings of the National Academy of Sciences, USA* **80**, 7080.

Emr, S.D., Vassarotti, A., Garrett, J., Geller, B.L., Takeda, M. and Douglas, M.G. (1986). *Journal of Cell Biology* **102**, 523.

Erhart, E. and Hollenberg, C.P. (1983). *Journal of Bacteriology* **156**, 625.

Etcheverry, T. (1990). *Methods in Enzymology* **185**, 319.

Fabiani, L., Aragona, M. and Frontali, L. (1990). *Yeast* **6**, 69.

Fagan, M.C. and Scott, J.F. (1985). *Gene* **40**, 217.

Fasullo, M.T. and Davis, R.W. (1988). *Molecular and Cellular Biology* **8**, 4370.

Fields, S. and Song, O.-K. (1989). *Nature* **340**, 245.

Fitzgerald-Hayes, M., Buhler, J.-M., Cooper, T.G. and Carbon, J. (1982a). *Molecular and Cellular Biology* **2**, 82.

Fitzgerald-Hayes, M., Clarke, L., and Carbon, J. (1982b). *Cell* **29**, 235.

Fleig, U.N., Pridmore, R.D., and Philippsen, P. (1986). *Gene* **46**, 237.

Futcher, A. B. (1986). *Journal of Theoretical Biology* **119**, 197.

Garfinkel, D.J. and Strathern, J.N. (1991). *Methods in Enzymology* **194**, 342.

Garfinkel, D.J., Mastrangelo, M.F., Sanders, N.J., Shafer, B.K. and Strathern, J.N. (1988). *Genetics* **120**, 95.

Gatignol, A., Baron, M. and Tiraby, G. (1987). *Molecular and General Genetics* **207**, 342.

Gent, M.E., Crowley, P., Ludwig, J.R., Anwar, R., Sugden, D.A., Sims, P.F.G. and Oliver, S.G. (1985). *Current Genetics* **10**, 29.

Gerbaud, C., Fournier, P., Blanc, H. Aigle, M., Heslot, H. and Guerineau, M. (1979). *Gene* **5**, 233.

Gietz, R.D. and Sugino, A. (1988). *Gene* **74**, 527.

Gleeson, M.A., Ortori, G.S. and Sudbery, P.E. (1986). *Journal of General Microbiology* **132**, 3459.

Glumoff, V., Käppeli, O., Fiechter, A. and Reiser, J. (1989). *Gene* **84**, 311.

Gmünder, H. and Kohli, J. (1989). *Molecular and General Genetics* **220**, 95.

Goldstein, A. and Lampen, J.O. (1975). *Methods in Enzymology* **42**, 504.

Goursot, R., Goze, A., Niaudet, B. and Ehrlich, S.D. (1982). *Nature* **298**, 488.

Green, R., Schaber, M.D., Shields, D. and Kramer, R. (1986). *Journal of Biological Chemistry* **261**, 7558.

Grimm, C. and Kohli, J. (1988). *Molecular and General Genetics* **215**, 87.

Grimm, C., Kohli, J., Murray, J. and Maundrell, K. (1988). *Molecular and General Genetics* **215**, 81.

Gritz, L. and Davies, J. (1983). *Gene* **25**, 179.

Guarente, L. and Ptashne M. (1981). *Proceedings of the National Academy of Sciences, USA* **78**, 2199.

Guarente, L., Yocum, R.R. and Gifford, P. (1982). *Proceedings of the National Academy of Sciences, USA* **79**, 7410.

Gunge, N. (1988). *In* "Viruses of Fungi and Simple Eukaryotes" (Y. Koltin and M.J. Leibowitz, eds), pp. 265–282. Marcel Dekker, Inc., New York.

Hadfield, C., Cashmore, A.M. and Meacock, P.A. (1986) *Gene* **45**, 149.

Hahnenberger, K.M., Baum, M.P., Polizzi, C.M., Carbon, J. and Clarke, L. (1989). *Proceedings of the National Academy of Sciences, USA* **86**, 577.
Hanes, S.D. and Brent, R. (1989). *Cell* **57**, 1275.
Hartley, J.L. and Donelson, J.E. (1980). *Nature* **286**, 860.
Hartwell, L.H., Dutcher, S.K., Wood, J.S. and Garvik, B. (1982). *Recent Advances Yeast Molecular Biology* **1**, 28.
Hashimoto, H., Morikawa, H., Yamada, Y. and Kimura, A. (1985). *Applied Microbiology and Biotechnology* **21**, 336.
Heinemann, J.A. and Sprague, G.F., Jr (1989). *Nature* **340**, 205.
Henikoff, S. (1984). *Gene* **28**, 351.
Herskowitz, I., and Jensen, R.E. (1991). *Methods in Enzymology* **194**, 132.
Heusterspreute, M., Oberto, J., Ha-Thi, V. and Davison, J. (1985). *Gene* **34** 363.
Heyer, W.-D., Sipiczki, M. and Kohli, J. (1986). *Molecular and Cellular Biology* **6**, 80.
Hicks, J.B., Hinnen, A. and Fink, G.R. (1978). *Cold Spring Harbor Symposium on Quantitative Biology* **43**, 1305.
Hieter, P., Pridmore, D., Hegemann, J.H., Thomas, M., Davis, R.W. and Philippsen, P. (1985). *Cell* **42**, 913.
Hill, J.E., Myers, A.M., Koerner, T.J. and Tzagoloff, A. (1986). *Yeast* **2**, 163.
Hinnen, A, Hicks, J.B. and Fink, G.R. (1978). *Proceedings of the National Academy of Sciences, USA* **75**, 1929.
Hitzeman, R.A., Hagie, F.E., Levine, H.L., Goeddel, D.V., Ammerer, G. and Hall, B.D. (1981). *Nature* **293**, 717.
Hitzeman, R.A., Chang, C.N., Matteucci, M., Perry, L.J., Kohr, W.J.,Wulf, J.J., Swartz, J.R., Chen, C.Y. and Singh, A. (1986). *Methods in Enzymology* **119**, 424.
Hoekstra, M.F., Seifert, H.S., Nickoloff, J. and Heffron, F. (1991). *Methods in Enzymology* **194**, 329.
Hoffman, C.S. and Winston, F. (1989). *Gene* **84**, 473.
Hohn, B. and Hinnen, A. (1982). In "Genetic Engineering: Principles and Methods", Vol. 2 (J.K. Setlow and A. Hollaender, eds), pp. 169–183. Plenum Press, New York.
Horinouchi, S. and Weisblum, B. (1982a). *Journal of Bacteriology* **150**, 804.
Horinouchi, S. and Weisblum, B. (1982b). *Journal of Bacteriology* **150**, 815.
Huisman, O., Raymond, W., Froehlich, K.-U., Errada, P., Kleckner, N., Botstein, D. and Hoyt, M.A. (1987). *Genetics* **116**, 191.
Hwang, Y.-I., Harashima, S. and Oshima, Y. (1988). *Applied Microbiology and Biotechnology* **28**, 155.
Innis, M.A., Holland, M.J., McCabe, P.C., Cole, G.E., Wittman, V.P., Tal, R., Watt, K.W.K., Gelfand, D.H., Holland, J.P. and Meade, J.H. (1985). *Science* **228**, 21.
Ito, H., Fukuda, Y., Murata, K. and Kimura, A. (1983). *Journal of Bacteriology* **153**, 163.
Jacobs, E., Dewerchin, M. and Boeke, J.D. (1988). *Gene* **67**, 259.
Jacobson, M.A., Forma, F.M., Buenaga, R.F., Hofmann, K.J., Schultz, L.D., Gould, R.J. and Friedman, P.A. (1989). *Gene* **85**, 511.
Jagadeeswaran, P. and Haas, P. (1990). *Gene* **86**, 279.
Jayaram, M., Li, Y.-Y. and Broach, J.R. (1983). *Cell* **34**, 95.
Jensen, R., Sprague, G.F., Jr and Herskowitz, I. (1983). *Proceedings of the National Academy of Sciences, USA* **80**, 3035.
Johnston, M. and Davis, R.W. (1984). *Molecular and Cellular Biology* **4**, 1440.
Johnston, S.A., Anziano, P.Q., Shark, K., Sanford, J.C. and Butow, R.A. (1988). *Science* **240**, 1538.
Jones, J.S. and Prakash, L. (1990). *Yeast* **6**, 363.
Kaiser, C.A. and Botstein, D. (1990). *Molecular and Cellular Biology* **10**, 3163.
Kaiser, C.A., Preuss, D., Grisafi, P. and Botstein, D. (1987). *Science* **235**, 312.
Kang, Y.-S., Kane, J., Kurjan, J., Stadel, J.M. and Tipper, D.J. (1990). *Molecular and Cellular Biology* **10**, 2582.
Karube, I., Tamiya, E. and Matsuoka, H. (1985). *FEBS Letters* **182**, 90.
Khan, M.I., Ecker, D.J., Butt, T., Gorman, J.A. and Crooke, S.T. (1987). *Plasmid* **17**, 171.
Kikuchi, Y. (1983). *Cell* **35**, 487.

Kingsman, A.J., Clarke, L., Mortimer, R.K. and Carbon, J. (1979). *Gene* **7**, 141.

Kingsman, S.M., Cousens, D., Stanway, C.A., Chambers, A., Wilson, M. and Kingsman, A.J. (1990). *Methods in Enzymology* **185**, 329.

Klein, R.D. and Favreau, M.A. (1988). *Journal of Bacteriology* **170**, 5572.

Kolodziej, P.A. and Young, R.A. (1991). *Methods in Enzymology* **194**, 508.

Koshland, D. and Hieter, P. (1987). *Methods in Enzymology* **155**, 351.

Kramer, R.A. and Andersen, N. (1980). *Proceedings of the National Academy of Sciences, USA* **77**, 6541.

Kramer, R.A., DeChiara, T.M., Schaber, M.D. and Hilliker, S. (1984). *Proceedings of the National Academy of Sciences, USA* **81**, 367.

Kranz, J.E. and Holm, C. (1990). *Proceedings of the National Academy of Sciences, USA* **87**, 6629.

Kudla, B., Persuy, M.-A., Gaillardin, C. and Heslot, H. (1988). *Nucleic Acids Research* **16**, 8603.

Kurtz, M.B., Kelly, R. and Kirsch, D.R. (1990). *In* "The Genetics of Candida" (D.R. Kirsch, R. Kelly and M.B. Kurtz, eds), pp. 21–73. CRC Press, Boston.

Kwak, J.W., Kim, J., Yoo, O.J. and Han, M.H. (1987). *Biochemical and Biophysical Research Communications* **149**, 846.

Lagosky, P.A., Taylor, G.R. and Haynes, R.H. (1987). *Nucleic Acids Research* **15**, 10355.

Lee, M.G. and Nurse, P. (1987). *Nature* **327**, 31.

Lemontt, J.F., Wei, C.-M., and Dackowski, W.R. (1985). *DNA* **4**, 419.

Liu, H., Krizek, J. and Bretscher, A. (1992). *Genetics* **132**, 665.

Losson, R. and Lacroute, F. (1983). *Cell* **32**, 371.

Ma, H., Kunes, S., Schatz, P.J. and Botstein, D. (1987). *Gene* **58**, 201.

Ma, J. and Ptashne, M. (1987). *Cell* **51**, 113.

Macreadie, I.G. (1990). *Nucleic Acids Research* **18**, 1078.

Maine, G.T., Surosky, R.T. and Tye, B.-K. (1984). *Molecular and Cellular Biology* **4**, 86.

Mann, C. and Davis, R.W. (1986). *Molecular and Cellular Biology* **6**, 241.

Marchuk, D. and Collins, F.S. (1988). *Nucleic Acids Research* **16**, 7743.

Maundrell, K. (1990). *Journal of Biological Chemistry* **265**, 10857.

Maundrell, K., Hutchison, A. and Shall, S. (1988). *EMBO Journal* **7**, 2203.

Maundrell, K., Wright, A.P.H., Piper, M. and Shall, S. (1985). *Nucleic Acids Research* **13**, 3711.

McDonnell, D.P., Pike, J.W., Drutz, D.J., Butt, T.R. and O'Malley, B.W. (1989). *Molecular and Cellular Biology* **9**, 3517.

McLeod, M., Stein, M. and Beach, D. (1987). *EMBO Journal* **6**, 729.

McNeil, J.B. and Friesen, J.D. (1981). *Molecular and General Genetics* **184**, 386.

Miyajima, A., Miyajima, I., Arai, K.-I. and Arai, N. (1984) *Molecular and Cellular Biology* **4**, 407.

Miyajima, A., Bond, M.W., Otsu, K., Arai, K.-I. and Arai, N. (1985). *Gene* **37**, 155.

Moir, D.T. and Dumais, D.R. (1987). *Gene* **56**, 209.

Moreno, S., Klar, A. and Nurse, P. (1991). *Methods in Enzymology* **194**, 795.

Murray, A.W. and Szostak, J.W. (1983a). *Cell* **34**, 961.

Murray, A.W. and Szostak, J.W. (1983b). *Nature* **305**, 189.

Murray, J.A.H. (1987). *Molecular Microbiology* **1**, 1.

Murray, J.A.H., Scarpa, M., Rossi, N. and Cesareni, G. (1987). *EMBO Journal* **6**, 4205.

Murray, J.A.H., Cesareni, G. and Argos, P. (1988). *Journal of Molecular Biology* **200**, 601.

Myers, A.M., Tzagoloff, A., Kinney, D.M. and Lusty, C.J. (1986). *Gene* **45**, 299.

Mylin, L.M., Hofmann, K.J., Schultz, L.D. and Hopper, J.E. (1990). *Methods in Enzymology* **185**, 297.

Orr-Weaver, T.L., Szostak, J.W. and Rothstein, R.J. (1981). *Proceedings of the National Academy of Sciences, USA* **78**, 6354.

Orr-Weaver, T.L., Szostak, J.W. and Rothstein, R.J. (1983). *Methods in Enzymology* **101**, 228.

Oshima, Y. (1982). *In* "The Molecular Biology of the Yeast *Saccharomyces*: Metabolism and

Gene Expression" (J.N. Strathern, E.W. Jones and J.R. Broach, eds), pp. 159-180. Cold Spring Harbor Press, Cold Spring Harbor.

Panzeri, L., Landonio, L., Stotz, A. and Philippsen, P. (1985). *EMBO Journal* **4**, 1867.

Parent, S.A., Fenimore, C.M. and Bostian, K.A. (1985). *Yeast* **1**, 83.

Parent, S.A., Tait-Kamradt, A.G., Levitre, J., Lifanova, O. and Bostian, K.A. (1987). *In* "Phosphate Metabolism and Cellular Regulation in Microorganisms" (A. Torriani-Gorini, F.G. Rothman, S. Silver, A. Wright and E. Yagil, eds), pp. 63-70. American Society for Microbiology, Washington, DC.

Picard, D., Schena, M. and Yamamoto, K.R. (1990). *Gene* **86**, 257.

Polumienko, A.L., Grigor'eva, S.P., Lushnikov, A.A. and Domaradskij, I.V. (1986). *Biochemical and Biophysical Research Communications* **135**, 915.

Price, V.L., Taylor, W.E., Clevenger, W., Worthington, M. and Young, E.T. (1990). *Methods in Enzymology* **185**, 308.

Purvis, I.J., Chotai, D., Dykes, C.W., Lubahn, D.B., French, F.S., Wilson, E.M. and Hobden, A.N. (1991). *Gene* **106**, 35.

Reynolds, A.E., Murray, A.W. and Szostak, J.W. (1987). *Molecular and Cellular Biology* **7**, 3566.

Rine, J. (1991). *Methods in Enzymology* **194**, 239.

Roberts, T.M., Swanberg, S.L., Poteete, A., Riedel, G. and Backman, K. (1980). *Gene* **12**, 123.

Rogers, D.T., Lemire, J.M., and Bostian, K.A. (1982). *Proceedings of the National Academy of Sciences, USA* **79**, 2157.

Roggenkamp, R., Hansen, H., Eckart, M., Janowicz, Z. and Hollenberg, C.P. (1986). *Molecular and General Genetics* **202**, 302.

Romanos, M.A., Scorer, C.A. and Clare, J.J. (1992). *Yeast* **8**, 423.

Rose, A.B. and Broach, J.R. (1990). *Methods in Enzymology* **185**, 234.

Rose, M., Grisafi, P. and Botstein, D. (1984). *Gene* **29**, 113.

Rose, M. and Winston, F. (1984). *Molecular and General Genetics* **193**, 557.

Rose, M.D. and Broach, J.R. (1991). *Methods in Enzymology* **194**, 195.

Rose, M.D., Misra, L.M. and Vogel, J.P. (1989). *Cell* **57**, 1211.

Rose, M.D., Novick, P., Thomas, J.H., Botstein, D. and Fink, G.R. (1987). *Gene* **60**, 237.

Rosenberg, S., Coit, D. and Tekamp-Olson, P. (1990). *Methods in Enzymology* **185**, 341.

Rothstein, R.J. (1983). *Methods in Enzymology* **101**, 202.

Russell, P. and Nurse, P. (1986). *Cell* **45**, 145.

Rymond, B.C., Zitomer, R.S., Schümperli, D. and Rosenberg, M. (1983). *Gene* **25**, 249.

Sakaguchi, J. and Yamamoto, M. (1982). *Proceedings of the National Academy of Sciences, USA* **79**, 7819.

St John, T.P., Scherer, S., McDonell, M.W. and Davis, R.W. (1981). *Journal of Molecular Biology* **152**, 317.

Schena, M., Picard, D. and Yamamoto, K.R. (1991). *Methods in Enzymology* **194**, 389.

Scherer, S. and Davis, R.W. (1979). *Proceedings of the National Academy of Sciences, USA* **76**, 4951.

Schild, D., Brake, A.J., Kiefer, M.C., Young, D. and Barr, P.J. (1990). *Proceedings of the National Academy of Sciences, USA* **87**, 2916.

Schultz, L.D., Hofmann, K.J., Mylin, L.M., Montgomery, D.L., Ellis, R.W. and Hopper, J.E. (1987). *Gene* **61**, 123.

Seifert, H.S., Chen, E.Y., So, M. and Heffron, F. (1986). *Proceedings of the National Academy of Sciences, USA* **83**, 735.

Sengstag, C. and Hinnen, A. (1987). *Nucleic Acids Research* **15**, 233.

Shampay, J., Szostak, J.W. and Blackburn, E.H. (1984). *Nature* **310**, 154.

Shortle, D., Haber, J.E. and Botstein, D. (1982). *Science* **217**, 371.

Sidhu, R.S. and Bollon, A.P. (1987). *Gene* **54**, 175.

Sikorski, R.S. and Hieter, P. (1989). *Genetics* **122**, 19.

Sledziewski, A.Z., Bell, A., Yip, C., Kelsay, K., Grant, F.J. and MacKay, V.L (1990). *Methods in Enzymology* **185**, 351.

Smith, R.A., Duncan, M.J. and Moir, D.T. (1985). *Science* **229**, 1219.
Snyder, M., Elledge, S. and Davis, R.W. (1986). *Proceedings of the National Academy of Sciences, USA* **83**, 730.
Snyder, M., Elledge, S., Sweetser, D., Young R.A. and Davis, R.W. (1987). *Methods in Enzymology* **154**, 107.
Som, T., Armstrong, K.A., Volkert, F.C. and Broach, J.R. (1988). *Cell* **52**, 27.
Sreekrishna, K., Webster, T.D. and Dickson, R.C. (1984). *Gene* **28**, 73.
Stearns, T., Ma, H. and Botstein, D. (1990). *Methods in Enzymology* **185**, 280.
Stinchcomb, D.T., Struhl, K. and Davis, R.W. (1979). *Nature* **282**, 39.
Stinchcomb, D.T., Thomas, M., Kelly, J., Selker, E. and Davis, R.W. (1980). *Proceedings of the National Academy of Sciences, USA* **77**, 4559.
Stinchcomb, D.T., Mann, C. and Davis, R.W. (1982). *Journal of Molecular Biology* **158**, 157.
Struhl, K. (1983). *Gene* **26**, 231.
Struhl, K. (1985). *Nucleic Acids Research* **13**, 8587.
Struhl, K., Stinchcomb, D.T., Scherer, S. and Davis, R.W. (1979). *Proceedings of the National Academy of Sciences, USA* **76**, 1035.
Szostak, J.W. and Blackburn, E.H. (1982). *Cell* **29**, 245.
Tait-Kamradt, A.G., Turner, K.J., Kramer, R.A., Elliott, Q.D., Bostian, S.J., Thill, G.P., Rogers, D.T. and Bostian, K.A. (1986). *Molecular and Cellular Biology* **6**, 1855.
Tikhomirova, L.P., Ikonomova, R.N. and Kuznetsova, E.N. (1986). *Current Genetics* **10**, 741.
Tipper, D.J. and Bostian, K.A. (1984). *Microbiological Reviews* **48**, 125.
Toh-e, A. and Utatsu, I. (1985). *Nucleic Acids Research* **13**, 4267.
Tschopp, J.F., Sverlow, G., Kosson, R., Craig, W. and Grinna, L. (1987). *Bio/Technology* **5**, 1305.
Tschumper, G. and Carbon, J. (1980). *Gene* **10**, 157.
Tschumper, G. and Carbon, J. (1982). *Journal of Molecular Biology* **156**, 293.
Uno, I., Fukami, K., Kato, H., Takenawa, T. and Ishikawa, T. (1988). *Nature* **333**, 188.
Utatsu, I., Utsunomiya, A. and Toh-e, A. (1986). *Journal of General Microbiology* **132**, 1359.
Utatsu, I., Sakamoto, S., Imura, T. and Toh-e, A. (1987). *Journal of Bacteriology* **169**, 5537.
Volkert, F.C. and Broach, J.R. (1986). *Cell* **46**, 541.
Vollrath, D., Davis, R.W., Connelly, C. and Hieter, P. (1988). *Proceedings of the National Academy of Sciences, USA* **85**, 6027.
Webster, T.D. and Dickson, R.C. (1983). *Gene* **26**, 243.
West, R.W. (1988). In "Vectors: A Survey of Molecular Cloning Vectors and Their Uses" (R.L. Rodriguez and D.T. Denhardt, eds), pp. 387–404. Butterworths, Boston.
Wing, R.A. and Ogrydziak, D.M. (1985). In "Molecular Genetics of Filamentous Fungi" (W.E. Timberlake, ed.), pp. 367–381. Alan R. Liss, New York.
Winston, F., Chumley, F. and Fink, G.R. (1983). *Methods in Enzymology* **101**, 211.
Wright, A., Maundrell, K., Heyer, W.-D., Beach, D. and Nurse P. (1986a). *Plasmid* **15**, 156.
Wright, A.P.H., Maundrell, K. and Shall, S. (1986b). *Current Genetics* **10**, 503.
Yanisch-Perron, C., Vieira, J. and Messing, J. (1985). *Gene* **33**, 103.
Young, R.A. and Davis, R.W. (1983). *Science* **222**, 778.
Zealey, G.R., Goodey, A.R., Piggott, J.R., Watson, M.E., Cafferkey, R.C., Doel, S.M., Carter, B.L.A. and Wheals, A.E. (1988). *Molecular and General Genetics* **211**, 155.
Zimmermann, F.K., Khan, N.A. and Eaton, N.R. (1973). *Molecular and General Genetics* **123**, 29.
Zvonok, N.M., Horváth, E. and Bajszár, G. (1988). *Gene* **66**, 313.

5 Yeast Genome Structure

David B. Kaback

Department of Microbiology and Molecular Genetics, University of Medicine and Dentistry—New Jersey Medical School, 185 South Orange Avenue, Newark, NJ 07103, USA

I. INTRODUCTION

The genomes of yeasts have been the subject of intense investigations during the last three decades. Even the largest yeast chromosomes are an order of magnitude smaller than human chromosomes and therefore provide simple

The Yeasts Vol. 6, 2nd edition
ISBN 0-12-596416-1

models for investigating both chromosome structure and function. The range of knowledge about yeast genomes extends from physical maps for many different chromosomes in several different yeasts to the nucleotide sequence of several entire chromosomal DNA molecules from *Saccharomyces cerevisiae*. From these studies, several generalities have emerged that can be briefly summarized. Genome sizes are relatively constant and range from 10 to 20 megabases (Mb), presumably encoding 5000–10 000 genes. Haploid chromosome numbers vary from three to 16 and individual yeast chromosomes vary in size by more than an order of magnitude from approximately 0.2 to 7.0 Mb. It appears that yeast chromosomes are tightly packed with genes. In addition, each chromosome contains specialized structures that enable it to replicate and segregate properly at each mitotic or meiotic division. Besides chromosomal DNA, yeast contains several species of extrachromosomal DNA. These include the mitochondrial genome, which is 19–75 kb long, depending on the organism, and several small circular and linear plasmids that reside in the nucleus. Finally, several intracellular RNA viruses have been found. Some of these viruses represent the RNA stage of retrotransposable elements, while others appear to have simple double-stranded RNA life cycles. This review focuses specifically on structure and organization of chromosomes from nuclei of several different yeasts. Extrachromosomal genetic elements are reviewed elsewhere in this volume (Chapter 9). In addition, several recent reviews are available that deal in depth with specific topics and these are referenced in the text.

II. THE GENOME OF *SACCHAROMYCES CEREVISIAE*

A. Introduction

The most widely investigated yeast is *Sacch. cerevisiae*, which contains 16 haploid chromosomes (Mortimer and Schild, 1980, 1985). Based on chemical determinations and renaturation kinetics, its haploid nuclear genome size was estimated to be 13.5 ± 3.0 Mb, which included approximately 1 Mb of 9.1 kb tandem repeats of rDNA (Ogur *et al.*, 1952; Schweizer *et al.*, 1969; Lauer *et al.*, 1977). Pulsed-field gel electrophoresis has been used to separate chromosomes of *Sacch. cerevisiae* by size (Schwartz and Cantor, 1984; Carle and Olson, 1984, 1985) and the size of each chromosome determined by physical mapping using the rare-cutting restriction endonucleases *Not*I and *Sfi*I (Kaback *et al.*, 1989; Link and Olson, 1991). Adding up chromosome sizes confirmed earlier estimates and indicated that the nuclear genome size of haploid strain AB972, a close relative of the widely utilized strain S288C, was 12.49 Mb plus 100–200 tandem 9.1 kb rDNA repeats (Link and Olson, 1991). However, different strains exhibit naturally occurring chromosome-length polymorphisms (isoforms) (Carle and Olson, 1985; Johnston and Mortimer, 1986; Ono and Ishino-Arao, 1988; Adams et al., 1992) so that the amount of DNA in each haploid genome could vary by several per cent depending on the combination of chromosomal isoforms present. The NotI-SfiI mapping studies indicated that the smallest chromosome (I) in strain AB972 has 0.24 Mb, while the largest exclusive of rDNA (Chromosome IV) has 1.64 Mb. Chromosome XII contains approximately 1.1 Mb of "unique" DNA together with 1–2 Mb of tandemly repeated rDNA. Genome structure and organization in Sacch. cerevisiae has been reviewed recently (Newlon, 1988, 1989; Olson, 1991). The following review is designed to cover briefly many of the main points previously addressed and update those areas in which there has been recent progress.

B. Electrophoretic karyotyping

The display of intact chromosomes separated by pulsed-field gel electrophoresis is known as an electrophoretic karyotype (Carle and Olson, 1985). In strain AB972, the 16 chromosomes were separated into 14 distinct bands. Chromosomes V + VIII and VII + XV co-migrated under most conditions. Since different strains exhibit chromosome-length polymorphisms, it has been possible to breed yeast strains, such as YPH80, in which one of the two AB972 doublets is resolved and only two chromosomes are not separated as distinct gel bands (Gerring et al., 1991). Blot hybridization of separated chromosomes (termed chromo-blots) has been useful for assigning the chromosomal location of cloned sequences (Mortimer et al., 1989, 1992). Prepared gel samples from these strains are commercially available. However, caution should be exercised in interpreting chromo-blots, since separation between adjacent chromosomal bands can be minimal, leading to errors in chromosomal assignments (Barton et al., 1992). Suspected chromosomal locations should be confirmed by tetrad analysis or some other reliable mapping technique.

Electrophoretic karyotyping has been valuable in determining the nature of chromosomal re-arrangements, such as translocations and large deletions (Fasullo and Davis, 1988; Vollrath et al., 1988; Kaback et al., 1992). Careful measurement of relative band intensities also enables detection of aneuploidy (deJonge et al., 1986; Guacci and Kaback, 1991). These analyses have been useful for assessing the genetic makeup of industrial strains of

Sacch. cerevisiae and for investigating taxonomy and lineages (Johnston and Mortimer, 1986; Ono and Ishino-Arao, 1988).

C. Genome mapping

Efforts are currently underway to clone and physically map the entire genome of *Sacch. cerevisiae*. Several "chromosome walks" were reported where large (40–100 kb) regions of several chromosomes were isolated, and their genetic and physical structures analysed (Chinault and Carbon, 1979; Clarke and Carbon, 1980; Coleman *et al.*, 1986; Steensma *et al.*, 1987; Kaback *et al.*, 1989). In addition, a circular 200 kb derivative of chromosome III was physically isolated by density-gradient centrifugation, its restriction map determined and its cloning completed (Strathern *et al.*, 1979; Newlon *et al.*, 1991). These efforts were followed by construction of lambda or cosmid clone banks using partial *Sau*3A digests of DNA isolated from pulsed-field or field-inversion gel purified individual chromosomes (Yoshikawa and Isono, 1990, 1991; Oliver *et al.*, 1992; Tanaka and Isono, 1992; Tanaka *et al.*, 1992; A. Thierry and B. Dujon, personal communication). For chromosome III, analysis of 344 lambda clones by restriction endonucleases indicated that the bacteriophage inserts represented two contiguous stretches of DNA (*contigs*) covering virtually the entire chromosomal DNA molecule (Yoshikawa and Isono, 1990). Single *contigs* were obtained for chromosomes VI and I and two *contigs* were obtained for chromosomes V and VIII. (Iwasaki *et al.*, 1991; Yoshikawa and Isono, 1991; Tanaka and Isono, 1992; Tanaka *et al.*, 1992). Concurrently, progress was made towards cloning the entire yeast genome by rapidly analysing large numbers of randomly selected lambda-bacteriophage clones (Olson *et al.*, 1986; Riles *et al.*, 1993). Cloned inserts were subjected to digestion with a combination of *Eco*RI and *Hin*dIII leading to a characteristic gel-electrophoretic fingerprint for each cloned region. The fingerprints were compared to find overlapping regions, which were pieced together into contigs. Previously characterized cloned regions and the *Not*I-*Sfi*I map (Link and Olson, 1991) were used to optimize the alignment. Continuous physical maps were obtained for most chromosomes even though several chromosomes were represented by more than one *contig*. Cloning and physical mapping of the six smallest chromosomes is 97% complete, with the missing DNA being largely from the ends of each chromosome. Chromosomes I, VI, III and V are represented by single *contigs*, while chromosomes VIII and IX have single internal gaps in their maps. The missing regions are probably attributable to difficulties in stably cloning certain sequences in bacteriophage or plasmid vectors (Riles *et al.*, 1993). Substantial progress towards cloning and mapping most of the larger yeast chromosomes has also been made so that ordered clone banks representing more than 90% of the yeast genome are currently available from the American Type Culture Collection (Jong and Edwards, 1990). In addition, these clone banks have been gridded to nitrocellulose filters that also are available from the American Type Culture Collection. The nitrocellulose filters can be hybridized to cloned probes enabling genomic localization of the cloned sequence to a specific bacteriophage insert. In most studies, this information can be combined with data on additional restriction endonuclease-cleavage sites and used to position the cloned insert on the physical map of a specific chromosome.

D. Gene organization

Transcribed regions on large segments of the genome of *Sacch. cerevisiae* have been analysed by R-loop electron microscopy (Kaback *et al.*, 1979). These studies showed that most of the genome was transcribed and that large regions devoid of genes were either rare or non-existent. Northern-blot hybridization using cloned probes has been used to generate transcription maps for all of chromosomes III and VI, for large parts of chromosome I and for many regions contiguous to various cloned genes. For chromosome III, 156 poly(A)-containing transcripts were distributed over 315 kb of DNA (Yoshikawa and Isono, 1990). Transcript density on chromosome VI was slightly lower; 97 transcripts were found distributed over 270 kb of DNA (Yoshikawa and Isono, 1991). The 160 kb so far examined on chromosome I contained 57 poly(A)-containing transcripts (Coleman *et al.*, 1986; Steensma *et al.*, 1987; Diehl and Pringle, 1991; Barton and Kaback, 1991). The average transcript size was 1.9 kb in all three studies. Extrapolating these data (310 transcripts for 745 kb of DNA) to the entire 12.49 Mb genome indicates that there should be approximately 5000 genes expressed in exponentially growing cells. This number is in close agreement with the previously obtained estimate examining global transcription by R-loop electron microscopy (Kaback *et al.*, 1979). Assuming transcribed regions are non-overlapping, it appears that at least 80% of the DNA from *Sacch. cerevisiae* is transcribed. In most experiments, mapping was carried out with poly(A)-containing RNA isolated from cells growing exponentially in a rich medium. Genes that are not expressed under these conditions would have been missed. Several gene-size regions that did not correspond to a vegetative-cell transcript were found. It is known that some genes are only expressed under special physiological or genetic conditions. Therefore, it is likely that large regions devoid of transcripts contain genes expressed during different circumstances. Furthermore, most of these studies did not include non-polyadenylated RNA species such as tRNAs and snRNAs. There are approximately 350 tRNA genes (Schweitzer *et al.*, 1969) and at least two dozen snRNA genes (Riedel *et al.*, 1986) that appear randomly dispersed throughout the genome (Beckmann *et al.*, 1977). Therefore, it is likely that a somewhat greater fraction of the genome of *Sacch. cerevisiae* genome is transcribed and that the total number of genes could be as high as 6500.

Transcript abundance was also measured for chromosomes III and VI (Yoshikawa and Isono, 1990, 1991). With several exceptions, transcript levels were highest in the middle of each chromosome arm. However, the reason for this correlation has yet to be investigated. Within 15–20 kb of each telomere, transcripts were found at low but detectable levels. Low rates of gene expression in subtelomeric regions may be characteristic of some general repression mechanism involving the ends of chromosomes (Gottschling *et al.*, 1990; Aparicio *et al.*, 1991; Gottschling, 1992).

Analysis of many cloned genes revealed that introns are relatively rare. In addition, no more than a single intron has ever been found in a gene from *Sacch. cerevisiae*. When introns were found they were relatively small, in the range of 200–500 bp, and usually located near the 5′ end of the gene (Fink, 1987). Introns were first found in tRNA genes (Goodman *et al.*, 1977; Valenzuela *et al.*, 1978), the actin gene (Gallwitz and Sures, 1980; Ng and Abelson, 1980) and several of the genes encoding ribosomal proteins (reviewed in Woolford and Warner, 1991). A description of currently known intron-containing genes is available (reviewed in Rymond and Rosbash, 1993). There is no doubt that several more will be discovered. However, based on the frequency at which introns have been found so far, it is likely that no more than 2–4% of all genes in *Sacch. cerevisiae* will contain them.

The position and paucity of introns led to the suggestion that *Sacch. cerevisiae* had largely cleansed itself of these sequences during evolution. It was proposed that reverse transcription of mRNA without introns produced double-stranded cDNA copies. These cDNA copies recombined with genomic copies using either double cross-overs or double-stranded break recombination, and replaced the intron-bearing genomic copy with a cDNA without introns (Fink, 1987). Model systems that reproduced these events have been constructed in *Sacch. cerevisiae* making this mechanism plausible (Derr *et al.*, 1991).

E. Genome sequencing

The most accurate look at genome structure has come from determining the entire nucleotide sequence of a single chromosome from *Sacch. cerevisiae* (Oliver *et al.*, 1992). Chromosome III was divided up by a consortium of 35 laboratories each responsible for 3–10% of the sequence. The sequence has been placed in the EMBO Laboratory (EMBL) database (aquisition number X59720) and was analysed for open-reading frames and other salient features. In addition, individual groups have reported independently on regions that they sequenced (see sequence reports in the journal *Yeast*, 1990–1993).

The nucleotide sequence revealed that chromosome III was approximately 9% shorter than the 335 kb estimate obtained by restriction-endonuclease analysis. The sequence was 315 357 bp long but excluded approximately 0.4 kb from the right end and telomeric repeats, which have various lengths in the range 0.5–3 kb (see p. 201). Whether or not these restriction endonuclease-determined size inaccuracies will be found in other chromosomes has yet to be reported.

The 315 kb sequence revealed 182 open-reading frames (ORFs) for proteins longer than 100 amino-acid residues. Numbering these ORFs was carried out on a systematic basis that is dependent on its position on the chromosome. This system is applicable to all chromosomes in *Sacch. cerevisiae* and appears to be in the process of being adopted universally. Each ORF was labelled by a three-letter prefix followed by a number and a W (Watson) or a C (Crick) depending upon strand. The first letter of the prefix is Y, standing for the yeast *Sacch. cerevisiae*; the second letter corresponds to the letter alphabetically assigned to each chromosome, A for chromosome I, B for chromosome II, C for chromosome III, and so on; the third letter is either L or R corresponding to the left or right arm of the chromosome, respectively. Following the prefix is a number designating the relative position of an ORF from the centromere. For example, YCR23C is the 23rd ORF from the centromere on the right arm of chromosome III and is oriented in the same direction as other genes denoted by a W.

The map of ORFs was in close agreement with the transcription map, except that it contained 26 additional potential ORFS that did not correspond to poly (A) containing RNA. The 26 genes presumably represent that fraction of the genome, which is either not expressed or expressed at levels too low for detection in cells used to isolate the poly(A)-containing RNA. Many of these genes were small and would have been hard to detect by blot hybridization. It also is likely that there are several additional genes that encode proteins smaller than 100 amino-acid residues. In total, it appears that more than 90% of chromosome III has the potential to be transcribed. Thus, no more than 10% of the genome, or on average 190 bp for each gene is available to serve as upstream control sequences for gene expression. Indeed, many of the already characterized upstream-control regions appear to be of that size.

Of the 182 ORFs on chromosome III, 145 represent newly discovered genes whose function is unknown. Clearly, determining the function of these unidentified genes represents an important step in completely understanding the life cycle of *Sacch. cerevisiae.*

In addition to protein-encoding genes, 10 potential tRNA genes and several transposons and transposon-terminal repeats were precisely located during sequencing of chromosome III. In agreement with the previously noted association, all 10 tRNA genes were within 500 bp of a transposon or a transposon terminal repeat (see below).

Complete elucidation of the chromosome III sequence led to a concerted international effort to sequence the rest of the genome of *Sacch. cerevisiae* (Vassarotti *et al.*, 1990). Several large-scale projects aimed at sequencing additional chromosomes are already underway. All of chromosomes I, II, VIII and XI are sequenced (H. Bussey, H. Feldmann, M. Johnston and B. Dujon, personal communication). Genomic sequencing of *Sacch. cerevisiae* is expected to have a great impact upon understanding both the structure and function of chromosomes, and the function of genes contained in an entire organism. To date (January 1995), there are approximately 4000 data entries in either GENBANK or EMBL databases representing approximately 50% of the entire nuclear genome in *Sacch. cerevisiae*. Half of this data comes from genomic sequencing. The rest represents incidental sequencing of genes and their flanking sequences. The aim is to complete the entire genomic sequencing project by the end of 1996.

F. Comparison of genetic and physical maps

The 11th edition of the genetic map has 857 markers that have been mapped genetically and an additional 189 genes whose chromosomal position has been determined by physical studies (Mortimer *et al.*, 1992). Thus, approximately 20% of the total number of genes in *Sacch. cerevisiae* have been defined by one of a number of either classical or molecular criteria and placed on the map (Mortimer and Schild, 1980, 1985; Mortimer *et al.*, 1989, 1992). The remaining genes are either totally uncharacterized or have just been defined by a transcribed region or ORF of unknown function.

The genetic map of *Sacch. cerevisiae* contains 16 *bona-fide* linkage groups. Each linkage group contains a centromere, no less than 15 genetically linked markers and corresponds to a specific DNA molecule observable in an electrophoretic karyotype. The existence of a 17th linkage group has been suggested based on the apparent first-division meiotic segregation of the *KRB1* gene and the failure to find linkage to any of the known centromere markers in *Sacch. cerevisiae* (Wickner *et al.*, 1983). However, its physical nature and chromosomal whereabouts remain enigmatic (Mortimer *et al.*, 1989).

The genetic map of *Sacch. cerevisiae* is currently 4300 cM long and has not been increased substantially in length for more than 15 years, despite a tripling in the number of mapped genes. Thus, it is likely that genetic map lengths for each chromosome are very close to complete (Mortimer *et al.*, 1989, 1992).

Since many genes have been physically mapped, it has been possible to compare genetic and physical maps for several chromosomes (Kaback *et al.*, 1989; Mortimer *et al.*, 1989, 1992; Yoshikawa and Isono, 1990, 1991; Oliver *et al.*, 1992; Riles *et al.*, 1993; D.B. Kaback, unpublished results). With few exceptions, the two maps are colinear. The exceptions could be due to small inversions in strains used either for mapping or cloning or to inaccuracies inherent in placing genes on genetic maps using independent data sets.

A comparison of genetic and physical maps describes relative amounts of meiotic cross-over activity on each segment of a chromosomal DNA molecule. This activity is expressible as centimorgans per kilobase. The most extensive study of this activity utilized chromosome I, which has many markers that are both genetically and physically mapped over most of its length (Kaback *et al.*, 1989; Steensma *et al.*, 1989; Y. Su and D. Kaback, unpublished results). These markers are distributed from 8 kb from the left telomere to 4 kb from the right telomere. A plot of centimorgans per kilobase over 95% of the physical length of this chromosome indicated that rates of recombination were depressed at the centromere and in both subtelomeric regions. The recombination rate over the rest of the physical length of the chromosome was relatively constant except for a recombination hot spot located in the *CDC24-PYK1* interval (Coleman *et al.*, 1986). Similar results were observed or could be calculated from the available data for chromosomes II, III, V and VI, which also showed relatively uniform levels of recombination over most of their physical lengths. Regions adjacent to centromeres and to three telomeres had low recombination rates (Clarke and Carbon, 1980; Lambie and Roeder, 1986; Mortimer *et al.*, 1989, 1992; Yoshikawa and Isono, 1990, 1991; D.B. Kaback, unpublished results). Unfortunately, genetically mapped telomere proximal markers were not available at each end of these four other chromosomes. Translocation experiments that moved centromeres to new locations indicated that the depressions in rates of recombination found near centromeres were dependent upon the presence of functional *CEN* DNA (Lambie and Roeder, 1986). Similar experiments suggest that telomeric DNA sequences can also inhibit meiotic recombination directly in subtelomeric regions (D. Kaback, Y. Su, D. Barber and J. Mahon, unpublished results).

A few small regions that showed high recombination rates were observed on other chromosomes. (Nag *et al.*, 1989; Cao *et al.*, 1990; Malone *et al.*, 1992). High-resolution analysis of the recombinational behaviour of a 23 kb region from chromosome III indicated that different segments of this region underwent both meiotic recombination and gene conversion at different rates (Symington and Petes, 1988a,b; Symington *et al.*, 1991).

Comparing total genetic and physical map lengths of each chromosome indicated that on average, smaller chromosomes undergo reciprocal recombination at higher rates than larger chromosomes (Kaback *et al.*, 1989; Mortimer *et al.*, 1989, 1992; Riles *et al.*, 1993). The rate of recombination on chromosome I was $0.60 \, \text{cM} \, \text{kb}^{-1}$ while the rate on chromosome IV was $0.29 \, \text{cM} \, \text{kb}^{-1}$. Intermediate values were obtained for other chromosomes and the average for the entire genome was $0.34 \, \text{cM} \, \text{kb}^{-1}$ (Kaback *et al.*, 1989; Mortimer *et al.*, 1989, 1992). It also was demonstrated that the rate of recombination was directly dependent on chromosome size (Kaback *et al.*, 1992). Small chromosomes made larger showed lower rates of reciprocal recombination while large chromosomes made smaller showed higher rates (Kaback *et al.* 1992 and unpublished results). In addition, the rate of recombination at the *CDC24-PYK1* recombination hot spot was substantially lowered when chromosome size was increased. These experiments demonstrated that chromosome size-dependent control of recombination was superimposable on hot spots making it unlikely that *cis* acting sequences were responsible for enhancing rates of recombination on small chromosomes. It was suggested that chromosome size-dependent control of recombination was important for ensuring that all chromosomes undergo recombination (Kaback *et al.*, 1992). In addition, it is possible that this mechanism limits the amount of crossing-over on the larger chromosomes.

G. Repetitive DNA

1. Ribosomal RNA genes

Saccharomyces cerevisiae contains 100–200 copies of a 9.1 kb tandem rRNA gene (rDNA) repeat located on the right arm of chromosome XII (Schweitzer *et al.*, 1969; Harbitz and Oyen, 1974; Cramer *et al.*, 1977; Petes and Botstein, 1977; Petes, 1979a,b; Kaback and Davidson, 1980; G.F. Carle and M.V. Olson, unpublished observation, cited in Olson, 1991). The structure of each rRNA gene and control of rRNA synthesis have been reviewed recently (Woolford and Warner, 1991). Briefly, each repeat contains genes for 18S, 26S, 5.8S and 5S ribosomal RNA (rRNA) and a non-transcribed spacer (Rubin and Sulston, 1973; Kaback *et al.*, 1976; Nath and Bollon, 1976; Bell *et al.*, 1977; Cramer *et al.*, 1977; Valenzuela *et al.*, 1977; Kramer *et al.*, 1978; Philippsen *et al.*, 1978). A 6.6 kb 35S rRNA precursor molecule is transcribed from one strand of the rDNA by RNA polymerase I. This precursor is sequentially processed to produce 18S, 26S and 5.8S rRNAs (Udem and Warner, 1972). The 5S species is transcribed by RNA polymerase III from the opposite strand of rDNA (Valenzuela *et al.*, 1977). This gene is located in the non-transcribed spacer region between adjacent 35S precursor genes. Interspersion of 5S rRNA genes between the 35S rRNA precursor sequences has been observed in several other fungi but is not universal for the class (see p. 205). Indeed, most 5S genes from higher eukaryotes are found in clusters separated from the main rDNA cluster. The non-transcribed spacer contains an enhancer for rRNA transcription (Elion and Warner, 1984, 1986). Ribosomal DNA from *Sacch. cerevisiae* can be purified by density-gradient centrifugation, where it forms a dense satellite, or by pulsed-field gel electrophoresis (Cramer *et al.*, 1972; Cramer and Rownd, 1980; G.F. Carle and M.V. Olson, unpublished results, cited in Olson, 1991).

Initial cloning and Southern-blot analysis suggested that rDNA was organized in 9.1 kb tandem repeats (Bell *et al.*, 1977; Cramer *et al.*, 1977; Valenzuela *et al.*, 1977; Kramer *et al.*, 1978; Philippsen *et al.*, 1978). Electron-microscope observations of R-loop-containing DNA indicated that virtually all of the rDNA was localized on simple tandem repeats with very few rDNA–chromosomal junctions (Kaback and Davidson, 1980). These results suggested there was a single cluster of rDNA repeats. There have been reports of limited-length heterogeneity in an rDNA cluster due to 50 bp insertions within several non-transcribed spacers (Jemptland *et al.*, 1986). Besides these, failure to find significant restriction-fragment heterogeneity, except due to rDNA–chromosomal junctions, supported the idea of a single tandem repeat. One rDNA–chromosome junction has been cloned and characterized, and it was found that different strains could have different single-copy sequences at this junction (Zamb and Petes, 1982). Pulsed-field gel electrophoresis of genomic DNA that had been cleaved with enzymes that do not cut within the 9.1 kb repeat, revealed that the rDNA cluster runs as a single high molecular-weight DNA species providing further evidence for a single rDNA gene cluster. The high molecular-weight species varies from 1 to 2 Mb corresponding to approximately 100–200 rRNA genes (G.F. Carle and M.V. Olson, unpublished observation, cited in Olson, 1991).

Both quantitative hybridization and pulsed-field gel electrophoresis indicated that different strains and even different isolates of the same strain had different numbers of rDNA repeats (Oyen, 1973; Harbitz and Oyen, 1974; G.F. Carle and M.V. Olson, unpublished observation, cited in Olson, 1991). Furthermore, strains with low numbers of rDNA repeats underwent spontaneous rDNA magnification, which increased rDNA gene numbers significantly (Kaback et al., 1976; Kaback and Halvorson, 1977, 1978). There is now evidence for unequal mitotic recombination and unequal meiotic sister chromatid exchange between rDNA repeats (Petes, 1980; Szostak and Wu, 1980). These types of recombination could cause the observed alterations in the number of rDNA repeats. In addition, many strains have extrachromosomal circles containing rDNA (Meyerink et al., 1979; Clark-Walker and Azad, 1980; Larionov et al., 1980; Devenish and Newlon, 1982). These circles presumably arise by intrachromosomal recombination between tandem repeats. Therefore, rDNA cluster-length changes observable by pulsed-field gels could be due to excision or re-integration of circular rDNA species. However, the fact that quantitative hybridization experiments show that the total number of repeats actually changes, make it likely that molecular-weight determinations faithfully reflect actual expansion and contraction of the rDNA cluster.

Restriction fragment-length polymorphisms were used to map the rDNA to chromosome XII (Petes and Botstein, 1977; Petes, 1979a,b). These experiments also showed that meiotic recombination between homologous chromosomes within the rDNA repeat was suppressed. Consistent with this suppression, meiotic double-strand breaks were rare in the rDNA repeat (Hogst and Oyen, 1984). The mechanism for suppression of recombination is not known. However, it appears to require topo-isomerases I and II and the product of the SIR2 gene (Christman et al., 1988; Gottlieb and Esposito, 1989). In addition, the nucleolus of Sacch. cerevisiae does not dissociate during meiosis. Therefore, it is possible that rDNA is inaccessible to the recombinational machinery. Irrespective of which mechanism operates, suppression of recombination could minimize the potential for unequal crossing-over between homologues leading to production of spores with little or no rDNA.

2. Subtelomeric regions

Chromosomal regions adjacent to simple telomeric repeat sequences (see p. 201) are largely or totally comprised of repeated sequences and genes. With the possible exception of the right arm of chromosome VI in strain S288C, all subtelomeric regions contain what are termed X sequences and about half contain what are termed Y′ sequences (Button and Astell, 1986; Zakian and Blanton, 1988; Jager and Philippsen, 1989; Link and Olson, 1991). The X and Y′ sequences are separated from each other by simple telomeric repeat sequences, while Y′ sequences are found closer to the end of the chromosome (Chan and Tye, 1983a,b; Walmsley et al., 1984; Louis and Haber, 1990). X and Y′ sequences are unrelated but both contain autonomously replicating sequence elements (see below) (Chan and Tye, 1980, 1983a, b; Walmsley et al., 1984). The X sequences are heterogeneous in size, ranging from 0.3 to 4.0 kb (Chan and Tye, 1983a,b). They are also heterogeneous in composition and do not always cross-hybridize with each other (Zakian and Blanton, 1988; Jager and Philippsen, 1989; Link and Olson, 1991). Y′ sequences are more highly conserved and are either 5.2 kb or 6.7 kb long depending upon the presence or absence of a 1.5 kb sequence. Otherwise, 6.7 and 5.2 kb sequences appear identical (Chan and Tye, 1983b; Louis and Haber, 1990). The number of Y′ sequences at each end varies from 0 to 4. When multiple Y′ elements are present, they are found in the same orientation and are separated by short segments of simple telomeric repeat sequences. The total number of Y′ sequences appears to vary from strain to strain (Chan and Tye, 1983a,b; Zakian and Blanton, 1988; Jager and Philippsen, 1989; Louis and Haber, 1990). The Y′ sequences have short internal repeats reminiscent of satellite DNA in higher eukaryotes (Horowitz and Haber, 1984). Finally, sequence studies hint that both X and Y′ sequences have retroviral origins (Louis and Haber, 1992; Voytas and Boeke, 1992).

Neither X nor Y′ sequences appear to be required for normal mitotic chromosome function. A derivative of chromosome III that has neither functions perfectly well during mitosis (Murray and Szostak, 1986). While it is not known whether these sequences are required for normal meiosis, chromosome I lacks Y′ sequences (Zakian and Blanton, 1988; Jager and Philippsen, 1989; Steensma et al., 1989; Link and Olson, 1991) and chromosome I fragments lacking both X and Y′ sequences on one arm are fully functional during meiosis (Guacci and Kaback, 1991; Kaback et al., 1992).

Subtelomeric regions often contain copies of repetitive functional gene families. Members of the *SUC*, *MAL* and *PHO* gene family have all been found imbedded in or adjacent to subtelomeric X or Y′ sequences (Carlson *et al.*, 1985; Charron *et al.*, 1989; Steensma *et al.*, 1989; Ventor and Horz, 1989; Michels *et al.*, 1992). Different chromosomes have different combinations of these genes and the combinations can vary from strain to strain. Transcription of most subtelomeric genes appears to be regulated by external stimuli. With the exception of *MAL63*, these genes are all abundantly transcribed under conditions that promote expression but are otherwise, silent or near silent. The *MAL63* gene is furthest from the telomere and shows a 2–3-fold difference in transcript levels in induced compared with uninduced cells (Needleman *et al.*, 1984). It was also shown that transcript levels were low for most of the genes located near the ends of both chromosomes III and VI (Yoshikawa and Isono, 1990, 1991). In addition, translocation of a *URA3* to a subtelomeric region largely prevented its transcription (Gottschling *et al.*, 1990; Aparicio *et al.*, 1991). These observations suggest that there could be a general mechanism for lowering gene expression, near telomeres, and that this mechanism can be overcome under appropriate physiological conditions, which induce specific gene expression. Thus, it is possible that the ends of chromosomes provide a convenient site that permits efficient repression of genes that are only required under certain conditions.

3. Other repetitive genes

There have been many reports of repeated protein-encoding genes. There are two or more copies of genes encoding cytochrome *c*, each of the four core histones, many ribosomal proteins and glycolytic enzymes, G1 and G2 cyclins, α-tubulin, β-hydroxy-β-methylglutaryl-SCoA (HMG-CoA) reductase, copper chelatin (metallothionein) and the Ras proteins, to name just a few (Hereford *et al.*, 1979; Holland *et al.*, 1981; Fogel and Welch, 1982; DeFeo-Jones *et al.*, 1983; Fogel *et al.*, 1983; Smith and Murray, 1983; Abovich and Rosbash, 1984; Powers *et al.*, 1984; Schatz *et al.*, 1986; Basson *et al.*, 1987; Hadwiger *et al.*, 1989; Woolford and Warner, 1991). With the exception of copper chelatin genes, gene-family members are not clustered. Most repeated genes found so far were detected by cross-hybridization to cloned probes. Accordingly, they are at least 60% identical at the nucleic-acid level. There are certainly additional repeated genes that lack the identity to be detected by cross-hybridization. Use of oligonucleotides encoding conserved protein motifs has allowed isolation of putative gene homologues using the polymerase chain reaction (PCR) (Surana *et al.*, 1991). In addition, genetic screens for unlinked non-complementing mutations, and for synthetic lethal genes that utilize plasmid-shuffling and aphenotypic mutants deleted for one copy of the gene of interest, will no doubt reveal additional duplicated genes (Botstein, 1988; also see Chapter 3 by M.D. Rose, this volume).

There are two examples of large duplicated genomic regions. At least 25 kb from the right end of chromosome I are repeated 10 kb from the right end of chromosome VIII (Steensma et al., 1989; Ventor and Horz, 1989). Restriction maps of these two regions are virtually identical except for a Ty1 transposon present on chromosome VIII. Both contain genes encoding acid phosphatase as well as X sequences. A 7.5 kb sequenced region on chromosome X, containing genes CYC1, OSM1 and RAD7, appears to be less closely but nevertheless related to a region on chromosome V containing genes CYC7, ANB1 and RAD23 (Melnick and Sherman, 1990). Both CYC genes encode cytochrome c. However, the two loci are regulated differently (Laz et al., 1984). Functional relationships between OSM1 and ANB1, and RAD7 and RAD23, have yet to be definitively demonstrated.

In some strains, it is believed that gene duplication, such as that found for ribosomal protein genes, has evolved simply to produce higher levels of that gene product (Woolford and Warner, 1991). However, it is also common to find differentially regulated members of gene families. Besides the CYC1 and CYC7 genes, histone genes and members of the 70 kDa heat-shock protein gene family show different patterns of regulation (Norris and Osley, 1987; Cross and Smith, 1988; Craig, 1989; Werner-Washburne et al., 1989). In addition, mutations in the two different RAS genes have different phenotypes (Tatchell et al., 1984, 1985). These data suggest that different family members may respond differently under a variety of physiological conditions enabling the organism to perform vital functions. Furthermore, different gene-family members may play distinct roles in various parts of the yeast life cycle. For example, different members of the glucoamylase gene family are expressed during vegetative growth and sporulation (Pugh and Clancy, 1990). The sporulation enzyme is believed to be primarily involved in degrading internal glycogen, while the vegetative form is secreted in order to degrade starch for use as a carbon source.

H. Transposons

The discovery of transposons in yeast sparked a great deal of research into mechanisms of transposition. Details of these mechanisms are spelled out in several recent reviews (Boeke and Sandmeyer, 1991; Sandmeyer, 1992). Standard laboratory strains of Sacch. cerevisiae contain between 2 and 30 copies of at least five different retrotransposons, labelled Ty1-Ty5. The number of copies for each retrotransposon is highly variable depending on the strain examined. Transposons in *Sacch. cerevisiae* are widely dispersed in the genome, appear to undergo transposition through an RNA intermediate and are all related to retroviruses. Each retrotransposon consists of 5.3-5.7 kb of internal DNA surrounded by ~350 bp long terminal repeats (LTRs). Five nucleotides of the chromosomal target DNA are duplicated at each end of the transposon. So far, we know of three different types of LTR, called sigma, delta and tau. In addition, subtelomeric X sequences may represent a fourth class (Voytas and Boeke, 1992). Different LTR sequences appear unrelated. Retrotransposons Ty1 and Ty2 are flanked by delta sequences, Ty3 by sigma sequences, Ty4 by tau and Ty5 by X. The central region of the transposon encodes components of virus-like particles including a reverse transcriptase, integrase, protease and nucleocapsid (*gag*) proteins. Based on the organization of the virus-like particle genes, Ty1, Ty2 and Ty4 appear to belong to the *copia*-like family of transposons, while Ty3 belongs to the *gypsy*-like family. Transposons Ty1 and Ty2 are closely related and are almost identical for more than half of their lengths. With the exception of their *copia*-like organization and conservation of a few functional protein motifs, they do not share significant homology with Ty4 (reviewed in Boeke and Sandmeyer (1991) and Sandmeyer (1992); see also, Stucka *et al.* (1992)).

Since identification of several retrotransposons was based solely on homology, it is possible that many of those found so far are defective and incapable of transposition (Boeke *et al.*, 1988). For example, the first sequenced Ty3 element (Ty3-2) contained a frameshift mutation that rendered it inactive. Correction of this mutation *in vitro* produced a transpositionally active copy of this retrotransposon (Hansen *et al.*, 1988; Hansen and Sandmeyer, 1990). In addition, the recently discovered Ty5-1 element is believed to have accumulated a large number of mutations that obscured its identification as a retrotransposon (Voytas and Boeke, 1992).

There are many more copies of LTRs dispersed throughout the genome than there are copies of the central sequence (Cameron *et al.*, 1979; Chisholm *et al.*, 1984; Genbauffe *et al.*, 1984; Sandmeyer *et al.*, 1988; Stucka *et al.*, 1989). The solo LTRs are believed to be the product of homologous recombination between the two LTRs flanking a retrotransposon. This recombination event expels the central sequence and one LTR, leaving behind a solo LTR (Cameron *et al.*, 1979). Clusters of delta elements have been observed. In one cluster, recombination between solo deltas appears to cause frequent deletions and re-arrangements of DNA between the cluster (Liebman *et al.*, 1981; Rothstein *et al.*, 1987).

Retrotransposons or solo LTRs are frequently found near the 5' end of tRNA genes. Retrotransposon Ty3 elements or sigma sequences are almost always found within 20 nucleotides of a tRNA gene (Sandmeyer *et al.*, 1988; Chalker and Sandmeyer, 1990). In addition, 11 out of 13 randomly selected tRNA genes contained either delta, sigma or tau elements located nearby (Hauber *et al.*, 1988). Delta and tau LTRs are usually located within 500 nucleotides of the tRNA gene (Eigel and Feldmann, 1982; Gafner *et al.*, 1983; Stucka *et al.*, 1989). However, there are no obvious sequence similarities that suggest a specific transposon-insertion site. It has been proposed that the transposition machinery might interact with factors involved in tRNA transcription (Olson, 1991).

It is likely that additional retrotransposons remain to be discovered. Sequence studies suggest the presence of at least one additional transposon. A sequence related to LTRs was found near the gene for a tRNA^{Glu} on chromosome X. However, this sequence fails to cross-hybridize with other genomic sequences (Melnick and Sherman, 1990). Perhaps surveying genomes of independently maintained yeast strains might produce the rest of this particular Ty element.

I. DNA sequences involved in chromosome function

A great deal has been learned about functional elements on eukaryotic chromosomes using *Sacch. cerevisiae*. The ability to transform this yeast with chimeric plasmid DNA rapidly led to the discovery of origins of DNA replication, centromeric DNA sequences and telomeric DNA sequences. These functional elements are the subject of intense investigation and have all been reviewed recently in great detail (Bloom and Yeh, 1989; Bloom *et al.*, 1989; Newlon, 1989; Zakian, 1989; Clarke, 1990; Murphy and Fitzgerald-Heyes, 1990; Zakian *et al.*, 1990; Blackburn, 1991; Campbell and Newlon, 1991; Schulman and Bloom, 1991; Fangman and Brewer, 1992). However, several salient features of functional elements in *Sacch. cerevisiae* chromosomes are discussed.

The first of these elements, termed autonomously replicating sequences or *ARS*, increased the efficiency of DNA transformation by several orders of magnitude because they enabled plasmid DNA to replicate autonomously in the nucleus (Hsiao and Carbon, 1979; Kingsman *et al.*, 1979; Stinchcomb *et al.*, 1979; Struhl *et al.*, 1979; Chan and Tye, 1980; also reviewed in Campbell and Newlon, 1991). These *ARS* elements are believed to correspond to origins of DNA replication and appear to be equivalent to *ori* sequences from prokaryotes (Brewer and Fangman, 1987; Huberman *et al.*, 1987). Many *ARS* sequences have been cloned and studied. They are about 100 bp in length and occur about once every 20–40 kb in the genome (Chan and Tye, 1980; Newlon *et al.*, 1991). Their genomic distribution is consistent with fibre autoradiographic evidence on distribution of chromosomal origins of DNA replication (Petes and Williamson, 1975). Studies using chromosome III suggest that most but not all *ARS* elements correspond to origins of DNA replication that initiate during every round of DNA replication during both mitotic and meiotic cell cycles (Greenfeder and Newlon, 1992; I. Collins and C.S. Newlon, personal communication). One *ARS* sequence is known to be used infrequently whereas two that occur near the *HML* silent mating-type locus do not appear to function as origins during normal S-phase DNA synthesis (Dubey *et al.*, 1991; Greenfeder and Newlon, 1992; Rivier and Rine, 1992). While the role of these *ARS* elements is not known, it is conceivable that they function only under special physiological conditions. Not all *ARS* elements initiate replication simultaneously. An *ARS* located near the telomere of chromosome V was found to initiate replication late during S phase (Ferguson *et al.*, 1991). However, this effect may be due to its existence near the end of a chromosome rather than to a specific sequence found within that particular *ARS* (Ferguson and Fangman, 1992).

All *ARS* sequences contain the core consensus:

5'A/T TTTA T/C A/G TTT A/T T/C/G3'

also known as domain A (Stinchcomb *et al.*, 1981; Broach *et al.*, 1983; Van Houten and Newlon, 1990). Analysis of mutants within this region have shown that the core consensus is essential for *ARS* function (Kearsey, 1984; Van Houten and Newlon, 1990; Deshpande and Newlon, 1992) and it has been suggested that this region serves as a binding site for an as yet unidentified DNA-initiation protein (Campbell and Newlon, 1991). Conserved sequences located at the 3' end of the core-consensus sequence have also been shown to be required for *ARS* function. These sequences, termed domain B, are more variable in both length and sequence context than the core consensus. However, their precise function in replication is not agreed upon yet. In contrast to the sequences located at the 3' end of the core consensus, there is no general requirement for sequences at the 5' end of the core consensus. Nevertheless, in plasmid assays, a few *ARS* elements appear to be affected by deletions in this region (reviewed in Campbell and Newlon, 1991).

Circular plasmids containing only an *ARS* do not segregate efficiently into progeny cells. In most cells, both copies of the replicated plasmid remain in the mother cell leaving the daughter cell plasmid-free. Therefore, *ARS* plasmids are frequently lost unless selection for the plasmid marker is used to maintain only the plasmid-containing population of cells. As a result of this transmission inefficiency, cells transformed with *ARS*-containing plasmids frequently acquire many copies of the plasmid. In addition, these cells often produce small slow-growing colonies on selective media, since not every progeny cell retains plasmid and is capable of reproduction (Stinchcomb *et al.*, 1979; Hsiao and Carbon, 1979; Clarke and Carbon, 1980; Murray and Szostak, 1983a).

Centromeres are the sites of kinetochore formation and chromosome attachment to mitotic and meiotic spindles. Centromeric DNA (*CEN*), the second *cis*-acting functional sequence element that was isolated, was first obtained by cloning the region near the *CDC10* gene, which genetically maps very close to the chromosome III centromere (Clarke and Carbon, 1980). One of the isolated sequences enabled *ARS*-containing plasmids to be inherited stably. Plasmids containing this sequence still transformed cells at high efficiency; however, transformants contained only 1–3 plasmid copies in each cell and plasmid loss was infrequent. As a result, transformants containing these sequences grew with near-normal doubling times and gave rise to normal-size colonies. In addition, plasmids containing this sequence exhibited proper first- and second-division segregation in approximately half of the meiotic divisions (Clarke and Carbon, 1980). Similar results were found using a sequence that was close to the centromere of chromosome IV (Stinchcomb *et al.*, 1982). Genetic properties exhibited by plasmids containing these sequences, combined with the close proximity of the sequences to the genetically defined centromere, made it apparent that *CEN* DNA had been cloned. Deletion of these sequences from the chromosome, or conditional inactiviation of their function, caused rapid chromosome loss (Clarke and Carbon, 1983; Hill and Bloom, 1987). These experiments confirmed that these DNA sequences were functional centromeres.

The ability of this sequence to impart mitotic stability and enable rapid growth of transformants led to cloning and characterization of 12 out of the 16 centromeric DNA sequences from *Sacch. cerevisiae* (Hsiao and Carbon, 1981; Panzeri and Philippsen, 1982; Maine *et al.*, 1984; Hieter *et al.*, 1985; Neitz and Carbon, 1985; Huberman *et al.*, 1986). Sequencing all of the cloned *CEN*s and mutational analysis of several of them have indicated that all *CEN* DNAs are closely related but not identical and are 111–119 nucleotides long (Fitzgerald-Hayes *et al.*, 1982; Hieter *et al.*, 1985; Huberman *et al.*, 1986).

Centromere-swapping experiments indicated that different centromeres were interchangable and that orientation with respect to the rest of the chromosome was not important for proper mitotic and meiotic function (Clarke and Carbon, 1983; Carbon and Clarke, 1984). Meiotic competition experiments in trisomes for chromosome I, where two homologues contained one *CEN* while a third homologue contained either a different or an inverted *CEN*, gave equivalent results. All of the three copies behaved identically confirming *CEN* DNA equivalence (V. Guacci and D.B. Kaback, unpublished results). The equivalence of centromere DNA suggests that these sequences do not play a role in homologue identification during meiosis I.

Sequencing indicated that each *CEN* contains three adjacent conserved DNA sequence elements (CDE) named *CDEI*, *CDEII* and *CDEIII*. The consensus *CDEI* is 8 bp long with the sequence: 5'PuTCACPuTG3'. The *CDEI* sequence is not exclusive to centromeres and has been found to be a functional element in promoters of many genes (Bram and Kornberg, 1987; Thomas *et al.*, 1989; Vogel *et al.*, 1989; Mellor *et al.*, 1990; Ogawa and Oshima, 1990). The *CDEII* consists of a 78–86 bp highly variable sequence that is 90% A + T while the consensus *CDEIII* is 25 bp long with the sequence:

$$5'\text{TGT A/T T A/T TGNNTTCCGAANNNNNAAA}3'$$

(Fitzgerald-Hayes *et al.*, 1982; Hieter *et al.*, 1985; Huberman *et al.*, 1986). A 125 bp sequence composed of the three *CDE* elements is sufficient to produce complete mitotic and meiotic *CEN* function (Cottarel *et al.*, 1989).

Mutational analysis of each sequence motif indicated that *CDEII* and *CDEIII* are essential to maintain both plasmid and chromosome stability during mitotic growth. Deletion of either one of these elements eliminates centromere function (Carbon and Clarke, 1984; Hegemann *et al.*, 1986; Panzeri *et al.*, 1985; McGrew *et al.*, 1986; Gaudet and Fitzgerald-Hayes, 1987). In addition, point mutations in *CDEIII* have been found that completely eliminate *CEN* function (McGrew *et al.*, 1986; Gaudet and Fitzgerald-Hayes, 1989; Jehn *et al.*, 1991). This result combined with nuclease protection and gel-shift experiments suggested that *CDEIII* forms a specific protein complex (Ng and Carbon, 1987). The *CDEII* sequence can be replaced with other A + T-rich sequences and appears to be made of bent DNA that may have some functional significance (Ng *et al.*, 1986; Cumberledge and Carbon, 1987; Murphy *et al.*, 1991). Finally, the orientation of *CDEIII* with respect to the other elements is critical for proper centromere function (Murphy *et al.*, 1991).

The *CDEII* and *CDEIII* sequences are believed to be essential for meiotic function. Unfortunately, chromosomes carrying *CDEIII* mutations are not stable mitotically and cannot be easily tested for meiotic function. However, plasmids and chromosomes carrying *CDEII* alleles with partial mitotic defects showed precocious meiosis-I segregation suggesting that *CDEII* is required to keep sister chromatids together during meiosis I (Cumberledge and Carbon, 1987; Murphy *et al.*, 1991).

The role of *CDEI* is more enigmatic, since mutations in this region, including deletions, have little or no effect on either stability of plasmids or mitotic segregation of chromosomes (Panzeri *et al.*, 1985; Cumberledge and Carbon, 1987; Hegemann *et al.*, 1988; Gaudet and Fitzgerald-Hayes, 1989). It has been reported that *CDEI* mutations cause a mitotic delay in the cell cycle (Spencer and Hieter, 1992). Furthermore, mutations in the non-essential *CBF1* gene (also known as *CEP1* and *CPF1*), which encodes a helix-loop-helix protein that binds to *CDE1*, cause chromosome instability (Baker and Masison, 1990; Cai and Davis, 1990; Mellor *et al.*, 1990). These results indicate that *CDE1* is indeed required for optimal centromere function. It has been suggested that *CDE1* is involved in the efficient recruitment of the microtubule to the kinetochore (Spencer and Hieter, 1992). However, a definitive mitotic role for this sequence has not yet been found.

The role of *CDEI* during meiosis is also not clearly defined. *CEN* plasmids with *CDEI* mutations exhibit high levels of precocious disjunction at meiosis I. However, chromosomes containing *CDEI* mutations did not exhibit this behaviour (Panzeri *et al.*, 1985; Cumberledge and Carbon, 1987; Hegemann *et al.*, 1988; Gaudet and Fitzgerald-Hayes, 1989). The absence of an effect on meiotic chromosomes has been attributed to other *CDEI* sequences located elsewhere on the same chromosome that can substitute for the centromere *CDEI*. However, further experimentation is required to verify this hypothesis and to determine precisely the role of *CDEI* in meiosis.

There is a large effort underway aimed at isolating proteins that interact with *CEN* DNA to form a functional kinetochore. Chromatin studies have shown that the core 120 together with approximately 50 nucleotides on each side are largely nuclease-resistant. Nucleosomes are precisely positioned on both sides of this nuclease-resistant core (Bloom and Carbon, 1982; reviewed in Bloom *et al.* (1989) and Bloom and Yeh (1989)). In addition, centromeres produce a characteristic *in vivo* footprint that suggests the presence of specific protein-binding sites (Denzmore *et al.*, 1991). Coincidentally, the length of the nuclease-resistant core is equal to the diameter of a microtubule. However, experiments using a different centromere suggested that the nuclease-resistant region was 20% smaller (Funk *et al.*, 1989). Mutations that destroy mitotic *CEN* function alter the chromatin structure and the observed *in vivo* "footprint" (Saunders *et al.*, 1988; Denzmore *et al.*, 1991). A genetically engineered centromere flanked by restriction-endonuclease linkers can be excised from chromatin using the enzyme *Bam*HI (Kenna *et al.*, 1988). Isolation of native centromere complexes using this construct may provide a novel approach to characterizing centromere-binding proteins from *Sacch. cerevisiae*.

As already stated, a *CDEI*-binding protein, namely Cbf1, and its corresponding gene, *CBF1*, have been reported (Bram and Kornberg, 1987;

Jiang and Philippsen, 1989; Baker and Masison, 1990; Cai and Davis, 1990; Mellor *et al.*, 1990). In addition, a *CDEIII*-binding complex, called Cbf3, comprising three proteins has been purified by *CEN* DNA affinity chromatography (Lechner and Carbon, 1991). The genes that encode the Cbf3 complex have been isolated (W. Jiang, J. Lechner and J. Carbon, personal communication). The genes for two of the three proteins (*CBF2* and *CBF3C*) have been analysed and were found to be essential. The *CBF2* gene encodes a 110 kDa protein previously called Cbf3a and is identical with the *NDC10* gene defined by a thermosensitive mutant defective in nuclear division (Goh and Kilmartin, 1993; Jiang *et al.*, 1993). The *CBF3C* gene encodes a 58 kDa protein and is identical with the *CTF13* gene defined by a chromosome-segregation defect (Doheny *et al.*, 1993; J. Carbon and J. Lechner, personal communication). The CBF3 protein complex appears to link centromere DNA to microtubules and has an *in vitro* microtubule motor activity. The motor appears to propel itself towards the minus end of a microtubule, consistent with current models for anaphase movement of chromosomes (Hyman *et al.*, 1992). Thus, it is likely that this complex is an important component of the yeast kinetochore.

In vitro binding of *CEN*-containing plasmids to microtubules has also been reported. Complex formation was dependent upon the presence of cell extracts from *Sacch. cerevisiae*. Furthermore, the ability of the extract to promote microtubule binding was dependent on the stage of the cell cycle from which the cell extract was made (Kingsbury and Koshland, 1991). Whether or not this activity is the same as that observed with *CEN* DNA affinity-purified complex has yet to be determined.

The third class of functional elements isolated from yeast chromosomes were the telomeres (*TEL*), specialized sequences that enable complete replication of the ends of chromosomes and prevent their degradation within the cell. Telomeres were first isolated from the macronucleus of the ciliate *Tetrahymena thermophila* (Blackburn and Gall, 1978). Short restriction fragments containing these sequences, when ligated to both ends of linearized ARS-containing plasmid DNA molecules, produced linear plasmids. These linear plasmids replicated and were maintained as linear DNA when transformed into *Sacch. cerevisiae* (Szostak and Blackburn, 1982). Normally linear DNA introduced into *Sacch. cerevisiae* is either integrated into a chromosome or lost. These first linear constructs were then used to isolate yeast telomeres by finding *Sacch. cerevisiae* DNA sequences that would substitute for the ciliate telomeres. The yeast sequences were shown to come from the ends of chromosomes (Szostak and Blackburn, 1982). It was also discovered that the ciliate telomeres initially introduced into yeast were replaced in progeny chromosomes by yeast telomeric sequences, suggesting that telomere formation might occur *de novo* (Shampay *et al.*, 1984).

Characterization of the yeast sequences indicated that yeast *TEL* DNA was comprised of simple repeated sequences that are similar but not identical to the ciliate sequences (Walmsley *et al.*, 1984). The *TEL* sequence in *Sacch. cerevisiae* is [C(1–3)A]n, where n is approximately 100. However, n can vary depending on the strain background. In addition, several mutations are known that have considerable effects on telomere length (Carson and Hartwell, 1985; Lustig and Petes, 1986; Lundblad and Szostak, 1989; Conrad *et al.*, 1990; Lustig *et al.*, 1990; Hardy *et al.*, 1992; Kyrion *et al.*, 1992).

Telomeres do not appear to be packaged into conventional nucleosomes but, instead, show a diffuse region of nuclease protection limited to the C(1–3)A repeat (Wright *et al.*, 1992). Gel-retardation studies and DNA-affinity chromatography identified an abundant protein, which binds to C(1–3)A. This protein, called Rap1, copurifies with the nuclear scaffold and also binds *in vitro* to other sites including transcriptional control sequences on some ribosomal protein genes, and the sequences needed to repress silent mating-type loci (Berman *et al.*, 1986; Shore and Nasmyth, 1987; Buchman *et al.*, 1988; Hoffmann *et al.*, 1989; Conrad *et al.*, 1990; Lustig *et al.*, 1990; Moehle and Hinnebusch, 1991). Immunofluorescent studies indicate that, *in vivo*, Rap1 is most abundant near the ends of chromosomes. Furthermore, it appears that chromosome ends are attached to the nuclear periphery (Klein *et al.*, 1992). The *RAP1* gene has been cloned and *rap1* mutations shown to affect telomere length (Shore and Nasmyth, 1987; Buchman *et al.*, 1988; Hoffmann *et al.*, 1989; Conrad *et al.*, 1990; Lustig *et al.*, 1990; Kyrion *et al.*, 1992). Based on these data, Rap1 is likely to be important for telomere structure and function. Interactions between Rap1 and telomeres also may be important in organizing chromosomes within the nucleus.

Several genes that affect telomere length have been characterized. The first, *CDC17*, encodes a subunit of DNA polymerase I (Carson, 1987). Mutations in this gene cause many phenotypes including longer telomeres at the permissive temperature as well as the more familiar thermosensitive block in DNA synthesis (Carson and Hartwell, 1985). The second, *EST1*, has the mutant phenotypes of telomere shortening and early cell senescence (Lundblad and Szostak, 1989). Based on the proposed mechanism for telomere synthesis, *EST1* is a likely candidate for encoding an enzyme directly involved in telomere synthesis. The third, *RIF1*, was identified using the two-hybrid screen of Fields and Song (1989), as a gene that produced a protein that interacted with Rap1. Mutants of *RIF1* appear to have phenotypes similar to *RAP1* mutants, suggesting that their two gene products interact (Hardy *et al.*, 1992). Finally, *TEL1* and *TEL2* are two genes defined by mutants that produce shorter telomeres but otherwise have no additional obvious phenotypic defects (Lustig and Petes, 1986).

Isolation of *ARS*, *CEN* and *TEL* DNA enabled the construction of synthetic *Sacch. cerevisiae* chromosomes. Incorporation of a *CEN*, an *ARS* and a selectable marker such as the *URA3* gene on an appropriately sized linear bacteriophage lambda DNA molecule bounded by telomeric repeat sequences yielded a DNA molecule capable of acting as a functional chromosome when introduced into vegetatively growing cells (Murray and Szostak, 1983b). When these artificial chromosomes were 45 kb or larger, they exhibited high mitotic stability and were transmitted normally in at least 98% of all cell divisions. When below 45 kb in length, the artificial chromosomes showed mitotic instability. Minichromosomes made of lambda multimers that were approximately 150 kb in length exhibited mitotic stability, which approached that of normal chromosomes (Murray *et al.*, 1986). Similar results were obtained using chromosome fragments 100 kb long or greater (Surosky and Tye, 1985; Surosky *et al.*, 1986; Zakian *et al.*, 1986).

During meiosis, lambda-based artificial chromosomes recombine at lower than normal rates and segregate at meiosis I by distributive disjunction (Dawson *et al.*, 1986; Mann and Davis, 1986), that is they segregated from either homologous or non-homologous chromosomes in the absence of crossing-over. Distributive disjunction occurred 80–90% of the time and was not a feature distinct to synthetic chromosomes, since native yeast chromosomes were also shown to segregate by distributive disjunction (Guacci and Kaback, 1991). Incorporation of 12-kb stretches of yeast DNA enhanced the ability of two homologous lambda-based artificial chromosomes to recombine and segregate during meiosis (Ross *et al.*, 1992). In these experiments, the recombinant lambda-based artificial chromosomes segregated from each other nearly 100% of the time at meiosis I. However, the total fraction of recombinant minichromosomes was still low and non-recombinant molecules presumably segregated from each other by distributive disjunction at the lower efficiency. It is not known why artificial chromosomes comprised primarily of lambda-DNA functioned like native chromosomes during mitosis but not during meiosis. However, two artificial chromosomes containing greater than 225 kb of homologous human DNA appear to recombine and segregate from each other in most meioses (Sears *et al.*, 1992). Furthermore, functional chromosome fragments as small as 135 kb also recombine and segregate from each other close to 100% of the time (Murray and Surosky, 1986; Zakian *et al.*, 1986; Surosky and Tye, 1988; Guacci and Kaback, 1991; Kaback *et al.*, 1992). Therefore, it is possible that the bacteriophage DNA sequences impede normal meiotic functioning of these artificial chromosomes.

Construction of linear artificial chromosomes led to the development of yeast artificial chromosome vectors or YACs (Burke *et al.*, 1987). These vectors enable large pieces (50–2000 kb) of chromosomal DNA from higher eukaryotes to be introduced into *Sacch. cerevisiae* and manipulated experimentally. These vectors have been used extensively to clone large segments of genomes of many different organisms including *Schizosaccharomyces pombe* (see p. 204), nematodes, fruit flies and humans (reviewed in Schlessinger, 1990).

In summary, the genome of *Sacch. cerevisiae* has proved to be incredibly amenable to molecular analyses and has led to characterization of elements that permit proper mitotic function. While it is not yet clear whether the nature of every element will be conserved in higher organisms, it has provided an important paradigm for examining chromosome function in all organisms.

III. THE GENOME OF *SCHIZOSACCHAROMYCES POMBE*

A. Introduction

The nuclear genome of the fission yeast *Schizosaccharomyces pombe* is also well characterized and is almost the same size as that of *Sacch. cerevisiae*. Electrophoretic karyotyping indicates that *Schiz. pombe* contains three chromosomes of 5.7, 4.6 and 3.5 Mb, corresponding to chromosomes I, II and III, respectively (Smith *et al.*, 1987; Fan *et al.*, 1989, 1991). Unlike the chromosomes of *Sacch. cerevisiae*, those in *Schiz. pombe* are large enough and condense sufficiently to be visible during mitosis and meiosis. Thus, *Schiz. pombe* presents a cytogenetic bridge for examining chromosome structure and function that is not found in *Sacch. cerevisiae*.

B. Genome mapping and cloning

A physical map using the rare-cutting endonucleases *Not*I and *Sfi*I has been generated for each of the three chromosomes of *Schiz. pombe* and many genetic markers have been physically located on these maps (Fan *et al.*, 1989, 1991). In addition, the entire genome of *Schiz. pombe* has been placed in an ordered library of 1248 YAC clones in *Sacch. cerevisiae* with an average insert size of 535 kb. A minimal subset of 26 YAC clones was selected, which was sufficient to cover the entire genome (Maier *et al.*, 1992). Efforts at placing the genome in ordered sets of cosmid and P1 clones also are in progress (Maier *et al.*, 1992; Hoheisel *et al.*, 1993; Mizukami *et al.*, 1993).

C. The genetic map

The genetic map of *Schiz. pombe* currently contains more than 500 loci dispersed on three linkage groups that total about 2000 cM of recombinational map distance (Gygax and Thuriaux, 1984; Kohli, 1987; Lennon and Lehrach, 1992). A database containing defined genetic loci is available (Lennon and Lehrach, 1992).

The map of the largest chromosome is at least 800 cM long indicating that this chromosome undergoes an average of 16 cross-overs in each meiosis. The map of the smallest chromosome is at least 400 cM long indicating it undergoes an average of eight cross-overs during each meiosis (Gygax and Thuriaux, 1984). The total number of cross-overs for each chromosome is very large compared with other organisms and is 2–3-fold higher than for chromosomes of *Sacch. cerevisiae*. Cross-over (chiasma) interference, the mechanism that prevents further cross-overs in regions that have already undergone crossing-over, is not detected in *Schiz. pombe* (Snow, 1979; Munz *et al.*, 1989). Therefore, it is possible that the large number of cross-overs for each chromosome could be partially due to lack of cross-over interference. Interestingly, *Schiz. pombe* and several other fungi that also do not exhibit cross-over interference do not appear to produce meiotic synaptonemal complexes (Olson *et al.*, 1978; Egel-Mitani *et al.*, 1982; Bähler *et al.*, 1993). This correlation has led to the suggestion that cross-over interference is somehow linked to the appearance of a synaptonemal complex (Egel, 1978).

Two interesting *cis*-acting sequences, which affect meiotic recombination, have been described. A single-point mutation in the *ade6* gene, termed M26, causes 5–10-fold higher levels of gene conversion (Gutz, 1971; Ponticelli *et al.*, 1988; Schuchert and Kohli, 1988). The molecular mechanism of the enhanced recombination is currently under investigation (Ponticelli and Smith, 1992). A meiotic recombination cold spot has also been found. The 15 kb region between *mat2* and *mat3* undergoes virtually no meiotic recombination (Egel, 1984). *Trans*-acting mutations that decrease homothallic interconversion of mating-type genes restore meiotic recombination to this region (Klar and Bonaduce, 1991). These results have been used to suggest that the mating-type region is packaged in the nucleus in a different conformation that prevents the normal interaction of homologous chromosomes (Klar, 1992).

D. Genome organization

Examination of several sequences cloned from *Schiz. pombe* suggests that the distribution of genes resembles that in *Sacch. cerevisiae*. Unlike *Sacch. cerevisiae*, it has been estimated that a large fraction of the genes in *Schiz. pombe* have introns. Of about 100 genes surveyed, 36 contained introns. On average, these introns were significantly smaller than those found in *Sacch. cerevisiae* and averaged only 100 bp in length (reviewed in Prabhala *et al.*, 1992). In fact, *Schiz. pombe* contains one of the smallest known naturally occurring introns (35 bases; Azuma *et al.*, 1991) Introns in *Schiz. pombe* are scattered throughout genes and multiple introns are not rare. The branch-point sequence that is almost completely invariant in *Sacch. cerevisiae* is somewhat more variant. Thus, with the exception of their size, introns in *Schiz. pombe* more closely resemble those found in higher eukaryotes than *Sacch. cerevisiae* (Prabhala *et al.*, 1992).

Approximately 1 Mb of chromosome III is made up of about 100 10.4 kb rDNA repeats that contain genes for 18S, 26S and 5.8S RNA (Umesono *et al.*, 1983; Toda *et al.*, 1984). Cytology indicated that this rDNA was divided between both ends of the chromosomal DNA molecule. One of the two rDNA gene clusters was genetically mapped by integrating the *leu1* gene from *Schiz. pombe* into the rDNA cluster and demonstrating tight linkage to the *ade5* gene (Toda *et al.*, 1984). A few copies of the rDNA have also been detected by hybridization to YAC clones from chromosomes I and II (Maier *et al.*, 1992). However, it is not known whether these represent functional rRNA genes.

Organization of the 5S RNA genes in *Schiz. pombe* has also been investigated. In contrast to *Sacch. cerevisiae* and many other fungi, where 5S genes are regularly interspersed with the genes encoding the other rRNA species (Rubin and Sulston, 1973; Valenzuela *et al.*, 1977; Garber *et al.*, 1988; Howlett *et al.*, 1992), 5S RNA genes in *Schiz. pombe* are located in clusters unlinked to other rRNA genes (Shaak *et al.*, 1982). This arrangement is more typical of higher organisms but has also been found in the ascomycete, *Neurospora crassa* (Metzenberg *et al.*, 1985).

Two putative transposons, *Tf1* and *Tf2*, have been identified in *Schiz. pombe* (Levin *et al.*, 1990). These transposons possess canonical terminal repeats and homologies to retrovirus proteins. Their gene organization puts them both in the *gypsy* family. These transposons are found in various locations in different strains and induction of *Tf1* transcription from heterologous promotors induces transposition (Levin and Boeke, 1992). These results are consistent with their being retrotransposons.

E. DNA sequences involved in chromosome function

DNA fragments from *Schiz. pombe* DNA have been isolated, which confer high-frequency transformation to plasmids and thus resemble *ARS* sequences from *Sacch. cerevisiae* (Beach and Nurse, 1981; Losson and Lacroute, 1983; Sakaguchi and Yamamoto, 1982). However, it is not known if these sequences serve as origins of DNA replication. Evidence for a chromosomal origin has been found using two-dimensional gel electrophoresis (Zhu *et al.*, 1992). Initiation of DNA synthesis did not appear to occur at a discrete site but was distributed over a distance of a few kilobases. Similar models have been presented for initiation sites in higher eukaryotes. However, at the present time, it should be emphasized that the true nature of origins of DNA synthesis in higher eukaryotes remains controversal.

Autonomously replicating sequences (*ARS*) in *Schiz. pombe* occur at approximately 20 kb intervals, are A + T rich and have a core consensus sequence that is similar but not identical with the *ARS* core consensus sequence from *Sacch. cerevisiae* (Wright *et al.*, 1986; Maundrell *et al.*, 1988). It appears that this core consensus region is not required for *ARS* function as measured by high-frequency transformation. Since the consensus is found in all sequenced *ARS* elements, it is possible that it still plays a role if and when *ARS* elements function in the chromosome. Finally, a sequence has been isolated that stabilizes *ARS*-containing plasmids in *Schiz. pombe* by improving mitotic partitioning (Heyer *et al.*, 1986). The mechanism of this stabilization remains to be determined.

Centromeric DNA sequences from all three chromosomes of *Schiz. pombe* have been characterized (Clarke *et al.*, 1986; Nakeseko *et al.*, 1986). They impart mitotic stability and predictable meiotic segregation to both *ARS*-containing circular minichromosomes and linear YACs transformed into *Schiz. pombe*. However, these sequences are much larger and have a more complex structure than centromeres of *Sacch. cerevisiae* (Clarke *et al.*, 1986; Nakaseko *et al.*, 1986, 1987; Chikashige *et al.*, 1989; Hahnenberger *et al.*, 1989; reviewed in Clarke (1990) and Schulman and Bloom (1991)). Centromeric DNA of *Schiz. pombe* ranges in size from 33 kb to 100 kb and comprises a 5–15 kb central core surrounded by one half of a large inverted repeat segment on each side. The central cores are unique to each centromere. The inverted repeat is made up of groups of three or four different repeated sequences. These repeated sequences are largely confined to the three centromeres. The composition of repeated sequences varies between centromeres and between the same centromere in different strains. The presence of these repeated sequences suggests that *Schiz. pombe* centromeres have elements that closely resemble the heterochromatic repeated DNA found in centromeres of higher eukaryotes.

Centromere deletion experiments suggest that the core contains nonessential segments or functional redundancy, since half of it can be removed without impairing its ability to stabilize circular and linear minichromosomes (Clarke and Baum, 1990; Hahnenberger *et al.*, 1991). Alternatively, the segregation of minichromosomes might require attachment of fewer microtubules than full-length chromosomes, enabling a much smaller central core to perform adequately in this assay. Deletion experiments involving a large part of the long inverted repeat sequences indicated that these sequences are not essential for proper mitotic function but are required for maintenance of sister-chromatid attachment during meiosis I (Clarke and Baum, 1990; Hahnenberger *et al.*, 1991).

Sequence analysis of a large fraction of *cen1* and *cen2* indicated that one of the repeats was composed of clusters of tRNA genes (Kuhn *et al.*, 1991). The DNA of *cen1* is surrounded by six tRNA genes, while *cen2* is surrounded by 22 tRNA genes. For both *cen* sequences, the tRNA genes directly flank or extend into the central core. There are no genes encoding poly(A)-containing RNA within 35 kb of the central region of *cen2* (Fishel *et al.*, 1988). Indeed, insertion of the *ura4* gene within the region prevents normal expression of that gene (R. Allshire, personal communication). As with *CEN* DNA in *Sacch. cerevisiae*, meiotic recombination is also repressed close to *Schiz. pombe* centromeres (Clarke *et al.*, 1986). Finally, chromatin structure in the central core but not the flanking repeats is different from the nucleosomal repeat of the bulk chromatin (Pollizi and Clarke, 1991).

Telomeres have been characterized from *Schiz. pombe*. They have a simple repeat element that is different from that in DNA of *Sacch. cerevisiae* with the more complex repeat unit C(1-6)G(0-1)T(0-1)GTA(1-2) (Sugawara and Szostak, 1986, cited in Matsumoto *et al.*, 1987).

IV. OTHER YEAST GENOMES

A. Introduction

Genomes of many different yeasts, including members of the ascomycetous, basidiomycetous and deuteromycetous families, have been investigated using pulsed-field gel electrophoresis (deJonge *et al.*, 1986; Johnston *et al.*, 1988; Stoltenburg *et al.*, 1992; Passoth *et al.*, 1992). Yeast species that are not inbred showed large variations in their electrophoretic karyotypes making generalizations difficult for certain species. In general, genome sizes range from 10 to 20 Mb. Since many of these yeasts are not characterized genetically, genomes that are 20 Mb long may simply be due to diploidy combined with isomorphic species of most chromosomes. However, it is still possible that genomes of some yeast species are significantly larger than others. Most yeasts have 5–15 chromosomal bands that are between 1 and 4 Mb long. Occasionally, one or two small chromosome bands of 50–250 kb can be observed. In cases where haploidy and diploidy have been established genetically, it is clear that many homologous chromosomes have polymorphic lengths. In cases where DNA per cell is significantly greater than the masses of all chromosomes combined, it is possible the yeast being examined is indeed diploid. Pulsed-field karyotypes also provide a convenient yeast-typing marker that should prove important in both phylo-genetic and clinical studies. The genomes of several of the more widely investigated pathogenic and industrial yeasts are described below. Each was selected because it has provided important new insights in genome research. Descriptions of several important yeasts are clearly omitted. This omission is a reflection of my own bias rather than a measure of the usefulness of these organisms.

B. Pathogenic yeasts

1. *Candida albicans*

The dimorphic yeast *Candida albicans* is part of the normal human flora and an important opportunistic pathogen. Accordingly, it has been the subject of increased investigation over the past few years. Its genetics have been reviewed recently (Scherer and Magee, 1990; Poulter, Chapter 8, this volume). Since this organism is an imperfect fungus, much of what is known about its genetics comes from molecular analysis. However, para-sexual genetics have been used to define several linkage groups (Poulter *et al.*, 1982; Hilton *et al.*, 1985). Electrophoretic karyotypes combined with blot hybridizations using many cloned probes have led to the conclusion that *C. albicans* is normally diploid and contains eight pairs of chromosomes ranging from 1.2 to 3 Mb (Magee and Magee, 1987; Lasker *et al.*, 1989; Wickes *et al.*, 1991). As a result of chromosome-size polymorphisms, most strains produce more than seven electrophoretic bands (Magee and

Magee, 1987; Magee *et al.*, 1988; Lasker *et al.*, 1989; Wickes *et al.*, 1991). Some of these chromosome-length polymorphisms have been attributed to translocations (Thrash-Bingham and Gorman, 1992). Totalling the average molecular weight for each chromosome indicates that the haploid genome size is 16 Mb, which includes 40–80 units of the 12–14 kb tandem rDNA repeat. Polymorphisms present in single isolates account for the size range of the rDNA repeat (Magee *et al.*, 1987). In addition, some strains contain a very small chromosome. Karyotype instability has also been observed in laboratory strains (Kelly *et al.*, 1987; Suzuki *et al.*, 1989).

Many genes have been isolated from *C. albicans* by virtue of their ability to complement mutations in *Sacch. cerevisiae* (reviewed in Scherer and Magee, 1990). Cloned DNA sequences appear to integrate homologously when transformed into *C. albicans* (Kurtz *et al.*, 1986). In addition, sequences have been isolated from *C. albicans* that impart high-frequency transformation to plasmids and may be analogous to *ARS* elements in *Sacch. cerevisiae*. However, unlike *Sacch. cerevisiae*, plasmids transformed into *C. albicans* replicate as multimers (Kurtz *et al.*, 1987). Finally, a retrotransposon-like element designated "alpha" has also been described (Chen and Fonzi, 1992).

2. Cryptococcus neoformans

Cryptococcus neoformans, also known as *Filobasidiella neoformans*, is a pathogenic encapsulated yeast that can cause systemic mycoses commonly involving the respiratory and central nervous systems. Some classical genetic analysis has been carried out on determinants of the capsule and a few auxotrophic mutations and cloned genes have been reported. Three studies reported characterization of the electrophoretic karyotype of this basidiomycete (deJonge *et al.*, 1986; Perfect *et al.*, 1989; Polachek and Lebens, 1989). These studies indicated that *C. neoformans* contains 10–12 chromosomes and that the total genome is 15–17 Mb long. The distribution of electrophoretic bands was different in the various serotypes analysed. Thus, electrophoretic karyotyping may be useful in taxonomic and epidemiological studies.

A system for transforming *C. neoformans* has been described that uses the homologous *URA5* gene and electroporation (Varma *et al.*, 1992; Edman, 1992). While these plasmids transform efficiently, sequences providing the ability to replicate autonomously have not been characterized. Interestingly, linearized vectors transform more efficiently than circular ones. Investigation of the transformants revealed that telomere-like structures are added *in vivo* to the transforming DNA. This added DNA has been characterized. Based on a limited sample, it appears that the telomere repeat is AG(3–5)T(2). Similar *de novo* addition of telomeres has been observed in the pathogenic yeast, *Histoplasma capsulatum* (Woods and Goldman, 1992).

C. Industrial yeasts

1. *Kluyveromyces* spp.

The genomes of several species from the genus *Kluyveromyces* have been investigated. Most is known about *K. marxianus* and *K. lactis*. It has been suggested that these two species are homothallic and heterothallic variants of the same species (Sor and Fukuhara, 1989). However, Steensma *et al.* (1988) believe them to be distinct species. DNA from more than 25 species of the genus *Kluyveromyces* were examined on pulsed-field gels and were found to contain an average of seven bands (deJonge *et al.*, 1986; Johnston *et al.*, 1988; Steensma *et al.*, 1988; Sor and Fukuhara, 1989). Several strains had as many as 13 or as few as three electrophoretic bands. Estimates of genome size based on pulsed-field gels indicate that, on average, members of the genus *Kluyveromyces* contain 12 ± 3.5 Mb of DNA. *Kluyveromyces waltii* appeared to contain the smallest genome with only 7 Mb while *K. marxianus* strain CBS1553 appeared to contain the largest genome with 18 Mb of DNA (Sor and Fukahara, 1989). The apparent small genome size of *K. waltii* might be due to the presence of several chromosomes of similar size not separating under the electrophoretic conditions used. The large genome size of *K. marxianus* CBS1553 genome may be due either to diploidy or to partial aneuploidy. In addition, *Candida pseudotropicalis*, which may be a member of the genus *Kluyveromyces*, appeared to contain 22–27 Mb of DNA in each cell (Sor and Fukuhara, 1989). This high content of genomic DNA may be a property of polyploidy or aneuploidy.

   The arrangement of the rRNA gene cluster in *K. lactis* is very similar to that in *Sacch. cerevisiae*. The 5S rRNA genes are interspersed between the the genes encoding the large precursor that is processed into 18S, 26S and 5.8S rRNA (Verbeet *et al.*, 1984). Slow-growing mutants lacking half of the rDNA have been obtained (Maleszka and Clark-Walker, 1989). Examination of fast-growing revertants indicated that rDNA had reverted to the wild-type level (Maleszka and Clark-Walker, 1990). This phenomenon closely resembled the rDNA magnification first observed in fruit-fly *minute* mutants (Ritossa, 1972), in *Sacch. cerevisiae* (Kaback *et al.*, 1976; Kaback and Halvorson, 1977) and in *Neurospora crassa* (Russell and Rodland, 1986).

In addition to chromosomal DNA, *K. drosophilarum* contains a 1.6 μm plasmid termed pKD1. This plasmid has a functional organization similar to 2 μm circular DNA in *Sacch. cerevisiae* and has been used to construct vectors for transformation of *K. lactis* (Bianchi *et al.*, 1987). *ARS*-containing DNA has also been isolated from the linear killer plasmids found in certain strains of *K. lactis* (Thompson and Oliver, 1986; Fujimura *et al.*, 1987; reviewed in Stark *et al.*, 1990).

Chromosomal elements in *Kluyveromyces* spp. bear a striking resemblance to those found in *Sacch. cerevisiae* Chromosomal *ARS* elements from *K. lactis* were isolated by obtaining restriction fragments that imparted a high frequency of transformation to plasmids (Fabiana *et al.*, 1990). Some of these sequences act as *ARS* elements in both *K. lactis* and *Sacch. cerevisiae* while others are specific to *K. lactis*. Analysis of one *ARS* from *K. lactis* indicated that it contained the same core consensus sequence as found in *Sacch. cerevisiae*. However, deletion of this 12 bp sequence did not eliminate *ARS* function in *K. lactis*, although it did prevent it from functioning in *Sacch. cerevisiae*. Sequence studies now underway are designed to determine the properties that cause the specificity shown (L. Fabiani, personal communication).

The DNA from *K. lactis* centromeres has been isolated on the basis of its ability to stabilize *ARS* plasmids. The *CEN*s of *K. lactis* are remarkably similar to those found in *Sacch. cerevisiae*. However, they differ in sequence and cannot function heterologously. Those in *K. lactis* have two conserved sequences of nine (*CDEI*) and 26 bp (*CDEIII*) flanking a 161–164 bp A + T-rich region (*CDEII*). In addition, these *CEN*s contain a second A + T-rich sequence located upstream of the 9 bp *CDEI* sequence. Counting the extra conserved sequence and the two-fold larger *CDEII* analogue, it appears that the *K. lactis CEN* DNA is considerably longer than the approximately 120 bp *Sacch. cerevisiae* minimal *CEN* (Heus *et al.*, 1990; J. Heus and H.Y. Steensma, personal communication).

2. *Yarrowia lipolytica*

So far five genetic linkage groups have been defined by tetrad analysis for this lipophilic yeast (Ogrydziak *et al.*, 1982; D. Ogrydziak, personal communication). Unfortunately, spore viability in *Y. lipolytica* is insufficient to demonstrate first-division segregation of markers. Therefore, it has not been possible to demonstrate that that each linkage group is associated with a centromere and represents a single chromosome. However, coordinate loss of linked markers when chromosomes were lost from diploids supports the notion that *Y. lipolytica* contains five chromosomes (D. Ogrydziak, personal communication). Pulsed-field gels have revealed as few as two chromosomal bands (Johnston and Mortimer, 1986; deJonge *et al.*, 1986) and as many as six (Fournier *et al.*, 1993; D. Ogrydziak, personal communication). However, larger chromosomes would not have been resolved by the pulsed-field gel conditions where only two bands were found. In experiments revealing six bands, the possibility that different isoforms of the same chromosome were responsible for more than one band was not eliminated. Thus, chromosome number for this organism is still not certain.

Isolation of *ARS* sequences from *Y. lipolytica* has been problematical but nevertheless extremely interesting. Despite repeated attempts, only two different sequences displaying high-frequency transformation were isolated from genomic clone banks (Fournier *et al.*, 1991). Sequencing indicated that the two sequences share several regions of homology and several near matches to the *ARS* core consensus in *Sacch. cerevisiae* (Fournier *et al.*, 1993). However, it is not known whether these sequences are of functional significance. It was suggested that they contain both *ARS* elements as well as centromeric DNA since only 1–3 copies of these were present per cell. In addition, markers homologously integrated into chromosomes at sites adjacent to the putative *ARS* sequences showed tight centromere linkage during meiosis. Integration of these *ARS* sequences at other chromosomal loci in *Y. lipolytica* led to chromosome breakage. These results suggested that the integration event may have created dicentric chromosomes, which break during subsequent mitotic divisions. Plasmids containing deletions for part of these sequences lost their ability to promote high-frequency transformation but still could cause chromosome breakage (Fournier *et al.*, 1993). Thus, *ARS* and *CEN* functions on this DNA may be separable. Nevertheless, at the present time, it is not possible to conclude that these sequences represent *bona fide* centromeric DNA, since plasmids in *Y. lipolytica* did not exhibit first-division meiotic segregation. Overall, these experiments suggest that an *ARS* sequence is not sufficient to produce plasmid transformation in *Y. lipolytica* and that additional DNA such as a centromere is required for either transmission or stabilization of this DNA.

V. CONCLUSIONS

Studies on the genomes of several yeasts have produced a huge amount of information regarding placement of genes and molecular characteristics of the functional elements present. The sequences of several yeast chromosomes have recently been completed and we are nearing the point where we will have the entire nucleotide sequence of one and perhaps two different yeast genomes. A great deal has been learned about sequences involved in chromosome function so that synthetic chromosomes that work during mitotic growth have been produced. This technology has enabled the use of one yeast to become a suitable host for cloning and mapping both mammalian and other eukaryotic genomes. Nevertheless, there is still a tremendous amount that needs to be learned about chromosome structure and function. For example, it has been suggested that chromosomes are arranged in a specific manner within the nucleus. If this is true, we need to discover the arrangement, and learn what sequences or structures are involved. Furthermore, mechanisms for homologous chromosome pairing and meiotic segregation remain a mystery. Perhaps further analysis will lead to the discovery of specific sequences or chromosomal structures involved in these processes. There is little doubt that yeast chromosomes will continue to be an important research focus for many scientists. Their small size combined with the genetic malleability of both their component DNA molecules and the yeast which they inhabit assure that yeast chromosomes will remain important models for understanding the biology of these organelles from all organisms.

VI. ACKNOWLEDGEMENTS

I am most grateful to my friends and colleagues for sharing their data and providing manuscripts far in advance of publication. Particular thanks go to Rod Rothstein, Amar Klar, Carol Newlon, Marjorie Brandriss, Yde Steensma, Linda Riles, Corrinne Michels, Pete Magee, David Ogrydziak, Yuping Su and Susan Forsburg for reading parts of the manuscript, discussing their ideas, or providing either references or copies of papers from their extensive collections. I also wish to thank the members of my laboratory for many stimulating discussions. Finally, I am grateful to the office staff of the Department of Microbiology and Molecular Genetics for help with the references. Work in my laboratory is supported by grants from the National Science Foundation and the National Center for Human Genome Research of the National Institutes of Health.

References

Abovich, N. and Rosbash, M. (1984). *Molecular and Cellular Biology* **4**, 1871.
Adams, J., Puska-Rosza, S. Simlar, J. and Wilke, C.M. (1992). *Current Genetics* **22**, 13.

214   D.B. Kaback

Aparicio, O.M., Billington, B.L. and Gottschling, D.E. (1991). *Cell* **66**, 1279.
Azuma, Y., Yanagishi, M., Ueshima, R. and Ishihama, A. (1991). *Nucleic Acids Research* **19**, 461.
Bähjer, J., Wyler, T., Loidlg, J. and Kohli, J. (1993). *Journal of Cell Biology* **121**, 241.
Baker, R.E. and Masison, D.C. (1990). *Molecular and Cellular Biology* **10**, 2458.
Barton, A.B. and Kaback, D.B. (1994). *Journal of Bacteriology* **176**, 1872.
Barton, A.B., Davies, C.J., Hutchison C.V., III and Kaback, D.B. (1992). *Gene* **117**, 137.
Basson, M.E., Moore, R.L., O'Rear, J and Rine, J. (1987). *Genetics* **117**, 645.
Beach, D. and Nurse P. (1981). *Nature* **290**, 140.
Beckmann, J., Johnson, P. and Abelson, J. (1977). *Science* **196**, 205.
Bell, G.K., De Gennaro, L.J., Gelfund, D.H., Bishop, R.J., Valenzuela, P. and Rutter, W.J. (1977). *Journal of Biological Chemistry* **252**, 8118.
Berman, J., Tachibana, C.Y. and Tye, B.-K. (1986). *Proceedings of the National Academy of Sciences, USA* **83**, 3713.
Bianchi, M.M., Falcone, C., Chen, X.J., Wesolowski-Louvel, M., Frontali, L. and Fukuhara, H. (1987). *Current Genetics* **12**, 185.
Blackburn, E.A. (1991). *Nature* **30**, 569.
Blackburn, E.H. and Gall, J.G. (1978). *Journal of Molecular Biology* **120**, 33.
Bloom, K. and Carbon, J. (1982). *Cell* **29**, 305.
Bloom, K.S., Hill, A., Kenna, M. and Saunders, M. (1989). *Trends in Biochemical Science* **14**, 223.
Bloom, K. and Yeh, E. (1989). *Current Opinion in Cell Biology* **1**, 526.
Boeke, J.B. and Sandmeyer, S.B. (1991). *In* "The Molecular and Cellular Biology of the Yeast *Saccharomyces*" (J.R. Broach, J.R. Pringle and E.W. Jones, eds), Vol. I, pp. 193-261. Cold Spring Harbor Laboratory Press, Cold Spring Harbor.
Boeke, J.D., Eichinger, D., Castrillon, D. and Fink, G.R. (1988). *Molecular and Cellular Biology* **8**, 1432.
Botstein, D. (1988). "*The Harvey Lectures*", Series 82, 1986-87, pp. 157-167. Alan R. Liss, New York.
Bram, R.J. and Kornberg, R.D. (1987). *Molecular and Cellular Biology* **7**, 403.
Brewer, B.J. and Fangman, W.L. (1987). *Cell* **51**, 463.
Broach, J.R., Li, Y.-Y. Feldman, J., Jayaram, M., Abraham, J., Nasmyth, K.A. and Hicks, J.B. (1983). *Cold Spring Harbor Symposium in Quantitative Biology* **47**, 1165.
Buchman, A.R., Kimmerly, W.J., Rine, J. and Kornberg, R.D. (1988). *Molecular and Cellular Biology* **8**, 210.
Burke, D.T., Carle, G.F. and Olson, M.V. (1987). *Science* **236**, 806.
Button, L.L. and Astell, C.R. (1986). *Molecular and Cellular Biology* **6**, 1352.
Cai, M. and Davis, R.W. (1990). *Cell* **61**, 437.
Cameron, J.R., Loh, E.Y. and Davis, R.W. (1979). *Cell* **16**, 739.
Campbell, J.L. and Newlon, C.S. (1991) *In* "The Molecular and Cellular Biology of the Yeast *Saccharomyces*" (J.R. Broach, J.R. Pringle and E.W. Jones, eds), Vol. I, pp. 41-146. Cold Spring Harbor Laboratory Press, Cold Spring Harbor.
Cao, L., Alani, E. and Kleckner, N. (1990). *Cell* **61**, 1089.
Carbon, J. and Clarke, L. (1984). *Journal of Cell Science*, Suppl. **1**, 43.
Carle, G.F. and Olson, M.V. (1984). *Nucleic Acids Research* **12**, 5647.
Carle, G.F. and Olson, M.V. (1985). *Proceedings of the National Academy of Sciences, USA* **82**, 3756.
Carlson, M., Celenza, J.L. and Eng, F.J. (1985). *Molecular and Cellular Biology* **5**, 2894.
Carson, M.J. (1987). PhD Thesis: University of Washington, Seattle.
Carson, M.J. and Hartwell, L. (1985). *Cell* **42**, 249.
Chan, C.S.M. and Tye, B.-K. (1980). *Proceedings of the National Academy of Sciences, USA* **77**, 6329.
Chan, C.S.M. and Tye, B.-K. (1983a). *Cell* **33**, 563.
Chan, C.S.M. and Tye, B.-K. (1983b). *Journal of Molecular Biology* **168**, 505.
Charron, M.J., Read, E., Haut, S.R. and Michels, C.A. (1989). *Genetics* **122**, 307.
Chalker, D.L. and Sandmeyer, S.B. (1990). *Genetics* **126**, 837.
Chen, J.-Y. and Fonzi, W.A. (1992). *Journal of Bacteriology* **174**, 5624.

Chikashige, Y., Kinoshita, N., Nakaseko, Y., Matsumoto, T., Murakami, S., Niwa, O. and Yanagida, M. (1989). *Cell* **57**, 739.

Chinault, A.C. and Carbon, J. (1979). *Gene* **5**, 111.

Chisholm, G.E., Genbauffe, F.S. and Cooper, T.G. (1984). *Proceedings of the National Academy of Sciences, USA* **81**, 2965.

Christman, M.F., Dietrich, F.S. and Fink, G.R. (1988). *Cell* **55**, 413.

Clark-Walker, G.D. and Azad, A.A. (1980). *Nucleic Acids Research* **8**, 1009.

Clarke, L. (1990). *Trends in Genetics* **6**, 150.

Clarke, L. and Baum, M.P. (1990). *Molecular and Cellular Biology* **10**, 1863.

Clarke, L. and Carbon, J. (1980). *Nature* **287**, 504.

Clarke, L. and Carbon, J. (1983). *Nature* **305**, 23.

Clarke, L., Amstutz, H., Fishel, B. and Carbon, J. (1986). *Proceedings of the National Academy of Sciences, USA* **83**, 8253.

Coleman, K.G., Steensma, H.Y., Kaback, D.B. and Pringle, J.R. (1986). *Molecular and Cellular Biology* **6**, 4516.

Conrad, M.N., Wright, J.H., Wolf, A.J. and Zakian, V.A. (1990). *Cell* **63**, 739.

Cottarel, G., Shero, J.H., Hieter, P. and Hegemann, J.H. (1989). *Molecular and Cellular Biology* **9**, 3342.

Craig, E. (1989). *BioEssays* **11**, 48.

Cramer, J.H. and Rownd, R.H. (1980). *Molecular and General Genetics* **177**, 199.

Cramer, J.H., Bhargava, M.M. and Halvorson, H.O. (1972). *Journal of Molecular Biology* **71**, 11.

Cramer, J.H., Farrelly, F.W., Barnitz, J.T. and Rownd, R.H. (1977). *Molecular and General Genetics* **151**, 229.

Cross, S.L. and Smith, M.M. (1988). *Molecular and Cellular Biology* **8**, 945.

Cumberledge, S. and Carbon, J. (1987). *Genetics* **117**, 203.

Dawson, D.S., Murray, A.W. and Szostak, J.W. (1986). *Science* **234**, 713.

DeFeo-Jones, D., Scolnick, E.M., Koller, R. and Dhar, R. (1983). *Nature* **306**, 707.

deJonge, P., DeJongh, F.C.M., Meijers, R., Steensma, H.Y. and Scheffers, W.A. (1986). *Yeast* **2**, 193.

Denzmore, L., Payne, W.E. and Fitzgerald-Hayes, M. (1991). *Molecular and Cellular Biology* **11**, 154.

Derr, L.K., Strathern, J.M. and Garfinkel, D.J. (1991). *Cell* **67**, 355.

Desphande, A.M. and Newlon, C.S. (1992). *Molecular and Cellular Biology* **12**, 4305.

Devenish, R.J. and Newlon, C.S. (1982). *Gene* **8**, 277.

Diehl, B.E. and Pringle, J.R. (1991). *Genetics* **127**, 837.

Doheny, K.F., Sorger, P.K., Hyman, A.A., Tugenreich, S., Spencer, F. and Hieter, P. (1993). *Cell* **73**, 761.

Dubey, D.D., Davis, L.R., Greenfeder, S.A., Ong, L.Y., Zhu, J., Broach, J.R., Newlon, C.S. and Huberman, J.A. (1991). *Molecular and Cellular Biology* **11**, 5346.

Edman, J.C. (1992). *Molecular and Cellular Biology* **12**, 2777.

Egel, R. (1978). *Heredity* **41**, 233.

Egel, R. (1984). *Current Genetics* **8**, 199.

Egel-Mitani, M., Olson, L.W. and Egel, R. (1982). *Hereditas* **97**, 179.

Eigel, A. and Feldmann, H. (1982). *EMBO Journal* **1**, 1245.

Elion, E.A. and Warner, J.R. (1984). *Cell* **39**, 663.

Elion, E.A. and Warner, J.R. (1986). *Molecular and Cellular Biology* **6**, 2089.

Fabiani, L., Aragona, M. and Frontali, L. (1990). *Yeast* **6**, 69.

Fan, J.B., Chikashige, Y., Smith, C.L., Niwa, O., Yanagida, M. and Cantor, C.R. (1989). *Nucleic Acids Research* **17**, 2801.

Fan, J., Grothues, D. and Smith, C.L. (1991). *Nucleic Acids Research* **19**, 6289.

Fangman, W.L. and Brewer, B.J. (1992). *Cell* **71**, 363.

Fasullo, M.T. and Davis, R.W. (1988). *Molecular and Cellular Biology* **8**, 4370.

Ferguson, B.M. and Fangman, W.L. (1992). *Cell* **68**, 333.

Ferguson, B.M., Brewer, B.J., Reynolds, A.E. and Fangman, W.L. (1991). *Cell* **65**, 507.

Fields, S. and Song, O.K. (1989). *Nature* **340**, 245.

Fink, G.R. (1987). *Cell* **49**, 5.

216   D.B. Kaback

Fishel, B., Amstutz, H., Baum, M., Carbon, J. and Clarke, L. (1988). *Molecular and Cellular Biology* **8**, 754.
Fitzgerald-Hayes, M., Clarke, L. and Carbon, J. (1982). *Cell* **29**, 235.
Fogel, S. and Welch, J.W. (1982). *Genetics* **79**, 5342.
Fogel, S., Welch, J.W., Cathala, G. and Karin, M. (1983). *Current Genetics* **7**, 347.
Fournier, P., Guyaneux, L., Chasles, M. and Gaillardin, C. (1991). *Yeast* **7**, 25.
Fournier, P., Abbas, A., Chasles, M., Kudla, B., Ogrydziak, D.M., Yaver, D., Xuan, J.-W., Peito, A., Ribet, A.-M., Feynerol, C., He, F. and Gaillardin, C. (1993). *Proceedings of the National Academy of Sciences, USA* **90**, 4912.
Fujimura, H., Hishinuma, F. and Gunge, N. (1987). *Current Genetics* **12**, 99.
Funk, M., Hegemann, J.H. and Philippsen, P. (1989). *Molecular and General Genetics* **219**, 153.
Gafner, J., DeRobertis, E.M. and Philippsen, P. (1983). *EMBO Journal* **2**, 583.
Gallwitz, D. and Sures, I. (1980). *Proceedings of the National Academy of Sciences, USA* **77**, 2546.
Garber, R.C., Turgeon, B.G., Selker, E.V. and Yoder, O.C. (1988). *Current Genetics* **14**, 573.
Gaudet, A. and Fitzgerald-Hayes, M. (1987). *Molecular and Cellular Biology* **7**, 68.
Gaudet, A. and Fitzgerald-Hayes, M. (1989). *Genetics* **121**, 477.
Genbauffe, F.S., Chisholm, G.E. and Cooper, T.G. (1984). *Journal of Biological Chemistry* **259**, 10518.
Gerring, F., Connelly, C. and Hieter, P. (1991). *Methods in Enzymology* **194**, 57.
Goh, P.-V. and Kilmartin, J.V. (1993). *Journal of Cell Biology* **121**, 503.
Goodman, H.M., Olson, M.V. and Hall, B.D. (1977). *Proceedings of the National Academy of Sciences, USA* **74**, 5423.
Gottlieb, S. and Esposito, R.E. (1989). *Cell* **56**, 771.
Gottschling, D.E. (1992). *Proceedings of the National Academy of Sciences, USA* **89**, 4062.
Gottschling, D.E., Aparicio, O.M., Billington, B.L. and Zakian, V.A. (1990). *Cell* **63**, 751.
Goyon, C. and Lichten, M. (1993). *Molecular and Cellular Biology* **13**, 373.
Greenfeder, S.A. and Newlon, C.S. (1992). *Molecular Biology of the Cell* **3**, 999.
Guacci, V. and Kaback, D.B. (1991). *Genetics* **127**, 485.
Gutz, H. (1971). *Genetics* **69**, 317.
Gygax, A. and Thuriaux, P. (1984). *Current Genetics* **8**, 85.
Hadwiger, J.A., Wittenberg, C., Richardson, H.E., de Barros Lopes, M. and Reed, S.I. (1989). *Proceedings of the National Academy of Sciences, USA* **86**, 6255.
Hahnenberger, K.M., Baum, M.P., Polizzi, C.M., Carbon, J. and Clarke, L. (1989). *Proceedings of the National Academy of Sciences, USA* **86**, 577.
Hahnenberger, K.M., Carbon, J. and Clarke, L. (1991). *Molecular and Cellular Biology* **11**, 2206.
Hansen, L.J. and Sandmeyer, S.B. (1990). *Journal of Virology* **64**, 2599.
Hansen, L.J., Chalker, D.L. and Sandmeyer, S.B. (1988). *Molecular and Cellular Biology* **8**, 5245.
Harbitz, I. and Oyen, T.B. (1974). *Biochemical and Biophysical Acta* **366**, 374.
Hardy, C., Sussel, L. and Shore, D. (1992). *Genes and Development* **6**, 801.
Hauber, J., Stucka, R., Krieg, R. and Feldmann, H. (1988). *Nucleic Acids Research* **16**, 10623.
Hegemann, J.H., Pridmore, R.D., Schneider, R. and Philippsen, P. (1986). *Molecular and General Genetics* **205**, 305.
Hegemann, J.H., Spero, J.H., Cottarel, G., Philippsen, P. and Hieter, P. (1988). *Molecular and Cellular Biology* **8**, 2523.
Hereford, L., Fahrner, K., Woolford, J., Rosbash, M. and Kaback, D.B. (1979). *Cell* **18**, 1261.
Heus, J.J., Zonneveld, G.J.M., Steensma, H.Y. and Van den Berg, J.A. (1990). *Current Genetics* **18**, 517.
Heyer, W.D., Sipiezki, M. and Kohli, J. (1986). *Molecular and Cellular Biology* **6**, 80.
Hieter, P., Pridmore, D., Hegemann, J.H., Thomas, M., Davis, R.W. and Philippsen, P. (1985). *Cell* **42**, 913.
Hill A. and Bloom, K. (1987). *Molecular and Cellular Biology* **7**, 2397.

Hilton, C., Markie, D., Corner, B., Rikkerink, E. and Poulter, R.T. (1985). *Molecular and General Genetics* **200**, 162.
Hoffmann, J.F., LaRoche, T., Brand, A. and Gasser, S.M. (1989). *Cell* **57**, 725.
Hoheisel, J.G., Maier, E., Mott, R., McCarthy, L., Grigoriev, A.V., Schwalkwyk, L.C., Nizetic, D., Francis, F. and Lehrach, H. (1993). *Cell* **73**, 109.
Hogst, A. and Oyen, T.B. (1984). *Nucleic Acids Research* **18**, 7199.
Holland, M.J., Holland, J.P., Thill, G.P. and Jackson, K.A. (1981). *Journal of Biological Chemistry* **256**, 1385.
Horowitz, H. and Haber, J.E. (1984). *Nucleic Acids Research* **12**, 7105.
Howlett, B.J., Brownlee, A.G., Guest, D.I., Adcock, G.J. and McFadden, G.I. (1992). *Current Genetics* **22**, 455.
Hsiao, C.-L. and Carbon, J. (1979). *Proceedings of the National Academy of Sciences, USA* **76**, 3829.
Hsiao, C.-L. and Carbon, J. (1981). *Proceedings of the National Academy of Sciences, USA* **78**, 3760.
Huberman, J.A., Pridmore, R.D., Jaeger, D., Zonneveld, B. and Philippsen, P. (1986). *Chromosoma* **94**, 162.
Huberman, J.A., Spotila, L.D., Nawotka, K.A., El-Assoult, S.M. and Davis, L.R. (1987). *Cell* **51**, 473.
Hyman, A.A., Middleton, K., Centola, M., Mitchison, T.J. and Carbon, J. (1992). *Nature* **359**, 533.
Iwasaki, T., Shirahige, K., Yoshikawa, H. and Ogasawara, N. (1991). *Gene* **109**, 81.
Jaeger, D. and Philippsen, P. (1989). *Molecular and Cellular Biology* **9**, 5754.
Jehn, B., Niedenthal, R. and Hegemann, J.H. (1991). *Molecular and Cellular Biology* **11**, 5212.
Jemptland, R., Maehluin, E., Gabrielson, O.S. and Oyen, T.B. (1986). *Nucleic Acids Research* **14**, 5145.
Jiang, W. and Philippsen, P. (1989). *Molecular and Cellular Biology* **9**, 5585.
Jiang, W., Lechner, J. and Carbon, J. (1993). *Journal of Cell Biology* **121**, 513.
Johnston, J.R. and Mortimer, R.K. (1986). *International Journal of Systematic Bacteriology* **36**, 569.
Johnston, J.R., Contopoulou, C.R. and Mortimer, R.K. (1988). *Yeast* **4**, 191.
Jong, S.C. and Edwards, M.J. (eds) (1990). "ATCC-Catalogue of Yeasts". American Type Culture Collection, Rockville.
Kaback, D.B. and Davidson, N. (1980). *Journal of Molecular Biology* **138**, 745.
Kaback, D.B. and Halvorson, H.O. (1977). *Proceedings of the National Academy of Sciences, USA* **74**, 1177.
Kaback, D.B. and Halvorson, H.O. (1978). *Journal of Bacteriology* **134**, 237.
Kaback, D.B., Halvorson, H.O. and Rubin, G.M. (1976). *Journal of Molecular Biology* **107**, 385.
Kaback, D.B., Angerer, L.A. and Davidson, N. (1979). *Nucleic Acids Research* **6**, 2499.
Kaback, D.B., Guacci, V., Barber, D. and Mahon, J. (1992). *Science* **256**, 485.
Kaback, D.B., Steensma, H.Y. and deJonge, P. (1989). *Proceedings of the National Academy of Sciences, USA* **86**, 3694.
Kearsey, S. (1984). *Cell* **37**, 299.
Kelly, R., Miller, S.M., Kurtz, M.B. and Kirsch, D.R. (1987). *Molecular and Cellular Biology* **7**, 199.
Kenna, M., Amaya, A. and Bloom, R. (1988). *Journal of Cell Biology* **107**, 9.
Kingsbury, J. and Koshland, D. (1991). *Cell* **66**, 483.
Kingsman, A.J., Clarke, L., Mortimer, R.K. and Carbon, J. (1979). *Gene* **7**, 141.
Klar, A.J.S. (1992). *Trends in Genetics* **8**, 208.
Klar, A.J.S. and Bonaduce, M.M. (1991). *Genetics* **129**, 1033.
Klein, F., LaRoche, T., Cardena, M.E., Hoffmann, J.F.X., Schweizer, D. and Gasser, S.M. (1992). *Journal of Cell Biology* **117**, 935.
Kohli, J. (1987). *Current Genetics* **11**, 575.
Kramer, R.A., Philippsen, P. and Davis, R.W. (1978). *Journal of Molecular Biology* **123**, 405.

Kuhn, R.M., Clarke, L. and Carbon, J. (1991). *Proceedings of the National Academy of Sciences, USA* **88**, 1306.

Kurtz, M.B., Cortelyou, M.W. and Kirsch, D.R. (1986). *Molecular and Cellular Biology* **6**, 142.

Kurtz, M.B., Cortelyou, M.W., Miller, S.M., Lai, M. and Kirsch, D.R. (1987). *Molecular and Cellular Biology* **7**, 209.

Kyrion, G., Boakye, K.A. and Lustig, A.J. (1992). *Molecular and Cellular Biology* **12**, 5159.

Lambie, E.J. and Roeder, G.S. (1986). *Genetics* **114**, 769.

Larionov, V.L., Grishin, A.V. and Smirnov, M.N. (1980). *Gene* **12**, 41.

Lasker, B.A., Carle, G.F., Kobayashi, G.S. and Medoff, G. (1989). *Nucleic Acids Research* **17**, 3783.

Lauer, G.D., Roberts, T.M. and Klotz, L.C. (1977). *Journal of Molecular Biology* **114**, 507.

Laz, T.M., Pietras, D.F. and Sherman, F. (1984). *Proceedings of the National Academy of Sciences, USA* **81**, 4475.

Lechner, J. and Carbon, J. (1991). *Cell* **64**, 717.

Lennon, G.G. and Lehrach, H. (1992). *Current Genetics* **21**, 1.

Levin, H.L. and Boeke, J.D. (1992). *EMBO Journal* **11**, 1145.

Levin, H.L., Weaver, D.C. and Boeke, J.D. (1990). *Molecular and Cellular Biology* **10**, 6791.

Liebman, S., Shalit, P. and Picologlou, S. (1981). *Cell* **26**, 401.

Link, A.J. and Olson, M.V. (1991). *Genetics* **127**, 681.

Losson, R. and Lacroute, F. (1983). *Cell* **32**, 371.

Louis, E.J. and Haber, J.E. (1990). *Genetics* **124**, 533.

Louis, E.J. and Haber, J.E. (1992). *Genetics* **131**, 559.

Lundblad, V. and Szostak, J.W. (1989). *Cell* **57**, 633.

Lustig, A., Kurtz, S. and Shore, D. (1990). *Science* **250**, 549.

Lustig, A.J. and Petes, T.D. (1986). *Proceedings of the National Academy of Sciences, USA* **83**, 1398.

Magee, B.B. and Magee, P.T. (1987). *Journal of General Microbiology* **133**, 425.

Magee, B.B., D'Souza, T.D. and Magee, P.T. (1987). *Journal of Bacteriology* **169**, 1639.

Magee, B.B., Koltin, Y., Gorman, J. and Magee, P.T. (1988). *Molecular and Cellular Biology* **8**, 4721.

Maier, E., Hoheisel, J.D., McCarthy, L., Mott, R., Grigoriev, A.V., Monaco, A.P., Larin, Z. and Lehrach, H. (1992). *Nature Genetics* **1**, 273.

Maine, G.T., Surosky, R.T. and Tye, B.-K. (1984). *Molecular and Cellular Biology* **4**, 86.

Malone, R.E., Bullard, S., Lundquist, S., Kim, S. and Tarkowski, T. (1992). *Nature* **359**, 154.

Mann, G. and Davis, R.W. (1986). *Proceedings of the National Academy of Sciences, USA* **83**, 6017.

Maleszka, R. and Clark-Walker, G.D. (1989). *Current Genetics* **16**, 429.

Maleszka, R. and Clark-Walker, G.D. (1990). *Molecular and General Genetics* **223**, 342.

Matsumoto, T., Fukui, K., Niwa, O., Sugawara, N., Szostak, J.W. and Yanagida, M. (1987). *Molecular and Cellular Biology* **7**, 4424.

Maundrell, K., Hutchison, A. and Shall, S. (1988). *EMBO Journal* **7**, 2203.

McGrew, J., Diehl, B. and Fitzgerald-Hayes, M. (1986). *Molecular and Cellular Biology* **6**, 530.

Mellor, J., Jiang, W. and Funk, M. (1990). *EMBO Journal* **9**, 4017.

Melnick, L. and Sherman, F. (1990). *Nucleic Acids Research* **87**, 157.

Metzenberg, R.L., Stevens, J.N., Selker, E.V. and Morzycka-Wroblewska, E. (1985). *Proceedings of the National Academy of Sciences, USA* **82**, 2607.

Meyerink, J.A., Klootwijk, J., Planta, R.J., van der Ende, A. and van Bruggen, E.F.J. (1979). *Nucleic Acids Research* **7**, 69.

Mizukami, T., Chang, W.I., Gargartzer, I., Kaplan, N., Lombardi, D., Matsumoto, T., Osami, N., Kounosu, A., Yanagida, M., Marr, T.G. and Beach, D. (1993). *Cell* **73**, 121.

Michels, C.A., Read, E., Nat, K. and Charron, M. (1992). *Yeast* **8**, 655.

Moehle, C.M. and Hinnebusch, A.G. (1991). *Molecular and Cellular Biology* **11**, 2723.

Mortimer, R.K. and Schild, D. (1980). *Microbiology Review* **44**, 519.

Mortimer, R.K. and Schild, D. (1985). Microbiology Review **49**, 181.

Mortimer, R.K., Schild, D., Contopoulou, C.R. and Kans, J.A. (1989). *Yeast* **5**, 321.
Mortimer, R.K., Cantopoulou, R.C. and King, J.S. (1992). *Yeast* **8**, 817.
Munz, P., Wolf, K., Kohli, J. and Leopold, V. (1989). *In* "Molecular Biology of Fission Yeast" (A. Nasim, P. Young and B. Johnson, eds), pp. 1-30. Academic Press, San Diego.
Murphy, M.R. and Fitzgerald-Hayes, M. (1990). *Molecular Microbiology* **4**, 329.
Murphy, M.R., Fowlkes, D.M. and Fitzgerald-Hayes, M. (1991). *Chromosoma* **101**, 189.
Murray, A.W. and Szostak, J.W. (1983a). *Nature* **305**, 189.
Murray, A.W. and Szostak, J.W. (1983b). *Cell* **34**, 961.
Murray, A.W. and Szostak, J.W. (1986). *Molecular and Cellular Biology* **6**, 3166.
Murray, A.W., Schultes, N.P. and Szostak, J.W. (1986). *Cell* **5**, 529.
Nag, D.N., White, M.A. and Petes, T.D. (1989). *Nature* **340**, 318.
Nakaseko, Y., Adachi, Y., Funahashi, S., Niwa, O. and Yanagida, M. (1986). *EMBO Journal* **5**, 1011.
Nath, K. and Bollon, A.P. (1976). *Molecular and General Genetics* **147**, 153.
Needleman, R.B., Kaback, D.B., Dubin, R., Perkins, E.L., Rosenberg, N.G., Sutherland, K.A., Forrest, D.B. and Michels, C.A. (1984). *Proceedings of the National Academy of Sciences, USA* **81**, 2811.
Neitz, M. and Carbon, J. (1985). *Cellular and Molecular Biology* **5**, 2887.
Newlon, C.S. (1988). *Microbiological Reviews* **52**, 586.
Newlon, C.S. (1989). *In* "The Yeasts" (A.H. Rose and J.S. Harrison, eds), Vol. III, pp. 57-116. Acaemic Press, San Diego.
Newlon, C.S., Lipchitz, L.R., Collins, I., Deshpande, A., Devenish, R.J., Green, R.P., Klein, H.L., Palzkill, T.G., Ren, R., Synn, S. and Woody, S.T. (1991). *Genetics* **129**, 343.
Ng, R. and Abelson, J. (1980). *Proceedings of the National Academy of Sciences, USA* **77**, 3912.
Ng, R. and Carbon, J. (1987). *Molecular and Cellular Biology* **7**, 4522.
Ng, R., Cumberledge, S. and Carbon, J. (1986). *In* "Yeast Cell Biology" (J. Hicks, ed.), pp. 225-239. Alan R. Liss, Inc., New York.
Nicolas, A., Treco, D., Schultes, N.P. and Szostak, J.W. (1989). *Nature* **338**, 35.
Norris, D. and Osley, M.A. (1987). *Molecular and Cellular Biology* **7**, 3473.
Ogawa, N. and Oshima, Y. (1990). *Molecular and Cellular Biology* **10**, 2224.
Ogur, M., Minckler, S., Lindegren, G. and Lindegren, C.C. (1952). *Archives of Biochemistry and Biophysiology* **40**, 175.
Ogrydziak, D., Bassel, J. and Mortimer, R. (1982). *Molecular and General Genetics* **188**, 179.
Oliver, S.G. *et al.* (1992). *Nature* **357**, 38.
Olson, L.W., Eden, U., Egel-Mitani, M. and Egel, R. (1978). *Hereditas* **89**, 189.
Olson, M.V. (1991). *In* "The Molecular Biology of the Yeast *Saccharomyces*" (J.R. Broach, J.R. Pringle and E.W. Jones, eds), pp. 1-39. Cold Spring Harbor Laboratory Press, Cold Spring Harbor.
Olson, M.V., Dutchik, J.E., Graham, M.Y., Brodeur, G.M., Helms, C., Frank, M., MacCollin, M., Scheinman, R. and Frank, T. (1986). *Proceedings of the National Academy of Sciences, USA* **83**, 7826.
Ono, B. and Ishino-Arao, I. (1988). *Current Genetics* **14**, 413.
Oyen, T.B. (1973). *FEBS Letters* **30**, 53.
Panzeri, L. and Philippsen, P. (1982). EMBO Journal **1**, 1605.
Panzeri, L., Landonico, L., Stotz, A. and Philippsen, P. (1985). *EMBO Journal* **4**, 1867.
Passoth, V., Hansen, M., Klinner, U. and Emeis, C.C. (1992). *Current Genetics* **22**, 429.
Perfect, J.R., Magee, B.B. and Magee, P.T. (1989). *Infection and Immunity* **57**, 2624.
Petes, T.D. (1979a). *Journal of Bacteriology* **138**, 185.
Petes, T.D. (1979b). *Proceedings of the National Academy of Sciences, USA* **76**, 410.
Petes, T.D. (1980). *Cell* **19**, 765.
Petes, T.D. and Botstein, D. (1977). *Proceedings of the National Academy of Sciences, USA* **74**, 5091.
Petes, T.D. and Williamson, D.H. (1975). *Experimental Cell Research* **95**, 103.
Philippsen, P., Kramer, R.A. and Davis, R.W. (1978). *Journal of Molecular Biology* **123**, 371.
Polacheck, I. and Lebens, G.A. (1989). *Journal of General Microbiology* **135**, 65.

Pollizi, C. and Clarke, L. (1991). *Journal of Cell Biology* **112**, 191.
Ponticelli, A.S. and Smith, G.R. (1992). *Proceedings of the Natural Academy of Sciences, USA* **89**, 227.
Ponticelli, A.S., Sena, E. and Smith, G.R. (1988). *Genetics* **119**, 491.
Poulter, R.T. Hanrahan, V., Jeffrey, K., Markie, D., Shepherd, M.G. and Sullivan, P.A. (1982). *Journal of Bacteriology* **152**, 969.
Powers, S., Kataska, T., Fasano, O., Goldfarb, M., Broach, J.R. and Wigler, M. (1984). *Cell* **36**, 607.
Prabhala, G., Rosenberg, G.H. and Kaufer, N.F. (1992). *Yeast* **8**, 171.
Pugh, T.A. and Clancy, M.J. (1990). *Molecular and General Genetics* **222**, 87.
Riedel, N., Wise, J., Swerdlow, H., Mak, A. and Guthrie, C. (1986). *Proceedings of the National Academy of Sciences, USA* **83**, 8097.
Riles, L., Dutchik, J.E., Baktha, A., McCauley, B.K., Thayer, E.C., Leckie, M.P., Braden, V.V., Depke, J.E. and Olson, M.V. (1993). *Genetics* **134**, 81.
Ritossa, F.M. (1972). *Nature New Biology* **204**, 109.
Rivier, D.H. and Rine, J. (1992). *Science* **256**, 659.
Ross, L.O., Treco, D., Nicolas, A., Szostak, J.W. and Dawson, D. (1992). *Genetics* **131**, 541.
Rothstein, P., Helms, C. and Rosenberg, N. (1987). *Molecular and Cellular Biology* **7**, 1198.
Rubin, G.M. and Sulston, J.E. (1973). *Journal of Molecular Biology* **79**, 521.
Russell, P.J. and Rodland, K.D. (1986). *Chromosoma* **93**, 337.
Rymond, B.C. and Rosbash, M. (1993). *In* "The Molecular and Cellular Biology of the Yeast *Saccharomyces*" (J.R. Broach, J.R. Pringle and E.W. Jones, eds), Vol. II, p. 143. Cold Spring Harbor Laboratory Press, Cold Spring Harbor.
Sakaguchi, J. and Yamamoto, M. (1982). *Proceedings of the National Academy of Sciences, USA* **79**, 7819.
Sandmeyer, S.B. (1992). *Current Opinion in Genetics and Development* **2**, 705.
Sandmeyer, S.B., Bilanchone, V.W., Clark, D.J., Morcos, P., Carle, G.F. and Brodeur, G.M. (1988). *Nucleic Acids Research* **16**, 1499.
Saunders, M. Fitzgerald-Hayes, M. and Bloom, K. (1988). *Proceedings of the National Academy of Sciences, USA* **85**, 175.
Schatz, P.J., Pillus, L., Grisafi, P., Solomon, F. and Botstein, D. (1986). *Molecular and Cellular Biology* **6**, 3711.
Scherer, S. and Magee, P.T. (1990). *Microbiological Reviews* **54**, 226.
Schlessinger, D. (1990). *Trends in Genetics* **6**, 248.
Schuchert, P. and Kohli, J. (1988). *Genetics* **119**, 507.
Schulman, I. and Bloom, K.S. (1991). *Annual Review of Cell Biology* **7**, 311.
Schwartz, D.C. and Cantor, C.R. (1984). *Cell* **37**, 67.
Schweitzer, E., MacKechnie, C. and Halvorson, H.O. (1969). *Journal of Molecular Biology* **40**, 261.
Sears, D.D., Hegemann, J.H. and Hieter, P. (1992). *Proceedings of the National Academy of Sciences, USA* **89**, 5296.
Shaak, J., Mao, J. and Soll, D. (1982). *Nucleic Acids Research* **10**, 2851.
Shampay, J., Szostak, J.W. and Blackburn, E.H. (1984). *Nature* **310**, 154.
Shore, D. and Nasmyth, K. (1987). *Cell* **51**, 721.
Smith, C.L., Matsumoto, T., Osami, O., Klco, S., Fan, J.-B., Yanagida, M. and Cantor, C.R. (1987). *Nucleic Acids Research* **15**, 4481.
Smith, M.M. and Murray, K. (1983). *Journal of Molecular Biology* **169**, 641.
Snow, R. (1979). *Genetics* **92**, 231.
Sor, F. and Fukuhara, H. (1989). *Yeast* **5**, 1.
Spencer, F. and Hieter, P. (1992). *Proceedings of the National Academy of Sciences, USA* **89**, 8908.
Stark, M.J.R., Boyd, A., Mileham, A.J. and Romanos, M.A. (1990). *Yeast* **6**, 1.
Steensma, H.Y., Crowley, J.C. and Kaback, D.B. (1987). *Molecular and Cellular Biology* **7**, 410.
Steensma, H.Y., de Jonge, P., Kaptein, A. and Kaback, D.B. (1989). *Current Genetics* **16**, 131.

Steensma, H.Y., de Jongh, F.C.M. and Linnekamp, M. (1988). *Current Genetics* **14**, 311.
Stinchcomb, D.T., Struhl, K. and Davis, R.W. (1979). *Nature* **282**, 39.
Stinchcomb, D.T., Mann, C. and Davis, R.W. (1982). *Journal of Molecular Biology* **158**, 157.
Stinchcomb, D.T., Mann, C., Selker, E. and Davis, R.W. (1981). *ICN-UCLA Symposium on Molecular and Cellular Biology* **22**, 473.
Stoltenburg, R., Klinner, U., Ritzerfeld, P., Zimmermann, M. and Emeis, C. (1992). *Current Genetics* **22**, 441.
Strathern, J.N., Newlon, C.S., Herskowitz, I. and Hicks, J.B. (1979). *Cell* **18**, 309.
Struhl, K., Stinchcomb, D.T., Scherer, S. and Davis, R.W. (1979). *Proceedings of the National Academy of Sciences, USA* **76**, 1035.
Stucka, R., Lochmueller, H. and Feldmann, H. (1989). *Nucleic Acids Research* **17**, 4993.
Stucka, R., Schwarzlose, C., Lochmüller, H., Häcker, U. and Feldmann, H. (1992). *Gene* **122**, 119.
Sugawara, N. and Szostak, J.W. (1986). *Yeast* **2** (Supplement), 373.
Surana, T., Robitsch, H., Price, C., Schuster, T., Fitch, I., Futcher, A.B. and Nasmyth, K. (1991). *Cell* **65**, 145.
Surosky, R.T., Newlon, C.S. and Tye, B.-K. (1986). *Proceedings of the National Academy of Sciences, USA* **83**, 414.
Surosky, R.T. and Tye, B.-K. (1985). *Proceedings of the National Academy of Sciences, USA* **82**, 2106.
Surosky, R.T. and Tye, B.-K. (1988). *Genetics* **119**, 273.
Suzuki, T., Kobayashi, I., Kanbe, T. and Tanaka, K. (1989). *Journal of General Microbiology* **135**, 425.
Symington, L.S. and Petes, T.D. (1988a). *Molecular and Cellular Biology* **8**, 595.
Symington, L.S. and Petes, T.D. (1988b). *Cell* **52**, 237.
Symington, L.S., Brown, A., Oliver, S.G., Greenwell, P. and Petes, T.D. (1991). *Genetics* **128**, 717.
Szostak, J.W. and Blackburn, E.H. (1982). *Cell* **29**, 245.
Szostak, J.W. and Wu, R. (1980). *Nature* **284**, 426.
Tanaka, S. and Isono, K. (1992). *Nucleic Acids Research* **20**, 3011.
Tanaka, S., Yoshikawa, A. and Isono, K. (1992). *Journal of Bacteriology* **174**, 5985.
Tatchell, K., Chaleff, D.T., DeFeo-Jones, D.T. and Scolnick, E.M. (1984). *Nature* **309**, 523.
Tatchell, K., Robinson, L.C. and Breitenbach, M. (1985). *Proceedings of the National Academy of Sciences, USA* **82**, 3785.
Thompson, A. and Oliver, S.G. (1986). *Yeast* **2**, 179.
Thomas, D., Cherest, H. and Surdin-Kerjan, J. (1989). *Molecular and Cellular Biology* **9**, 3292.
Thrash-Bingham, C. and Gorman, J.A. (1992). *Current Genetics* **22**, 93.
Toda, T., Nakaseko, Y., Osami, N. and Yanagida, M. (1984). *Current Genetics* **8**, 93.
Udem, S.A. and Warner, J.R. (1972). *Journal of Molecular Biology* **65**, 227.
Umesono, K., Hiraoka, Y., Toda, T. and Yanagida, M. (1983). *Current Genetics* **7**, 123.
Valenzuela, P., Bell, G.I., Venegas, A., Sewell, E.T., Masiarz, F.R., DeGennaro, L.J., Weinberg, F. and Rutter, W.J. (1977). *Journal of Biological Chemistry* **252**, 8126.
Valenzuela, P., Venegas, A., Weinberg, F., Bishop, R. and Rutter, W.J. (1978). *Proceedings of the National Academy of Sciences, USA* **75**, 190.
Vassarotti, A., Goffeau, A., Magnien, E., Loder, B. and Fasella, P. (1990). *Biofutur* **94**, 84.
Van Houten, V.J. and Newlon, C.S. (1990). *Molecular and Cellular Biology* **10**, 3917.
Varma, A., Edman, J.C. and Kwon-Chung, K.J. (1992). *Infection and Immunity* **60**, 1101.
Ventor, U. and Hörz, W. (1989). *Nucleic Acids Research* **17**, 1313.
Verbeet, M.P., vanHeerikhuizen, H., Klootwijk, J., Fontijn, R.D. and Planta, R.J. (1984). *Molecular and General Genetics* **195**, 116.
Vogel, K., Horz, W. and Hinnen, A. (1989). *Molecular and Cellular Biology* **9**, 2050.
Vollrath, D., Davis, R.W., Connelly, C. and Hieter, P. (1988). *Proceedings of the National Academy of Sciences, USA* **85**, 6027.
Voytas, D.F. and Boeke, J.D. (1992). *Nature* **358**, 717.
Walmsley, R.W., Chan, C.S.M., Tye, B.-K. and Petes, T.D. (1984). *Nature* **310**, 157.

Werner-Washburne, M., Becker, J., Kosic-Smithers, J. and Craig, E.A. (1989). *Journal of Bacteriology* **171**, 2680.

Wickes, B., Staudinger, J., Magee, B.B., Kwon-Chung, K.-J., Magee, P.T. and Scherer, S. (1991). *Infection and Immunity* **59**, 2480.

Wickner, R.B., Boutelet, F. and Hilger, F. (1983). *Molecular and Cellular Biology* **3**, 415.

Woods, J.P. and Goldman, W.E. (1992). *Molecular Microbiology* **6**, 3603.

Woolford, J.L. and Warner, J.R. (1991). *In* "The Molecular and Cellular Biology of the Yeast *Saccharomyces*" (J.R. Broach, J.R. Pringle and E.W. Jones, eds), Vol. 1, pp. 587–626. Cold Spring Harbor Laboratory Press, Cold Spring Harbor.

Wright, A.P.H., Maundrell, K. and Shall, S. (1986). *Current Genetics* **10**, 503.

Wright, J.W., Gottschling, D.E. and Zakian, V. (1992). *Genes and Development* **6**, 197.

Yoshikawa, A. and Isono, K. (1990). *Yeast* **6**, 401.

Yoshikawa, A. and Isono, K. (1991). *Nucleic Acids Research* **19**, 1189.

Zakian, V. (1989). *Annual Review of Genetics* **23**, 579.

Zakian, V. and Blanton, H.M. (1988). *Molecular and Cellular Biology* **8**, 2257.

Zakian, V., Blanton, H.M., Witzel, L. and Dani, G.M. (1986). *Molecular and Cellular Biology* **6**, 925.

Zakian, V.A., Runge, K. and Wang, S.S. (1990). *Trends in Genetics* **6**, 12.

Zamb, T.J. and Petes, T.D. (1982). *Cell* **28**, 355.

Zhu, J., Brun, C., Kurooka, H., Yanagida, M. and Huberman, J.A. (1992). *Chromosoma* **102**, S7–S16.

6 Genetics of Brewing Yeasts

Morten C. Kielland-Brandt*, Torsten Nilsson-Tillgren†, Claes Gjermansen*,‡,
Steen Holmberg† and Mogens Bohl Pedersen‡

*Department of Yeast Genetics, Carlsberg Laboratory, Gamle Carlsberg Vej 10, DK-2500
Copenhagen Valby, Denmark, †Department of Genetics, Institute of Molecular Biology, Univer-
sity of Copenhagen, Øster Farimagsgade 2A, 1353 Copenhagen K, Denmark and ‡Carlsberg
Research Laboratory, Gamle Carlsberg Vej 10, DK-2500 Copenhagen Valby, Denmark

I. INTRODUCTION

The conversion of brewers' wort into beer is a fermentation in which yeast
has always played the key role. Almost all beer production world-wide
today is carried out with pure cultures, i.e. single-cell cultures (Hansen,
1888), of yeasts belonging to the genus *Saccharomyces*. There are exceptions. Some types of beer obtain their characteristics only if produced by
mixed fermentation (Verachtert and Dawoud, 1990). Belgium is known for
its beers of these types (van Oevelen *et al.*, 1977). It is even possible to
make beer with yeasts that are not members of the genus *Saccharomyces*.

223

The Yeasts Vol. 6, 2nd edition
ISBN 0-12-596416-1

Schizosaccharomyces pombe has its species-name from swahili; *pombe* is the word for beer. These exceptions will not be discussed further in the present chapter.

Not all *Saccharomyces* spp. are suitable for beer production. A simple way to realize the industrial importance of yeast genetics is to carry out test fermentations with existing strains. In this way, brewing yeasts can be compared with wild *Saccharomyces* spp. or with the many closely related strains of *Saccharomyces cerevisiae* which, because of their regular behaviour in sexual reproduction, are used all over the world for basic biological studies. It is generally found that only the brewing yeasts yield beer of a desired quality and behave satisfactorily in terms of indicators, such as fermentation rate and sedimentation. In addition, a brewing yeast suitable for a certain fermentation plant and a certain type of beer may be inappropriate in other brewing situations. Typical faults of non-brewing *Saccharomyces* spp. include production of phenolic off-flavour, inability to utilize maltotriose and low fermentation rate at the temperature optimal for the desired aroma.

Even for the existing brewing yeasts, the characteristics can be improved. In some markets, there may be a demand for new types of beer, for which yeasts with new characteristics may be expedient. In the past and present, most brewers have found that altering the characteristics of the yeast can provide a more reproducible process or a lowering of production costs, with no change in the resulting beer. Such altered characteristics may include a high rate of fermentation, a decrease in maturation time, better flocculation at elevated temperatures, and increased tolerance to high-gravity wort and alcohol, i.e. often characteristics that minimize the need for investment in fermentation tanks.

There are two major factors that currently limit progress in the breeding of brewing yeasts. First, it is often complicated to translate the desired change in yeast behaviour into biochemical and genetic terms. Much research has been devoted to this area for several decades and significant progress is steadily being made, often guided by model systems in non-brewing yeasts. The second difficulty is that brewing yeasts generally have deficiencies in their sexual reproduction. As a consequence, it is difficult or impossible to carry out many of the breeding steps and procedures in genetic analysis that are trivial with the non-brewing yeasts used as genetic reference strains in academic studies.

In principle, there are two strategies for dealing with these deficiencies. One is rather radical. A genetically well-behaved non-brewing yeast is modified to meet the requirements of the brewer. While this strategy is interesting and deserves attention, it is generally regarded as overambitious if traditional flavour characteristics are to be maintained. The other strategy is more conservative, the principle being to breed on existing brewing yeasts by circumventing or correcting some of their deficiencies in sexual reproduction.

If the potential of breeding of brewing yeasts is to be utilized to any reasonable degree, it is essential to know their genetic structure. It has long been clear that they have a close relationship with genetic reference strains of *Sacch. cerevisiae*, but molecular details of this relationship and its significance for genetic exchange have only been known for slightly more than a decade. Knowledge of the genetic structure of brewing yeasts is evidently quite modest when compared with the enormous amount of information available on the genetics of the academic reference strains.

The present chapter describes yeast used for production of lager beer, that is, bottom-fermenting yeast often referred to as *Sacch. carlsbergensis*. The main reason is that relatively little information is available on the genetics of top-fermenting yeasts, used for production of ales. Also, these yeasts constitute, on the one hand, a more diverse group than lager yeasts, thereby making generalizations difficult; on the other hand, they are more closely related to the genetic reference yeasts (Pedersen, 1986a).

II. TAXONOMY

Brewing strains of *Saccharomyces*, including those belonging to the taxon *Sacch. carlsbergensis* (Hansen, 1908), are currently all placed under the species name *Sacch. cerevisiae* (Kreger-van Rij, 1984). From the point of view of industrial microbiology, the virtues and shortcomings of this and similar conventions have been discussed by various authors including Gilliland (1971), Campbell (1987), van der Walt (1987), Martini and Martini (1989) and Barnett (1992). In this review, we refer to lager production yeast as *Sacch. carlsbergensis*; more specifically we shall use terms such as "the *Sacch. carlsbergensis* specific chromosome" for a chromosome found in lager yeast but not in genetic reference strains of *Sacch. cerevisiae*. We do this for practical reasons and it is not our present intention to insist that "*Sacch. carlsbergensis*" be regarded as a species name. Should, however, taxonomists wish to reconsider this question, we would like to draw their attention to a few important facts. As will be clear from the following, lager yeast is alloploid with a genetic structure reminiscent of species hybrids in higher plants. It has chromosomes in which the divergence in nucleotide sequence is correlated with a severe drop in meiotic recombination. It is likely that this low recombination causes fertility barriers to an extent, which is often used as a criterion in discrimination of species. Yeast molecular taxonomy is largely based on DNA association data (Kurtzman and Phaff, 1987), which do not clearly reveal alloploidy, although it may be suggested (Martini and Kurtzman, 1985).

As will be discussed later, it should be noted that Pedersen (1986a,b) found distinct nucleotide sequence differences between modern lager-brewing yeasts and the type strain of *Sacch. carlsbergensis*. Considering their close relationship, however, we use the name *Sacch. carlsbergensis* for both. A source of confusion is the existence of strains, which, because of their ability to ferment melibiose, have been named *Sacch. carlsbergensis* without being alloploid. Ten Berge (1972) investigated the genetics of strain NCYC 74 and found that it is diploid, and that its meiotic progeny can be crossed with genetic reference strains of *Sacch. cerevisiae*, yielding diploids with high spore viability, suggesting little divergence from *Sacch. cerevisiae*. Nevertheless, this strain and its progeny, e.g. CB11, have in recent literature retained the name *Sacch. carlsbergensis* (Federoff *et al.*, 1982) or been named *Sacch. uvarum* (Séraphin *et al.*, 1987). Several other yeast strains existing in collections under the name *Sacch. carlsbergensis* exhibit finger-prints resembling *Sacch. cerevisiae* in restriction fragment-length analysis (Pedersen, 1986a). Strain CBS 2354, synonymous with both NCYC 74 and ATCC 9080, exhibits chromosome lengths that are clearly different from those of industrial strains of *Sacch. carlsbergensis* (Takata *et al.*, 1989; M.B. Pedersen, unpublished data). We suggest that strain NCYC 74 (and therefore CBS 2354 and ATCC 9080) are not classified as *Sacch. carlsbergensis*.

The large number of interbred, genetically well-behaving *Sacch. cerevisiae* strains used world-wide for studies of basic biological phenomena will in this chapter mostly be referred to as "genetic reference strains".

III. GENES OF BREWING YEASTS

Identification of a gene associated with a given phenotype obviously involves the demonstration that the presence or absence in the cell of some allele can result in the phenotype. In academic strains of *Sacch. cerevisiae*, this has for the large majority of genes been carried out by mutagenesis of a haploid strain, screening or selection of a mutant of interest, and sub-sequent crosses to test whether segregation of the phenotype is Mendelian, indicating that it can, in a given genetic background, be ascribed to a single gene located on a chromosome. The success and importance of this approach are well known (see, e.g. Mortimer *et al.*, 1989).

For brewing yeasts, this approach has not had general success. The first problem appears in the search for mutants. Haplo-diplontic strains of *Saccharomyces* spp. seem to have a preference for the diploid phase and, since brewing yeasts have no or poor mating ability, they are not haploid, as expected. Indeed, direct searches for mutant types presumed to be recessive have failed with lager yeasts, even after heavy mutagenesis (Kielland-Brandt *et al.*, 1979). A report to the contrary (Molzahn, 1977) describes mutants of interesting phenotypes, but the mutants were not characterized at the biochemical or the genetic level. Mutant types expected to be dominant, on the other hand, appear readily. They include mutations in structural genes for biosynthetic enzymes, conferring resistance of enzymic activity to physiological inhibitors, such as isoleucine (Kielland-Brandt *et al.*, 1979), or non-physiological inhibitors such as sulphometuron methyl (Gjermansen and Sigsgaard, 1986; Galván *et al.*, 1987; Gjermansen *et al.*, 1988). However, only a minority of genes can be identified by dominant mutations.

The next obvious step is, therefore, an attempt to isolate haploids by sporulation of brewing yeasts. However, most brewing yeasts exhibit poor sporulation and low spore viability (Winge, 1944; Thorne, 1951; Johnston, 1965; Fowell, 1969; Anderson and Martin, 1975). Over a century ago, Hansen (1888) noted that lager yeasts exhibited poor ascus formation, although he could not know the consequences in terms of their life cycle.

To isolate meiotic progeny of lager yeasts, it is necessary to deviate from standard procedures used with *Sacch. cerevisiae* (Gjermansen and Sigsgaard, 1981; Bilinski *et al.*, 1987a). Notably, the temperature should be lower. Because of low spore viability, microdissection is not expedient. Mass-spore isolation yields colonies exhibiting a pronounced diversity in morphology and size. Some of the clones are temperature-sensitive for growth (Gjermansen and Sigsgaard, 1986). This character can be used as a selective marker in hybridizations. Some of the meiotic offspring express mating type but conjugation, as detected in the microscope, does not always lead to proper genetic hybridization. In a study of 134 zygotes from a cross between two meiotic segregants of *Sacch. carlsbergensis*, only 10% of the resulting colonies appeared as regular hybrids (Gjermansen and Sigsgaard, 1986). The remaining colonies consisted either of parental cells or were sectored, with **a**, α, and hybrid sectors. This observation suggests a high frequency of failure in nuclear fusion. Although the meiotic progeny of the lager yeast are not generally haploid (Nilsson-Tillgren *et al.*, 1986; Kielland-Brandt *et al.*, 1989), some of the strains yield recessive mutants after mutagenesis (Gjermansen, 1983). However, even though mutants can be obtained and genetic variation among existing strains can be utilized, it is generally hopeless to carry out tetrad analysis with brewing yeasts or their progeny. The study by Thorne (1951) on the genetics of flocculence of

Sacch. cerevisiae is illustrative. Out of 20 top-fermenting English brewing strains, he found 12 that sporulated adequately, although their spore viability averaged only 12%. Out of 372 different attempted crosses of the spores, only eight hybrids were obtained, none of which was amenable to tetrad analysis.

Faced with these difficulties, it has proved fruitful to resort to a primary identification of genes of brewing yeasts by assaying their phenotype, not in the brewing yeast but fully or partially in the genetic background of academic strains of *Sacch. cerevisiae*. Thus, in order to study inheritance of the flocculation character of a brewing strain, Thorne (1951) crossed some of the spores to non-brewing strains. This approach is probably rather generally applicable. After one or more backcrosses to a suitable non-brewing parent, spore viability may be adequate for tetrad analysis (Thorne, 1951; Lewis *et al.*, 1976). A more straightforward utilization of the same principle is construction in plasmid vectors of gene libraries from brewing yeasts, and cloning of the gene of interest by selection or screening for the relevant phenotype after transformation of an appropriate genetic reference strain (Casey, 1986a; Casey and Pedersen, 1988; Gjermansen *et al.*, 1988). It may sometimes be expedient to supplement this approach by screening the vector library for other characteristics, such as molecular hybridization with known probes (Nilsson-Tillgren *et al.*, 1986). When the phenotype conferred by a gene is influenced by polymorphisms at other loci, such as with genes for flocculence (Lewis *et al.*, 1976; Holmberg and Kielland-Brandt, 1978), molecular cloning of a gene from a brewing yeast, such as *FLO1* (Watari *et al.*, 1989), is the most reliable identification of the gene, since it allows comparison of isogenic strains. Furthermore, molecular cloning is an obvious step in the analysis of the structure and function of the gene. In general, genes from brewing yeasts have functional counterparts in the genetic reference strains of *Sacch. cerevisiae*. So far, only a few genes have been found in brewing yeasts, which give phenotypes clearly different from those usually found in genetic reference strains.

As already mentioned, several genes for flocculence from brewing yeasts have been identified by Mendelian segregation in *Sacch. cerevisiae* (Gilliland, 1951; Thorne, 1951; Lewis *et al.*, 1976), obviously implying that, for these genes, the alleles present in the tester strains were different from those identified in the brewing yeasts. It is not known whether there is a general polymorphism for these particular genes among genetic reference yeasts, but it is the experience of most geneticists working with *Saccharomyces* spp. that there is polymorphism for several genes affecting flocculence. A number of genes affecting flocculence have been identified in more recent years (see Stewart and Russell, 1987). The best studied gene for flocculence, *FLO1* (Watari *et al.*, 1994), was derived from a brewing yeast (Lewis *et al.*, 1976).

Another example is the *pof1* gene (Goodey and Tubb, 1982; Meaden and Taylor, 1991), for which brewing yeasts generally seem to be homozygous, while standard academic yeasts have the functional allele, *POF1*. The latter is responsible for the ability of the strain to decarboxylate derivatives of cinnamic acid, such as ferulic acid. As this compound occurs in brewers' wort, yeasts containing *POF1* will by decarboxylation produce 4-vinylguaiacol (Ryder *et al.*, 1978), which imparts a phenolic off-flavour to the beer. Brewing yeasts, lacking the *POF1* allele, do not produce 4-vinylguaiacol.

IV. NON-MENDELIAN ELEMENTS

The most important non-Mendelian genetic element is mitochondrial DNA. It is essential for respiration, while other mitochondrial functions, such as biosynthesis of isoleucine and valine (Ryan and Kohlhaw, 1974), can proceed without it. Mitochondrial DNA can also be important for functions that are not connected with respiration in very obvious ways. Thus, Evans and Wilkie (1976) found that defects in the mitochondrial DNA in *Sacch. cerevisiae* could impede utilization of maltose and galactose.

An extremely frequent mutation seen in *Saccharomyces* spp. is loss of some or other part of the mitochondrial DNA, with a concomitant amplification of the remainder. The resulting respiratory-deficient mutants are called "petites". The frequency of petites in cultures of *Sacch. cerevisiae* is typically of the order of one to several per cent. The rate of the mutation at the molecular level is hard to estimate. Together with mitotic segregation among the many copies of mitochondrial DNA, it establishes a directly observable rate of mutation at the cellular level. Experiments with genetic reference yeasts have shown that this rate is under the control of several nuclear genes (Devin and Koltovaya, 1981) as well as mitochondrial DNA. Thus, deletion of non-coding sequences in mitochondrial DNA can in some cases decrease the rate of petite formation (Clark-Walker *et al.*, 1985; Piškur, 1989). In addition, environmental factors can be important; certain drugs, like ethidium bromide, dramatically increase the rate.

Petite mutants of brewing yeasts are undesirable in beer production. Important deficiencies have been reported (Šilhánková *et. al.*, 1970), such as increased diacetyl concentrations (Morrison and Suggett, 1983) and inability to utilize maltotriose (Gyllang and Martinson, 1972), although the latter deficiency is not found in petites of all brewing yeasts (Morrison and Suggett, 1983). Uptake of maltose and glucose are affected (Stewart *et. al.*, 1990). Altered flocculence has been reported by several authors (Lewis *et al.*, 1976; Stewart and Russell, 1977; Holmberg and Kielland-Brandt, 1978; Hinrichs *et al.*, 1988).

There is extensive conservation of the functions of structural genes in mitochondrial DNA. The organization of introns and intergenic regions, on the other hand, is quite divergent and differences are also seen among *Saccharomyces* spp. including brewing yeasts (Aigle *et al.*, 1984). To what extent these differences are responsible for differences in fermentation behaviour and flavour of beer (Hammond and Eckersley, 1984) is not easy to assess. The *karl-1* mutation isolated by Conde and Fink (1976) in *Sacch. cerevisiae* provides a convenient tool for studies of the effects of various cytoplasmic genetic markers, while maintaining a reasonably constant nuclear genetic background. A mating in which at least one parent carries *karl-1* proceeds through cytoplasmic fusion but karyogamy does not occur in most cases. What looks like a zygote is actually a heterokaryon, which gives rise to mostly haploid progeny having the nuclear genotype of one parent or the other, but often cytoplasmic markers from both parents. Studies based on this technique, which may be called cytoduction (Zakharov and Yarovoy, 1977), have suggested the possibility that the mitochondrial genome influences the concentrations in beer of 4-vinylguaiacol (Conde and Mascort, 1981; Tubb *et al.*, 1981). One mechanism by which mitochondrial DNA could affect general cell metabolism in a population is by its control of the rate of formation of petites.

There may be a difference between mitochondrial DNA of genetic reference strains of *Sacch. cerevisiae* and that of some brewing yeasts concerning its maintenance or expression. C. Gjermansen (unpublished observation) has observed that cytoduction of respiratory-proficient mitochondria readily occurs from academic strains of *Sacch. cerevisiae* to petites of meiotic progeny of *Sacch. carlsbergensis*, while the reverse experiment was unproductive. In crosses between *Sacch. douglasii* and *Sacch. cerevisiae*, Claisse *et al.* (1987) observed a similar phenomenon. They were able to identify a function, provided by chromosome IX in *Sacch. douglasii*, necessary for *cox1* mRNA maturation in mitochondria in this yeast.

The only other non-Mendelian elements that have been investigated in brewing yeasts are 2 μm DNA (Tubb, 1980; Aigle *et al.*, 1984) and the double-stranded RNA elements, L and M (Young, 1981; Aigle *et al.*, 1984). They seem to be quite similar to the corresponding elements found in academic strains of *Sacch. cerevisiae* (Wickner, 1986; Volkert *et al.*, 1989). Mead *et al.* (1986) found that strains of *Sacch. cerevisiae* that have spontaneously lost their 2 μm DNA have a slight competitive advantage over their plasmid-containing parents, and Xiao and Rank (1990) suggest this as a breeding step. Most brewing yeasts are non-killers (Kreil *et al.*, 1975), i.e. they lack the M element. It has been suggested that infection of beer fermentations with wild killer yeasts would be less serious if the brewing yeast had the killer character, or at least were killer tolerant or killer resistant. Strain modification based on this suggestion has been carried out (Young, 1981, 1983; Hammond and Eckersley, 1984).

V. CHROMOSOMES OF BREWING YEASTS

Having established the presence of genes of specific interest as well as various marker genes, the next level of organization of the genetic material to be studied is the chromosome. However, the low spore viability of brewing yeasts becomes an even more serious problem here. Repeated back-crosses and molecular cloning, which can be used to study segregation, expression and structure of a gene from a brewing yeast, are not by themselves efficient for studying the genetic structure and function of a full chromosome. Development of the single-chromosome transfer technique changed the situation. The basic idea of this technique is the same as already described for the study of genes of brewing yeasts, namely to place the entity to be investigated in a strain with regular genetic behaviour. The difference is that now this entity is a full chromosome.

A. The single-chromosome transfer technique

This technique, like cytoduction already described (Section IV), makes use of the *kar1-1* mutation isolated by Conde and Fink (1976). It takes advantage of the fact that not only cytoplasmic genetic elements but also nuclear elements from one haploid strain can be combined with the nuclear genome of another. Livingston (1977) found that 2 μm DNA is transferred from one haploid genotype to the other, and, in *kar1-1* produced hetero-karyons, with an efficiency as high as about half that with which mito-chondrial DNA is transferred, even though 2 μm DNA is likely to have a nuclear location (Kielland-Brandt *et al.*, 1980; Broach, 1981). A plasmid 2–3 times larger, constructed by insertion of a chromosomal fragment into 2 μm DNA, was transferred with a much lower frequency of 10^{-4}–10^{-3} (Kielland-Brandt *et al.*, 1981). Importantly, even a whole chromosome can be transferred (Nilsson-Tillgren *et al.*, 1980; Dutcher, 1981). Nilsson-Tillgren *et al.* (1980) crossed two haploid strains of *Sacch. cerevisiae* with the genotypes *MATα his4 ade2 can1 kar1* [rho⁻] and *MATa* with one another and plated the mixture on minimal medium containing adenine and canavanine in order to select for clones of the genotype of the first parent that had received the *HIS4* wild-type gene of the second parent. The first parent would not grow since it requires histidine and, because of canavanine sensitivity, the second parent would not grow, nor would normal diploid products of mating, the gene for canavanine resistance (*can1*) being recessive. As *ade2* cells accumulate a red pigment, colony colour provided an additional screen. Red colonies appeared at low frequency. They proved to be respiratory sufficient, histidine prototrophic, adenine requiring, canavanine-resistant, non-sporulating and non-mating. This indicates that they had the genotype of the first parent, except that they had received from the second parent functional mitochondrial DNA and chromosome III (with *MATa* and *HIS4*), but few or no other chromosomes. Chromosome V (with the *CAN1* wild-type gene) and chromosome XV (with *ADE2*) were clearly not transferred. Although these clones were non-maters, they segregated mating cells at a low frequency, allowing recovery of prototrophs by mating the clones to appropriate haploid tester strains. Sporulation and tetrad analysis of prototrophs from five different crosses confirmed the disomic nature of the clones that were crossed and the rare mitotic events that gave rise to mating cells were identified. Chromosome III, as identified by four markers distributed on both arms (*HIS4 LEU2 MATa THR4*), was transferred at a frequency of one out of several hundred thousand *ade2 can1* [RHO$^+$] progeny strains from the cross. In that study, cotransfer of five other chromosomes was looked for but was not detected. In an independent study of crosses involving *kar1-1*, Dutcher (1981) discovered transfer of several different chromosomes. The frequency of transfer was inversely correlated with the known map size. Chromosome III was transferred 100–1000 times more frequently than the rate found by Nilsson-Tillgren *et al.* (1980), probably as a result of strain differences. Dutcher (1981) found that transfer of more than one chromosome to the same cell was much rarer than transfer of one chromosome but more frequent than that calculated for independent events. In crosses where a strain of *Sacch. carlsbergensis* was a donor (see the following sections), Nilsson-Tillgren *et al.* (1981) observed some progeny that may be explained by multiple chromosome transfer. The mechanism of transfer, however, is unknown.

B. Chromosome III of Saccharomyces carlsbergensis

Transfer of single chromosomes into haploid strains of *Sacch. cerevisiae* can be used for analysis of the gene content of chromosomes of *Saccharomyces* strains in general. However, a prerequisite for the method is that the donor strain is able to mate. Lager yeast strains of *Sacch. carlsbergensis* do not mate but, as already described, it is possible to derive meiotic progeny with mating ability from at least some of them. These strains proved to be excellent tools for genetic characterization of the lager yeast. First, the chromosome in *Sacch. carlsbergensis* carrying the *HIS4* gene was analysed. Nilsson-Tillgren *et al.* (1981) used a donor with mating type **a**, which was crossed to a haploid strain of *Sacch. cerevisiae* with the genotype *MATα his4 ade2 can1 kar1*. Histidine-independent progeny that were canavanine-resistant and red (expressing *can1* and *ade2*) were selected. The strains were non-maters, indicating that *MATa* was cotransferred with *HIS4*, suggesting linkage between these two genes as in the recipient strain of *Sacch. cerevisiae* (see Fig. 1). Genetic analysis of the transferred chromosome from *Sacch. carlsbergensis* proceeded as already described for chromosome III from *Sacch. cerevisiae*. Rare mating to haploid strains of *Sacch. cerevisiae* with a mutant *his4* gene, permitting selection for histidine independence, allowed recovery of the chromosome from *Sacch. carlsbergensis* in a strain that could be analysed by tetrad analysis. In several cases, it was found that the mating cell had become a mater by loss of the entire chromosome III from *Sacch. cerevisiae*. Fortunately, the products of these crosses gave high spore viability, so that tetrad analysis was straightforward. The chromosome from *Sacch. carlsbergensis* turned out to be functionally equivalent to chromosome III of *Sacch. cerevisiae* and could fully substitute for this chromosome in the haploid progeny, providing all information for functions essential for haploid viability as well as mating. By analogy with the terminology used by wheat geneticists, we use the term "chromosome-addition strain" for a *Sacch. cerevisiae* strain that is disomic by having an extra chromosome from *Sacch. carlsbergensis*. Likewise, the term "chromosome-substitution strain" refers to the situation where a chromosome from *Sacch. carlsbergensis* has replaced one in *Sacch. cerevisiae*, resulting in a euploid strain. In spite of the functional equivalence of the two chromosomes, they exhibit a substantial genetic difference. In

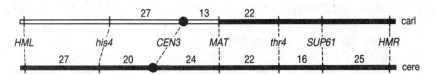

Fig. 1. Genetic map of a chromosome III from *Saccharomyces carlsbergensis* (carl; data of Nilsson-Tillgren *et al.*, 1981; Skaanild, 1985; Resnick *et al.*, 1989; T. Nilsson-Tillgren, unpublished data) compared with a chromosome III from *Saccharomyces cerevisiae* (cere; Mortimer *et al.*, 1989). The chromosome of *Sacch. carlsbergensis* is composed of a part that is divergent (open bar) from that in *Sacch. cerevisiae* and a part that is highly homologous (closed bar). Numbers refer to map distances (cM).

a meiosis involving the two chromosomes, no recombination between *his4* and *MAT* was observed (Nilsson-Tillgren *et al.*, 1981). Later experiments demonstrated that meiotic recombination is not completely repressed, although it was decreased approximately 100-fold. The region of decreased recombination was later extended (T. Nilsson-Tillgren, unpublished data) to the *CHA1* gene, which maps close to the left telomere (Bornæs *et al.*, 1992). In contrast, normal recombination frequencies were observed between the *MAT* locus and *thr4*. Suppression of crossing-over can be caused by inversions, making the recombinants non-viable. However, the spore viability did not suggest this explanation. Because of the degeneracy of the genetic code, identical proteins can be encoded by nucleotide sequences that differ from each other, and intergenic sequences may vary considerably without effect on essential functions. In yeast, small differences in nucleotide sequence are known to affect recombination (Smolik-Utlaut and Petes, 1983; Borts and Haber, 1987). To investigate the nature of the difference between the two chromosomes III, a restriction-length analysis of selected genes in the chromosome-substitution strain as well as the donor strain of *Sacch. carlsbergensis* was performed, using cloned genes from *Sacch. cerevisiae* as hybridization probes (Nilsson-Tillgren *et al.*, 1981; Holmberg, 1982). For both strains analysed, the genes *SUP-RL1* (allelic to *SUP61*, Fig. 1) and *HMR* in the recombining region gave a restriction-fragment pattern very similar to that found in genetic reference strains of *Sacch. cerevisiae*, indicating that the nucleotide sequences were very similar or identical. However, for the genes *HIS4*, *MAT* and *HML*, from the non-recombining region, not only were the lengths of the restriction fragments observed in the chromosome-substitution strain generally different, but the hybridization intensities were strongly affected by nucleotide sequence non-homology between genes in *Sacch. carlsbergensis* and those in *Sacch. cerevisiae*. Thus, the absence of recombination already described correlates with regions of general nucleotide-sequence non-homology. When the *LEU2* gene was used as a probe at high stringency, no hybridization to DNA of *Sacch. carlsbergensis* was detected (Holmberg, 1982). However, Pedersen (1985) detected clearly two different *LEU2* genes in *Sacch. carlsbergensis*, one less like that in *Sacch. cerevisiae* than the other, and Casey and Pedersen (1988) have cloned a *LEU2* gene from the brewing yeast having a strongly deviating restriction-endonuclease map. Interestingly, the results from the parental *Sacch. carlsbergensis* strain showed that it was structurally heterozygous in and around the *HIS4* and the *MAT* loci, suggesting that this strain was a hybrid, also containing a chromosome III, which more closely resembled the chromosome in *Sacch. cerevisiae*, but which has not been analysed by single-chromosome transfer.

In diploids formed between chromosome III substitution strains and normal *Sacch. cerevisiae* strains, all genes in the non-recombining region will segregate during the first meiotic division and therefore appear as tightly linked to the centromere. For this reason, genetic distances cannot be determined from tetrad analysis of such diploids, but have to be analysed in diploids formed by crossing two chromosome-substitution strains with the opposite mating type. Diploids of this nature are perfectly viable and able to go through meiosis, forming asci with viable spores. Originally, the only available genetic marker on the chromosome from *Sacch. carlsbergensis* was the *MAT* locus and its distance from the centromere (see Fig. 1) could be determined in crosses that included heterozygosity for a centromere marker, such as *trp1*, on one of the chromosome pairs in *Sacch. cerevisiae*. With the isolation of *his4* mutants in chromosome-substitution strains (Skaanild, 1985; Resnick *et al.*, 1989), it became possible to obtain crosses that allowed positioning of the centromere between the *MAT* locus and *HIS4*. As already indicated, rare crossing-over is observed in the region of chromosome III, which is divergent between *Sacch. cerevisiae* and *Sacch. calsbergensis*. The occurrence of such rare reciprocal recombination placed *CHA1* distal to *his4* on the left arm of the chromosome (T. Nilsson-Tillgren, unpublished data). A map of the analysed chromosome III from *Sacch. carlsbergensis* is shown in Fig. 1, from which it is apparent that not only the gene order but also the gene distances are remarkably conserved in the two types of chromosome III. The close relatedness between them was further illustrated by the isolation of *his4* mutations in chromosome III substitution strains (Skaanild, 1985). Seven independent mutant alleles were found, six of which were able to complement specific alleles in *Sacch. cerevisiae*. It was possible to construct a complementation map of the *HIS4* locus in *Sacch. carlsbergensis* from the data obtained by crossing the mutants to known alleles of *Sacch. cerevisiae* (Fig. 2).

The analysis already described was based on the lager production strain referred to earlier as strain 244 of *Sacch. carlsbergensis*. Pedersen (1983, 1985, 1986a,b) has carried out an extensive analysis of *Sacch. carlsbergensis* strains and related strains directed at several chromosomes, but with an emphasis on chromosome III. Electrophoretic karyotyping (Carle and Olson, 1985) of strains of recent or current use in lager-beer production gave a very uniform pattern with a single chromosome III band smaller than that observed in academic yeasts of *Sacch. cerevisiae* (Pedersen, 1986b; Casey and Pedersen, 1988; Fig. 3), which means that the two types of chromosome III present in these yeasts (Pedersen, 1985) must have the same size. From another brewing strain of *Sacch. carlsbergensis*, Pedersen (1986b) analysed a chromosome III by single chromosome

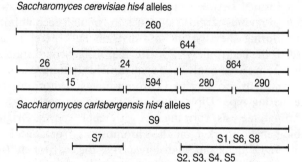

Fig. 2. Complementation pattern of nine *his4* mutations induced in the chromosome III of *Saccharomyces carlsbergensis* presented in Fig. 1. The mutants of *Saccharomyces cerevisiae* and their data are from Fink and Styles (1974) and a coherent complementation pattern was obtained by including the *his4* mutants of *Sacch. carlsbergensis* (Skaanild, 1985; Resnick *et al.*, 1989; T. Nilsson-Tillgren, unpublished data). Numbers refer to the allele designations.

transfer with results identical with those obtained by Nilsson-Tillgren *et al.* (1981).

C. Chromosome V of *Saccharomyces carlsbergensis*

Analogous studies were performed for chromosome V (Nilsson-Tillgren *et al.*, 1986). As for chromosome III, the chromosome V donor strains were meiotic progeny strains of *Sacch. carlsbergensis*, expressing mating type **a** or α, derived from a lager production strain. The previously used selectable marker, *can1*, is located on chromosome V, and therefore another marker had to be used to counterselect against the donor strain and products of complete nuclear fusion. The recipient strains were *Sacch. cerevisiae* haploids constructed to carry the *kar1-1* mutation, a recessive gene for resistance to cycloheximide (*cyh2*) located on chromosome VII, the chromosome XV marker *ade2* for red colour, as well as several chromosome V markers (*can1*, *ura3*, *ilv1*, *his1* and *rad3*). After mating of the recipient and donor strains, red cycloheximide-resistant colonies were selected on complete medium lacking histidine, uracil and isoleucine, singly or in combinations. The products of chromosome transfer were found to be histidine, uracil, and isoleucine independent, no matter which medium they were selected on, indicating that *HIS1*, *URA3* and *ILV1* were indeed linked on

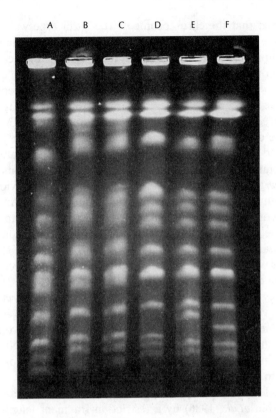

Fig. 3. Electrophoretic karyotypes of strains of *Saccharomyces carlsbergensis* and *Saccharomyces cerevisiae*. Lane A is for *Sacch. cerevisiae* top-fermenting strain 300; lane B, *Sacch. carlsbergensis* lager production strain 224; lane C, *Sacch. carlsbergensis* lager production strain 244; lane D, chromosome substitution strain 290986J-11, containing a chromosome III from *Sacch. carlsbergensis* strain 244 while the other chromosomes are from *Sacch. cerevisiae* strain K5-5A; lane E, *Sacch. cerevisiae* haploid genetic reference strain K5-5A; lane F, *Sacch. cerevisiae* haploid genetic reference strain X2180-1A, in which bands from below are known to correspond to chromosomes I, VI, III and IX, respectively, after which double bands appear.

the transferred chromosome of *Sacch. carlsbergensis*. The disomic nature of the presumed chromosome-addition strains was revealed by crossing them to genetically marked haploid strains for subsequent sporulation and tetrad analysis. A large fraction of the chromosome-addition strains were canavanine–resistant, a fact defining two different types of chromosome V in *Sacch. carlsbergensis*, one carrying a wild-type arginine-permease gene, *CAN1*, and the other carrying a defective allele or missing the gene. Genetic

238   M.C. Kielland-Brandt, T. Nilsson-Tillgren, C. Gjermansen, S. Holmberg and M.B. Pedersen

analysis showed that the chromosome of the latter type was homologous to chromosome V of *Sacch. cerevisiae*, i.e. genetic recombination between them was unimpaired, yielding gene distances almost identical with those found in standard crosses of *Sacch. cerevisiae* (Fig. 4). In contrast, the chromosome with the functional *CAN1* gene did not recombine with a chromosome V of *Sacch. cerevisiae* in any part of the tested region, spanning the *can1* locus close to the left telomere to *rad3* at the right telomere. The disomic chromosome V-addition strains with this chromosome exhibited some mitotic instability. After 8–10 generations in non-selective medium, about 1% of the cells had lost the chromosome from *Sacch. carlsbergensis*, as revealed by the appearance of ultraviolet-sensitive cells with requirements for isoleucine, uracil and histidine. An assumption that also the chromosome from *Sacch. cerevisiae* might become lost turned out to be true, so that chromosome-substitution strains could easily be isolated. It was found that both types of chromosome V from *Sacch. carlsbergensis* were functionally homologous to that from *Sacch. cerevisiae* and, as in the case of chromosome III, they could provide the cells with all functions essential for viability. Spontaneous *can1* and *ura3* mutations, as well as *in vitro* constructed *ilv1* mutations, have made it possible to map these genes on the homologous, non-recombining chromosome, by crossing two substitution strains of opposite mating type (Nielsen, 1991). As presented in Fig. 4, the gene order and gene distances closely resemble those found in *Sacch. cerevisiae* strains.

The physical nature of the two types of chromosome V was investigated (Nilsson-Tillgren *et al.*, 1986) by a restriction-fragment length analysis of four loci on chromosome V (*CAN1, CYC7, URA3* and *ILV1*). The genes on the recombining chromosome gave patterns identical or nearly so to the patterns in *Sacch. cerevisiae*. In contrast, the non-recombining chromosome harboured alleles of the genes that, at high stringency, hybridized very poorly with probes from *Sacch. cerevisiae* and, at lower stringency, gave

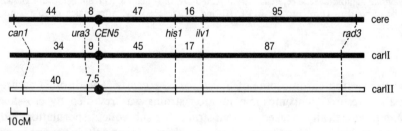

Fig. 4. Genetic maps of two chromosomes V from *Saccharomyces carlsbergensis* (carlI and carlII; data of Nilsson-Tillgren *et al.*, 1986; T. Nilsson-Tillgren, unpublished data) compared with chromosome V of *Sacch. cerevisiae* (cere; Mortimer *et al.*, 1989). Closed bars refer to *Sacch. cerevisiae* sequences or *Sacch. cerevisiae*-like sequences, while the open bar refers to divergent sequences. Numbers refer to map distances (cM).

very different restriction-fragment patterns. Thus, as was observed for chromosome III, the decreased recombination in the homoeologous chromosome V correlates with diverged nucleotide sequences. Genetic and molecular analyses of the chromosome-substitution strains clearly demonstrate the presence of two different types of chromosome V in *Sacch. carlsbergensis*. Restriction-fragment length analyses (Pedersen, 1994) have revealed the occurrence of at least four types of chromosome V among brewing yeasts, as defined by *URA3* diagnostic fragment patterns. The Carlsberg lager-brewing yeast contains two *Sacch. cerevisiae*—like chromosomes V (*URA3* fragment patterns I and III) together with the homoeologous chromosome V already described.

D. Chromosome X of *Saccharomyces carlsbergensis*

Casey (1986b) performed a detailed study of chromosome X. The results were analogous with and extended those obtained with chromosomes III and V. Using the chromosome-transfer technique, it was possible to isolate two different chromosomes X from *Sacch. carlbergensis*. Both types harboured wild-type alleles of *ARG3*, *MET3*, *ILV3*, *CDC11* and *HOM6*, but only one type (I) complemented *rad7*, i.e. it had a functional *RAD7* gene. Recombination with chromosome X of tester strains of *Sacch. cerevisiae* was limited to certain regions, as for chromosome III. The type I chromosome X (Casey, 1986b) recombined in the left arm between the centromere and *arg3*, while type II recombined in the right arm between *ilv3* and *hom6* (Fig. 5). In the short region between the centromere marker *met3* and *ilv3* on the right arm, neither type I nor type II recombined. The observed recombination frequencies were decreased to almost 50% of the standard values (Mortimer *et al.*, 1989), in contrast to what is observed in the homologous regions of chromosomes III and V. Casey (1986b) observed that considerably higher recombination frequencies in the right arm could be calculated from asci with three living spores compared with what was calculated from those with four spores. He concluded that recombination between chromosomes X from *Sacch. carlsbergensis* and *Sacch. cerevisiae* in this region might result in a non-viable product. This would lead to an overrepresentation of parental ditype asci among asci with four viable spores. A similar decreased viability of the products of recombination between the chromosomes might also explain the map reduction observed in the left arm.

Using probes from the *ILV3* and *CYC1* genes, Casey (1986a,b) performed a restriction-fragment length analysis of these two loci on types I and II

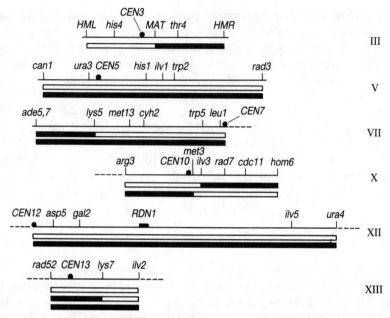

Fig. 5. Structures of chromosomes of *Saccharomyces carlsbergensis* recovered in *Saccharomyces cerevisiae* by single-chromosome transfer. Bars are aligned with the linkage groups of *Sacch. cerevisiae* identified with Roman numbers (Mortimer *et al.*, 1989) to indicate the segments that have been analysed at the genetic and/or molecular levels. Closed bars indicate segments that recombine with *Sacch. cerevisiae*, while open bars indicate those that do not. One type of chromosome III and three types of chromosome XIII have been recovered, while two types have been recovered for each of the other four analysed linkage groups.

of chromosome X, which confirmed the earlier observation that genes mapping in recombining regions are closely homologous to those of *Sacch. cerevisiae*, while genes from the non-recombining regions are divergent in DNA sequence and specific to *Sacch. carlsbergensis* in the present context. With the aid of the restriction-fragment length analysis, it was possible to determine whether the product of chromosome transfer was disomic or euploid in the cases where chromosome X type I had been transferred. In contrast to the situation with chromosomes III and V, Casey (1986b) found a majority of spontaneous substitution strains. However, for chromosome X type II, only chromosome-addition strains could be isolated, as confirmed by tetrad analysis and pulse-field electrophoretic separation of chromosomes.

The genetic marker for the *RAD7* gene in this work was *cyc1-1*, a large deletion in the *CYC1* region, which also includes the neighbouring *RAD7* gene. The presence of a functional *CYC1* gene on the type II chromosome X of *Sacch. carlsbergensis* was not tested by Casey (1986b). However, the restriction-fragment length analysis, where the probe was a fragment from the *CYC1* promoter of *Sacch. cerevisiae*, indicates that this gene is present on chromosome X.

E. Other chromosomes of *Saccharomyces carlsbergensis*

Single-chromosome transfer experiments have also been performed for analysis of chromosomes VII (T. Nilsson-Tillgren, unpublished data), XII and XIII (Petersen *et al.*, 1987) since the reference to Fig. 5 not only concerns chromosomes XII and XIII, but also, e.g. chromosome VII. In these cases, a detailed genetic analysis was hampered by low spore viability in diploids formed by crossing chromosome-addition strains to tester strains of *Sacch. cerevisiae*. Some chromosome-addition strains with these chromosomes exhibited considerable mitotic instability, losing the acquired chromosome from *Sacch. carlsbergensis* at a high rate. However, in spite of considerable effort, it was not possible to isolate chromosome-substitution strains for these three chromosomes.

For chromosome VII, only the left arm has been analysed by complementation and recombination (T. Nilsson-Tillgren, unpublished data) and, as illustrated in Fig. 5, two different chromosomes were found. One was freely recombining with, and thus homologous to, chromosome VII in *Sacch. cerevisiae*. The other was, like chromosomes III and X, mosaic in nature, with a non-recombining, homoeologous region from the centromere to *lys5* and a homologous region from *lys5* to *ade5*. The structure of the right arm is unknown.

Petersen *et al.* (1987) were able to isolate two types of chromosome XII. One was found to be homoeologous to chromosome XII in *Sacch. cerevisiae*, as it did not recombine with this chromosome in three intervals tested (*asp5–gal2*, *gal2–ilv5*, *ilv5–ura4*) on the right arm of the chromosome. The other type was homologous to chromosome XII of *Sacch. cerevisiae* and exhibited a considerable recombination in all three intervals. There is no information about the structure of the left arm, which makes it impossible to determine whether the former chromosome XII is entirely homoeologous, like chromosome V, or of a mosaic nature. A restriction-fragment length analysis of the *ILV5* gene showed the presence in the lager yeast of two different alleles of this gene. As expected, one allele closely resembled the one found in *Sacch. cerevisiae*, and was present on the recombining chromosome, while the other was divergent and found on the non-recombining chromosome. In this respect, the situation is similar to that reported for the structure of homologous and homoeologous regions of other chromosomes. The ribosomal-RNA locus of *Sacch. cerevisiae*, *RDN1*, is located on chromosome XII (Petes, 1979). The *RDN1* locus has been found to be polymorphic for restriction sites and has been classified into two different types, namely I and II, by Petes *et al.* (1978). Type I is common to most genetic reference strains of *Sacch. cerevisiae*, while lager strains of *Sacch. carlsbergensis* have type II (Pedersen, 1985). By molecular probing, Petersen *et al.* (1987) demonstrated that the *Sacch. cerevisiae*-like chromosome had an *RDN1* gene cluster, while *RDN1* could not be detected on the transferred chromosome XII specific to *Sacch. carlsbergensis*. Since, however, the parental lager yeast strain and a few analysed progeny strains in chromosome-separation gels exhibit several *ILV5* hybridizing bands, all of which contain *RDN1* sequences (M.B. Pedersen, unpublished observation), it cannot be precluded that the transferred chromosome was atypical because of size selection. Deletion of ribosomal-RNA genes has been observed in alloploid species of plants. For example, in hexaploid wheat, rDNA could not be detected in the A genome (Appels *et al.*, 1980).

Petersen *et al.* (1987) isolated three different chromosomes XIII by single-chromosome transfer. Genetic analysis showed the presence of: (i) a chromosome with homology to that in *Sacch. cerevisiae* allowing meiotic recombination between all three markers tested (*RAD52, LYS7, ILV2*); (ii) a non-recombining homoeologous chromosome; and (iii) a mosaic-type chromosome exhibiting recombination with *Sacch. cerevisiae* between *RAD52* and *LYS7*, but not between *LYS7* and *ILV2*. The mitotic stabilities of the three types of chromosome XIII in the chromosome-addition strains varied widely. In contrast to other chromosomes of *Sacch. carlsbergensis* studied, where the mitotic stability is rather high, the homoeologous chromosome XIII could not be retained in *Sacch. cerevisiae* under non-selective conditions. The homologous chromosome was found to be stable, while the mosaic type was maintained with intermediate stability. The *ILV2* locus of chromosome XIII was analysed by molecular hybridization and, as for the other genes of *Sacch. carlsbergensis* that have been analysed, two different alleles were found. The *ILV2* allele similar to that in *Sacch. cerevisiae* was associated with the homologous chromosome and deviated from that in *Sacch. cerevisiae* at a single *Xba*I site in the 5' region. The other allele was divergent in sequence and has been cloned (Gjermansen *et al.*, 1988) and sequenced (Gjermansen, 1991).

VI. GENOME STRUCTURE

In all of the chromosomes investigated from *Sacch. carlsbergensis*, the general picture that emerges is that chromosomal regions, and sometimes full chromosomes, are represented in the genome of lager yeast in two versions, one closely related to that present in genetic reference strains of *Sacch. cerevisiae*, while the other has a different albeit related nucleotide sequence. Several strains were detected where genes were found, or suggested to be missing or defective, in chromosomes of *Sacch. carlsbergensis* when compared with the chromosomes of *Sacch. cerevisiae*, but this was not very frequent nor was it limited to chromosomal regions specific to *Sacch. carlsbergensis*. In general, functionality and linkage have been highly conserved. Chromosome III of *Sacch. cerevisiae* has been sequenced and contains 182 open-reading frames for proteins longer than 100 amino-acid residues (Oliver *et al.*, 1992). It is noteworthy that it can be substituted with the analysed chromosome III of *Sacch. carlsbergensis* with no apparent phenotypic consequence. On the other hand, the divergence between the two versions of a given region is so large that most fragment sizes for commonly used restriction endonucleases are not conserved and a lowering of the stringency of hybridization from the standard conditions is needed to obtain a normal response in Southern analysis. The divergence is correlated with a severe drop in genetic recombination.

A. Nucleotide-sequence comparisons

The divergence between the nucleotide sequences in *Sacch. cerevisiae* and the chromosomal regions specific to *Sacch. carlsbergensis* has been analysed directly in two selected loci by Gjermansen (1991). In *ILV1* on chromosome V, the open-reading frames are of identical lengths and the alignment suggests deletion/insertion only outside the reading frames. An amino-acid sequence identity of 95.7% of the encoded polypeptides was derived from nucleotide sequences of 86.4% identity. The encoded enzyme is an anabolic threonine deaminase, an allosteric enzyme with a mitochondrial location (Ryan and Kohlhaw, 1974). It is hypothesized to have a transit peptide (Kielland-Brandt *et al.*, 1984; Holmberg *et al.*, 1985). This would be in accordance with the finding that conservation of the amino-acid sequence is much lower in the most N-terminal, 40 amino-acid residue long, non-acidic part of the encoded polypeptide than in the remainder of the protein.

The other locus is *ILV2* on chromosome XIII, encoding acetohydroxyacid synthase. The sequence in *Sacch. cerevisiae* was determined by Falco *et al.* (1985). Gjermansen (1991) found nucleotide-sequence and derived amino-acid sequence homologies quite similar to those found for *ILV1*. For *ILV2* he found 85.4% and 92.3%, respectively. In both loci, the nucleotide-sequence homology decreases outside the reading frames. Conservation of putative transcriptional control sites supports notions concerning their importance (Gjermansen, 1991). Hansen and Kielland-Brandt (1994) have cloned two *MET2* alleles from *Sacch. carlsbergensis*. One is virtually indistinguishable in its restriction endonuclease map from the *MET2* gene from *Sacch. cerevisiae* (located on chromosome XIV), while the sequence of the other deviates from the sequence of the *MET2* gene from *Sacch. cerevisiae*. Nucleotide-sequence identity with the sequence of the latter, published by Langin *et al.* (1986), is 84% in the coding region, while the derived amino-acid sequences are 94% identical. Gjermansen (1991) has reported partial sequence data that correspondingly suggest a homology in the *URA3* coding region of 79% with a derived amino-acid sequence homology of 93%.

B. Ploidy and origin

Having established the alloploid nature of *Sacch. carlsbergensis*, the important question of ploidy can be posed as follows: "What is the copy number of each version of a given chromosome or chromosomal region?" This question has been tackled for the *ILV2* locus in a lager production strain (Gjermansen *et al.*, 1988; Kielland-Brandt *et al.*, 1989, unpublished data). The experiments were part of a study of the consequences of eliminating *ILV2* gene function in the brewing yeast. In each of the two wild-type alleles, cloned in bacterial plasmids, a deletion was constructed *in vitro*, leaving enough of the sequences from the *ILV2* locus on both sides of the deletion to allow use of the two-step gene-replacement technique (Scherer and Davis, 1979) also referred to as pop-in/pop-out replacement (Rothstein, 1991). The vector carried a gene conferring resistance to the antibiotic geneticin (G418) on the yeast (Yocum, 1985), allowing selection of transformants as well as screening for subsequent spontaneous loss of vector sequences. The geneticin-sensitive segregants were screened by Southern analysis to reveal strains in which a wild-type gene, rather than the mutant gene, had looped away with the vector. It was found that the brewing yeast contained two copies of each of the two versions of the *ILV2* region. Generalization of this finding would mean that the brewing yeast is allotetraploid. A hypothesis of allotetraploidy in *Sacch. carlsbergensis* is consistent with the finding by Casey (1986b) of chromosomes of three sizes that hybridize with a chromosome X probe. Also, the strain of *Sacch. carlsbergensis* named "bottom fermenting yeast no. I" by Hansen (1908), which he isolated from a single cell in 1882–83, has chromosome III-hybridizing bands corresponding to three sizes (Pedersen, 1986b), in agreement with tetraploidy or triploidy, band intensities possibly favouring the former. Although this yeast can also be distinguished in other ways from the very uniform group of modern lager brewing yeasts, it is closely related to these (Pedersen, 1986b). Tetraploid *Sacch. cerevisiae* can exhibit high spore viability, yielding diploid spores with only occasional aneuploidy (Roman *et al.*, 1955). Being allotetraploid is no reason by itself for low spore viability; on the contrary, bivalent formation should if anything be more efficient than in a normal tetraploid. There are several reasons, however, why allotetraploidy and low spore viability in the brewing yeast are compatible. Firstly the alloploidy appears to be irregular, with many chromosomes exhibiting mosaic patterns of homology with *Sacch. cerevisiae*. Secondly, the finding of missing functions on some chromosomes, when they are compared to the homologues or homoeologues (*CAN1*, *RAD7*) in *Sacch. cerevisiae*, suggests that essential genes may also in some cases be defective or missing. Indeed, the failure of attempts to derive substitution strains from certain of the addition strains supports this notion. Finally, a low spore viability can obviously be caused by physiological factors that may be influenced by the environment (Bilinski *et al.*, 1987a).

One may speculate on the origin of *Sacch. carlsbergensis*. It seems likely that it is the result of hybridization between *Sacch. cerevisiae* and another species of *Saccharomyces*. Since the observed divergence is correlated with a drop in meiotic recombination expected to cause fertility barriers, we prefer to regard this hypothetical hybridization as being interspecific. Although the parent (which is not *Sacch. cerevisiae*) is not known, there are obvious candidates. In addition to the bottom fermenting yeast no. I, Hansen (1908) described bottom fermenting yeast no. II, which he named *Sacch. monacensis*. Pedersen (1986a,b) analysed this strain for restriction-fragment lengths at several loci, including Ty1 elements, and we see nothing that argues against the hypothesis proposed by Pedersen (1986b) that it represents the non-*Sacch. cerevisiae* parent species of *Sacch. carlsbergensis*. An alternative view is that the closely related *Sacch. bayanus* (Martini and Kurtzman, 1985; Martini and Martini, 1989), but the data of Pedersen (1986a,b), Casey and Pedersen (1988) and Pedersen (1993) point to *Sacch. monacensis* as being a more obvious candidate. Hybridization to chromosome-separation blots and restriction-fragment length analysis of *HIS4* and *LEU2* (chromosome III), as well as *URA3* and *ILV1* (chromosome

V), in *Sacch. monacensis* suggested that the strain is polyploid and homozygous for the alleles, which are unlike those in *Sacch. cerevisiae* but are
found in lager-brewing yeast. Furthermore Hansen and Kielland-Brandt
(1994) determined nucleotide sequences of a 330 base-pair *MET2* fragment
(located on chromosome XIV) recovered by polymerase chain reaction
from several *Saccharomyces* species. They found that the sequence in
Sacch. monacensis (CBS 1503) is identical with the non-*Sacch. cerevisiae*-
like *MET2* sequence from *Sacch. carlsbergensis* strains (a lager-brewing
yeast and the type strain CBS 1513), but only 93% identical with *MET2*
sequences found in *Sacch. bayanus* (type strain CBS 380) and *Sacch.
uvarum* (type strain CBS 395) (which are 100% identical).

Fragment-length analysis of *Ava*I-digested mitochondrial DNA from
the same *Saccharomyces* yeast strains (M.B. Pedersen, unpublished data)
support the hypothesis that *Sacch. monacensis* is one of the parental strains
of *Sacch. carlsbergensis*. The pattern of mtDNA from *Sacch. monacensis*
(CBS 1503) is very similar to mtDNA molecules from *Sacch. carlsbergensis*
(lager-brewing yeast and type strain CBS 1513) as well as *Sacch. pastoria-
nus* (CBS 1538). *Saccharomyces bayanus* (CBS 380) and *Sacch. uvarum*
(CBS 395) have nearly identical fragment patterns and are substantially different from *Sacch. monacensis*. Molecular hybridization of *ori1*-containing
sequences with Southern blots of the *Ava*I digested mtDNA gave results
consistent with this general picture. The size of the wild-type mtDNA from
Sacch. monacensis is 68 kb, in agreement with the data of Good *et al.* (1993)
for *Sacch. carlsbergensis* lager-brewing yeast.

The appearance of chromosomes of a mosaic type in analysis after
single-chromosome transfer may raise a few questions. Firstly, in terms of
evolutionary origin, it could in principle be the chromosome of *Sacch.
cerevisiae* that is the mosaic, rather than the given analysed one in *Sacch.
carlsbergensis*. However, the occurrence of a mosaic type of chromosome
XIII as well as the two non-mosaics argues against this view, as does the
occurrence of two types of mosaics of chromosome X. It is also more
obvious to think of mosaic chromosomes arising in an alloploid genetic
background that is known to exist rather than in hypothetical hybrids
that would have to be invoked. Another obvious question is whether the
recovered mosaic chromosomes exist as such in the brewing yeast or
whether they were formed in the meiotic divisions giving rise to the mating
chromosome donor strains. In fact, we do not know but we find the first
alternative the more likely one. Unless the constraints on meiotic recombination conferred by sequence divergence are dramatically relaxed in the
brewing yeast as compared with *Sacch. cerevisiae*, the second alternative
can explain only occasional recovery of mosaic chromosomes. The repeated
recovery of a single pattern of mosaicism in chromosome III (Nilsson-

Tillgren *et al.*, 1981; Pedersen, 1986b) also argues for the first alternative. We thus favour the idea that mosaic chromosomes exist in the brewing yeasts and that they have arisen in the alloploid situation by rare recombinational events. One way these events could be brought about is recombinational repair of double-strand breaks. Although this repair is strongly affected by the sequence divergence between homoeologues of *Sacch. cerevisiae* and *Sacch. carlsbergensis* (Resnick *et al.*, 1989), it seems to be generally less affected by this divergence than is meiotic recombination (Tullin, 1990; Resnick *et al.*, 1992).

VII. CONSEQUENCES FOR BREEDING

Various aspects of the breeding of brewing yeasts have been reviewed recently by Tubb and Hammond (1987), Bilinsky and Casey (1989), Kielland-Brandt *et al.* (1989) and Hinchliffe (1991). We here mainly concern ourselves with the consequences of the genetic structure of *Sacch. carlsbergensis* already described.

The difficulties we are confronted with when genetic crosses and mutagenesis are applied in genetic analysis of brewing yeast obviously also meet us when we try to put the same techniques to work in breeding. This is one of the reasons why these "classical" techniques have received relatively little attention, especially after the introduction of other techniques for genetic change. In our view, they still deserve much attention, but obviously their use needs to take the genetic structure of the yeast into account.

The work by Gjermansen and Sigsgaard (1981) illustrates the potential for cross-breeding of *Sacch. carlsbergensis*. Pairwise crosses were performed between some of the maters derived by sporulation of a single production strain and the resulting hybrids were screened. Colony size and aerobic growth rate were initial criteria, and subsequently the fermentation behaviour was tested in brewer's wort in 2-litre cultures under conditions designed to simulate reasonably fermentation in large tanks. Most hybrids that reached this stage showed inferior performance, but the one that most resembled the parental production strain turned out to perform well in larger experimental-scale and in production trials. This shows that, although the mating meiotic segregants are probably alloploid with an irregularity at least as bad as their parent, it is possible to find among them a pair that by mating yields a good production strain.

The maters were originally thought to be haploid or near haploid, since recessive mutations can be induced in them (Gjermansen, 1983), but alloploidy has been observed in them for chromosomes III (S. Holmberg, unpublished observation), V (Nilsson-Tillgren *et al.*, 1986), X (Casey, 1986b) and XII and XIII (Petersen *et al.*, 1987). They may well be mostly alloploid and frequently aneuploid, as might be expected for meiotic segregants of an irregular allotetraploid. With this in mind, it is not surprising that the hybrids were a heterogeneous population, but this finding also suggests itself as a basis for conservative cross-breeding. The hybrids performing well are expected to differ slightly in various ways, such as in flocculence, to make them individually better suited for certain tank sizes, or other features of the process or production plant. Whether the maters are derived from different yeasts (outbreeding; Bilinski *et al.*, 1987b) or a single production yeast (inbreeding), the obtained genetic variation will be much richer than the variation obtainable from screening a mitotic population. Obviously, cross-breeding can be applied to combine known desired features of parental strains (see, for example Emeis, 1971; Janderová *et al.*, 1990).

However, the most important potential of pairs of maters that can yield good production strains lies in their lower ploidy, enabling the breeder to induce recessive mutations (Gjermansen, 1983). When the desired mutants cannot be recovered in a single step, consideration of the possible allodiploid nature of the maters is helpful, as illustrated by the work now to be described (C. Gjermansen, unpublished data).

A notable example showing that recessive mutants can be desired for breeding purposes is shown by the effect of *ilv2* mutations on the time required for beer maturation. The *ILV2* gene codes for acetolactate synthase, which catalyses the first step on the biosynthetic pathway from pyruvate to valine. The bulk of the first intermediate, acetolactate, normally goes to produce valine but, during the primary fermentation in beer production, some acetolactate leaks out of yeast cells to be slowly converted to diacetyl. Unless the yeast is allowed during the secondary fermentation (lagering, maturation) to reduce the diacetyl, excessive amounts of this compound will give the beer a buttery off-flavour. Removal of diacetyl is one of the main functions of lagering, since maturation of beer can be effected enzymically in 24 hours by addition of α-acetolactate decarboxylase (Godtfredsen and Ottesen, 1982). Various approaches have been taken by several laboratories to construct yeast strains that leak less acetolactate to permit a decrease in maturation time (Gjermansen *et al.*, 1988; Fujii *et al.*, 1990; Suihko *et al.*, 1990; Villanueba *et al.*, 1990). The approach that was first investigated consisted in a decrease in the activity of acetolactate synthase (Ramos-Jeunehomme and Masschelein, 1977). Complete elimination of acetolactate synthase activity in *Sacch. carlsbergensis* can be carried out by *in vitro* deletion and recombinational replacement of *ILV2* genes with the mutant alleles as already described (C. Gjermansen, unpublished observation). However, completely blocking this biosynthetic step has not yielded brewing yeasts with good characteristics, presumably because the parental brewing yeast does not take up isoleucine and valine as efficiently as *Sacch. cerevisiae* (Tullin, 1990; Kielland-Brandt *et al*, 1990). A slight decrease in acetolactate synthase activity in polyploid brewing yeasts can be obtained by dominant mutation. Sulphonyl urea herbicides, such as sulphometuron methyl, induce isoleucine and valine requirements in *Sacch. cerevisiae* by inhibiting acetolactate synthase, and dominant mutations in *ILV2* can cause resistance by encoding an acetolactate synthase with lowered binding affinity to the herbicide (Falco and Dumas, 1985). In some mutants, the enzyme has become less active. Since spontaneous dominant mutations are easily obtained in polyploids, this fact has been used in pratical breeding but, as also expected in polyploids, only moderate results can be obtained in this way (Gjermansen and Sigsgaard, 1986; Galván *et al.*, 1987).

However, C. Gjermansen (unpublished results, see also Kielland-Brandt *et al.*, 1989) devised a procedure for producing recessive mutations in genes, such as *ILV2*, in which dominant mutations can be obtained. The basis is a pair of presumed allodiploid maters, which by mating can yield a good production strain (Gjermansen and Sigsgaard, 1981). Firstly, spontaneous herbicide-resistant mutants of these strains are isolated. In a diploid, such a mutant will expectedly contain one wild-type *ILV2* gene and one *ILV2* gene mutated to resistance. Mutagenic treatment with, for example, ultraviolet radiation, and subsequent screening for reversion to sensitivity can now yield secondary mutants in which the resistance gene has been inactivated. In diploid strains in *Sacch. cerevisiae*, this procedure would yield strains with two wild-type *ILV2* genes, just like the original diploid, since mitotic recombination induced by the mutagenic treatment would result in a high rate of homozygotization of the wild-type *ILV2* gene present in the primary resistant mutant. However, knowing that we are dealing with alloploidy, we can expect that the decreased frequency of mitotic recombination (Skaanild, 1985; Resnick *et al.*, 1989) allows isolation of maters with just one active *ILV2* gene. Indeed, C. Gjermansen (unpublished data) has found that further mutagenesis of these maters, and screening for slow growth on medium devoid of isoleucine and valine, yields strains with very low acetolactate synthase activity. Mating of these strains yields hybrids that grow well in wort, produce little diacetyl and give an acceptable beer (C. Gjermansen, K. Erdal and P. Sigsgaard, unpublished observations).

In strain construction, one is often interested in providing the yeast with the potential of synthesizing an extra protein by introducing the corresponding gene. If it is a yeast protein, it may be expedient to use the yeast gene with its own signals for expression (such as promoter, terminator, signal sequence codons). If it is a heterologous protein, it is generally advantageous to flank the coding region with a yeast promoter and sometimes also other signals.

In commercial production of a purified heterologous protein, one usually utilizes the enhanced production potential provided by the high copy number of some self-replicating plasmids. Self-replicating plasmids can also be employed in breeding of industrial yeasts for traditional purposes (see, for example, Hinchliffe, 1991; Suihko *et al.*, 1990; Villanueba *et al.*, 1990). However, the required production of an extra protein may for many purposes be lower and be accomplished with a single copy of the gene in question. In this situation, the breeder has the option of integrating the gene into a yeast chromosome with the stable presence of the gene as a result. With gene-replacement procedures (Scherer and Davis, 1979; Rothstein, 1991), it is straightforward to obtain full stability in a haploid strain. In polyploid yeasts, it may be expedient to carry out the integration at a given locus in two or more homologous chromosomes in order to render loss of the gene by mitotic recombination impossible. In this way, high stability is attained. The stringency of the requirement for a low rate of loss of the introduced gene depends obviously on whether there is significant selection against cells harbouring the gene. In the alloploid situation, it may well be sufficient to integrate the gene into highly homologous chromosomes, since loss by mitotic recombination with diverged chromosomes will take place at a lower rate. Yocum (1985, 1986) constructed vectors useful for integration of genes into the *HO* locus of prototrophic yeasts. A *Sacch. carlsbergensis* strain that he investigated turned out to harbour three copies of an *HO* or *ho* gene, like that present in *Sacch. cerevisiae*, while he did not look for a diverged gene using low stringency hybridization. In order to obtain sufficient stability of the presence of an amyloglucosidase gene from *Aspergillus niger* in the *HO* locus of *Sacch. carlsbergensis* to allow practical use, it was necessary to integrate the gene in two homologues.

References

Aigle, M., Erbs, D. and Moll, M. (1984). *Journal of the American Society of Brewing Chemists* **42**, 1.
Anderson, E. and Martin, P.A. (1975). *Journal of the Institute of Brewing* **81**, 242.
Appels, R., Gerlach, W.L., Dennis, E.S., Swift, H. and Peacock, W.J. (1980). *Chromosoma* **78**, 293.
Barnett, J.A. (1992). *Yeast* **8**, 1.
Bilinski, C.A. and Casey, G.P. (1989). *Yeast* **5**, 429.

Bilinski, C.A., Russell, I. and Stewart, G.G. (1987a). *Journal of the Institute of Brewing* **93**, 216.

Bilinski, C.A., Russell, I. and Stewart, G.G. (1987b). "Proceedings of the 21st European Brewery Convention Congress", Madrid, pp. 497–504. IRL Press, Oxford.

Bornæs, C., Petersen, J.G.L. and Holmberg, S. (1992). *Genetics* **131**, 531.

Borts, R.H. and Haber, J.E. (1987). *Science* **237**, 1459.

Broach, J.R. (1981). *In* "The Molecular Biology of the Yeast *Saccharomyces*. Life Cycle and Inheritance" (J.N. Strathern, E.W. Jones and J.R. Broach, eds), pp. 445–470. Cold Spring Harbor Laboratory, Cold Spring Harbor.

Campbell, B. (1987). *In* "Brewing Microbiology" (F.G. Priest and I. Campbell, eds), pp. 1–13. Elsevier Applied Science, London, New York.

Carle, G.F. and Olson, M.V. (1985). *Proceedings of the National Academy of Sciences, USA* **82**, 3756.

Casey, G.P. (1986a). *Carlsberg Research Communications* **51**, 327.

Casey, G.P. (1986b). *Carlsberg Research Communications* **51**, 343.

Casey, G.P. and Pedersen, M.B. (1988). *Carlsberg Research Communications* **53**, 209.

Claisse, M., Michel, F. and Hawthorne, D. (1987). *In* "Plant Mitochondria. Structural, Functional, and Physiological Aspects" (A.L. Moore and R.B. Beechey, eds), pp. 283–292. Plenum Press, New York.

Clark-Walker, G.D., Evans, R.J., Hoeben, P. and McArthur, C.R. (1985). *In* "Achievements and Perspectives of Mitochondrial Research" (E. Quagliariello, E.C. Slater, F. Palmieri, C. Saccone and A.M. Kroon, eds), Vol. II, pp. 71–78. Biogenesis, Elsevier, Amsterdam.

Conde, J. and Fink, G.R. (1976). *Proceedings of the National Academy of Sciences, USA* **73**, 3651.

Conde, J. and Mascort, J.L. (1981). *Proceedings of the 18th European Brewery Convention Congress*, Copenhagen, pp. 177–186. IRL Press, London.

Devin, A.B., and Koltovaya, N.A. (1981). *Mutation Research* **91**, 451.

Dutcher, S.K. (1981). *Molecular and Cellular Biology* **1**, 245.

Emeis, C.C. (1971). *American Society of Brewing Chemists Proceedings*, pp. 58–62.

Evans, I.H. and Wilkie, D. (1976). *Genetical Research, Cambridge* **27**, 89.

Falco, S.C. and Dumas, K.S. (1985). *Genetics* **109**, 21.

Falco, S.C., Dumas, K.S. and Livak, K.J. (1985). *Nucleic Acids Research* **13**, 4011.

Federoff, H.J., Cohen, J.D., Eccleshall, T.R., Needleman, R.B., Buchferer, B.A., Giacalone, J. and Marmur, J. (1982). *Journal of Bacteriology* **149**, 1064.

Fink, G.R. and Styles, C.A. (1974). *Genetics* **77**, 231.

Fowell, R.R. (1969). *In* "The Yeasts" (A.H. Rose and J.S. Harrison, eds), Vol. 1, pp. 303–383. Academic Press, London.

Fujii, T., Kondo, K., Shimizu, F., Sone, H., Tanaka, J.-I. and Inoue, T. (1990). *Applied and Environmental Microbiology* **56**, 997.

Galván, L., Pérez, A., Delgado, M. and Conde, J. (1987). *Proceedings of the 21st European Brewery Convention Congress*, Madrid, pp. 385–392. IRL Press, Oxford.

Gilliland, R.B. (1951). *Proceedings of the European Brewery Convention Congress, Brighton*, pp. 35–58. Elsevier Publishing Company, Amsterdam.

Gilliland, R.B. (1971). *Journal of the Institute of Brewing* **77**, 276.

Gjermansen, C. (1983). *Carlsberg Research Communications* **48**, 557.

Gjermansen, C. (1991). PhD Thesis: University of Copenhagen.

Gjermansen, C. and Sigsgaard, P. (1981). *Carlsberg Research Communications* **46**, 1.

Gjermansen, C. and Sigsgaard, P. (1986). *E.B.C.-Symposium on Brewers' Yeast*, Vuoranta, Finland, pp. 156–168. Verlag Hans Carl, Nürnberg.

Gjermansen, C, Nilsson-Tillgren, T., Petersen, J.G.L., Kielland-Brandt, M.C., Sigsgaard, P. and Holmberg, S. (1988). *Journal of Basic Microbiology* **28**, 175.

Godtfredsen, S.E. and Ottesen, M. (1982). *Carlsberg Research Communications* **47**, 93.

Good, L., Dowhanick, T.M., Ernandes, J.E., Russell, I. and Stewart, G.G. (1993). *Journal of the American Society of Brewing Chemists* **51**, 35.

Goodey, A.R. and Tubb, R.S. (1982). *Journal of General Microbiology* **128**, 2615.

252   M.C. Kielland-Brandt, T. Nilsson-Tillgren, C. Gjermansen, S. Holmberg and M.B. Pedersen

Gyllang, H. and Martinson, E. (1972). *Proceedings of the 13th European Brewery Convention Congress*, 1971, Estoril, pp. 265–271. Elsevier Publishing Company, Amsterdam.
Hammond, J.R.M. and Eckersley, K.W. (1984). *Journal of the Institute of Brewing* **90**, 167.
Hansen, E.C. (1888). *Meddelelser fra Carlsberg Laboratorium* **2**, 257.
Hansen, E.C. (1908). *Meddelelser fra Carlsberg Laboratorium* **7**, 166.
Hansen, J. and Kielland-Brandt, M.C. (1994). *Gene* **140**, 33.
Hinchliffe, E. (1991). In "Applied Molecular Genetics of Fungi" (J.F. Peberdy, C.E. Caten, J.E. Ogden and J.W. Bennett, eds), pp. 129–145. Cambridge University Press, Cambridge.
Hinrichs, J., Stahl, U. and Esser, K. (1988). *Applied Microbiology and Biotechnology* **29**, 48.
Holmberg, S. (1982). *Carlsberg Research Communications* **47**, 233.
Holmberg, S. and Kielland-Brandt, M.C. (1978). *Carlsberg Research Communications* **43**, 37.
Holmberg, S., Kielland-Brandt, M.C., Nilsson-Tillgren, T. and Petersen, J.G.L. (1985). *Carlsberg Research Communications* **50**, 163.
Janderová, B., Cvrčková, F. and Bendová, O. (1990). *Journal of Basic Microbiology* **30**, 499.
Johnston, J.R. (1965). *Journal of the Institute of Brewing* **71**, 130.
Kielland-Brandt, M.C., Petersen, J.G.L. and Mikkelsen, J.D. (1979). *Carlsberg Research Communications* **44**, 27.
Kielland-Brandt, M.C., Wilken, B., Holmberg, S., Petersen, J.G.L. and Nilsson-Tillgren, T. (1980). *Carlsberg Research Communications* **45**, 119.
Kielland-Brandt, M.C., Nilsson-Tillgren, T., Petersen, J.G.L. and Holmberg, S. (1981). In "Molecular Genetics in Yeast, Alfred Benzon Symposium" (D. von Wettstein, J. Friis, M. Kielland-Brandt and A. Stenderup, eds), pp. 369–382. Munksgaard, Copenhagen.
Kielland-Brandt, M.C., Holmberg, S., Petersen, J.G.L. and Nilsson-Tillgren, T. (1984). *Carlsberg Research Communications* **49**, 567.
Kielland-Brandt, M.C., Gjermansen, C., Nilsson-Tillgren, T. and Holmberg, S. (1989). *Proceedings of the 22nd European Brewery Convention Congress*, Zürich, pp. 37–47. Oxford University Press, Oxford.
Kielland-Brandt, M.C., Gjermansen, C., Tullin, S., Nilsson-Tillgren, T., Sigsgaard, P. and Holmberg, S. (1990). *Proceedings of the 6th International Symposium on Genetics of Industrial Microorganisms*, Strasbourg, (H. Heslot, J. Davies, J. Florent, L. Bobichon, G. Durand and L. Penasse, eds), pp. 877–885. Société Française de Microbiologie.
Kreger-van Rij, N.J.W. (1984). "The Yeasts, a Taxonomic Study". Elsevier Biomedical Press, Amsterdam.
Kreil, H., Kleber, W. and Teuber, M. (1975). *Proceedings of the 15th European Brewery Convention Congress*, Nice, pp. 323–329. Elsevier Scientific Publishing Company, Amsterdam.
Kurtzman, C.P. and Phaff, H.J. (1987). In "The Yeasts" (A.H. Rose and J.S. Harrison, eds), Vol. 1, pp. 63–94. Academic Press, London.
Langin, T., Faugeron, G., Goyon, C., Nicolas, A. and Rossignol, J.-L. (1986). *Gene* **49**, 283.
Lewis, C.W., Johnston, J.R. and Martin, P.A. (1976). *Journal of the Institute of Brewing* **82**, 158.
Livingston, D.M. (1977). *Genetics* **86**, 73.
Martini, A.V. and Kurtzman, C.P. (1985). *International Journal of Systematic Bacteriology* **35**, 508.
Martini, A.V. and Martini, A. (1989). In "Biotecnology Applications in Beverage Production" (C. Cantarelli and G. Lanzarini, eds), pp. 1–16. Elsevier Applied Science, London.
Mead, D.J., Gardner, D.C.J. and Oliver, S.G. (1986). *Molecular and General Genetics* **205**, 417.
Meaden, P.G. and Taylor, N.R. (1991). *Journal of the Institute of Brewing* **97**, 353.
Molzahn, S.W. (1977). *Journal of the American Society of Brewing Chemists* **35**, 54.
Morrison, K.B. and Suggett, A. (1983). *Proceedings of the 19th European Brewery Convention Congress*, London, pp. 489–496. IRL Press, Oxford.
Mortimer, R.K., Schild, D., Contopoulou, C.R. and Kans, J.A. (1989). *Yeast* **5**, 321.
Nielsen, T.L. (1991). PhD Thesis: University of Copenhagen.
Nilsson-Tillgren, T., Petersen, J.G.L., Holmberg, S. and Kielland-Brandt, M.C. (1980). *Carlsberg Research Communications* **45**, 113.

Nilsson-Tillgren, T., Gjermansen, C., Kielland-Brandt, M.C., Petersen, J.G.L. and Holmberg, S. (1981). *Carlsberg Research Communications* **46**, 65.
Nilsson-Tillgren, T., Gjermansen, C., Holmberg, S., Petersen, J.G.L. and Kielland-Brandt, M.C. (1986). *Carlsberg Research Communications* **51**, 309.
Oliver, S.G. *et al.* (1992). *Nature* **357**, 38.
Pedersen, M.B. (1983). *Carlsberg Research Communications* **48**, 485.
Pedersen, M.B. (1985). *Carlsberg Research Communications* **50**, 263.
Pedersen, M.B. (1986a). *Carlsberg Research Communications* **51**, 163.
Pedersen, M.B. (1986b). *Carlsberg Research Communications* **51**, 185.
Pedersen, M.B. (1994). *Journal of the American Society of Brewing Chemists* **52**, 23.
Petersen, J.G.L., Nilsson-Tillgren, T., Kielland-Brandt, M.C., Gjermansen, C. and Holmberg, S. (1987). *Current Genetics* **12**, 167.
Petes, T.D. (1979). *Proceedings of the National Academy of Sciences, USA* **76**, 410.
Petes, T.D., Hereford, L.M. and Skryabin, K.G. (1978). *Journal of Bacteriology* **134**, 295.
Piškur, J. (1989). *Gene* **81**, 165.
Ramos-Jeunehomme, C. and Masschelein, C.A. (1977). *Proceedings of the 16th European Brewery Convention Congress*, Amsterdam, pp. 267–283. European Brewery Convention, Zoeterwoude.
Resnick, M.A., Skaanild, M. and Nilsson-Tillgren, T. (1989). *Proceedings of the National Academy of Sciences, USA* **86**, 2276.
Resnick, M.A., Zgaga, Z., Hieter, P., Westmoreland, J., Fogel, S. and Nilsson-Tillgren, T. (1992). *Molecular and General Genetics* **234**, 65.
Roman, H., Phillips, M.M. and Sands, S.M. (1955). *Genetics* **40**, 546.
Rothstein, R. (1991). *In* "Methods in Enzymology" (C. Guthrie and G.R. Fink, eds), Vol. 194, pp. 281–301. Academic Press, San Diego.
Ryan, E.D. and Kohlhaw, G.B. (1974). *Journal of Bacteriology* **120**, 631.
Ryder, D.S., Murray, J.P. and Stewart, M. (1978). *Master Brewers Association of the Americas Technical Quarterly* **15**, 79.
Scherer, S. and Davis, R.W. (1979). *Proceedings of the National Academy of Sciences, USA* **76**, 4951.
Séraphin, B., Simon, M. and Faye, G. (1987). *Journal of Biological Chemistry* **262**, 10146.
Šilhánková, L., Moštek, J., Šavel, J. and Šolínová, H. (1970). *Journal of the Institute of Brewing* **76**, 289.
Skaanild, M.T. (1985). MSc Thesis: University of Copenhagen.
Smolik-Utlaut, S. and Petes, T.D. (1983). *Molecular and Cellular Biology* **3**, 1204.
Stewart, G.G. and Russell, I. (1977). *Canadian Journal of Microbiology* **23**, 441.
Stewart, G.G. and Russell, I. (1987). *E.B.C.-Symposium on Brewers' Yeast, 1986*, Vuoranta, Finland, pp. 53–70. Verlag Hans Carl, Nürnberg.
Stewart, G.G., d'Amore, T., Novak, S. and Russell, I. (1990). *Proceedings of the 6th International Symposium on Genetics of Industrial Microorganisms*, Strasbourg (H. Heslot, J. Davies, J. Florent, L. Bobichon, G. Durand and L. Penasse, eds), pp. 887–896. Société Française de Microbiologie.
Suihko, M.-L., Blomqvist, K., Penttilä, M., Gisler, R. and Knowles, J. (1990). *Journal of Biotechnology* **14**, 285.
Takata, Y., Watari, J., Nishikawa, N. and Kamada, K. (1989). *Journal of the American Society of Brewing Chemists* **47**, 109.
Ten Berge, A.M.A. (1972). *Molecular and General Genetics* **115**, 80.
Thorne, R.S.W. (1951). *Comptes Rendus des Travaux du Laboratoire Carlsberg, Série Physiologique* **25**, 101.
Tubb, R.S. (1980). *Journal of the Institute of Brewing* **86**, 78.
Tubb, R.S. and Hammond, J.R.M. (1987). *In* "Brewing Microbiology" (F.G. Priest and I. Campbell, eds), pp. 47–82. Elsevier Applied Science, London.
Tubb, R.S., Searle, B.A., Goodey, A.R. and Brown, A.J.P. (1981). *Proceedings of the 18th European Brewery Convention Congress*, Copenhagen, pp. 487–496. IRL Press, London.
Tullin, S. (1990). MSc Thesis: University of Copenhagen.

Van Oevelen, D., Spaepen, M., Timmermans, P. and Verachtert, H. (1977). *Journal of the Institute of Brewing* **83**, 356.

van der Walt, J.P. (1987). *In* "The Yeasts" (A.H. Rose and J.S. Harrison, eds), 2nd edn, Vol. 1, pp. 95-121. Academic Press, London.

Verachtert, H. and Dawoud, E. (1990). *Louvain Brewing Letters* **3** (1/2), 15.

Villanueba, K.D., Goossens, E. and Masschelein, C.A. (1990). *Journal of the American Society of Brewing Chemists* **48**, 111.

Volkert, F.C., Wilson, D.W. and Broach, J.R. (1989). *Microbiological Reviews* **53**, 299.

Watari, J., Takata, Y., Ogawa, M., Nishikawa, N. and Kamimura, M. (1989). *Agricultural and Biological Chemistry* **53**, 901.

Watari, J., Takata, Y., Ogawa, M. *et al.* (1994). *Yeast* **10**, 211.

Wickner, R.B. (1986). *Annual Review of Biochemistry* **55**, 373.

Winge, Ö. (1944). *Comptes Rendus des Travaux du Laboratorie Carlsberg, Série Physiologique* **24**, 79.

Xiao, W. and Rank, G.H. (1990). *Journal of the American Society of Brewing Chemists* **48**, 107.

Yocum, R.R. (1985). European patent application 85303625.9

Yocum, R.R. (1986). *In* "BioExpo 86 Proceedings", pp. 171-180. Butterworth Publishers, Stoneham.

Young, T.W. (1981). *Journal of the Institute of Brewing* **87**, 292.

Young, T.W. (1983). *Proceedings of the 19th European Brewery Convention Congress*, London, pp. 129-136. IRL Press Ltd, Oxford.

Zakharov, I.A. and Yarovoy, B.Ph. (1977). *Molecular and Cellular Biochemistry* **14**, 15.

7 Genetics of Industrial Yeasts other than *Saccharomyces cerevisiae* and *Schizosaccharomyces pombe*

Peter E. Sudbery

Department of Molecular Biology and Biotechnology, University of Sheffield, Sheffield S10 2TN, UK

The Yeasts Vol. 6, 2nd edition
ISBN 0-12-596416-1

I. INTRODUCTION

The importance of *Saccharomyces cerevisiae*, not only in fundamental research, but also as an industrial yeast is well known. Much of this series is devoted to chronicling the large body of knowledge concerning its biochemistry, physiology, genetics and molecular biology that has accumulated over the last few decades. Its use in the production of alcoholic beverages and industrial ethanol, baking, and its use as a foodstuff, is ancient. More recently, it has become one of the major vehicles for production of recombinant proteins.

Nevertheless, other species of yeast are either in use or have the demonstrated potential for a wide variety of industrial processes. These include production of single-cell protein, organic acids, industrial enzymes, lipids, ethanol, fine biochemicals, vitamins and recombinant proteins. An important advantage is the ability of some of these yeasts to utilize unusual growth substrates, which are often either industrial waste products or inexpensive. These include methanol, alkanes and fuel oils, milk whey, wood sugars, sulphite-waste liquor and xylose. This list is not intended to be exhaustive; a more extensive review on the applications of industrial yeasts and the variety of potential growth substrates may be found in Cartledge (1987), and Burden and Eveleigh (1990).

Many of these yeasts have been the subject of genetic analysis reviewed recently by Ogrydziak (1988). In a recent report on a conference of the genetics of non-conventional yeasts, Weber (1988) listed 53 species for which genetic data were presented. Extensive analysis is limited to a much smaller group of organisms; indeed at the same conference, Ogrydziak (1988) listed only six species for which linkage data have been obtained from tetrad analysis. Molecular transformation systems are, however, available for a growing number of these other yeasts (Table I). The scope of this chapter is to review the progress made in this field. The basic genetic approaches available will be introduced, and the problems and strengths of each discussed. This will be followed by brief reviews of those organisms where extensive genetic systems have been developed or where the organism is of particular industrial importance.

II. TECHNIQUES OF GENETIC ANALYSIS

A. Classical techniques

Classical techniques of genetic analysis encompass the capability to cross genetically marked haploid strains to form a diploid, and to study allelic

Table I. Reports of transformation in yeasts other than *Saccharomyces cerevisiae* and *Schizosaccharomyces pombe*

Species	Selectable marker	Fate of DNA	References
Candida albicans	*ADE1*	Integrating	Kurtz et al. (1986)
	ADE1	Replicating[b]	Kurtz et al. (1987)
Candida guilliermondii	*ARG4* (S.c.)[a]	Replicating[b]	Kunze et al. (1985a,b)
Candida maltosa	*ARG4* (S.c.)	Replicating[b]	Kunze et al. (1985a,b)
	LEU2 (S.c.)	Replicating	Takagi et al. (1986)
	HIS5	Replicating	Hikiji et al. (1989)
Candida tropicalis	*URA3* (S.c.)	Integrating	Haas et al. (1990)
Hansenula polymorpha	*URA3* (S.c.)	Replicating	Roggenkamp et al. (1986)
	LEU2 (S.c.)	Replicating[b]	Gleeson et al. (1986)
	LEU2 (S.c.)	Replicating	Tikhomirova et al. (1986)
	CUP1[r] (S.c.)	Replicating	Tikhomirova et al. (1988)
	G418[r]	Integrating	Sudbery et al. (1988)
Klyuveromyces lactis	*G418*[r],	Replicating	Das and Hollenberg (1982)
	TRP1 (S.c.),		
	β-gal		
	URA3 (S.c.),	Replicating	Bianchi et al. (1987)
	G418[r]		
Klyuveromyces fragilis	*G418*[R]	Replicating	Das et al. (1984)
Pichia pastoris	*HIS4*[c]	Replicating/ integrating	Cregg et al. (1985)
Schwanniomyces occidentalis	*TRP5* (S.c.)	Replicating	Dohmen et al. (1989)
	ADE2	Replicating	Klein and Faureaum (1988)
Trichosporon cutaneum	Hygromycin B[r],	Replicating	Glumoff et al. (1989)
	Phleomycin[r]		
Yarrowia lipolytica	*LEU2*	Integrating	Davidow et al. (1985)
	LYS2 (S.c.)	Integrating	Gaillardin et al. (1985)
	Phleomycin[r],	Integrating	Gaillardin and Ribet (1987)
	β-gal		
	Invertase (S.c.)	Integrating	Nicaud et al. (1989)

[a] "S.c." denotes that the sequence in the selection originated in *Sacch. cerevisiae*. The symbols R (dominant) and r (recessive) are used to indicate the resistance conferred by gene products.
[b] Replication is not programmed by specific ars-like sequences but fortuitously from undefined sequences within the transforming DNA.
[c] Both homologous and *Sacch. cerevisiae HIS4* sequences were used.

relationships of mutations by complementation and linkage relationships by meiotic recombination. In addition, parasexual phenomena such as mitotic recombination and haploidization may be utilized. The use of these techniques allows the basic genetic constitution of a yeast to be elaborated, is a powerful tool in elucidation at the molecular level of the basis and regulation of all aspects of the yeast's biology, and allows strain improvement by recombination of favourable traits from different strains.

Most perfect yeast species exhibit a similar life cycle with a bipolar mating-type system and sporulation of the diploid, producing unordered four-spored asci. Despite the bipolar mating-type system, mating is often homothallic. In *Sacch. cerevisiae*, this comes about through mating-type switching during cell division. The switching event is controlled by the *HO* locus which, upon mutation, results in the yeast displaying an apparently heterothallic system. Because of the general similarity in the life cycles of yeasts, similar genetic techniques can be applied. This results in programmes of genetic analysis in new yeasts species borrowing heavily from the methodology first elaborated in *Sacch. cerevisiae*. In order for such programmes to be successful, two requirements must be fulfilled. Firstly, the yeast must be perfect, that is the sexual stage must be known. Secondly, mutant isolation must be possible.

With regard to the life cycle, the conditions necessary to induce mating in haploids and sporulation in diploids must be known and easily manipulable. Efficiencies of mating, sporulation and spore germination should be reasonably high. Low mating efficiencies make it difficult to carry out the large-scale complementation tests necessary to assign a large number of mutants with the same phenotype to complementation groups. Low-spore germination makes tetrad analysis difficult and extremely inefficient as only a small minority of tetrads will be complete. A related point is that asci can often be difficult to dissect, which again hinders tetrad analysis.

A knowledge of whether mating systems are homothallic or heterothallic is clearly necessary. Although heterothallic systems are usually thought to be preferable, homothallic systems are in some ways very convenient, since all haploid strains can be intercrossed immediately. An important proviso is that the haplophase can be stabilized, that is, mating is not constitutive but is under some controllable environmental condition such as growth on a particular carbon source. Often homothallic strains can be converted to the heterothallic state by mutation of the locus controlling mating-type switching. This is what occurred accidentally in the development of the genetics of *Sacch. cerevisiae* and has been deliberately engineered into *Pichia pinus* (Benevolenskii and Tolstorukov, 1980).

The ability to generate mutations requires a stable haplophase, since diploidy, aneuploidy or polyploidy will result in recessive mutations being masked by remaining wild-type alleles. Since auxotrophic mutations are the commonest form of genetic marker, a chemically defined axenic medium is extremely useful. Other factors that may prevent recovery of mutants are the inability to utilize exogenous supplements, which results in nutritional mutants being unconditionally lethal, and abnormally efficient repair systems, which prevent mutations from becoming fixed. Often the reason for a failure to recover mutants is unclear, since basic knowledge about the genetic properties of a yeast is, to some extent, dependent on the success of genetic analysis. An example of this difficulty is *Candida albicans*, where it has only recently been established that the organism is diploid, accounting for the previous difficulty in mutant recovery (see Poulter, Chapter 8 of this volume).

A number of other types of mutation in addition to auxotrophy may be used. Temperature—sensitive mutations may be useful where a defined medium cannot be used or uptake systems prevent isolation of nutritional markers. They are sometimes useful for analysis of particular systems, for example, the assembly of organelles or macromolecular structures is often blocked by cold-sensitive mutations. Drug-resistant mutations are valuable for two reasons. Firstly, they are often easy to isolate, since positive selection may be used. Secondly, they may be dominant, which allows genetic markers to be introduced into yeasts that are refractory to recessive mutations for the reasons already stated. Lastly, some yeasts are capable of producing respiratory-deficient petite mutations. Such mutations may be nuclear or mitochondrial, the latter again offering a route for introducing genetic markers into yeasts that are not amenable to other types of mutation. They have been used to effect interspecific hybridizations by protoplast fusion (de Richard and de van Broock, 1984).

Attempts to isolate mutants usually include mutagenesis. This may be through the use of chemicals such as ethylmethane sulphonate (EMS) or *N*-ethyl-*N'*-nitro-*N*-nitrosoguanidine (NTG), or the use of ultraviolet or X-ray irradiation. In general, irradiation is likely to cause chromosomal re-arrangements and should be avoided. With all mutagenesis, a balance must be struck between the frequency of mutants recovered, usually related to the degree of killing, and induction of multiple and undefined mutations. Very often, the end products of the programme of genetic improvements will be used in fermenters, where undefined mutations will cause dramatic changes in the properties of the yeast, which are often not apparent on Petri-dish cultures or batch cultures grown in shake flasks. Overenthusiastic use of mutagenesis may be a common reason for genetically manipulated strains performing disappointingly in an industrial situation.

The frequency of mutant recovery may be dramatically improved by use of suicide selection regimes. The nystatin-based technique (Snow, 1966; Sanchez and Demain, 1977) has successfully been used (Gleeson, 1986; Veale *et al.* 1992) to enrich over 100 000-fold for recovery of mutants of *Hansenula polymorpha*. So powerful is this technique that it allows recovery of mutants without the intervention of mutagenesis. Other examples of the use of suicide selection discussed later include the use of snail glusulase to recover mutants of *Candida utilis* (Delgardo *et al.*, 1979) and nystatin selection to recover mutants of *Candida tropicalis* (Haas *et al.*, 1990).

Before embarking on any programme of genetic analysis, care should be taken over the choice of strain. Many strains with the same species designation will have different characteristics with respect to mating, sporulation and spore-germination efficiencies. These characteristics may even vary among different samples of the same strain according to storage regimes used in different laboratories. Initial attempts at genetic analysis of *H. polymorpha* were hindered by the choice of a strain, which, although providing mutants, proved to be infertile when mating was attempted. Among other yeasts, the mating efficiency of *Yarrowia lipolytica* is reported to vary among strains and also to respond to selection (Ogrydziak *et al.*, 1978, 1982; Barth and Weber, 1984, 1985). Clearly, a survey of different strains available will minimize later difficulties.

Another consideration is whether a strain is already in use either for biochemical and physiological studies, industrial use or indeed genetic work. The use of different strains can result in strain effects complicating interpretation of data at all levels of research. Research on *Sacch. cerevisiae* has suffered considerably from this problem resulting from use of different isolates in initial studies and the largely undocumented hybridization among pairs of them. In contrast, the *Schiz. pombe* research community has benefited considerably from the isogenicity of strains in use.

B. Protoplast fusion

Protoplast fusion allows interspecific hybridization and intercrossing of strains not amenable to genetic analysis. It involves fusion of protoplasts using polyethylene glycol and selection of hybrids usually through complementary genetic markers. Drug-resistant and respiratory-deficient mutations are particularly useful in this respect as they can often be generated in strains in which recessive nutritional markers have been difficult or impossible to isolate. Hybrids generated by protoplast fusion are usually unstable, and this has been a major problem in the generation of commercially useful strains by this route.

C. Pulsed-field gel electrophoresis

The various pulsed-field gel electrophoresis techniques (for a review, see Anand, 1986) allow separation of chromosome-size DNA molecules. They are particularly useful in the genetics of yeasts as they provide a rapid method of karyotype determination. In yeasts, where genetic analysis is reasonably advanced, the number of chromosomes may thus be compared to the number of linkage groups determined by recombination. In addition, strains and species can be compared rapidly, giving information on their relatedness and evolutionary relationships. It is clear from studies using pulsed-field gel electrophoresis (de Jonge *et al.*, 1986; Sor and Fukuhara, 1989) that most yeasts have fewer, longer chromosomes compared with *Sacch. cerevisiae* (which has 16). Potentially, using cloned genes as probes, genomes can be mapped by hybridization to the large fragments generated when nuclear DNA is digested with rare cutting enzymes, such as *Not*I. Pulsed-field gel electrophoresis techniques may thus offer an alternative to recombination mapping.

D. Molecular transformation

The advantages and uses of molecular transformation are well known. It allows introduction of genes into a strain independently of classical routes, in a form where the expression of that gene may be precisely regulated both with respect to the strength and pattern of transcription. Moreover, the location (chromosomal/replicating) and dosage of the introduced gene is also controllable. Finally, engineered forms of a gene may be introduced to replace the native form. The utility of such techniques pervades every aspect of fundamental research at the molecular level. Industrial applications include production of recombinant proteins and improvement of strains by the introduction of an ability to catabolize novel substrates, such as dextrins, cellulose and xylose Table I lists 11 yeast species, other than *Sacch. cerevisisae* and *Schiz. pombe*, for which transformation systems have been reported.

As with classical genetics, the methodology is heavily dependent on *Sacch. cerevisiae*. Introduction of DNA requires both a selectable marker and a suitable recipient strain. The most common form of marker is one derived from a biosynthetic pathway equivalent to the *LEU2*, *URA3*, *TRP1*, *HIS4* genes in *Sacch. cerevisiae*. The sequence may be derived from the homologous system, but often it is found that the sequence derived from *Sacch. cerevisiae* is effective. The homologous gene may be cloned through complementation of the corresponding mutation in *E. coli* or by probing a gene library with the sequence from *Sacch. cerevisiae*. The recipient is marked with the appropriate mutation, which must be sufficiently stable to prevent revertants occurring at a similar or higher rate than transformants.

The mutant strain is usually identified by enzymic assay of mutants with the appropriate phenotype.

Maintenance of the introduced sequence demands either autonomous replication or chromosomal integration. Autonomous replication is usually based on ARS-like sequences isolated in the same way as they were originally isolated in *Sacch. cerevisiae* (Stinchcombe *et al.*, 1979). Although sequences with ARS activity in *Sacch. cerevisiae* can be isolated from the genome of a second yeast, these do not usually act in the same manner when returned to that yeast. Active ARS sequences must be selected for in the same yeast in which they will be used. However, heterologous sequences may have elements that fortuitously promote replication; for examples of this, see Table I. The origin of replication of the 2 μm plasmid of *Sacch. cerevisiae* does not function in other yeasts; indeed its presence has been reported to be inhibitory to transformation in *Kluyveromyces fragilis* (Das *et al.*, 1984).

Integration can be brought about by transformation with vectors lacking an origin of replication. In many yeasts, targeted integration or gene disruption can also be used. In the former technique, a circular plasmid is linearized within a region of chromosomal homology. In the latter, a linear fragment is used in which a homologous sequence has been disrupted with a selectable marker. The stringency of the recombination is variable so that, in some cases, integration may take place at non-homologous sites or complex events may occur.

III. METHYLOTROPHIC YEASTS

A. Introduction

The ability of certain yeasts to grow on methanol, a cheap and microbiologically clean substrate, has resulted in their use as a source of single-cell protein. However, the price of methanol relative to other inexpensive bulk substrates, such as molasses, has resulted in a decline in the use of these yeasts for this purpose. More recently, there has been considerable interest in their use as a vehicle for production of recombinant proteins using the alcohol oxidase promoter to drive expression of the heterologous gene.

Alcohol oxidase catalyses oxidation of methanol to hydrogen peroxide and formaldehyde, the first step in methanol utilization. During growth on methanol, the enzyme accounts for some 30–40% of total cell protein, sequestered in peroxisomes, which can occupy up to 80% of cell volume. During growth on glucose, the enzyme activity is decreased to very low levels. Regulation of expression is brought about by a combination of glucose or ethanol represson and methanol induction. In *H. polymorpha*, this binary control means that, under derepressing conditions, such as glucose limitation or growth on suboptimal carbon sources, such as glycerol, significant levels of alcohol oxidase activity is observed. In *Pichia pastoris*, there is an absolute requirement for methanol in order for any significant expression to occur.

Methylotrophic yeasts are found in at least four different genera, namely *Hansenula*, *Pichia*, *Torulopsis* and *Candida*. The taxonomic studies of Lodder (1970) reported that, while species of *Pichia* and *Hansenula* are haploid ascosporogenic yeasts, no sexual phase is evident for species of *Candida* and *Torulopsis*. Genetic studies have therefore been limited to species of the former two genera, although physiological and biochemical studies have been more widespread. For a general review of methlylotrophic yeasts, the reader is referred to Gleeson and Sudbery (1988a) and, for a recent extensive review of genetic studies of these organisms, to Sudbery and Gleeson (1989).

B. Classical genetic analysis

1. Pichia pinus

The first methylotrophic yeast to be studied at a genetic level was *Pichia pinus*. Germination of four-spored asci of this yeast gives rise to haploid ascospores displaying a bipolar mating type with two α and two **a** cells in each ascus (Tolstorukov and Benvolenskii, 1978). The separated ascospores are stable on a rich medium. Nutritional deprivation in the form of nitrogen starvation gives rise to mating-type switching followed by conjugation to produce diploids (Tolstorukov and Benvolenskii, 1980). Switching is random, occurring in both mother and daughter cells. Genetic analysis based on mutants, which are unable to complete the life cycle, showed that switching is controlled by a single locus, namely *HTH1* analogous to the *HO* locus of *Sacch. cerevisiae* (Benvolenskii and Tolstorukov, 1980). Mutation at this locus results in heterothallic strains. The diploid progeny produced are also stable but are induced to sporulate by nutritional deprivation giving rise to brown-coloured colonies.

This life cycle lends itself readily to genetic analysis. Auxotrophic mutants can be generated in the haploid phase. These can be readily mated to allow complementation tests between pairs of alleles to define genes. After sporulation of the diploid, recombinant progeny provide data for construction of linkage maps. In addition, natural and radiation-induced loss of chromosomes provide a parasexual cycle allowing assignment of genes to linkage groups (Tolstorukov et al., 1979, 1982, 1983).

Tolstorukov et al. (1977) isolated 84 auxotrophic mutants after NTG and ultraviolet light mutagenesis. There was a bias towards an excess of Ade⁻, Met⁻, Lys⁻ and Cys⁻ phenotypes. Complementation tests assigned the mutants to 35 separate loci. A combination of chromosome loss studies and tetrad analysis allowed construction of a genetic map defining the position of 25 loci on four separate linkage groups and two additional fragments. Centromere-linked markers were identified for each of the linkage groups allowing the position of the centromeres to be defined (Tolstorukov and Efremov, 1984).

These genetic techniques have been employed in a detailed investigation into mechanisms controlling alcohol oxidase expression in *P. pinus* (Sibirny et al., 1988). Glucose repression was shown to be mediated by a locus designated *GCR1*, and methanol repression by a locus designated *ECR1*, mutations in these genes resulting in insensitivity of alcohol oxidase expression to glucose and ethanol, respectively. Mutants designated *gcr1* have a low level of phosphofructokinase activity, while *ecr1* mutants have a lowered level of 2-oxyglutarate dehydrogenase activity, but the significance of these observations is unclear nor is it known whether the mutations affect structural genes for these enzymes.

2. Hansenula polymorpha

The life cycle of *H. polymorpha* is similar to that of *P. pinus*, it being an ascosporogenic yeast displaying homothallic mating based on a bipolar mating-type system. Haploid and diploid phases are stable when grown on rich media, with mating and sporulation being induced by growth on malt extract. In order to elaborate techniques for genetic analysis, a collection of 218 auxotrophic and temperature-sensitive mutants was produced by NTG mutagenesis (Gleeson et al., 1985; Gleeson and Sudbery, 1988b). As with *P. pinus*, there was a preponderance of Ade⁻ and Met⁻ phenotypes. The mutant alleles were assigned by complementation to 57 separate loci. This study included an analysis of 106 Ade⁻ mutants, which defined 12 complementation groups. Of particular interest was the discovery of intragenic complementation in a locus designated *ADE2* defining two subgroups *ADE2.1* and *ADE2.2*. This closely parallels the *ADE5 ADE7* complex in *Sacch. cerevisiae* (Roman, 1956; Costello and Bevan, 1964).

Sporulation of heteroallelic diploids allowed recombination studies either by random-spore analysis or by tetrad dissection. The former was subjected to distortion by viability effects, while the latter was physically difficult due to the small size of the spores and their physical attachment to each other by a thread-like structure. Nevertheless, linkage has been demonstrated but a map as detailed as that for *P. pinus* has not been produced.

It is clear that different strains of *H. polymorpha* have different propensities for genetic analysis. Our initial studies were made using a sample of strain CBS 4732. While mutant isolation was straightforward, mating only occurred rarely (at a frequency less than 10^{-6}), which seriously hampered complementation studies. Another *H. polymorpha* strain, designated NCYC 495, was found to be fully fertile and we carried out studies using this strain. Subsequently we have found that another isolate of CBS 4732 is fertile. In addition, Bodunova *et al.* (1986) found irregular segregation of markers in a strain VKM-U-1397, which originated in the mitotic progeny of a heterozygous hybrid produced by protoplast fusion.

C. Molecular transformation

1. *Pichia pastoris*

Techniques for molecular transformation were developed by Cregg *et al.* (1985). Deoxyribonucleic acid was introduced by the protoplast method. The selectable marker used was the sequence homologous to the *HIS4* gene in *Sacch. cerevisiae* complementing a *P. pastoris* mutant defective in histidinol dehydrogenase. Either the homologous *HIS4* sequence from *P. pastoris* or the heterologous sequence from *Sacch. cerevisiae* would function. Sequences, called PARS, acting in the same manner as the ARS sequences of *Sacch. cerevisiae* were isolated. These resulted in a high frequency of transformation (greater than 10^5 transformants for each microgram of DNA) and in unstable autonomous replication of the plasmid for over 50 generations. Fragments carrying the *LEU2* or *HIS4* genes from *Sacch. cerevisiae* also replicated autonomously showing that they contain ARS activity in *P. pastoris* (but not in *Sacch. cerevisiae*). A similar situation with respect to the *LEU2* gene in *Sacch. cerevisiae* is found in *H. polymorpha*.

Integration is most effectively achieved by transforming with linear molecules produced by digestion of plasmids carrying the *HIS4* gene from *P. pastoris* with *Stu*I for which the *HIS4* contains a unique site. Alternatively, linear fragments carrying the *HIS4* gene between the the 5′ and 3′ ends of the locus for alcohol oxidase result in replacement of the resident alcohol oxidase locus (Cregg *et al.*, 1987). In the latter situation, the coding sequence for alcohol oxidase is replaced by a heterologous gene (see later), so that the alcohol oxidase function is lost in the replacement event. Since *P. pastoris* contains a second alcohol oxidase locus, this does not result in loss of capacity to grow on but it does slow down the growth rate. Multiple integration events leading to increased gene dosage elevate product yield without lowering stability.

A number of heterologous proteins have been expressed in *P. pastoris* using the alcohol oxidase promoter and terminator. These are listed in Table II. Generally high cell densities (greater than 100 g dry wt l^{-1}) are produced by growth on glycerol (i.e. in the absence of expression) followed by a production phase induced by addition of methanol. Product yields were high, at least as good or better than the comparable product in *Sacch. cerevisiae*. With hepatitis B surface antigen, the yield was 50–100-fold higher, and the product naturally aggregated into 22 nm virus-like particles necessary for immunogenicity. Expression can be internal or secreted through the leader sequence of the invertase from *Sacch. cerevisiae*. In this case, glycosylation occurs. In a study of invertase from *Sacch. cerevisiae* expressed in *P. pastoris*, Grinna and Tschopp (1989) showed that glycosylation occurred at nine out of the possible 14 N-asparagine-linked glycosylation sites in the native protein in *Sacch. cerevisiae*. In contrast to *Sacch. cerevisiae*, however, the product from *P. pastoris* contains much shorter and more uniformly sized chains, which, it is claimed, more closely resemble the enzyme in animal cells.

2. Hansenula polymorpha

Transformation systems for *H. polymorpha* have been developed by several groups (Gleeson *et al.*, 1986; Roggenkamp *et al.*, 1986; Tikhomirova *et al.*,

Table II. Heterologous proteins expressed in *Pichia pastoris*

Protein	Yield	Reference
Hepatitis B surface antigen	400 mg l^{-1}	Cregg *et al.* (1987)
Invertase from *Saccharomyces cerevisiae*	2.5 g l^{-1}	Tschopp *et al.* (1987)
Streptokinase from *Streptococcus equisimilis*	250 mg l^{-1}	Hagenson *et al.* (1989)
Human tumour necrosis factor	10 g l^{-1}	Sreekrishna *et al.* (1989)
Bovine lysozyme	550 mg l^{-1}	Digan *et al.* (1989)
Human tissue plasminogen-activating factor	25 mg l^{-1}	Ratner (1989)
Fragment C of tetanus (*Clostridium tetani*) toxin	10 g l^{-1}	Clare *et al.* (1991a)
Whooping cough (*Bordetella pertussis*) p69 antigen	5 g l^{-1}	Romanos *et al.* (1990)
Mouse epidermal growth factor	0.45 g l^{-1}	Clare *et al.* (1991b)

1986, 1988). Roggenkamp *et al.* (1986) made use of the *URA3* sequence in *Sacch. cerevisiae* to complement the equivalent mutation in *H. polymorpha*. Replication was from an ARS-like sequence, called HARS, isolated from the *H. polymorpha* genome. Gleeson *et al.* (1986) used the *LEU2* sequence from *Sacch. cerevisiae* present on YEp13 (Broach *et al.*, 1979). This is a 2 μm-based plasmid commonly used in *Sacch. cerevisiae*. On transformation into *H. polymorpha*, this plasmid complements a Leu⁻ mutant defective in β-isopropylmalate dehydrogenase, the enzyme coded for by *LEU2* in *Sacch. cerevisiae*. The plasmid replicated autonomously. Tikhomirova *et al.* (1986) also used a system based on the *LEU2* gene from *Sacch. cerevisiae* and fragments from mitochondrial DNA of *C. utilis* to programme replication. This resulted in concatamerization of the plasmids *in vivo*.

Besides auxotrophic markers, dominant selectable markers have also been used for transformation. Tikhomirova *et al.* (1988) used the copper resistance *CUP1ʳ* gene to select copper resistance, and I and my colleagues have used resistance to the aminoglycoside G418, coded for by the *NPRT II* gene of Tn5 (Sudbery *et al.*, 1988; Veale, 1989). In the case of copper resistance, Tikhomirova *et al.* (1988) showed that, although it was an inefficient way of selecting transformants, it was possible to increase copy number by growing transformed strains on successively increasing concentrations of copper ions. They inserted the *NPRT II* gene between the promoter and terminator for alcohol oxidase and transformed the construct on a linear fragment with the free ends homologous to the the two flanking regions of the locus for alcohol oxidase. This was effectively the one-step gene-disruption procedure of Rothstein (1983) commonly used in *Sacch. cerevisiae*. The transformants had a disrupted gene for alcohol oxidase and were thus unable to grow on methanol. They were resistant to very high concentrations of G418 (greater than 20 mg ml⁻¹). This procedure allows stable integration of a heterologous gene into the genome. Alternatively, prolonged growth (about 80 generations) on selective medium can also be used to effect integration.

Recombinant proteins have also been expressed in *H. polymorpha* from the promoter and terminator for alcohol oxidase. In this way my colleagues and I have expressed α-galactosidase from the guar plant (*Cyamposis tetragonobla*; Veale, 1989; Fellinger *et al.*, 1991). In my laboratory, glucose oxidase from *Aspergillus niger*, a 21 kDa seed-storage protein from *Theobroma cacao* and a human lipase have also been expressed (unpublished observations). Glycosylation has not been studied in detail but, at least in the case of guar α-galactosidase, it has been shown to occur. Janowicz *et al.* (1991) have demonstrated simultaneous expression of the S and L surface antigens of hepatitis B, which could form composite subviral particles. The level of the S antigen was reported to be 5-8% of the total cell protein. They also demonstrated that these proteins could effectively be expressed using the promoter from the formate dehydrogenase gene. Finally, Gellisen *et al.* (1991) have reported efficient expression of glucoamylase from *Schwanniomyces occidentalis*.

IV. CANDIDA SPECIES

A. Introduction

The genus *Candida* contains some important yeasts of industrial and medical relevance, such as *Candida utilis* and *Candida albicans*. Yet, since by definition they contain no sexual stage (Barnett *et al.*, 1983), they present the severest problems in elaborating techniques of genetic analysis. Furthermore, both *C. utilis* and *C. albicans* are diploid so that mutant isolation is difficult for reasons already discussed. Nevertheless, in the case of *C. albicans*, considerable progress has been made both in terms of genetic analysis employing a parasexual cycle and through the development of molecular transformation technology. This work is reviewed in detail elsewhere in this volume (Chapter 8) and will not be considered further in this review. This section reviews progress with industrially important *Candida* species, such as *C. utilis, Candida maltosa, Candida tropicalis* and *Candida guillermondii* (*Pichia guillermondii*).

B. Formal genetic analysis

Delgardo *et al.* (1979) used enrichment with snail-gut glusulase to isolate auxotrophic mutants of *C. utilis*. Pairs of strains containing complementing sets of double auxotrophic mutants were induced to mate by mixing stationary-phase cells on selective medium (Pérez-Bolaños and Herrera, 1981). The resulting hybrid was induced to produce ascus-like structures by incubation on a glucose–acetate medium. Upon germination in a glucose-rich medium, recombinants were detected by replica-plating the resulting colonies. It was concluded that, although mechanisms such as mitotic recombination were not formally excluded, it seemed most likely that they arose through meiotic recombination.

When the DNA contents of the two parent strains were compared with those of the hybrids, they were found to be similar. Moreover, the parent strains formed asci when incubated in the acetate-containing medium. It was concluded that the parent strains themselves were diploid and that mating could only occur during the transitory existence of haploids produced by low-frequency sporulation in the parents. This conclusion is supported by the observation that, upon first isolation, the auxotrophic mutants readily formed hybrids and had a DNA content half of that of the parent strain. However, upon growth, the DNA content increased and the ability to form hybrids was lost. Thus, *C. utilis* is probably a diploid yeast with a transitory haploid phase, mutant isolation and mating only occurring in this phase. However, inspection of the DNA contents reported by Pérez-Boloños and Herrera (1981) shows that the wild-type DNA content is twice that of a diploid strain of *Sacch. cerevisiae* and four times that of a haploid strain of this yeast. So *C. utilis* may be tetraploid or polyploid, further increasing the difficulties in establishing formal genetic analysis in this organism. An interesting observation made in these studies was that spores recovered do not resemble those of *Hansenula jadinii*, which Kurtzman *et al.* (1979) proposed was conspecific with *C. utilis*, the latter being the imperfect form of the former.

In an attempt to circumvent these difficulties, Pérez *et al.* (1984) attempted to transfer chromosomes from *C. utilis* into *Sacch. cerevisiae*. so that they could be analysed by surrogate genetics in that organism. Marked strains of *Sacch. cerevisiae* and *C. utilis* were hybridized by protoplast fusion (Delgardo and Herrera, 1981; Pérez *et al.*, 1984). Most of the prototrophic hybrids were unstable, producing auxotrophic segregants. One stable strain was recovered, which, upon treatment with ultraviolet radiation to produce mitotic recombination or treatment with *p*-fluorophenylalanine to induce chromosome loss, produced auxotrophic segregants. Both the spontaneous and induced segregants always included markers from both parents. The recombinant fraction was always in excess, suggesting that nuclear fusion had taken place. There was, however, a bias towards recovery of markers originating with the *C. utilis* parent, suggesting that the chromosomes of *Sacch. cerevisiae* were preferentially lost. The stable hybrid was unable to sporulate but, when mass-mated with α-type *Sacch. cerevisiae* (the opposite mating type to the *Sacch. cerevisiae* parent) and placed on sporulation medium, a high level of sporulation was observed, indicating that the hybrid retained the mating specificities of the *Sacch. cerevisiae* parent. The asci were dissected and segregation of markers in the progeny followed. Markers from both parents were recovered but, with the exception of the mating-type alleles, segregation of alleles was complex. It was hoped that genes in *C. utilis* could be mapped by crossing to standard laboratory strains of *Sacch. cerevisiae*. However, genetic and cytological data showed that, when this was attempted, karyogamy did not occur and there was internuclear transfer of genetic material (Pérez and Benítez, 1986).

Protoplast-fusion experiments have also been carried by de Richard and de van Broock (1984). They fused petite cells of *C. utilis* with respiratory-competent but auxotrophic cells of *Sacch. cerevisiae*. In this case, the hybrids retained the properties of the *C. utilis* parent with the exception of respiratory competence, which was transferred from the *Sacch. cerevisiae* parent. It seems that nuclei from *Sacch. cerevisiae* were rapidly lost from the hybrids, and that some form of recombination had occurred between the two sets of mitochondria.

A parasexual cycle has been developed for genetic analysis of the methylotrophic yeast *Candida pelliculosa* (Lahtchev and Tuneva, 1986). Hybrids were formed by protoplast fusion of auxotrophic mutants. Mitotic chromosome loss and mitotic recombination resulted in segregation of markers. The data suggested the presence of at least four chromosomes. A parasexual cycle, based on protoplast fusion and mitotic segregation, has also been developed for *C. maltosa* (Klinner *et al.*, 1984). Linkage was detected between two auxotrophic markers, *ade26* and *pro1*, and the centromere, the order being *CEN–ade26–pro1*.

C. Molecular transformation

Despite reports of the *LEU2* gene from *C. utilis* being cloned and sequenced (Zhang and Reddy, 1986; Hamasawa *et al.*, 1987) and of the recovery of *C. utilis* sequences, which have ARS activity in *Sacch. cerevisiae* (Hsu *et al.*, 1983; Tikhomirova *et al.*, 1983), no system of molecular transformation has been developed in this organism. However, systems have been developed for other *Candida* species.

Kunze *et al.* (1985a,b) used the *ARG4* gene from *Sacch. cerevisiae* to transform the *n*-alkane-utilizing yeasts *C. maltosa* and *P. guilliermondii* (*C. guilliermondii*). The plasmid used contained only the *ARG4* gene cloned into pBR322 yet, in both species, the plasmid replicated extrachromosmally as judged by stability, Southern hybridization and the ability to rescue plasmids by transformation of *E. coli* with total DNA preparations. This shows again that, in many yeasts, heterologous DNA often contains sequences that will fortuitously direct replication. Both β-lactamase and arginosuccinate lyase activities were detectable in the transformants. An interesting feature of this work is that the most effective method of introducing the DNA was that of Imura *et al.* (1983), which uses intact cells treated with polyethylene glycol.

Transformation of *C. maltosa* has also been demonstrated by Takagi *et al.* (1986). This system is based on mutants of *C. maltosa* defective in β-isopropylmalate dehydrogenase, equivalent to *leu2* mutants of *Sacch. cerevisiae*. The DNA used for transformation contained the *LEU2* sequence from *Sacch. cerevisiae*, together with sequences from genomic DNA of *C. maltosa*, which would direct autonomous replication (called TRA). The homologous *LEU2* sequence, called C-*LEU2*, has also been isolated and sequenced (Kawamura *et al.*, 1983; Takagi *et al.*, 1987).

A problem with this system is that the Leu⁻ host (*C. maltosa* J288) has an impaired growth rate when grown on alkanes compared with the wild type (Hikiji *et al.*, 1989). This probably came about through the vigorous mutagenesis necessary to isolate Leu⁻ mutants (Chang *et al.*, 1984), thus causing additional mutations. Other mutant phenotypes such as His⁻ and Ade⁻ were more commonly recovered and their isolation did not require the same degree of mutagenesis. On the basis of DNA measurements, Chang *et al.* (1984) suggested that the genome of *C. maltosa* is at least in part diploid, while Hikiji *et al.* (1989) proposed that the differential recovery of mutants comes about through aneuploidy with two copies of the C-*LEU2* gene but only one of the C-*HIS5* (see later) and that this would explain the different availability of Leu⁻ compared to His⁻ and Ade⁻ phenotypes. However, as discussed in Section II, such a biased recovery of mutants is commonly observed when working with yeasts even in species that are known to be haploid. In the diploid *C. albicans*, similar observations have been reported where it is thought that heterozygosity for mutant alleles rather than aneuploidy is responsible.

In order to overcome the problem caused by the poor growth of J288, Hikiji *et al.* (1989) made used of a His⁻ mutant (CH1), which was recovered after light mutagenesis, and had a growth rate on alkanes nearer to the wild type. The chromosomal copy of the affected gene was isolated by complementation from a genomic library and by complementation of mutants of *Sacch. cerevisiae* shown to correspond to the *HIS5* locus. It was thus called C-*HIS5*. Vectors containing this sequence together with TRA elements were effective in transforming CH1, thus providing an alternative system overcoming the growth-rate problem encountered with the J228 strain. Gene C-*HIS5* was sequenced and shown to be homologous to the gene in *Sacch. cerevisiae*.

Haas *et al.* (1990) developed molecular transformation in the alkane- and fatty acid-utilizing yeast *C. tropicalis*. Their system is based on the gene coding for orotidine monophosphate decarboxylase (OMD) coded for by the *URA3* gene in *Sacch. cerevisiae*. Mutants defective in OMD (*ura3*) were isolated in *C. tropicalis* by successive application of nystatin enrichment and selection for 5-fluoro-orotic acid resistance. The *URA3* gene in

C. tropicalis was isolated using a library of *Sau*3a fragments cloned into a replicating vector of *Sacch. cerevisiae* and using this to complement *ura3* mutants in this yeast. A 5.8 kb fragment was recovered which contained the *URA3* gene from *C. tropicalis*. Hybridization of this gene to genomic DNA from *C. tropicalis* showed that two separate sequences were present. The reason for this is uncertain, but it could be due to a restriction-fragment length polymorphism being present as an allelic pair in the presumptive diploid.

The *URA3* sequence in *C. tropicalis* was subcloned into pUC18. Mutants of *ura3* in *C. tropicalis* were then transformed with circular or linearized forms of the resulting plasmid, using either the protoplast or the cation-treatment method. The frequency was between 1.7 and 4.0×10^4 transformants $(\mu g \ DNA)^{-1}$. A mixture of small and large colonies was recovered with linearized DNA but only small colonies when circular DNA was used. It was shown that the small colonies contain unstable extrachromosomal forms of the DNA whereas the large colonies were stable. Some of these were the result of gene conversion, but others arose through integration of a single copy at the chromosomal *URA3* locus or integration of multiple tendem copies at that locus. Linearization of the plasmid clearly promotes targeted integration.

Interestingly, the *URA3* sequence in *Sacch. cerevisiae* was ineffective at complementing *ura3* mutants of *C. tropicalis*. If integration is the only route by which transformation could come about in the absence of a homologous ARS, then the *URA3* sequence in *Sacch. cerevisiae* presumably lacks sufficient homology to allow such integration. No attempt was made to isolate an ARS sequence in *C. tropicalis* but, if one were available, it would be interesting to see if the heterologous *URA3* could operate on a replicating plasmid.

V. YARROWIA LIPOLYTICA

A. Introduction

Yarrowia lipolytica is a dimorphic yeast of some industrial importance. It has been used for production of single-cell protein as it can utilize a range of growth substrates, such as lipids, proteins and hydrocarbons. It has also been used for production of organic acids, such as citric acid. It can secrete a number of extracellular enzymes, such as ribonucleases, lipases and both acid and alkaline proteases. Lastly, some strains have a dsRNA genome responsible for a killer phenotype (Groves *et al.*, 1983; Jewers *et al.*, 1983). The alkaline protease is of particular interest, since its production is strongly regulated by pH value and the presence of proteins in the medium (Ogrydziak and Sharf, 1982). The structural gene has been cloned (Davidow *et al.*, 1987; Matoba *et al.*, 1988; Nicaud *et al.*, 1989) and the process of secretion studied (Matoba and Ogrydziak, 1989), showing the involvement of KEX2-like and dipeptidylaminopeptidase proteolytic-cleavage steps. The latter was unusual in that it was apparently an early step in processing, occurred 120 residues upstream of the N-terminal residue of the mature protein and was not involved in enzyme activation. The leader sequences of this protein can be used to secrete heterologous proteins (Nicaud *et al.*, 1989).

B. Formal genetics

Wickerham *et al.* (1970) reported that *Candida lipolytica*, which up to then was thought to be imperfect, did have a sexual stage. They showed that most wild isolates were haploid with a heterothallic mating system consisting of two mating types, subsequently designated as A and B (Bassel and Mortimer, 1973). One rare strain was diploid, and could be induced to sporulate producing 1–4 ascopores in each ascus with a wide range of morphology. Subsequently, a diploid strain was constructed, which produced uniform spores shaped like a "shallow bowl". Strains of *C. lipolytica* exhibiting sexuality were renamed *Saccharomycopsis lipolytica* (Yarrow, 1972), which has subsequently been changed to *Yarrowia lipolytica*.

Bassel *et al.* (1971) demonstrated mutant isolation, complementation between pairs of auxotrophic strains and genetic recombination in this yeast. Subsequently, a full genetic system was developed (Bassel and Mortimer, 1973; Gaillardin *et al.*, 1973; Esser and Stahl, 1976; Ogrydziak *et al.*, 1978, 1982). Collections of mutants have been accumulated exhibiting auxotrophic, temperature-sensitive and alkane-defective phenotypes. These have been assigned to individual loci by complementation, while a genetic map based on tetrad analysis has been produced (Ogrydziak *et al.*, 1978, 1982). The latest map based on 81 characterized genes shows the position of 29 of them on five linkage fragments but with only one example of possible centromere linkage. The relative lack of centromere linkage led to the suggestion that only a few long chromosomes were present. This is supported by later pulsed-field gel electrophoresis studies of de Jonge *et al.* (1986) who were able to resolve only two large chromosomes in this organism. Kurischko (1986) reported that genes can be assigned to complementation groups by spontaneous haploidization occurring shortly after conjugation. This technique may prove helpful in establishing the number of linkage groups.

Gaillardin and Heslot (1979) carried out a detailed study of lys⁻ mutants and identified the structural gene for homocitrate synthase. Mutant alleles showed intragenic complementation, which could be correlated with altered properties of the enzyme *in vitro*. A fine-structure map of the alleles was constructed by γ-ray-induced gene conversion. Mutants affecting hydrocarbon utilization have been characterized by Bassel and Mortimer (1982).

A common problem reported by all workers is low mating, sporulation and spore-germination frequencies. The last of these problems can be remedied by inbreeding strains starting with spores from complete tetrads. However, introduction of markers from other strains immediately causes a drop in the frequency of sporulation and spore germination (Ogrydziak, 1988). Strain heterogeneity has clearly been a problem in this field. Strains with higher mating (Barth and Weber, 1984) and sporulation (Barth and Weber, 1985) efficiencies have been produced. Dissection of tetrads is difficult (Ogrydziak, 1988).

C. Molecular transformation

Techniques for molecular transformation have been developed independently by Davidow *et al.* (1985) and by Gaillardin *et al.* (1985). Davidow *et al.* (1985) cloned the *Y. lipolytica* homologue of the *LEU2* gene in *Sacch. cerevisiae* by complementation of a *leuB6* mutant of *E. coli* with a *Y. lipolytica* gene library in strain pBR322 of *Y. lipolytica*. A complementing plasmid was then used to transform a *leu2* mutant of *Y. lipolytica* using the lithium acetate method. The transformation frequency was increased 100-fold by linearization of the plasmid within the region of chromosomal homology. After optimization of experimental parameters, transformation frequencies of 10^4 transformants µg DNA^{-1} were obtained. The transforming DNA always integrated into the region of chromosomal homology and was completely stable. The *LEU2* gene of *Y. lipolytica* would complement a *leu2* mutant of *Sacch. cerevisiae* but a plasmid containing the *URA3* gene of *Sacch. cerevisiae* would not function in *Y. lipolytica*.

Gallardin *et al.* (1985) made use of the observation that the *LYS2* gene of *Sacch. cerevisiae* complements a *lys2* mutant of *Y. lipolytica* in a protoplast-fusion experiment. Accordingly, they used the already available *LYS2* sequence of *Sacch. cerevisiae* cloned in pBR322. A library of DNA fragments from *Y. lipolytica* was then cloned 5' to the *LYS2* sequence. The purpose of this was two-fold. Firstly, it was hoped that an ARS-like sequence may be recovered, which would allow autonomous replication of the transforming plasmid; secondly, that sequences may be recovered, which would promote expression of the heterologous gene. In the event, integrative transformation occurred, which, was dependent on the presence of genomic fragments in the 5' (but not the 3') position relative to the *LYS2* sequence. Multiple tandem integration events were revealed by Southern hybridization, which were shown to be due to *in vivo* ligation of cotransforming plasmids followed by targeted integration into a region of chromosomal homology. So there was no correlation between activity of the *LYS2* gene, as revealed by enzyme assay and copy number. The function of the fragments in the 5' position was thought to be as promoters.

Both of these systems depend on the presence of suitable mutations in the host strain so that transformation of a new strain requires introduction by genetic crossing of the selectable marker. As already discussed, crossing of non-isogenic strains results in significant changes in strain characteristics including a decrease in sporulation and spore-germination efficiency. It would clearly be advantageous to have a system based on a dominant selectable marker, which could be used on all strains. Unfortunately, *Y. lipolytica* is resistant to the antibiotics commonly used with *Sacch. cerevisiae*, including chloramphenicol and G418.

Gaillardin and Ribet (1987) used the homologous *LEU2* promoter to express β-galactosidase as a colour marker and the phleomycin resistance gene from Tn5 to select for phleomycin-resistant transformants. However, when the latter was used, half of the colonies that appeared were found to have arisen through spontaneous resistance to the drug. While this problem was prevented by inclusion of an expression phase in the protocol, this caused a 60-fold decrease in transformation frequency. There were also fears that phleomycin may act as a mutagen.

A more satisfactory alternative was developed by Nicaud *et al.* (1989), who exploited the inability of *Y. lipolytica* to utilize sucrose. The invertase gene from *Sacch. cerevisiae* was expressed from the promoter and signal sequence of the *XPR2* gene, which codes for the alkaline extracellular protease. The heterologous protein was efficiently expressed, retaining physiological control of the *XPR2* promoter and was exported predominantly to the periplasmic space. Its expression resulted in the yeast being able to grow on sucrose, and it provided a selection system as efficient as the *LEU2* system but which could be used on all strains.

VI. *KLUYVEROMYCES* SPECIES

A. Introduction

Kluyveromyces fragilis and *K. lactis* are closely related strains used industrially because of their ability to utilize lactose and thus grow on milk whey. This ability is dependent on production of β-galactosidase. They differ from each other in that *K. fragilis* can grow over a wider range of growth temperatures, by their chromosome banding pattern on OFAGE, mitochondrial banding patterns after restriction digestion and ability to support replication of the pKD1 plasmid (Sor and Fukuhara, 1989).

Kluyveromyces lactis is notable for its killer system specified by two linear DNA genomes. These have been the subject of considerable research. This has been recently reviewed in detail (Stark *et al.*, 1990) and will not be considered here. *Kluveromyces drosophilarum* contains a 1.6 μm autonomously replicating plasmid (Falcone *et al.*, 1986). This has a similar organization (but not sequence homology) to the 2 μm plasmid of *Sacch. cerevisiae*, namely three genes equivalent to those of the 2 μm plasmid in *Sacch. cerevisiae* and a pair of inverted repeats. Isomeric forms of the plasmid interconvert by a flip-flop mechanism. This plasmid replicates stably when introduced into *K. lactis* (Bianchi *et al.*, 1987) but not *K. fragilis* (Sor and Fukuhara, 1989).

Formal genetic analysis is poorly developed but controlled mating, sporulation and tetrad analysis have been used by Brunner *et al.* (1977, 1987) during the course of studies on mutations leading to resistance to drugs that affect mitochondria in *K. lactis*. They demonstrated involvement of both nuclear and mitochondrial genes from the tetrad ratios recovered.

B. Molecular transformation

Transformation systems were developed at an early stage for both species of *Kluyveromyces*. Das and Hollenberg (1982) transformed *K. lactis* using three types of selection. The kanamycin-resistant determinant of Tn601 was used to select for G418-resistant cells, and the β-galactosidase gene from *E. coli* to complement a mutation in the *LAC4* gene, the structural gene for β-galactosidase. The *TRP1* gene in *Sacch. cerevisiae* was used to complement a corresponding mutation in *K. lactis*. Transformation frequencies were greatly increased by the use of *KARS* sequences in the transforming plasmid, which promoted unstable autonomous replication.

However, the *ARS1* in *Sacch. cerevisiae* present on YRp7 did not function in *K. lactis*. In the absence of a functioning origin of replication, low-frequency transformation occurred by chromosomal integration. The 2 μm DNA from *Sacch. cerevisiae* allowed replication but transformation frequencies were very low. The isolated origin from the 2 μm plasmid did not function even though it does in cir° strains of *Sacch. cerevisiae*. Das *et al.* (1984) transformed *K. fragilis* using G418 resistance as a selection and the *KARS2* from *K. lactis* to promote replication. This sequence also acted to promote replication in *Sacch. cerevisiae*. Thus, the *KARS2* element is functional in all three yeasts.

Bianchi *et al.* (1987) used the pKD1 plasmid from *K. drosophilarum* to transform *K. lactis* using G418 or the *URA3* sequence from *Sacch. cerevisiae* as selectable markers. The plasmid allowed high-frequency transformation and stable maintenance of the transformed plasmid. The killer DNA genomes were shown to interfere with transformation, decreasing transformation frequencies 3–5-fold when present. They also investigated the role of various regions of the plasmid genomes. Deletions generally had little effect on transformation frequencies but did lead to increased instability. This could be complemented by the presence of an intact pKD1 plasmid in the host presumably supplying missing functions in *trans*.

VII. *Pachysolen tannophilus*

A. Introduction

Pachysolen tannophilus has the ability to ferment hexose and pentose sugars to ethanol. Since lignocellulase hydrolysates contain mixtures of such sugars, it has the potential to convert plant biomass, particularly woody tissue, into ethanol. While molecular systems are as yet poorly developed, formal genetic manipulation has proved useful in removal of unwanted characteristics and strain improvement.

B. Formal genetic analysis

The species was first described by Boiden and Adzet (1957) who reported that vegetative reproduction occurred by budding and sporulation by ascospore formation, whereby a cell projects a long tube at the tip of which a four-spore ascus appears. Of the four strains initially studied, three sporulated at a low frequency; the other with a much higher frequency. Wickerham (1970) concluded that this strain was diploid, the others haploid. Clones derived from isolated ascospores sporulate at the same low frequency as haploids, so that mating is homothallic.

Genetic analysis was established by James and Zahab (1982, 1983). Conjugant cells entering the sexual cycle appeared on malt-extract medium. They showed that diploidization does not result from mating of two cells but is apogamic, that is, the two products of a mitosis fuse to form the diploid nucleus. This observation and the transient nature of the diplophase were not conducive to genetic analysis. Nevertheless, they showed that, when pairs of complementary mutants were mixed on minimal medium, it was possible to select hybrids. Furthermore, the diploid phase could be stabilized, if conjugant cells were transferred by micromanipulation from malt extract–yeast extract medium to a richer medium. The clones derived from such cells were larger than haploids, and sporulated more rapidly and more profusely when placed on malt extract. Tetrad dissection was possible and normal segregation patterns for the mutant markers were observed, including a pair of alleles showing linkage.

Several examples of strain improvement using genetic manipulation have been reported. James and Zahab (1983) reasoned that, as most industrial strains of *Sacch. cerevisiae* used for ethanol production were either polyploid or aneuploid, the same might be true of *Pa. tannophilus*. Accordingly they constructed triploids and tetraploids by the forced mating of diploids with haploids and diploids with diploids, respectively. Confirmation of the polyploid state was made by tetrad analysis, which also provided aneuploids. From the ratio of complete tetrads in the segregation of triploids, they were able to estimate the number of linkage groups to be 5–7. This was confirmed by later observations of Maleszka and Skrzypek (1990) who observed seven bands and one possible doublet after electrophoretic separation of chromosomes. The yield of ethanol from D-xylose and D-galactose was significantly higher in triploid and tetraploid strains.

When growing on lignocellulose hydrolysates, which contain a mixture of glucose and D-xylose, *Pa. tannophilus* will preferentially ferment glucose to ethanol and, because the yeast has a low ethanol tolerance, will not then fully utilize the xylose. Wedlock *et al.* (1989) addressed this problem by isolating a series of mutants, which separately inactivated each of three enzymes with glucokinase activity (hexokinase A and B, and glucokinase). When the mutations were combined in the same strain, which was then grown on glucose–xylose mixtures, the yeast preferentially utilized the xylose, although it would only grow on the xylose in yeast extract–peptone medium not in minimal medium. A strain defective in hexokinase A and glucokinase simultaneously utilized glucose and xylose. These studies open the way for fermentation of hydrolysates with mixed yeast cultures with a first phase, in which *Pa. tannophilus* ferments the xylose and a second with *Sacch. cerevisiae* fermenting glucose.

In a separate study Clark *et al.* (1986) combined two different mutations, which each led to increased ethanol yield. The first (*eth2-1*), isolated by Lee *et al.* (1986), was selected as a mutant unable to utilize ethanol and involves a defect in malate dehydrogenase activity. The second strain, NO_3-NO_3-4, was selected by Jeffries (1984) by increased growth on a xylitol–nitrate medium, the rationale being that nitrate utilization requires NADPH produced by the pentose phosphate pathway. Increased activity in this pathway should result in faster fermentation of xylose and therefore production of more ethanol. The second strain did indeed produce 32% more ethanol. It was found to harbour multiple mutations, so that the hybrid between this strain and the *eth2-1* strain was backcrossed three times to the NO_3-NO_3-4 parent. The resultant strain produced even more ethanol from xylose after growth into the stationary phase. Compared to the NO_3-NO_3-4 strain, the hybrid gave the same peak levels of ethanol but, in the former, levels declined towards the end of the fermentation as ethanol was consumed. This decline was not observed in the hybrid.

C. Molecular techniques

Molecular transformation of *Pa. tannophilus* has been reported by Wedlock and Thornton (1989) using the lithium-ion method. A hexokinase-defective mutant was complemented with a plasmid carrying a sequence encoding the hexokinase PII enzyme from *Sacch. cerevisiae*. Maleska and Skrzypek (1990) cloned several genes from *Pa. tannophilus* by complementation in *Sacch. cerevisiae*. Chromosome blots produced by pulsed-field gel electrophoresis were then probed with these cloned sequences together with other sequences from *Sacch. cerevisiae*. Ethidium bromide gels showed the presence of seven bands, one of which is probably a doublet giving a total of eight chromosomes. The location of 11 genes among these was determined by hybridization.

VIII. CONCLUDING REMARKS

The introduction of techniques of molecular manipulation have allowed rapid advances in the utilization of new yeast species. Methylotrophic yeasts already at least match the performance of *Sacch. cerevisiae* and in many ways are superior. They are already being used in full-scale production of recombinant proteins. Rapid advances are also being made in the *Kluyveromyces* species and *Y. lipolytica*. One major disappointment is the relative lack of progress on the non-pathogenic *Candida* species, especially *C. utilis*. Considerable resources are now being invested in work on the *Candida* species of medical importance, especially *C. albicans*. The problems are essentially similar, so these studies may facilitate progress in other *Candida* species.

References

Anand, R. (1986). *Trends in Genetics* 2, 278.
Barnett, J.A., Payne, R.W. and Yarrow, D. (1983). "Yeasts Characteristics and Identification". Cambridge University Press, Cambridge.
Barth, G. and Weber, H. (1984). *Zeitschrift für Allgemeine Mikrobiologie* 24, 403.
Barth, G. and Weber, H. (1985). *Antonie van Leeuwenhoek* 51, 167.
Bassel, J. and Mortimer, R. (1973). *Journal of Bacteriology* 114, 894.
Bassel, J. and Mortimer, R. (1982). *Current Genetics* 5, 77.
Bassel, J., Warfel, J. and Mortimer, R. (1971). *Journal of Bacteriology* 108, 609.
Benevolenskii, S.V. and Tolstorukov, I.I. (1980). *Genetika* 16, 1342.
Bianchi, M.M., Falconi, C. Chen, X.J., Weslowski-Louvel, Frontali, L. and Fukuhara, H. (1987). *Current Genetics* 12, 185.
Bodunova, E.N., Donich, V.N. and Nesterova, G.F. (1986). *Genetika* 22, 939.
Boidin, J. and Adzet, J. (1957). *Bulletin de la Société Mycologique de France* 73, 331.
Broach, J.R., Strathern, J.N. and Hicks, J.B. (1979). *Gene* 8,121.
Brunner, A., de Cobas, A.T. and Griffiths, D.E. (1977). *Molecular and General Genetics* 152, 183.
Brunner, A., Mendoza, L.V. and de Cobas, A.T. (1987). *Current Genetics* 11, 475.
Burden, D.W and Eveleigh, D.E. (1990). *In* "Yeast Technology" (J.F.T. Spencer and D.M. Spencer, eds), pp. 199–227. Springer Verlag, Berlin, Heidelberg, New York.
Cartledge, T.G. (1987). *In* "Yeast Biotechnology" (D.R. Berry, I. Russell and G.G. Stewart, eds), pp. 311–344 Allen and Unwin, London.
Chang, M.C., Jung, H.K., Suzuki, T., Takagi, M. and Yano, K. (1984). *Journal of General and Applied Microbiology* 30, 489.
Clark, T. Wedlock, N. James, A.P., Deverell, K. and Thornton, R.J. (1986). *Biotechnology Letters* 8, 801.
Clare, J.J., Rayment, F.B., Ballintine, S.P., Sreekrishna, K. and Romanos, M.A. (1991a). *Bio/Technology* 9, 455.
Clare, J.J., Romanos, M.A., Rayment, F.B., Rowedder, J.E., Smith, M.A., Payne, M.E., Sreekrishna, K. and Henwood, C.A. (1991b). *Gene* 105, 205–212.
Costello, W.P. and Bevan E.A. (1964). *Genetics* 50, 1219.
Cregg, J.M., Barringer, K.J., Hessler, A.Y. and Madden, K.R. (1985). *Molecular and Cellular Biology* 5, 3376.
Cregg, J.M., Tschopp, J.F., Stillman, C., Siegel, R., Akong, M., Craig, W.S., Buckholtz, R.G., Madden, K.R., Kellaris, P.A. Davis, G.R., Smiley, B.L., Cruze, J., Terragrossa, R., Velicelebi, G. and Thill, G. (1987). *Bio/Technology* 5, 479–485.
Das, S. and Hollenberg, C.P. (1982). *Current Genetics* 6, 123.
Das, S., Ellerman, E. and Hollenberg, C.P. (1984). *Journal of Bacteriology* 158, 1165.

Davidow, L.S., Opostolakos, D., O'Donnell, M.M., Proctor, A.R., Ogrydziak, D.M., Wing, R.A., Stasko, I. and DeZeeuw, J.R. (1985). *Current Genetics* 10, 39.
Davidow, L.S., O'Donnell, M.M., Kacymarek, F.S., Pereira D.A., DeZeeuw, J.R. and Franke, A.E. (1987). *Journal of Bacteriology* 169, 4621.
de Jonge, P., de Jonge, F.C.M., Meijers, R., Yde Steensma, H. and Scheffers, W. (1986). *Yeast* 2, 193.
de Richard, M.S. and de van Broock, M.R. (1984). *Current Microbiology* 10, 117.
Delgardo, J.M. and Herrera, L.S. (1981). *Acta Microbiologica of the Academy of Sciences Hungary* 28, 339.
Delgardo, J.M., Perez, C., Herrera, L.S. and Lopez, R. (1979). *Canadian Journal of Microbiology* 25, 486.
Digan, M.E., Lair, S.V., Brierley, R.A., Siegel, R.S., Williams, M.E., Ellis, S.B., Kellaris, P.A., Provow, S.A., Craig, W.S., Velicelbeli, G., Harpold M.M. and Thill G.P. (1989). *Bio/Technology* 7, 160.
Dohmen, R.J., Strasser, A.W.M., Zitomer, R.S. and Hollenberg C.P. (1989). *Current Genetics* 15, 319.
Esser, K. and Stahl, U. (1976). *Molecular and General Genetics* 146, 101.
Falcone, C., Saliola, M., Chen, X.J., Frontali, L. and Fukihara, H. (1986). *Plasmid* 15, 248.
Fellinger, A.J., Veale, R.A., Sudbery, P.E., Bom, I.M., Overbeke, N., Verbakel, J.M.A. and Verrips, C.T. (1991). *Yeast* 7, 463.
Gaillardin, C.M. and Heslot, H. (1979). *Molecular and General Genetics* 172, 185.
Gaillardin, C.M. and Ribet, A.M. (1987). *Current Genetics* 11, 377.
Gaillardin, C.M., Charoy, V. and Heslot, H. (1973). *Archives of Microbiology* 92, 69.
Gaillardin, C.M., Ribet, A.M. and Heslot, H. (1985). *Current Genetics* 10, 49.
Gellisen, G., Janowicz, Z.A., Merckelbach, A., Piontek, M., Keup, P., Weydemann, U., Hollenberg, C.P. and Strasser, A.W.M. (1991). *Bio/Technology* 9, 291.
Gleeson, M.A.G. (1986). PhD Thesis: University of Sheffield.
Gleeson, M.A.G. and Sudbery, P.E. (1988a). *Yeast* 4, 1.
Gleeson, M.A.G. and Sudbery, P.E. (1988b). *Yeast* 4, 293.
Gleeson, M.A.G., Waites, M.J. and Sudbery, P.E. (1985). *In* "Microbial Growth on C1 Compounds" (R.L. Crawford and R.S. Hanson, eds) pp. 228–235. American Society for Microbiology, Washington, DC.
Gleeson, M.A.G., Ortori, G.S. and Sudbery, P.E. (1986). *Journal of General Microbiology* 132, 3459.
Glumoff, V., Kappeli, O., Fiechter, A. and Reiser, J. (1989). *Gene* 84, 311.
Grinna, L.S. and Tschopp, J.F. (1989). *Yeast* 5, 106.
Groves, D.P., Clare, J.J. and Oliver, S.G. (1983). *Current Genetics* 7, 185.
Haas, L.O.C., Cregg, J.M. and Gleeson, M.A.G. (1990). *Journal of Bacteriology* 8, 4571.
Hagenson, M.J., Holden, K.A., Parker, K.A., Wood, P.J., Cruze J.A., Fuke, M., Hopkins, T.R. and Stromen, D.W. (1989). *Enzyme and Microbial Technology* 11, 650.
Hamasawa, K., Kobayashi, Y., Hamada, S., Yoda, K., Yamasaki, M. and Tamura, G. (1987). *Journal of General Microbiology* 133, 1089.
Hikiji, T., Ohkuma, M., Takagi, M. and Yano, K. (1989). *Current Genetics* 16, 261.
Hsu, W.H., Magee, P.T., Magee, B.B. and Reddy, C.A. (1983). *Journal of Bacteriology* 154, 1033.
Imura, Y., Gotoh, K., Ouchi, K. and Nishiya, T. (1983). *Agricultural and Biological Chemistry* 47, 897.
James, A.P. and Zahab, D.M. (1982). *Journal of General Microbiology* 128, 2297.
James, A.P. and Zahab, D.M. (1983). *Journal of General Microbiology* 129, 2489.
Janowicz, Z.A., Melber, K., Merckelbach, A., Jacobs, E., Harford, N., Comberbach, M. and Hollenberg, C.P. (1991). *Yeast* 7, 431.
Jeffries, T.W. (1984). *Enzyme and Microbial Technology* 6, 254.
Jewers, R.J., El Shebani, M., Mountain, H.A., Bostian K.A., Bevan, E.A. and Mitchell, D.J. (1983). *Heredity* 52, 458.
Kawamura, M., Takagi, M. and Yano, K. (1983). *Gene* 24 ,157.
Klein, R.D. and Faureaum, A. (1988). *Journal of Bacteriology* 170, 5572.

282   P.E. Sudbery

Klinner, U., Samsanova, I.A. and Böttcher, F. (1984). *Current Microbiology* **11**, 241.
Kunze, G., Petzoldt, C., Bode, R., Samasonova, I.A., Böttcher, F. and Birnbaum, D. (1985a). *Journal of Basic Microbiology* **25**, 141.
Kunze, G., Petzoldt, C., Bode, R., Samasonova, I.A., Hecker, M. and Birnbaum, D. (1985b). *Current Genetics* **9**, 205.
Kurischko, C. (1986). *Current Genetics* **10**, 709.
Kurtz, M.B., Cortelyou, M.W. and Kirsch, C.R. (1986). *Molecular and Cellular Biology* **6**, 142.
Kurtz, M.B., Cortelyou, M.W., Miller, S.M., Lai, M. and Kirsch, C.R. (1987). *Molecular and Cellular Biology* **7**, 209.
Kurtzman, C.P., Johnson, C.J. and Smiley, M.J. (1979). *Mycologia* **71**, 844.
Lahtchev, K. and Tuneva, D. (1986). *Current Microbiology* **14**, 121.
Lee, H., James, A.P., Zahab, D.M., Mahmourides, G., Maleska, R. and Schneider, H. (1986). *Applied Environmental Microbiology* **44**, 909.
Lodder, J. (1970). "The Yeasts, A Taxonomic Study". North Holland, Amersterdam.
Maleszka, R. and Skrzypek, M. (1990). *FEMS Microbiology Letters* **69**, 79.
Matoba, S. and Ogrydziak, D.M. (1989). *Journal of Biological Chemistry* **264**, 6037.
Matoba, S., Fukayama, J., Wing, R. and Ogrydziak, D.M. (1988). *Molecular and Cellular Biology* **8**, 4903.
Nicaud, J.M., Fabre, E. and Gaillardin, C. (1989). *Current Genetics* **16**, 253.
Ogrydziak, D. (1988). *Journal of Basic Microbiology* **28**, 185.
Ogrydziak, D. and Sharf, S.J. (1982). *Journal of General Microbiology* **128**, 1225.
Ogrydziak, D., Bassel, R., Contopoulou, R. and Mortimer, R. (1978). *Molecular and General Genetics* **163**, 229.
Ogrydziak, D., Bassel, R. and Mortimer, R. (1982). *Molecular and General Genetics* **188**, 179.
Pérez, C. and Benitez, J. (1986). *Current Genetics* **10**, 639.
Pérez, C., Vallin, C. and Benitez, J. (1984). *Current Genetics* **8**, 575.
Pérez-Bolaños, C. and Herrera, L.S. (1981). *Experimental Mycology* **5**, 15.
Ratner, R.M. (1989). *Bio/Technology* **7**, 1129.
Roggenkamp, R., Hansen, H., Eckart, M., Janowitz, Z. and Hollenberg, C.P. (1986). *Molecular and General Genetics* **202**, 302.
Roman, H. (1956). *Comptes Rendues du Laboratoratoire Carlsbergensis, Séries Physiology* **26**, 299.
Romanos, M.A., Rayment, F., Beesley, K.M. and Clare, J.J. (1990). *Yeast* **6**, S248.
Rothstein, R.J. (1983). *Methods in Enzymology* **101**, 167.
Sanchez, S. and Demain, A.L. (1977). *European Journal of Applied Microbiology* **4**, 45.
Sibirny, A.A., Titorenko, V.I., Gonchar, M.V., Ubyvovk, V.M., Ksheminskaya, G.P. and Vivitskaya, O.P. (1988). *Journal of Basic Microbiology* **28**, 293.
Snow, R. (1966). *Nature* **211**, 206.
Sor, F. and Fukuhara, H. (1989). *Yeast* **5**, 1.
Sreekrishna, K., Nelles, L., Potencz, R., Cruze, J., Mazzaferro, P., Fish, W., Fuke, M., Holden, K., Phelps, D. and Wood, P. (1989). *Biochemistry* **28**, 4117.
Stark, M.J.R., Boyd A., Mileham A.J. and Romanos, M.A. (1990). *Yeast* **6**, 1.
Stinchcombe, D.T., Struhl, K. and Davies, R.W. (1979). *Nature* **282**, 39.
Sudbery, P.E. and Gleeson, M.A.G. (1989). In "Molecular and Cell Biology of Yeasts" (E.F. Walton and G.T. Yarranton, eds), pp. 304–329. Blackie, Glasgow, London.
Sudbery, P.E., Gleeson, M.A., Veale, R.A., Ledeboer, A.M. and Zoetmulder, M.C.M. (1988). *Transactions of the Biochemical Society* **16**, 1081.
Takagi, M., Kawai, S., Chang, M.C., Shibuya, I. and Yano, K. (1986). *Journal of Bacteriology* **167**, 551.
Takagi, M., Kobayashi, N., Sugimoto, M., Fujii, T., Wateri, J. and Yano, K. (1987). *Current Genetics* **11**, 451.
Tikhomirova, L.P., Kryukov, V.M., Strizhov, N.I. and Bayev, A.A. (1983). *Molecular and General Genetics* **189**, 479.
Tikhomirova, L.P., Ikonomova, R.N. and Kutzetsova, E.N. (1986). *Current Genetics* **10**, 741.

Tikhomirova, L.P., Ikonomova, R.N., Kutzetsova, E.N., Fodor, I.I., Bystrykh, L.V., Aminova, L.R. and Trotsenko, Y.A. (1988). *Journal of Basic Microbiology* **28**, 343.
Tolstorukov, I.I. and Benevolenskii, S.V. (1978). *Genetika* **14**, 519.
Tolstorukov, I.I. and Benevolenskii, S.V. (1980). *Genetika* **16**, 1335.
Tolstorukov, I.I. and Efremov, B.D. (1984). *Genetika* **20**, 1099.
Tolstorukov, I.I., Dutova, T.A., Benevolenskii, S.V. and Soom, Y.O. (1977). *Genetika* **13**, 322.
Tolstorukov, I.I., Bliznik, K.M. and Korogodin, V.I. (1979). *Genetika* **15**, 2140.
Tolstorukov, I.I., Bliznik, K.M. and Korogodin, V.I. (1982). *Genetika* **18**, 1276.
Tolstorukov, I.I., Efremov, B.D. and Bliznik, K.M. (1983). *Genetika* **19**, 897.
Tschopp, J.F., Sverlow, G., Kosson, R., Craig, W. and Grinna, L. (1987). *Bio/Technology* **5**, 1305.
Veale, R.A. Giuseppin, M.F., Van Gijk, H.M.S., Sudbery, P.E. and Verrips, C.T. (1992) Development of a strain of *Hansenula polymorpha* for the efficient expression of guar α-galactosidase. *Yeast* **8**, 361–372.
Weber, H. (1988). *Yeast* **4**, 235.
Wedlock, D.N. and Thornton, R.J. (1989). *Biotechnology Letters* **9**, 601.
Wedlock, D.N., James, A.P. and Thornton, R.J. (1989). *Journal of General Microbiology* **135**, 2019.
Wickerham, L.J. (1970). *In* "Yeasts: A Taxonomic Study" (J. Lodder, ed.), 2nd edn, pp. 448–454, North Holland Publishing Company, Amsterdam.
Wickerham, L.J., Kurtzman, C.P. and Herman, A.I. (1970). *Science* **167**, 1141.
Yarrow, D. (1972). *Antonie van Leeuwenhoek* **3570**, 360.
Zhang, Y.Z. and Reddy, C.A. (1986). *Current Genetics* **10**, 571.

8 Genetics of *Candida* Species

Russell T.M. Poulter

Department of Biochemistry, University of Otago, Dunedin, New Zealand

I. INTRODUCTION

The genetics of *Candida albicans* has developed rapidly in the last 10 years. The purpose of this chapter is to summarize the present state of analysis. There have been a number of recent reviews of this field (Kurtz *et al.*, 1988;

The Yeasts Vol. 6, 2nd edition
ISBN 0-12-596416-1

Magee *et al.*, 1988). Any review of this topic must owe a great debt to the most recent review (Kirsch *et al.*, 1990). The analysis of other *Candida* spp. is less developed but is now progressing rapidly (Kirsch *et al.*, 1990).

It is reasonable to ask why there should be this developing interest in the genetics of *Candida* spp. There are two possible responses. First, *C. albicans* is the most common and one of the most serious fungal pathogens. Its association with human immunodeficiency virus (HIV) infection has increased the clinical significance of candidosis. The yeast continues to be a serious problem in such superficial infections as vaginitis. Other *Candida* spp. have specific significance. Secondly, the genetics of *Candida* spp. would be of interest even if they were not clinically or economically significant. *Candida* spp. are, by definition, imperfect. Those species that have been analysed have been found to be diploid. This diploid imperfect status presents technically interesting problems to the geneticist. Molecular genetics and parasexual genetics have been applied to circumvent these problems. The natural diploid imperfect status is, however, of great intrinsic interest. The evolutionary problems faced by such organisms are distinct but they may offer insights into the general processes of evolution.

II. PHYSICAL CHARACTERIZATION OF THE GENOME OF *CANDIDA ALBICANS*

A. DNA content

The determination of the DNA content of *C. albicans* has been used in the analysis of ploidy. There is, in general, close agreement between the various DNA values reported in the last decade. These values fall in the range of 35–40 fg per cell. This value resembles the DNA content of diploid *Saccharomyces cerevisiae*. The values reported are: 35.5–39.0 fg (Whelan *et al.*, 1980), 35.2 fg (Whelan and Magee, 1981), 39.6–39.9 fg (Sarachek *et al.*, 1981), 36.5–39.0 fg (Riggsby *et al.*, 1982), 39 ± 2.4 fg (Rhoads and Sarachek, 1984), 37.7 ± 1.9 fg (Dvorak *et al.*, 1987), 38.3–41.9 (Kwon-Chung *et al.*, 1987). There is one obviously discordant value (Buckley *et al.*, 1982). These DNA values are derived by dividing the total DNA extracted from a culture by the number of cells in the culture. The DNA extraction must be quantitative without any significant degradation. The cells should be in the stationary phase of growth and without significant bud formation.

The DNA of *C. albicans* has been shown (Wills *et al.*, 1984) to have a G + C content of 32.8% (main nuclear band), 40.3% (satellite nuclear band) and 38.2% (mitochondrial). Most of the repeated DNA of *C.*

albicans consists of mitochondrial DNA or rDNA. There are at least two repeat families. One repeat family contains multiple copies of a sequence, which has been termed Ca3 by one group (Soll *et al.*, 1989) and 27A by another group (Scherer and Stevens, 1987). This is present at about 10 copies per genome and has a repeat size of about 15 kb. The other repeat family contains the Ca7 sequence (Soll *et al.*, 1989) and is telomeric.

B. Ploidy

The conclusion that all (or most) *C. albicans* strains are diploid is derived from three independent types of evidence, namely molecular genetics (Section III), parasexual genetics (Section IV) and physical evidence. The physical evidence for ploidy comes from measurement of the re-association kinetics of DNA from *C. albicans* together with information on the total DNA content of each cell. There is so little repetitious DNA that it does not interfere with this analysis. Re-association has been measured by hydroxyapatite chromatography, optical hypochromicity and S1 nuclease digestion (Riggsby *et al.*, 1982). Relative to a control renaturation of DNA from *E. coli*, performed under identical conditions, it was found that the complexity of DNA from *C. albicans* corresponds to an unrepeated genome size of 18 fg. This number is approximately half the total DNA content in each cell (35–40 fg). It is apparent from a comparison of complexity (18 fg) and total cell content (35–40 fg) that *C. albicans* is diploid.

The literature contains some reports of other ploidy states for *C. albicans*. It has been suggested (Olaiya *et al.*, 1980) that some strains, originally isolated as non-pathogenic, were in fact haploid. Others (Buckley *et al.*, 1982; Fleischman and Howard, 1988) have subsequently reported that these strains are normal diploids. It has been claimed (Suzuki *et al.*, 1982) that several other *C. albicans* isolates are non-diploid. There is however (Riggsby, 1985), some doubt as to whether some of these strains should be accepted as *C. albicans*.

C. Karyotype

Electrophoretic karyotyping has found considerable use in studying the genetics of *C. albicans*. Originally, orthogonal field agarose gel electrophoresis (OFAGE) and field inversion electrophoresis (FIGE) were employed, but most recent studies have used clamped homogeneous field gel electrophoresis (CHEF). Some reports using transverse alternating field electrophoresis (TAFE) have been made.

There is now emerging a consensus as to the electrophoretic karyotype (Riggsby, 1990). Problems with the analyses include a general lack of understanding of the theoretical basis of the electrophoretic protocols. Different chromosome bands may migrate in a different sequence if slightly different protocols are employed. In addition, any analysis must take into account the diploid nature of *C. albicans*. In specific strains, different homologues may be electrophoretically very different or apparently identical. The earliest electrophoretic separations used OFAGE. In one early study (Magee and Magee, 1987), as many as ten bands were resolved but the authors noted that this might reflect divergence of homologues in different strains. In a study (Merz *et al.*, 1988) of 17 clinical isolates, ten strains were found to have seven bands, five strains had eight bands and two strains had nine bands. FIGE analysis (Lott *et al.*, 1987; Johnston *et al.*, 1988) generally resolves rather fewer bands than OFAGE. CHEF analysis (Vollrath and Davis, 1987) usually resolves seven or eight bands. Other reports (Magee *et al.*, 1988) show similar resolution with CHEF.

It is clear from the appearance of gels presented in these analyses that chromosomes of *C. albicans* are smaller and more numerous than those of *Schizosaccharomyces pombe*. They are also clearly larger and less numerous than those of *Sacch. cerevisiae* (which has 16 chromosomes). Given that the genome complexity of *C. albicans* and *Sacch. cerevisiae* are similar (as is the DNA content, if *C. albicans* is compared with diploid *Sacch.* cerevisiae), it follows from these gel analyses that a "generalized" or "simplified" *C. albicans* strain should have about half as many chromosomes as *Sacch. cerevisiae* and that these should be larger than those of *Sacch. cerevisiae*. In the absence of homologue divergence, a band number of 6–8 would be widely accepted as reasonable. Homologue divergence will increase this band number. It was suggested at the American Society of Microbiology *Candida* Conference held in Palm Springs, California, in 1987 that the correct number of bands is six, but subsequent analyses have suggested that the number may be seven or eight.

One method of identifying diverged chromosome homologues is to use Southern blotting with defined probes. This procedure should permit the eventual analysis of the molecular basis of homologue divergence (including translocation, deletion and duplication). The first thorough study using this approach (Magee *et al.*, 1988) probed OFAGE, FIGE and CHEF gels with 14 gene-specific probes for *C. albicans* and three unidentified cloned sequences. In several cases, a probe was found to hybridize to two bands, a result that might suggest homologue divergence. These workers suggested that the correct chromosome number is seven. A subsequent analysis (Lasker *et al.*, 1989) using OFAGE and FIGE reported similar results but suggested several significant modifications. The most important modification was the suggestion that the largest band detected by Magee *et al.* (1988) was in fact two bands, bringing the chromosome number to eight. In addition, it was further suggested that the *TRP1* gene hybridized to one of these large bands, whereas Magee *et al.* (1988) had reported the *TRP1* probe as hybridizing to a small band.

A brief (and selective) summary of the present consensus electrokaryotype and Southern hybridization patterns is presented in Table I. The chromosomes are described according to the number system (Magee *et al.*, 1988) and the letter system.

The origin and characterization of some of these probes is given later in the section on molecular genetics (Section III).

Many examples of homologue divergence have been found in various strains. Such divergence would act as a barrier to a meiotic cycle, because of the difficulty of generating balanced haploids. It is probably incorrect to see the divergence as the reason why there is no sexual cycle. More probably the homologue divergence is a consequence of the absence of a sexual cycle and the constraints that meiosis applies.

D. Mitochondrial DNA

The mitochondrial DNA of *C. albicans* (Wills *et al.*, 1984; Shaw *et al.*, 1989) has a G + C content of 38.2% and is 40 kb in length. It is circular on the basis of electron-microscope studies and detailed restriction-fragment analysis. This size is in the middle range of the mitochondrial

Table I. Electrokaryotype and probe analysis

Chromosome number	Chromosome letter	Probe
1	H and G	Actin, β-tubulin, rDNA, *GAL1 SOR9* (*TRP1*, Lasker)
2	F	*HIS3*
3	E	*URA3 ADE2 SOR2*
4	D	pCHR4 (anonymous)
5	C	*LYS2*
6	B	*BEN1* (benlate resistance)
7	A	pCHR7 (anonymous)

DNA of fungi (*Torulopsis glabrata* 18 kb; *Sacch. cerevisiae* 78 kb). There is a 5 kb inverted repeat in the genome (Shaw *et al.*, 1989). The inverted repeat serves as a site for recombination so that the mitochondrial genome exists in two forms or isomers. The inverted repeats include the gene for subunit III of cytochrome *c* oxidase, which is therefore present twice in the genome. The inverted repeats divide the molecule into two domains. The larger domain (just over 20 kb) includes the genes for the small rRNA, cytochrome *b*, subunit 1 of cytochrome *c* oxidase, and subunits 6 and 9 of ATPase. The smaller domain (less than 10 kb) includes subunits 2 of cytochrome *c* oxidase and the large rRNA. These assignments were made using petite mitochondrial DNA from defined strains of *Sacch. cerevisiae* as probes (Riggsby, 1990). There have been no reports of mitochondrial petite strains of *C. albicans* despite numerous attempts to isolate them. It is probable therefore that *C. albicans* is a true petite-negative yeast. It is likely that the petite state is lethal in this and other petite-negative yeasts but it is not obvious what the basis of this lethality is.

E. Genomic variation and strain recognition

There are a number of possible ways for attempting strain discrimination with genomic DNA analysis. Several of these can be discarded as of limited value. For example, electrokaryotyping is time-consuming and there are difficulties in resolving many strains. Similarly, the analysis is of restriction-fragment length polymorphisms (RFLP analysis) with unique sequence probes, shows limited usefulness. It has, however, been used to support the belief that *C. albicans* is diploid (Kurtz *et al.*, 1988). Restriction analysis of mitochondrial DNA reveals some strain variation when four basepair restriction site endonucleases (such as *Hae*III and *Taq*I) are used. The variation, is however, limited (Olivo *et al.*, 1987). Restriction analysis of mitochondrial DNA has proved useful (Kwon-Chung *et al.*, 1989) in separating *Candida stellatoidea* strains into two groups — one (Type II) considered to be simply sucrose-negative variants of *C. albicans* and the other group of strains (Type I) perhaps only distantly related to *C. albicans*. Ribosomal DNA has been employed (Scherer and Stevens, 1987) in strain recognition by restricting DNA and observing the intense (repeat sequence) bands on ethidium bromide agarose gels. Alternatively (Magee *et al.*, 1987), rDNA has been visualized using Southern blotting with a rDNA probe. A large number of strains cannot be resolved by these procedures as they give similar or identical band patterns.

Moderately repetitious probes show greatest promise for strain discrimination. Probe 27A is a species-specific probe present in about 10 copies. The repeat elements are dispersed throughout the genome. In one study (Scherer and Stevens, 1988), probe 27A Southern analysis was used to establish clonal identity between strains isolated from different sites in a number of patients (urine, anus, vagina). In another study using probe 27A (Fox *et al.*, 1989), it proved possible to discriminate 60 strains isolated from 63 patients. In this study it was found that, when 57 isolates from 20 patients were analysed, all strains from a particular patient were the same regardless of the site of isolation. Strains isolated from patients over a period of 2–18 months remained unchanged (Fox *et al.*, 1989). In summary, it appears that moderately repetitive probes, such as probe 27A, have an overwhelming advantage for epidemiological analysis of *C. albicans*.

III. MOLECULAR GENETICS OF *CANDIDA ALBICANS*

A. Gene isolation

Specific gene isolation is a prerequisite for such procedures as DNA sequencing and site-specific gene disruption. There are two broad types of gene isolation, those relying on function and those relying on DNA homology and hybridization.

A number of DNA libraries have been constructed for *C. albicans* (Kirsch *et al.*, 1990). These can be employed to isolate a specific gene of interest by functional complementation of a suitable recipient cell. The libraries have been constructed in λ vectors (Kurtz *et al.*, 1988; Scherer and Stevens, 1988), *Sacch. cerevisiae/E. coli* shuttle vectors such as YEp13 (Gillum *et al.*, 1984; Rosenbluh *et al.*, 1985; Kurtz *et al.*, 1986) and in the *C. albicans* vector pCARS1 (Kelly *et al.*, 1988). Libraries have been constructed with DNA, which has been completely digested, for example, with *Hind*III or *Eco*RI (Kurtz *et al.*, 1988; Scherer and Stevens, 1988). Alternatively, and preferably, libraries have been constructed using partial digestion with the four basepair restriction enzyme *Sau*3A (Gillum *et al.*, 1984; Rosenbluh *et al.*, 1985; Kurtz *et al.*, 1986). Even in a partial-digest library, a desired sequence may be absent because of the statistical chance of it being unrepresented or restricted. In addition, some sequences may be lethal in the recipient system if they are expressed. When no suitable recipient cell exists, it may be possible to detect the cloned gene because of a high plasmid copy number leading to overexpression and resultant resistance to a specific inhibitor (Kurtz *et al.*, 1987, 1988).

The limited amount of sequence information available suggests that *C. albicans* (like *Sacch. cerevisiae* but unlike *Aspergillus nidulans, Neurospora crassa* or *Schiz. pombe*) has few genes with introns. This means that there is a reasonable chance of a gene carried on an *E. coli* vector being expressed in *E. coli*, although the expression may not be very efficient in such a heterologous host. It has been estimated that 20–30% of genes in *Sacch. cerevisiae* can be expressed in *E. coli*. The frequency of gene expression in *C. albicans* is uncertain but may be similar.

A more promising strategy is the cloning of genes into "shuttle" vectors for *Sacch. cerevisiae* such as YEp13. In such systems, the libraries may be amplified in *E. coli* but screened for expression in *Sacch. cerevisiae*. Given that both *C. albicans* and *Sacch. cerevisiae* lack introns, it is likely that a reasonably high proportion of genes will be expressed in this heterologous system (Kurtz *et al.*, 1990). There are some genes that cannot be isolated in such heterologous systems (morphology genes, for example) and, for this reason, there is particular value in having a homologous expression system. This topic will be discussed later, but such a system has been developed for *C. albicans*.

Cloning of genes by DNA homology and hybridization is independent of expression. Unfortunately many genes in *C. albicans* lack significant homology with sequences in *Sacch. cerevisiae* (Gillum *et al.*, 1984; Kurtz *et al.*, 1986, 1987; Kelly *et al.*, 1988; Kurtz and Marrinan, 1989). The exceptions to this general rule are the tightly conserved structural genes such as actin (*ACT1*; Lasker *et al.*, 1989) and β-tubulin (*TUB2*; Gorman *et al.*, 1988), which have been cloned by screening *C. albicans* libraries with *Sacch. cerevisiae* probes. There are, in addition to the structural genes, some enzymes that are sufficiently conserved to permit isolation by hybridization with *Sacch. cerevisiae* probes. The lanosterol 14α demethylase gene from *C. albicans* was cloned using this procedure (Kirsch *et al.*, 1988). The sequence was shown to complement a mutant of *Sacch. cerevisiae* if present on a high copy number vector (Kirsch *et al.*, 1988). The sequence conferred resistance to imidazole antifungal compounds on *C. albicans* recipients when present on a multicopy plasmid.

B. Cloned genes

1. *URA3*. Orotidine-5′-phosphate decarboxylase

This was the first gene to be cloned in *C. albicans* (Gilllum *et al.*, 1984) and was isolated by heterologous expression of a YEp13 library in an *ura3* strain of *Sacch. cerevisiae*. The *C. albicans URA3* clones also complemented an *E. coli pyrF* mutant. It was subsequently shown that the sequence could be disrupted to yield a *ura3 C. albicans* auxotroph, which could be complemented by the *C. albicans URA3* sequence. There was no hybridization of the sequence to *Sacch. cerevisiae*. The minimum functional sequence that was subcloned was a 1.5 kb *Xba*I–*Sca*I fragment. One of the original clones derived from the library was not colinear with the genome, presumably due to a multiple cloning event. It is important that the sequence is proved to be colinear, if the clone is to be employed for gene disruption.

2. *LEU2*. 3-Isopropylmalate dehydrogenase

The *C. albicans LEU2* gene has been isolated (Kelly *et al*., 1988) using a *Candida* library constructed in the homologous shuttle vector pCARS1. The *E. coli* recipient strain C600 (*leub*) was used and it showed functional complementation. The minimum functional sequence was a 2.2 kb *Bam*HI–*Eco*RI fragment. The sequence was subcloned to a *Sacch. cerevisiae* vector, where it complemented the *leu2* auxotrophy of *Sacch. cerevisiae* strain AH22. The *LEU2* sequence was disrupted and an attempt was made to generate a *C. albicans leu2* auxotroph. This experiment will be discussed further in the section on gene disruption but the attempt produced unexpected results, which are difficult to interpret without further analysis.

3. *HIS3*. Imadazole-glycerolphosphate dehydrogenase

This gene was isolated by heterologous complementation in *Sacch. cerevisiae* (Rosenbluh *et al*., 1985). The gene has been used as a probe in electrokaryotyping.

4. *ADE2*. Phosphoribosylaminoimadazole carboxylase

This gene was cloned by heterologous complementation in *Sacch. cerevisiae* (Kurtz *et al*., 1986) The essential region is a 2.3 kb *Sau*3A–*Eco*RV segment. The particular value of this system is the ease with which visual selection and analysis can be used. Colonies of strain *ADE2* are white while *ade2* colonies are red. Sectoring (unstable) colonies are easily detected.

5. ADE1. Phosphoribosylaminoimadazole succino-carboximide synthetase

This gene is the second "red" adenine gene. There is some contradiction in the literature concerning this locus. It was reported (Magee *et al.*, 1988) that the *ADE1* sequence only complemented some strains of *Sacch. cerevisiae*. As the authors point out, this could suggest a cloned suppressor sequence. Subsequent work (Kelly *et al.*, 1988) supported this suggestion and described the isolation of a different sequence carrying what is probably the structural *ADE1* gene. This has been shown to complement *ade1* strains of *Sacch. cerevisiae* and *C. albicans*. This sequence has also been shown to complement a Type II *C. stellatoidea ade1* strain (Kurtz *et al.*, 1990).

6. TRP1. Phosphoribosyl anthranilate isomerase

This gene was isolated (Rosenbluh *et al.*, 1985) by complementation of a *trp1* mutant of *Sacch. cerevisiae*.

7. DRF1. Dihydrofolate reductase

The isolation of this gene (Kurtz *et al.*, 1987) was unusual in that it made use of high level expression in the heterologous *E. coli* system. The recipient *E. coli* became resistant to the prokaryote dihydrofolate reductase inhibitor trimethoprim.

8. HEM3. Uroporphyrinogen I synthase

A sequence that contains the *HEM3* structural gene was isolated (Kurtz and Marrinan, 1989) by complementation of a *hem3* strain of *Sacch. cerevisiae*. The minimum sequence was 2 kb. The sequence was disrupted and used to form a *hem3* auxotroph of *C. albicans*. This auxotroph shows different auxotrophic requirements to the corresponding strain of *Sacch. cerevisiae*.

9. TMP1. Thymidylate synthase

This sequence has been isolated (Singer *et al.*, 1989) by complementation of a *tmp1* mutant of *Sacch. cerevisiae*. The enzyme is of considerable interest as a target for chemotherapeutic agents. The gene has been sequenced (Singer *et al.*, 1989), and the 5' and 3' sequences characterized. The coding region shows 70% identity at the nucleotide level and 74% at the amino-acid level, when compared with the *TMP1* gene from *Sacch. cerevisiae*.

10. *SOR2* and *SOR9*. Sorbitol utilization

Sequences which enable *Sacch. cerevisiae* to grow on media containing sorbitol as a sole carbon source have been described (J.A. Gorman, communication). The *SOR2* 2.8 kb fragment has been sequenced. It is not certain what these genes do, but one or other may represent a sorbitol dehydrogenase. A curious observation is that exogenous glucose regulates the level of expression of *SOR2* and *SOR9* both in *Sacch. cerevisiae* and *C. albicans*, suggesting that glucose regulation must act by some common pathway in both of these yeasts.

11. *GAL1*. Galactokinase

The galactokinase gene in *C. albicans* has been cloned by complementation of a *gal1* mutant of *Sacch. cerevisiae*. The complete nucleotide sequence has been determined (Kurtz *et al.*, 1990). It has been demonstrated that the *GAL1* gene in *C. albicans* resembles the *Sacch. cerevisiae* in that it is positively regulated by the *GAL4* gene product of *Sacch. cerevisiae*.

12. Ribosomal DNA

Ribosomal DNA sequences in *C. albicans* show strong homology with the homologous sequences in *Sacch. cerevisiae*. The arrangement of genes for the various ribosomal RNAs is also conserved (Kurtz *et al.*, 1990).

13. *SUC*. α-Glucosidase

A putative α-glucosidase gene has been isolated (Kurtz *et al.*, 1990) from *C. albicans* by complementation of a *suc2* (invertase) mutant of *Sacch. verevisiae*. The gene enables the mutant to utilize both sucrose and maltose.

14. Others

A number of other genes have been isolated from *C. albicans*. These include: *ERG7*, 2,3,-oxidosqualene cyclase, which is involved in ergosterol biosynthesis (Kelly *et al.*, 1990); *ERG16*, lanosterol 14α-demethylase (Kirsch *et al.*, 1988); *CHS1*, chitin synthetase (J. Au-Young and P. Robbins, personal communication); *TUB2*, β tubulin (Smith *et al.*, 1988); and *ACT1*, actin (Lasker *et al.*, 1989).

Although only a small number of *C. albicans* genes have been sequenced, it is of interest to suggest what general features are apparent (Kurtz *et al.*,

1990). The genes show considerable similarity in their molecular organization to those in *Sacch. cerevisiae*. This may be a general pattern or may reflect the selection systems used to isolate most of the genes in *C. albicans*. Most genes do not contain introns (the exceptions are *ACT1* and *TUB2*). Most genes have TATA sequences. Codon bias is found only in the highly expressed genes, as in *Sacch. cerevisiae*. The 3' transcription termination and polyadenylation sites resemble those in *Sacch. cerevisiae*. There are therefore no obvious or distinctive features of genes in *C. albicans*. It is to be hoped, therefore, that heterologous transformation with genes from *Sacch. cerevisiae* can be used to characterize *C. albicans* mutants. Although the reverse procedure has been extensively employed, there are few examples of using sequences from *Sacch. cerevisiae* to characterize *C. albicans* mutants. The *LEU2* gene from *Sacch. cerevisiae* is one of the few that has been shown to function in *C. albicans*.

C. Transformation

The first *C. albicans* strains identified as carrying specific genes (Poulter and Rikkerink, 1983) were *ade1* and *ade2* auxotrophic strains. One of these *ade2* strains, hOG300, was used as a recipient in the development of transformation. The corresponding *ADE2* sequence was isolated on the YEp13 shuttle vector by complementation of an *ade2 Sacch. cerevisiae* strain (Kurtz *et al.*, 1986). Transformation was performed with *C. albicans* protoplasts. The frequency of transformants was 0.5–5 transformants per μg DNA. The frequency of stable transformants was 3–6 per 10^6 viable spheroplasts. Southern hybridization demonstrated that transformation was integrative and that it was homologous. The transformant carried, as an integrated sequence, the whole transforming plasmid.

For many purposes, transformation is facilitated by using plasmids which have an autonomously replicating sequence (ARS). For example, the frequency of transformation is higher with ARS plasmids as they are not dependent on homologous integration. Such transformation systems have the additional property that the transformants are unstable on non-selective media. An ARS sequence was isolated (Kurtz *et al.*, 1987) by screening a library of *Rsa*I-digested DNA from *C. albicans* inserted into an *ADE2* plasmid. Colonies were isolated which were *ADE2* but unstable. These were easily detected because of the white/red sectoring. Such unstable colonies give rise, at a low frequency, to stable integrated derivatives. Only one ARS plasmid, namely p56, was isolated from the library, suggesting that the frequency of ARS sequences may be somewhat lower than in *Sacch.*

cerevisiae. Unlike the ARS vectors from *Sacch. cerevisiae*, the CARS vector from *C. albicans* is present in transformed cells as a high molecular weight concatamer (> 40 kb). The average copy number was 10. The CARS from *C. albicans* does not function in *Sacch. cerevisiae*. Neither the 2 μm ori from *Sacch. cerevisiae* nor ARS1 function in *C. albicans*.

D. Gene disruption

Gene disruption by site-specific mutagenesis provides an alternative to parasexual mutant selection. *Candida albicans* displays a high frequency of homologous recombination with transforming sequences, a prerequisite for gene disruption. The yeast does, however, present the unusual problem that it is diploid and, therefore, the initial product of gene disruption will be a heterozygous strain, which must subsequently be brought to homozygosity by induced mitotic recombination. The general procedures resemble those developed in *Sacch. cerevisiae* and employ linearized genes, which have been disrupted internally, usually with a selectable marker. The disrupted sequence is exchanged for the natural prototrophic sequence by homologous recombination. The procedure requires that the gene to be mutated has been cloned, and at least a partial restriction map, revealing internal cut sites, must be available. Three genes have been disrupted in this way, namely *URA3* (Kelly *et al.*, 1987), *HEM3* (Kurtz and Marrinan, 1989) and *LEU2* (Kelly *et al.*, 1988). The *LEU2* analysis has a number of unclear aspects. The only *leu2* auxotroph isolated remained heterozygous for the disrupted sequence. The other allele showed a normal restriction pattern and this suggests that it had suffered a point mutation. It is unclear why the disrupted sequence could not be brought to homozygosity. It may be that the disruption included an overlap into another (vital) gene adjacent to the *LEU2* sequence.

IV. PARASEXUAL GENETIC ANALYSIS OF *CANDIDA ALBICANS*

A. Introduction

Despite being diploid and imperfect, the yeast *C. albicans* can be analysed by the application of parasexual techniques. The three basic procedures (mutant selection, complementation testing and recombination analysis) are required for classical parasexual analysis to be performed with reasonable efficiency in this organism. Methods for directed site-specific mutagenesis have been described in Section III. There is at present a vigorous effort aimed at integrating the genetic map with the electrokaryotype described in Section II. Many metabolic pathways of *C. albicans* closely resemble those of *Sacch. cerevisiae*. This has proved useful in isolating and characterizing *C. albicans* genes. The genes of *C. albicans* are, where possible, named on the basis of homology with those of *Sacch. cerevisiae*.

B. Mutants

Given that *C. albicans* is diploid, it follows that a recessive mutation must be homozygous before it is expressed. The creation of a *de novo* mutant is considered to involve two steps, namely generation of the mutation (represented initially heterozygously) followed by a mitotic recombination event bringing the mutation to homozygosity. Mitotic recombination is a repair response and it is generally true that procedures that are mutagenic will also be recombinogenic. The nature of the mitotic recombination events will be discussed further in Section IV.D. In practice, it has proved reasonably easy to isolate auxotrophic mutants from *C. albicans*.

In this review, it is impossible to describe all of the systems that have been analysed. Three will be considered because of their significance and because they illustrate general principles. Genetic markers that can be selected visually or by antibiotics are especially valuable in this genetically difficult organism.

The most useful auxotrophs so far employed in *C. albicans* are the red adenine auxotrophs *ade1* and *ade2*. These mutants were used in the original analysis of protoplast fusion (Poulter *et al.*, 1981; Sarachek *et al.*, 1981) during the exploration of the ploidy question (Poulter *et al.*, 1982) and extensively in mapping (Poulter *et al.*, 1982; Poulter and Rikkerink, 1983). In addition, *ade2* auxotrophs produced by parasexual genetics were used in the establishment of transformation procedures (Kurtz *et al.*, 1986, 1987). These *ade1* and *ade2* auxotrophs can be selected visually because, on media containing limiting concentrations of adenine, they grow as red colonies. An additional advantage is that the red adenine auxotrophs contain representatives of only two genetic classes namely *ade1* and *ade2*. Both genes have been mapped and are unlinked.

There are two pyrimidine biosynthesis auxotrophs of particular interest, namely, *ura3* and *ura5*. Both auxotrophic types are resistant to 5-fluoro-orotic acid and they can therefore be readily selected. Transformation systems using *ura3* recipients have been developed (see Section III). Both the *ura3* and *ura5* genes have been mapped, and are unlinked.

In *C. albicans*, sulphite reductase mutants can be visually selected. *Candida albicans* has an unusual response to media containing bismuth ammonium citrate and sulphite. On such media, prototrophic colonies of *C. albicans* are black. The black pigment, bismuth sulphide, does not interfere with viability. Sulphite reductase-deficient colonies appear white on such media (Glasgow, 1985). A modification of this medium has been used to distinguish by colour "opaque" and "white" colonies in switching strains of *C. albicans* (Rikkerink *et al.*, 1988).

C. Protoplast fusion and complementation

Protoplasts can be formed from *C. albicans* using Zymolyase (Poulter *et al.*, 1981; Kakar and Magee, 1982; Whelan *et al.*, 1986) or glucuronidase (Sarachek *et al.*, 1981; Evans *et al.*, 1982; Pesti and Ferenczy, 1982). Osmotic buffering of the protoplasts can be achieved with magnesium sulphate, potassium chloride, mannitol or sorbitol. Protoplast fusion is achieved by mixing protoplasts in the presence of polyethylene glycol and calcium salts.

Protoplast hybridization is achieved by crossing two diploid parental strains each of which is auxotrophic. The selection of hybrid fusion products is performed on minimal regeneration medium on which the hybrid will grow but on which neither parental strain will grow. In order to avoid false positives, stable auxotrophs must be used. It is frequently useful to have two independent auxotrophies in each parental strain. Protoplast fusions typically give rise to two types of colonies on regeneration plates. At a low frequency, fast-growing stable colonies arise. Such colonies have been shown (Whelan *et al.*, 1985) to be uninucleate. This study (Whelan *et al.*, 1985) also determined the DNA content of seven diploid parental auxotrophic strains and six hybrids formed between these strains. The parental diploid strains had DNA values between 36.7 and 38.8 fg per cell, in good agreement with other published values. The DNA content of five of the hybrids was ($\pm$12%) the sum of the parental values suggesting that they are stable tetraploids. The sixth hybrid strain was apparently hexaploid.

Slow-growing colonies arise on regeneration plates at a higher frequency. Cells from these slow-growing colonies are multinucleate (Sarachek *et al.*, 1981) and they are presumably therefore heterokaryons. These slow-growing colonies are unstable and, on non-selective media, give rise to the two parental types. When subcultured on selective media, slow-growing cultures give rise to colonies with faster growing papillae and sectors. An analysis of the DNA content (Saracheck *et al.*, 1981) of seven fast-growing derivatives of one heterokaryon suggested that some derivatives were probably tetraploids while others were probably aneuploids. Evidence suggests (Sarachek and Weber, 1984) that the aneuploid derivatives may have arisen by a process of chromosome transfer or partial karyogamy analogous to the process described in *kar1* mutants of *Sacch. cerevisiae* (Dutcher, 1981).

Protoplast fusion on selective media can be used to test whether phenotypically similar auxotrophs complement. The most extensive such analysis (Poulter and Rikkerink, 1983) was performed on 23 independent red adenine mutants. Of these strains, nine fell into a simple non-complementing goup deduced to be *ade1*. The remaining 14 strains complemented with *ade1* strains. In addition, strains of this group complemented in some combinations with other members of the group in a complex pattern. This complementation behaviour, suggestive of intragenic complementation, indicated that the group of strains corresponded to the *ade2* gene, since this gene is known to display intragenic complementation in *Sacch. cerevisiae*. It should be possible to perform such complementation testing among any reasonably stable auxotrophic alleles. Complementation between non-auxotrophic characters (such as morphology mutants) can be performed by using auxotrophic markers for selection and treating the morphology phenotype as an "unselected" marker (Pomes *et al.*, 1985).

D. Mitotic recombination and mapping

Protoplast fusion results in two auxotrophic parental strains giving rise to a tetraploid hybrid. The hybrid is a duplex tetraploid, having two defective and two functional alleles at each locus. While it is possible to produce homozygous recombinant derivatives from such tetraploids, this occurs at a very low frequency. It is more efficient first to reduce the hybrid tetraploid to the diploid state by induced chromosome loss. Various procedures will induce such diploidization, but the most effective procedure (Hilton *et al.*, 1985) is brief heat-shock (51°C for 90 seconds). Heterozygous para-sexual diploids (paradiploids) give rise to homozygous derivatives with a high frequency following induction of mitotic recombination by ultraviolet irradiation.

It is probable that ultraviolet irradiation induces various responses including chromosome loss (rarely), mitotic crossing-over and mitotic gene conversion. In contrast, heat shock appears to induce chromosome loss but no other type of mitotic recombination. Use can be made of these systems to map genetic markers. This is facilitated if at least one of the markers in the analysis can be selected, visually or using antibiotics. Mitotic recombination is an uncommon event even following ultraviolet induction and some system of selection of recombinant colonies is therefore valuable. Mapping of genes is further facilitated if the recessive mutant alleles are present in *cis* in one of the parental strains. In this situation, they are likely to cosegregate and therefore come to expression together.

If two genes are present on different chromosomes, they will only very rarely cosegregate from a paradiploid. Such cosegregation would require independent events at the two sites. If two genes frequently cosegregate, then they are assumed to be on the same chromosome. If two genes are on opposite arms (transcentromeric), then they will invariably cosegregate following heat shock but only rarely following ultraviolet irradiation. If two genes are on the same chromosome arm, they will invariably cosegregate following heat shock and frequently cosegregate following ultraviolet irradiation as a result of mitotic crossing-over. It should be possible to map the relative position of genes on the same chromosome arm. Mitotic crossing-over will bring to homozygosity that part of the chromosome arm distal to the site of crossing-over, while that part of the chromosome arm proximal to the crossing-over site will remain heterozygous. If a group of recombinants are analysed, then the patterns of cosegregation should reveal the relative positions. The most distal gene will be able to come to homozygosity on its own, whereas the proximal genes will only come to homozygosity as cosegregants. This theoretical approach is complicated by the presence of mitotic gene conversion. Mitotic gene conversion will result in proximal genes being able to come to homozygosity in isolation. It should be possible to obtain absolute or relative frequencies of mitotic recombinants but this is made difficult by strain variation in ultraviolet sensitivity. In addition, there is variation in the ease of detection and purification of auxotrophic clones; for example, red adenine auxotrophs are especially easy to find and purify.

A number of linkage groups have been analysed (Table II). There are therefore six linkage groups so far described and 23 genes placed on these linkage groups (Poulter, 1990).

E. Natural heterozygosity

Some strains of *C. albicans* (perhaps 10–20%) carry recessive mutant alleles. Such heterozygous strains give rise to a high frequency of a specific type of property (for example, auxotrophy) following ultraviolet

Table II. Linkage groups of *Candida albicans*

Linkage group number	Relative position[a] of mapped genes[b]
1	(leu)*ino*–cen–*ura5*–*tsm2*–*met1*–*arg4*–*ade1*
2	(his,ilv)*pro*–cen–*lys*
3	cen–*met2*–*ade2*–*ura3*
4	(hom)cen–*let2*–*let1*–*sfi1*
5	*sfi2*–cen–*asn*
6	*thr*–cen–*fcy1*–*his3*

[a] Loci in brackets are mapped to the linkage group but the position is undefined. The *leu* gene on linkage group 1 is very close to *ade1*.
[b] The loci are named following conventions used in *Saccharomyces cerevisiae*; *tsm2* is a temperature-sensitive locus as is *let1*, *let2* is a recessive lethal, *sfi1* and *sfi2* are involved in sulphite utilization, *fcy1* is the fluorocytosine resistance gene, and cen represents the centromere position.

irradiation. This behaviour was one of the first lines of evidence that suggested that *C. albicans* was diploid (Whelan *et al.*, 1980). It is surprising, on theoretical grounds, that only a small percentage of strains are heterozygous for auxotrophic mutations. It is also surprising that the strains that carry such heterozygous genes usually only carry a single such heterozygosity (Poulter, 1990). It might have been expected that this imperfect diploid yeast would have accumulated more such mutations. There is one further unexpected feature of this phenomenon. While about 50% of auxotrophic heterozygosities represent a wide range of genes, the remaining 50% all involve a single gene, *sfi1* (Poulter, 1990). This gene is part of the sulphite reductase complex. Protoplast fusion has been employed to show that 11 out of 12 natural *sfi1* heterozygosities were non-complementing. Natural heterozygosities for *fcy1* (fluorocytosine resistance, uracil phosphoribosyltransferase) are also frequent and are responsible for clinical resistance to 5-fluorocytosine. Natural heterozygosities have been used for strain recognition for epidemiological purposes.

F. Switching

Some strains of *C. albicans* display a spontaneous high frequency of reversible morphological change (Soll, 1990). The frequency is higher than would be expected from classical mutation. The "white–opaque" transition shows a spontaneous white-to-opaque change of 5×10^{-4}, the reverse process being less frequent. "White" colonies are morphologically normal

(opaque, smooth and white), while "opaque" colonies are flat and trans-
lucent. Incubation of "opaque" cells at 34°C for 5 hours results in all cells
undergoing transition to the "white" type. Ultraviolet irradiation also
influences these frequencies (Rikkerink *et al.*, 1988). It is not clear what
process underlies this phenomenon. It is possible that a transposon-like
element is involved but as yet there is no evidence to support this suggestion.

V. GENETICS OF OTHER *CANDIDA* SPECIES

A. Introduction

By definition yeast of the genus *Candida* are imperfect. Genetic analysis
of these yeasts is therefore limited to parasexual and molecular approaches.
These approaches are broadly similar to those employed in analysis of
C. albicans. Some aspects of this field have recently been reviewed
(Rachubinski, 1990).

Parasexual protocols employ protoplast fusion between suitable auxotrophic parental strains and subsequent selection on minimal media. Auxotrophs are reasonably readily isolated from *Candida* spp. even though
species are diploid.

Molecular approaches can be broadly subdivided into a direct study of
the genome and indirect study employing manipulation of the genome. Of
particular importance in manipulation of the genome is the characterization
of an efficient transformation system. Such transformation systems usually
require plasmids carrying suitable prototrophic alleles together with an
effective autonomously replicating sequence. In addition, transformation
will require a suitable auxotrophic recipient strain corresponding to the
cloned prototrophic gene. It cannot be assumed that a prototrophic gene
from one species will necessarily function in a different species nor that an
ARS active in one species will be active in another. In the absence of a
procedure for transforming a species, it is still possible to isolate and
analyse genes of interest in another suitable yeast (such as *Sacch.
cerevisiae*). Once such genes are isolated, they may prove useful in
producing auxotrophic mutants by gene-disruption techniques.

B. *Candida tropicalis*

Candida tropicalis is considered to be second only in importance to
C. albicans as a human fungal pathogen. The yeast is also of interest for

the study of alkane and fatty-acid degradation in peroxisomes. Parasexual techniques have been described for this diploid yeast. Protoplast fusion generates stable hybrids and from these hybrids parental markers may be recovered following induced recombination (Fournier *et al.*, 1977; Vallin and Ferenczy, 1978). Recently, intraspecific and inter-specific (with *C. albicans*) hybrids were described between red adenine auxotrophs (Corner and Poulter, 1989). Red adenine auxotrophs of *C. tropicalis* fall into two complementation classes, which correspond to the *ade1* and *ade2* mutants of *Sacch. cerevisiae* and *C. albicans*. These auxo-trophs may be of use in establishing a transformation system, although as yet there has been no report of transformation of *C. tropicalis*.

A number of genes in *C. tropicalis* have been cloned and characterized. Five peroxisome protein genes have been described (Kamiryo and Okazaki, 1984; Okazaki *et al.*, 1986, 1987; Murray and Rachubinski, 1987; Szabo *et al.*, 1989). Peroxisome protein genes do not contain introns and have typical TATA box sequences 5′ to the coding region together with typical yeast polyadenylation sites. Codon usage is highly biased in these five peroxisome genes and resembles the usage found in highly expressed genes in *Sacch. cerevisiae*. Two cytochrome P450 systems are known in *C. tropicalis*; one is involved in sterol biosynthesis and the other in the initial step in alkane degradation. The genes for both P450 lanosterol 14 α demethylase and the alkane-induced P450 have been cloned and charac-terized (Chen *et al.*, 1988; Sanglard and Loper, 1989). The characteristics of these P450 genes closely resemble those described for peroxisome genes.

C. *Candida utilis*

This diploid yeast is important industrially. Protoplast fusion has been described (Delgado and Herrera, 1981) between *C. utilis* auxotrophs, and between *C. utilis* and *Sacch. cerevisiae*.

The *LEU2* gene of *C. utilis* has recently been sequenced (Hamasawa *et al.*, 1987). The *LEU2* gene shows 76.2% nucleotide-residue identity and 85.4% amino-acid residue identity with the *LEU2* gene of *Sacch. cerevisiae*. The 370 nucleotide-residue pairs of sequence 5′ to the *LEU2* coding region do not seem to include either the leucine residue-rich peptide or the tRNA^{Leu} gene found in *Sacch. cerevisiae*. The 5′ region does not possess any sequence homology with the large open-reading frame found 420 nucleotide residues 5′ to the *LEU2* sequence in *C. maltosa* and 5′ to the *LEU2* sequence of *C. albicans*. The cloned *LEU2* gene may prove useful in establishing transformation in this organism.

D. *Candida maltosa*

Candida maltosa is a useful industrial yeast and resembles *C. tropicalis* in that it can grow on *n*-alkanes as the sole carbon and energy source. Several transformation systems have been described for *C. maltosa*. Auxotrophs carrying a *lys2* allele can be transformed and complemented by plasmids of *Sacch. cerevisiae* carrying *LYS2* and ARS sequences (Kunze *et al.*, 1987). A similar system (Kunze *et al.*, 1985) has been described for the *arg4* locus. Recently an ARS has been isolated from *C. maltosa*, which is functional in both *C. maltosa* and *Sacch. cerevisiae*. The ARS has been sequenced and, in a region of 200 basepairs, five 11-basepair sequences closely homologous to the ARS consensus in *Sacch. cerevisiae* were found (Kawai *et al.*, 1987).

The *LEU2* gene of *C. maltosa* has recently been isolated (Takagi *et al.*, 1987). This gene, like that of *C. utilis*, shows close similarity to the *LEU2* gene in *Sacch. cerevisiae*, there being 84% nucleotide sequence and 94% amino-acid sequence homology. The sequence 5' to the LEU2 coding region does not show any evidence of homology to the leucine residue-rich leader peptide and tRNA^{Leu} found 5' to the *LEU2* coding region in *Sacch. cerevisiae*. The region 5' to the *LEU2* gene in *C. maltosa* does reveal one or two long open-reading frames of unknown function (R. Poulter, unpublished observation). The published sequence describes 2193 bases (from −859 to 1334). The sequence from −859 to −626 is one open-reading frame, which is followed (−628 to −492) by another open-reading frame. It is possible that these two are in fact one continuous open-reading frame (with a sequencing error at position −627). If this is so, then the open-reading frame extends from before −859 (the start of the sequence) to −492. This is of particular interest because a closely homologous sequence is found 5' to the *LEU2* gene of *C. albicans* (R. Cannon, personal communication). In *C. albicans*, there is one continuous-reading frame, which shows 94% amino-acid-residue homology and 84% nucleotide-residue homology to that in *C. maltosa*. The function of this gene, and the reason for its conservation in sequence and position 5' to the *LEU2* gene deserve further analysis.

A gene coding for one of the peroxisome proteins (*POX4*) has recently been isolated from *C. maltosa* (Hill *et al.*, 1988). It closely resembles the analogous gene from *C. tropicalis*; for example, there is 83% identity at the amino-acid-residue level. The physical characteristics of mitochondrial DNA from *C. maltosa* have recently been studied (Kunze *et al.*, 1986). The DNA is circular and 52 kb in length.

306   R.T.M. Poulter

References

Buckley, H.R., Price, M.R. and Daneo-Moore, L (1982). *Infection and Immunity* **37**, 1209.
Chen, C., Kalb, V.F., Turi, T.G. and Loper, J.C. (1988). *DNA* **7**, 617.
Corner, B.E. and Poulter, R.T.M. (1989). *Journal of Bacteriology* **171**, 3586.
Delgado, J.M. and Herrera, L.S. (1981). *Acta Microbiologica Academiae Scientiarum Hungaricae* **28**, 339.
Dutcher, S.K. (1981). *Molecular and Cellular Biology* **1**, 245.
Dvorak, J.A., Whelan, W.L., McDaniel, J.P., Gibson, C.C. and Kwon-Chung, K.J. (1987). *Infection and Immunity* **5**, 1490.
Evans, K.O., Adeniji, A. and McClary, D.O. (1982). *Antonie van Leeuwenhoek* **48**, 169.
Fleischman, J. and Howard, D.H. (1988). *Nucleic Acids Research* **16**, 765.
Fournier, P., Provost, A., Bourgignon, C. and Heslot, H. (1977). *Archives of Microbiology* **115**, 143.
Fox, B.C., Mobley, H.L.T. and Wade, J.C. (1989). *Journal of Infectious Diseases* **159**, 488.
Gillum, A.M., Tsay, E.Y.H. and Kirsch, D.R. (1984). *Molecular and General Genetics* **198**, 179.
Glasgow, B. (1985). MSc Thesis: University of Otago, Dunedin.
Gorman, J.A., Smith, H.E., Koltin, Y. and Gorman, J.W. (1988). *Yeast* **4**, S149.
Hamasawa, K., Kobayashi, Y., Harada, S., Yoda, K., Yamasaki, M. and Tamura, G. (1987). *Journal of General Microbiology* **133**, 1089.
Hilton, C., Markie, D., Corner, B., Rikkerink, E. and Poulter, R. (1985). *Molecular and General Genetics* **200**, 162.
Hill, D.E., Boulay, R. and Rogers, D. (1988). *Nucleic Acids Research* **16**, 365.
Johnston, J.R., Contopoulou, C.R. and Mortimer, R.K. (1988). *Yeast* **4**, 191.
Kakar, S.N. and Magee, P.T. (1982). *Journal of Bacteriology* **151**, 1247.
Kamiryo, T. and Okazaki, K. (1984). *Molecular and Cellular Biology* **4**, 2136.
Kawai, S., Hwang, C.W., Sugimoto, M., Takagi, M. and Yano, K. (1987). *Agricultural and Biological Chemistry* **51**, 1587.
Kelly, R., Miller, S.M., Kurtz, M.B. and Kirsch, D.R. (1987). *Molecular and Cellular Biology* **7**, 199.
Kelly, R. Miller, S.M. and Kurtz, M.B. (1988). *Molecular and General Genetics* **214**, 24.
Kelly, R., Miller, S.M., Lai, M.H. and Kirsch, D.R. (1990). *Gene* **87**, 177.
Kirsch, D.R., Lai, M.H. and O'Sullivan, J. (1988). *Gene* **68**, 229.
Kirsch, D.R., Kelly, R. and Kurtz, M.B. (1990). "Genetics of *Candida*". CRC Press, Alma Ata.
Kunze, G., Petzoldt, C., Bode, R., Samsonova, I., Hecker, M. and Birnbaum, D. (1985). *Current Genetics* **9**, 205.
Kunze, G., Bode, R. and Birnbaum, D. (1986). *Current Genetics* **10**, 527.
Kunze, G., Bode, R., Schmidt, H., Samsonova, I.A. and Birnbaum, D. (1987). *Current Genetics* **11**, 385.
Kurtz, M.B. and Marrinan, J. (1989). *Molecular and General Genetics* **217**, 47.
Kurtz, M.B., Cortelyou, M.W. and Kirsch, D.R. (1986). *Molecular and Cellular Biology* **6**, 142.
Kurtz, M.B., Cortelyou, M.W., Miller, S.M., Lai, M. and Kirsch, D.R. (1987). *Molecular and Cellular Biology* **7**, 209.
Kurtz, M.B., Kirsch, D.R. and Kelly, R. (1988). *Microbiological Sciences* **5**, 58.
Kurtz, M.B., Kelly, R. and Kirsch, D.R. (1990). *In* "The Genetics of *Candida*" (D.R. Kirsch, R. Kelly and M.B. Kurtz, eds) pp. 21–74. CRC Press, Alma Ata.
Kwon-Chung, K.J., Wickes, B.L. and Whelan, W.L. (1987). *Infection and Immunity* **55**, 3207.
Kwon-Chung, K.J., Riggsby, W.S., Uphoff, R.A., Hicks, J.B., Whelan, W.L., Reiss, E., Magee, B.B. and Wickes, B.L. (1989). *Infection and Immunity* **57**, 527.
Lasker, B.L., Carle, G.F., Kobayashi, G.S. and Medoff, G. (1989). *Nucleic Acids Research* **17**, 3783.

Lott, T.J., Boiron, P. and Reiss, E. (1987). *Molecular and General Genetics* **209**, 170.

Magee, B.B. and Magee, P.T. (1987). *Journal of General Microbiology* **133**, 425.

Magee, B.B., D'Souza, T.M., Magee, P.T. (1987). *Journal of Bacteriology* **169**, 1639.

Magee, B.B., Koltin, Y., Gorman, J.A. and Magee, P.T. (1988). *Molecular and Cellular Biology* **8**, 4721.

Merz, W.G., Connelly, C. and Heiter, P. (1988). *Journal of Clinical Microbiology* **26**, 842.

Murray, W.W. and Rachubinski, R.A. (1987). *Gene* **51**, 119.

Okazaki, K., Takechi, T., Kambara, N., Fukui, S., Kubota, I. and Kamiryo, T. (1986). *Proceedings of the National Academy of Sciences, USA* **83**, 1232.

Okazaki, K., Tan, H., Fukui, S., Kubota, I. and Kamiryo, T. (1987). *Gene* **58**, 37.

Olaiya, A.F., Steed, J.R. and Sogin, S.J. (1980). *Journal of Bacteriology* **141**, 1284.

Olivo, P.D., McManus, E.J., Riggsby, W.S. and Jones, J.M. (1987). *Journal of Infectious Diseases* **156**, 214.

Pesti, M. and Ferenczy, L. (1982). *Journal of General Microbiology* **128**, 123.

Pomes, R., Gil, C. and Nombela, C. (1985). *Journal of General Microbiology* **131**, 2107.

Poulter, R. (1990). *In* "The Genetics of *Candida*" (D.R. Kirsch, R. Kelly and M.B. Kurtz, eds) pp. 75-124. CRC Press, Alma Ata.

Poulter, R., Jeffery, K., Hubbard, M.J., Shepherd, M.G. and Sullivan, P.A. (1981). *Journal of Bacteriology* **146**, 833.

Poulter, R., Hanrahan, V., Jeffery, K., Markie, D., Shepherd, M.G. and Sullivan, P.A. (1982). *Journal of Bacteriology* **152**, 969.

Poulter, R.T.M. and Rikkerink, E.H.A. (1983). *Journal of Bacteriology* **156**, 1066.

Rachubinski, R.A. (1990). *In* "The Genetics of *Candida*" (D.R. Kirsch, R. Kelly and M.B. Kurtz, eds) pp. 177-186. CRC Press, Alma Ata.

Rhoads, D.D. and Sarachek, A. (1984). *Mycopathologia* **87**, 35.

Riggsby, W.S. (1985). *Microbiological Sciences* **2**, 258.

Riggsby, W.S. (1990). *In* "The Genetics of *Candida*" (D.R. Kirsch, R. Kelly and M.B. Kurtz, eds) pp. 125-146. CRC Press, Alma Ata.

Riggsby, W.S., Torres-Bauza, L.J., Wills, J.W. and Townes, T.M. (1982). *Molecular and Cellular Biology* **2**, 853.

Rikkerink, E.H.A., Magee, B.B. and Magee, P.T. (1988). *Journal of Bacteriology* **170**, 895.

Rosenbluh, A., Mevarech, M., Koltin, Y. and Gorman, J.A. (1985). *Molecular and General Genetics* **20**, 500.

Sanglard, D. and Loper, J.C. (1989). *Gene* **76**, 121.

Sarachek, A. and Weber, D.A. (1984). *Current Genetics* **8** 181.

Sarachek, A., Rhoads, D.D. and Schwarzhoff, R.H. (1981). *Archives of Microbiology* **129**, 1.

Scherer, S. and Stevens, D.A. (1987). *Journal of Clinical Microbiology* **25**, 675.

Scherer, S. and Stevens, D.A. (1988). *Proceedings of the National Academy of Sciences, USA* **85**, 1452.

Shaw, J.A., Troutman, W.B., Lasker, B.A., Mason, M.M. and Riggsby, W.S. (1989). *Journal of Bacteriology* **171**, 6353.

Singer, S.C., Richards, C.A., Ferone, R., Benedict, D. and Ray, P. (1989). *Journal of Bacteriology* **171**, 1372.

Smith, H.A., Allaudeen, H.S., Whitman, M.H., Koltin, Y. and Gorman, J.A. (1988). *Gene* **63**, 53.

Soll, D.R. (1990). *In* "The Genetics of *Candida*" (D.R. Kirsch, R. Kelly and M.B. Kurtz, eds), pp. 147-176. CRC Press, Alma Ata.

Soll, D.R., Galask, R., Isley, S., Rao, T.V., Stone, D., Hicks, J., Schmid, J., Mac, K. and Hanna, C. (1989). *Journal of Clinical Microbiology* **27**, 681.

Suzuki, T., Nishibayashi, S., Turoiwa, T., Kanbe, T. and Tanaka, K. (1982). *Journal of Bacteriology* **152**, 893.

Szabo, L.J., Small, G.M. and Lazarow, P.B. (1989). *Gene* **75**, 119.

Takagi, M., Kobayashi, N., Sugimoto, M., Fujii, T., Watari, J. and Yano, K. (1987). *Current Genetics* **11**, 451.

Vallin, C. and Ferenczy, L. (1978). *Acta Microbiologica Academiae scientarum Hungarica* **25**, 209.

Vollrath, D. and Davis, R.W. (1987). *Nucleic Acids Research* **15**, 7865.
Whelan, W.L., Partridge, R.M. and Magee, P.T. (1980). *Molecular and General Genetics* **180**, 107.
Whelan, W.L. and Magee, P.T. (1981). *Journal of Bacteriology* **145**, 896.
Whelan, W.L., Markie, D.M., Simkin, K.G. and Poulter, R.M (1985). *Journal of Bacteriology* **161**, 1131.
Whelan, W.L., Markie, D. and Kwon-Chung, K.J. (1986). *Antimicrobial Agents and Chemotherapy* **29**, 726.
Wills, J.W., Lasker, B.A., Sirotkin, K. and Riggsby, W.S. (1984). *Journal of Bacteriology* **157**, 918.

9 Non-Mendelian Genetic Elements in *Saccharomyces cerevisiae*. RNA Viruses, 2 μm DNA, ψ, [URE3], 20S RNA and Other Wonders of Nature

Reed B. Wickner

Section on Genetics of Simple Eukaryotes, LBP, National Institutes of Diabetes, Digestive and Kidney Diseases, NIH Bethesda, MD 20892, USA

The Yeasts Vol. 6, 2nd edition
ISBN 0-12-596416-1

I. INTRODUCTION

Since most traits of *Saccharomyces cerevisiae* and other yeasts are, of course, determined by chromosomal genes, why do such non-chromosomal genomes as mitochondrial DNA, dsRNA viruses, 2 μm DNA, the linear DNA killer system of *Kluyveromyces lactis* and others receive so much attention out of proportion to the amount of information they encode? One reason is that they provide models for viruses (both DNA and RNA) and plasmids of larger eukaryotes. Another is their usefulness as vectors for production of foreign proteins in yeast cells. These elements also provide examples of different methods by which nucleic acids can be replicated and expressed. The study of mitochondria, in particular has shown us the flexibility of the genetic code, self-splicing introns, intron-encoded maturases and transposases, and numerous other important phenomena and mechanisms. Previously, most genetic elements were discovered by their phenotypes and later identified with a non-chromosomal nucleic-acid replicon. More recently, however, a number of molecules unassociated with a phenotype have been found.

In this chapter, I review selected aspects of these non-Mendelian genetic elements and non-chromosomal replicons. Mitochondrial genetics, the first non-Mendelian system studied in yeast, and the *Ustilago maydis* dsRNA-based killer system are reviewed elsewhere (Koltin and Leibowitz, 1988; Newlon, 1989; Guérin, 1991; Slonimski, 1993) and will not be covered here for lack of space, but certainly not for lack of interest or importance. The retroviruses of yeast, called Ty1 to Ty5 (so far), are likewise important genetic elements that violate Mendelian rules by their ability to hop from place to place (reviewed by Boeke, 1989). Common themes in the genetics of non-Mendelian elements will be emphasized along with basic molecular mechanisms that have been worked out with their aid.

II. GENETIC CRITERIA FOR A NON-MENDELIAN GENETIC ELEMENT

Several criteria can be used to distinguish those traits which are chromosomally encoded from those determined by non-Mendelian genetic elements (Table I). In a cross of a strain with the trait (+) and one lacking it (−), 4 + :0 segregation in meiosis is the classical sign of "cytoplasmic" inheritance. Of course, the same pattern can be seen if the (+) parent has multiple unlinked chromosomal loci any one of which is sufficient for the phenotype and the (−) parent has none. Successive crosses of the (+) offspring with the (−) parent can eventually rule out this possibility.

Cytoduction is a second criterion for non-chromosomal inheritance. Mating of two yeast strains normally results in a low frequency of non-fusion of the parental nuclei. In the next cell division, these unfused nuclei separate but both daughter cells receive the mixed cytoplasm of the zygote. Transmission of a trait from one parent to the other in this type of cross is strong evidence of a non-chromosomal gene (Wright and Lederberg, 1957; Aigle and Lacroute, 1975). This method was dramatically simplified by the *kar1* mutant (Conde and Fink, 1976), which is defective for nuclear fusion (karyogamy). Cytoplasmic transfer using the *kar1* mutant has been used extensively in the killer system to study dsRNAs, and their interactions with each other and with chromosomal genes and other cytoplasmic elements (Toh-e *et al.*, 1978; Toh-e and Wickner, 1979; Wickner, 1980) and in the study of ψ^- mutants (Cox *et al.*, 1980). The *kar1* strains make it possible to move cytoplasm from strain to strain quickly and easily. This method is equivalent to the heterokaryon transfer or transfer by hyphal anastomosis used in fungi in which nuclear fusion is not a necessary accompaniment of cell fusion. While dsRNAs, ψ, [URE3] and mitochondrial DNA are efficiently transferred by cytoduction, $2 \mu m$ DNA, because it is largely confined to the nucleus, is transferred at only 50% efficiency even though 100 copies are present in each cell (Livingston, 1977).

Mitotic segregation is a third hallmark of non-Mendelian genes. When strains with distinguishable alleles for some chromosomal trait are mated,

Table I. Methods for distinguishing chromosomal and non-chromosomal traits in yeasts

	Chromosomal traits	Non-chromosomal traits
Meiotic segregation	2+: 2−	4 +: 0 *or* irregular
Mitotic segregation	No: heterozygosis maintained	Yes: alleles separate
Transfer by cytoduction (*kar1* crosses)	Very rare	50–100% efficient
Curing	No	Yes: various agents

the resulting diploids may have the phenotype of one or the other parent, or some intermediate phenotype, but this phenotype is largely stable. Continued subcloning will result in only a very low frequency of segregation of the parental traits due to mitotic recombination. In contrast, most non-Mendelian elements segregate relative to the same replicon carrying a distinguishable allele. This was clearly shown for antibiotic resistance markers on mitochondria (Coen et al., 1970), and is also true for an M_1 dsRNA determining the $K^+ R^+$ phenotype (killer toxin-secreting and resistant to the toxin) segregating from an M_1 dsRNA with a defective toxin gene ($K^- R^+$) (Somers and Bevan, 1969). Within a few subclonings, nearly all of the cells have only one or the other parental allele.

Efficient curability by treatments that are only modestly mutagenic for nuclear genes is a fourth trait typical of non-Mendelian elements. Ethidium bromide for mitochondrial DNA (Goldring et al., 1970), cycloheximide (Fink and Styles, 1972) or elevated temperature (Wickner, 1974a) for M_1 dsRNA, and high osmotic strength (Singh et al., 1979) or low concentrations of guanidine hydrochloride for ψ (Tuite et al., 1981b), η (Leibman and All-Robyn, 1984) or [URE3] (M. Aigle, cited in Cox et al., 1988) can all produce complete or nearly complete elimination of the respective element.

While each of these criteria has its exceptions and its pitfalls, they serve together to provide an unambiguous distinction between traits determined by chromosomal genes and those determined by other elements, whether nuclear or cytoplasmic. While all non-Mendelian elements of yeast whose molecular basis has been resolved have proven to be extrachromosomal replicons, there remains the possibility of other "epigenetic" phenomena, a hypothetical example of which is discussed in connection with the ψ element.

III. COMMON FEATURES OF NON-MENDELIAN ELEMENTS OF *SACCHAROMYCES CEREVISIAE*

All known yeast viruses, including the dsRNA viruses L-A, L-BC, M and the Ty retroviruses, have as their sole known route of infection transmission by cell–cell fusion. Sphaeroplasts transformed with a DNA plasmid in the presence of dsRNA viruses often take up and become stably infected with the viruses (El-Sherbeini and Bostian, 1987), but this is not a natural infection cycle. The absence of an extracellular route of infection is common to all fungal viruses (Koltin and Leibowitz, 1988), but such viruses are nonetheless widespread. Probably, because cell–cell fusion is a very frequent event among fungi, extracellular spread is unnecessary for these viruses. Thus, the L-A and L-BC viruses are present in a majority of strains of *Saccharomyces cerevisiae* and strains lacking Ty1 are unknown. Viruses of larger eukaryotes have not neglected this route of transmission. All herpes viruses and human immunodeficiency viruses (HIV) are known to be transmitted by direct cell contact or cell fusion but, since cell–cell fusion is a relatively rare event in animals, an animal virus that could not leave one cell and get into another would not be likely to become widespread enough to become an object of study.

Non-Mendelian elements, being generally small molecules and limited in their coding capacity, must depend heavily on their host for their expression and replication. An extreme example of this from other systems is the viroids (reviewed by Owens and Hammond, 1988; Bruening *et al.*, 1988), which are small circular single-stranded RNA molecules only a few hundred bases in length. They encode no proteins themselves and, except for the self-cleavage and self-ligation activities that the RNAs themselves display, they depend entirely on host enzymes for their replication. Studies of bacteriophage indicate that there is an inverse relationship between the size of a phage genome and the number of host genes on which it depends. The very large bacteriophage T4 encodes nearly all of its own replication proteins, while the medium-sized phage λ encodes only two and the small ϕX174 encodes only one and uses many host functions, which T4 encodes itself (DNA polymerase, DNA ligase, and single-stranded DNA binding protein to mention a few of the most familiar examples).

So it is not surprising that studies on mitochondrial DNA, the various dsRNAs, 2 μm DNA, ψ, and [*URE3*] have revealed in each case chromosomal genes essential for maintenance or replication of these different elements (Table II). Likewise, chromosomal genes contribute to expression of the genetic information carried on these elements. Most of the proteins involved in mitochondrial protein-synthesis machinery are encoded by the nucleus (reviewed by Pon and Schatz, 1991), the *KEX* genes encode proteases necessary for secretion of the killer toxin encoded by M_1 dsRNA, antisuppressor and allosuppressor genes modify the activity of ψ and 2 μm DNA requires RNA polymerase II for its transcription.

While mitochondrial DNA, and viral DNAs and RNAs of higher cells are as fully accessible for analysis as are those in *Sacch. cerevisiae*, it is the possibility of carrying out detailed host genetics, which is, among eukaryotes, unique to yeasts and other lower eukaryotes like *Drosophila* spp. The precedents from the studies of plasmids and viruses of bacteria tell us that study of the interactions of chromosomal and non-chromosomal genes in yeast will give us unique insights not easily obtained in higher organisms. Moreover, since evolution generally proceeds by accretion

Table II. Non-Mendelian elements of *Saccharomyces cerevisiae*

Genetic element	Nucleic acid	Phenotype	Host-replication genes	Host-expression genes	Curing
[Rho]	Mitochondrial DNA	Respiration	*PET18*, ...	Many *PET* genes	Ethidium bromide
[cir]	2 μm DNA	"Nib"-nibbled colonies if host is *nib1*⁻	*CDC* genes, *MAP1*	RNA polymerase II and others	Transform with 2 μm DNA-based vector
[HOK], [NEX], [EXL], [B]	L-A dsRNA	Needed for killer phenotype	*MAK3*, *MAK10*, *PET18*		Growth at 39°C
L-BC dsRNA	L-BC dsRNA		*CLO1*		
[KIL-k_1], [KIL-k_2], [KIL-k_{28}]1, ...	M_1,M_2, M_{28}, ... dsRNAs	$K_1 R_1$, ... killer phenotypes	*MAK1*, ... , *MAK31*	KEX1 and KEX2 proteases for toxin secretion	Cycloheximide, growth at 39°C
[D]	?	*ski*⁻ M strain sicker	*MAD*		Growth at 39°C
20S RNA = W dsRNA	20S circular ssRNA = W dsRNA	?	?		
T dsRNA	T dsRNA	?			
ψ	?	Increased ochre suppr.	*PNM* genes	Allo- and anti-suppressors	Guanidine hydrochloride, high osmolarity
η	?	Increased omnipotent suppression	?		Guanidine hydrochloride
[URE3]	?	Ureidosuccinate uptake in ammonia-containing medium	*URE2*		Guanidine hydrochloride

rather than by revolution, what is true of yeast(s) is likely to also be true (in some form) of man.

IV. L-A AND M dsRNA VIRUSES

In 1963, Makower and Bevan discovered that certain strains of *Sacch. cerevisiae* secrete a toxin lethal to other strains (the K_1^+ phenotype) but to which the secreting strains were immune (the R_1^+ phenotype). Crossing K_1^+ R_1^+ strains with K^- R^- strains produced only K_1^+ R_1^+ diploids, which, on sporulation, produced 4 K_1^+ R_1^+: 0 segregation (Somers and Bevan, 1969). This is classical non-Mendelian segregation and led to the demonstration that the toxin was encoded by a 1.8 kb dsRNA molecule, M_1, which replicates as a satellite virus of the cytoplasmic dsRNA virus, L-A (4.6 kb) (Bevan *et al.*, 1973; Buck *et al.*, 1973; Herring and Bevan, 1974; Vodkin *et al.*, 1974; Bostian *et al.*, 1980a,b; Sommer and Wickner, 1982b; Dihanich *et al.*, 1989). We are beginning to learn some of the details of replication of these viruses, and we are getting some hints as to the nature of their interactions with each other and with their host. The occurrence of killer phenomena in various genera and applications of these phenomena in brewing, epidemiology and biotechnology have been reviewed elsewhere (Young, 1987; Wickner, 1991).

A. Viral replication cycles: *in vivo* and in isolated virions (= *in viro*)

Figure 1 shows the replication cycles of L-A dsRNA and its satellites, M, S and X dsRNAs. Viral particles containing one L-A dsRNA molecule each carry out conservative transcription, that is, the parental strands stay together while the newly synthesized (+) strand is single-stranded (Newman *et al.*, 1981; Sclafani and Fangman, 1984; Fujimura *et al.*, 1986; Williams and Leibowitz, 1987) and the (+) strand product is extruded from the particle (Esteban and Wickner, 1986). These (+) strands then serve as mRNA for synthesis of both the major coat protein (76 kDa, analogous to the retroviral *gag* protein) (Hopper *et al.*, 1977; Icho and Wickner, 1989) and a 170 kDa *gag-pol*-type fusion protein, described below (Sommer and Wickner, 1982b; Fujimura and Wickner, 1988b; Icho and Wickner, 1989). The (+) strands are also the species encapsidated to form new viral particles (Fujimura *et al.*, 1986). These new (+) strand-containing particles then carry out the replication step, that is synthesis of the (−) strand on the (+) strand template to complete the cycle.

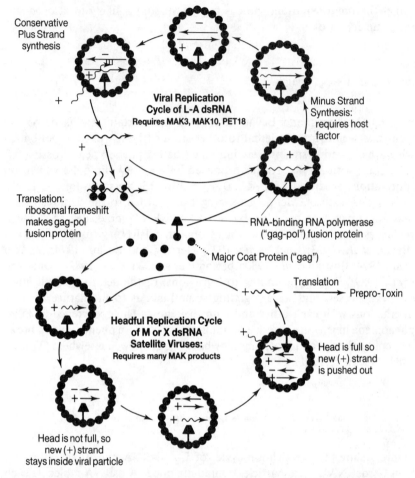

Fig. 1. Replication cycles for L-A dsRNA (above) and its satellites, M, S (a deletion mutant of M) and X (a deletion mutant of L-A dsRNAs (below). Replication of all of these genomes is intraviral, meaning that both (+) and (−) strands are made within the particles, conservative, meaning parental strands stay together, and asynchronous, meaning (+) and (−) strands are made at different points in the replication cycle. The smaller genomes show headful replication, meaning that a single (+) strand is packaged and replicated, and transcribed to fill the head; then all of the new (+) strands are extruded from the head. Wavy lines represent single-stranded RNA and straight lines strands in double-stranded RNA.

The M dsRNAs depend for their replication and maintenance on the viral coat proteins provided by L-A (Hopper *et al.*, 1977; Bostian *et al.*, 1980b; Sommer and Wickner, 1982b; Wickner *et al.*, 1991). The virus L-A can also support replication of deletion mutants of M or those derived from L-A itself (Somers, 1973; Vodkin *et al.*, 1974; Fried and Fink, 1978; Esteban and

Wickner, 1988). These satellite genomes follow a replication cycle similar to L-A with one exception. While all L-A (+) strand transcripts are extruded from the viral particles, those of M and the deletion mutants of M and L-A are largely retained within the particles, apparently because the particles are not full (Esteban and Wickner, 1986, 1988). These retained (+) strands can be replicated to form a second dsRNA molecule within the same particle. In fact, X, a deletion mutant of L-A whose length is only 530 bp (one eighth the size of L-A), is found in viral particles with 1–8 X dsRNA molecules in each particle. Those viral particles with two M dsRNA molecules in each particle (or eight X dsRNA molecules per particle) extrude all of their new (+) strand transcripts. This is referred to as "headful replication" (Esteban and Wickner, 1986, 1988). The model predicted that only a single (+) strand is initially packaged in each particle and this has been directly verified (Fujimura *et al.*, 1990; see below). However, it may then replicate several times within the head until the head is full. Then all new (+) strands must go out into the world to make their own way. The headful replication model may also apply to the dsRNA virus system from *Ustilago maydis*, where 1–3 M dsRNA segments have been found encapsidated in a single particle (Bozarth *et al.*, 1981).

B. *In vitro* replication, transcription and packaging

Though it was possible to determine the viral replication cycle from *in vivo* studies and studies on the activities of isolated viral particles with their endogenous template (that is, *in viro*), detailed understanding of the replication, transcription and packaging mechanisms cannot be obtained without development of template-dependent *in vitro* systems. Treatment of purified L-A dsRNA-containing particles with solutions of low ionic strength results in particles bursting and releasing their dsRNA. These opened empty particles can then be re-isolated and they now specifically bind viral (+) strands, a reaction that is part of the packaging process, as discussed below. On addition of NTPs and a host-factor fraction (a rather crude preparation from a strain lacking L-A), they convert these (+) strands to dsRNA form by synthesis of a complementary (−) strand (the replication step) (Fujimura and Wickner, 1988a,b; Esteban *et al.*, 1988). These opened empty particles can also transcribe added viral dsRNA to form (+) ssRNA (Fujimura and Wickner, 1989). The latter reaction requires a high concentration of added polyethylene glycol, probably to increase the effective concentration of template dsRNA.

C. Recognition sites for replication, transcription and packaging

Using the template-dependent *in vitro* systems described above, the sites on viral RNA necessary and sufficient for these processes have been determined (Esteban *et al.*, 1988, 1989; Fujimura *et al.*, 1990). The site necessary for binding of (+) strands of X (a deletion mutant of L-A) was found to be a 24-base region about 400 bases from the 3′ end. This region was predicted by computer analysis to have the secondary structure shown in Fig. 2(b) and detailed experimental analysis using compensatory mutations has shown that this is the structure of the region, and that the structure is necessary for binding. The sequence of the stem is not important, only that it be a stem. The A residue protruding on the 5′ side of the stem is essential and no other base will suffice. The sequence of the loop is also important. This region is also sufficient for the binding, as is a region of similar structure found by computer analysis of the M_1 (+) strands (Georgopoulos *et al.*, 1986; Esteban *et al.*, 1989; Fujimura *et al.*, 1990).

That this binding site is the packaging signal was shown by inserting the synthetic site in a yeast expression vector and showing that the *in vivo* transcripts were packaged inside L-A viral particles (Fujimura *et al.*, 1990). As predicted by the headful replication model, these transcripts were found in particles by themselves, not in the majority of particles containing an L-A dsRNA molecule. Thus, a single (+) strand is packaged in each particle (Fujimura *et al.*, 1990).

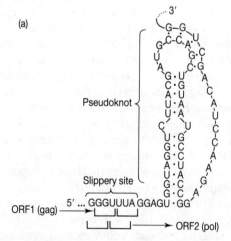

Fig. 2. *Cis*-acting sites on L-A and M_1 (+) strands. (a) The region determining the −1 ribosomal frameshift that fuses ORF1 and ORF2 (Icho and Wickner, 1989; Dinman *et al.*, 1991). The ribosomes translating the "slippery site" bump into the pseudoknot structure and slip back one base on the mRNA, changing the reading frame from that of ORF1 to that

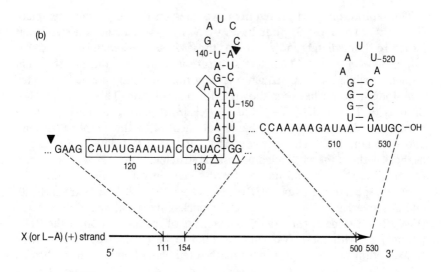

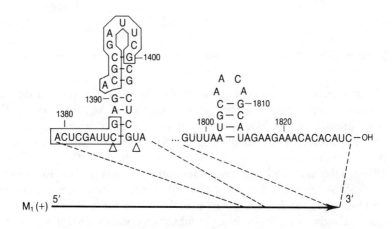

of ORF2. This results in the synthesis of a "gag-pol"-type fusion protein. (b) The packaging signals and sites needed for replication of viral (+) strands (Esteban *et al.*, 1989). The packaging signal is the stem-loop structure between the open triangles (△). Replication requires the 3′ end four bases and the adjacent stem-loop structure. The internal replication enhancer (IRE) is necessary for optimal template activity in the replication reaction and is the region between the closed triangles (▲). Imperfect direct repeats in both the L-A and M_1 sequences (boxed) in this region have been noted, but their role has not been examined. The L-A 3′ end sequences shown are necessary for the replication reaction. The M_1 3′ end sequences shown can substitute for the L-A 3′ end. ▼-▼, limits of the IRE; △-△, viral particle binding site = packaging signal.

The replication reaction requires two sites on the (+) strand template (Fig. 2(b)). The most 3' four bases of X are essential, as is a stem-loop structure immediately adjacent to those bases (Esteban *et al.*, 1989; Fujimura *et al.*, 1990). The stem structure is important, but not the stem sequence, while the loop sequence is important for template activity. This stem-loop structure has been shown to form in solution (Thiele *et al.*, 1984). Strands of L-A and M_1 (+) and (−) each are known to have an extra adenine residue at the 3' end that is not encoded by the template (Bruenn and Brennan, 1980; Thiele *et al.*, 1984), and such templates are as active as those without the extra adenine residue (Esteban *et al.*, 1989). The second region of the (+) strand needed for optimal template activity, called the internal replication enhancer (IRE), is located about 400 bases from the 3' end and overlaps with the binding site-packaging signal already discussed (Fig. 2(b)) (Esteban *et al.*, 1989). The nature of the interaction of the RNA polymerase with these two sites remains to be elucidated. The *in vitro* transcription system is specific for dsRNA templates known to be replicated by the L-A system but the precise signals recognized are not known (Fujimura and Wickner, 1989).

D. Expression of L-A: ribosomal frameshifting makes a *gag-pol* fusion protein

The sequence of several complete cDNA clones of L-A showed that it has two long open-reading frames (Fig. 3) (Icho and Wickner, 1989). ORF1 (*gag*) encodes the 76 kDa major coat protein (Icho and Wickner, 1989). ORF2 is expressed only as an ORF1–ORF2 170 kDa fusion protein, which is present in about 1–2 copies in each particle (Fujimura and Wickner, 1988b; Icho and Wickner, 1989; Dinman *et al.*, 1991). This fusion protein has ssRNA-binding activity but the major coat protein does not (Fujimura and Wickner, 1988b). The ORF2 protein, produced in *Escherichia coli* also has this activity (T. Fujimura, unpublished observation). It has been suggested that this activity may be responsible for packaging of the (+) strands to form new virions (Fujimura and Wickner, 1988b). Although this ssRNA-binding activity is not specific for the packaging signal, this may be because the activity is assayed on North-Western blots after treatment with sodium dodecyl sulphate and urea.

The sequence of ORF2 (Icho and Wickner, 1989) matches the consensus for RNA-dependent RNA polymerases of (+) strand RNA viruses first described by Kamer and Argos (1984) (Fig. 4). Moreover, all dsRNA viruses appear to share this consensus, suggesting that these groups of viruses may be more closely related than had previously been thought.

That the L-A *gag-pol* fusion protein is made by ribosomal frameshifting was first suggested (Icho and Wickner, 1989) by the resemblance of a region

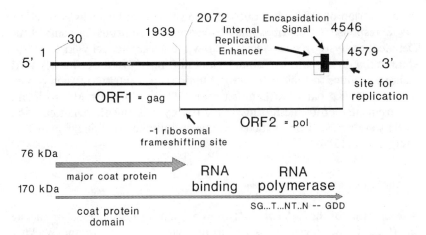

Fig. 3. Structure and expression of the L-A genome. *ORF1* encodes the major coat protein and *ORF*1 + *ORF*2 together encode the minor 170 kDa protein. These two *ORF*s are fused by a −1 ribosomal frameshift whose mechanism is much like that of retroviruses. *ORF*2 has an amino-acid sequence pattern typical of the RNA-dependent RNA polymerases of (+) strand and double-stranded RNA viruses. It also has a ssRNA binding activity that is thought to be involved in packaging of (+) strands. Sites essential for replication ((−) strand synthesis on (+) strand template) are at the extreme 3′ end and 400 bases from the 3′ end (see Fig. 2). The packaging signal (see Fig. 2) is about 400 bases from the 3′ end overlapping the internal replication enhancer.

(+) Strand RNA Viruses:
```
BromMV      ADLSKFDK..FQRRTGDAFTYFGNTLVtMAMI..AIFSGDDSLII..DPLREI
TobMV       IDFSKFDK..YQRKSGDVTTFIGNTVIiAACL..GAFCGDDSLLY..DPLKLI
AlfalfMV    IDFSKFDK..FQRRTGDALTYLGNTIVtLACL..VVASGDDSLIG..NPLKLL
CowpeaMV    CDYSSFDG..CGIPSGFPMTVIVNSIFNEILI..LVTYGDDNLIS..DFLKRT

Sinbis      TDIASFDK..AMMKSGMFLTLFVNTVLNVVIA..AAFIGDDNIIH..DPLKRL

FootMDV     VDYDAFDA..GGMPSGCSATSIINTILNNIYV..MISYGDDIVVA..VFLKRH
EMC         VDYSNFDS..GGLPSGCAATSMLNTIMNNIII..VLSYGDDLLVA..VFLKRK
Polio       FDYTGYDA..GGMPSGCSGTSIFNSMINNLII..MIAYGDDVIAS..TFLKRF
                *    *    **   *   **   *         ***        ***
```

ssCircular RNA:
```
20S RNA     RDTLKGDF..RGILmGLPTTWAIlnlmhLWCW..CRVCGDDLIGV..IPLKGL
```

dsRNA Viruses:
```
L-A         PDVAVVDQ..GTLLSGWRLTTFMNTVLNWAYM..SVHNGDDVMIS
BTV         IDYSEYDT..DTHLSGENSTLIANSMHNMAIG..EQYVGDDTLFY
ROTA V      TDVSQWDS..GAVASGEKQTKAANSIANLALI..IRVDGDDMYAV
REOVIRUS    IDISACDA..TTFPSGSTATSTEhTANNSTMM  YVCQGDDGLMI..PFLKMV
PHI6        GDPSNPDL..VGLSSGQGATDLMgTLLmSITY..QISKsDDAILG
IBDV        YGQGSGNAATFINNhLLsTLVL..IERSiDDIRGK
```

Fig. 4. RNA-dependent RNA polymerase consensus sequence. This consensus was first recognized by Kamer and Argos (1984) for (+) strand RNA viruses. It is clearly present in the L-A ORF2 (Icho and Wickner, 1989) and probably present in 20S RNA (= W dsRNA) (Matsumoto and Wickner, 1991; Rodriguez-Cousino *et al.*, 1991).

in the overlap of ORF1 and ORF2 to the site shown by Jacks *et al.* (1988) to be responsible for ribosomal frameshifting in Rous sarcoma virus. Detailed analysis of this region (shown diagrammatically in Fig. 2(a)) showed that it promotes ribosomal frameshifting at just the efficiency sufficient to produce the observed ratio of major coat protein to fusion protein in isolated viral particles (Dinman *et al.*, 1991). Moreover, the structural requirements for frameshifting promoted by this region (Dinman *et al.*, 1991) were precisely as predicted by the "simultaneous slippage model" of Jacks *et al.* (1988).

1. Why make a gag-pol fusion protein?

The structure of the L-A *gag-pol* fusion protein suggests a model of packaging (Fig. 5) that may apply as well to retroviruses (Fujimura and Wickner, 1988b). It is suggested that the *pol* domain's ssRNA-binding activity holds the viral (+) strand while the N-terminal major coat protein domain primes

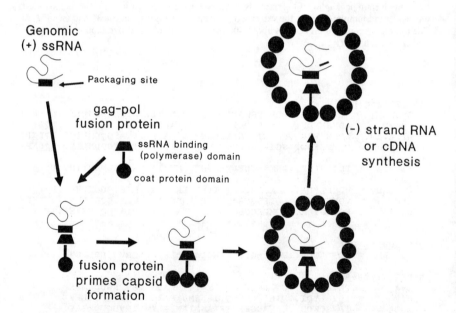

Fig. 5. Packaging model for L-A and its satellite viruses (Fujimura and Wickner, 1988b). It is proposed that the *gag-pol* fusion protein holds on to the viral (+) strand (the species to be packaged) with its RNA-binding *pol* domain and then its *gag* domain primes polymerization of the coat. This naturally results in packaging of one viral (+) ssRNA in each particle. This model may also apply to retroviruses where *gag-pol* fusion proteins are made.

polymerization of the coat. Thus, covalent attachment of the *pol* domain to the *gag* domain ensures both the packaging of the RNA polymerase and the viral genome itself (Fig. 5). The standard model of packaging in retroviruses is that the ssRNA-binding activity of the mature nucleocapsid protein (or the equivalent domain of the *gag* precursor protein) binds to the retrovirus "psi" packaging signal, thus packaging the genome. However, while the nucleocapsid region of *gag* is clearly important in packaging (Fu *et al.*, 1988), the few critical experiments to test this model suggest that the *pol* domain may, indeed, also be necessary (Haffar *et al.*, 1990). Further work will be needed to determine whether the model proposed by Fujimura and Wickner applies to either L-A or to retroviruses.

2. Why make the *gag-pol* fusion protein by ribosomal frameshifting?

One possible answer to this question has been proposed by Icho and Wickner (1989). Retroviruses, (+) strand RNA viruses and dsRNA viruses all use their (+) strands for three functions: as mRNA, as the species packaged to make new viral particles, and as a template for viral replication. If viruses of any of these groups modify their (+) strands for the purpose of translation, such as by splicing or by RNA editing, these modified (+) strands could become mutant genomes unless the modification destroys a site necessary for packaging or replicating the RNA. Indeed, retroviruses regularly carry out splicing in order to synthesize the *env* protein and the minor regulatory proteins made by HIV. However, all these splicing events result in the removal of the "psi" packaging site. Thus the altered (+) strands cannot become mutant viruses. Among the many studied (+) strand RNA and dsRNA viruses, splicing is completely unknown, although it is known in a number of (−) strand RNA viruses and is, of course, widespread in DNA viruses. A form of RNA editing has recently been described in the para-influenza viruses, which are also (−) strand RNA viruses (Paterson and Lamb, 1990).

Synthesis of the *gag-pol* fusion protein, perhaps for the purpose of packaging as already discussed, is carried out by ribosomal frameshifting in the case of L-A and in most retroviruses. Murine leukaemia virus fuses the *gag* and *pol* open-reading frames by read-through of a termination codon, a mechanism also used by the (+) strand animal viruses, Sindbis virus and Middelburg virus (Strauss *et al.*, 1984), and by Tobacco Mosaic virus (Pelham, 1978), Tobacco Rattle virus (Pelham, 1979), Beet Necrotic Yellow Vein virus (Ziegler *et al.*, 1985) and Carnation Mottle virus (Harbison, 1985) (reviewed by Valle and Morch, 1988). Both ribosomal frameshifting and termination codon read-through are mechanisms that do not modify the mRNA (viral (+) strands) but do modify the outcome of the translation.

E. Chromosomal genes affecting L-A and M replication and expression

1. MAK genes: replication and maintenance

Mutants in over 30 chromosomal genes lead to loss of M_1 dsRNA (Wickner, 1979). Something of the host role of some of these genes is known: MAK1 encodes the nucleolar enzyme DNA topoisomerase I (Thrash et al., 1984), MAK8 encodes ribosomal protein L3 (Wickner et al., 1982), MAK11 encodes a membrane-associated protein that is essential for cell growth (Icho and Wickner, 1988); the MAK16 protein is necessary for progression through the G1 phase of the cell cycle and is a nuclear protein (Wickner, 1988), and the GCD genes are probably translational factors (Harashima and Hinnebusch, 1986). However, the role of these proteins in the replication or maintenance of M_1 remains completely obscure.

Only MAK3, MAK10 and PET18 are necessary for replication or maintenance of L-A dsRNA (Sommer and Wickner, 1982b). Dihanich et al. (1989) found that mak10 and mak3 mutants grow slowly on glycerol, and elimination of the mitochondrial genome partially suppresses a mak10 mutation (Wickner, 1977). Viral particles isolated from strains carrying pet18^{ts} and mak10^{ts} mutations are structurally unstable, suggesting that these gene products are involved in viral packaging (Fujimura and Wickner, 1986, 1987). The pet18 mutants are all large deletions of several open-reading frames probably due to homologous recombination between two direct repeats of Ty1 on chromosome III (Toh-e and Sahashi, 1985). This deletion results in an absolute inability to maintain the mitochondrial genome, and a temperature–sensitive growth defect and temperature–sensitive maintenance of L-A and M dsRNAs (Leibowitz and Wickner, 1978; Sommer and Wickner, 1982b). The regions needed for cell growth and dsRNA maintenance were distinct from that needed for mitochondrial DNA replication (Toh-e and Sahashi, 1985). This temperature-dependent maintenance of L-A and M in pet18 strains, the heat curability of M and L-A (Wickner, 1974a; Sommer and Wickner, 1982a), and the similar temperature-dependent requirement for the MKT1 gene (Wickner, 1987; see below) of M_2 indicate that dsRNA replication differs somewhat as a function of temperature.

The requirement of M_1 for MAK10 can be by-passed by providing the L-A-encoded information from a cDNA clone of L-A, showing that M_1

does not itself need the *MAK10* product, except insofar as L-A needs it and M_1 needs L-A-encoded proteins. Mutants in the *MAK3* gene are not so suppressed, so that M_1 replication or maintenance apparently needs *MAK3* more directly (Wickner *et al.*, 1991).

Unexpectedly, while L-A only needs *MAK3*, *MAK10* and *PET18*, X dsRNA, a 530 bp deletion mutant of L-A requires all of the *MAK* genes tested (*MAK4, 6, 16, 18, 21, 26, 27*) like M_1 (Esteban and Wickner, 1988). This may be a clue to the role of the many *MAK* genes in M replication. Both M and X use the L-A-encoded proteins for their own replication and must appropriate them from L-A. In doing so, each lowers the copy number of L-A ten-fold (Ball *et al.*, 1984). They also differ from L-A in their following the headful replication pattern already discussed. Perhaps the *MAK* genes needed by X and by M_1 have some role in these two processes. Alternatively, since mutations in the antiviral genes, *SKI2, SKI3, SKI4, SKI6, SKI7* and *SKI8* (see below) all suppress the group of *mak* mutations that result in M loss but not L-A loss, it is possible that these *MAK* genes regulate one or more of the *SKI* genes, and their mechanism of action is to be sought by a study of the *SKI* products (Toh-e and Wickner, 1980).

2. *SKI* genes: an antiviral system

Mutations in six chromosomal genes result in derepressed replication of M (hence the gene name — superkiller), L-A, L-BC and 20S RNA (Toh-e *et al.*, 1978; Ball *et al.*, 1984; Matsumoto *et al.*, 1990). Since deletion mutations in the *SKI3* and *SKI8* genes do not show a growth defect in the absence of M dsRNA, it has been suggested that these genes are an antiviral system dedicated to preventing the cytopathology produced by excessive viral reproduction (Sommer and Wickner, 1987; Rhee *et al.*, 1989). The *SKI3* product is a 163 kDa nuclear protein (Rhee *et al.*, 1989) but its function is completely unknown. Further discussion of the *SKI* genes and their possible function will be found in Sections IV.F and V.

3. Mitochondrial functions and dsRNA replication control

Recently, mutants in two mitochondrial components were found to affect L-A copy number. Deletion of the major mitochondrial outer-membrane protein, known as porin, thought to function as a diffusion pore allowing hydrophilic metabolites access across the mitochondrial outer membrane, results in delayed growth on glycerol, as expected, but also in accumulation of L-A-containing viral particles (Dihanich *et al.*, 1987, 1989). This accumulation is much greater than that previously documented for wild-type strains grown on ethanol (Oliver *et al.*, 1977; Liu and Dieckmann, 1989), and occurs in spite of the known independence of the L-A and mitochondrial replicons. Defects in *mak10* or *mak3* were further found to cause slow growth on glycerol (Dihanich *et al.*, 1989) but this effect for *mak10* is independent of its loss of L-A (Y.-J. Lee and R.B. Wickner, unpublished observation). Making cells ρ^0 partially suppresses the requirement of L-A for the *MAK10* product (Wickner, 1977), again connecting the mitochondrial genome and control of L-A replication.

Like the porin gene, mutation of the *NUC1* gene, which encodes the major non-specific nuclease in mitochondria (Vincent *et al.*, 1988) results in overproduction of L-A dsRNA and viral particles (Liu and Dieckmann, 1989). As with the *ski* mutants that overproduce L-A (Ball *et al.*, 1984), the overproduction in *nuc1⁻* strains is eliminated by the introduction of M (Liu and Dieckmann, 1989), which represses L-A replication (Ball *et al.*, 1984) perhaps by competing for some factor. Liu and Dieckmann (1989) point out, however, that the porin-deficient and *nuc1* strains define a new group of chromosomal genes controlling L-A replication, since neither gives the superkiller phenotype. This evidence points to an effect of mitochondrial function and/or products on L-A replication, rather than an effect of L-A on mitochondria. The nature of this effect is as yet unclear.

4. KEX2 protease can replace similar mammalian prehormone proteases

Mammalian prehormones, such as proinsulin and pro-opiomelanocortin, are processed to their mature form — insulin in the former case and ACTH, β-lipotropin, the melanocyte stimulating hormones, enkephalin, endorphin and probably others in the latter case — by specific peptidases. These enzymes cleave first after pairs of basic residues, and then remove these pairs to yield the mature hormones (Fig. 6). While the cleavages made have been known for some time, the enzymes responsible them have been elusive. A study of the chromosomal genes responsible for processing of the killer toxin has recently led to the discovery of several candidates for these hormone-processing genes.

The mechanisms of processing, secretion and action of toxin have been the subject of a series of elegant studies mostly by Bussey, Bostian, Tipper, Thomas, Thorner, Fuller and their colleagues. This body of knowledge has been recently reviewed (Tipper and Bostian, 1984; Bussey, 1988; Fuller *et al.*, 1988; Wickner, 1991). Among the first chromosomal mutants found, which affect the killer phenotype, were those that produced the K^-R^+ (neutral) phenotype and the opposite K^+R^- (suicide) phenotype (Wickner, 1974b, 1976). The former mutants involved two genes, called *KEX1* and

PreproInsulin

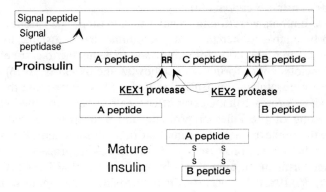

Pro-opiomelanocortin

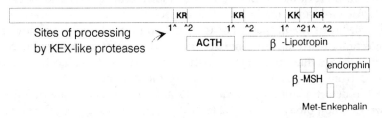

Prepro-Killer Toxin

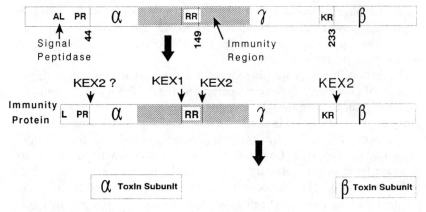

Fig. 6. Processing of the K_1 killer toxin by the KEX1 and KEX2 proteases resembles prohormone processing in mammals.

KEX2 (for killer expression), while the latter single mutant was designated *REX1* (for resistance expression).

The *kex2* mutants were found to have an α mating-type specific mating defect, to be unable to secrete the α pheromone that is a signal to **a** cells to arrest in G1 in preparation for mating and to be unable to carry out later stages of meiotic spore formation (Leibowitz and Wickner, 1976). These phenomena were explained when Julius *et al.* (1984) showed that the KEX2 gene encodes a protease that cleaves after a pair of basic residues, carrying out processing of the killer preprotoxin and α pheromone that closely resembles the prehormone processing that occurs in animals (Fig. 6). Furthermore, the KEX2 protease could substitute in processing of pro-opiomelanocortin in an animal cell-line deficient in the KEX2-type step (Thomas *et al.*, 1988). The next step in killer-toxin, α pheromone or prehormone processing is removal of two basic residues from the carboxy-terminus. This step is carried out by the KEX1 protease (Dmochowska *et al.*, 1987; Wagner and Wolf, 1987; Cooper and Bussey, 1989). The reader is referred to reviews by Bussey (1988), Fuller *et al.* (1988) and Wickner (1991) for more information on this subject.

Knowledge of the properties of the KEX2 protease in yeast has now led to detection of two human proteases with similar sequences, including conservation of residues known to be critical for activity. One, a cDNA clone isolated from a human insulinoma using the polymerase chain reaction (PCR) to detect a clone with sequences similar to the active site of KEX2, shows 49% homology with KEX2 (Smeekens and Steiner, 1990). The other, called furin, shares 50% homology with the catalytic domain of KEX2 and so is, likewise, a possible human prohormone-processing protease (Roebroek *et al.*, 1986; Fuller *et al.*, 1989). Two mouse genes, isolated using PCR and oligonucleotides based on the furin–KEX2 protease active-site sequence, show extensive homology to KEX2 and furin (Seidah *et al.*, 1990). The tissue distribution of their mRNA suggests that they may be prehormone-processing enzymes.

Overproducing prepro-α-pheromone in a *kex2* strain results in some production of properly processed α-pheromone (Egel-Mitani *et al.*, 1990). The aspartyl protease responsible for this processing is called YAP3 (yeast aspartyl protease 3), and is similar to the PEP4 and BAR1 proteases (Egel-Mitani *et al.*, 1990). The effect of this gene on processing of the killer protoxin was not reported. Loh *et al.* (1985) previously described an aspartyl protease from bovine pituitary that cleaves between or after paired basic residues in pro-opiomelanocortin to produce mature hormones. This purified enzyme makes the same cleavages as occur in the pituitary (Loh *et al.*, 1985; Estivariz *et al.*, 1989). This suggests that yeast may have two protease systems similar to mammalian prehormone processing enzymes.

F. L-A genetics: [HOK], [EXL], [NEX], [B]

The discovery of several non-Mendelian genetic elements affecting M_1 and M_2 replication and maintenance led to the discovery of L-A and L-BC as distinct systems (see below), and a definition of the interactions of L-A with M and with several chromosomal genes.

A natural variant of L-A, called L-A-E, was found to be able to support M_1 in a ski⁻ host, but not in a wild-type host. The non-Mendelian element [HOK] (helper of killer) is the ability of "normal" L-As (e.g. L-A-HN or L-A-HNB) to supply the function(s) deficient in L-A-E so that M_1 may be stable in a wild-type host (Sommer and Wickner, 1982b; Wickner and Tohe, 1982; Ridley *et al.*, 1984). Expression of ORF1 alone from a cDNA clone of L-A is sufficient to supply [HOK] activity (Wickner *et al.*, 1991). These results suggest that the major coat protein of L-A (encoded by ORF1) functions to counter the antiviral activity of the SKI system. If the SKI system encodes an activity that attacks the viral RNA itself or the coat protein, then these results could be easily explained.

There exists an element, [EXL], that can dramatically lower the copy number of a particular L-A (called L-A-H because it does have the [HOK] activity described above) so that an M that was supported by the L-A-H is lost (excluded) (Wickner, 1980; Sommer and Wickner, 1982b; Ridley *et al.*, 1984; Hannig *et al.*, 1985). The [EXL] element is carried on the L-A-E variant of L-A (hence the E) and so, as already mentioned, is itself unable to maintain M in a wild-type strain. Most L-A variants are not reduced in copy number by L-A-E and so are said to make M non-excludable ([NEX]; L-A-HN or L-A-HNB).

The designation [B] (bypass) is given to the ability of some L-As (L-A-HNB) to support M_1 in certain *mak* mutants from which it would be lost if it were depending on L-A-HN (the most common variety) (Uemura and Wickner, 1988). For example, *mak11-1* L-A-HN M_1 clones rapidly and completely loses M_1, but *mak11-1* L-A-HNB M_1 clones stably maintain M_1 at the same copy number as if they were MAK^+. As with [HOK], expression of ORF1 was necessary and sufficient to express [B] activity from a cDNA clone of an L-A-HNB, indicating an important relation between the major coat protein gene and the *MAK* genes (Wickner *et al.*, 1991).

V. THE L-BC SYSTEM

The first hint that there existed more than one species of dsRNA of 4.6 kb was the finding that *mak3* mutants cosegregated with a low copy number of L dsRNA (Wickner and Leibowitz, 1979). When later work (Wickner and Toh-e, 1982) showed that *pet18* and *mak10* mutants had the same effect, the existence of a second kind of L dsRNA was hypothesized. This was then confirmed by T1 RNAse fingerprint analysis (Sommer and Wickner, 1982b) and by hybridization (Field *et al.*, 1982). That it was L-A on which M depends was shown by curing experiments (Sommer and Wickner, 1982a) and by identification of L-A as the species carrying the [HOK] gene (see above). The T1 fingerprint experiments clearly distinguished two L species (in different strains) that were distinct from L-A, and these were named L-B and L-C, respectively (Sommer and Wickner, 1982b). However, since L-B and L-C showed some cross-hybridization (Sommer and Wickner, 1982b) and behaved biologically as members of a family, they have been referred to collectively as L-BC.

In a wild-type host, L-BC seems to be unaffected by the presence or absence of dsRNAs from L-A or M (Wickner and Toh-e, 1982; Ball *et al.*, 1984). The L-BC system is independent of all of the *MAK* genes for its replication, but is lost at low temperatures from strains carrying the chromosomal *clo* mutation, a defect that does not affect L-A or M (Wesolowski and Wickner, 1984). The copy number for L-BC is repressed 3–4-fold by the *SKI* products. It is present in intracellular viral particles, which, like those containing L-A, have an RNA-dependent RNA polymerase (Sommer and Wickner, 1982a; Thiele *et al.*, 1984). Moreover, L-BC seems to follow a replication cycle like that shown for L-A in Fig. 1 (Fujimura *et al.*, 1986) but it has not been studied in as great detail as has L-A.

VI. 20S RNA, A CIRCULAR SINGLE-STRANDED RNA REPLICON (= W dsRNA)

20S RNA was first described as a species of RNA, migrating slightly slower on acrylamide gels than 18S rRNA, and whose synthesis was induced on shifting a culture to potassium acetate medium, the same conditions that induce meiosis and sporulation in *Sacch. cerevisiae* (Kadowaki and Halvorson, 1971a,b). Under these conditions, 20S RNA synthesis accounted for up to 15% of stable RNA synthesized. Although this was at first thought to be a form of rRNA whose processing differed in sporulation from vegetative conditions, Garvik and Haber (1978) clearly showed that some strains lacked the ability to amplify 20S RNA, and that this ability was inherited as a non-Mendelian genetic element. Ability to amplify 20S RNA segregated 4 + : 0 in meiosis and could be transferred efficiently by cytoduction. They also showed that haploids and other strains unable to carry out meiosis or sporulation could often amplify 20S RNA, while some strains unable to make 20S RNA were nonetheless sporulation competent. Evidence was presented that the 20S RNA element was distinct from mitochondrial DNA, L-A or M dsRNA or 2 μm DNA (Garvik and Haber, 1978). Moreover, amplification was not inhibited by actinomycin D, indicating that it was not produced by transcription.

More recently, Matsumoto *et al.* (1990) showed that 20S RNA is an independent replicon. Clones of 20S RNA showed no homology to cellular DNA, chromosomal or otherwise, and both (+) and (−) strands of 20S RNA could be detected, although (−) strands were about ten-fold less abundant than (+) strands. Strains unable to amplify 20S RNA were shown to lack it completely. Vegetative cells have about 5–20 copies in each cell and transfer to potassium acetate medium results in a 10 000-fold amplification. The copy number of 20S RNA is also controlled about 3–5-fold by *SKI* products so that a *ski* strain grown on acetate will have more 20S RNA than 18S rRNA. Molecules of 20S RNA were proposed to be circular based on: (i) electron microscopy of purified 20S RNA (Fig. 7); (ii) the inability to label either 3′ ends using RNA ligase or 5′ ends using 5′ polynucleotide kinase; and (iii) migration on two-dimensional gel electrophoresis characteristic of non-linear ssRNAs (Matsumoto *et al.*, 1990). However, this study did not rule out unusual linkage of the 3′ and 5′ ends of a fundamentally linear molecule.

Most of the 20S RNA genome has now been cloned and sequenced (Matsumoto and Wickner, 1991). There is a long open-reading frame extending throughout the region cloned that appears to encode an RNA-dependent RNA polymerase (Fig. 4), although the match of the sequence to the consensus is not as clear as for L-A. Potential RNA-binding sequences were also detected. The presence of a site for cAMP-dependent phosphorylation in the sequence of the putative RNA-dependent RNA polymerase suggests a possible mechanism for amplification in cells grown on acetate medium. A drop in the cellular content of cAMP is the signal to the meiosis-sporulation system that cells are on sporulation medium and, conceivably, the same signal results in a change in the phosphorylation state of the RNA polymerase resulting in its activation.

W dsRNA was described as a low copy species of about 2.25 kb present in strains lacking L-A, L-BC and M dsRNAs (Wesolowski and Wickner, 1984). Its copy number was amplified ten-fold by growing cells at 37°C. Most of W dsRNA has likewise been cloned and sequenced (Rodriguez-Cousino *et al.*, 1991), and, unexpectedly, the sequences of W and 20S RNA are essentially identical (Matsumoto and Wickner, 1991; Rodriguez-Cousino *et al.*, 1991). The two RNA species are clearly different forms of

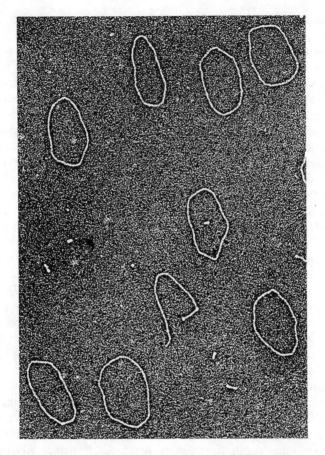

Fig. 7. Electron micrograph of 20S RNA spread after treatment with bacteriophage T4 gene 32 protein and glutaraldehyde cross-linking. The micrograph is reproduced by courtesy of Richard Fishel of the National Cancer Institute (Matsumoto *et al.*, 1990).

the same replicon. The conformation (circular or linear) of W has not yet been reported, but Rodriguez-Cousino *et al.* (1991) have argued that W and 20S RNA are both linear molecules. Both studies have failed to sequence the same region of the W and 20S molecules, suggesting the presence of some unusual structure whose elucidation will need further study.

Wejksnora and Haber (1978) reported that 20S RNA is present in 32S particles containing 18–20 copies in each particle of an acidic 23 kDa protein that lacked methionine residues. Production of 23 kDa protein was, like 20S RNA itself, induced by transfer of cells to acetate medium. A strain lacking 20S RNA failed to induce production of the 23 kDa protein

(Wejksnora and Haber, 1978). Based on its size, amino-acid composition and pattern of induction, Susek and Lindquist (1989) suggested that the 23 kDa protein might be the product of the *HSP26* gene (Bossier *et al.*, 1989; Susek and Lindquist, 1989), and that this gene might be providing a sort of viral coat protein to 20S RNA. The sequence of 20S RNA (representing about 90% of the total 20S RNA) does not contain a region that could encode a protein with the amino-acid composition of the 23 kDa protein, consistent with the hypothesis of Susek and Lindquist (1989). In fact, the 23 kDa protein has now been shown to be the Hsp26 protein (Widner *et al.*, 1991). However, there appears to be no clear functional or physical association between 20S RNA and Hsp26. A disruption mutant of *HSP26* amplifies 20S RNA to the same extent as an isogenic wild-type strain and this 20S RNA has the same sedimentation rate in sucrose gradients as that from the wild type. Moreover, a strain lacking 20S RNA still produced the Hsp26 protein. Thus, 20S RNA appears to be essentially naked in the cell, much like viroids and other circular RNA replicons.

VII. T dsRNA

A minor species of about 2.7 kb, T dsRNA, was found by Wesolowski and Wickner (1984) in strains lacking L-A, L-BC and M dsRNAs. The species is not homologous to any cellular DNAs or other dsRNAs. Like W, its copy number is induced at least ten-fold by growing of cells at 37°C. All strains known to carry T also carry W but many strains with W lack T, findings consistent with a dependence of T on W. Carrying T dsRNA was not associated with any phenotype.

VIII. DISEASE [D] AS A GENETIC ELEMENT IN KILLER STRAINS

Derepressed replication of M dsRNA, as a result of a *ski* mutation, results in cells being cold sensitive for growth (Ridley *et al.*, 1984). Growth is normal in the range 20–37°C, but there is no growth at 8–12°C and cells die with a half-life of less than a day. This cold sensitivity requires both the *ski⁻* mutation and M dsRNA. It is not due to the toxin or immunity protein encoded by M, since deletion mutants of M_1 (called S dsRNA for small) lacking essentially all of the coding sequences also make these strains cold sensitive. Nor is the cold sensitivity due to the total load of dsRNA being too great for the cell. M dsRNA represses L-A copy number

5–10-fold, so that eliminating M from a strain increases the L-A copy number dramatically and results in the M-o derivative having a higher dsRNA content than the original killer strain. Nonetheless, loss of M from the strain results in disappearance of the cold-sensitive phenotype (Ridley et al., 1984).

A new non-Mendelian genetic element, called [D] for disease, makes ski^- killer strains grow poorly or die at any temperature (Esteban and Wickner, 1987). The ski^- [D] M_1 strains become healthy on losing M_1, or on losing [D] or on becoming SKI^+. The disease genetic element is distinct from L-A, M, L-BC or W dsRNAs. It is also different from mitochondrial DNA, 2 μm DNA or ψ. Several mutants which lose [D] (called mad for maintenance of [D]) were isolated (Esteban and Wickner, 1987). The mad mutants retained M, L-A, W, mitochondrial DNA and 2 μm DNA. At least one mad mutation was shown to be in a single chromosomal gene. Curing of L-A by growing cells at high temperatures was accompanied by loss of [D] but, since several lines of evidence including transfection of L-A viral particles by the method of El-Sherbeini and Bostian (1987) showed that [D] is not a property of L-A, it seems that [D] depends on L-A for its replication or maintenance (Esteban and Wickner, 1987). There is at present no positive evidence concerning the molecular identity of [D].

IX. GENETIC ELEMENTS [Psi] AND [Eta]

A non-Mendelian genetic element, termed ψ, was originally detected by its ability to increase the efficiency of a weak ochre suppressor, namely SUQ5 (Cox, 1965). The effects of ψ have been explored in detail by biochemical and genetic studies, and much is known about factors affecting its inheritance. The molecular nature of ψ, however, remains unknown. An excellent detailed review by Cox et al. (1988) should be consulted by anyone interested in pursuing this fascinating puzzle. Eta (η) is closely related as a genetic element but has somewhat different effects on translation (Liebman and All-Robyn, 1984).

A. What do [Psi] and [Eta] do?

[Psi] augments the efficiency of the relatively weak serine-inserting ochre suppressors SUQ5 (=SUP16), SUP17 and SUP19 (Cox, 1965; Liebman et al., 1975; Ono et al., 1979; Palmer et al., 1979a; Waldron et al., 1981).

Tyrosine-inserting ochre suppressors are quite efficient in the absence of ψ and several become lethal in its presence (Cox, 1971). However, the activity of amber and opal (UGA) suppressors is not affected by ψ (Cox, 1971). Paromomycin can induce suppression of amber, ochre and opal codons *in vivo* and this effect requires the presence of ψ (Palmer *et al.*, 1979b). The frameshift suppressors *SUF1*, *SUF4* and *SUF6* are mutations in glycyl-tRNAs and their efficiency is also increased by ψ (Cummins *et al.*, 1980). [Psi] does not affect the activity of the omnipotent suppressors *sup35* and *sup45* (Liebman and Cavanagh, 1980). [Psi] alone, without an iden-tifiable chromosomal ochre suppressor, can suppress an ochre mutation in *cyc1* to a 1% level and the easily suppressed *trp5-48* ochre mutation (Liebman and Sherman, 1979), but the existence of low-level endogenous suppressors (reviewed by Cox *et al.*, 1988) leaves the possibility that ψ cannot act alone.

[Eta] does not affect the efficiency of the ochre suppressor alleles of *SUP11*, *SUP3* or *SUP53*, but it causes lethality in combination with some alleles of the omnipotent suppressors *sup35* and *sup45* (Leibman and All-Robyn, 1984). Other alleles are made stronger suppressors by η and some new genes able to mutate to omnipotent suppressors have been defined using this effect (All-Robyn *et al.*, 1990).

B. How does [Psi] work?

In vitro studies of the mechanism of ψ action have used the *in vitro* protein synthesis system of *Sacch. cerevisiae* (Gasior *et al.*, 1979; reviewed by Tuite and Plesset, 1986). Tuite *et al.* (1981a) showed that, in this system, extracts of ψ$^+$ strains could efficiently read through amber, ochre and UGA termination codons. This result differs from the *in vivo* result, which is specific for ochre codons. Extracts of ψ$^-$ strains were unable to do so, even if the extract contained suppressor tRNAs whose activity *in vivo* was independent of ψ. Also unexpected was the result of mixing ψ$^+$ and ψ$^-$ extracts; addition of only 20% of the ψ$^-$ extract eliminated read-through activity of the ψ$^+$ extract (Tuite *et al.*, 1981a, 1983, 1987). Fractionation of extracts indicated that the factor present in ψ$^-$ extracts that prevented read-through was loosely ribosome associated but it has not yet been precisely identified (Tuite *et al.*, 1987). Thus, ψ probably prevents synthesis of this ribosome-associated fidelity-enhancing factor, although why the *in vivo* and *in vitro* results differ is not clear (Tuite *et al.*, 1983).

C. Chromosomal genes affecting replication of [Psi]

Both dominant and recessive chromosomal mutants unable to maintain ψ have been isolated (Young and Cox, 1971; Cox *et al.*, 1980). These define genes (called *PNM* for [Psi] no more), which have not yet been characterized but are at least formally analogous to the *MAK* genes needed to replicate or maintain L-A and M dsRNAs, some of the *PET* genes such as *PET18*, which is needed to replicate mitochondrial DNA, and *ure2*, which is needed to maintain *[URE3]*. The mutants lose ψ and, when it is reintroduced, fail to maintain it.

D. Efficient conversion to ψ⁻ by guanidine hydrochloride or high osmolarity

Singh *et al.* (1979) found that medium containing 2.5 M potassium chloride or 1.8 M ethylene glycol efficiently induced conversion of ψ⁺ cells to ψ⁻. The same effect was seen by growth of cells on 5 mM guanidine hydrochloride (Tuite *et al.*, 1981b). However, while ψ⁻ mutants induced by high osmolarity are revertible to ψ⁺, those induced by guanidine hydrochloride are not (Lund and Cox, 1981). This suggests that the latter are deletions of all or part of the ψ genome and so should be especially useful in identifying what is ψ. Eta is also cured by guanidine hydrochloride suggesting that, although its phenotypic effects are clearly different from those of ψ, it may be a replicon very similar to ψ (All-Robyn *et al.*, 1990).

E. What is [Psi]?

It is clear that ψ is not related to mitochondrial DNA, to killer RNAs (L-A or M), or to 2 μm DNA (Young and Cox, 1972; Leibowitz and Wickner, 1978; Singh *et al.*, 1979; Tuite *et al.*, 1982). Evidence exists that ψ is distinct from [URE3], 20S RNA and η (summarized by Cox *et al.*, 1988).

Dai *et al.* (1986) have shown that 3 μm plasmid DNA (Clark-Walker and Azad, 1980) from yeast has the ability to transform yeast from [Psi⁻] to [Psi⁺]. Three-micrometre DNA is the circular form of rDNA presumably formed by homologous recombination between two of the 100 tandem repeats on chromosome XII. While this seemed to have solved the problem of "What is ψ?", Cox *et al.* (1988) have argued that the mystery remains. They point out that the stability of ψ is inconsistent with the low copy of

3 μm DNA and its lack of a centromere, that only a few copies of rDNA in plasmid form would not be expected to affect translational fidelity substantially, and that [Psi⁻] strains continue to retain 3 μm DNA. It seems clear that, while the results of Dai *et al.* (1986) cannot be less than an important hint, more work will be needed to demystify ψ.

Another type of model for ψ, not involving any non-chromosomal replicon, has been proposed by Strathern (cited by Cox, *et al.*, 1988; Fig. 8). The read-through product of the ochre terminator codon of a recessive *pnm* gene (*long* Pnm) is supposed to promote ochre read-through, but the product of the same gene, if terminated at the ochre codon (*short*, Pnm), has no such activity. The presence of *long* Pnm promotes its own synthesis, but some non-mutagenic agents, like guanidine hydrochloride or high osmotic strength, might be expected to inactivate *long* Pnm. Thereafter, only *short* Pnm would be made. This clever model explains much but not all of the known ψ phenomena, as discussed at length by Cox *et al.* (1988). For example, the presence in a [Psi⁻] strain of a normal ochre suppressor would be expected to result regularly in the strain becoming [Psi⁺], although this is not observed. Perhaps as important as whether this model is correct or not is that it provides a very clear example of the sort of phenomenon that can behave just like a non-chromosomal replicon genetically, and yet be based solely on the behaviour of chromosomal genes.

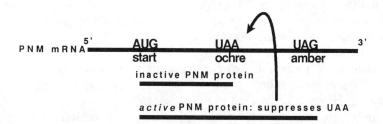

Fig. 8. Strathern's clever model of ψ (cited in Cox *et al.*, 1988) may or may not be true (see text), but illustrates how phenomena involving only chromosomal genes can masquerade as non-chromosomal replicons. The fundamental assumption of the model is that the short form of Pnm is inactive while the long form promotes ochre-codon read-through. A cell in which only the short form of Pnm is being expressed cannot make the long form because it cannot read-through the UAA termination codon. Thus, this [Psi⁻] state is stable. A cell that is initially making the long form of Pnm (the [Psi⁺] state) is able to continue doing so because this form is active in suppressing ochre mutations, thus allowing its own synthesis to continue. Mating a [Psi⁺] with a [Psi⁻] produces cells in which all copies of the Pnm mRNA are experiencing read-through and so all offspring are [Psi⁺].

X. [URE3] ALLOWS YEAST TO TAKE UP UREIDOSUCCINATE

The URA2 gene encodes aspartate carbamoyltransferase, whose product is ureidosuccinate (carbamoyl-L-aspartate). Wild-type yeast cannot take up ureidosuccinate when growing on ammonium sulphate or glutamine as a nitrogen source, but mutants in the chromosomal ure1 gene can do so when growing in an ammonium sulphate medium. This uptake is inhibited by ammonia, and selection of strains able to take up ureidosuccinate in the presence of high concentrations of ammonia produced mutants in another chromosomal gene, ure2, and in [URE3], a non-Mendelian mutation (Lacroute, 1971). While ammonia still inhibits of ureidosuccinate uptake in ure2 and [URE3] mutants, the level of uptake is high enough to by-pass the ura2 mutation. Only the [URE3] mutation is dominant. The ure1, ure2 and [URE3] mutants all also secrete uracil when supplied with ureidosuccinate on ammonium acetate medium. Spreading such a test plate with an inoculum of a uracil auxotroph produces a halo of growth around the ure1, ure2 or [URE3] strains (Lacroute, 1971; Drillien and Lacroute, 1972). This assay is particularly convenient in that it allows assay of the presence of [URE3] even in a strain that is not ura2.

[URE3] and ure2, in addition to their effects on uptake of ureidosuccinate, also produce derepressed levels of the degradative glutamate dehydrogenase (the enzyme that uses NAD^+) and somewhat reduced levels of the biosynthetic enzyme (that uses $NADP^+$) (Drillien et al., 1973). The connection of this phenotype to the effects on uptake of ureidosuccinate are not clear. The URE2 gene encodes a 40 kDa protein with homology to glutathione transferases (Coschigano and Magasanik, 1991). The URE2 protein regulates nitrogen metabolism through the GLN3 protein, but not by altering the transcription or translation of GLN3 (Courchesne and Magasanik, 1988).

That [URE3] was a non-chromosomal mutation was first suggested by 4[URE3]:0 segregation in crosses of the type ura2 [URE3] × ura2 [ure3+]. However, compared to other elements like L-A or M_1 dsRNA, ψ or [Rho], the mitochondrial genome, [URE3], is not completely consistent in segregating 4[URE3]:0. Aigle and Lacroute (1975) then showed that the [URE3] element could be transferred by cytoduction. These experiments were done before the kar1 mutants of Conde and Fink (1976) were available, and so relied on the low level of natural non-fusion of nuclei first described by Wright and Lederberg (1957).

Aigle and Lacroute (1975) also showed, by both meiotic crosses and by cytoduction experiments, that the ure2 mutants were unable to maintain the [URE3] element. How can ure2 produce the same mutant phenotype as [URE3] and yet be unable to maintain the [URE3] element? This recalls the relations among the suppressive mutants of M_1, called S mutants, and the *mak* mutants. In that case, if one equates an S mutant dsRNA (a deletion mutant of M_1 that excludes the normal M_1 replicon when both are introduced into the same cell) with the [*URE3*] mutant, and equates the wild-type [*ure3$^+$*] replicon with the wild-type M_1, then the *mak* mutants are analogous to the *ure2* mutants. This suggests that both the *ure2* mutants and the [*URE3*] mutants produce their common phenotype by eliminating (losing) the wild-type [*ure3$^+$*] genome. By this interpretation, [*URE3*] is, like the S mutants, dominant because it excludes the wild-type [*ure3$^+$*] genome, not because it is expressing a product that does something. This model suggests that one might look for the molecular correlate of [*URE3*] or [*ure3$^+$*], using the absence of the molecule in a *ure2* mutant as a control. [*URE3*] is, like ψ and η, reported to be efficiently cured by growth of cells in the presence of guanidine hydrochloride (M. Aigle, cited in Cox *et al.*, 1988; Leibman and All-Robyn, 1984), suggesting that the [*URE3*] replicon may be the same as, or related to those of ψ and η.

XI. TWO-MICRON DNA

The molecular biology of 2 μ DNA and its use in constructing cloning vectors has been reviewed recently (Futcher, 1988; Newlon, 1989; Rose and Broach, 1990; Parent and Bostian, Chapter 4 of this volume), and this section will only briefly summarize those aspects in order to focus on genetical aspects of 2 μ DNA and to compare it with other non-Mendelian elements. Two-micron DNA has also been used to develop a simple genetic mapping procedure (Wakem and Sherman, 1990).

Two-micron DNA was first discovered by electron microscopy, and was named for its contour length and closed circular conformation (Sinclair *et al.*, 1967; Clark-Walker, 1973) as 2 μ circles. Two-micron DNA is 6318 bp in length and has two 599 bp identical regions present in inverted orientation separating unique 2774 and 2346 bp domains (Hartley and Donelson, 1980). Frequent recombination between these inverted repeats, catalysed by the plasmid-encoded FLP protein, results in cells carrying a mixture of two forms of 2 μ DNA differing only in the relative orientation of the unique regions (Guerineau *et al.*, 1976; Hollenberg *et al.*, 1976; Gubbins *et al.*, 1977; Livingston and Klein, 1977). Two-micrometre DNA replicates as θ structures (Petes and Williamson, 1975) and is expressed in the nucleus as a histone-covered molecule (Livingston and Hahne, 1979; Nelson and Fangman, 1979). It is quite stably maintained at a copy number of around 60, slowing the growth of its host by about 1% (Futcher and Cox, 1983; Mead *et al.*, 1986).

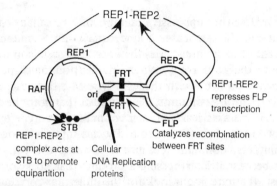

Fig. 9. The 2 μ DNA plasmid of *Sacch. cerevisiae*, showing genes, sites and structure. The extent and direction of transcription of the open-reading frames is indicated, along with sites of action of the REP1–REP2 complex and the FLP protein.

It encodes four proteins (Fig. 9) (Hartley and Donelson, 1980). These are *REP1* and *REP2*, which are necessary for proper plasmid partition (Kikuchi, 1983) and repress production of the FLP protein (Murray *et al.*, 1987), *FLP* (for flip) that catalyses site-specific intramolecular recombination between the inverted repeats (Vetter *et al.*, 1983; Babineau *et al.*, 1985; Meyer-Leon *et al.*, 1987), and *RAF* (for REP antagonizing factor or recombinase activating factor) that modulates repression of *FLP* by *REP1* and *REP2* (Murray *et al.*, 1987). The critical sites on 2 μ DNA include a replication origin (*ORI*; Broach and Hicks, 1980; Brewer and Fangman, 1987), the site at which *FLP* acts called *FRT* (Broach *et al.*, 1982), and a partition site called *STB* for stability (or *REP3*) that is acted on by *REP1* and *REP2* (Kikuchi, 1983) (Fig. 9).

Related plasmids, with similar overall structure, gene organization and mechanism of replication and partition, but only rather limited homology with 2 μ DNA, have been found in *Zygosaccharomyces bailii* (pSB1 and pSB2; Toh-e *et al.*, 1984; Utatsu *et al.*, 1987), *Zygosacch. rouxii* (pSR1; Toh-e *et al.*, 1982; Araki *et al.*, 1985), *Zygosacch. bisporus* (pSB3, Toh-e *et al.*, 1984), *Zygosacch. fermentati* (pSM1; Utatsu *et al.*, 1987), and *Kluyveromyces drosophilarum* (pKD1; Chen *et al.*, 1986).

A. Partition of the 2 μ DNA

The instability of ARS plasmids and of 2 μ DNA-derived vectors in a *REP⁻* environment (Broach and Hicks, 1980) is due to asymmetric segregation of the plasmid copies at mitosis (Kikuchi, 1983; Murray and

Szostak, 1983). Even though each cell may have 60 copies of the vector, in the absence of *REP1*, *REP2* or *STB* function, all of the copies will stay with the mother cell in most cell divisions. Surprisingly, these positive functions are needed to prevent asymmetric distribution of plasmid copies to the mother and daughter cells. The mechanism by which the *REP* genes and *STB* site promote distribution of plasmid copies is not known, but the finding that REP1 is a nuclear matrix protein (Wu *et al.*, 1987) suggests that attachment of $2\,\mu$ DNA molecules to the nuclear matrix is needed to move it into the daughter cells. The *STB cis* site consists of a partition site and a transcription termination site, the latter being required because transcription interferes with partition (Kikuchi, 1983; Murray and Cesareni, 1986).

B. Flp-mediated $2\,\mu$ DNA amplification

Since each $2\,\mu$ DNA molecule replicates exactly once in each cell cycle, it would seem difficult to understand how a single molecule transformed into a cell could reach high copy number. A clever solution (Futcher, 1986) proposes that Flp-catalysed recombination between the inverted repeat sequences of a single molecule that has begun replication would generate a molecule in which the two replication forks would be proceeding in the same direction around the circle instead of opposite directions (the usual situation). These two forks, instead of meeting each other at the other side of the molecule and thus terminating the round of replication, would follow each other around, making multiple copies after a single initiation event (Fig. 10) (Futcher, 1986). Strong support for this model was then supplied by Volkert and Broach (1986) who showed that a single copy of $2\,\mu$ DNA excised by Flp from a chromosomal site could amplify to normal copy number only if its *FRT* site was intact. The model (described in greater detail by Futcher (1988)) has the virtue of explaining why $2\,\mu$ DNA and all related plasmids devote so much of their genome to *FLP* and inverted repeats, the consistent placement of *ORI* near or in the inverted repeats, and a mechanism for amplification without violating the one initiation in each cell cycle rule.

This mechanism also makes sense of the *REP*-mediated repression of FLP. When the copy number of $2\,\mu$ DNA is low, the concentration of *REP* products is low and FLP is derepressed, leading to amplification. When the copy number of $2\,\mu$ DNA increases, so does that of the REP proteins and FLP is shut off, so that $2\,\mu$ DNA replicates only once in each cycle thereafter.

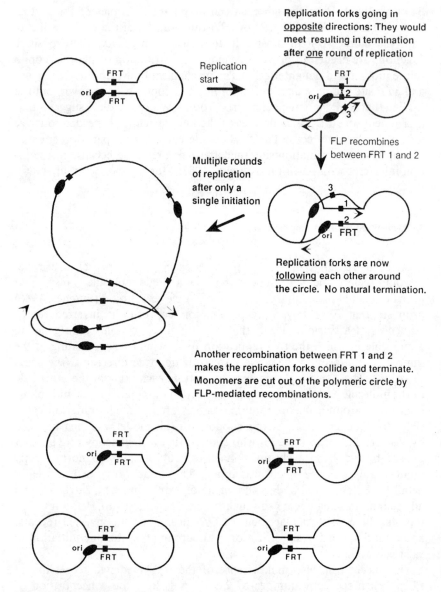

Fig. 10. *FLP*-mediated amplification of 2 μm DNA plasmid copy number according to the model of Futcher (1986).

C. Mechanism of *FLP* action

The *FLP–FRT* reaction has been studied in great detail by groups led by Sadowski, Cox and Jayaram (reviewed by Sadowski, 1986; Cox, 1989). The FLP protein uses no ATP or other energy source to carry out the reaction. Rather, the FLP protein forms phosphotyrosine linkages with a DNA substrate, maintaining the high energy in these bonds (Gronostajski and Sadowski, 1985). The *FRT* site (Broach *et al.*, 1982; Andrews *et al.*, 1985; Jayaram, 1985; Senecoff *et al.*, 1985) includes three 13 bp repeats, all of which are protected by FLP protein, but only two of which are important for binding of FLP protein. These two repeats are in inverted orientation and surround an 8 bp core region whose sequence is not recognized by FLP, but which must be identical between the two FRT sites between which recombination is carried out. FLP makes staggered cuts at each side of the 8 bp core sequence and the requirement for identity in the core sequences implies that the single-stranded regions produced must pair. The normal core sequence is asymmetrical, so that the pairing can only yield inversion of the segment in between the inverted repeats, not excision of the interval between the repeats. However, if the cores are both changed to be a symmetrical sequence, the efficiency of the FLP reaction is not affected, but now either flipping or excision is possible (Senecoff and Cox, 1986).

The sequence of events (see Fig. 11) is as follows. (a) Binding: one FLP monomer binds to each of the 13 bp inverted repeats in the *FRT* site. (b) Bending: the FLP monomers interact to bend the *FRT* site by 144° (Schwartz and Sadowski, 1989, 1990). (c) First nicking: one FLP protein on each *FRT* site nicks next to the core sequence attaching the 3′ end formed to a tyrosine residue of FLP and freeing the 5′ ended core to pair (Qian *et al.*, 1990). (d) Synapsis: two *FRT* sites are associated by protein–protein interactions of FLP monomers. (e) Strand exchange: 5′ ended cores hybridize with the intact strand of the other core and are ligated with the tyrosyl-3′ end there. (f) Isomerization: a conformational change of the complex occurs to prepare for nicking the second strand. (g) Second nicking: the as yet intact parental strand in each *FRT* site is nicked. (h) Resolution: the resealing of the recombined second strands to end the reaction.

The FLP protein is homologous to other site-specific recombinases (Argos *et al.*, 1986) and detailed studies of partial reactions of mutant proteins may lead to an understanding of the role of conserved domains in this family.

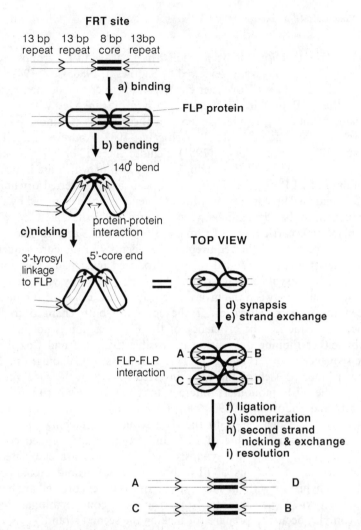

Fig. 11. The mechanism of FLP-catalysed recombination at FRT sites. This model is a simplification of the models of Schwartz and Sadowski (1990) and P.D. Sadowski (personal communication).

D. Genetics of 2 μ DNA

Two-micron DNA is a typical non-Mendelian element in that mating strains with [cir⁺] and without [cir°] 2 μ DNA produces diploids stably maintaining 2 μ DNA and meiotic segregants, all of which carry 2 μ DNA

(Livingston, 1977). However, cytoduction transfers 2 μ DNA at only 50% efficiency, consistent with its nuclear location. Mating strains with distinguishable 2 μ DNA genomes leads to zygote clones, and mitotic and meiotic progeny, all of which carry a mixture of the two varieties (Livingston, 1977). Nearly all strains have 2 μ DNA, and it is only rarely lost in mitotic growth (Futcher and Cox, 1983). It can be eliminated, however, by simply growing cells mitotically for prolonged periods and screening by colony hybridization (Futcher and Cox, 1983) or by trans-forming cells with a 2 μ DNA plasmid containing a defective *LEU2* gene, which only makes cells Leu$^+$ when in high copy (Dobson *et al.*, 1980; Toh-e and Wickner, 1981; Erhart and Hollenberg, 1983; Harford and Peeters, 1987). This gene is found on pJDB219 (Beggs, 1978) and deriva-tives thereof. While selection for high copy and consequent competition with the normal 2 μ DNA is clearly part of the explanation, the phe-nomenon seems to be more complicated (see discussion by Rose and Broach, 1990).

Two-micron DNA replication is controlled by the cell cycle such that, like a chromosome, each 2 μ DNA molecule replicates once during each S phase (Zakian *et al.*, 1979). Thus, 2 μ DNA replication requires products of the *CDC28*, *CDC4*, *CDC7*, and *CDC8* genes (Petes and Williamson, 1975; Livingston and Kupfer, 1977) and is inhibited by treatment of a cells with α factor, but is unaffected by the *CDC13* gene, which is blocked in nuclear division (Livingston and Kupfer, 1977). The *ORI* of 2 μ DNA is a typical chromosomal *ORI* (Broach *et al.*, 1983) so its activation only once per cell cycle is as expected.

The *nib1* mutant results in very high 2 μ DNA copy number and a nibbled-colony morphology due to cell death (Holm, 1982a,b). Presum-ably, the *NIB1* gene product is a host component involved in the 2 μ DNA partition process or, perhaps, in repressing 2 μ DNA replication. This phenomenon is apparently a combination of the effects of 2 μ DNA and another element (Sweeney and Zakian, 1989).

Mutations in a single chromosomal gene, *MAP1*, affect replication of 2 μm DNA and all *ARS* and *CEN* vectors tested, but not chromosomal DNA (Kikuchi and Toh-e, 1986). Plasmids able to overcome a *map1* mutation were found to have acquired either a second *ARS* sequence or to have acquired a second inverted repeat (*FLP* site). This suggests that the *MAP1* gene affects primarily replication of *ARS* sequences rather than plasmid segregation (Kikuchi and Toh-e, 1986). Presumably chromosomal *ARS*s are similarly affected but, because there are many *ARS*s on each chromosome, they need not all fire in order to replicate the chromosome. Acquisition of the 2 μ DNA inverted repeat containing the *FLP* site should, by the Futcher (1986) mechanism, allow the plasmid to carry out multiple rounds of replication by a rolling-circle mechanism after only a single replication initiation event.

XII. LINEAR DNA KILLER PLASMIDS OF *KLUYVEROMYCES LACTÎS*

Some strains of *K. lactis* carry linear double-stranded DNA plasmids (Gunge *et al.*, 1981; Wesolowski *et al.*, 1982a; reviewed by Stark *et al.*, 1990). The smaller 8.9 kb plasmid (called pGKl1 or k1) encodes the three subunits of a secreted killer toxin and an immunity protein, and the 13.4 kb plasmid (pGKl2 or k2) is required for maintenance of k1 in a manner superficially similar to the L-A and M dsRNAs in *Sacch. cerevisiae*. Each of these plasmids has terminal inverted repeat sequences, and proteins covalently attached to their 5′ termini such as phage ϕ29 or adenovirus. They replicate in the cytoplasm of either *K. lactis* or *Sacch. cerevisiae*. This is in contrast to 2 μ DNA and results in their having to encode their own replication and transcription functions, since the nuclear apparatus is unavailable. Their ability to undergo homologous recombination has recently led to a method for study and manipulation of their genomes. Similar linear DNA plasmids have been found in *Pichia inositovora* (Ligon *et al.*, 1989), at least one of which appears to be associated with a killer phenotype (Hayman and Bolen, 1990).

A. Genome structure and replication

The 8874 bp k1 has 202 bp terminal inverted repeats, while the 13 457 bp k2 has completely unrelated 182 bp inverted repeats (Sor *et al.*, 1983; Hishinuma *et al.*, 1984; Stark *et al.*, 1984; Sor and Fukuhara, 1985; Tommasino and Galeotti, 1988). The 5′-terminal covalently-attached proteins are 28 kDa and 36 kDa, respectively (Kikuchi *et al.*, 1984; Stam *et al.*, 1986). In adenovirus and phage ϕ29, these proteins are primers for DNA replication and the same is presumed in this case, although there is no direct evidence on this point.

Based on both microscopy and subcellular fractionation experiments, the k1 and k2 plasmids are located in the cytoplasm of either *K. lactis* or *Sacch. cerevisiae* (Gunge *et al.*, 1982; Stam *et al.*, 1986). The k1 and k2 segments each encode a protein with clear homology to family-B DNA polymerases, which includes those encoded by phage ϕ29, herpes simplex virus, adenovirus, vaccinia and varicella viruses, phage T4 and others (Bernad *et al.*, 1987; Fukuhara, 1987; Jung *et al.*, 1987; Tommasino and Galeotti, 1988). It is likely that some of the other k2-encoded proteins are polymerase accessory proteins. The k1 and k2 plasmids can be stably maintained in *Sacch. cerevisiae*, but only in strains that are ρ^0 (Gunge and Yamane, 1984) and only if the strain is haploid (Gunge *et al.*, 1990). The reasons for these effects are not yet entirely clear.

B. Expression mechanisms

Their location in the cytoplasm requires these plasmids to make provision for a transcription apparatus. Plasmid k2 encodes a protein with striking homology to RNA polymerase in the large subunits of prokaryotes and eukaryotes (Wilson and Meacock, 1988). Unlike other RNA polymerases, this protein has homology in its central region to three domains conserved in β subunits, and in its C-terminal region, homology to two domains conserved in β′ subunits (Wilson and Meacock, 1988; Stark *et al.*, 1990).

The near absence of intergenic spaces, and the high AT content of the k1 and k2 genomes set constraints on the promotor structures, but these have not yet been defined due to difficulties of the system. The fact that k1 and k2 genes are transcribed by their own RNA polymerase is reflected in the fact that insertion of these genes into nuclear plasmid vectors results in little or no expression (reviewed by Stark *et al.*, 1990). Furthermore, the unique structure of the ends of k1 and k2 results in difficulties in transforming cells with modified linear plasmids.

C. Recombination and genome manipulation

In spite of its location in the cytoplasm, recombination of defective k1 genomes to give wild-type recombinants was demonstrated by Wesolowski *et al.* (1982b). Recently, this phenomenon was used to produce a directed modification of k1 by the method routinely used for gene disruptions (Kamper *et al.*, 1989). As reviewed by Stark *et al.* (1990), previous efforts to use these plasmids as vectors for expression or secretion of foreign proteins had met with little success. The *in vivo* recombination method promises to provide a means for modifying the promoter sequences, to insert foreign proteins to be expressed, and otherwise alter k1 and k2 in order to study and use them. A kanamycin-resistance gene has been expressed from k1 by this method (Tanguy-Rougeau *et al.*, 1990). It is as yet unclear what enzymes are carrying out this recombination, and whether there exists a special plasmid-encoded system or a host system.

D. Toxin structure, processing and action

A more detailed review of the toxin can be found in Stark *et al.* (1990). The k1-encoded toxin consists of three distinct subunits, α, β and γ (Stark and Boyd, 1986). Subunits α and β are encoded by a single ORF (Stark and Boyd, 1986) (Fig. 12) and are processed by proteolytic cleavages after pairs of basic residues by an enzyme called *KEX1*, which is homologous to and can be substituted by the product of the *KEX2* gene in *Sacch. cerevisiae* (Wesolowski *et al.*, 1982a; Tanguy-Rougeau *et al.*, 1988; Wesolowski-Louvel *et al.*, 1988). The toxin was thought to act by an effect on cAMP but this has been questioned (Stark *et al.*, 1990). The γ subunit is responsible for the toxic effect, since its expression alone from a cDNA clone is sufficient to kill cells from inside, and toxin-immune cells are also immune to this effect (Tokunaga *et al.*, 1988). Presumably, α and β subunits are involved in getting γ into the cell.

XIII. OTHER POSSIBLE NON-MENDELIAN ELEMENTS

Wild-type *Sacch. cerevisiae* carrying one of the *MAL* loci can use maltose as a carbon source whether they are ρ⁺ or ρ⁻. Schamhart *et al.* (1975) found that crosses of a *MAL4 Sacch. cerevisiae* strain 1403-7A with a *MAL6 Sacch. carlsbergensis* strain CB-6 produced *MAL6* segregants, all

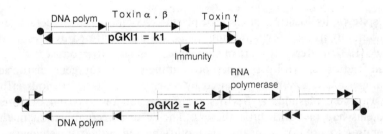

Fig. 12. Structure of the pGKL1, i.e. k1 (above) and pGKL2, i.e. k2 (below) linear DNA killer plasmids of *Kluyveromyces lactis*. The filled circles represent the 5'-linked proteins. The arrowheads at the termini represent the terminal inverted repeat sequences. The approximate locations and functions of open-reading frames are indicated.

of which were unable to grow or were able to grow only very poorly on maltose when they were made $\rho°$ with ethidium bromide, this although both parents were able to grow on maltose in the $\rho°$ state. The effect was specific for *MAL6*, since *MAL4* segregants were normal. The genetic element producing this effect was not eliminated by treatment of cells with ethidium bromide and so is not on mitochondrial DNA. This single report has not, to my knowledge, been pursued.

Wild-type yeast cannot grow on non-fermentable carbon sources in the presence of glucosamine. Mutants resistant to this effect include one, GR10, which showed tetrad ratios in meiosis that varied from 4 resistant:0 to 0:4 sensitive, in various crosses with sensitive strains. The phenotype was unaffected by ethidium bromide (mitochondrial genome) or by curing the killer trait using cycloheximide or heat. No relationship with ψ or [*URE3*] was detected (Kunz and Ball, 1977).

Mention has already been made of an element different from 2 μ DNA, which affects lethal sectoring produced by the *nib1* mutation (Sweeney and Zakian, 1989) and [D], the non-Mendelian factor(s) affecting the cytopathology produced by M dsRNA in a *ski⁻*-host (Esteban and Wickner, 1987). Each of these elements has been the subject of only a single report and each needs to be pursued more thoroughly. With a substantial number of both molecular orphans and genetic orphans, it seems likely that some of each may find homes in the near future.

XIV. UPDATE

Evidence that [URE3] and [Psi] are prions. A prion is an infectious protein, a concept originating in studies of scrapie of sheep, and kuru and Creutzfeldt–Jacob disease of man (reviewed in Prusiner, 1994). Based on their genetic properties, it has been proposed that both [URE3] and [Psi] are prions of yeast (Wickner, 1994). Both elements can be cured, but the cured strains can reacquire the element without its introduction from another cell, unlike nucleic acid replicons (Singh *et al.*, 1979; Tuite *et al.*, 1981; Wickner, 1994). [URE3] depends on the chromosomal *URE2* gene for its propagation (Aigle and Lacroute, 1975; Wickner, 1994), and [Psi] similarly depends on *SUP35* (Cox, 1993; TerAvanesyan *et al.*, 1994; Doel *et al.*, 1994). In each case, the presence of the dominant non-Mendelian element confers the *same* phenotype as does a recessive mutation in the chromosomal gene necessary for its propagation. In contrast, strains carrying M_1 dsRNA have a phenotype *opposite* to that of *mak* mutants, and strains with the mitochondrial genome have a phenotype *opposite* to that of mutants in chromosomal genes that cannot maintain it. Finally, over-producing Ure2p or Sup35p induce the appearance of [URE3] (Wickner, 1994) and [Psi] (Chernoff *et al.*, 1993), respectively. These three proper-ties all point to [URE3] and [Psi] being prions, rather than nucleic acid replicons (Wickner, 1994).

These developments have supported the still controversial prion concept, and suggested that prions can be the basis of inheritance as well as infection. It also indicates that the occurrence of prions is not restricted to the rare mammalian spongiform encephalopathies.

XV. ACKNOWLEDGEMENTS

I am grateful to my many colleagues for their sharing of information, criticism of the manuscript, and the part many have had in the work of my own laboratory. Particular thanks are due to Paul Sadowski, Y. Peng Loh, Rosaura Valle for their contributions to this manuscript, and to Stephen Ball, Jonathan Dinman, Rosa Esteban, Tsutomu Fujimura, Patricia Guerry-Kopecko, Tateo Icho, Hyun-Sook Lee, Yang-Ja Lee, Michael Leibowitz, Yutaka Matsumoto, Sang-Ki Rhee, Juan Carlos Ribas, S. Porter Ridley, Steve Sommer, Juan Carlos Tercero, Akio Toh-e, Hiroshi Uemura, Rosaura Valle, Micheline Wesolowski, and William R. Widner, for their efforts in my laboratory over the last 17 years.

References

Aigle, M. and Lacroute, F. (1975). *Molecular and General Genetics* **136**, 327.
All-Robyn, J.A., Kelley-Geraghty, D., Griffin, E., Brown, N. and Liebman, S.W. (1990). *Genetics* **124**, 505.
Andrews, B.J., Proteau, G.A., Beatty, L.G. and Sadowski, P.D. (1985). *Cell* **40**, 795.
Araki, H., Jearnpipatkul, A., Tatsumi, H., Sakurai, T., Ushio, K., Muta, T. and Oshima, Y. (1985). *Journal of Molecular Biology* **182**, 191.
Argos, P., Landy, A., Abremski, K., Egan, J.B., Haggard-Ljungquist, E., Hoess, R.H., Kahn, M.L., Kalionis, B., Narayana, S.V.L., Pierson, L.S., Sternberg, N. and Leong, J.M. (1986). *EMBO Journal* **5**, 433.
Babineau, D., Vetter, D., Andrews, B.J., Gronostajski, R.M., Proteau, G.A., Beatty, L.G. and Sadowski, P.D. (1985). *Journal of Biological Chemistry* **260**, 12313.
Ball, S.G., Tirtiaux, C. and Wickner, R.B. (1984). *Genetics* **107**, 199.
Beggs, J.D. (1978). *Nature* **275**, 104.
Bernad, A., Zaballos, A., Salas, M. and Blanco, L. (1987). *EMBO Journal* **6**, 4219.
Bevan, E.A., Herring, A.J. and Mitchell, D.J. (1973). *Nature* **245**, 81.
Boeke, J.D. (1989). *In* "Mobile DNA" (D. Berg and M. Howe, eds), pp. 335–374. ASM Publications, Washington, DC.
Bossier, P., Fitch, I.T., Boucherie, H. and Tuite, M.F. (1989). *Gene* **78**, 323.

Bostian, K.A., Hopper, J.E., Rogers, D.T. and Tipper, D.J. (1980a). *Cell* **19**, 403.

Bostian, K.A., Sturgeon, J.A. and Tipper, D.J. (1980b). *Journal of Bacteriology* **143**, 463.

Bozarth, R.F., Koltin, Y., Weissman, M.B., Parker, R.L., Dalton, R.E. and Steinlauf, R. (1981). *Virology* **113**, 492.

Brewer, B.J. and Fangman, W.L. (1987). *Cell* **51**, 463.

Broach, J.R. and Hicks, J.B. (1980). *Cell* **21**, 501.

Broach, J.R., Guarascio, V.R. and Jayaram, M. (1982). *Cell* **29**, 227.

Broach, J.R., Li, Y.-Y., Feldman, J., Jayaram, M., Abraham, J., Nasmyth, K. and Hicks, J.B. (1983). *Cold Spring Harbor Symposia on Quantitative Biology* **47**, 1165.

Bruening, G., Buzayan, J.M., Hampel, A., and Gerlach, W.L. (1988). *In* "RNA Genetics" (E. Domingo, J.J. Holland and P. Ahlquist, eds), pp. 127-145. CRC Press, Boca Raton.

Bruenn, J.A. and Brennan, V.E. (1980). *Cell* **19**, 923.

Buck, K.W., Lhoas, P. and Street, B.K. (1973). *Biochemical Society Transactions* **1**, 1141.

Bussey, H. (1988). *Yeast* **4**, 17.

Chen, X.J., Saliola, M., Falcone, C., Bianchi, M.M. and Fukuhara, H. (1986). *Nucleic Acids Research* **14**, 4471.

Chernoff, Y.O., Derkach, I.L. and Inge-Vechtomov, S.G. (1993). *Current Genetics* **24**, 268.

Clark-Walker, G.D. (1973). *European Journal of Biochemistry* **32**, 263.

Clark-Walker, G.D. and Azad, A.A. (1980). *Nucleic Acids Research* **8**, 1009.

Coen, D., Netter, P., Petrochilo, E. and Slonimski, P. (1970). *Symposium of the Society for Experimental Biology* **24**, 449.

Conde, J. and Fink, G.R. (1976). *Proceedings of the National Academy of Sciences, USA* **73**, 3651.

Cooper, A. and Bussey, H. (1989). *Molecular and Cellular Biology* **9**, 2706.

Coschigano, P.W. and Magasanik, B. (1991). *Molecular and Cellular Biology* **22**, 822-832.

Courchesne, W.E. and Magasanik, B. (1988). *Journal of Bacteriology* **170**, 708.

Cox, B.S. (1965). *Heredity* **20**, 505.

Cox, B.S. (1971). *Heredity* **26**, 211.

Cox, B.S. (1993). *In* "The Early Days of Yeast Genetics", pp. 219-239. Cold Spring Harbor Press, Cold Spring Harbor Laboratory.

Cox, B.S., Tuite, M.F. and Mundy, C.J. (1980). *Genetics* **95**, 589.

Cox, B.S., Tuite, M.F. and McLaughlin, C.S. (1988). *Yeast* **4**, 159.

Cox, M.M. (1989). *In* "Mobile DNA" (D.E. Berg and M.M. Howe, eds), pp. 661-670. American Society for Microbiology, Washington, DC.

Cummins, C.M., Gaber, R.F., Culbertson, M.R., Mann, R. and Fink, G.R. (1980). *Genetics* **95**, 855.

Dai, H., Tsay, S.-H., Lund, P.M. and Cox, B.S. (1986). *Current Genetics* **11**, 79.

Dihanich, M., Suda, K. and Schatz, G. (1987). *EMBO Journal* **6**, 723.

Dihanich, M., Van Tuinen, E., Lambris, J.D. and Marshallsay, B. (1989). *Molecular and Cellular Biology* **9**, 1100.

Dinman, J.D., Icho, T. and Wickner, R.B. (1991). *Proceedings of the National Academy of Sciences, USA* **88**, 174.

Dmochowska, A., Dignard, D., Henning, D., Thomas, D.Y. and Bussey, H. (1987). *Cell* **50**, 573.

Dobson, M.J., Futcher, A.B. and Cox, B.S. (1980). *Current Genetics* **2**, 210.

Doel, S.M., McCready, S.J., Nierras, C.R. and Cox, B.S. (1994). *Genetics* **137**, 659.

Drillien, R. and Lacroute, F. (1972). *Journal of Bacteriology* **109**, 203.

Drillien, R., Aigle, M. and Lacroute, F. (1973). *Biochemical and Biophysical Research Communication* **53**, 367.

Egel-Mitani, M., Flygenring, H.P. and Hansen, M.T. (1990). *Yeast* **6**, 127.

El-Sherbeini, M. and Bostian, K.A. (1987). *Proceedings of the National Academy of Sciences, USA* **84**, 4293.

Erhart, E. and Hollenberg, C.P. (1983). *Journal of Bacteriology* **156**, 625.

Esteban, R. and Wickner, R.B. (1986). *Molecular and Cellular Biology* **6**, 1552.

352   R.B. Wickner

Esteban, R. and Wickner, R.B. (1987). *Genetics* **117**, 399.
Esteban, R. and Wickner, R.B. (1988). *Journal of Virology* **62**, 1278.
Esteban, R., Fujimura, T. and Wickner, R.B. (1988). *Proceedings of the National Academy of Sciences, USA* **85**, 4411.
Esteban, R., Fujimura, T. and Wickner, R.B. (1989). *EMBO Journal* **8**, 947.
Estivariz, F.E., Birch, N.P. and Loh, Y.P. (1989). *Journal of Biological Chemistry* **264**, 17796.
Field, L.J., Bobek, L., Brennen, V., Reilly, J.D. and Bruenn, J. (1982). *Cell* **31**, 193.
Fink, G. R. and Styles, C.A. (1972). *Proceedings of the National Academy of Sciences, USA* **69**, 2846.
Fried, H.M. and Fink, G.R. (1978). *Proceedings of the National Academy of Sciences, USA* **75**, 4224.
Fu, X.D., Katz, R.A., Skalka, A.M. and Leis, J. (1988). *Journal of Biological Chemistry* **263**, 2140.
Fujimura, T. and Wickner, R.B. (1986). *Molecular and Cellular Biology* **6**, 404.
Fujimura, T. and Wickner, R.B. (1987). *Molecular and Cellular Biology* **7**, 420.
Fujimura, T. and Wickner, R.B. (1988a). *Journal of Biological Chemistry* **263**, 454.
Fujimura, T. and Wickner, R.B. (1988b). *Cell* **55**, 663.
Fujimura, T. and Wickner, R.B. (1989). *Journal of Biological Chemistry* **264**, 10872.
Fujimura, T., Esteban, R. and Wickner, R.B. (1986). *Proceedings of the National Academy of Sciences, USA* **83**, 4433.
Fujimura, T., Esteban, R., Esteban, L.M. and Wickner, R.B. (1990). *Cell* **62**, 819.
Fukuhara, H. (1987). *Nucleic Acids Research* **15**, 10046.
Fuller, R.S., Stearne, R.E. and Thorner, J. (1988). *Annual Reviews of Physiology* **50**, 345.
Fuller, R.S., Brake, A.J. and Thorner, J. (1989). *Science* **246**, 482.
Futcher, A.B. (1986). *Journal of Theoretical Biology* **119**, 197.
Futcher, A.B. (1988). *Yeast* **4**, 27.
Futcher, A.B. and Cox, B.S. (1983). *Journal of Bacteriology* **154**, 612.
Garvik, B. and Haber, J.E. (1978). *Journal of Bacteriology* **134**, 261.
Gasior, E., Herrera, F., Sadnik, I., McLaughlin, C.S. and Moldave, K. (1979). *Journal of Biological Chemistry* **254**, 3965.
Georgopoulos, D.E., Hannig, E.M. and Leibowitz, M.J. (1986). *In* "Extrachromosomal Elements in Lower Eukaryotes" (R.B. Wickner, A. Hinnebusch, A.M. Lambowitz, I.C. Gunsalus and A. Hollaender eds), pp. 203-213. Plenum Press, New York.
Goldring, E.S., Grossman, L.I., Krupnick, D., Cryer, D.R. and Marmur, J. (1970). *Journal of Molecular Biology* **52**, 323.
Gronostajski, R.M. and Sadowski, P.D. (1985). *Molecular and Cellular Biology* **5**, 3274.
Gubbins, E.J., Newlon, C.S., Kann, M.D. and Donelson, J.E. (1977). *Gene* **1**, 185.
Guérin, B. (1991). *In* "The Yeasts", vol. 4 (Rose and Harrison, eds). pp. 541-600. Academic Press, London.
Guerineau, M., Grandchamp, C. and Slonimski, P. (1976). *Proceedings of the National Academy of Sciences, USA* **73**, 3030.
Gunge, N. and Yamane, C. (1984). *Journal of Bacteriology* **159**, 533.
Gunge, N., Murakami, K., Takesato, T. and Moriyama, H. (1990). *Yeast* **6**, 417.
Gunge, N., Murata, K. and Sakaguchi, K. (1982). *Journal of Bacteriology* **151**, 462.
Gunge, N., Tamaru, A., Ozawa, F. and Sakaguchi, K. (1981). *Journal of Bacteriology* **151**, 462.
Haffar, O., Garrigues, J., Travis, B., Moran, P., Zarling, J. and Hu, S.-L. (1990). *Journal of Virology* **64**, 2653.
Hannig, E.M., Leibowitz, M.J. and Wickner, R.B. (1985). *Yeast* **1**, 57.
Harashima, S. and Hinnebusch (1986). *Molecular and Cellular Biology* **6**, 3990.
Harbison (1985). *Journal of General Virology* **66**, 2597.
Harford, M.N. and Peeters, M. (1987). *Current Genetics* **11**, 315.
Hartley, J.L. and Donelson, J.E. (1980). *Nature* **286**, 860.
Hayman, G.T. and Bolen, P.L. (1990). *Yeast* **6**, S548.

Herring, A.J. and Bevan, A.E. (1974). *Journal of General Virology* **22**, 387.
Hishinuma, F., Nakamura, K., Hirai, K., Nishizawa, R., Gunge, N. and Maeda, T. (1984). *Nucleic Acids Research* **12**, 7581.
Hollenberg, C.P., Degelmann, A., Kustermann-Kuhn, B. and Royer, H.D. (1976). *Proceedings of the National Academy of Sciences, USA* **73**, 2072.
Holm, C. (1982a). *Molecular and Cellular Biology* **2**, 985.
Holm, C. (1982b). *Cell* **29**, 585.
Hopper, J.E., Bostian, K.A., Rowe, L.B. and Tipper, D.J. (1977). *Journal of Biological Chemistry* **252**, 9010.
Icho, T. and Wickner, R.B. (1988). *Journal of Biological Chemistry* **263**, 1467.
Icho, T. and Wickner, R.B. (1989). *Journal of Biological Chemistry* **264**, 6716.
Jacks, T., Madhani, H.D., Masiarz, F.R. and Varmus, H.E. (1988). *Cell* **55**, 447.
Jayaram, M. (1985). *Proceedings of the National Academy of Sciences, USA* **82**, 5875.
Julius, D., Brake, A., Blair, L., Kunisawa, R. and Thorner, J. (1984). *Cell* **36**, 309.
Jung, G., Leavitt, M.C. and Ito, J. (1987). *Nucleic Acids Research* **15**, 9088.
Kadowaki, K. and Halvorson, H.O. (1971a). *Journal of Bacteriology* **105**, 826.
Kadowaki, K. and Halvorson, H.O. (1971b). *Journal of Bacteriology* **105**, 831.
Kamer, G. and Argos, P. (1984). *Nucleic Acids Research* **12**, 7269.
Kamper, J., Meinhardt, F., Gunge, N. and Esser, K. (1989). *Nucleic Acids Research* **17**, 1781.
Kikuchi, Y. (1983). *Cell* **35**, 487.
Kikuchi, Y. and Toh-e, A. (1986). *Molecular and Cellular Biology* **6**, 4053.
Kikuchi, Y., Hirai, K. and Hishinuma, F. (1984). *Nucleic Acids Research* **12**, 5685.
Koltin, Y. and Leibowitz, M.J. (1988). "Viruses of Fungi and Simple Eukaryotes". Marcel Dekker, Inc., New York.
Kunz, B.A. and Ball, A.J.S. (1977). *Molecular and General Genetics* **153**, 169.
Lacroute, F. (1971). *Journal of Bacteriology* **106**, 519.
Leibowitz, M.J. and Wickner, R.B. (1976). *Proceedings of the National Academy of Sciences, USA* **73**, 2061.
Leibowitz, M.J. and Wickner, R.B. (1978). *Molecular and General Genetics* **165**, 115.
Liebman, S.W. and All-Robyn, J.A. (1984). *Current Genetics* **8**, 567.
Liebman, S.W. and Cavanagh, M. (1960). *Genetics* **95**, 49.
Liebman, S.W. and Sherman, F. (1979). *Journal of Bacteriology* **139**, 1068.
Liebman, S.W., Stewart, J.W. and Sherman, F. (1975). *Journal of Molecular Biology* **94**, 595.
Ligon, J.M., Bolen, P.L., Hill, D.S., Bothast, R.J. and Kurtzman, C.P. (1989). *Plasmid* **21**, 185.
Liu, Y. and Dieckmann, C.L. (1989). *Molecular and Cellular Biology* **9**, 3323.
Livingston, D.M. (1977). *Genetics* **86**, 73.
Livingston, D.M. and Hahne, S. (1979). *Proceedings of the National Academy of Sciences, USA* **76**, 3727.
Livingston, D.M. and Klein, H.L. (1977). *Journal of Bacteriology* **129**, 472.
Livingston, D.M. and Kupfer, D.M. (1977). *Journal of Molecular Biology* **116**, 249.
Loh, Y.P., Parish, D.C. and Tuteja, R. (1985). *Journal of Biological Chemistry* **260**, 7194.
Lund, P.M. and Cox, B.S. (1981). *Genetical Research* **37**, 173.
Makower, M. and Bevan, E.A. (1963). *Proceedings of the XIth International Congress of Genetics* **1**, 202.
Matsumoto, Y. and Wickner, R.B. (1991). *Journal of Biological Chemistry* **266**, 12779.
Matsumoto, Y., Fishel, R. and Wickner, R.B. (1990). *Proceedings of the National Academy of Sciences, USA* **87**, 7628.
Mead, D.J., Gardner, D.C.J. and Oliver, S.G. (1986). *Molecular and General Genetics* **205**, 417.
Meyer-Leon, L., Gates, C.A., Attwood, J.M., Wood, E.A. and Cox, M.M. (1987). *Nucleic Acids Research* **16**, 6469.
Murray, A.W. and Szostak, J.W. (1983). *Cell* **34**, 961.
Murray, J.A.H. and Cesareni, G. (1986). *EMBO Journal* **5**, 3391.
Murray, J.A.H., Scarpa, M., Rossi, N. and Cesareni, G. (1987). *EMBO Journal* **6**, 4205.

Nelson, R.G. and Fangman, W.L. (1979). *Proceedings of the National Academy of Sciences, USA* **76**, 6515.

Newlon, C.S. (1989). In "The Yeasts," 2nd edn, (A.H. Rose and J.S. Harrison, eds) Vol. 3, pp. 57-116. Academic Press, London.

Newman, A.M., Elliott, S.G., McLaughlin, C.S., Sutherland, P.A. and Warner, R.C. (1981). *Journal of Virology* **38**, 263.

Oliver, S.G., McCready, S.J., Holm, C., Sutherland, P.A., McLaughlin, C.S. and Cox, B.S. (1977). *Journal of Bacteriology* **130**, 1303.

Ono, B.I., Stewart, J.W. and Sherman, F. (1979). *Journal of Molecular Biology* **128**, 81.

Owens, R.A. and Hammond, R.W. (1988). In "RNA Genetics" (E. Domingo, J.J. Holland and P. Ahlquist, eds) pp. 107-125. CRC Press, Boca Raton.

Palmer, E., Wilhelm, J. and Sherman, F. (1979a). *Journal of Molecular Biology* **128**, 107.

Palmer, E., Wilhelm, J. and Sherman, F. (1979b). *Nature* **277**, 148.

Paterson, R.G. and Lamb, R.A. (1990). *Journal of Virology* **64**, 4137.

Pelham, H.R.B. (1978). *Nature* **272**, 469.

Pelham, H.R.B. (1979). *Virology* **97**, 256.

Petes, T.D. and Williamson, D.H. (1975). *Cell* **4**, 249.

Pon, L. and Schatz, G. (1991). In "The Molecular Biology of the Yeast *Saccharomyces*: Genome Dynamics, Protein Synthesis and Energetics," (J.R. Breach, J.R. Pringle and E.W. Jones, eds), pp. 333-406. Cold Spring Harbor Laboratory, Cold Spring Harbor.

Prusiner, S.B. (1994). *Annual Review of Microbiology* **48**, 655.

Qian, X.-H., Inman, R.B. and Cox, M.M. (1990). *Journal of Biological Chemistry* **265**, 21779.

Rhee, S.-K., Icho, T. and Wickner, R.B. (1989). *Yeast* **5**, 149.

Ridley, S.P., Sommer, S.S. and Wickner, R.B. (1984). *Molecular and Cellular Biology* **41**, 761.

Rodriguez-Cousino, N., Esteban, L.M. and Esteban, R. (1991). *Journal of Biological Chemistry* **266**, 12772.

Roebroek, A.J., Schalken, J.A., Leunissen, J.A., Onnekink, C., Bloemers, H.P., Van de Ven, W.J. (1986). *EMBO Journal* **5**, 2197.

Rose, A.B. and Broach, J.R. (1990). *Methods in Enzymology* **185**, 234.

Sadowski, P.D. (1986). *Journal of Bacteriology* **165**, 341.

Schamhart, D.H.J., Ten Berge, A.M.A. and Van de Poll, K.W. (1975). *Journal of Bacteriology* **121**, 747.

Schwartz, C.J.E. and Sadowski, P.D. (1989). *Journal of Molecular Biology* **205**, 647.

Schwartz, C.J.E. and Sadowski, P.D. (1990). *Journal of Molecular Biology* **216**, 289.

Sclafani, R.A. and Fangman, W.L. (1984). *Molecular and Cellular Biology* **4**, 1618.

Seidah, N.G., Gaspar, L., Mion, P., Marcinkiewicz, M., Mbikay, M. and Chretien, M. (1990). *DNA and Cell Biology* **9**, 415.

Senecoff, J.F. and Cox, M.M. (1986). *Journal of Biological Chemistry* **261**, 7380.

Senecoff, J.F., Bruckner, R.C. and Cox, M.M. (1985). *Proceedings of the National Academy of Sciences, USA* **82**, 7270.

Sinclair, J.H., Stevens, B.J., Sanghavi, P. and Rabinowitz, M. (1967). *Science* **156**, 1234.

Singh, A.C., Helms, C. and Sherman, F. (1979). *Proceedings of the National Academy of Sciences, USA* **76**, 1952.

Slonimski, X. (1993).

Smeekens, S.P. and Steiner, D.F. (1990). *Journal of Biological Chemistry* **265**, 2997.

Somers, J. and Bevan, E.A. (1969). *Genetical Research* **13**, 71.

Somers, J.M. (1973). *Genetics* **74**, 571.

Sommer, S.S. and Wickner, R.B. (1982a). *Journal of Bacteriology* **150**, 545.

Sommer, S.S. and Wickner, R.B. (1982b). *Cell* **31**, 429.

Sommer, S.S. and Wickner, R.B. (1987). *Virology* **157**, 252.

Sor, F. and Fukuhara, H. (1985). *Current Genetics* **9**, 147.

Sor, F., Wesolowski, M. and Fukuhara, H. (1983). *Nucleic Acids Research* **11**, 5037.

Stam, J.C., Kwakman, J., Meijer, M. and Stuitje, A.R. (1986). *Nucleic Acids Research* **14**, 6871.

Stark, M.J.R. and Boyd, A. (1986). *EMBO Journal* **5**, 1995.
Stark, M.J., Boyd, A., Mileham, A.J. and Romanos, M.A. (1990). *Yeast* **6**, 1.
Stark, M.J.R., Mileham, A.J., Romanos, M.A. and Boyd, A. (1984). *Nucleic Acids Research* **12**, 6011.
Strauss, E.G., Rice, C.M. and Strauss, J.H. (1984). *Virology* **133**, 92.
Susek, R.E. and Lindquist, S.L. (1989). *Molecular and Cellular Biology* **9**, 5265.
Sweeney, R. and Zakian, V.A. (1989). *Genetics* **122**, 749.
Tanguy-Rougeau, C., Wesolowski-Louvel, M. and Fukuhara, H. (1988). *FEBS Letters* **234**, 464.
Tanguy-Rougeau, C., Chen, X.J., Wesolowski-Louvel, M. and Fukuhara, H. (1990). *Gene* **91**, 43.
TerAvanesyan, A., Dagkesamanskaya, A.R., Kushnirov, V.V. and Smirnov, V.N. (1994). *Genetics* **137**, 671.
Thiele, D.J., Hannig, E.M. and Leibowitz, M.J. (1984). *Molecular and Cellular Biology* **4**, 92.
Thomas, G., Thorne, B.A., Thomas, L., Allen, R.G., Hruby, D.E., Fuller, R. and Thorner, J. (1988). *Science* **241**, 226.
Thrash, C., Voelkel, K., DiNardo, S. and Sternglanz, R. (1984). *Journal of Biological Chemistry* **259**, 1375.
Tipper, D.J. and Bostian, K.A. (1984). *Microbiological Reviews* **48**, 125.
Toh-e, A. and Sahashi, Y. (1985). *Yeast* **1**, 159.
Toh-e, A., Araki, H., Utatsu, I. and Oshima, Y. (1984). *Journal of General Microbiology* **130**, 2527.
Toh-e, A., Tada, S. and Oshima, Y. (1982). *Journal of Bacteriology* **151**, 1380.
Toh-e, A. and Wickner, R.B. (1981). *Journal of Bacteriology* **145**, 1421.
Toh-e, A. and Wickner, R.B. (1980). *Proceedings of the National Academy of Sciences, USA* **77**, 527.
Toh-e, A. and Wickner, R.B. (1979). *Genetics* **91**, 673.
Toh-e, A., Guerry, P. and Wickner, R.B. (1978). *Journal of Bacteriology* **136**, 1002.
Tokunaga, M., Kawamura, A. and Hishinuma, F. (1988). *Nucleic Acids Research* **17**, 3435.
Tommasino, M. and Galeotti, C.L. (1988). *Nucleic Acids Research* **16**, 5863.
Tuite, M.F. and Plesset, J. (1986). *Yeast* **2**, 35.
Tuite, M.F., Cox, B.S. and McLaughlin, C.S. (1981a). *Journal of Biological Chemistry* **256**, 7298.
Tuite, M.F., Mundy, C.R. and Cox, B.S. (1981b). *Genetics* **98**, 691.
Tuite, M.F., Lund, P.M., Futcher, A.B., Dobson, M.J., Cox, B.S. and McLaughlin, C.S. (1982). *Plasmid* **8**, 103.
Tuite, M.F., Cox, B.S. and McLaughlin, C.S. (1983). *Proceedings of the National Academy of Sciences, USA* **80**, 2824.
Tuite, M.F., Cox, B.S. and McLaughlin, C.S. (1987). *FEBS Letters* **225**, 205.
Uemura, H. and Wickner, R.B. (1988). *Molecular and Cellular Biology* **8**, 938.
Utatsu, I., Sakamoto, S., Imura, T. and Toh-e, A. (1987). *Journal of Bacteriology* **169**, 5537.
Valle, R.P.C. and Morch, M.-D. (1988). *FEBS Letters* **235**, 1.
Vetter, D., Andrews, B.J., Roberts-Beatty, L. and Sadowski, P.D. (1983). *Proceedings of the National Academy of Sciences, USA* **80**, 7284.
Vincent, R.D., Hofmann, T.J. and Zassenhaus, H.P. (1988). *Nucleic Acids Research* **16**, 3297.
Vodkin, M., Katterman, F. and Fink, G.R. (1974). *Journal of Bacteriology* **117**, 681.
Volkert, F.C. and Broach, J.R. (1986). *Cell* **46**, 541.
Wagner, J.-C. and Wolf, D.H. (1987). *FEBS Letters* **221**, 423.
Wakem, L.P. and Sherman, F. (1990). *Genetics* **125**, 333.
Waldron, C., Cox, B.S., Wills, N., Gesteland, R.F., Piper, P.W., Colby, D. and Guthrie, C. (1981). *Nucleic Acids Research* **9**, 3077.
Wejksnora, P.J. and Haber, J.E. (1978). *Journal of Bacteriology* **134**, 246.
Wesolowski, M., Algeri, A., Goffrini, P. and Fukuhara, H. (1982a). *Current Genetics* **5**, 191.
Wesolowski, M., Algeri, A. and Fukuhara, H. (1982b). *Current Genetics* **5**, 205.
Wesolowski-Louvel, M., Tanguy-Rougeau, C. and Fukuhara, H. (1988). *Yeast* **4**, 71.

356   R.B. Wickner

Wesolowski, M. and Wickner, R.B. (1984). *Molecular and Cellular Biology* **4**, 181.
Wickner, R.B. (1974a). *Journal of Bacteriology* **117**, 1356.
Wickner, R.B. (1974b). *Genetics* **76**, 423.
Wickner, R.B. (1976). *Genetics* **82**, 429.
Wickner, R.B. (1977). *Genetics* **87**, 441.
Wickner, R.B. (1979). *Journal of Bacteriology* **140**, 154.
Wickner, R.B. (1980). *Cell* **21**, 217.
Wickner, R.B. (1987). *Journal of Bacteriology* **169**, 4941.
Wickner, R.B. (1988). *Proceedings of the National Academy of Sciences, USA* **85**, 6007.
Wickner, R.B. (1991). *In* "The Molecular and Cellular Biology of the Yeast *Saccharomyces*" (J. Broach, E. Jones and J. Pringle, eds), pp. 263–296. Cold Spring Harbor Press, Cold Spring Harbor, New York.
Wickner, R.B. (1994). *Science* **264**, 566.
Wickner, R.B. and Leibowitz, M.J. (1979). *Journal of Bacteriology* **140**, 154.
Wickner, R.B. and Toh-e, A. (1982). *Genetics* **100**, 159.
Wickner, R.B., Ridley, S.P., Fried, H.M. and Ball, S.G. (1982). *Proceedings of the National Academy of Sciences, USA* **79**, 4706.
Wickner, R.B., Icho, T., Fujimura, T. and Widner, W.R. (1991). *Journal of Virology* **65**, 155.
Widner, W.R., Matsumoto, Y. and Wickner, R.B. (1991). *Molecular and Cellular Biology* **11**, 2905.
Williams, T.L. and Leibowitz, M.J. (1987). *Virology* **158**, 231.
Wilson, D.W. and Meacock, P.A. (1988). *Nucleic Acids Research* **16**, 8097.
Wright, F. and Lederberg, J. (1957). *Proceedings of the National Academy of Sciences, USA* **43**, 919.
Wu, L.C., Fisher, P.A. and Broach, J.R. (1987). *Journal of Biological Chemistry* **262**, 883.
Young, C.S.H. and Cox, B.S. (1971). *Heredity* **26**, 413.
Young, C.S.H. and Cox, B.S. (1972). *Heredity* **28**, 189.
Young, T.W. (1987). *In* "The Yeasts" (A.H. Rose and J.S. Harrison, eds) Vol. 2, pp. 131–164. Academic Press, London.
Zakian, V.A., Brewer, B.J. and Fangman, W.L. (1979). *Cell* **17**, 923.
Ziegler, V., Richards, K., Guilley, H., Jonard, G. and Putz, C. (1985). *Journal of General Virology* **66**, 2079.

10 Double-strand Breaks and Recombinational Repair: The Role of Processing, Signalling and DNA Homology

Michael A. Resnick, Craig Bennett, Ed Perkins, Greg Porter and Scott D. Priebe

Laboratory of Molecular Genetics, National Institute of Environmental Health Sciences, Box 12233, Research Triangle Park, NC 27709 USA

The Yeasts Vol. 6, 2nd edition
ISBN 0-12-596416-1

I. THE RELEVANCE OF DOUBLE-STRAND BREAK RECOMBINATIONAL REPAIR: AN OVERVIEW

Systems that repair chromosomal damage protect cells from external and internal DNA-damaging events. These systems may also contribute to normal chromosomal functions throughout the life cycle of an organism (for a review see Bennett *et al.*, 1991). For *Saccharomyces cerevisiae*, several genes involved in repair are also essential for mitotic growth and meiotic development (Resnick, 1987). In light of the similarity between molecular structures generated during repair as compared with those occurring in recombination and replication, it is not surprising that repair systems are involved with normal cellular DNA metabolic process. Thus, just as membrane-signalling proteins are considered part of an integrated system required for growth and cell development, the various repair genes function in a network of processes that co-ordinate chromosomal metabolism with the cell cycle.

An important DNA lesion in *Sacch. cerevisiae* is the DNA double-strand break (DSB). It can arise naturally during the life cycle or in response to a DNA-damaging agent. Attention has been drawn to this lesion and its repair because of the biological relevance of DSBs and the intricate machinery required for DSB recombinational repair. Examples of the importance of DSBs in other organisms include V(D)J recombination in mouse thymocytes (Roth *et al.*, 1992), P-element transposition in *Drosophila melanogaster* (Engels *et al.*, 1990) and transposition in *Escherichia coli* (Bainton *et al.*, 1991; Haniford *et al.*, 1991). The processing and repair of DSBs brings together DNA-lesion recognition, cell signalling and mechanisms of recombination. The genes involved in DSB repair in *Sacch. cerevisiae* have important roles in chromosomal metabolism and many are essential for meiotic recombination.

The consequences of DSBs include loss of chromosomal integrity and blockage of DNA replication. The ends of DSBs can be considered as highly reactive in that they undergo recombinational repair or lead to gross chromosomal changes such as deletions, duplications, inversions and translocations. The capability for repair implies that, following DNA damage, a cell is able to detect the damage and to recruit repair systems that can restore continuity of chromosomal DNA. The recruitment may be part of a signalling system between the damage and induction of the appropriate repair enzymes. This signalling system may also interact with the components involved in cell cycling (see Section V). For the case of DSBs, there is an inhibition of cell-cycle progression thereby assuring adequate time for repair.

As will be discussed, a study of DSB recombinational repair has provided insight into the mechanisms of recombination (Sections II–IV) and associated cell signalling (Section V). This chapter will emphasize cell signalling associated with DSBs, the processing of breaks, the search for and the role of homology in DSB repair, the consequences of unrepaired breaks to cell progression and recombination between diverged DNAs. As discussed in Section IV, the latter type of recombination can lead to novel genetic information as well as chromosomal alterations.

Since the development of the DSB recombinational-repair model (Resnick, 1976), there have been several modifications that have depended, in part, upon the system being studied (summarized in Thaler and Stahl, 1988; Petes *et al.*, 1991). The common features (Fig. 1(a)) include induction of a break, 5′ to 3′ nucleolytic degradation of an end (see Section II, interaction between pre-existing homologous DNAs and replication across the break or gap. Subsequent events include features common to the Holliday model for recombination (Holliday, 1964). Repair can be accomplished through recombination between sister chromatids in G-2 haploid cells or homologous chromosomes in G-1 diploid cells (see Fig. 1; Resnick, 1975; Resnick and Martin, 1976; Luchnik *et al.*, 1977; Brunborg *et al.*, 1980). The absence of opportunities for recombinational repair in G-1 haploid cells results in sensitivity to ionizing radiation (Beam *et al.*, 1954).

Cellular recognition that a DSB has occurred and a search for homology must be included in any model of DSB recombinational repair (Fig. 1(b)). As discussed in Section IV and V, it is possible to develop model systems in yeast to address specifically the signalling effect of a DSB, the importance of homology and consequences of reduced or no homology.

There is also a non-conservative mode of recombination between repeat DNAs that is similar to that previously reported in mammalian cells and bacteria (see Section II, Fig. 2). While 5′ to 3′ exonucleolytic digestion has been proposed for both conservative and non-conservative recombinational repair of DSBs, the role of a 3′ single-strand tail in recombination remains to be established. Originally, it was proposed to function in both strand-invasion and initiation of replication to fill in information across a break (Resnick, 1976).

Recombinational repair has been shown for ionizing radiation-induced DSBs (Ho, 1975; Resnick and Martin, 1976; summarized in Frankenberg-Schwager and Frankenberg, 1991), double-strand cut plasmids (Orr-Weaver and Szostak, 1983), and DSBs occurring naturally during the life cycle by

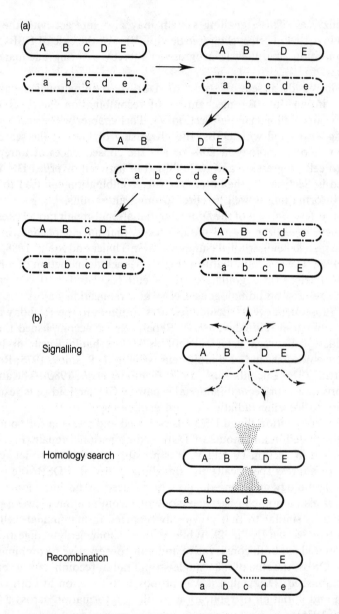

Fig. 1. A summary model of double-strand break (DSB) recombinational repair, the genetic consequences and proposed intermediate steps of repair. (a) A DSB or double-strand gap is produced in one of two copies of a chromosome (or other homologous DNA such as a sister chromatid or repeated DNA). A common feature of most models is 5′ to 3′ degradation of the broken ends by a deoxyribonuclease. As shown in Fig. 1(b), there is an interaction (strand invasion) between the two DNAs and a subsequent copying of information across the break

mating-type switching or during meiosis (see Section II). Gene targeting (Rothstein, 1983) may also be considered a special case of DSB recombinational repair. Although events induced by radiation damage are rare, it is possible to induce specific DSBs using mating-type locus sequences in combination with an inducible nuclease that cuts DNA specifically at the *MAT* YZ junction (discussed in Sections II and V). The opportunity to induce DSBs at precise sites has rendered this lesion particularly convenient for probing mechanisms of recombinational repair as well as biological consequences of DSBs and their repair.

Judging by the extent of DSB repair after low doses of ionizing radiation (summarized in Frankenberg-Schwager and Frankenberg, 1991), cells possess an efficient mechanism(s) for recognizing a rare disruption in chromosomal DNA, finding the homologous sequence, and repairing the lesion. This may not be surprising given the excellent capability for recombination between homologous sequences found in meiotic yeast cells, even when the sequences are on different chromosomes (ectopic recombination; reviewed in Petes *et al.*, 1991). While strand exchange and auxiliary proteins that facilitate interactions between homologous DNAs (see Section II) have been identified in yeast, *in vivo* accuracy and efficiency suggest an elaborate system that accomplishes a lesion-induced search and subsequent recombination interaction. For enzymically-generated lesions using the mating-type switching system, repair between homologous DNAs appears to be completely efficient for both intrachromosomal and interchromosomal events in mitotic and meiotic cells (discussed in Section II).

The only clear example of recombinational repair in yeast is for DSBs. Recombinational repair appears to deal with lesions induced by methylmethanesulphonate (Chlebowicz and Jachymczyk, 1979), bleomycin (Moore, 1978, 1989) and DNA cross-links induced by psoralens (Jachymczyk *et al.*, 1981; Magana-Schwencke *et al.*, 1982), which may be processed to DSBs. Post-replication repair, which accounts for damage-induced mutagenesis (see Section V), may also involve recombinational repair. The term post-replication repair refers to the capability of cells to deal with DNA lesions during and following replication. Frequently, it is considered in terms of cellular processing of pyrimidine dimers induced by ultraviolet radiation. This repair may include replication past a lesion, formation of

---

or gap. This newly synthesized DNA enables hybridization of the two fragments. The copying of information from the original intact chromosome (indicated by dashes) can lead to gene conversion. Subsequent recombinational steps that can lead to reciprocal exchange (Resnick, 1976; Szostak *et al.*, 1983) are not presented. (b) A DSB break can lead to a signal that can interact with cell-cycle processes (as described in Section V, p. 000). There must also be a search for homologous sequences, presumably mediated by strand-exchange proteins.

single-strand gaps in the vicinity of the lesion with subsequent filling-in of the gapped region and/or recombinational repair of the gapped region using the sister chromatid as occurs in *E. coli* (Ganesan, 1974). Although the ability to replicate past DNA lesions and genetic control of post-replication repair have been demonstrated in yeast, evidence for post-replication recombinational repair of single-strand damage is lacking for cells growing vegetatively (Prakash, 1981; Resnick *et al.*, 1981) or undergoing meiosis (DNA damage decreases meiotic recombination; Resnick *et al.*, 1983).

This chapter focuses on DSB recombinational repair systems because of their importance in maintaining genome stability and a relationship between DNA damage and cell-signalling processes that are revealed by studies of DSBs and repair. Relatively little emphasis has been placed on other repair systems in light of recent reviews (Moustacchi, 1987; Friedberg *et al.*, 1991). There are several in-depth reviews on various types of recombination in yeast including DSB-induced recombination (Game, 1983; Thaler and Stahl, 1988; Petes *et al.*, 1991). It should be noted that the various repair pathways have been divided into three epistasis groups (Cox and Game, 1974); these are excision repair (identified as the *RAD1* pathway), mutational repair (i.e. the *RAD6* pathway) and ionizing damage/DSB repair (i.e. the *RAD52* pathway).

II. RECOMBINATIONAL REPAIR OF DOUBLE-STRAND BREAKS

A. Genetics of DSB recombinational repair

Numerous genes have been identified in *Sacch. cerevisiae* that are responsible for repair of DNA damage. Many of these can be categorized into epistasis groups or genetic pathways of repair (Game, 1983). A more complicated picture emerges, however, upon examination of the roles of these genes in other processes such as mitotic and meiotic recombination, vegetative growth, meiotic development, genome stability and chromosome segregation. In many processes, a gene ascribed to one particular repair pathway has an overlapping or different function in another pathway (Game, 1983).

When a particular repair pathway is blocked by mutation of a necessary gene, potentially lethal DNA damage can often be shunted to another repair pathway. For example, a small proportion of ultraviolet damage is repaired by the error-prone *RAD6* pathway. Mutations of some genes in the *RAD6* epistasis group result in decreased ultraviolet-induced mutation and increased sensitivity to ultraviolet radiation (Cox and Parry, 1968). Apparently, lesions partially processed by the *RAD6* pathway can neither be eliminated nor repaired by other repair systems and are lethal. However, if the *SRS2/HPR5/RADH* gene is also inactivated, the increase in ultraviolet-sensitivity is eliminated and there is a corresponding rise in ultraviolet-induced recombination (Lawrence and Christensen, 1979a; Aboussekhra *et al.*, 1989; Schiestl *et al.*, 1990; Rong *et al.*, 1991). Lesions that cannot be repaired by excision repair or defective error-prone repair pathways can be corrected through the *RAD52* recombinational repair pathway. If the recombinational repair pathway is also inactivated, ultraviolet-sensitivity is again observed (Schiestl *et al.*, 1990). Interestingly, the *SRS2/HPR5* gene encodes a protein with DNA helicase activity (Aboussekhra *et al.*, 1989); its action apparently renders chromosomal DNA near an ultraviolet-induced lesion unsuitable for repair by recombination. Contrary to the foregoing, the only pathway available in yeast for repair of DSBs involves recombination. If the recombinational repair pathway is not available, repair of a DSB cannot be shunted into other pathways, and the unrepaired DSB is nearly always lethal. Consequently, assignment of genes to the DSB-repair pathway is straightforward; mutations that confer sensitivity to ionizing radiation and a defect in mating-type switching are likely to be involved in DSB repair.

Mutation of any of the genes in the *rad50* epistasis group (summarized in Haynes and Kunz, 1981; Game, 1983) confers sensitivity to ionizing radiation. The *rad52* mutants are deficient in many types of recombination (Game, 1983); these effects are described more extensively in Section III. Mutants *rad50*, *rad55* and *rad57* are defective in meiosis and exhibit reduced levels of damage-induced mitotic recombination. *RAD51* has recently been shown to encode a DNA strand-transfer protein with homology to the *recA* protein of *E. coli* (Shinohara *et al.*, 1992), which catalyses transfer of strands during bacterial recombination. Deletion of *RAD51* results in defects in meiotic recombination, in spore inviability and in processing of DSBs formed at hotspots for recombination during meiosis. Another gene, *DMC1*, has recently been identified and appears to be related to *RAD51* and *recA*, but is abundantly transcribed only during meiosis (Bishop *et al.*, 1992). Deletion of the gene results in a phenotype comparable to *RAD51*. A *dmc1* mutant is not sensitive to methylmethane-sulphonate, a compound that mimics ionizing radiation (Snow and Korch, 1970); the sensitivity to ionizing radiation has not been reported.

Two other recently identified genes, appear to affect DSB-induced recombination. A temperature-sensitive mutation in the *REC1* gene, *rec1-1*, results in sensitivity to ionizing radiation at non-permissive temperatures, and in a lack of mating-type switching (Esposito and Brown, 1990). The

REC1 product appears to regulate expression of strand-exchange proteins (M. Esposito, personal communication). In extracts prepared from *REC1/ REC1* diploids, the major strand-exchange activity is the 170 kDa product of the *KEM1/SEP1/DST2* locus (Kim *et al.*, 1990; Dykstra *et al.*, 1991; Johnson and Kolodner, 1991). At the permissive temperature, *rec1-1/rec1-1* diploids contain an immunologically unrelated protein of molecular weight 43 kDa. Extracts from *rec1-1/rec1-1* diploids grown at the non-permissive temperature appear to contain no strand-exchange activity, even though both proteins are active *in vitro* at the non-permissive temperature. Expression of strand-exchange proteins within the mutant at the permissive temperature may mimic meiosis or other cellular conditions for which special strand-exchange proteins are required. *REC1* may be identical to the essential gene *KEM3*, which affects nuclear fusion (Kim *et al.*, 1990).

The *hrr25-1* mutation confers sensitivity to ionizing radiation and methylmethanesulphonate, and is deficient in proper completion of meiosis (Hoekstra *et al.*, 1991). Null alleles of *HRR25* also exhibit a severe growth defect and aberrant cellular morphologies. The predicted protein product of the *HRR25* locus contains domains characteristic of serine–threonine protein kinases, with additional similarity to the Raf, PKS, *mos* subgroup. Protein kinases are critical components of cell-cycle control. The similarity of predicted *HRR25* gene product to protein kinases suggests that DNA-repair systems are interrelated with systems that control the cell cycle. The products of the genes already described along with proteins that have essential roles in DNA metabolism such as DNA ligase and DNA polymerases, are likely to comprise most of the participants in DSB repair.

B. Processing of broken ends and genetic control

Genetic and physical consequences of a DSB are best examined at a defined DSB. Double-strand breaks caused by external agents such as ionizing radiation are difficult to study, since they are randomly distributed and infrequent at biologically meaningful doses (Resnick and Martin, 1976). Studies of DSB metabolism in yeast have frequently employed components of the endogenous mating-type system for producing a DSB at a defined location.

During mating-type switching in *Sacch. cerevisiae*, the product of the *HO* (homothallic) gene, an endonuclease, catalyses formation of a DSB at a specific location in the mating-type locus (Strathern *et al.*, 1982; Kostriken *et al.*, 1982; Kostriken and Heffron, 1984). The *HO* endonuclease gene can be fused to an inducible promoter, allowing experimental control of its expression (Jensen and Herskowitz, 1984; Kostriken and Heffron, 1984). The YZ junction target site can be cleaved by the HO nuclease when introduced elsewhere in the genome. Induction of an HO-mediated DSB and subsequent events may be monitored at the DNA level. The DNAs adjacent to a cleaved YZ junction have been examined for nucleolytic degradation using strand-specific probes on native and denaturing Southern blots. Physical monitoring revealed that 3′ single-stranded DNA (ssDNA) tails are produced by 5′ to 3′ resection at HO-induced breaks located at the MAT locus, at other chromosomal locations, and in plasmids. Such processing has also been detected at DSBs formed during meiosis at hotspots for meiotic recombination.

Induction of a DSB between directly repeated homologous DNAs on a chromosome (Sugawara and Haber, 1992) or within one of two repeats on a plasmid (Fishman-Lobell *et al.*, 1992) leads to extensive degradation of each 5′ end, sometimes extending greater than 2 kb. Degradation is under the control of *RAD50* and *RAD52* genes. Single-strand tails are longer in *rad52* mutants and there is eventual degradation of single-strand tails; no recombinant products are formed. On the other hand, single-strand tail formation is slower in a *rad50* mutant, as is formation of recombination products. In a *rad52 rad50* double mutant, the rate of single-strand tail formation is similar to that in the *rad50* mutant; however, no recombinant products are formed (Sugawara and Haber, 1992). On the basis of these results, the *rad52* defect in recombination does not appear to relate to levels of degradation.

Processing of ends formed at a natural YZ junction, during mating-type switching appears different from that already described. A single-strand tail is created at the double-strand end distal to *MAT* by 5′ to 3′ degradation; however, there is no degradation at the proximal end (White and Haber, 1990). Further analysis suggests that the 3′ tail invades sequences at the donor (*HML*) locus and serves as a primer for DNA synthesis. Asymmetric degradation is controlled by the transcriptionally silent MAT information at *HML* and *HMR*, since the proximal end is also processed if both donor loci are removed. In *rad52* strains, the 3′ ssDNA tails are more extensive. Both *rad52* and haploid strains lacking the donor loci die in an attempt to switch, presumably due to the effects of unrepaired DSBs.

Double-strand breaks and associated nucleolytic processing also appear important in meiotic recombination. Transient DSBs have been detected in regions previously identified as hotspots for meiotic recombination (Sun *et al.*, 1989, 1991; Cao *et al.*, 1990). These breaks occur at specific sites irrespective of whether the sequence is in the chromosome or in a plasmid (Sun *et al.*, 1989), but the appearance of the breaks can be modified by changing the sequence environment of the hotspot, in some cases without changing the level of meiotic recombination. This suggests that these sites are preferred but not absolutely required for meiotic recombination (Wu and Lichten, 1992). Sun *et al.* (1991) demonstrated that 3′ ssDNA tails are formed at the DSBs that appear during meiosis.

Formation of meiotic DSBs requires the *SPO11* gene, which is required for early steps in meiosis, and the *RAD50* gene (Cao *et al.*, 1990). In cells harbouring the *rad50S* allele, DSBs appear during meiosis at a frequency greater than in *RAD*+ cells; unlike the situation in *RAD*+ cells, the DSBs are not processed (Cao *et al.*, 1990; Sun *et al.*, 1991). Double-strand breaks are not produced in a *rad50S spo11* double mutant. The ssDNA tails formed in strains deficient in either of the Rec-A-like gene products, Rad51 or Dmc1, are longer and, in the *dmc1* mutant, they are more persistent, similar to the effect of *rad52* on HO-induced DSBs. The Rad51 protein has been shown to bind to the Rad52 protein (Shinoharai *et al.*, 1992). This suggests that the *RAD51* and *RAD52* gene products may act as a complex or, in a concerted manner, catalyse strand transfer and completion of recombination. Thus 5′ to 3′ degradation seems to be a common pathway for processing of DSBs. In accordance with the original predictions for DSB repair (Resnick, 1976), the single-strand ends generated have the predicted 3′ polarity that would allow them to serve as invasive ends and primers for DNA synthesis.

C. The conservative recombination pathway

Except for mating-type switching, most repair of DSBs is likely to occur between sequences on sister chromatids or homologous chromosomes, and is expected to involve conservative recombination (Fig. 1; see Resnick, 1979). Some reciprocal exchange of outside markers is anticipated with recombination (Resnick, 1976; Szostak *et al.*, 1983). In one study of repair of X-ray-induced recombination, 12% of gene conversion events exhibited reciprocal exchange (Wildenberg, 1970). In meiosis, the frequency with which reciprocal exchange is associated with gene conversion varies depending on the interval monitored (0.18–0.66) (Fogel *et al.*, 1981). However, the contribution of DSBs to initiation of recombination remains to be established.

Double-strand breaks induced by HO have been used to examine recombination between homologous chromosomes. Depending on the efficiency of HO induction, 10–100% of the cells exhibit recombination, compared to less than 10^{-3} for spontaneous recombination (Ray *et al.*, 1988; J.N. Strathern, personal communication). J.N. Strathern (personal communication) has also examined reciprocal exchange associated with gene conversion of a mutation next to an HO-induced DSB. Reciprocal recombination could be ascribed to no less than 6% and no more than 13% of the gene conversions. Ray *et al.* (1988) observed 6% association of reciprocal exchange with intrachromosomal gene conversion induced by a DSB within one of two directly repeated sequences. The recombinogenic effects of DSBs appear to be transmissible over a long distance. Ray *et al.* (1989) demonstrated that an induced DSB can stimulate recombination between a locus 8.6 kb from the DSB and an unlinked homologue.

Systems involving inverted repeat sequences are promising model systems for studying DSB-induced repair and associated reciprocal exchange. There does not appear to be a competing non-conservative pathway such as exists for direct repeats, as discussed in the next section. Physical monitoring of recombination induced by a DSB within one of two inverted repeats on a plasmid has demonstrated the presence of the expected exchange and non-exchange products (Rudin *et al.*, 1989). Surprisingly, one of the two products diagnostic of reciprocal exchange appears earlier than the other. Exchange accompanies repair in approximately 50% of the surviving plasmids.

D. A non-conservative recombination pathway

Studies involving DSB-induced recombination between direct repeats has unexpectedly revealed that most putative gene conversions with associated reciprocal exchange are not truly reciprocal. Instead they are the result of a non-conservative deletion of the sequences between the repeats. For the case of a break within a repeat, gene conversion without reciprocal exchange results in replacement of the YZ sequences and surrounding region with information from the other repeat; the final product has two complete repeats. Reciprocal recombination is predicted to lead to a single repeat in the chromosome and an excised circle comprised of one repeat and the sequences originally between the repeats. However, in many studies only the former of the expected reciprocal products is observed; the frequency of these simple deletion events often exceeds that of gene conversions (Nickoloff *et al.*, 1989; Rudin *et al.*, 1989; Ozenberger and Roeder, 1991; Fishman-Lobell *et al.*, 1992). Similar results are obtained if the DSB is induced in the non-homologous DNA between direct repeats (Rudin and Haber, 1988; Nickoloff *et al.*, 1989; Sugawara and Haber, 1992). In one study, however, all deletion events were accompanied by formation of a small circle and thus appeared to be truly reciprocal (Ray *et al.*, 1988). In this study, gene conversions were in great excess over reciprocal events. The high frequency of deletions appears to be due to a non-conservative recombination pathway that acts specifically to repair a DSB within or between nearby direct repeats. It has been proposed that deletion products in yeast can be produced from the 3′ tails described above by annealing of complementary sequences followed by DNA replication and ligation to fill gaps (Fig. 2) (Ozenberger and Roeder, 1991; Fishman-Lobell et al., 1992). This model for non-conservative repair previously has been proposed for short tracts of homology in E. coli (Conley et al., 1986) and for mammalian cells (Lin et al., 1984, 1990; Maryon and Carroll, 1991). The distinction between this non-conservative pathway and gene conversion, which is also characterized by asymmetric transfer of information, is that a single-strand annealing pathway involves loss of an entire repeat, as well as any spacer DNA between repeats, while the conservative pathway leading to gene conversion involves only a localized loss of information within one repeat.

It appears that a DSB can be processed by either the conservative recombination pathway or the non-conservative annealing pathway, and the likelihood of repair by the former increases with distance between repeats. Gene-conversion products in a plasmid system increase at the expense of deletion events as the amount of intervening DNA between the repeats is increased from zero up to 4.4 kb. The time at which deletion products first appear increases with increasing size of the intervening DNA, whereas the first appearance of gene-conversion products is not delayed by additional spacer DNA (Fishman-Lobell et al., 1992). The delay seen for deletions is consistent with the rate of nucleolytic degradation observed with other constructs (White and Haber, 1990; Sugawara and Haber, 1992).

Most, but not all, of the presumed deletion events require RAD52 function. In a rad52 mutant, deletion formation resulting from repair of a DSB in a direct repeat on a plasmid was 13% of the wild-type level (Fishman-Lobell et al., 1992). Efficient RAD52-independent repair was identified in ribosomal repeat DNA, or in a CUP1 tandem gene array (18 repeats) (Ozenberger and Roeder, 1991). When CUP1 was present as three tandem genes, RAD52-independent repair was inefficient (80% cell inviability). Similarly, repair of plasmid-borne lacZ repeats was largely, but not completely, abolished in a rad52 background, and gene-conversion events were completely absent (Fishman-Lobell et al., 1992).

These results may be explained if RAD52 is required for timely catalysis of strand-transfer or annealing prior to a slow degradation of the 3′ ssDNA tail. A RAD52 would be necessary for two repeats but, in larger arrays, each repeat would represent a separate opportunity for uncatalysed annealing and repair. Such a view is supported by physical data. In rad52 mutants, single-strand tail formation is more extensive before eventually

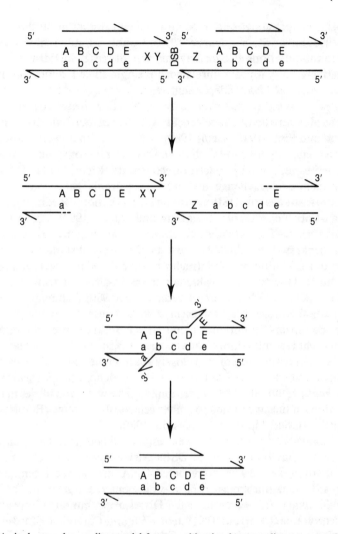

Fig. 2. A single-strand annealing model for recombination between direct repeats (Sugawara and Haber, 1992). The thick arrows above the top panel indicate each repeat. A–E indicate sequences within the repeat; lower case letters correspond to the complementary sequences. X, Y, and Z indicate sequences between the direct repeats. Nucleolytic processing 5′ to 3′ into the repeats eventually exposes complementary single strands (second panel). These can subsequently anneal (third panel); the annealing could be facilitated by proteins associated with the *RAD52* pathway of recombination (see the text for an explanation). Resolution of the annealed products leaves one copy of the repeat while the X, Y and Z sequences are lost (Panel 4).

disappearing, presumably due to low-level degradation (Fishman-Lobell *et al.*, 1992; Sugawara and Haber, 1992). Double-strand breaks in cells with *RAD51* null mutations also display this phenotype (Shinohara *et al.*, 1992). Alternatively, a pathway unrelated to single-strand annealing may be responsible for *RAD52*-independent repair.

Studies of spontaneous mitotic recombination between direct repeats have revealed a paucity of reciprocal exchange associated with conversion (Jackson and Fink, 1981; Klein, 1984). This led to the suggestion that there are topological constraints that inhibit intrachromosomal reciprocal exchange between closely spaced direct repeats (Klein, 1984). The absence of any reciprocal exchange attending measurable gene conversion in a number of studies of DSB-induced recombination (Rudin *et al.*, 1989; Fishman-Lobell *et al.*, 1992; Sugawara and Haber, 1992) is consistent with this hypothesis. The observation of reciprocal exchange between *ADE4* repeats separated by 5.43 kb suggests that this distance is above the threshold for inhibition of intrachromosomal reciprocal exchange (Ray *et al.*, 1988). However, this value is not vastly different from the distance between repeats in other studies (Rudin *et al.*, 1989; Fishman-Lobell *et al.*, 1992) and, therefore, may represent a special case.

Inverted repeats do not appear to suffer from the same proximity effect on reciprocal recombination that appears to exist for direct repeats. Induction of a DSB within one of two closely spaced inverted *lacZ* repeats on the chromosome leads to a low but measurable incidence of reciprocal exchange (Rudin *et al.*, 1989). Similarly, recombination between *his3* alleles in inverted orientation on plasmids leads to some reciprocal exchange (Embretson and Livingston, 1984; Ahn and Livingston, 1986).

The biological importance of the presumed single-strand annealing pathway in yeast may be in repair of DSBs arising within the ribosomal repeat array, as there are few other repeated DNAs in the yeast genome. Non-conservative deletion seems to be a prominent mechanism of DSB repair in higher eukaryotes, where repeated DNAs are common (Lin *et al.*, 1984, 1990; Maryon and Carroll, 1991; Jeong-Yu and Carroll, 1992). In light of the strong dependence on *RAD52*, single-strand annealing may not be a separate pathway, but instead it may be an alternative resolution of the conservative recombination pathway in situations where formation of the second reciprocal product is not feasible. Possibly there is a competition between the annealing reaction by single-strand tails and the presumed strand-invasion step of conservative recombination.

III. THE *RAD52* GENE: A KEY PLAYER IN RECOMBINATION, REPAIR AND POSSIBLY OTHER PROCESSES

The *RAD52* gene is required for DSB repair and appears to function in other cellular and genetic processes. A *rad52* mutant was originally identified on the basis of ionizing radiation sensitivity (Resnick, 1969). Subsequently, it was shown that the sensitivity was due to an inability to carry out DSB repair (Ho, 1975; Resnick and Martin, 1976). As previously stated, *RAD52* and other members of this epistasis group are responsible for DSB repair. Furthermore, *RAD52* is a necessary component for completion of mating type interconversion (Malone and Esposito, 1980; Weiffenbach and Haber, 1981; Raveh *et al.*, 1989) and for successful completion of meiosis (Resnick, 1987), both of which require recombinational repair. During meiotic recombination, the *RAD52* gene appears to function in an intermediate step (Resnick *et al.*, 1981, 1984), whereas, during mating-type interconversion, it appears to function at an early step in processing of DSBs at the *MATYZ* junction (see Section II for further discussion) (Connolly *et al.*, 1988; White and Haber, 1990). The *RAD52* gene may also affect mitotic chromosome segregation (Mortimer *et al.*, 1981); its role, however, is not clear since different mutations (point mutations as distinct from deletions) and strain backgrounds appear to have different effects on growth and chromosome loss (E. Perkins and M.A. Resnick, unpublished observation; V.L. Larionov, unpublished observation). The effects of point mutations in *RAD52* must be reconsidered in light of the finding that *RAD52* and *RAD51* appear to interact in a protein–protein complex (Shinohara *et al.*, 1992). In addition to the events already discussed *RAD52* also appears to influence spontaneous mitotic recombination (discussed below).

The *RAD52* gene has been emphasized because of its role in many chromosomal metabolic processes, particularly DSB repair. Historically, identification of a *RAD52*-dependent genetic process was taken as *prima facie* evidence that a DSB intermediate was involved. However, while *RAD52* is required for DSB repair, an event that requires *RAD52* may not necessarily indicate involvement of DSBs (see below and Fig. 3).

A. Interaction of *RAD52* and DNA replication systems

An initial indication that *RAD52* may be functioning in cellular processes other than DSB repair comes from analysis of the enhanced mutator effect observed in some *rad52* mutants (Kunz *et al.*, 1989). Spontaneous *SUP4-o*

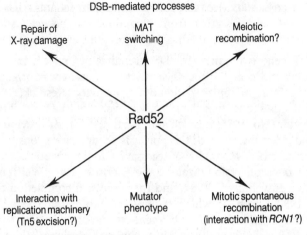

Fig. 3. Roles of *RAD52* in DNA metabolic processes. Some are initiated by DSBs; others may not involve DSBs.

mutations were analysed in strains carrying a disrupted *RAD52* gene, and the mutational spectrum was compared with that from a wild-type strain. The increase in mutator activity of a *rad52* mutation is associated with an increase in frequency of base-pair substitutions. While all types of substitutions were observed, there was an increase in G–C to C–G transversions over that of A–T to G–C transitions. Interestingly, Ty element insertions and multiple base-pair deletions were only detected in *RAD*⁺ strains. The authors suggested that the increased mutator activity may be due to altered replication fidelity or channelling of spontaneous lesions produced during mitotic growth through other mutagenic repair pathways.

Further evidence that *RAD52* interacts with the DNA-replication machinery comes from the study of bacterial Tn5 excision from the yeast *LYS2* gene (Gordenin *et al.*, 1988). This novel system provides a model for examining the manner in which small repeat sequences may influence genome stability, since repeated sequences can serve as a substrate for genomic re-arrangements such as duplications and deletions (Berg, 1989). Precise excision of Tn5 in bacteria occurs between the 9 bp direct repeats of the transposon target sequence and this event appears to be stimulated by 1.5 kb inverted repeats between the short direct repeats. It has been proposed that these deletions are a result of illegitimate recombination between the short repeats, possibly due to slippage during replication (Egner and

Berg, 1981; Berg, 1989; Kleckner, 1989). A similar mechanism has been suggested for deletions resulting from recombination between short direct repeats (Efstratiadis *et al.*, 1980). Therefore, Tn5 excision in yeast might prove useful for investigating various DNA metabolic events such as repair, replication and recombination.

Initially, the transposon Tn5 was introduced into the yeast *LYS2* gene (generating lysine auxotrophs), and a low excision rate was observed (10^{-9}) (Gordenin *et al.*, 1988). Mutants were isolated in which the excision rate was elevated 100-fold. One mutation isolated, *tex1*, was found to be allelic to *POL3*, the structural gene for DNA polymerase δ (Gordenin *et al.*, 1992). Mutations in *POL1* (DNA polymerase α) but not *POL2* (DNA polymerase ε) also led to increased Tn5 excision. In fact, *pol2* mutations appear to block the increased excision observed for *tex1* strains (Gordenin *et al.*, 1992; D. Godenin, E. Perkins and M.A. Resnick, unpublished observation). These results indicate that some components involved in DNA replication play a role in the excision process.

In addition to various DNA polymerase mutations, other mutations representing the three epistasis groups involved in DNA repair (see Sections I and II) were also examined for their effects on Tn5 excision. Only the *rad50* and *rad52* mutations (other genes in this epistasis group were not examined) decreased the high excision levels observed in *pol3* strains (Gordenin *et al.*, 1992). Furthermore, *rad52 pol3* double mutants have a reduced semi-permissive temperature for growth and enhanced lethality at non-permissive temperatures (in contrast to *pol3* alone) suggesting an incompatible interaction between the two mutations (Hartwell and Smith, 1985; Gordenin *et al.*, 1992). It has been suggested that the *pol3 rad52* defect is due to a lack of recombinational repair (Hartwell and Smith, 1985). However, it is possible that both *RAD52* and *RAD50* carry out some other DNA metabolic function (i.e. a non-DSB repair function) in relation to DNA replication.

As proposed by Morrison *et al.* (1990) the function of DNA polymerase α is to synthesize RNA primers and possibly short DNA sequences at the replication origins on the leading strand, and at the beginning of Okazaki fragments on the lagging strand. Polymerase δ is proposed to finish synthesis on the lagging strand while polymerase ε performs leading strand synthesis. The rates of leading and lagging strand synthesis during vegetative growth would be tightly co-ordinated such that both rates are equal. Gordenin *et al.* (1992) suggested that in *pol3* strains the leading strand synthesis may outpace that of the lagging strand, such that the lagging strand will contain longer-lived single-stranded regions. This could allow basepairing between Tn5 inverted repeats, generating a hairpin secondary structure on the lagging strand. To by-pass this hairpin and finish replication on the lagging strand, replicative slippage may be necessary. The genes *RAD50* and *RAD52* could play a role(s) in stabilizing pairing of inverted repeats, thereby facilitating by-pass of the hairpin or in transferring the 3' stalled end at the base of the hairpin to the next repeat thus allowing continued replication. It is interesting that the *RAD50* and *RAD52* genes may also have a role in interaction between complementary single strands in the non-conservative recombination pathways (see Section II).

Though the presence of a DSB during Tn5 excision cannot be ruled out, it is unlikely that DSBs are responsible for the *pol3 rad52* incompatibility (E. Perkins, D.A. Gordenin and M.A. Resnick, unpublished observation). This has been examined in a strain containing a circular chromosome III. During gel electrophoresis to separate whole chromosomes (OFAGE analysis), the unbroken circular chromosome does not migrate into the gel; if it contains a DSB it can then enter the gel (Game *et al.*, 1989). When a *pol3 rad52* double mutant is grown at non-permissive temperatures, the cells arrest and eventually die; however, DSBs in the circular chromosome III are not detectable (no migration into the gel) (E. Perkins, D.A. Gordenin and M.A. Resnick, unpublished observation). As a control, linear chromosome III can be detected when cells (*pol3 rad52*) are exposed to levels of ionizing radiation that yield less than one DSB in circular chromosome III in each cell. The role of *RAD52* in mutagenesis, Tn5 excision and the incompatibility of *rad52* with *pol3* suggests that it and other genes in this epistasis group may be involved with replication machinery as well as DSB recombinational repair.

B. Spontaneous mitotic recombination: roles for *RAD52* and *RNC1*

Many of the genes responsible for DNA damage repair and/or meiotic development also appear to influence spontaneous and damage-induced mitotic recombination (Petes *et al.*, 1991; Roeder and Stewart, 1988; Orr-Weaver and Szostak, 1985; Esposito and Klapholz, 1981). Rates of spontaneous mitotic recombination are typically 10^{-5} or less for each cell generation (orders of magnitude less than during meiosis).

When investigating the role of *RAD52* in DSB repair events (e.g. repair of X-ray-induced damage, HO catalysed DSBs, plasmid gap repair), the Rad52 protein was clearly established as a necessary component of these repair processes. However, analyses of the role *RAD52* plays in spontaneous mitotic recombination events has yielded a myriad of results. Spontaneous recombination is often examined with haploid strains disomic for a particular chromosome bearing heteroalleles or with a strain that has naturally occurring or artificially constructed repeats. Initially, Jackson and Fink (1981) examined the role of *RAD52* in recombination between repeats flanking plasmid sequences integrated at the *HIS4* locus. While the wild-type and *rad52* displayed similar levels of reciprocal recombination (as measured by plasmid pop-out), gene conversion was drastically decreased in *rad52* strains. Recombination leading to loss of a *SUP4-o* allele integrated into a cluster of repeated sequences was decreased in *rad52* mutants (Rothstein *et al.*, 1987). For integration of plasmids following transformation, Orr-Weaver *et al.* (1981) found no effect of *rad52* on integration of a plasmid containing a *rRNA* gene, whereas Malone *et al.* (1988) demonstrated a decrease in integration for a plasmid containing a gene that was singly represented in the genome. In a screen for mutants exhibiting elevated mitotic recombination between tandem heteroallelic genes, Aguilera and Klein (1988) identified mutations in eight different genes exhibiting a hyper-rec phenotype. Among these were mutations in *POL1* and *POL3* as well as DNA topo-isomerase genes. Furthermore, all the mutations were dependent on *RAD52* for their hyper-rec phenotype (Aguilera and Klein, 1988).

While the influence of *RAD52* on mitotic recombination varies, recombination can still occur when it is inactivated suggesting that an alternative pathway (*RAD52*-independent) exists for mitotic recombination (Haber and Hearn, 1985; Hoekstra *et al.*, 1986). The alternative may be an excision-repair pathway, which includes the *RAD1* gene. Mutations in *rad1* have been shown to affect mitotic recombination stimulated by *HOT1*, the Pol1 rDNA promoter (Keil and Roeder, 1984; Roeder *et al.*, 1986). Furthermore, *RAD1* plays a role in direct-repeat recombination and homologous integration of linear plasmids (Schiestl and Prakash, 1988). Thomas and Rothstein (1989a) observed synergism between *rad1* and *rad52* mutations with respect to recombination (i.e. the rate of recombination in the double mutant was much less than anticipated based on decreased recombination in the single mutants), but not with respect to survival after exposure to ultraviolet or gamma rays (Thomas and Rothstein, 1989b). Interestingly, recombination still occurs in the double mutant, although drastically decreased, indicating that there is another pathway for mitotic recombination. This proposed pathway appears to be influenced by transcription.

An insight into the role that *RAD52* plays in mitotic recombination comes from examination of a DNAse whose presence appears to be influenced by the *RAD52* gene (Chow and Resnick, 1987, 1988). Characterization of yeast nucleases will be important for elucidating mechanisms of recombination, since recombination and repair require nuclease activities either for processing of ends or for resolution of recombination/repair intermediates. Previously, a nuclease activity for *RAD52* had been reported based on its ability to complement T4 phage genes *46* and *47* (Chen and Bernstein, 1988). However, these findings must be seriously questioned, since repetition of the results has proved unsuccessful (E. Perkins, J. Drake and M.A. Resnick, unpublished results) and the original strains and plasmids are not available (D. Chen, personal communication). Utilizing an antibody raised against an endonuclease/exonuclease from *Neurospora crassa*, Chow and Resnick (1987, 1988) isolated a Mg^{2+}-dependent 72 kDa DNAse with single-stranded endonuclease/exonuclease and double-strand exonuclease activity. The levels of antibody cross-reactivity were dependent upon a functional *RAD52*. The protein has 5′ to 3′ double-strand exonuclease activity (T.-Y.K. Chow, personal communication), suggesting it may function in processing of DSB ends (see Section II). In *rad52* strains, the level of the nuclease is 10% of that in wild-type strains, and there is no meiotic induction unlike in wild-type strains.

A gene referred to as *RNC1* was subsequently cloned using the *N. crassa* antibody described above to identify sequences that when expressed in a λ library led to protein capable of cross-reacting with the polyclonal antibody (Chow *et al.*, 1992). The gene was mapped to chromosome XI, and subsequently the precise location was determined as part of the yeast genome sequencing effort (van Vliet-Reedijk and Planta, 1993) (see reference in Note Added in Proof). An apparant sequencing error by Chow (personal communication) led to the conclusion that *RNC1* coded for a chimeric protein that included a G-protein rather than two genes as was subsequently determined. The G-protein was found to be *RHO4* (Matsui and Toh-e, 1992). (see reference in Note Added in Proof) and was determined to be independent from the closely linked gene, *NUD1* (van Vliet-Reedijk and Planta, 1993; Lewis, Perkins and Resnick, unpublished). The function of *NUD1*, the putative gene coding for the *N. crassa* identified nuclease, is currently under investigation (Perkins and Resnick, unpublished).

Nucleases that may be similar to *RNC1* are beginning to be identified in yeast as well as other species. Using the antibody derived from *N. crassa*, a mammalian nuclease has been identified and purified to homogeneity (Couture and Chow, 1992; see note added in proof) and a similar sized cross-reacting protein (approximately 70 kDa) has been identified in *Schizosaccharomyces pombe* (E. Perkins and M.A. Resnick, unpublished observation).

IV. DNA HOMOLOGY, DIVERGENCE AND RECOMBINATIONAL REPAIR

Homology between interacting DNAs is implicit in recombination. During recombinational repair, there must be a searching process to locate DNA with which to interact following induction of a lesion (see Fig. 1(b)). This searching process must be highly efficient in yeast, given the extent of DSB repair (Resnick and Martin, 1976; summarized in Frankenberg-Schwager and Frankenberg, 1991).

The requirements for DNA homology in recombinational repair, as well as the genetic consequences of decreased homology, can be addressed uniquely in yeast and several approaches have been taken. Recombination is decreased if the DNA available for homologous interaction is short. Estimates for the minimum length of DNA required for homologous recombination, either spontaneous or associated with DSB recombinational repair, vary from 50 to 100 bp (Ahn *et al.*, 1988; Sugawara and Haber, 1992) and are comparable to those reported for *E. coli* and mammalian cells (summarized in Sugawara and Haber, 1992). For yeast transformation, rare illegitimate integrants can be obtained when cells are exposed to linear DNA in which there are only four basepairs of chromosomal homology at the ends (Schiestl and Petes, 1991) or to oligonucleotides (with less than 50 bases) (Moerchell *et al.*, 1988). The consequences of removing homology, thereby preventing recombinational repair in yeast, can have profound effects on cell development and viability (see Section V).

Decreasing the level of homology between interacting DNAs may affect recombination. Mismatches in heteroduplex DNA could influence the initial steps in recombination or subsequent processing of DNAs so as to lower the extent of recombination. This has been demonstrated for meiotic recombination between diverged chromosomes in that DNA divergence greatly decreases recombination (see below). As discussed later, DNA divergence can have major effects on recombinational repair and its genetic outcomes, both for recombination between chromosomes and during transformation. In the following section, the consequences of there being no opportunity for recombinational repair of DSBs are discussed.

A. DNA divergence decreases DSB recombinational repair

1. Chromosomal interactions

Since many organisms contain related repeat DNA sequences, DNA divergence may be a factor in recombinational interactions. In mammalian species, repeat DNAs exhibiting up to 20% divergence can account for nearly 30% of the genome (discussed further in Section VI.A). In *Sacch. cerevisiae*, there are several examples of related genes that arose by duplication and subsequent divergence from a common progenitor. Rare spontaneous recombinants are detected between pairs of diverged genes *SAM1* and *SAM2* (Bailis and Rothstein, 1990) and also the *CYC1* and *CYC7* pair (Ernst *et al.*, 1981). Recombination is ectopic, since the genes are located on different chromosomes. Such recombination between these genes accounts for a unique category of mutation reversions (Ernst *et al.*, 1981). Naturally occurring chromosomes, which appear to be diverged (homoeologous) over their entire length, are present within *Sacch. carlsbergensis* (Nilsson-Tillgren *et al.*, 1981; Holmberg, 1982). The biological consequence of recombination between diverged DNAs may be significant, since it could be a source of translocations and novel chimeric genes.

Recently, the homology requirement for recombinational repair between chromosomes in *Sacch. cerevisiae* was examined using homoeologous chromosomes from *Sacch. carlsbergensis*. Several chromosomes in *Sacch. carlsbergensis* are functionally comparable to those on *Sacch. cerevisiae* even though they exhibit considerable DNA divergence (Nilsson-Tillgren *et al.*, 1981; Holmberg, 1982). For chromosomes III and V, the level of divergence for coding sequences is approximately 10-20% and is even higher in non-coding regions. Since pairs of homoeologous chromosomes in *Sacch. cerevisiae* exhibit almost no meiotic recombination (Nilsson-Tillgren *et al.*, 1981, 1986; Holmberg, 1982), decreased homology clearly influences some aspect(s) of recombination. This is a well-established observation for many genetic systems. For example, divergence prevents recombination between chromosomes of related bacteria; however, mutations in mismatch repair genes lower recombinational incompatibility (see Section IV.B; Rayssiguier *et al.*, 1989).

The role of homology in recombinational repair has been addressed by examining the response of individual homoeologous *Sacch. carlsbergensis/Sacch cerevisiae* chromosome pairs to low doses of ionizing radiation where the fate of each chromosome could be followed genetically. Since the other chromosome pairs were homologous, DSBs in the rest of the genome would be repaired by homologous interactions.

Following exposure of G-1 diploid cells to non-lethal doses of ionizing radiation, there was considerable chromosome loss among homoeologous pairs of chromosomes (III and V were examined). The high level of aneuploidy (in up to 10% of the cells) was concluded to be due to a lack of opportunity for recombinational repair, since there was little radiation-induced aneuploidy in cells containing only homologous chromosomes. Considering that recombination and recombinational repair must rely to some extent on homology, it was not surprising that ionizing radiation-induced repair between divergent chromosomes was low compared to homologous chromosomes (Resnick *et al.*, 1989). Even so, the extent of the decrease was much less than would be predicted from the effects of sequence divergence on meiotic recombination. Lack of recombination could also be due to effects of divergence on chromosome pairing prior to recombination.

The decreased homology, which prevents meiotic recombination, does not appear to prevent DSB repair completely. For example, the efficiency of chromosome loss for each estimated DSB induced in either of the homoeologous chromosomes is approximately 0.4, suggesting that DSB repair could still occur (Resnick *et al.*, 1989). This was confirmed by demonstrating induction by ionizing radiation of recombination between *HIS4* or *ILV1* heteroalleles on homoeologous chromosomes (Resnick *et al.*, 1992). Among the recombinants, there was no chromosome loss as would be expected, if recombinational repair were complete. Thus, it appears that the requirement for homology in meiotic recombination may be greater than for recombinational repair in mitotic cells. The observation of recombinational repair between diverged sequences suggests that this pathway could be a novel source of mutations or altered genes.

2. Plasmid/chromosome interactions and targeting

Spontaneous or damage-induced recombination is difficult to study because it occurs randomly at low frequency. Experimental systems that rely on plasmid transformation circumvent this problem and have been used to study recombination induced by DSBs (Orr-Weaver and Szostak, 1983; Petes *et al.*, 1991). Plasmids capable or incapable of replication in yeast are used to examine non-reciprocal or reciprocal recombination, respectively. These plasmids contain yeast chromosomal DNA and are cleaved within the yeast DNA prior to transformation. Recombination efficiency is inferred from transformation frequency. Recombined plasmid or chromosomal DNA from individual transformants can then be analysed to determine the nature of the recombinational event. Replacement of plasmid-borne sequences with chromosomal information is formally equivalent to gene conversion.

Plasmid transformation has been used to study recombination induced by defined DSBs (Priebe *et al.*, 1994; see note added in proof). Haploid strains of *Sacch. cerevisiae*, which carry either their normal chromosome III or a substitute, diverged chromosome III from *Sacch. carlsbergensis* as already described, have been transformed with plasmids that carry the

HIS4 region from chromosome III of *Sacch. cerevisiae*. Replicating or non-replicating plasmids were cleaved within *HIS4* to yield a double-strand break or gap. Both types of plasmids transformed the recipients with the diverged chromosome 20-fold less frequently than the recipients with the homologous chromosome. Therefore, homoeologous DNAs can undergo DSB-induced reciprocal and non-reciprocal recombination, although at lower frequencies than homologous DNA.

Analysis of replicating plasmids following transformation into strains with the homoeologous chromosome gave interesting results. A small gap introduced into the plasmid-borne *HIS4* was filled with DNA homologous to the chromosomal *HIS4*. In half of the transformants, no chromosomal sequences beyond the original gap were detected in the recombinant plasmids. In the remaining transformants, the recombinant plasmids carried additional chromosomal information (over a kilobase in some transformants) but extending from only one side of the original gap. Of the models for DSB-induced recombination discussed by Thaler and Stahl (1988), these results support a model in which one end of a DSB (or gap) interacts with homologous DNA (Resnick, 1976), rather than a model in which both ends interact with the homologous DNA (Szostak *et al.*, 1983).

Gene conversion associated with plasmid integration into the diverged chromosome was also examined. Gene conversion of plasmid sequences adjacent to the original DSB was detected either at both sides, at one side only, or at neither side. The results suggest that integration-associated gene conversion for a given plasmid end was independent of associated conversion at the other end, and that the probability of associated gene conversion was 0.5 for each end. Conversion of chromosomal sequences was observed in only one of 25 transformants. Thus DNA divergence does not prevent recombination. Plasmid sequences adjacent to a DSB or gap appear to undergo some type of processing that leads to gene conversion. In current models for recombination, the processing would be the result of mismatch repair.

B. Recombinational as against mismatch repair

1. General considerations

Recombinational interactions between DNAs that differ at one or more sites generate heteroduplex DNA (hDNA) containing basepair mismatches, which may account for the decrease in recombinational repair already described. The decrease might be due to either an inhibition of hDNA formation or mismatch-specific processing of the hDNA intermediates. Most of our knowledge of DNA mismatch repair comes from studies with prokaryotes. As will be discussed, the key features are the ability to recognize basepair mismatches and to direct repair to a specific strand.

Mismatch repair systems for DNA have been described in prokaryotes, and they not only affect mutation arising from replication errors but also recombination due to mismatches in hDNA (Claverys and Lacks, 1986; Radman and Wagner, 1986; Radman, 1988; Modrich, 1991). Recombination intermediates in some systems are exquisitely sensitive to destruction by DNA-mismatch repair. In *Streptococcus pneumoniae*, the transformation frequency for a given marker can be decreased ten-fold due to mismatch repair (Claverys and Lacks, 1986). Mismatch repair for DNA has such a profound effect on lowering the frequency of Hfr conjugation between *E. coli* and *Salmonella typhimurium* that Rayssiguier *et al.* (1989) proposed that mismatch repair might be a significant factor in speciation by promoting recombination fidelity. Mismatch repair in yeast, therefore, might be expected to influence the outcome of recombination involving mismatches, particularly when the recombining DNAs are considerably diverged. The mismatch repair systems could act by limiting formation of a heteroduplex between diverged DNA, or repair may lead to destruction of recombination intermediates, and overlapping repair events might even lead to secondary DSBs (Borts *et al.*, 1990). The net result would be to decrease recombination efficiency, presumably as a function of sequence divergence.

The non-reciprocal recombination phenomena of gene conversion and post-meiotic segregation (PMS) in yeast and other fungi have been attributed to correction or lack of correction, respectively, of mismatches within hDNA formed during meiotic recombination (Fogel *et al.*, 1981). Post-meiotic segregation results when hDNA containing a basepair mismatch persists in a meiotic product, such as a spore. The hDNA is then resolved by DNA replication during the first mitotic cycle producing two daughter cells that differ genetically at the site of the mismatch. Three putative DNA-mismatch repair genes, *PMS1* (Williamson *et al.*, 1985; Kramer *et al.*, 1989), and *PMS2* and *PMS3* (Kramer *et al.*, 1989), have been identified in *Sacch. cerevisiae*. Mutations in these genes lead to increased levels of PMS, spontaneous mutation and heteroallelic recombination. The putative *PMS1* gene product exhibits homology to components of bacterial long-patch, mismatch repair systems, MutL of *Sal. typhimurium* and HexB of *Strep. pneumoniae* (Kramer *et al.*, 1989; Mankovich *et al.*, 1989; Prudhomme *et al.*, 1989). Earlier it had been shown that MutS of *Sal. typhimurium* and HexA of *Strep. pneumoniae* were homologous (Haber *et al.*, 1988; Priebe *et al.*, 1988), and homologues of these have since been identified in the mouse (Linton *et al.*, 1989) and human (Fujii and Shimada, 1989). The homology of *PMS1* and other eukaryotic proteins to components of bacterial long-patch repair systems implies a functional or mechanistic similarity.

In general, the repair specificity for various types of mismatches in yeast is very similar to that observed for bacterial long-patch, mismatch repair systems. Base-substitution mismatches, with the exception of C/C, and small insertion/deletion mismatches are corrected efficiently. This repair specificity is observed in meiotic recombination (White *et al.*, 1985; Detloff *et al.*, 1991), for repair of heteroduplex plasmids (made *in vitro*) during transformation (Bishop and Kolodner, 1986; Bishop *et al.*, 1987, 1989; Kramer *et al.*, 1989) and in cell-free extracts made from mitotic cells (Muster-Nassal and Kolodner, 1986). Generally, *pms1* (Bishop *et al.*, 1989; Kramer *et al.*, 1989), *pms2* or *pms3* (Kramer *et al.*, 1989) mutations lower the repair efficiency of all mismatches to the level of the C/C mismatch in wild-type. A denaturant-gel electrophoresis method revealed that, of possible C/C or G/G mispairs formed in recombination with an allele expressing high PMS, only the C/C mismatch persisted in post-meiotic DNA of spores from *PMS1* cells while, in spores from *pms1* cells, both the C/C and G/G mispairs persisted (Lichten *et al.*, 1990).

Excision and resynthesis tracts are thought to extend a kilobase or more during long-patch repair in bacteria (Claverys and Lacks, 1986; Modrich, 1991). Yeast alleles exhibiting high PMS yield lower PMS frequencies when tightly linked to low PMS alleles (Fogel *et al.*, 1981). This is similar to pneumococcal transformation in which high-efficiency markers have lower transformation frequencies when linked to low-efficiency markers (Claverys and Lacks, 1986). In both situations, the magnitude of the effect is related to the distance separating the alleles, suggesting that repair of mismatches in hDNA involves long-patch repair.

An important feature of DNA-mismatch repair, which has not been addressed adequately in yeast, is how repair is directed to one specific strand. A single-strand break or end is thought to serve as the signal that directs repair to the broken strand during bacterial mismatch repair (Claverys and Lacks, 1986; Modrich, 1991). Single-strand breaks on plasmid substrates direct mismatch repair to the broken strand following transformation in *E. coli* (Langle-Rouault *et al.*, 1987) and in cell-free extracts of *E. coli* (Lahue *et al.*, 1989), *Drosophila* spp. and man (Holmes *et al.*, 1990). Single-strand breaks are present in newly synthesized DNA at replication forks and in recombination intermediates at the end of an invading DNA single-strand, as in pneumococcal transformation. In addition to normal DNA metabolic activity, breaks in newly synthesized DNA can be provided by repair components, such as MutH of *E. coli* and *Sal.*

typhimurium that exploit transient undermethylation of newly synthesized DNA to nick the unmodified (synthesized) strand (Welsh *et al.*, 1987). Excision and resynthesis tracts are thought to extend from the breaks to the mismatch, leading to long patches of repair (Modrich, 1991).

The mutator phenotype of *pms1*, *pms2* and *pms3* suggests that mismatch repair in yeast is involved in mutation avoidance during replication. Morrison *et al.* 1993; see note added in proof, observed that mutations in the editing exonucleases from DNA polymerases increase mutation rates. If the strains are also *pms1*, there is a synergistic increase in mutation rates. Mismatch repair must be directed to the newly synthesized strand to function in mutation avoidance and is therefore strand-specific. Since yeast appears to lack DNA modification systems that could be used to direct repair (Hattman *et al.*, 1978; Proffit *et al.*, 1984), strand-specificity may depend upon single-strand ends present during replication (Claverys and Lacks, 1986; Modrich, 1991). Whether or not mismatch repair is strand-specific during recombination will be critical in elucidating mechanisms of recombination.

In some current models for recombination, repair of mismatches in hDNA formed during recombination is without strand bias, and leads to gene conversion in half of the cases (Holliday, 1964; Meselson and Radding, 1975). If true, repair of independent multiple mismatches in hDNA formed by recombination might lead to overlapping repair tracts and therefore to secondary DSBs. Borts and Haber (1987) and Borts *et al.* (1990) have reported observations that appear to support this view. They examined meiotic recombination patterns over a 9 kb interval flanked by *MAT* loci. As the number of heterologous sites introduced into this interval increased, there was a decrease in reciprocal recombination within the interval and an increase in non-reciprocal recombination involving *MAT* loci. This effect was partially reversed by *pms1*. They proposed that DSBs resulted from overlapping mismatch-repair tracts and lead to secondary recombination events.

A hypothesis of mismatch repair without strand bias during recombination predicts that recombination between homoeologous DNAs, as compared to homologous DNAs, should be decreased by mismatch repair. The large number of mismatches in hDNA should lead to overlapping repair tracts and then secondary DSBs, many of which would not be repaired (Resnick *et al.*, 1989). Therefore, a defect in DNA-mismatch repair, e.g. *pms1*, should lead to a higher recombination efficiency between homoeologous DNAs by lowering the number of recombinants lost. However, this has not been observed in the *pms1* mutation.

2. Interactions between diverged DNAs

Recombination between homoeologous DNAs from *E. coli* and *Sal. typhimurium* during Hfr conjugation is 10^6-fold lower than during conjugation between bacteria of the same species. However, if the recipient carries a mismatch-repair defect in *mutL* or *mutS*, the homoeologous recombination frequency increases 10^3-fold, demonstrating that mismatch repair poses a significant barrier to recombination between homoeologous DNAs (Rayssiguier *et al.*, 1989). A similar outcome might be expected for recombination between homoeologous DNAs in yeast having a mutation in *PMS1* which is homologous to *mutL* (Kramer *et al.*, 1989; Mankovich *et al.*, 1989; Prudhomme *et al.*, 1989).

Resnick *et al.* (1992) examined recombinational repair of ionizing-radiation damage in G-1 cells of *Sacch. cerevisiae* between homoeologous chromosomes in *PMS1* and *pms1* backgrounds. Chromosome loss frequencies, which indicate a lack of repair, were the same in both backgrounds, indicating that mismatch repair did not affect the frequency of repair events. However, *pms1* cells exhibited a 2–3-fold increased frequency of intragenic recombination as compared to *PMS1* cells. Taken together, these results suggest that mismatch repair might have an effect on intermediates or products of recombination between homoeologous DNAs, but not on the efficiency.

Bailis and Rothstein (1990) examined spontaneous recombination in *Sacch. cerevisiae* between the homoeologous *SAM1* and *SAM2* genes, which are on different chromosomes. They found that *pms1* cells had a 4.5-fold higher non-reciprocal recombination frequency, but this increase was the same as that observed for ectopic recombination between homologous genes. Also, the same length distribution for gene conversion tracts was observed in recombinants generated in wild-type and in a *pms1* mutant.

Approaches using plasmid transformation have been used to examine the effect of mismatch repair on recombination between homoeologous DNAs of *Sacch. cerevisiae* and *Sacch. carlsbergensis* (Priebe *et al.* 1994; see note added in proof). The frequency of transformation was slightly higher for *pms1*-recipient cells, both for repair of a double-strand gap in an episomal plasmid (non-reciprocal recombination) and for integration of a non-replicating plasmid (reciprocal recombination). Physical analysis of the recombinants did not reveal any obvious effect of mismatch repair on gene-conversion tract length for recombination between the homoeologous DNAs. These results are consistent with those already described for spontaneous recombination (Bailis and Rothstein, 1990).

If mismatch repair is defective, DNA replication will resolve basepair mismatches in hDNA formed during recombination. Some transformant colonies (20%) of the *pms1* recipient contained a mixture of cells differing in the extent of gene conversion associated with plasmid integration. For gapped, episomal plasmids, a PMS⁺ recipient yielded one mixed transformant colony out of 18, and a pms⁻ recipient yielded no mixed colonies out of 16. The majority of transformant colonies were pure, suggesting that either hDNA formation during either reciprocal or non-reciprocal recombination is not extensive or yeast has another mismatch-repair system.

In summary, mismatch repair in yeast, as defined by *PMS1*, appears to have little effect on the overall efficiency of recombinational repair between homoeologous DNAs. This contrasts with the influence of mismatch repair on homoeologous recombination in bacteria. Although gene-conversion frequencies in *pms1* mutants are higher for spontaneous recombination (Bailis and Rothstein, 1990) and for recombination induced by ionizing radiation (Resnick *et al.*, 1992), examination of recombinant DNAs resulting from spontaneous recombination (Bailis and Rothstein, 1990) and during transformation (Priebe *et al.* 1994; see note added in proof) offers no obvious explanation. Finally, gene conversion during homoeologous recombination occurs in *pms1* mutants. Therefore, either a *PMS1*-independent mismatch-repair system exists in yeast or there is another mechanism for gene conversion.

V. GLOBAL SIGNALLING RESPONSES TO DOUBLE-STRAND BREAKS AND OTHER DNA LESIONS

The consequences of unrepaired DSBs are poorly understood. These as well as other types of lesion can have both direct and indirect effects on metabolism of the eukaryotic chromosome. Among the direct effects are blockage of replication and loss of information. Indirect effects that are more global include enhanced expression of repair/recombination proteins and signalling of cell-cycle arrest. Inhibition of cell-cycle progression following induction of DNA damage provides a greater opportunity for repair prior to mitosis.

The inducible SOS system of *E. coli* (Walker, 1984; Witkin, 1991) has served as a model for understanding damage-inducible responses in both prokaryotes and eukaryotes. Yeast-inducible systems are described in the following paragraphs that illustrate similarities between yeast and *E. coli* in responses to DNA-damaging agents. Summarized in Fig. 4 are features of signalling systems in eukaryotes that appear to be comparable with the SOS system.

A. SOS repair in *Escherichia coli*: a model for damage-inducible cell signalling

Activation of the SOS regulatory response in *E. coli* following induction of cellular DNA damage results in co-ordinated derepression of a number of unlinked damage inducible (*din*) genes (for a review, see Walker, 1984). While the effect of DSBs is not known, many agents that produce bulky DNA lesions that block DNA synthesis also delay cell-cycle progression. *De novo* induced expression of these *din* genes is required for a diverse set of phenomena. These include enhanced error-free and error-prone DNA repair, enhanced chromosomal alterations resulting from mutagenic replication and increased recombination, enhanced survival (referred to as Weigle re-activation) and enhanced mutagenesis (Weigle mutagenesis) of DNA-damaged phage, lysogenic induction of latent prophage and cellular filamentation due to cessation of cell cycling.

In undamaged *E. coli*, the *din* genes of the SOS regulon are repressed to varying degrees by the LexA repressor (Witkin, 1991). Among the 16 *din* genes of the SOS regulon are those coding for the repressor itself, LexA, and the multifunctional protein RecA. The *recA* gene product is essential for recombination and is also a major regulator of the SOS response. Following DNA damage in *E. coli*, a signal is generated such that RecA becomes activated (RecA*), which promotes cleavage of the LexA repressor (Fig. 4). This results in derepression of the SOS genes.

There appears to be an indirect role for single-strand DNA (ssDNA) in signalling as well as a direct role in recombination (Sassanfar and Roberts, 1990; see Section II). Although the exact nature of the SOS inducing signal is unclear, activation of RecA to RecA* requires ssDNA and a nucleotide triphosphate (dATP, ATP or ATP γ-S). Furthermore, cleavage of the LexA repressor *in vivo* after ultraviolet irradiation requires post-irradiation DNA synthesis (Witkin, 1991). Ongoing replication has been proposed to leave ssDNA gaps in a daughter strand at blocked replication forks such that RecA can bind ssDNA and be activated to RecA* in the presence of a nucleoside triphosphate. Alternatively, exonuclease digestion of one strand at a DSB (i.e. 5' to 3'; Section II) producing ssDNA tails would allow binding by RecA and subsequent activation to RecA*.

B. SOS-like damage-inducible systems in yeast

Derepression of the SOS regulon and expression of SOS phenotypes depend on a signal transduction mechanism or mechanisms. The existence of SOS-

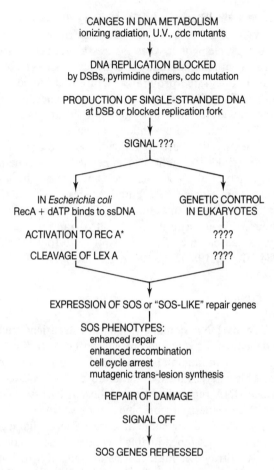

CANGES IN DNA METABOLISM
ionizing radiation, U.V., cdc mutants

DNA REPLICATION BLOCKED
by DSBs, pyrimidine dimers, cdc mutation

PRODUCTION OF SINGLE-STRANDED DNA
at DSB or blocked replication fork

SIGNAL???

IN *Escherichia coli*   GENETIC CONTROL
RecA + dATP binds to ssDNA   IN EUKARYOTES

ACTIVATION TO REC A*   ????

CLEAVAGE OF LEX A   ????

EXPRESSION OF SOS or "SOS-LIKE" repair genes

SOS PHENOTYPES:
enhanced repair
enhanced recombination
cell cycle arrest
mutagenic trans-lesion synthesis

REPAIR OF DAMAGE

SIGNAL OFF

SOS GENES REPRESSED

Fig. 4. Comparison of general features of pathways for the recognition, processing and signalling resulting from DNA damage induced in *Escherichia coli* and eukaryotes.

like phenomena in yeast and other eukaryotes suggests similar damage-dependent processes. Several SOS-like phenomena occur in yeast, which suggest that signal transduction may be an important feature of repair of DNA damage, particularly DSBs, and may influence overall genome stability.

1. Ty transposition in yeast

Activation of the SOS pathway by ultraviolet irradiation results in λ prophage induction, since RecA* induces autocleavage of the λ repressor.

Similarly, ultraviolet irradiation of mammalian cells results in induction of latent integrated viruses, such as SV40 and retroviruses (Defais *et al.*, 1983). The transposable Ty element in yeast shares a number of structural and replicative features with mammalian retroviruses (Boeke *et al.*, 1985). These features include an integrated proviral structure that includes tandem long terminal repeats (LTRs) flanking *pol* and *env*-like genes, and the ability to translocate and integrate into new genomic locations using an RNA intermediate. There is a dose-dependent increase in Ty insertions at the *ADH2* locus following exposure to ionizing radiation (Morawetz, 1987). Furthermore, ultraviolet irradiation or exposure to ethylmethane sulphonate results in induction of Ty transcription and transposition (Rolfe *et al.*, 1986; Morawetz and Hagen, 1990) and suggests that similarities exist between signalling in *E. coli*, mammalian cells and yeast. The mechanism of damage-induced transposition may be at a transcriptional level, since mutagenic agents can enhance Ty transcription 20-fold (McClanahan and McEntee, 1984; Rolfe, 1985).

2. Enhancement of DNA repair

Induction of SOS by DNA damage results in a transient enhancement of repair that is dependent on *de novo* protein synthesis. Indirect evidence for inducible DNA repair in mammalian cells and yeast has been derived from split-dose protocols whereby a small initial dose of a DNA-damaging agent is used to induce DNA repair functions followed by a second larger dose at a later time. Mutagenesis, recombination or survival is then compared with controls that received the same amount of DNA damage in a single dose. In yeast, a small but reproducible increase (2–5-fold) occurs in repair endpoints such as ultraviolet mutagenesis or gene conversion, for a split dose as compared with a single dose (Eckhardt *et al.*, 1978; Siede and Eckhardt, 1984). These increases are transient and are not observed when the inducible repair system is blocked by the protein-synthesis inhibitor cycloheximide (Eckhardt *et al.*, 1978).

3. Damage-inducible (*DIN*) genes in yeast

The existence of repair genes, which are derepressed following DNA damage, suggests that yeast has a damage-inducible signalling system. Putative SOS-like genes have been cloned from mammalian cells (Fornace *et al.*, 1989; Boothman *et al.*, 1990) and yeast (Friedberg *et al.*, 1991) based on the damage-inducible nature of the transcripts from these genes.

Using a molecular genetic approach in yeast, DNA-damaging agents have been shown to induce expression of at least 20 repair genes (Rolfe, 1985; Ruby and Szostak, 1985; Cole *et al.*, 1987; Madura *et al.*, 1990; Friedberg *et al.*, 1991). Although a number of these *DIN* genes have been characterized in yeast, a common co-ordinate derepression, as observed in prokaryotes utilizing LexA repressor cleavage, has not been found. Furthermore, unlike SOS regulation, deletion in yeast of the damage-responsive control element (inducible promoter of *RAD54*) does not increase cellular lethality in response to DNA damage (Cole and Mortimer, 1989). In addition, some yeast *DIN* genes are induced by a broad range of DNA-damaging agents, while others are induced by only a few agents (Ruby and Szostak, 1985), suggesting differential regulation among sets of *DIN* genes. Although yeast *DIN* genes appear to be regulated differently from the SOS system, their existence is indicative of a signal transduction pathway.

4. Mutagenic *trans*-lesion DNA synthesis

The SOS-mediated mutagenic by-pass of replication-blocking lesions in *E. coli* requires direct interaction of SOS proteins RecA*, to inhibit polymerase proof-reading functions, and UmuDC to increase polymerase processivity in the vicinity of the lesion. Mutagenic by-pass of replication-blocking lesions in eukaryotes requires direct participation of the replicative apparatus and therefore is restricted to the S-phase of the cell cycle (for a review, see Lawrence, 1982). Blockage of DNA replication during S-phase results in formation of discontinuities in daughter strands, which are eliminated by post-replication repair. Although details of post-replication repair are not entirely clear. In eukaryotes, it may involve either a recombinational repair pathway (as seen in *E. coli*, see Section I) or a mutagenic by-pass of the lesion that initially resulted in daughter-strand discontinuity (for a review, see Kaufmann, 1989). It is interesting that both direct and indirect mechanisms of DNA-synthesis inhibition have been observed in S-phase of mammalian cells. Bulky lesions can directly inhibit chain elongation by polymerase complexes and can also indirectly suppress initiation events at undamaged replicons (Kaufmann, 1989). This implies that indirect DNA damage can globally influence replication.

In yeast, direct participation of the replicative apparatus in mutagenic processing (*trans*-lesion synthesis) is suggested by the sequence of the *REV3* gene, because it is required for ultraviolet mutagenesis (Lawrence and Christensen, 1979b; Lawrence *et al.*, 1984). The *REV3* gene is non-essential but bears striking sequence homology to the sequence of a number of DNA polymerases including those from adenovirus and herpesvirus (Morrison *et al.*, 1989). It has been suggested that *REV3* is a SOS polymerase that is induced following DNA damage and enables mutagenic by-bass of potentially lethal replication blocks such as thymine dimers.

5. Enhanced repair indirectly induced by DNA damage in yeast

In *E. coli*, episomal-DNA damage can indirectly enhance chromosomal mutagenesis utilizing the SOS-repair pathway (Walker, 1984, 1985). In yeast, enhanced recombination has been observed in an unirradiated heteroallelic diploid when mated to an ultraviolet or X-irradiated haploid, where the nuclei of the two cells are prevented from fusing (Fabre and Roman, 1977; discussed in Resnick, 1979). This result implies that a *trans*-acting factor, generated in the haploid, is able to induce enhanced recombination in the diploid nucleus.

C. Cell-cycle arrest induced by double-strand breaks and DNA lesions

Saccharomyces cerevisiae progresses through four morphologically distinct stages (G-1, S, G-2 and M) during a single cell cycle (Fig. 5). In eukaryotes, DNA-damaging agents have been shown to cause a transient arrest in cell cycling similar to that observed in *E. coli* (Burns, 1956; Hittleman and Rao, 1974; Tobey, 1975; Brunborg and Williamson, 1978; Busse *et al.*, 1978; Lau and Pardee, 1982; Vindelov *et al.*, 1982; Wennerberg *et al.*, 1984; Kupiec and Simchen, 1985; Fingert *et al.*, 1986; Weinert and Hartwell, 1988, 1990; Schiestl *et al.*, 1989). Recently, cell-cycle arrest in G-1 in response to DNA damage has also been described for mammalian cells (Kastan *et al.*, 1991 Kuerbitz *et al.*, 1992).

It has been proposed that a surveillance mechanism operates in G-2 that can detect the presence of DNA damage and prevent the cell from entering mitosis (Fig. 5; Weinert and Hartwell 1988; Hartwell and Weinert, 1989). This has been suggested in part from experiments with mammalian cells treated with DNA-damaging agents together with caffeine or methylxanthine. Caffeine and methylxanthine treatment prevents G-2 arrest and allows immediate progression from G-2 into mitosis even though chromosomal-DNA damage has not been repaired (Lau and Pardee, 1982; Fingert *et al.*, 1986). This results in enhanced lethality from chromosomal aberrations.

1. Genetic control of cell-cycle arrest

Several studies have demonstrated that chromosomal-DNA damage can induce a G-2 delay in *Sacch. cerevisiae* and that this process is mediated by the *RAD9* gene product (Weinert and Hartwell, 1988, 1990; Hartwell and Weinert, 1989; Schiestl *et al.*, 1989). The *RAD9* gene is required for mutagenic ultraviolet repair and chromosome stability, but is not essential for normal mitotic growth (Schiestl *et al.*, 1989; Weinert and Hartwell, 1990). Arrest dependent on *RAD9* occurs by a post-translational mechanism; both induction of G-2-mediated arrest and recovery from arrest can occur in the presence of cycloheximide (Weinert and Hartwell, 1990). Furthermore, *RAD9* does not appear to code for a DNA-repair protein, since an increase in X-ray-induced lethality of *rad9* mutants can be compensated by treatment with the microtubule poison methyl-2-yl-carbamate, that arrests cells in G-2 (Weinert and Hartwell, 1988). Moreover, temperature-sensitive mutations in some DNA-replication genes such as *cdc2* (DNA pol δ), *cdc9* (DNA ligase) and *cdc17* (DNA pol α) result in a G-2 arrest at the non-permissive temperature that is dependent on *RAD9* (Hartwell and Weinert, 1989). It has therefore been proposed that there are genetically controlled checkpoints, such as the one controlled by *RAD9*, that enforce a dependence of a late cell-cycle event (mitosis) on completion of earlier events such as DNA replication (Weinert and Hartwell, 1988; Hartwell and Weinert, 1989).

In *Schiz. pombe*, the gene products Cdc2 (a protein kinase) and Cdc25 (a tyrosine phosphatase) are required for cells to delay in G-2 in response to inhibition of DNA synthesis (Nurse and Thuriaux, 1980). Activation of Cdc2 and entry into mitosis (G-2 → M) requires the activity of Cdc25 (Russell and Nurse, 1986). Inactivation of Cdc2 and cell-cycle delay require the protein tyrosine kinase Wee1 (Russell and Nurse, 1987). Surprisingly, a functional Wee1 is required for ionizing radiation-induced mitotic delay in G-2 but does not involve the Cdc25 pathway (Rowley *et al.*, 1992). Therefore, unlike *Sacch. cerevisiae*, mitotic delay in response to DNA damage is separate from that induced by replication arrest in *Schiz. pombe*.

The *RAD9*-dependent cell-cycle arrest in response to DNA damage is, at best, indirect evidence for a global signal induced by DNA lesions in eukaryotes. Studies with the two yeasts have relied on procedures that produce DNA damage directly to the genome. This makes it impossible to separate the indirect (signalling) from the direct (chromosome loss) effects of damage. Furthermore, the physical or chemical agents used to damage DNA produce a wide spectrum of lesions.

(a)

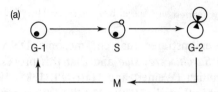

(b)

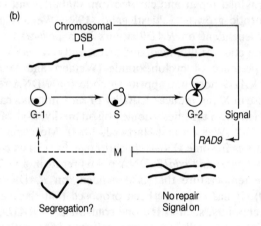

(c)

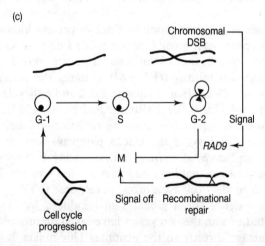

Fig. 5. Proposed summary of cell-cycle arrest in G-2 following production of a DSB within a chromosome in haploid *Saccharomyces cerevisiae* and dependence on the *RAD9* gene. (a) Cellular morphology during cell-cycle progression of undamaged yeast cells. (b) A chromosomal DSB induced by ionizing radiation before replication is completed (G-1 or early S-phase of the cell cycle) causes a *RAD9*-dependent arrest at the end of G-2. Because the damage occurred before replication, the sister chromatid (following S-phase) also possesses a break.

2. Arrest of cell progression and lethality caused by a single double-strand break in a plasmid

To overcome problems associated with a wide spectrum of chromosomal damage induced by ionizing radiation, a novel yeast system has been developed (Bennett *et al.*, 1993) that allows induction of a single defined lesion, namely a DSB, in non-essential plasmid DNA that is not subject to recombinational repair (see Fig. 6 legend for details). This has allowed objective examination of indirect global effects of a DSB on cellular metabolism and genome stability.

A single persistent non-chromosomal DSB has severe consequences on cellular growth, decreasing survival to 2% in wild-type cells. Lethality in a RAD^+ strain can be divided into two categories based on examination of single cells from a rapidly cycling, unsynchronized population. An unrepairable DSB prevents growth in about one-half of the population, both in non-budded (G-1) and budded, (G-2) cells. The second type of lethality is seen as a transient G-2 delay followed by clonal death within 4–5 generations (microcolonies comprised of dead cells; Fig. 7). Neither type of lethality is due to loss of genetic information, since selection for the YZ-CEN plasmid information was removed at the time of DSB induction. Thus, this system represents the first true assay for indirect, DNA damage-induced signalling.

The unexpectedly low survival exhibited by wild-type cells suggests that an unrepaired DSB can cause dominant lethality, even without loss of essential genetic information. Possible reasons for cell lethality may include repeated rounds of cell-cycle delays due to persistence of the unrepaired lesion, unbalanced growth in subsequent generations or consequences to other aspects of DNA/chromosome metabolism (such as chromosome segregation, recombination or replication).

Recombinational repair cannot occur due to lack of an intact homologue. This results in persistence of the DSB-inducing signal and a lethal $RAD9$-dependent arrest at the G-2/M boundary. Any cells that escape the $RAD9$-dependent checkpoint and progress to mitosis risk loss of essential chromosomal material. The broken ends could also be subject to deoxyribonucleolytic degradation. Cells deleted for $RAD9$ continue to progress past G-2/M but die as microcolonies, presumably due to loss of essential genetic material during mitosis (see text). (c) A chromosomal DSB induced by ionizing radiation after DNA replication (late S- or G-2), results in a $RAD9$-dependent arrest at the G-2/M boundary. Because the chromosome is duplicated as sister chromatids, DSB recombinational repair occurs, thereby abrogating the signal. Cells can then move into mitosis and segregate repaired chromosomal DNA.

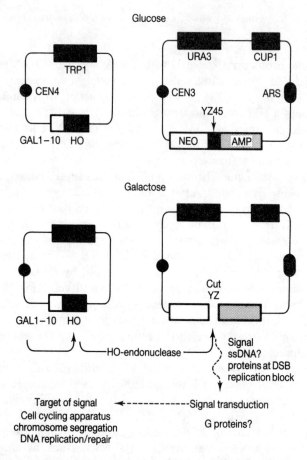

Fig. 6. Experimental design for production of a persistent irreparable DSB in a dispensable single-copy (CEN) yeast plasmid. A non-switching (*MAT* Δ) haploid strain containing a selectable (*TRP1*) CEN plasmid (pGALHOT) with the HO-endonuclease under the control of the inducible *GAL1-10* promoter was transformed with a second selectable (*URA3*) plasmid (YZ-CEN) containing a YZ junction (from *MAT*). Both plasmids are maintained when cells are grown in glucose-containing medium lacking uracil and tryptophan. Transfer to medium containing galactose and uracil leads to transcriptional induction of the HO endonuclease resulting in a DSB at the YZ junction in the target plasmid. The YZ junction in the target plasmid is located between flanking non-yeast sequences (*neo* and *amp*) to prevent homologous recombination with the yeast chromosome. As discussed in the text (also see Fig. 7) the persistent unrepaired DSB results in arrest of cell progression and lethality even when the plasmid is dispensible (+uracil).

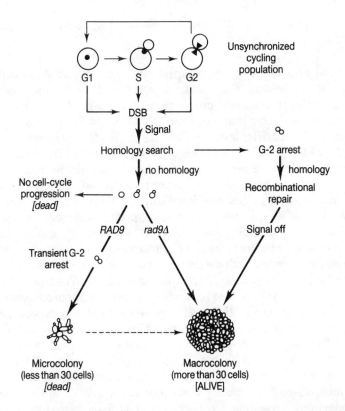

Fig. 7. Cell progression and lethality in response to a persistent unrepaired DSB in a dispensable yeast plasmid. In RAD^+ cells, an unrepaired DSB on a plasmid generates a signal that results in cells that either do not progress or, alternatively, pause transiently at G-2 and proceed for a limited number of divisions resulting in a microcolony (containing less than 30 cells and elongated non-viable cells. Only a small percentage (~2%) go on to produce macrocolonies (more than 30 cells). In $rad9\Delta$ cells, the unrepaired DSB results in either non-viable cells that do not progress or cells that continue to divide, producing macrocolonies of viable cells with normal morphology. A DSB between flanking direct repeats or in the presence of an undamaged homologue in G-2 undergoes recombinational repair. It is hypothesized that, in the presence of homology, the DSB is repaired and the growth-arrest signal is shut off allowing cells to continue in the cell cycle.

3. *RAD9*-independent and *RAD9*-dependent inhibition of cell progression by a double-strand break in a dispensable plasmid

A low dose of a DNA-damaging agent can cause *RAD9*-dependent cell-cycle arrest, suggesting a signalling interaction between the lesion and gene(s) controlling cell-cycle progression. The plasmid system described above has enabled examination of the role of *RAD9* in the global cellular response to a specific lesion. There is a *RAD9*-independent and a *RAD9*-dependent lethal response to the DSB (Fig. 7). The frequency of cells that do not exhibit cell progression and die (*RAD9*-independent lethality) is comparable for a *rad9* deletion mutant (*rad9Δ*) and wild-type cells. This implies that there may be an additional control(s) on cell growth in response to DNA damage that is not related to a specific stage of the cell cycle. Stage-independent lethality has also been observed for eukaryotic cells undergoing apoptosis (programmed cell death). Similarities to apoptosis include independence of cell-cycle stage and an irreversible commitment to death following appearance of chromosomal DSBs (Williams, 1991).

A role for the *RAD9* gene in response to a single extrachromosomal DSB was clearly manifested in those cells capable of cell progression (*RAD9*-dependent lethality). Rather than exhibiting limited growth and clonal death, *rad9Δ* cells with a DSB yield viable colonies at high frequency. This accounted for the large difference in survival (2% as against 32%) between RAD^+ and *rad9Δ* cells. Cellular and nuclear morphologies observed for *rad9Δ* mutants differed greatly from those of a RAD^+ strain soon after DSB induction, suggesting a role for the *RAD9* gene product in cellular metabolism immediately following induction of DNA damage. Among cells that were capable of cycling, the *RAD9* gene was responsible for a transient G-2/M delay (Fig. 7), similar to that observed in previous studies. By eliminating the G-2/M delay in *rad9* mutants, it has been proposed that there would be less opportunity for DNA repair of chromosomal lesions resulting in loss of essential genetic information. Increased survival of the *rad9*, as compared to wild-type strains, is consistent with this interpretation, since the lesion on the plasmid would not be expected to have direct genetic consequence.

4. Molecular origin of the damage-inducible growth-arrest signal

While there has been substantial progress in characterizing the molecular and genetic control of the yeast cell cycle (Reed, 1991; Pringle and Hartwell, 1981), little is known about the cell-cycle response to DNA damage or the signalling mechanisms. However, replication and/or processing of the ends of a DSB are likely to figure prominently. For example, a regulatory SOS-like signal could be generated by a replication block at a DSB. Interruptions in DNA replication result in SOS-mediated cessation of cell cycling in *E. coli* (Walker, 1984; Lutkenhaus, 1990), resulting in filamentation. This response is mediated by the SOS gene *sulA* (formerly *sfiA*), the product of which is thought to bind to the FtsZ protein. This complex in turn plays a central role in initiation of cell division (Huisman and D'Ari, 1981; Lutkenhaus, 1990). There are several possible sources for the signal in yeast. A replication-related regulatory signal could be induced by interaction of a replication complex with stable unrepaired double-strand broken ends. These ends are still present between 10 and 24 hours after HO induction (Bennett *et al.*, 1993), suggesting that there may be protection against extensive nucleolytic attack at the ends, possibly by a protein such as the Ku protein from humans (Mimori and Hardin, 1986).

A signal could arise in response to ssDNA being exposed. As described in Fig. 4, a signalling role for ssDNA has been proposed for activation of the SOS regulon in *E. coli*, resulting in cell-division arrest following DNA damage (Walker, 1984; Sassanfar and Roberts, 1990; Fig. 4). Nuclease 5'-3' digestion exposing a 3' ssDNA tail has been detected at DSBs induced at YZ sequences (see Section II.B). This 5' to 3' processing of double-stranded ends could act indirectly in DSB recombinational repair by signalling growth arrest thereby allowing time for repair prior to M phase, as well as acting directly by providing a recombinationally invasive 3' end (Resnick, 1976; Szostak *et al.*, 1983).

5. Implications of double-strand break-induced signalling to recombinational repair between diverged DNAs

The above system was designed to examine the effects of a DSB in yeast when there is no opportunity for recombinational repair. It is probable that the lethality observed results from persistence of the damage-induced growth-arrest signal. In the event of repair, the signal is turned off, the cell survives and progresses through the cell cycle (Figs. 4 and 7). What happens to this signal when a DSB is produced in DNA of reduced homology (homoeologous) is not clear. However, observation of chromosome-loss events in irradiated yeast strains containing a homoeologous pair of chromosomes (see Section IV), implies that a lethal growth-arresting signal is not produced under these conditions. Perhaps the ability to find a partially homologous template is enough to abrogate the lethal growth-arresting signal. Further experiments are now underway to produce site-specific DSBs, where the DSB will be induced between either homologous or homoeologous repeat DNAs. This system may provide opportunities for designing strategies to modulate the effects of DNA-damaging agents, including limitation of uncontrolled cell proliferation.

VI. NOVEL FEATURES OF RECOMBINATIONAL REPAIR

A. Yeast artificial chromosomes and cloning

Recombinational repair could lead to deletions and re-arrangements through recombination between repeated DNA sequences. Spontaneous recombination between repeats on plasmids and in chromosomes of yeast have been examined extensively (Petes and Hill, 1988). Conservative and non-conservative recombinational repair of DSBs can occur in repeat DNA (Section II). Repeat sequences are uncommon in yeast; however, they are frequent in many organisms, particularly mammals. Nearly 20–30% of mammalian chromosomal DNA is repetitive DNA (Britten and Kohne, 1968), including short repeats (i.e. less than 10–20 bp sequences), *Alu* sequences (approximately 300 bp, appearing on the average every 3–5 kb), LINE sequences (5–10 kb), pseudogenes and possibly amplified genes. As much as 20% divergence is observed for individual repetitive elements such as LINES and SINES (summarized in Hutchison *et al.*, 1989; Deininger, 1989). While investigations of repair and recombination in these diverged repeat DNAs within mammalian cells are lacking, it is now possible to study portions of higher chromosomes within using yeast artificial chromosomes (YACs) containing mammalian DNA. Such studies are particularly relevant to the international Human Genome Project.

One of the major goals of the Human Genome Project is the generation of accurate and comprehensive maps of the human genome. Physical analysis of the human genome is simplified by cloning human DNAs into YACs, because genomic DNA inserts isolated by this approach can be several hundred kilobases in length. In addition, YACs can be manipulated genetically and physically as yeast chromosomes, and those containing human DNA can be returned to mammalian cells (Pachnis *et al.*, 1990; Pavan *et al.*, 1990). It is essential for accurate characterization of the genome that DNA within YACs is an exact replica of that in human cells. However, errors can occur either during transformation of the human DNA-containing YACs into yeast or during subsequent propagation. Many errors are likely to result from recombination between the diverged repeats. Recombinational repair may play an important role as indicated by reduction of errors in *rad52* mutants (Section VI.B).

Recombination plays a prominent role in the use of human DNA-containing YAC libraries. While it provides a means for mapping, it is also a source of errors. The ability to target sequences to YACs has become a useful means for producing terminal and interstitial deletions (Pavan *et al.*, 1990; Campbell *et al.*, 1991). However, recombination may also be important in development of artefacts in YAC libraries (Albertsen *et al.*, 1990; Neil *et al.*, 1990). Neil *et al.* (1990) investigated the suitability of YAC vectors for cloning tandemly repeated human Y-chromosome sequences. Established YACs retransformed into yeast frequently (up to 60%) exhibited gross structural re-arrangements. Recombination was implicated, since the frequency of re-arrangements was lowered in a *rad52* mutant (Kouprina *et al.* 1994; note added in proof).

Recently, it was shown that human DNA-containing YACs are in many ways similar to yeast chromosomes in terms of recombination. They undergo meiotic recombination and exhibit levels of reciprocal exchange comparable with that for yeast chromosomes (Green and Olson, 1990; Sears *et al.*, 1992). Based on genetic evidence, recombinational repair between homologous human DNAs in yeast is as efficient as between homologous yeast chromosomes (Resnick *et al.*, 1992). This was shown by comparing YAC loss in response to ionizing radiation for cells containing one as compared with two copies of a human DNA-containing YAC. Following exposure to non-lethal doses, a single YAC that was frequently lost (approximately 6% after 10 krad) was stable when two homologous YACs were present. However, loss of the single YAC was less than expected based on size and the estimated frequency of radiation-induced DSBs. In addition, there was considerable variability in the response of individual, single-copy YACs. Differences in loss rates may be accounted for by differences in intrachromosomal recombinational repair between repeated, albeit diverged, DNAs. There is considerable variability in frequencies of repeats between YACs (Sainz *et al.*, 1992).

B. Transformation-associated recombination overcomes divergence

Both mutation and recombination can occur in DNA during transformation. In eukaryotes, transformation frequently results in DNA being altered. Transfection is highly mutagenic in mammalian cells (Calos *et al.*, 1983; Razzaque *et al.*, 1984). Similarly, transformation in yeast results in high levels of mutation in unselected markers, about 1% for 10 kb circular plasmids (Clancy *et al.*, 1984). Plasmid mutations are primarily deletions ranging from 1.5 to 3.0 kb.

Recombination during transformation of yeast can contribute substantially to production of changes within plasmids if they contain repeated DNAs. Tschumper and Carbon (1986) found that, during transformation with supercoiled plasmids containing the yeast Ty transposon, nearly 20% of transposons were excised. The high level of excision occurred by homologous recombination between direct terminal 0.3 kb repeats (deltas) of the Ty element. The extent of excision was dependent on the state of the DNA; there was more excision in supercoiled plasmids. The Ty excision frequency was greatly decreased in a *rad52* mutant, further implicating a role for recombination.

In addition to targeted recombination between plasmids and chromosomes (see Section IV.B), efficient recombination between plasmid molecules during transformation has also been demonstrated. Such recombination between plasmid fragments that share homology has been described in yeast and the frequencies can be high (Ma *et al.*, 1987). As few as 100 basepairs of homology are required for recombination.

Recombination between repeated DNAs during transformation has been investigated and compared with recombination during mitotic growth (Larionov *et al.* 1993; note added in proof). The transformation process leads to exceedingly high levels of recombination between homologous or diverged repeats. Once the plasmids are established, considerably lower frequencies are detected. Two plasmids (Fig. 8) were constructed to analyse recombination between repeats (E. Perkins, G. Porter and M.A. Resnick, unpublished observation) with the direct-repeat motif *URA3–ADE2–ura3** (*ura3** is a frameshift mutation). The 1 kb repeats were either homologous or diverged by 20%. A third control plasmid contained only a single *URA3*.

Frequencies of loss of the indicator colour-marker *ADE2* gene during transformation and subsequent mitotic propagation are summarized in Table I. There was a high rate of loss of *ADE2* during transformation. Both for homologous and diverged repeats, the frequency was over 100-fold greater during transformation than during mitotic growth. Subsequent restriction analysis of the plasmids after shuttling them to *E. coli* revealed that loss of the *ADE2* gene was due to recombination between repeats. The frequency of recombination between homologous direct repeats could be increased from less than 1% for supercoiled plasmids to 40% of the transformants by introduction of a specific single-strand nick in one of the repeats (Table I). Frequency of loss of *ADE2* in the control plasmid (one copy of *URA3*) was considerably lower (10^{-4}). For plasmids containing diverged repeats, 12% of the plasmids were recombined. A restriction enzyme-induced DSB increased the level of recombination a further 2–3-fold.

Systems involved in recombinational repair appear to be responsible for high levels of recombination based on stimulation by single-strand breaks

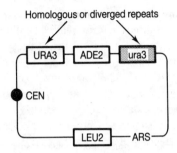

Homologous or diverged repeats

Fig. 8. Structure of a centromere-containing plasmid designed to detect recombination between direct repeats that are either homologous or diverged. Cells carrying the plasmid yield white colonies. Recombination between the repeats can lead to loss of *ADE2*, resulting in a red colony since the chromosomal copy of *ADE2* is mutated. Recombination can also be detected as conversion of *URA3* to the *ura3*.

or DSBs as well the considerable decrease (greater than 20-fold) in recombination in a *rad52* deletion mutant. Surprisingly, levels of recombination between diverged DNAs during transformation approach levels observed for homologous repeat DNAs.

The high level of recombination during transformation may relate to the nakedness of the DNA (Lin *et al.*, 1990). Transforming DNA, which lacks chromatin-associated proteins, might be more accessible to repair and recombination systems or to attack by DNA metabolic enzymes such as deoxyribonucleases.

The unexpectedly high frequency of recombination between diverged DNAs during transformation is consistent with the observations of Pompon and Nicolas (1989). These workers examined bimolecular recombination between much more highly diverged (approximately 40%) cytochrome P-450 DNAs from mouse and rabbit. A gap was created by restriction enzymes in one of the P-450 genes contained in a plasmid. This gapped plasmid was transformed into yeast along with a corresponding fragment from the other P-450 gene that covered the gapped region and extended to both sides of the gap. A small amount of gap-repair was detected. The observation of recombinational repair between diverged sequences suggests that this phenomenon could be a source of mutations or could result in mosaic genes.

Table I. Frequency of recombination between homologous and diverged *URA3* repeats leading to loss of the intervening *ADE2* gene during transformation or mitotic growth

	Homologous	Diverged
Transformation		
Covalently closed plasmid DNA[a]	$\sim 10^{-2}$	$\sim 10^{-3}$
Plasmid nicked in repeat[b]	$4.0 \cdot 10^{-1}$	$1.2 \cdot 10^{-1}$
Plasmid linearized in repeat[c]	$5.0 \cdot 10^{-1}$	$3.0 \cdot 10^{-1}$
Mitotic growth	$5.0 \cdot 10^{-4}$	$< 10^{-5}$

[a] The covalently closed plasmid DNA preparation contains a small amount (~ 1–3%) of relaxed DNA and this appears to contribute to the variable low of recombination.
[b] The nick was generated in the *URA3* sequence by appropriate restriction-enzyme treatment in the presence of ethidium bromide (Short and Botstein, 1983). Nearly 90% of the supercoiled plasmid was converted to a relaxed form and the remaining plasmid was linear.
[c] Plasmids were linearized by a restriction enzyme that cut in one of the repeats.

C. Double-strand breaks and biological relevance: evolutionary concerns

Repair of DSBs in yeast appears to occur mainly through processes involving recombination. Two mechanisms of *RAD52*-dependent recombinational repair have been identified, namely conservative and non-conservative. The latter may be a special case in that it appears to be related specifically to repair between repeat DNAs. Other mechanisms of repair, such as ligation, seem unlikely, particularly for radiation-induced DSBs. In a *rad52* mutant that lacks recombinational repair, there is nearly a 1:1 relationship between DSBs and lethality (Resnick and Martin, 1976). A similar relationship holds for haploid cells irradiated in G-1 (Luchnik *et al.*, 1977).

As discussed earlier in this chapter, DSBs are an important part of the biology and life cycle of yeast. They correspond to the initiating event in mating-type switching and may be responsible for much of meiotic recombination in *RAD*[+] strains, although single-strand breaks may also be important (Resnick *et al.*, 1984). To some extent they can also overcome homology barriers to recombination (Section IV).

A question remains as to the origin of a system(s) for repair of DSBs. A commonly held viewpoint is that it may have evolved as a result of environmental exposure to ionizing radiation. This seems unlikely in view of the low absorbed dose rate of ionizing radiation at the earth's surface. The maximum current level is less than 1 rad per year. This corresponds to less than 0.001 DSB in each non-dividing cell each year. Given the considerable amount of growth expected in natural yeast, the environmental threat by ambient radiation would appear to be insignificant even if dose rates were much higher in earlier stages of evolution.

There are other more likely sources of DSBs or double-strand damage that may have contributed to evolution of DSB recombinational-repair systems. Double-strand breaks could arise from overlapping excisional repair of damage on opposite strands. Such secondary DSBs have been observed for ultraviolet damage in mammalian cells (Bradley and Taylor, 1981) and in bacteria (Bonura and Smith, 1975). Surprisingly, pyrimidine dimers in close proximity on opposite strands are induced with near-linear kinetics (Lam and Reynolds, 1986). Both cross-linked DNA strands (Jachymczyk *et al.*, 1981; Magana-Schwencke *et al.*, 1982) and methyl-methanesulphonate damage (Chlebowicz and Jachymczyk, 1979) are processed to DSBs in yeast. While not yet demonstrated, it is possible that DSBs could also arise during chromosomal metabolism, possibly by the action of topo-isomerases or deoxyribonucleases. For example, during DNA synthesis the lagging strand may be open to single-strand endonucleases. Whatever the evolutionary pressure, DSBs represent a major threat to genomic stability of cells. As already discussed, there appears to be an elaborate mechanism for their recognition and repair, which includes participation of cell-cycle control processes.

VII. CONCLUSIONS

Over the past decade, it has been shown that systems responsible for DSB recombination repair are essential in mating-type switching, meiotic recombination and gene targeting in transformation. Using yeast, it has been demonstrated that the recombinational repair systems are interrelated with many aspects of cell development and genome stability. This chapter has addressed the current emerging themes in recombinational repair. The focus has been on DSBs because they are known to be repaired only by recombinational repair mechanisms and because they are one of the few lesions for which model systems are conveniently available to study their processing. Interactions between DNAs and consequences of diminished interactions are now being addressed. Approaches being taken provide for new understanding of the origins of translocations, novel genes and damage-induced aneuploidy. It is clear that recombinational repair of DNA lesions is integrated into control of the cell cycle. The RhoNUC rho-related nuclease and the HRR25 protein kinase appear to be the first proteins identified in eukaryotes that integrate repair with cell-cycle control. Future studies on recombinational repair should provide substantial insight into the intricate relationships between DNA metabolism, cell development and DNA repair.

VIII. ACKNOWLEDGEMENTS

We wish to thank Drs Jim Mason, Vladimer Larionov, Dmitry Gordenin, and Kevin Lewis for helpful and critical comments during the writing of this chapter. Dr Ed Perkins was supported by an NIH postdoctoral fellowship; Dr Scott D. Priebe was supported by a NRC postdoctoral fellowship.

NOTE ADDED IN PROOF

We would like to acknowledge the enormous contributions made by Dr Rose to yeast biology and genetics. His enthusiasm and concerns about the field will be missed.

This chapter was submitted in April 1992 and has not been modified since. While there has been considerable progress in the area of recombination, many of the ideas presented remain current. The following are references corresponding to material that was unpublished in the original manuscript. These are identified as "see Note Added in Proof . . ."

van Vliet-Reedijk, J.C. and Planta, R.J. (1993). *Yeast* **9**, 1139.
Matsui, Y. and Toh-e, A. (1992). *Gene* **144**, 43.
Couture, C. and Chow, T.Y.-K. (1992). *Nucleic Acids Research* **20**, 4355.
Morrison, A., Johnson, A.L., Johnston, L.H. and Sugino, A. (1993). *EMBO J.* **12**, 1463.
Priebe, S.D., Westmoreland, J., Nilsson-Tillgren, T. and Resnick, M.A. (1994). *Molec. Cell. Biol.* **14**, 4802.
Larionov, V., Kouprina, N., Eldarov, M., Perkins, E., Porter, G. and Resnick, M. (1993). *Yeast* **10**, 93.
Kouprina, N., Eldarov, M., Moyzis, Resnick, M. and Larionov, V. (1994). *Genomics* **21**, 7.

References

Aboussekhra, A., Chanet, R., Zgaga, Z., Cassier-Chauvat, C., Heude, M. and Fabre, F. (1989). *Nucleic Acids Research* **17**, 7211.
Aguilera, A. and Klein, H. (1988). *Genetics* **119**, 779.
Ahn, B.Y. and Livingston, D.M. (1986). *Molecular and Cellular Biology* **6**, 3685.
Ahn, B.Y., Dornfeld, K.J., Fagrelius, T.J. and Livingston, D.M. (1988). *Molecular and Cellular Biology* **8**, 2442.

Albertsen, H.M., Abderrahim, H., Cann, H.M., Dausset, J., Le Pastier, D. and Cohen, D. (1990). *Proceedings of the National Academy of Sciences, USA* **87**, 4256.
Bailis, A.M. and Rothstein, R. (1990). *Genetics* **126**, 535.
Bainton, R., Gamas, P. and Craig, N.L. (1991). *Cell* **65**, 805.
Beam, C.A., Mortimer, R.K., Wolfe, R.G. and Tobias, C.A. (1954). *Archives of Biochemistry and Biophysics* **49**, 110.
Bennett, C., Perkins, E. and Resnick, M.A. (1991). In "Modern Microbial Genetics" (U.N. Streips and R.E. Yasbin, eds), pp. 389–430. Wiley-Liss, New York.
Bennett, C.B., Lewis, A.L., Baldwin, K.K. and Resnick, M.A. (1993). *Proceedings of the National Academy of Sciences, USA* **90**, 5613–5617.
Berg, D.E. (1989). In "Mobile DNA" (D.E. Berg and M.M. Howe, eds), pp. 185–210. American Society for Microbiology, Washington, DC.
Bishop, D.K. and Kolodner, R. (1986). *Molecular and Cellular Biology* **6**, 3401.
Bishop, D.K., Williamson, M.S., Fogel, S. and Kolodner, R. (1987). *Nature* **328**, 362.
Bishop, D.K., Andersen, J. and Kolodner, R. (1989). *Proceedings of the National Academy of Sciences, USA* **86**, 3713.
Bishop, D.K., Park, D., Xu, L. and Kleckner, N. (1992). *Cell* **69**, 439.
Boeke, J., Garfinkle, D., Styles, C. and Fink, G. (1985). *Cell* **40**, 491.
Bonura, T. and Smith, K.C. (1975). *Journal of Bacteriology* **121**, 511.
Boothman, D.A., Lee, S., Trask, D.K., Dou, Q. and Hughs, E.N. (1990). In "Ionizing Radiation Damage to DNA: Molecular Aspects" (S. Wallace and R. Painter, eds), pp. 309–317. Wiley-Liss, New York.
Borts, R.H. and Haber, J.E. (1987). *Science* **237**, 1459
Borts, R.H., Leung, W.-Y., Kramer, W., Kramer, B., Williamson, M., Fogel, S. and Haber, J.E. (1990). *Genetics* **124**, 573.
Bradley, M.O. and Taylor, V.I. (1981). *Proceedings of the National Academy of Sciences, USA* **78**, 3619.
Britten, R.J. and Kohne, D.E. (1968). *Science* **161**, 529.
Brunborg, G. and Williamson, D.H. (1978). *Molecular and General Genetics* **162**, 277.
Brunborg, G., Resnick, M.A. and Williamson, D.H. (1980). *Radiation Research* **82**, 547.
Burns, V.W. (1956). *Radiation Research* **4**, 394.
Busse, P.M., Bose, S.K., Jones, R.W. and Tolmach, L.J. (1978). *Radiation Research* **76**, 292.
Calos, M.P., Lebkowski, J.S. and Botchan, M.R. (1983). *Proceedings of the National Academy of Sciences, USA* **80**, 3015.
Campbell, C., Gulati, R., Nandi, A.K., Floy, K., Hieter, P. and Kucherlapati, R.S. (1991). *Proceedings of the National Academy of Sciences, USA* **88**, 5744.
Cao, L., Alani, E. and Kleckner, N. (1990). *Cell* **61**, 1089.
Chen, D.S. and Bernstein, H. (1988). *Proceedings of the National Academy of Sciences, USA* **85**, 6821.
Chlebowicz, E. and Hachymczyk, W.J. (1979). *Molecular and General Genetics* **182**, 196.
Chow, T.-Y.K. and Resnick, M.A. (1987). *Journal of Biological Chemistry* **262**, 17659.
Chow, T.-Y.K. and Resnick, M.A. (1988). *Molecular and General Genetics* **211**, 41.
Chow, T.-Y.K., Perkins, E.L. and Resnick, M.A. (1992). *Nucleic Acids Research* **20**, 5215.
Clancy, S., Mann, C., Davis, R.W. and Calos, M.P. (1984). *Journal of Bacteriology* **159**, 1065.
Claverys, J.-P. and Lacks, S.A. (1986). *Microbiological Reviews* **50**, 133-5.
Cole, G.M. and Mortimer, R.K. (1989). *Molecular and Cellular Biology* **9**, 3314.
Cole, G.M., Schild, D., Lovett, S.T. and Mortimer, R.K. (1987). *Molecular and Cellular Biology* **7**, 1078.
Conley, E.C., Saunders, V.A., Jackson, V. and Saunders, J.R. (1986). *Nucleic Acids Research* **14**, 8919.
Connolly, B., White, C.I. and Haber, J.E. (1988). *Molecular and Cellular Biology* **8**, 2342.
Cox, B.S. and Game, J.C. (1974). *Mutation Research* **26**, 257.
Cox, B.S. and Parry, J.M. (1968). *Mutation Research* **6**, 37.
Defais, M., Hanawalt, P.C. and Sarasin, A.R. (1983). *Advances in Radiation Biology* **10**, 1.

Deininger, P.L. (1989). *In* "Mobile DNA" (D.E. Berg and M.M. Howe, eds), Vol. pp. 619–636. American Society for Microbiology, Washington, DC.

Detloff, P., Sieber, J. and Petes, T.D. (1991). *Molecular and Cellular Biology* **11**, 737.

Dykstra, C.C., Kitada, K., Clark, A.B., Hamatake, R.K. and Sugino, A. (1991). *Molecular and Cellular Biology* **11**, 2583.

Eckhardt, F., Moustacchi, E. and Haynes, R.H. (1978). *In* "DNA Repair Mechanisms" (P.C. Hanawalt, E.C. Friedberg and C.F. Fox, eds), pp. 421–423. Academic Press, New York.

Efstratiadis, A., Posokony, J.W., Maniatis, T., Lawn, R.M., O'Connell, C., Spritz, R.A., De Rail, J.K., Forget, B.C., Weissman, S.M., Slighton, J.L., Bleehe, A.E., Smithies, O., Baralle, F.E., Shoulders, C.C. and Proodfoot, N. (1980). *Cell* **21**, 653.

Egner, C. and Berg, D.E. (1981). *Proceedings of the National Academy of Sciences, USA* **78**, 459.

Embretson, J.E. and Livingston, D.M. (1984). *Gene* **29**, 292.

Engels, W.R., Johnson-Schlitz, D.M., Eggleston, W.B. and Sved, J. (1990). *Cell* **62**, 515.

Ernst, J.F., Stewart, J.W. and Sherman, F. (1981). *Proceedings of the National Academy of Sciences, USA* **78**, 6334.

Esposito, M.S. and Brown, J.T. (1990). *Current Genetics* **17**, 7.

Esposito, M.S. and Klapholz, S. (1981). *In* "The Molecular Biology of the Yeast *Saccharomyces*: Life Cycle and Inheritance" (J.N. Strathern, E.W. Jones and J.R. Broach, eds), Vol. 1, pp. 211–287. Cold Spring Harbor Laboratory, Cold Spring Harbor.

Fabre, F. and Roman, H. (1977). *Proceedings of the National Academy of Sciences, USA* **74**, 1667.

Fingert, H.J., Chang, J.D. and Pardee, A.B. (1986). *Cancer Research* **46**, 2463.

Fishman-Lobell, J., Rudin, N. and Haber, J.E. (1992). *Molecular and Cellular Biology* **12**, 1292.

Fogel, S., Mortimer, R.K. and Lusnak, K. (1981). *In* "The Molecular Biology of the Yeast *Saccharomyces*: Life Cycle and Inheritance" (J.N. Strathern, E.W. Jones and J.R. Broach, eds), Vol. 1, pp. 289–339. Cold Spring Harbor Laboratory, Cold Spring Harbor.

Fornace, A.J., Nebert, D.W., Hollander, C., Luethy, J.D., Papathanasiou, M., Fargnoli, J. and Holbrook, N.J. (1989). *Molecular and Cellular Biology* **9**, 4196.

Frankenberg-Schwager, M. and Frankenberg, D. (1991). *In* "Advances in Mutagenesis Research" (G. Obe, ed.), Vol. 3, pp. 1–27. Springer-Verlag, New York.

Friedberg, E.C., Siede, W. and Cooper, A.J. (1991). *In* "The Molecular and Cellular Biology of the Yeast *Saccharomyces*: Genome Dynamics, Protein Synthesis, and Energetics" (J.R. Broach, J.R. Pringle and E.W. Jones, eds), Vol. 1, pp. 147–192. Cold Spring Harbor Press, Cold Spring Harbour.

Fujii, H. and Shimada, T. (1989). *Journal of Biological Chemistry* **264**, 10057.

Game, J.C. (1983). *In* "Yeast Genetics: Fundamental and Applied Aspects" (J.F.T. Spencer, D.M. Spender and A.R.W. Smith, eds), pp. 109–119. Springer-Verlag, New York.

Game, J.C., Sitney, K.C., Cook, V.E. and Mortimer, R.K. (1989). *Genetics* **123**, 695.

Ganesan, A.K. (1974). *Journal of Molecular Biology* **87**, 103.

Gordenin, D.A., Trofimova, M.V., Shaburova, O.N., Pavlov, Y.I., Chernoff, Y.O., Chekuolene, Y.V., Proscyavichus, Y.Y., Sasnauskas, K.V. and Janulaitis, A.A. (1988). *Molecular and General Genetics* **213**, 388.

Gordenin, D.A., Malkova, A.L., Peterzen, A., Kulikov, V.N., Pavlov, Y.I., Perkins, E. and Resnick, M.A. (1992). *Proceedings of the National Academy of Sciences, USA* **89**, 3785.

Green, E.D. and Olson, M.V. (1990). *Science* **250**, 94.

Haber, J.E. and Hearn, M. (1985). *Genetics* **111**, 7.

Haber, L., Pang, P., Sobell, J.J.M. and Walker, G.C. (1988). *Journal of Bacteriology* **170**, 197.

Haniford, D.B., Benjamin, H.W. and Kleckner, N. (1991). *Cell* **64**, 171.

Hartwell, L.H. and Smith, D. (1985). *Genetics* **110**, 381.

Hartwell, L.H. and Weinert, T.A. (1989). *Science* **246**, 629.

Hattman, S., Kenny, C., Berger, L. and Pratt, K. (1978). *Journal of Bacteriology* **135**, 1156.

Haynes, R.H. and Kunz, B.A. (1981). *In* "The Molecular Biology of the Yeast *Saccharomyces cerevisae*: Life Cycle and Inheritance" (J.N. Strathern, E.W. Jones and J.R. Broach, eds), Vol. 1, pp. 371–414. Cold Spring Harbor Press, Cold Spring Harbour.

Hittleman, W.N. and Rao, P.N. (1974). *Cancer Research* **34**, 3433.

Ho, K.S.Y. (1975). *Mutation Research* **20**, 45.

Hoekstra, M.F., Naughton, T. and Malone, R.E. (1986). *Genetical Research* **48**, 9.

Hoekstra, M.F., Liskay, R.M., Ou, A.C., DeMaggio, A.J., Burbee, D.G. and Heffron, F. (1991). *Science* **253**, 1031.

Holliday, R. (1964). *Genetical Research* **5**, 282.

Holmberg, S. (1982). *Carlsberg Research Communications* **47**, 233.

Holmes, J., Clark, S. and Modrich, P. (1990). *Proceedings of the National Academy of Sciences, USA* **87**, 5837.

Huisman, O. and D'Ari, R. (1981). *Nature* **290**, 797.

Hutchison, C.A., Hardis, S.C., Loeb, D.D., Shehee, W.R. and Edgell, M.I.M.D. (1989). *In* "Mobile DNA" (D.E. Berg and M.M. Howe, eds), pp. 593–617. American Society for Microbiology, Washington, DC.

Jachymcyk, W.J., von Borstel, R.C., Mowat, M.R.A. and Hastings, P.J. (1981). *Molecular and General Genetics* **182**, 196.

Jackson, J.A. and Fink, J.R. (1981). *Nature* **292**, 306.

Jensen, R. and Herskowitz, I. (1984). *Cold Spring Harbor Symposia on Quantitative Biology* **49**, 97.

Jeong-Yu, S. and Carroll, D. (1992). *Molecular and Cellular Biology* **12**, 112.

Johnson, A.W. and Kolodner, R.D. (1991). *Journal of Biological Chemistry* **266**, 14046.

Kastan, M.B., Onyekwere, O., Sidransky, D., Vogelstein, B. and Craig, R.W. (1991). *Cancer Research* **51**, 6304.

Kaufmann, W.K. (1989). *Carcinogenesis* **10**, 1.

Keil, R. and Roeder, G.S. (1984). *Cell* **39**, 377.

Kim, J., Pjungdahl, P.O. and Fink, G.R. (1990). *Genetics* **126**, 799.

Kleckner, N. (1989). *In* "Mobile DNA" (D.E. Berg and M.M. Howe, eds), pp. 227–268. American Society for Microbiology, Washington, DC.

Klein, H.L. (1984). *Nature* **310**, 748.

Kostriken, R. and Heffron, F. (1984). *Cold Spring Harbor Symposia on Quantitative Biology* **49**, 89.

Kostriken, R., Strathern, J.N., Klar, A.J., Hicks, J.B. and Heffron, F. (1982). *Cell* **35**, 167.

Kramer, B., Kramer, W., Williamson, M.S. and Fogel, S. (1989). *Molecular and Cellular Biology* **9**, 4432.

Kramer, W., Kramer, B., Williamson, M. and Fogel, S. (1989). *Journal of Bacteriology* **171**, 5339.

Kuerbitz, S.J., Plunkett, B.S., Walsh, W.V. and Kastan, M.B. (1992). *Proceedings of the National Academy of Sciences, USA* **89**, 7491.

Kunz, B.A., Peters, M.G., Kohlami, S.E., Armstrong, J.D., Glattke, M. and Badiani, K. (1989). *Genetics* **122**, 535.

Kupiec, M. and Simchen, G. (1985). *Molecular and General Genetics* **201**, 558.

Lahue, R.S., Au, K.G. and Modrich, P. (1989). *Science* **245**, 160.

Lam, L.H. and Reynolds, R. (1986). *Biophysical Journal* **50**, 307.

Langle-Rouault, F., Maenhaut-Michel, G. and Radman, M. (1987). *EMBO Journal* **6**, 1121.

Lau, C.C. and Pardee, A.B. (1982). *Proceedings of the National Academy of Sciences, USA* **79**, 2942.

Lawrence, C.W. (1982). *Advances in Genetics* **21**, 173.

Lawrence, C.W. and Christensen, R.B. (1979a). *Journal of Bacteriology* **139**, 866.

Lawrence, C.W. and Christensen, R.B. (1979b). *Genetics* **92**, 397.

Lawrence, C.W., O'Brian, T. and Bond, J. (1984). *Molecular and General Genetics* **195**, 487.

Lichten, M., Goyon, C., Schultes, N.P., Treco, D., Szostak, J.W., Haber, J.E. and Nicholas, A. (1990). *Proceedings of the National Academy of Sciences, USA* **87**, 7653.

Lin, F.W., Sperle, K. and Sternberg, N. (1984). *Molecular and Cellular Biology* **4**, 1020.

Lin, F.W., Sperle, K. and Sternberg, N. (1990). *Molecular and Cellular Biology* **10**, 103.

Linton, J.P., Yen, J.-Y.J., Selby, E., Chen, Z., Chinsky, J.M., Liu, K., Kellems, R.E. and Crouse, G.F. (1989). *Molecular and Cellular Biology* 9, 3058.
Luchnik, A.M., Glaser, V.M. and Shestakov, S.V. (1977). *Molecular Biology Reports* 3, 437.
Lutkenhaus, J. (1990). *Trends in Genetics* 6, 22.
Ma, H., Kunes, S., Schatz, P.J. and Botstein, D. (1987). *Gene* 58, 201.
Madura, K., Prakash, S. and Prakash, L. (1990). *Nucleic Acids Research* 18, 771.
Magana-Schwencke, N., Averbeck, D., Henriques, J., Chanet, R. and Moustacchi, E. (1982). *Proceedings of the National Academy of Sciences, USA* 79, 1722.
Malone, R.E. and Esposito, R.E. (1980). *Proceedings of the National Academy of Sciences, USA* 77, 503.
Malone, R.E., Montelone, B., Edwards, C., Karney, K. and Hoekstra, M.F. (1988). *Current Genetics* 14, 211.
Mankovich, J.A., McIntyre, C.A. and Walker, G.C. (1989). *Journal of Bacteriology* 171, 5325.
Maryon, E. and Carroll, D. (1991). *Molecular and Cellular Biology* 11, 3278.
McClanahan, T. and McEntee, K. (1984). *Molecular and Cellular Biology* 4, 90.
Meselson, M. and Radding, C. (1975). *Proceedings of the National Academy of Sciences, USA* 72, 358.
Mimori, T. and Hardin, J.A. (1986). *Journal of Biological Chemistry* 261, 10375.
Modrich, P. (1991). *Annual Review of Genetics* 25, 229.
Moerchell, R.P., Tsunasawa, S. and Sherman, F. (1988). *Proceedings of the National Academy of Sciences, USA* 85, 524.
Moore, C.W. (1978). *Mutation Research* 51, 165.
Moore, C.W. (1989). *Cancer Research* 49, 6935.
Morawetz, C. (1987). *Mutation Research* 177, 53.
Morawetz, C. and Hagen, U. (1990). *Mutation Research* 229, 69.
Morrison, A., Christensen, R.B., Alley, J., Beck, A.K., Bernstine, E.G., Lemontt, J.F. and Lawrence, C.W. (1989). *Journal of Bacteriology* 171, 5659.
Morrison, A., Araki, H., Clark, A.B., Hamatake, R.K. and Sugino, A. (1990). *Cell* 62, 1143.
Mortimer, R.K., Contopoulou, R. and Schild, D. (1981). *Proceedings of the National Academy of Sciences, USA* 78, 5778.
Moustacchi, E. (1987). *In* "Advances in Radiation Biology" (J. Lett, ed.), Vol. 13, pp. 1–30. Academic Press, New York.
Muster-Nassal, C. and Kolodner, R. (1986). *Proceedings of the National Academy of Sciences, USA* 83, 7618.
Neil, D.L., Villasante, A., Fisher, R.B., Vetrie, D., Cox, B. and Tyler-Smith, C. (1990). *Nucleic Acids Research* 18, 1421.
Nickoloff, J.A., Singer, J., Hoekstra, M.F. and Heffron, F. (1989). *Journal of Molecular Biology* 207, 527.
Nilsson-Tillgren, T., Gjermansen, C., Kielland-Brandt, M.C., Petersen, J.G.L. and Holmberg, S. (1981). *Carlsberg Research Communications* 46, 65.
Nilsson-Tillgren, T., Gjermansen, C., Holmberg, S., Petersen, J.G.L. and Kielland-Brandt, M.C. (1986). *Carlsberg Research Communications* 51, 309.
Nurse, P. and Thuriaux, P. (1980). *Molecular and General Genetics* 96, 627.
Orr-Weaver, T.L. and Szostak, J.W. (1983). *Proceedings of the National Academy of Sciences, USA* 80, 4417.
Orr-Weaver, T.L. and Szostak, J.W. (1985). *Microbiological Reviews* 49, 33.
Orr-Weaver, T.L., Szostak, J.W. and Rothstein, R.J. (1981). *Proceedings of the National Academy of Sciences, USA* 78, 6354.
Ozenberger, B.A. and Roeder, G.S. (1991). *Molecular and Cellular Biology* 11, 1222.
Pachnis, V., Pevny, L., Rothstein, R. and Costantini, F. (1990). *Proceedings of the National Academy of Sciences, USA* 87, 5109.
Pavan, W.J., Hieter, P. and Reeves, R.H. (1990). *Proceedings of the National Academy of Sciences, USA* 87, 1300.
Petes, T.D. and Hill, C.W. (1988). *Annual Review of Genetics* 22, 147.
Petes, T.D., Malone, R.E. and Symington, L.S. (1991). *In* "The Molecular and Cellular

Biology of the Yeast *Saccharomyces*: Genome Dynamics, Protein Synthesis, and Energetics" (J.R. Broach, E.W. Jones and J.R. Pringle, eds), Vol. 1, pp. 407–521. Cold Spring Harbor Laboratory Press, Cold Spring Harbour.
Pompon, D. and Nicolas, A. (1989). *Gene* **83**, 15.
Prakash, L. (1981). *Molecular and General Genetics* **184**, 471.
Priebe, S.D., Hadi, S.M., Greenberg, B. and Lacks, S.A. (1988). *Journal of Bacteriology* **170**, 190.
Pringle, J.R. and Hartwell, L.H. (1981). In "The Molecular Biology of the Yeast *Saccharomyces*: Life Cycle and Inheritance" (J.N. Strathern, E.W. Jones and J.R. Broach, eds), Vol. 1, pp. 47–80. Cold Spring Harbor Laboratory, Cold Spring Harbour.
Proffitt, J.H., Davie, J.R., Swinton, D. and Hattman, S. (1984). *Molecular and Cellular Biology* **4**, 985.
Prudhomme, M., Martin, B., Mejean, V. and Claverys, J.-P. (1989). *Journal of Bacteriology* **171**, 5332.
Radman, M. (1988). In "Genetic Recombination" (R. Kucherlapati and G.R. Smith, eds), Vol. 20, pp. 169–192. American Society for Microbiology, Washington, DC.
Radman, M. and Wagner, R. (1986). *Annual Review of Genetics* **20**, 523.
Raveh, D.S., Hughes, S., Shafer, B. and Strathern, J.N. (1989). *Molecular and General Genetics* **220**, 33.
Ray, A., Siddiqi, I., Kolodkin, A.L. and Stahl, F.W. (1988). Journal of Molecular Biology **201**, 247.
Ray, A., Machin, N. and Stahl, F.W. (1989). *Proceedings of the National Academy of Sciences, USA* **86**, 6225.
Rayssiguier, C., Thaler, D.S. and Radman, M. (1989). *Nature* **342**, 396.
Razzaque, A., Chakrabati, S., Joffee, S. and Seidman, M. (1984). *Molecular and Cellular Biology* **4**, 435.
Reed, S.I. (1991). *Trends in Genetics* **7**, 95.
Resnick, M.A. (1969). *Genetics* **62**, 519.
Resnick, M.A. (1975). In "Molecular Mechanisms for Repair of DNA, Part B" (P.C. Hanawalt and R.B. Setlow, eds), Vol. pp. 549–556. Plenum Press, New York.
Resnick, M.A. (1976). *Journal of Theoretical Biology* **59**, 97.
Resnick, M.A. (1979). *Advances in Radiation Biology* **8**, 175.
Resnick, M.A. (1987). In "Meiosis" (P. Moens, ed.), Vol. pp. 157–209. Academic Press, New York.
Resnick, M.A. and Martin, P. (1976). *Molecular and General Genetics* **143**, 119.
Resnick, M.A., Boyce, J. and Cox, B. (1981). *Journal of Bacteriology* **146**, 285.
Resnick, M.A., Game, J.C. and Stasiewicz, S. (1983). *Genetics* **104**, 603.
Resnick, M.A., Chow, T.-Y.K., Nitiss, J. and Game, J. (1984). *Cold Spring Harbor Symposia on Quantitative Biology* **49**, 639.
Resnick, M.A., Skaanild, M. and Nilsson-Tillgren, T. (1989). *Proceedings of the National Academy of Sciences, USA* **86**, 2276.
Resnick, M.A., Zgaga, Z., Hieter, P., Westmoreland, J., Fogel, S. and Nilsson-Tillgren, T. (1992). *Molecular and General Genetics* **234**, 65.
Roeder, G.S. and Stewart, S.E. (1988). *Trends in Genetics* **4**, 263.
Roeder, G.S., Keil, R.L. and Voelkel-Meiman (1986). In "Current Communications in Molecular Biology: Mechanisms of Yeast Recombination" (A. Klar and J. Strathern, eds), pp. 29–33. Cold Spring Harbor Laboratory, Cold Spring Harbor.
Rolfe, M. (1985). *Current Genetics* **9**, 533.
Rolfe, M., Spanos, A. and Banks, G. (1986). *Nature* **319**, 339.
Rong, L., Palladino, F., Aguilera, A. and Klein, H. (1991). *Genetics* **127**, 75.
Roth, D.B., Nakajima, P.B., Menetski, J.P., Bosma, M.J. and Gellert, M. (1992). *Cell* **69**, 41.
Rothstein, R. (1983). *Methods in Enzymology* **101**, 202.
Rothstein, R., Helms, J.C. and Rosenberg, N. (1987). *Molecular and Cellular Biology* **7**, 1198.
Rowley, R., Hudson, J. and Young, P.G. (1992). *Nature* **356**, 353.
Ruby, S.W. and Szostack, J.W. (1985). *Molecular and Cellular Biology* **5**, 75.
Rudin, N. and Haber, J.E. (1988). *Molecular and Cellular Biology* **8**, 3918.

410   M.A. Resnick, C. Bennett, E. Perkins, G. Porter and S.D. Priebe

Rudin, N., Sugarman, E. and Haber, J.E. (1989). *Genetics* **122**, 519.
Russell, P. and Nurse, P. (1986). *Cell* **45**, 145.
Russell, P. and Nurse, P. (1987). *Cell* **49**, 559.
Sainz, J., Pevny, L., Wu, Y., Cantor, C.R. and Smith, C.L. (1992). *Proceedings of the National Academy of Sciences, USA* **89**, 1080.
Sassanfar, M. and Roberts, J.W. (1990). *Journal of Molecular Biology* **212**, 79.
Schiestl, R.H. and Petes, T.D. (1991). *Proceedings of the National Academy of Sciences, USA* **88**, 7585.
Schiestl, R.H. and Prakash, S. (1988). *Molecular and Cellular Biology* **8**, 3619.
Schiestl, R.H., Prakash, S. and Prakash, L. (1990). *Genetics* **124**, 817.
Schiestl, R.H., Reynolds, P., Prakash, S. and Prakash, L. (1989). *Molecular and Cellular Biology* **9**, 1882.
Sears, D.D., Hegemann, J.H. and Hieter, P. (1992). *Proceedings of the National Academy of Sciences, USA*, vol. **89**, pp. 5296–5300.
Shinohara, A., Ogawa, H. and Ogawa, T. (1992). *Cell* **69**, 457–470.
Shortl, D. and Botstein, D. (1983). *Methods in Enzymology* **100**, 457.
Siede, W. and Eckhardt, F. (1984). *Mutation Research* **129**, 3.
Snow, R. and Korch, C.T. (1970). *Molecular and General Genetics* **107**, 201.
Strathern, J.N., Klar, A.J.S., Hicks, J.B., Abraham, J.A., Ivy, J.M., Nasmyth, K.A. and McGill, C. (1982). *Cell* **31**, 183.
Sugawara, N. and Haber, J.E. (1992). *Molecular and Cellular Biology* **12**, 563.
Sun, H., Treco, D., Schultes, N.P. and Szostak, J.W. (1989). *Nature* **338**, 87.
Sun, H., Treco, D. and Szostak, J.W. (1991). *Cell* **64**, 1155.
Szostak, J.W., Orr, W.T., Rothstein, R.J. and Stahl, F.W. (1983). *Cell* **33**, 25.
Thaler, D. and Stahl, F. (1988). *Annual Review of Genetics* **22**, 169.
Thomas, B.J. and Rothstein, R. (1989a). *Cell* **56**, 619.
Thomas, B.J. and Rothstein, R. (1989b). *Genetics* **123**, 725.
Tobey, R.A. (1975). *Nature* **254**, 245.
Tschumper, G. and Carbon, J. (1986). *Nucleic Acids Research* **14**, 2989.
Vincent, R.D., Hofmann, T.J. and Zassenhaus, H.P. (1988). *Nucleic Acids Research* **16**, 3297.
Vindelov, L.L., Hansen, H.H., Gersel, A., Hirsch, F.R. and Nissen, N.I. (1982). *Cancer Research* **42**, 2499.
Walker, G.C. (1984). *Microbiological Reviews* **48**, 60.
Walker, G.C. (1985). *Annual Review of Biochemistry* **54**, 425.
Weiffenbach, B. and Haber, J.E. (1981). *Molecular and Cellular Biology* **1**, 522.
Weinert, T.A. and Hartwell, L.H. (1988). *Science* **241**, 317.
Weinert, T.A. and Hartwell, L.H. (1990). *Molecular and Cellular Biology* **10**, 6554.
Welsh, K.M., Lu, A.-L., Clark, S. and Modrich, P. (1987). *Journal of Biological Chemistry* **262**, 15624.
Wennerberg, J., Alm, P., Biorklund, A., Killande, D., Langstrom, E. and Trope, C. (1984). *International Journal of Cancer* **33**, 213.
White, C.I. and Haber, J.E. (1990). EMBO *Journal* **9**, 663.
White, J.H., Lusnak, K. and Fogel, S. (1985). *Nature* **315**, 350.
Wildenberg, J. (1970). *Genetics* **66**, 291.
Williams, G.T. (1991). *Cell* **65**, 1097.
Williamson, M.S., Game, J.C. and Fogel, S. (1985). *Genetics* **110**, 606.
Witkin, E.M. (1991). *Biochimie* **73**, 133.
Wu, T.-C. and Lichten, M. (1992). *In* "Meiosis II: Contemporary Approaches to the Study of Meiosis" (G.M. Cooper, F.P. Haseltine, S. Heyner and J.F. Strauss III, eds). AAAs, Washington, DC.

11 Mating and Mating-type Interconversion in *Saccharomyces cerevisiae* and *Schizosaccharomyces pombe*

George F. Sprague, Jr.

Institute of Molecular Biology and Department of Biology, University of Oregon, Eugene, Oregon 97403, USA

I. INTRODUCTION

The life cycles of *Saccharomyces cerevisiae*, referred to as budding yeast and *Schizosaccharomyces pombe*, referred to as fission yeast show several strong parallels, including the occurrence of two haploid mating types

The Yeasts Vol. 6, 2nd edition
ISBN 0-12-596416-1

determined by alleles of a single genetic locus, and a mating reaction initiated by response to secreted pheromones. In addition, both yeasts exhibit the remarkable ability to undergo mating-type interconversion. That is, a meiotic spore carrying one allele at the mating-type locus often gives rise, by virtue of a genetic switch, to mitotic progeny carrying the other allele. Cells of both mating types are therefore found in a clone produced by a single spore. Strains exhibiting this behaviour are referred to as homothallic and, in fact, wild-type strains of both yeasts are homothallic. Early in the development of these species as genetically tractable organisms, heterothallic strains in which mating type is stably inherited through mitosis were isolated as laboratory derivatives of wild-type strains. Naturally, most studies of mating-type switching use homothallic strains, and most studies of mating-type determination and the mating process itself use heterothallic strains.

In recent years, a great deal of progress has been made toward understanding the molecular mechanisms that underlie mating-type determination, pheromone response and mating-type switching in the two yeasts. In this article, I review this progress, highlighting when appropriate the similarities and differences between the yeasts. For each topic I first discuss the current understanding for budding yeast and then make comparisons to fission yeast. I adopt this approach in part because a deeper molecular understanding has often been achieved with Sacch. cerevisiae and, in part I confess, because my laboratory studies this yeast. Because both yeasts are only distantly related, a comparison of the mechanisms used by these two organisms may reveal fundamental molecular strategies that will apply to other species. Even in situations where molecular details are not precisely conserved, a general theme may emerge from the comparison. Quite a number of articles reviewing these topics have been published (Herskowitz and Oshima, 1981; Haber, 1983; Nasmyth and Shore, 1987; Cross et al., 1988; Egel, 1989; Herskowitz, 1989; Klar, 1989; Egel et al., 1990a; Fields, 1990; Blumer and Thorner, 1991; Dolan and Fields, 1991; Marsh et al., 1991; Rose, 1991; Sprague, 1990, 1991; Klar, 1992; Kurjan, 1992). Often these reviews provide different perspectives and emphasize particular aspects of mating or mating-type interconversion in one or the other yeast. Some of these reviews offer an historical perspective and discuss the experimental foundation for conclusions that serve as the starting point for this review, such as the $\alpha1-\alpha2$ hypothesis or the cassette hypothesis.

II. OVERVIEW OF MATING

Haploid cells of budding yeast exhibit either of two distinct phenotypes, **a** or α. These cell types proliferate by mitotic cell division and, hence, large pure populations of each cell type can be maintained. However, when co-cultured, **a** and α cells communicate their presence to each other, cease mitotic cell division, and carry out a mating process that leads to formation of a third cell type, an **a**/α diploid. Like the haploid cells, **a**/α cells can proliferate by mitosis, but unlike **a** and α cells, they cannot mate. Instead, **a**/α cells have a new property. When deprived of nutrients, especially nitrogenous ones, they undergo meiosis and sporulation, thereby regenerating haploid cell types.

The mating process requires that **a** and α cells first communicate their presence to each other and then interact physically so that cell and nuclear fusion can occur. Communication is achieved via cell type-specific production of secreted mating factors (pheromones) and receptors for those pheromones. In particular, only α cells secrete the 13 amino-acid-residue peptide, α-factor pheromone, which binds to a receptor present only on the surface of **a** cells. Likewise, **a** cells secrete a 12-amino-acid residue farnesylated lipopeptide, **a**-factor pheromone, which interacts with a receptor present on the surface of α cells. Binding of pheromone to receptor activates an intracellular signal-transduction pathway that is shared by **a** and α cells; that is, the intracellular pathway does not involve proteins present only in α or only in **a** cells (Bender and Sprague, 1986; Nakayama *et al.*, 1987). Propagation of a signal along this pathway elicits many physiological changes in the responding cell that allow it to mate. In essence, the role of the pheromones is to induce a transient developmental transformation of growing mitotically active cells into cells that express the characteristics of gametes. As a result of this cellular differentiation, cell and nuclear fusion can proceed efficiently to produce an **a**/α diploid.

Haploid *Schiz. pombe* also exhibits either of two mating types, designated *P* and *M*. Like their counterparts in *Sacch. cerevisiae*, these cells can grow vegetatively by mitosis but, unlike **a** and α cells, *P* and *M* cells must receive two environmental inputs in order to prepare for and execute mating. First, *P* and *M* cells must communicate via pheromones and pheromone receptors, which are similar to those produced by **a** and α cells of *Sacch. cerevisiae*. The pheromone secreted by *M* cells (*M*-factor) has recently been shown to be a nine amino-acid-residue lipopeptide, with the lipid moiety most likely to be a farnesyl group (Davey, 1992). The structure of *P* factor is not yet known. As discussed in more detail later in this review, the pheromone receptors are members of the same large receptor family to

which the budding-yeast pheromone receptors belong (Kitamura and
Shimoda, 1991; M. Yamamoto, personal comunication). Second, mating
of *P* and *M* cells occurs only if nutrients, particularly nitrogenous ones, are
limiting.

In the normal course of the life cycle of *Schiz. pombe*, the diploid phase
is transient. The zygote formed by mating of a *P* cell with an *M* cell on
a nitrogen-deficient medium immediately undergoes meiosis and sporulation. However, if the zygote is moved to a nitrogen-sufficient medium prior
to initiation of meiosis, it can grow vegetatively by mitosis. Thus, nutrient
limitation is required both for mating and for sporulation by fission yeast,
but only for sporulation by budding yeast. Because of this difference, the
usual vegetative state of a wild-type, homothallic fission yeast is as haploid
P and *M* cells, whereas the usual vegetative state of a wild-type, homothallic
budding yeast is as **a**/α diploid cells.

III. MATING-TYPE DETERMINATION

As the foregoing overview emphasized, the life cycles of both budding and
fission yeast involve three cell types, each with unique properties and
capabilities. How is this cell specialization achieved? In both species, cell
type is determined by alleles of a single genetic locus, namely the mating-
type locus, by convention symbolized *MAT* for budding yeast and *mat1* for
fission yeast. (For details of genetic nomenclature see Appendix 1 (p. 461).
For clarity, I will also use a minus (−) superscript to denote mutant alleles
in both fission and budding yeast. Budding-yeast proteins are designated
by upper case Roman letters and fission-yeast proteins by lower case Roman
letters.) Thus, in budding yeast, **a** cells contain *MAT***a**, α cells contain
*MAT*α, and **a**/α cells contain both alleles. Similarly, in fission yeast, *P* cells
contain *mat1-P*, *M* cells contain *mat1-M*, and *P/M*-cells contain both
alleles. A more explicit question, therefore, is: "What are the roles of the
products encoded by the mating-type locus alleles?" This question can be
answered precisely and in detail for budding yeast. The *MAT* products are
DNA-binding proteins that control transcription of structural genes for cell
type-specific proteins. For fission yeast, it seems likely that at least some
of the *mat1* products are also DNA-binding proteins, but this has not been
demonstrated experimentally. Potential target cell type-specific genes have
been identified only recently. Despite this potential similarity between
budding and fission yeasts, genetic analysis of *mat1* functions implies that
the molecular basis of their action will not be a simple replay of the
molecular strategy used by budding yeast.

A. Cell specialization in *Saccharomyces cerevisiae*: regulatory circuitry

The first insight into the role of the mating-type locus came from analysis of strains harbouring mutations at *MATα* or *MATa* (MacKay and Manney, 1974; Kassir and Simchen, 1976). This analysis led to the view that these alleles encode regulatory proteins that govern expression of four gene sets: α-specific genes, **a**-specific genes, haploid-specific genes and diploid-specific genes (Fig. 1, the α1–α2 hypothesis; Strathern *et al.*, 1981). This view has been verified using cloned structural genes for various cell type-specific proteins. These latter experiments also demonstrate that regulation

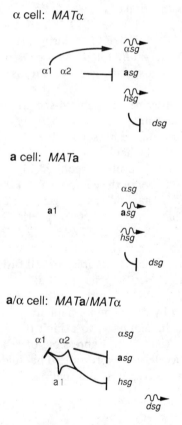

Fig. 1. The α1–α2 hypothesis. The mating-type locus products govern transcription of α-specific genes (α*sg*), **a**-specific genes (**a***sg*), haploid-specific genes (*hsg*) and diploid-specific genes (*dsg*). The pattern of transcription (wavy lines) of these gene sets is shown for the three cell types. Lines with arrowheads indicate stimulation of transcription. Lines with terminal bars indicate inhibition of transcription. See text for discussion of the circuitry.

is at the level of transcription. The regulatory circuitry by which the *MAT*-encoded mating-type locus products generate a unique pattern of transcription in each cell type is summarized below.

1. α cells

The locus *MATα* encodes two proteins, namely α1 and α2. The first of these proteins, α1, activates transcription of the α-specific genes set, which includes the a-factor receptor structural gene (*STE3*) and the α-factor pheromone structural genes (*MFα1* and *MFα2*) (Sprague *et al.*, 1983; Fields and Herskowitz, 1985; Inokuchi *et al.*, 1987; Flessel *et al.*, 1989). Conversely, α2 represses transcription of the a-specific gene set, which includes the α-factor receptor structural gene (*STE2*) and the a-factor pheromone structural genes (*MFA1* and *MFA2*) (Hartig *et al.*, 1986; Michaelis and Herskowitz, 1988; Dolan *et al.*, 1989). As a consequence of the action of these two regulators, the α-specific gene set is transcribed but the a-specific gene set is not.

Alpha cells also transcribe the haploid-specific gene set. As discussed below, a/α cells contain a unique repressor activity that blocks transcription of these genes. Because this repressor is absent from α and a cells, both of these cell types transcribe haploid-specific genes. As might be expected, many of these genes encode proteins required for mating by both a and α cells. For example, a number of the components of the pheromone-response pathway are specified by haploid-specific genes, including *GPA1 (SCG1)*, *STE4* and *STE18*, which encode a heterotrimeric G protein (Jahng *et al.*, 1988; Whiteway *et al.*, 1989), and *STE12*, which encodes a pheromone-responsive transcription factor (Fields and Herskowitz, 1987). In addition, proteins such as FUS1, which is required for cell fusion by the mating pair, are encoded by haploid-specific genes (McCaffrey *et al.*, 1987; Trueheart *et al.*, 1987). Finally, α cells fail to transcribe the diploid-specific gene set because it is repressed by one or more haploid-specific gene products (see a/α cells; see below). In summary, to exhibit the α cell phenotype, a cell must transcribe two gene sets. Expression of α-specific genes enables communication with a cells and expression of haploid-specific genes enables execution of pheromone response and mating.

2. a cells

The locus *MATa* encodes a single transcription regulator, namely a1, but it has no role in controlling expression of the α- or a-specific gene sets (Kassir and Simchen, 1976). Rather, the appropriate pattern of expression of these genes is the simple consequence of the absence of the two *MATα*-encoded regulators. Absence of α1 precludes transcription of α-specific genes, while absence of α2 permits transcription of a-specific genes. As expected given this view, a *matα1⁻matα2⁻* double mutant has the phenotype of an **a** cell; this was a key observation in formulation of the α1-α2 hypothesis (Strathern *et al.*, 1981).

Analysis of RNA reveals that *MATa* encodes a second transcribed gene, **a**2 (Klar *et al.*, 1981; Nasmyth *et al.*, 1981). However, mutation of **a**2 leads to no discernible phenotype (K. Tatchell, personal communication) and the sequence of the **a**2 transcription unit (deduced from the *MATa*2 DNA sequence) implies that it could encode only a very short protein. Whether this protein is actually synthesized has not been examined but at present no role in cell-type determination can be ascribed to this hypothetical gene product. For reasons already discussed for α cells, **a** cells transcribe haploid-specific genes but not diploid-specific genes.

3. a/α cells

These cells contain two repressor activities whose combined action prevents expression of α-, **a**- and haploid-specific genes, and permits expression of diploid-specific genes. First, the repressor α2 blocks transcription of **a**-specific genes, as it does in α cells. Second, in concert with **a**1, α2 exhibits a different DNA-binding specificity (Goutte and Johnson, 1988; Dranginis, 1990). This new activity, **a**1·α2, leads to repression of both α-specific and haploid-specific genes. Repression of α-specific genes is indirect. The combination **a**1·α2 binds to a site within the control region for the *MATα*1 gene and blocks its transcription, thereby preventing transcription of α-specific genes (Miller *et al.*, 1985). Repression of haploid-specific genes is direct in some cases, indirect in others. For example, the upstream control region of *STE5*, which encodes a component of the pheromone-response pathway, contains two presumptive **a**1·α2 binding sites (Miller *et al.*, 1985). On the other hand, *FUS1*, which is required for cell fusion, does not contain **a**1·α2 sites (McCaffrey *et al.*, 1987; Trueheart and Fink, 1989; Hagen *et al.*, 1991). Rather, *FUS1* fails to be transcribed in **a**/α cells because the transcription factor that activates its transcription, *STE12*, is itself repressed by **a**1·α2 (Fields and Herskowitz, 1987).

Transcriptional activation of diploid-specific genes is also an indirect consequence of the action of **a**1·α2. Genes such as *IME1* must be expressed to allow meiosis and sporulation to proceed in **a**/α diploids (Kassir *et al.*, 1988). Transcription of *IME1* requires that a cell contains both **a**1 and α2 and is also subject to nutrient limitation. The role of **a**1·α2 is to repress transcription of an inhibitor of meiosis, namely *RME1* (Mitchell and Herskowitz, 1986), apparently by binding to the promoter of *RME1* (Covitz *et al.*, 1991). Thus, *RME1* is another haploid-specific gene. In haploid cells, the presence of high levels of *RME1* prevents entry into meiosis. Repression of *RME1* mediated by a1·α2 in diploid cells, coupled with the appropriate nutritional signals, allows transcription of *IME1* and other genes required for sporulation and thereby permits initiation of meiosis.

B. Cell specialization in *Saccharomyces cerevisiae*: molecular basis for transcriptional regulation

The foregoing discussion provided partial insight as to how the mating-type locus products determine cell type. In molecular terms, how do these products bring about the transcriptional regulation that differentiates **a**, α and **a**/α cells? Given that α1, α2, and **a**1 regulate transcription, a reasonable supposition is that they are DNA-binding proteins. In fact, the sequences of **a**1 and α2 proteins (deduced from DNA sequences) support this supposition. Both of these proteins are members of a family of known DNA-binding proteins, the homeodomain family, first defined by proteins that play crucial roles in the development of *Drosophila* spp. (Laughon and Scott, 1984; Shepherd *et al.*, 1984). The sequence of α1, on the other hand, does not contain known DNA-binding motifs. The protein is very basic, however, especially in the N-terminal half, which exhibits a net charge of +21.

To determine how *MAT* products achieve transcription regulation, two complementary sets of studies have been carried out with representatives of each class of cell type-specific genes. On the one hand, mutational analysis of promoter regions has revealed minimum sequences that are necessary and sufficient for each type of regulation *in vivo*. On the other hand, analysis of protein–DNA complexes that form on these regions *in vitro* has shown that each *MAT* product participates in binding to the control regions of appropriate target genes and has revealed other proteins that are essential for protein–DNA complex formation. Together, these studies demonstrated that a combinatorial strategy generates the three cell types. Regulated transcription of α- and **a**-specific genes results from interaction of α1 and α2 with a general transcription factor/regulator, MCM1, which is present in all three cell types. The factor α1·MCM1 activates transcription of α-specific genes. Acting alone, MCM1 activates transcription of **a**-specific genes in **a** cells, while α2·MCM1 represses transcription of this gene set in α cells. Regulated transcription of haploid-specific genes results from a different combinatorial association. Together a1 and α2 form a heterodimer that has a different DNA-binding specificity than does α2 homodimer.

1. Activation of α-specific genes by α1·MCM1

Control regions for α-specific genes contain a 26 bp sequence element that functions as an upstream activation sequence (UAS) and imparts regulated transcription. The UAS is a composite of two elements. One part, termed P(asym), is an asymmetric version of a 16 bp symmetric dyad, P(PAL). The second part is an adjacent 10 bp Q sequence. Neither P(asym) nor Q alone has UAS activity but the composite QP(asym) element confers α-specific expression to reporter genes (Jarvis *et al.*, 1988). Symmetrical P elements such as P(PAL) also have UAS activity but in these cases the ability to act as a UAS is observed in all three cell types.

These observations led to the proposal that α-specific transcription is achieved by interaction and binding to QP(asym) of two proteins (Fig. 2; Bender and Sprague, 1987; Jarvis *et al.*, 1988), α1 and a general transcription factor, initially called PRTF but now known to be encoded by the *MCM1* gene (Bender and Sprague, 1987; Hayes *et al.*, 1988; Jarvis *et al.*, 1989; Keleher *et al.*, 1989; Passmore *et al.*, 1989; Ammerer, 1990). Support for this proposal has come from *in vitro* experiments that examined formation of protein–DNA complex on α-specific UAS elements. Individually, MCM1 and α1 bind poorly or not at all to QP(asym), but together they bind co-operatively and strongly (Bender and Sprague, 1987; Tan *et al.*, 1988). In contrast, MCM1 binds alone with high affinity to P(PAL). The interdependent binding of α1 and MCM1 to QP(asym) has three consequences that account for α-specific transcription. Firstly, a general transcription factor is recruited to bind to the elements. Secondly, α1 is thought to induce a conformational change in MCM1 that is important for its transcription activation function (Tan and Richmond, 1990). Thirdly, α1 itself has the ability to activate transcription if it is delivered to DNA as a LexA–α1 hybrid protein binding to LexA operator sites (Sengupta and Cochran, 1990). Presumably, α1 also contributes to transcription activation as part of the α1·MCM1 complex at α-specific genes. Thus, the interdependent binding of α1 and MCM1 may be viewed as an example of mutual recruitment by transcription factors.

420   G.F. Sprague

α cell

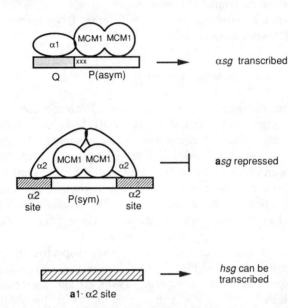

Fig. 2. Molecular model for cell type-specific transcription in α cells of *Saccharomyces cerevisiae*. The DNA-binding sites for proteins encoded by the mating-type locus are shown for α-specific, **a**-specific and haploid-specific genes. The protein α1 binds co-operatively with MCM1 to the QP (asym) sequence that functions as the upstream activation sequence (UAS) for α*sg* and, consequently, the gene set can be transcribed. The asymmetric nature of the P sequence is denoted by XXX. The protein α2 binds co-operatively with MCM1 to an operator sequence present in the upstream control regions of **a***sg*. The P segment of the operator is highly symmetrical (P (sym)). The α2·MCM1 complex brings about repression, apparently by recruiting SSN6 and TUP1. The **a**1·α2 sites within the control regions of *hsg* are not occupied because α cells do not contain **a**1 protein. Members of this gene set can therefore be transcribed by virtue of other sites that serve as UAS elements. Reprinted from Sprague (1992).

2. Regulation of **a**-specific genes by MCM1 and α2·MCM1

Control regions for **a**-specific genes also contain versions of the P sequence. At these genes, the sequence is highly symmetrical but not perfectly so, and MCM1 can bind well without interaction with coregulators (Bender and Sprague, 1987; Keleher *et al.*, 1988; Tan *et al.*, 1988). The MCM1/P sequence interaction contributes to transcription of the gene set in **a** cells (Fig. 3). In particular, it has been found that either mutation of the P site or depletion of MCM1 protein causes a drastic decrease (more than 50 times) in transcript levels (Keleher *et al.*, 1989; Kronstad *et al.*, 1987; Elble

a cell

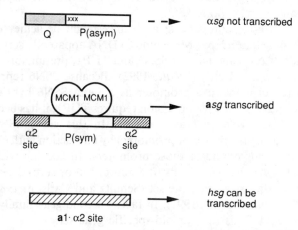

Fig. 3. Molecular model for cell type-specific transcription in **a** cells of *Saccharomyces cerevisiae*. The absence of α1 precludes interdependent binding of MCM1 and α1 to QP (asym) and hence α*sg* are not transcribed. In contrast, MCM1 binds well to the P (sym) sequence of **a***sg* and contributes to their transcription. Repression does not occur because α2 is not present. The *hsg* gene set can be transcribed because **a**1·α2 sites are not occupied. Reprinted from Sprague (1992).

and Tye, 1991; Hwang-Shum *et al.*, 1991). However, residual transcription is seen because the control regions of **a**-specific genes contain binding sites for STE12, a protein that mediates increased transcription in response to pheromone (see Section IV).

The MCM1/P sequence interaction has a second role in controlling expression of **a**-specific genes. It is essential for their repression in α cells (Keleher *et al.*, 1987). The P sequences of this gene set are flanked by binding sites for α2 protein. Proteins MCM1 and α2 bind co-operatively to the composite sequence (the α2·MCM1 operator) and formation of the ternary complex is required for repression *in vivo*. That is, mutation of either the P sequence or the α2 site at a particular **a**-specific gene allows transcription of that gene in α cells (Keleher *et al.*, 1989; Hwang-Shum *et al.*, 1991).

A remarkable feature of α2·MCM1-mediated repression is its ability to operate at a distance. This property is apparent at **a**-specific genes where α2·MCM1 prevents STE12 from bringing about transcription, but is more strikingly observed in artificial reporter-gene constructs. In these settings, an α2·MCM1 operator can repress transcription from UAS elements that are normally indifferent to cell type. Moreover, the operator functions when placed either upstream of the UAS or between the UAS and TATA element, although repression is more evident with the latter geometry (Johnson and Herskowitz, 1985).

Insight as to how repression at a distance might be achieved has come from the finding that α2·MCM1 bound to DNA apparently serves to recruit another pair of proteins, namely SSN6 and TUP1, presumably by protein–protein interactions (Keleher et al., 1992). Because SSN6 represses transcription when brought to a promoter as a LexA-SSN6 hybrid bound to LexA sites, and because this repression requires TUP1, it has been proposed that SSN6·TUP1 is a general repressor. In this view, SSN6·TUP1 is recruited to particular classes of promoters by interaction with class-specific regulatory proteins bound at those promoters. In keeping with this idea, mutation of SSN6 or TUP1 leads to derepression of several gene families, in addition to the a-specific gene set (Schultz and Carlson, 1987; Lemontt et al., 1980; Williams et al., 1991). In fact, a second gene family influenced by SSN6 and TUP1 is the haploid-specific gene set.

3. Repression of haploid-specific genes by a1·α2

Upstream regions from a number of haploid-specific genes contain a sequence element, the a1·α2 operator, which is similar in sequence to the α2·MCM1 operator (Fig. 4; Miller et al., 1985). The proteins a1 and α2 bind as a heterodimer to the a1·α2 operator sequences; α2 alone binds poorly, if at all (Goutte and Johnson, 1988; Draginis, 1990). Where it has been examined, mutation of the a1·α2 operator abolishes repression in a/α cells. Moreover, when inserted within the control regions of reporter genes, synthetic versions of the sequence bring the reporter genes under a1·α2 control (Miller et al., 1985). Because mutation of SSN6 or TUP1 leads to derepression of haploid-specific genes, it is likely that a1·α2 bound at its operator serves to recruit SSN6 and TUP1, just as α2·MCM1 bound at its operator within the promoters of a-specific genes is argued to do.

4. MCM1, α2 and a1 are members of large protein families

The foregoing discussion has highlighted the importance of combined associations among regulatory proteins for determining the α and a/α cell types. This strategy is made all the more intriguing by the realization that MCM1, α2 and a1 are members of larger protein families.

The MCM1 protein is part of a family that was first defined by a mammalian transcription factor, SRF (serum response factor), and that

a/α cell

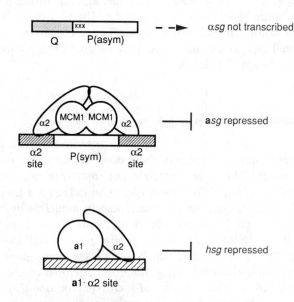

Fig. 4. Molecular model for cell type-specific transcription in a/α cells of *Saccharomyces cerevisiae*. Proteins a1 and α2 form a heterodimer that binds to sites present within the control regions of some *hsg* and also with the control region of the *MATα1* gene. The complex represses transcription from the UAS elements of these genes, perhaps by recruiting SSN6 and TUP1. Repression of *MATα1* precludes formation of the α1·MCM1 complex at QP (asym), and hence α*sg* are not transcribed. a*sg* are repressed as described in Fig. 3. Reprinted from Sprague (1992).

now includes proteins required for proper floral development in plants (Norman *et al.*, 1988; Yanofsky *et al.*, 1990; Sommer *et al.*, 1991). The similarity among these proteins can be quite striking; in a stretch of 80 residues, 55 are identical between SRF and MCM1. Moreover, the region of MCM1 homologous to SRF is the region responsible for interaction with coregulators α1 and α2 (Christ and Tye, 1991; Primig *et al.*, 1991; Bruhn *et al.*, 1992). The SRF region is known to interact with at least two coregulators, p62^{TCF} and SAP-1 (Shaw *et al.*, 1989; Dalton and Treisman, 1992). The SAP-1 protein does not show similarity to any of MCM1's coregulators but it will be interesting to learn whether p62^{TCF} does.

The α2 and a1 proteins are members of a different family, the homeodomain family, first defined by proteins that play crucial roles in the development of *Drosophila* spp. (Laughon and Scott, 1984; Shepherd *et al.*, 1984). As with **a**1 and α2, homeodomain proteins synthesized by *Dnosophila* spp. are thought to interact and form complexes with unique DNA-binding specificities. The ability of homeodomain proteins to interact with proteins in the SRF family, exemplified by the α2·MCM1 interaction, amplifies the combinatorial possibilities. Thus, the strategies used in determining cell type in budding yeast may be used on a grander scale in multicellular plants and animals.

C. Cell specialization in the fission yeast, *Schizosaccharomyces pombe*

The *mat1-P* and *mat1-M* alleles in the fission yeast, *Schiz. pombe* each specify two genes. The genes *mat1-Pc* and *mat1-Mc* are required for conjugation (Kelly *et al.*, 1988; Egel *et al.*, 1990b). They are transcribed at a basal level in vegetative cells and are modestly inducible by nutritional starvation. One attractive hypothesis is that *mat1-Pc* and *mat1-Mc* are required for transcription of *P*- and *M*-specific genes. Candidates for such genes include *mam2* and *map3*, which encode the pheromone receptors (Kitamura and Shimoda, 1991; M. Yamamoto, personal communication), and *mfm1* and *mfm2*, which encode *M*-factor pheromone (Davey, 1992). However, it is not known whether expression of these genes is controlled at the transcriptional level, nor is it known whether the Mc and Pc products are DNA-binding proteins. In support of this latter possibility, the Mc product shows considerable sequence similarity to the HMG class of DNA-binding proteins (Gubbay *et al.*, 1990). Among this class of proteins is a mammalian Y-chromosome-specific protein that may have a role in testes determination (Gubbay *et al.*, 1990). In addition to carrying out essential roles in conjugation, *mat1-Pc* and *mat1-Mc* are required for meiosis and sporulation.

The other genes specified by the *mat1* alleles, namely *mat1-Pm* and *mat1-Mm* (sometimes called *mat1-Pi* and *mat1-Mi*), are also required for meiosis and sporulation but, unlike Pc and Mc, they are not required for conjugation. The genes *mat1-Pm* and *mat-1Mm* are not transcribed in vegetative cells, but are highly transcribed under conditions that lead to mating (Kelly *et al.*, 1988). Induction of *mat1-Pm* requires both the pheromone and nutritional signals, whereas induction of *mat1-Mm* appears to require only a nutritional signal. The *mat1-Pm* product contains a homeodomain motif and is therefore likely to bind to DNA (Kelly *et al.*, 1988). Thus, it is tempting to speculate that the *mat1-Pm* and *mat1-Mm* gene products function in a manner analogous to the **a**1 and α2 products of budding yeast. That is, perhaps Pm and Mm form a heteromer that signals creation of the diploid state by binding to the promoter regions of genes that control meiosis and sporulation.

The analysis carried out thus far on fission-yeast cell-type determination reveals two departures from the strategies used by budding yeast. First, each of the fission-yeast *mat1* alleles encodes a function required for conjugation. The mutants *mat1-Pc⁻* and *mat1-Mc⁻* are mating defective, as are strains carrying a deletion of *mat1* (Kelly *et al.*, 1988). In contrast, the budding-yeast *MATa* locus is not required for mating. Deletion of the *MAT* locus, in particular the absence of α1 and α2 functions, creates a cell with the **a** phenotype. In effect, budding yeast exhibits a default haploid-cell type (the **a** phenotype) whereas fission yeast does not.

The second departure in regulatory strategy concerns expression of the mating-type locus genes. Transcription of budding-yeast *MAT* genes is unaffected by pheromone or nutritional condition whereas transcription of fission-yeast *mat1* genes is stimulated by both pheromone and nutrient limitation. Pheromone response is discussed later in this review. Here, I discuss the present understanding of induction of the *mat1* genes in response to nutrient limitation.

Genetic analysis has revealed two genes that are thought to have central roles in mediating this regulation. One of these, *ste11*, is presumed to encode a transcription factor required for expression of *mat1* genes (Sugimoto *et al.*, 1991). In particular, basal levels of *mat1* transcripts are greatly decreased in *ste11⁻* mutants and the levels are not increased upon starvation of nitrogenous nutrients. Conversely, unregulated overexpression of wild-type *ste11* causes cells to produce zygotes and asci in a rich medium. In keeping with its presumed role in transcription of *mat1*, the amino-acid-residue sequence of ste11 protein shows strong similarity with the HMG group of DNA-binding proteins (Sugimoto *et al.*, 1991) and ste11 binds to DNA *in vitro*. Binding appears to require a DNA sequence motif present within the promoter regions of *mat1-P* and *mat1-M*, and of other target genes. Interestingly, the *ste11* gene itself contains a version of this sequence, raising the possibility that *ste11* is subject to autoregulation.

How might nutrient deprivation control the activity or abundance of ste11? Nitrogen-nutrient starvation leads to a lowering of intracellular cAMP concentration. Although the pathway whereby cAMP concentrations are regulated is poorly understood, one consequence of lower concentrations is induction of *ste11* transcription (Sugimoto *et al.*, 1991). This finding, together with the possibility that *ste11* is autoregulated, has led to the following hypothesis. The ste11 protein is a transcription activator subject to phosphorylation by protein kinase A. When cAMP concentrations are high, ste11 is phosphorylated and therefore inactive. Upon nitrogen-nutrient starvation, cAMP concentrations fall, protein kinase A activity diminishes and, hence, active unphosphorylated ste11 can accumulate. Active ste11 stimulates transcription of target genes, including *mat1* and *mei2*, genes whose products promote meiosis. Because the *ste11* gene is also a target, a positive feedback loop is established, which may operate as a switch to commit the cell to sexual development.

The second gene thought to play a central role in regulating transcription of *mat1* genes is *pat1*, which encodes another protein kinase (Nielson and Egel, 1990). Loss of pat1 function leads to conjugation and meiosis even in media rich in nitrogenous nutrients, implying that pat1 is a negative regulator of these processes. Through use of a temperature-sensitive *pat1* allele, Nielson and Egel (1990) showed that partial inactivation of pat1 (achieved at a semi-permissive temperature) led to induction of *mat1* transcription and other events of conjugation. Induction of meiosis required total inactivation of pat1 at the restrictive temperature. This hierarchy of sensitivities to pat1-mediated inhibition may contribute to an orderly progression of events, allowing conjugation to occur before meiosis is induced. In any event, these data provoke two questions concerning the pat1 product. Firstly, how is its activity regulated and, secondly, what are its targets? Neither question can be answered yet with much sophistication. It is possible that pat1-kinase activity is sensitive to cAMP concentration and thereby sensitive to the nutritional state of the cell, but there is no direct evidence bearing on this. Activity of pat1 kinase may also be regulated by the pheromone-signalling pathway (see p. 438) as transcripts, such as *mat1-Pm*, which are inducible by pheromone, are induced in the *pat1$^-$* mutant. Targets of pat1 kinase are also unknown but may include ste11 and mei2, although the data are inconclusive. Mutation of *ste11* does suppress the phenotype of *pat1* mutations (Watanabe *et al.*, 1988; Sugimoto *et al.*, 1991). This finding could be taken as evidence that ste11 functions downstream of pat1, but is equally compatible with the possibility that ste11 and pat1 function on independent pathways. For example, in the absence of ste11-mediated transcription of *mat1*, the *pat1* mutation may not be able to exert its independent effect on transcription of the locus.

Two additional genes may have a role in controlling nutritional induction of *mat1* transcripts and other nutritionally sensitive genes. The *pac1* gene was identified as a multicopy plasmid suppressor of the *pat1$^-$* phenotype (Watanabe *et al.*, 1988). The second gene, *map1*, is required for induction of *mat1-Pc* transcript during nitrogen-nutrient starvation (Nielson and Egel, 1990). The *map1* gene appears to have a role only in *P* cells, as *M* cells carrying a *map1$^-$* mutation exhibit normal sexual differentiation. The molecular functions of *pac1* and *map1* are unknown.

IV. PHEROMONE RESPONSE

A. Overview of the signalling pathway in *Saccharomyces cerevisiae*

In response to pheromone, **a** and α cells undergo many physiological changes that allow mating to occur efficiently. This suite of changes includes: (i) increased transcription of genes whose products catalyse mating, either directly by promoting cell and nuclear fusion, or indirectly by contributing to alterations in cellular physiology necessary to accommodate these processes; (ii) arrest of the mitotic cell-division cycle in the G1 phase; (iii) preparation of the cell surface and nucleus for fusion with cognate organelles of the mating partner; (iv) alteration of cell polarity and morphology as the location of the mating partner is ascertained. As a result of these changes, the cell cycles of the mating pair are synchronized, and proteins that catalyse cell and nuclear fusion are poised to act. Mating to form an **a**/α diploid can proceed efficiently.

These events occur in both **a** and α cells. Although each cell type produces a unique pheromone and a unique cell-surface receptor that can be bound by the other cell type's pheromone, components of the intracellular signal-transduction pathway activated by the receptor–ligand are common to **a** and α cells. This conclusion follows from two types of observation. First, all mutations that effect response of **a** cells to α-factor also effect response of α cells to **a**-factor, except of course mutations in the receptor structural genes (Hartwell, 1980; Chan and Otte, 1982). Second, an **a** cell engineered to express the **a**-factor receptor, rather than the customary α-factor receptor, now responds to **a**-factor. Likewise, an **a** cell made to express the α-factor receptor responds to α-factor (Bender and Sprague, 1986; Nakayama *et al.*, 1987). Thus, the receptors are the only **a** and α-specific proteins required for response to the pheromones.

Genetic studies have been at the leading edge of efforts to understand the molecular basis of pheromone response. Two broad classes of mutants with altered pheromone response have been isolated. One class of mutants is non-responsive and defines elements necessary for pheromone response. These elements may be components of the pathway itself or may be functions required for synthesis or activity of pathway components. The other class of mutants display a phenotype that suggests that the pathway has become constitutively active, even in the absence of pheromone. These mutants may define elements that are required to hold the pathway in check until cells are exposed to pheromone. Formal genetic analysis, in particular, a study of double mutants created by combining mutations with contrasting phenotypes, has allowed the relative order of action of gene products to be deduced. In addition, cloning and sequencing of wild-type genes, coupled with physiological analyses of the mutants, has led to predictions for biochemical functions of the gene products. In some cases, these predictions have been verified by direct biochemical tests. Below, I provide a synopsis of the pathway and then discuss each of the identified components in detail.

The initial event in signal transduction involves binding of pheromone to receptor. Both receptors couple to the same heterotrimeric guanine nucleotide-binding protein (G protein) (Dietzel and Kurjan, 1987a; Jahng *et al.*, 1988; Fujimura, 1989). In the absence of a pheromone-receptor interaction, the G protein is inactive. When pheromone binds, the receptor presumably undergoes a conformational change that in turn activates the G protein. The immediate target of the activated G protein is not known, but five serine–threonine protein kinases that are required for signal transmission have been identified and the activities of at least three of these enzymes are increased when the pathway is stimulated.

What are the targets of the kinases? By analogy to other phosphorylation cascades that amplify an initial signal, some kinases may be substrates for others. Another likely target is STE12, the transcriptional activator that contributes to transcription of pheromone-responsive genes (Dolan *et al.*, 1989; Errede and Ammerer, 1989; Yuan and Fields, 1991). Upon stimulation of the signalling pathway, the ability of STE12 to promote transcription is increased and there is an attendant, perhaps causative, increase in its phosphorylation (Song *et al.*, 1991). Among the targets for this transcription factor are genes whose products participate in cell fusion, nuclear fusion, and control of the cell cycle.

Other targets of one or more of the kinases may be G1 cyclins or proteins that regulate cyclin activity or abundance. In mitotically active cells, G1 cyclins accumulate late in G1 and activate CDC28 protein kinase, itself required for progression from G1 to S in the cell cycle (Richardson *et al.*, 1990). Upon pheromone stimulation, the cyclins do not accumulate; consequently, CDC28 is inactive and cells arrest in G1.

In summary, it is now possible to trace the path of the signal from its initial generation at the cell surface to an identified transcription factor for pheromone-responsive genes. Increased transcription of target genes elevates intracellular concentrations of proteins that participate directly in cell fusion, nuclear fusion and cell-cycle arrest so, to a first approximation, the response pathway is one whereby the pattern of transcription is altered. Efficient completion of these events may require more than an increase in the quantity of appropriate gene products. For example, some products may be subject to pheromone-stimulated post-translational modifications that are important for maximal activity. Nonetheless, a molecular outline is now at hand for many of the physiological changes that characterize the pheromone response. The only changes not included in this outline are alterations of cell polarity and morphology, alterations for which there is as yet no molecular explanation.

B. Components of the pheromone signal-transduction pathway in *Saccharomyces cerevisiae*

The α-factor receptor is encoded by *STE2*, the a-factor receptor by *STE3* (Burkholder and Hartwell, 1985; Nakayama *et al.*, 1985; Hagen *et al.*, 1986). The first piece of evidence for this conclusion came from a study of mutants defective for mating and pheromone response. Mutation of *STE2* prevents mating and pheromone response by a cells but has no effect on the phenotype of α cells (MacKay and Manney, 1974; Hartwell, 1980). Similarly, mutation of *STE3* affects the mating and pheromone response only of α cells (MacKay and Manney, 1974; Hagen *et al.*, 1986). Since all other mutations that lead to a non-responsive phenotype affect both a and α cells, the cell-type specificity of *ste2* and *ste3* mutations suggests that they encode receptors. In keeping with the cell-type specificity of mutations, the genes exhibit cell-type specific transcription. The *STE2* gene is transcribed only in a cells, and *STE3* only in α cells (Sprague *et al.*, 1983; Hartig *et al.*, 1986). Compelling support for the suggestion that *STE2* and *STE3* encode receptors has come from three lines of experimentation. These experiments also provide important information about the receptors and insight as to how they function in transmembrane signalling.

First, ligand-binding studies reveal that α-factor binds specifically to a cells ($K_d \cong 10^{-9}$ M; 10^4 binding sites on each cell) and that a cells carrying temperature-sensitive mutations in *STE2* exhibit temperature-sensitive binding of α-factor (Jenness *et al.*, 1983, 1986). Moreover, bound α-factor can be cross-linked chemically to the *STE2* product (Blumer *et al.*, 1988). Binding experiments with a-factor have lagged behind those with α-factor, in part because of a-factor's extreme hydrophobicity. Nonetheless, preliminary experiments indicate that a-factor binds to α cells in a *STE3*-dependent fashion (J. Becker, personal communication).

Secondly, expression of *STE2* or *STE3* in cell types that normally do not express them allows those cells to respond to the cognate pheromone. As already described, transcription of *STE2* is limited to a cells, while transcription of *STE3* is limited to α cells. However, expression in other cell types can be achieved by placing the coding sequences under the control of a promoter whose activity is indifferent to cell type. By this device, it has been found that expression of *STE2* or *STE3* in *matα1* mutants, cells that do not express any **a**- or α-specific genes, allows mutants to respond to α-factor (*STE2*) or **a**-factor (*STE3*) (Bender and Sprague, 1986). Similarly, as alluded to in Section II, an **a** cell engineered to express *STE3* (rather than *STE2*) now responds to **a**-factor, and an α cell made to express *STE2* responds to α-factor (Bender and Sprague, 1986; Nakayama *et al.*, 1987).

Even more exotic expression studies give results that parallel those already summarized. For example, expression of *STE2* by oocytes of *Xenopus* spp. allows them to bind α-factor (Yu *et al.*, 1989). Similarly, expression in *Sacch. cerevisiae* of the *STE2* homologue from a related yeast, *Sacch. kluyveri*, allows the cells of *Sacch. cerevisiae* to respond strongly to the α-factor of *Sacch. kluyveri*, to which they normally respond only weakly (McCullough and Herskowitz, 1979; Marsh and Herskowitz, 1988).

Finally, the sequences of *STE2* and *STE3* predict that they encode integral membrane proteins and are members of a large family of receptor proteins including rhodopsin and the β-adrenergic receptor (Burkholder and Hartwell, 1985; Nakayama *et al.*, 1985; Hagen *et al.*, 1986). Both STE2 and STE3 are predicted to have an amino-terminal hydrophobic domain that spans the plasma membrane seven times, followed by a hydrophilic, cytoplasmic carboxy-terminal domain. Some facets of this presumed topology have been verified through use of gene fusions that tag carboxy-terminal deletion derivatives with reporter enzymes (Blumer *et al.*, 1988; Clark *et al.*, 1988; Cartwright and Tipper, 1991). However, despite the similar structural and topological organization of STE2 and STE3, the two proteins have essentially no sequence similarity. This is somewhat surprising given that they appear to couple to the same G protein. Perhaps, in three dimensions, the two receptors exhibit a structural similarity that is not evident from one-dimensional analysis, or alternatively the receptors may interact with different parts of the G protein (see p. 431).

The common structural and topological organization of yeast receptors to each other and to the larger family of receptors reflects the common functional organization as revealed by mutational analysis. Firstly, mutant receptors lacking the cytoplasmic tail are competent to bind ligand and initiate signal transduction (Konopka *et al.*, 1988; Blumer *et al.*, 1988; Davis *et al.*, 1993). Secondly, the cytoplasmic tail of both receptors is required for some aspects of desensitization. These mutant receptors are defective for endocytosis and confer a hypersensitive phenotype (Konopka *et al.*, 1988; Reneke *et al.*, 1988; Davis *et al.*, 1993). Thirdly, mutants altered in the cytoplasmic loop that connects transmembrane segments 5 and 6 (the so-called third cytoplasmic loop) exhibit a hypersensitive phenotype, perhaps due to alterations in interacting with the G protein (Kjelsberg *et al.*, 1992; C. Boone and G. Sprague, unpublished observation). Thus, as has been proposed for this family of receptors in general (Ross, 1989), perhaps the seven membrane-spanning segments of STE2 and STE3 form a barrel-like structure in the membrane. In such a structure, extracellular loops and perhaps parts of the hydrophobic barrel are proposed to interact with ligand, whereas the intracellular loops are proposed to interact with the G protein and other proteins that may regulate the receptor-G protein interaction. In this regard, it is interesting to note that properties of receptors encoded by chimeras of the *Sacch. cerevisiae* and *Sacch. kluyveri STE2* genes suggest that the α-factor-binding site includes extracellular loops and parts of transmembrane segments (Marsh, 1992). Similar conclusions have been reached through linker-insertion mutagenesis of *STE2* (Konopka and Jenness, 1991).

1. G proteins

Receptors of the rhodopsin family couple to G proteins and this is true for yeast receptors as well. The structural gene (*GPA1/SCG1/CDC70/DAC1*) for a yeast G_α subunit was identified by four independent approaches. These were hybridization to a mammalian G_α cDNA probe (Nakafuku *et al.*, 1987), selection for multicopy plasmids that confer resistance to pheromone (Dietzer and Kurjan, 1987a), isolation of suppressor mutations that restore mating competence to receptorless mutants (Jahng *et al.*, 1988), and isolation of a mutation that caused sterility in both **a** and α cells (Fujimura, 1989). The last three approaches suggest that G_α has a role in pheromone-mediated signal transduction (see also Miyajima *et al.*, 1987). Indeed, deletion of *GPA1* leads to constitutive activation of the pathway. Transcription of pheromone-sensitive genes is induced and cell division arrests in G1 phase. This result also implies that G_α is a negative regulator of the pathway, an initially surprising result because, in the archetypical mammalian systems, G_α has a positive role; that is, it is responsible for transmitting the signal to targets (Linder and Gilman, 1992).

The yeast G_α subunit is a membrane-associated myristoylated protein containing 472 amino-acid residues (Blumer and Thorner, 1990; Stone *et al.*, 1991). Myristoylization is required for activity of the yeast G_α, but apparently not for its membrane association (Stone *et al.*, 1991). The yeast protein shows significant similarity to mammalian G_α subunits. Overall, the greatest similarity of the yeast protein is with $G_\alpha i$ (45% identity, 65% similarity), while the least is with $G_\alpha s$ (33% identity, 51% similarity). The similarity is strong throughout the length of the proteins except that yeast G_α contains a 110 amino-acid insertion in a region that, for mammalian G_α subunits, has been proposed to interact with an effector (target) protein to propagate the signal. The role of this insertion is not known. Structural genes for the yeast G_β and G_γ subunits were identified by sequence analysis of genes required for pheromone response (Whiteway *et al.*, 1989). The STE4 protein shows similarity throughout its 423-residue length with mammalian G_β subunits; about 40% of the residues are identical. The STE18 protein is slightly longer at its amino- and carboxy-termini than mammalian G_γ subunits (110 residues compared with 74 for transducin γ), but about 37% of its residues are identical or similar to those of G_γ. Like mammalian G_γ, STE18 is isoprenylated near its carboxy-terminus, and prenylation is required for function (Nakayama *et al.*, 1988a; Finegold *et al.*, 1990). However, as for a-factor, yeast G_γ is modified by a farnesyl group (C_{15}), whereas mammalian G_γ subunit modification is by a larger geranylgeranyl group (C_{20}; Mumby *et al.*, 1990; Yamane *et al.*, 1990). The mammalian G_β and G_γ subunits purify as a complex. Because *ste14* and *ste18* mutants have similar phenotypes, it is reasonable to suppose that yeast subunits also form a complex and function as a unit (referred to as $G_{\beta\gamma}$).

The phenotype of strains lacking or G_β or G_γ is strikingly different from the phenotype of the G_α deletion mutant; *ste4* (G_β) and *ste18* (G_γ) null mutants are non-responsive to mating factor. Moreover, strains defective for both *GPA1* and *STE4*, or for *GPA1* and *STE18* are non-responsive (Miyajima *et al.*, 1988; Nakayama *et al.*, 1988b; Whiteway *et al.*, 1989). Together, these phenotypes indicate that the role of $G_{\beta\gamma}$ is to propagate the signal, whereas the role of G_α is to block activity of $G_{\beta\gamma}$ in the absence of a receptor–ligand interaction. Thus, by analogy with the biochemical properties of mammalian G proteins, the following events are presumed to occur in pheromone-signal transduction. In the unstimulated cell, G_α is bound by GDP and is in a complex with $G_{\beta\gamma}$, which is thereby rendered inactive. Binding of ligand to receptor presumably leads to a conformational change that is transmitted to G_α. As a result, GTP displaces GDP and $G_{\beta\gamma}$ is released to propagate the signal to downstream targets.

Although many of these biochemical predictions have not been tested, both biochemical and genetic experiments do point to a receptor–G_α interaction. (a) The guanine-nucleotide state of the G_α protein affects the affinity of receptor for its ligand, as is also observed for mammalian G-protein-receptor systems. Treatment of a-cell membranes with a non-hydrolysable GTP analogue, which presumably converts most of the G protein to the GTP-bound state, lowers receptor affinity for α-factor by about 10-fold (Blumer and Thorner, 1990). This G_α-receptor coupling also requires $G_{\beta\gamma}$, so perhaps $G_{\beta\gamma}$ interacts with receptor or makes G_α competent to associate correctly with receptor.

(b) Single amino-acid-residue substitutions near the carboxy-terminus of

G_α have different effects on pheromone response in **a** and α cells. A K468P mutant has a more severe defect in **a** cells than in α cells, whereas a K467P mutant has a more severe defect in α cells (Hirsch *et al.*, 1991). This differential effect suggests that different residues of G_α interact with the two receptors and, therefore, may explain how any one G_α type can interact with two dissimilar receptors. Mutant forms of G_α lacking the five carboxy-terminal residues (468–472) confer a non-responsive phenotype on both **a** and α cells. The phenotype is semi-dominant, presumably because mutant G_α molecules interact with and thereby sequester a limited supply of $G_{\beta\gamma}$. The importance of proper stoichiometry between G_α and G_β has also been revealed by other approaches. Overexpression of G_α leads to a non-responsive phenotype (Dietzel and Kurjan, 1987a), whereas over-expression of G_β leads to a constitutive phenotype (Cole *et al.*, 1990; Nomoto *et al.*, 1990; Whiteway *et al.*, 1990).

(c) Coupling of the β-adrenergic receptor to the yeast pathway requires co-expression of the receptor and mammalian G_αs. Coupling does not require mammalian G_β or G_γ, however. This finding implies that yeast G_α cannot interact with the β-adrenergic receptor but that mammalian G_αs can interact with yeast $G_{\beta\gamma}$ (King *et al.*, 1990). Similarly, expression of mammalian G_αi blocks constitutive signalling phenotype of a *gpa1* (G_α) mutant, indicating that G_αi can interact with yeast $G_{\beta\gamma}$. However, expression of G_αi does not allow response to pheromone, indicating that it cannot interact with the yeast receptor (Dietzer and Kurjan, 1987a). The properties of chimeras formed between G_αi and GPA1 support the idea that the carboxy-terminus of G_α interacts with the receptor (Kang *et al.*, 1990).

2. Downstream elements: STE5, five protein kinases and a transcription factor

One consequence of binding of pheromone to receptor is transcription induction of selected genes. Therefore, activated $G_{\beta\gamma}$ (liberated from G_α) must transmit a signal from the plasma membrane to the nucleus. How does this transmission of information occur? What might the targets of $G_{\beta\gamma}$ be? In addition to the receptor and G-protein structural genes, seven other genes are known to be required for pheromone response. Null alleles of *STE5*, *STE7*, *STE11*, *STE12* and *STE20* confer a non-responsive phenotype (Hartwell, 1980; Fields *et al.*, 1988; Clark and Sprague, 1989; Rhodes *et al.*, 1990; Leberer *et al.*, 1992). Similarly, *FUS3* and *KSS1* are required for response, although they are partially redundant in function. A *fus3 kss1* double mutant is non-responsive but single mutants retain some capacity to respond. Mutants of *fus3* show transcription induction but not cell-cycle arrest, whereas *kss1* mutants are normal for cell-cycle arrest but show a modest impairment for transcription induction (Courchesne *et al.*, 1989; Elion *et al.*, 1990, 1991b; Gartner *et al.*, 1992).

Each of these seven genes are argued, on the basis of double-mutant studies, to function after the G protein. Null alleles of *GPA1* or special dominant alleles of *STE4* lead to a constitutive phenotype, but double mutants formed with one of these alleles and null alleles of the *ste* loci or with a *fus3 kss1* double deletion are non-responsive (Nakayama *et al.*, 1988b; Blinder *et al.*, 1989; Elion *et al.*, 1990). A simple interpretation of these results is that the STE, FUS3 and KSS1 products act downstream of G$_\beta$, but more complicated possibilities cannot be excluded. For example, the G protein and STE products may function in parallel pathways, each of which is required for pheromone response. However, complementary double-mutant studies support the simple interpretation. Dominant alleles of *STE11* have been isolated that suppress *ste4* (G$_\beta$) deletion mutations (Stevenson *et al.*, 1992). These alleles lead to constitutive transcription of pheromone-sensitive genes, such that transcript levels are at or above values detected when wild-type cells are treated with pheromone. Thus, dominant *STE11* alleles are presumed to encode hyperactive or unregulated forms of STE11. Because the phenotype of any one *STE11* allele (non-responsive or constitutive) determines the phenotype of double mutants constructed with null or gain-of-function *STE4* alleles, the conclusion from these epistasis experiments is that STE11 functions after STE4 (G$_\beta$). Essentially identical experiments using a partially constitutive allele of *STE5* imply that STE5 also acts downstream of the G protein (J. Thorner, personal communication).

The dominant *STE11* mutations also permit ordering of STE11 with respect to other gene products. The *STE11* alleles suppress the phenotype of *ste5* mutants, but not of *ste7* or *ste12* mutants, nor of *fus3 kss1* double mutants, implying that STE11 acts after STE5 but before or in an interdependent fashion with STE7, FUS3, KSS1 and STE12. As discussed in detail later in this review, STE7 and FUS3 become phosphorylated in response to pheromone. Phosphorylation also takes place in *STE11* dominant mutants, even in the absence of pheromone, implying that this protein acts before STE7 and FUS3 (Cairn *et al.*, 1992; Gartner *et al.*, 1992; Stevenson *et al.*, 1992).

The ordering of gene function that emerges from this epistasis analysis has been reinforced and extended by other studies. For example, the hyperactive allele of *STE5* already discussed suppresses the mating defect of *ste4* and *ste20* mutants, but not of *ste7*, *ste11*, *fus3*, *kss1*, or *ste12* mutants (J. Thorner and M. Whiteway, personal communications). Similarly, overexpression of STE12 causes an increase in transcription of pheromone-responsive genes in *ste4*, *ste5*, *ste7* or *ste11* mutants, and restores some degree of mating competence to these mutants (Dolan and Fields, 1990). Finally, overexpression of *STE12*, but not of *STE4*, suppresses *ste20* mutations, further supporting the conclusion that STE20 acts after or at the G$_{\beta\gamma}$ step but before STE5 (Leberer *et al.*, 1992).

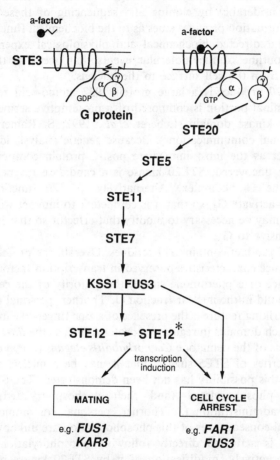

Fig. 5. A model for the pheromone response pathway in *Saccharomyces cerevisiae*. In α cells, binding of **a**-factor to the receptor encoded by *STE3* (or, in **a** cells, binding of α-factor to the receptor encoded by *STE2*) results in exchange of GTP for GDP on the α-subunit of the G protein and its dissociation from the β and γ subunits. The Gβγ dimer transmits the signal through the STE, FUS3 and KSS1 products, causing a change in modification of the transcription factor, STE12. Protein STE12* represents the phosphorylated form that appears to have transcriptional activity. Examples of inducible genes involved in mating and cell-cycle arrest are shown. Pheromone-mediated post-translation modification of FAR1 and FUS3, in addition to their increased abundance as a result of transcription induction, is likely to be essential to achieve cell-cycle arrest.

Altogether, these findings indicate that the order of gene function is: STE4 (G_β) → STE20 → STE5 → STE11 → STE7 → FUS3/KSS1 → STE12 (Fig. 5).

The formal genetic analysis already described identified genes whose products comprise part of the pheromone-response pathway or are required for activity of the pathway. This degree of understanding has been advanced considerably by cloning and sequencing of these genes. The sequence information provides clues as to the biochemical function of gene products, and directed biochemical and physiological experiments that yielded an outline of the molecular events that occur as the signal is transmitted from the cell surface to the nucleus.

The *STE20* product is a large protein (932 amino-acid residues). Its carboxy-terminal portion is composed of a presumptive serine/threonine-type protein-kinase domain (Leberer *et al.*, 1992; S. Ramer and R.W. Davis, personal communications). Because genetic analysis identifies this gene product as the most immediate post-G-protein component of the pathway yet uncovered, STE20 kinase is a candidate for being a direct target of the $G_{\beta\gamma}$ complex. Alternatively, *STE20* function may be necessary to activate $G_{\beta\gamma}$ so that it is competent to interact with its actual effector, or may be necessary to modify that effector so that is capable of being responsive to $G_{\beta\gamma}$.

The *STE5* product contains 917 residues. Overall, the protein has many of the sequence characteristics observed in transcription factors and, even in the absence of a pheromone signal, the majority of the cellular *STE5* protein is found in the nuclear fraction (J. Thorner, personal communication). One striking feature is the presence of a zinc finger-like motif, related to the Cys-rich domains in regulatory proteins such as the *lin-11* and *mec-3* gene products of the nematode *Caenorhabditis elegans* (Freyd *et al.*, 1990). These properties of STE5 suggest that it may be a nucleic acid-binding protein but this possibility has not been demonstrated. The STE5 protein is rapidly phosphorylated (and then dephosphorylated) following pheromone administration (J. Thorner, personal communication). The physiological consequences of this phosphorylation are unknown. Perhaps the protein is activated directly following phosphorylation by STE20 kinase. Alternatively, modification of it by STE20 kinase may make it competent to be activated by the $G_{\beta\gamma}$ complex, so the physiological role of STE5 protein in signal propagation remains elusive. Genetic analysis places STE5 early in the pathway, whereas biochemical analysis suggests that STE5 product is localized in the nucleus, like STE12 protein. Perhaps the STE5 protein shuttles from the membrane or cytosol to the nucleus, like the SW15 protein (Moll *et al.*, 1991), although there is no direct evidence to support this suggestion.

The *STE11* (Rhodes *et al.*, 1990), *STE7* (Teague *et al.*, 1986), *FUS3* (Elion *et al.*, 1990) and *KSS1* (Courchesne *et al.*, 1989) products all contain a domain that possesses all of the conserved consensus sequences diagnostic of the catalytic portion of known serine/threonine-specific protein kinases. Indeed, both the STE11 and STE7 products have been shown to possess protein kinase activity *in vitro* (Rhodes *et al.*, 1990; Z. Zhou and B. Errede, personal communication). In these proteins, the catalytic domain resides in the carboxy-terminal portion and is preceded by a long amino-terminal extension (over 400 residues in STE11 and nearly 200 residues in STE7). In STE11, mutations that activate the kinase either alter or remove the amino-terminal extension, suggesting that the amino-terminal sequence has a negative regulatory function (Cairns *et al.*, 1992; Stevenson *et al.*, 1992). In this sense, the STE11 protein resembles mammalian c-*raf* protein kinase and protein kinase C, but the yeast product lacks the cysteine-rich motifs found in the amino-terminal domains of these mammalian kinases.

Genes *FUS3* and *KSS1* encode relatively compact proteins (353 and 368 residues, respectively) that appear to consist only of a kinase catalytic domain. These two proteins are very similar to each other (56% identity) and are related most strongly (more than 50% identity) to the so-called MAP (or ERK) family of mammalian protein kinases that are activated during differentiation-inducing, as well as during growth-stimulatory, processes in animal cells (Pelech and Sanghera, 1992).

The realization that protein kinases are essential elements in pheromone response has prompted an examination of the phosphorylation state of other pathway components. Following treatment with pheromone, STE2 (Reneke *et al.*, 1988), STE4 (Cole and Reed, 1991), STE5 (J. Thorner, personal communication), FUS3 (Gartner *et al.*, 1992), KSS1 (J. Thorner, personal communication), STE7 (Cairns *et al.*, 1992; Stevenson *et al.*, 1992; Z. Zhou and B. Errede, personal communication), STE12 (Song *et al.*, 1991) and FAR1 (F. Chang and I. Herskowitz, personal communication) have been shown to become hyperphosphorylated. Phosphorylation of STE2 and STE4 apparently plays a role in desensitization of the initial signal but phosphorylation of the other proteins is likely to be important for execution of the response pathway.

Pheromone-stimulated hyperphosphorylation of the STE7, FUS3, and KSS1 proteins has been investigated in most detail, especially from a genetic perspective. Pheromone-stimulated phosphorylation of the STE7 enzyme is prevented in either a *ste11* mutant or a *fus3 kss1* double mutant, but not in a *ste12* mutant (Stevenson *et al.*, 1992; Z. Zhou and B. Errede, personal communication). Likewise, phosphorylation of FUS3 and KSS1 proteins requires *STE11* and *STE7* function, but not *STE12* function (Gartner *et al.*,

1992). In agreement with these findings, hyperphosphorylation of STE7 and FUS3 proteins is observed, even in the absence of pheromone, in strains carrying the dominant *STE11* mutations already discussed (Cairns *et al.*, 1992; Ganner *et al.*, 1992; Stevenson *et al.*, 1992; Z. Zhou and B. Errede, personal communication).

Phosphorylation of FUS3 and KSS1 occurs on tyrosine and threonine residues that are part of a -TEY- site situated near to the amino-terminal conserved -APE- motif that is an essential sequence element of the catalytic domain of all protein kinases (Hanks *et al.*, 1988). Such a -TEY- site is found in every MAP kinase examined to date and dephosphorylation of either residue eliminates catalytic activity *in vitro* (Anderson *et al.*, 1990; Payne *et al.*, 1991). Correspondingly, conversion of these amino-acid residues to similar but non-phosphorylatable residues inactivates the functions of both the FUS3 and KSS1 enzymes *in vivo* (Gartner *et al.*, 1992; J. Thorner, personal communication).

Recent reports involving co-immunoprecipitation of STE7 and FUS3 suggest that FUS3 is a direct target of STE7 (Z. Zhou, A. Gartner, K. Nasmyth, G. Ammerer and B. Errede, personal communication). This conclusion is in keeping with the finding that the MAP kinase activator is a protein kinase homologous to STE7 (Gomez and Cohen, 1991; Crews and Erickson, 1992; Matsuda *et al.*, 1992; Kosako *et al.*, 1992).

Taken together, the genetic and physiological findings imply that activity of the STE11 kinase increases in response to pheromone. The protein STE11, in turn, controls the phosphorylation state, and hence activity, of the STE7 and FUS3 kinases. Thus, there is reasonably compelling evidence that propagation and amplification of the initial pheromone signal involves a protein kinase cascade. Finally, STE12 protein is the most distal known gene product that operates downstream from the G-protein and is absolutely required for pheromone response. The STE12 protein is a transcription factor whose properties and function are discussed in greater detail later in this review.

C. Execution of pheromone response in *Saccharomyces cerevisiae*

1. Transcription induction

Many genes are known whose transcription is stimulated in response to pheromone. These genes include those required for response to pheromone, such as the receptor structural genes (Hagen and Sprague, 1984; Hartig *et al.*, 1986), as well as genes required for pheromone synthesis, such as the pheromone structural genes (Jarvis *et al.*, 1988; Achstetter, 1989; Dolan *et al.*, 1989; Flessel *et al.*, 1989). A second group of inducible genes encode products that facilitate mating or cell cycle arrest. These include *FUS1*, which is required for cell fusion and induced more than 100-fold (McCaffrey *et al.*, 1987; Trueheart *et al.*, 1987); *AGα1*, which encodes the α-agglutinin (Lipke *et al.*, 1989); *KAR3*, which is required for nuclear fusion (Meluh and Rose, 1990); *FUS3* (Elion *et al.*, 1990); and *CHS1*, which encodes a chitin synthetase (Appeltauer and Achstetter, 1989). A third group of inducible genes encode products that are involved in recovery from pheromone. They include *GPA1* (Jahng *et al.*, 1988), *SST2* (Dietzel and Kurjan, 1987b) and *BAR1* (Kronstad *et al.*, 1987). The magnitude of induction varies considerably from gene to gene. Many are induced 3–10-fold (e.g. *STE3* and *FUS3*), whereas a few are induced 20–100-fold (e.g. *KAR3* and *FUS1*). As discussed later, the different degrees of inducibility appear to reflect differences in the complexity of each genes' upstream regulatory regions.

Induction of pheromone-stimulated transcription is conferred by a DNA sequence, 5'-TGAAACA (termed the pheromone response element, PRE; Kronstad *et al.*, 1987; Van Arsdell and Thorner, 1987). Multiple tandem copies of this sequence are sufficient to confer inducibility to test promoters (Sengupta and Cochran, 1990; Hagen *et al.*, 1991; Davis *et al.*, 1992). Conversely, deletion of PREs from the upstream regions of *FUS1* or *BAR1* abolishes inducibility. With *FUS1*, deletion of PREs abolishes not only pheromone-stimulated transcription but also the low level of transcription that is normally detected in the absence of pheromone (Hagen *et al.*, 1991). Thus, PREs are the only elements within the *FUS1* upstream region that can act to drive transcription. In the absence of pheromone, they confer a low basal level of transcription; in the presence of pheromone they promote a greatly elevated level of transcription. With *BAR1* on the other hand, deletion of PREs abolishes induction, although substantial transcription of the gene still occurs (Kronstad *et al.*, 1987). The *BAR1* gene and many other pheromone-responsive genes contain a second sequence element, the P box, that serves as the binding site for MCM1. In the absence of pheromone, both the P box and the PRE contribute to basal transcription. Because only the PRE confers inducibility by pheromone, the magnitude of induction is more modest than observed with genes regulated by PRE elements only.

The STE12 protein binds to the PRE and mediates transcription induction (Dolan *et al.*, 1989; Errede and Ammerer, 1989; Yuan and Fields, 1991). Two lines of evidence indicate that activity of STE12 is limiting in non-stimulated cells and increases in response to pheromone. First, overexpression of STE12 by fusion of the coding sequence to the *GAL1* promoter increases transcription of *FUS1* in cells not exposed to pheromone and in cells mutant for *STE4, STE5, STE7, STE11* or *FUS3/KSS1* (Dolan and Fields, 1990; C. Adler and B. Errede, personal communication). Secondly, by the use of chimeras in which different segments of STE12 are fused to the DNA-binding domain of GAL4, it has been shown that the central one-third of STE12 is rapidly phosphorylated following treatment of cells with pheromone (Song *et al.*, 1991). These chimeras can activate transcription of reporter constructs containing GAL4-binding sites, but only after treatment of cells with pheromone. Thus, phosphorylation is correlated with ability of chimeras to stimulate transcription. It is therefore tempting to speculate that phosphorylation is causally related to the increase in transcriptional activity of STE12, and that STE12 is a direct target of one or more of the kinases that are part of the response pathway. In this view, STE12 is at the end of the signal-transduction pathway that leads to transcription induction.

The STE12 protein binds poorly to single PRE elements, but binds co-operatively if two or more elements are present on the DNA (Dolan *et al.*, 1989; Yuan and Fields, 1991). The protein can also bind co-operatively with other proteins, most notably MCM1 (Errede and Ammerer, 1989). Thus, in conjunction with binding of MCM1 to its site, binding of STE12 to single PRE elements is readily observed. The DNA-binding activities of STE12 therefore mirror *in vivo* transcriptional activities of PRE elements in different settings. A single PRE cannot serve alone to drive transcription, but two or more PREs in tandem can promote pheromone-stimulated transcription. However, a single PRE can confer pheromone-stimulated transcription if there is also an MCM1 site (P box) present (Errede and Ammerer, 1989; Sengupta and Cochran, 1990; Hagen *et al.*, 1991; Hwang-Shum *et al.*, 1991; F. Gimble and J. Thorner, personal communication).

It is worth noting that not all genes that exhibit pheromone-stimulated transcription contain PRE elements within their control regions. For example, the upstream activation sequence (UAS) from the (α-specific *STE3* gene (a-factor receptor structural gene) confers a modest degree of induction (around five-fold) to test promoters but lacks obvious PRE sequences (Hagen and Sprague, 1984). As already discussed, the *STE3* UAS serves as the binding site for α1 and MCM1. Although STE12 has not been shown to bind to this UAS even in the presence of α1 and MCM1, perhaps it does so *in vivo*. Alternatively, perhaps the pheromone-response pathway alters the transcriptional activity of α1 or MCM1, just as it does for STE12 (Sengupta and Cochran, 1991).

Finally, within the context of a natural promoter, the presence of PRE sequences is not necessarily sufficient to confer pheromone-stimulated transcription. For example, *STE7*, which contains multiple PREs, is not inducible (Z. Zhou and B. Errede, personal communication). Perhaps binding of other proteins to upstream regions of *STE7* occludes binding of STE12 or blocks STE12 activity once it is bound. Alternatively, perhaps other promoter elements for *STE7* become less active when cells are treated with pheromone, thereby counterbalancing PRE-mediated stimulation.

2. Cell-cycle arrest

In response to pheromone, cells arrest at a position in the cell cycle termed start, late in the G1 phase prior to the initiation of DNA synthesis. In mitotically dividing cells, the serine/threonine-specific protein kinase, CDC28, is required in order for cells to pass through start and initiate a new cell cycle. Activity of CDC28 is thought to be controlled by three G1 cyclins, namely CLN1, CLN2, and CLN3, whose abundance varies in the cell cycle (Cross, 1988; Nash *et al.*, 1988; Hadwiger *et al.*, 1989). The pheromone-response pathway apparently controls progression through start by preventing accumulation of the CLN proteins during G1.

Both transcriptional and post-transcriptional mechanisms operate to control CLN expression. Treatment of **a** cells with α-factor causes a decrease in concentration of RNA from *CLN1* and *CLN2* (Wittenberg *et al.*, 1990). Thus, there appears to be pheromone-mediated repression of these genes. In addition, a form of autogenous control almost certainly contributes to the lowering of levels of transcript. Specifically, CLN polypeptides and active CDC28 kinase are required for efficient transcription of *CLN1* and *CLN2* genes (Cross and Tinkelenberg, 1991; Dirick and Nasmyth, 1991). In mitotically cycling cells, this positive feedback loop presumably converts an initial gradual rise in CLN protein content during G1 into a sharp rise. Thus, the positive loop may serve as a switch and result in a firm commitment to initiate a new cell cycle. By the same token, the pheromone-mediated decreased in concentration of RNA from *CLN1* and *CLN2* may help throw the switch in the other direction and thereby contribute to G1 arrest.

Levels of RNA from *CLN3* are not influenced by pheromone so there must be other mechanisms that control CLN abundance following pheromone treatment (Nash *et al.*, 1988). In fact, CLN3 polypeptide is unstable and rapidly disappears when cells are treated with mating factor. Presumably, other CLN polypeptides are also degraded under these conditions. The CLN polypeptides contain PEST sequences, which have been implicated in protein turnover (Rogers *et al.*, 1986). The PEST sequences may therefore control CLN stability. In support of this possibility, a carboxy-terminal truncation of CLN3, which removes its

PEST sequence, creates a dominant, activated form of the protein. Cells producing this mutant protein are partially resistant to pheromone-mediated cell-cycle arrest (Cross, 1988).

Two genes, *FAR1* and *FUS3*, which contribute to inhibition of CLN function, have been identified by isolation of mutants that exhibit transcription induction but not cell-cycle arrest in response to pheromone. By virtue of the transcription-feedback loop already described, *FAR1* and *FUS3* products may contribute to inhibition of CLNI and CLN2 (Elion *et al.*, 1991a), but FAR1 appears to have a crucial role in inhibiting CLN2, and FUS3 a crucial role in inhibiting CLN3. This conclusion follows from the observation that *far1 cln2* and *fus3 cln3* double mutants arrest at start when exposed to pheromone, whereas the *far1* and *fus3* single mutants do not (Chang and Herskowitz, 1990; Elion *et al.*, 1990). A specific inhibitor of CLN1 has not been identified.

These studies, coupled with those on requirements for pheromone-stimulated transcription already described, suggest that FUS3 has two roles in pheromone response. One role, which can also be fulfilled by KSS1, is to allow transmission of the signal leading to transcription induction. The other role, which KSS1 cannot fulfil, is to promote cell-cycle arrest. Since FUS3 almost certainly has protein kinase activity, it may control the abundance of CLN3 by phosphorylating it or other proteins that control its degradation. Whether the difference in phenotype between *fus3* and *kss1* mutants arises from a qualitative or quantitative difference in kinase activities that they encode is not known.

Both the amount and activity of FAR1 and FUS3 are presumably increased by transmission of a signal through the pathway. Transcription of both genes is increased in response to pheromone, which presumably increases the abundance of the products (Chang and Herskowitz, 1990; Elion *et al.*, 1990). In addition, both proteins are rapidly phosphorylated when cells are treated with pheromone, which may increase their activities (Gartner *et al.*, 1992; F. Chang and I. Herskowitz, personal communication). Increased abundance and activity of FAR1 and FUS3 presumably ensure that CLN proteins do not accumulate and, consequently, CDC28 kinase is rendered inactive. In this way, mitosis is halted in G1.

3. Cell and nuclear fusion

Cell and nuclear fusion are poorly understood processes but both require pheromone-induced alterations in order to proceed efficiently. At least in part, these alterations result from transcription induction of genes whose products catalyse the processes.

Mutations in two genes, *FUS1* and *FUS2*, result in an interruption of the mating process just prior to cytoplasmic fusion. Transcription of both these genes is stimulated 50-fold or more by pheromone (McCaffrey *et al.*, 1987; Trueheart *et al.*, 1987; E. Elion, personal communication). In crosses of *fus1* or *fus2* mutants to wild-type cells, only a very modest defect is observed (McCaffrey *et al.*, 1987; Trueheart *et al.*, 1987). However, crosses involving two mutants, especially crosses involving two double mutant strains, show a severe defect in zygote formation. In these mutants, cell-wall fusion occurs but plasma membranes fail to fuse. The observation that the defect is most apparent in crosses involving double mutants has two implications. Firstly, *FUS1* and *FUS2* are partially redundant in function. Secondly, the presence of wild-type FUS1 or FUS2 protein in either cell of the mating pair is sufficient to bring about membrane fusion, albeit with decreased efficiency compared with crosses involving wild-type strains. The view that FUS1 and FUS2 have partially redundant activities is reinforced by the finding that either gene in multicopy partially overcomes the defect associated with mutation of the other (Trueheart *et al.*, 1987).

In accordance with the proposed role of FUS1 in membrane fusion, immunofluorescence studies using a FUS1–LacZ hybrid protein and antibodies to β-galactosidase reveal that hybrid protein is located at the morphological projection that emerges from the responding cell, that is, at the presumed site of fusion of the mating pair (Trueheart *et al.*, 1987). Moreover, *bona fide* FUS1 protein has been shown to span the plasma membrane (Trueheart and Fink, 1989). Biochemical experiments, coupled with information derived from the sequence of the gene, reveal that FUS1 contains a 71 amino-acid-residue extracellular amino-terminal domain, which shows O-linked glycosylation, followed by a 25-residue hydrophobic, membrane-spanning segment and a large (416 residue) cytoplasmic domain. The cytoplasmic domain contains two regions that show weak homology to myosin heavy chain, and a region with *src* homology 3 (SH3)-like domain near the carboxy-terminus. The SH3 domain, and at least one of the myosin homology regions, must be mutated before FUS1 function is significantly impaired. The amino-terminal domain is dispensable for function, so FUS1 may not directly catalyse cell fusion (J. Trueheart and J. Thorner, personal communication). Perhaps the cytoplasmic domain serves as a scaffold for other proteins that do catalyse the event. Immunofluorescence studies show that FUS2 is located in cortical dots near the intracellular face of the plasma membrane at the tip of the mating projection (E. Elion, personal communication).

Cell fusion, in particular membrane fusion, is a critical stage in mating and presumably must be controlled carefully to avoid cell death. A late event in pheromone response is an influx of Ca^{2+} ions that leads to a several-fold rise in the intracellular Ca^{2+} concentration (Iida et al., 1990). This influx appears to be important for maintaining viability of responding cells and of newly formed zygotes. Perhaps membrane changes associated with cell fusion are responsible for this Ca^{2+} sensitivity.

Three genes required for nuclear fusion have been identified (Conde and Fink, 1976; Polaina and Conde, 1982). One, KAR1, plays a role in spindle pole body duplication (Rose and Fink, 1987). Aberrant microtubular structures associated with the spindle pole body in kar1 mutants may interfere with karyogamy. A second gene, KAR2, encodes a homologue to mammalian BiP/GRP78, the HSP70 protein that resides in the endoplasmic reticulum (Normington et al., 1989; Rose et al., 1989). In yeast, the nuclear envelope is continuous with, and sometimes partially surrounded by, endoplasmic reticulum. In kar2 mutants, the nuclear envelope becomes hypertrophied, which may hinder karyogamy. Thus, the roles of KAR1 and KAR2 in karyogamy may be indirect.

The KAR3 product, on the other hand, may play a direct role in nuclear fusion. Firstly, KAR3 transcription is induced 20-fold or more in response to pheromone (Meluh and Rose, 1990). Secondly, the predicted KAR3 protein sequence suggests that it is a member of the kinesin protein family, which is characterized by a domain that serves as mechanochemical motor, connected to different protein-binding domains. In particular, KAR3 contains an amino-terminal globular domain that can associate with microtubules. This domain is connected by an extended coil domain to a carboxy-terminal globular domain that, by virtue of its homology to kinesin, is argued to bind ATP and to act as a device for moving microtubules. Mutation of the putative ATP-binding site confers a semi-dominant Kar^- phenotype. Thus, KAR3 is proposed to act during mating by crossbridging cytoplasmic microtubules emanating from spindle pole bodies of the two nuclei. The ATP-driven motor of KAR3 can then provide the force to draw the nuclei together (Meluh and Rose, 1990). In support of this possibility, KAR3 becomes relocalized following pheromone treatment; it moves from the nucleus to the cytoplasm, where it would be able to carry out its proposed function (P. Meluh, R. Monroe, and M. Rose, personal communication).

Three other genes required for karyogamy have been identified recently in a screen for mutants defective for spindle pole body and/or microtubule function (Page and Snyder, 1992). One of these genes, CIK1, encodes a protein with the potential to form an extended coil domain. As for KAR3, CIK1 transcription is induced by pheromone, and indirect immunofluorescence reveals that CIK1 protein is associated with the spindle pole body of α-factor-treated a cells (Page and Snyder, 1992). Null mutants of cik1 are defective for karyogamy and exhibit aberrant microtubule organization. These findings suggest that the spindle pole body performs a critical function in nuclear fusion.

D. Signalling pathway in *Schizosaccharemyces pombe*

Direct evidence that secreted pheromones play a crucial role in the mating process in fission yeast *Schiz. pombe* has been obtained only recently (Fukui *et al.*, 1986b; Leupold, 1987; Davey, 1991). Nonetheless, through cloning and sequencing of genes identified by the isolation of mating-deficient mutants and through cloning of homologues from this yeast to ras and G protein structural genes, insight into some of the molecular events that occur during pheromone response is emerging. The current picture of the pathway shows strong similarities to that in *Sacch. cerevisiae*, but there are intriguing differences as well (Fig. 6).

The initial event in pheromone signal transduction in *Schiz. pombe* presumably involves binding of the pheromones to unique receptors displayed on the surfaces of M and P cells. As noted in the discussion of the mating process, the structure of one of the pheromones, namely M-factor, has recently been deduced from the gene sequence (Davey, 1992). Like the a-factor of *Sacch. cerevisiae*, M-factor is a short lipo-peptide. The likely receptor for this pheromone has been identified through cloning and sequencing of *map3*, a gene required for mating only by P cells (M. Yamamoto, personal communication). The map3 product is a member of the rhodopsin/β-adrenergic receptor family and shows sequence similarity to the a-factor receptor (STE3) of *Sacch. cerevisiae*. Likewise, the sequence of the *mam2* gene, required for mating only by M cells, indicates that it encodes a seven-transmembrane receptor of the rhodopsin/β-adrenergic family, in this case exhibiting sequence similarity to the α-factor receptor (STE2) of *Sacch. cerevisiae* (Kitamura and Shimoda, 1991). Mutants defective for *mam2* do not respond to P-factor, adding support to the supposition that the mam2 product is a pheromone receptor.

In keeping with the finding that likely pheromone receptors are members of the seven-transmembrane receptor family, a G protein in fission yeast is argued to be part of the intracellular signal-transduction pathway. In particular, the structure a G_α homologue from *Schiz. pombe* has been identified by hybridization using mammalian G_α probes (Obara *et al.*, 1991). Inactivation of the *gpa1* gene in fission yeast leads to a mating-defective phenotype. Moreover, an alteration of gpa1 that is presumed to

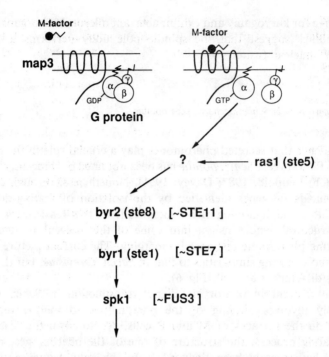

Fig. 6. A model for the pheromone-response pathway in *Schizosaccharomyces pombe*. Binding of M-factor to its receptor, map3, results in exchange of GTP for GDP on the α-subunit of the G protein and its dissociation from the β and γ subunits. The complex $G_\alpha \cdot$GTP transmits a signal, ultimately stimulating a protein kinase cascade. Homologues to the protein kinases of *Saccharomyces cerevisiae* are indicated in brackets. Targets for the kinases have not been identified. The ras1 protein is also required for propagation of the signal, perhaps by controlling the ability of G_α to function (see the text for further details).

activate G_α causes cells to exhibit mating responses even in the absence of pheromone stimulation and nitrogen starvation. Together, these pheno-types suggest that gpa1 in *Schiz. pombe* plays a positive role in signal transduction (Obara *et al.*, 1991), in contrast to the GPA1 in *Sacch. cerevisiae*, which plays a negative role. Structural genes for G_β and G_γ subunits in *Schiz. pombe* have not been identified, but presumably these subunits function to keep G_α inactive in the absence of pheromone stimulation. If so, the functions of G protein subunits in *Schiz. pombe* are similar to the prototypic mammalian G protein subunits.

The immediate target of the activated G_α is not known but, as for *Sacch. cerevisiae*, a protein-kinase cascade operates downstream from the G protein and is an essential feature of the pathway. The *byr2* (also called *ste8*), *byr1* (*ste1*), and *spk1* genes in fission yeast are homologous to *STE11*,

STE7 and *FUS3/KSS1* genes in *Sacch. cerevisiae*, respectively, with sequence identity ranging from 42% (*STE11* compared will *byr2*) to 60% (*KSS1* compared with *spk1*) (Nadim-Davis and Nasim, 1988, 1990; Toda *et al.*, 1991; Toda *et al.*, 1991; Wang *et al.*, 1991). Each of the genes in fission yeast is required for conjugation and sporulation, presumably because loss of these gene functions blocks pheromone response. To determine the relative order in which the protein kinases act, phenotypes of strains carrying a null allele of one gene and a plasmid that confers overexpression of a second gene have been investigated. Although not all combinations have yet been tested, thus far this epistasis analysis is consistent with the order: byr2 → byr1 → spk1. Additional support for this conclusion has come from experiments investigating whether genes in fission yeast and *Sacch. cerevisiae* can function when expressed in the heterologous host (Neiman *et al.*, 1993). In some strains, striking functional interchangeability is observed: *STE11* can complement *byr2⁻* from fission yeast, *byr1* can complement *ste7⁻* and *skp1* can complement *fus3⁻ kss1⁻*, both from *Sacch. cerevisiae*. In other strains expression of two kinases is required to observe complementation. For example, overexpression of both *byr2* and *byr1* is required to allow *ste11⁻* mutants of *Sacch. cerevisiae* to respond to pheromone. Nonetheless, the salient point is that complementation is observed, implying that pheromone-induced signal transduction employs a conserved set of protein kinases in these two yeasts.

Fission yeast has a functional requirement not seen for the pathway in *Sacch. cerevisiae*. In particular, the *ras1* (*ste5*) gene is required for pheromone response (Nielson *et al.*, 1992). Null alleles of *ras1* confer a non-mating phenotype, whereas the activated *ras1^{val17}* allele confers a hypersexual phenotype (Fukui *et al.*, 1986a; Nadim-Davis *et al.*, 1986). Functional relationships among gpa1, ras1 and the protein kinases have been investigated by epistasis experiments similar to those already described. Overexpression of either byr2 or byr1 suppresses the sporulation defects of *ras1⁻* and *gpa1⁻* mutants, while overexpression of byr1, but not byr2, weakly suppresses the conjugation defect of either strain (Wang *et al.*, 1991; Neiman *et al.*, 1993). Together, these observations imply that the kinase cascade functions downstream of both gpa1 and ras1. The relationship between gpa1 and ras1 is less clear but loss of gpa1 function suppresses the hypersexual phenotype of *ras1^{val17}*, suggesting that gpa1 functions downstream or at the same level as ras1 (Neiman *et al.*, 1993).

Thus, pheromone response in fission yeast occurs by a pathway similar to that established for *Sacch. cerevisiae* (Fig. 6). There are two important differences. Firstly, although both pathways involve a heterotrimeric G protein, in fission yeast the G_α subunit is the positive transducer of the signal rather than $G_{\beta\gamma}$. Secondly, the fission-yeast pathway depends on ras1 activity. This dependence on ras represents an interesting parallel with signal-transduction pathways activated by mammalian tyrosine kinases. These pathways depend on ras activity and transmit a signal to a protein-kinase cascade involving homologues to byr1 and spk1 (Thomas, 1992; Wood *et al.*, 1992).

V. MATING-TYPE INTERCONVERSION

A. Switching by *Saccharomyces cerevisiae*

As noted in Section I, yeast strains differ in the stability of their mating type. Strains with a stable mating type are referred to as heterothallic, whereas those with an unstable mating type are referred to as homothallic. These two types of strains differ in genetic content at a single locus. Homothallic strains carry the *HO* allele; heterothallic strains carry *ho*. The *HO* allele endows cells with the ability to undergo a genetic switch; an **a** cell can change its mating type information to *MAT*α and *vice versa* (Herskowitz and Oshima, 1981; Klar, 1989). This switching process requires not only *HO* but also two silent copies of mating-type information, namely *HML* and *HMR*, that reside on opposite arms of chromosome III, the same chromosome that carries the *MAT* locus (Fig. 7). In standard strains, *HML* contains α information and *HMR* contains **a** information. The *HML* and *HMR* loci are not expressed, and therefore do not influence the mating type of cells. Rather, these silent loci serve as donors of information during the switching process. Mating-type interconversion is initiated by the *HO* gene product, an endonuclease that makes a double-stranded cut at the *MAT* locus, near the junction of allele-specific DNA (**Ya** or Yα) and the so-called Z-sequence DNA shared by the three mating-type cassettes (Strathern *et al.*,

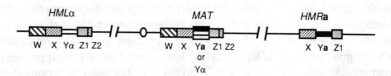

Fig. 7. Structure of mating-type information at *HML*, *MAT* and *HMR* on chromosome III of *Saccharomyces cerevisiae*. At all three loci, **a** and α information is distinguished by a non-homologous segment. Segment **Ya** has 642 bp, Yα 747 bp. These Y regions are flanked by sequences shared by *HML*, *MAT* and *HMR*. All three loci contain X (704 bp) and Z1 (239 bp). HML and MAT share additional sequences, namely W (723 bp) and Z2 (88 bp). The double-stranded cut that initiates mating-type interconversion occurs at *MAT* near the Y/Z1 junction.

1982; Kostriken *et al.*, 1983; Kostricken and Heffron, 1984). The double-strand break is then repaired by interaction with *HML* or *HMR*, which effects the genetic switch.

By examining switching events in cell-lineage pedigrees, rules that govern the process have been established. Firstly, α cells tend to switch using information at *HMR*, **a** cells using information at *HML* (Rine *et al.*, 1980; Klar *et al.*, 1982; Jensen and Herskowitz, 1984). Secondly, because yeast cells undergo mitosis by budding, the mother and daughter cells are readily distinguished. After each cell division, the mother cell is capable of switching whereas the daughter cell is not (Fig. 8; Strathern and Herskowitz, 1979). Thirdly, both products of any cell division are always of the same mating type, implying that the switching event occurred in the G1 phase of the cycle before replication of the *MAT* locus (Strathern and Herskowitz, 1979). Fourthly, **a**/α diploids cannot undergo mating-type interconversion. The molecular mechanisms that underlie the first rule are not understood. However, the other three rules reflect the pattern of transcriptional regulation of the *HO* gene. The *HO* gene is transcribed in **a** and α haploids but not in **a**/α diploids (Jensen *et al.*, 1983). Moreover, in haploids, it is transcribed only in mother cells and only in a narrow window of the cell cycle late in G1 (Nasmyth, 1983; Breeden and Nasmyth, 1987). Conversely, if daughter cells are made to express *HO*, these cells can then switch mating type (Jensen and Herskowitz, 1984; Nasmyth, 1987), implying that regulation of wild-type *HO* is sufficient to explain the pattern of switching. How then is regulation of *HO* achieved?

The upstream control region of *HO* is unusually large for a yeast gene, being about 1400 bp in length (Nasmyth, 1985; Russell *et al.*, 1986). Scattered throughout the 1400 bp region are 10 **a**1·α2 binding sites, which presumably account for the lack of transcription of *HO* in **a**/α cells (Miller *et al.*, 1985). The other two forms for regulation of *HO* have been ascribed to two broad segments, namely URS1 and URS2, that subdivide the 1400 bp

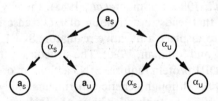

Fig. 8. Pattern of mating-type switching in budding yeast. An **a** cell, which is competent to switch (a_s), undergoes switching in late G1 and produces two daughter α cells, one that is competent to switch and one that is not (α_u). In the next cell-division cycle, the α_s daughter may switch, generating an a_s cell and an a_u cell. The α_u daughter cell does not switch, but generates one α_s and one α_u cell.

region. Segment URS1 has 600 bp and is responsible for mother–daughter control. However, within this segment, particular sequences sufficient to confer mother–daughter control to heterologous genes have not been identified. Segment URS2 has 800 bp and is responsible for cell-cycle control (Breeden and Nasmyth, 1987). This segment contains ten copies of a sequence motif, a cell-cycle box, that is sufficient to confer cell cycle-dependent transcription on heterologous genes (Breeden and Nasmyth, 1987; Andrews and Herskowitz, 1989a).

To identify genes whose products may operate through URS1 and URS2, mutants with altered transcription of *HO* have been isolated. Six *switch* (*SWI*) genes are required for *HO* transcription (Haber and Garvik, 1977; Stern *et al.*, 1984; Breeden and Nasmyth, 1987), so that formally the gene products behave as positive regulators of *HO*. However, the positive role of at least some of these gene products appears to result from their activity as antagonists of other gene products that serve as negative regulators of *HO*. This view emerged following isolation of suppressor mutations that restore transcription of *HO* in particular *swi⁻* mutants (Nasmyth *et al.*, 1987a; Sternberg *et al.*, 1987). Thus, *sin1⁻* (SWI-independent) and *sin2⁻* mutations suppress *swi1⁻*, *swi2⁻* or *swi3⁻* mutations, and *sin3⁻* and *sin4⁻* mutations suppress *swi5⁻* mutations. The sites of action of the *SWI* and *SIN* products have been determined by examining the effects of mutations in these genes on expression of URS1- or URS2-based reporter genes (Nasmyth *et al.*, 1987a; Sternberg *et al.*, 1987). These experiments show that SWI1, SWI2, SWI3, SIN1 and SIN2 act at both URS1 and URS2. On the other hand, the other SWI and SIN products appear to act at only one of the elements. Gene products SWI5, SIN3 and SIN4 work through URS1, whereas SWI4 and SWI6 work through URS2. This information, coupled with molecular and biochemical analysis of *SWI* and *SIN* genes, has led to the following specific model for transcriptional regulation of *HO* (Herskowitz, 1989).

Gene product SWI5 is a zinc-finger protein that binds to a site within URS1 (Nagai *et al.*, 1988; Stillman *et al.*, 1988), thereby setting in motion a chain of events that leads to activation of *HO* transcription. Binding of SWI5 is thought to displace a negative regulator, SDP1, which occupies a nearby site (Wang and Stillman, 1990). The structural gene for SDP1 is not known, but the SDP1 protein requires SIN3 activity in order to bind (Wang and Stillman, 1990). Although genetic experiments imply that SIN4 is also antagonized by SWI5, the molecular basis of SIN4 action is unknown. In keeping with the role of URS1 in mother–daughter control of *HO* transcription, overexpression of SWI5 (Nasmyth *et al.*, 1987b) or mutation of *SIN3* allows daughter cells to switch mating type. How SWI5 activity is confined to mother cells is as yet not understood.

Once bound to URSI, SWI5 is thought to interact with SWI1, SWI2 and SWI3. Because mutation of SWI1, SWI2 or SWI3 affects transcription of a number of genes in addition to *HO*, the gene products are thought to function as general activators of transcription (Peterson and Herskowitz, 1992). None of these three SWI proteins contains sequence motifs characteristic of DNA-binding proteins, so perhaps they are recruited to react with particular genes by protein–protein interaction involving site-specific gene-specific regulators such as SWI5. As a result of this recruitment, SWI1, SWI2 and SWI3 can antagonize SIN1 and SIN2, which are chromatin proteins (Kruger and Herskowitz, 1991). The SIN1 protein has significant sequence similarity to a mammalian non-histone chromatin protein, namely HMG1, while SIN2 is histone H3. Perhaps *HO* and other SWI1-, SWI2- and SWI3-sensitive genes have a particular chromatin structure that must be relieved before transcription can take place. In any event, SWI4 and SWI6 can now bind to cell-cycle boxes in URS2 and activate transcription of *HO* (Breeden and Nasmyth, 1987; Andrews and Herskowitz, 1989b; Taba *et al.*, 1991). The ability of these proteins to function is regulated by the CDC28 protein kinase (Breeden and Nasmyth, 1987; Andrews and Herskowitz, 1989b), which is itself active only late in the G1 phase.

In summary, genetic and molecular analysis of the SWI and SIN genes has led to an acceptable model as to how *HO* transcription is limited to a small window of the cell cycle and to particular cells in a mitotic population. Many features of the model remain to be tested but the dominant challenge for the future is to understand how SWI5 activity is restricted to mother cells.

B. Switching by *Schizosaccharomyces pombe*

Like haploid cells of *Sacch. cerevisiae*, *P* and *M* cells of fission yeast can undergo mating-type interconversion (Klar, 1989, 1992). Also, as with *Sacch. cerevisiae*, interconversion appears to be initiated by a double-stranded break at the mating-type locus (*mat1*), which is repaired by interaction with one of two silent mating-type cassettes, *mat2* or *mat3* (Fig. 9; Beach, 1983). However, despite the phenomenological similarity of the process in the two species, the mechanistic features of switching may be quite different.

Firstly, a fission-yeast gene equivalent to *HO* has yet to be identified. Ten genes required for switching have been identified and grouped based on the step in the process that appears to be defective in the *swi*⁻ mutants (Egel

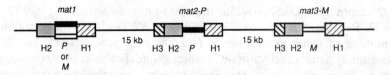

Fig. 9. Structure of mating-type information at *mat1*, *mat2* and *mat3* in *Schizosaccharomyces pombe*. At all three loci, *P* and *M* information is distinguished by non-homologous DNA, denoted *P* (1113 bp) or *M* (1127 bp). Non-homologous DNA is flanked by regions of homology, namely H1 (59 bp) and H2 (135 bp). The *mat2* and *mat3* loci share an additional homology region, H3 (57 bp). The double-stranded cut that is presumed to initiate mating-type interconversion is thought to occur at *mat1* near the junction of allele-specific DNA and the H1 homology region.

et al., 1984; Gutz and Schmidt, 1985; Schmidt, 1987). The *swi1⁻*, *swi3⁻* and *swi7⁻* mutants have decreased levels of the double-stranded cut at the *mat1* locus. These mutants are therefore considered to be defective for the initiation step in switching and, in principle, one of these genes could encode an endonuclease. The *swi2⁻*, *swi5⁻* and *swi6⁻* mutants exhibit normal levels of the double-stranded cut and are thought to be defective for utilization of the cut. The third group of mutants, *swi4⁻*, *swi8⁻*, *swi9⁻* and *swi10⁻*, are thought to be defective for resolution of the recombinational repair process that culminates the switching event, since re-arrangements of *mat1* are commonly seen in these mutants. Molecular analysis of these *swi* genes has not yet been reported, but it will certainly be interesting to learn the nature of the proteins they encode.

The more interesting distinctions between the switching process in *Sacch. cerevisiae* and fission yeast concerns two features of the pattern of mating-type switch events (Fig. 10(a); Miyata and Miyata, 1981). First, consider a haploid *P* cell that is competent to switch. After cell division, one of the cells will have switched to the *M* mating type but the other cell will still exhibit the *P* mating type. In the subsequent cell division, the newly generated *M* cell cannot switch, but the *P* cell can switch to yield one *M* and one *P* cell. Thus, after two cell-division cycles, a cell competent to switch will give rise to three granddaughters of one mating type and one granddaughter of the opposite type. This pattern suggests that the switching event occurs after DNA replication, rather than before, as in *Sacch. cerevisiae*.

The second distinction concerns the mechanism for asymmetrical inheritance of switching potential. By examining switching in diploid cells, Egel (1984) found that the competence of the two *mat1* loci to switch was independently determined. This finding suggests that the asymmetrical inheritance of switching potential is a property of the state of the *mat1*

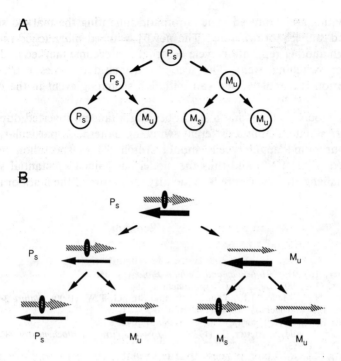

Fig. 10. Pattern of mating-type switching in fission yeast. (a) A fission-yeast P cell that is competent to switch (P_s) divides to generate one P_s cell and an M_u cell. The P_s cell can switch again in the next division, generating P_s and M_u cells. The M_u cell cannot switch but one of its daughters gains the competence to switch (M_s). (b) This switching pattern is interpreted to be the result of marking one strand of the DNA at the *mat1* locus. A cell that is competent to switch has one strand of the *mat1* locus marked by an unspecific modification (oval). Following DNA replication, one daughter inherits the marked strand, and is competent to switch in the next cell-division cycle. The other daughter will inherit a newly synthesized, unmarked version of this strand. This daughter is not competent to switch. Eventually, however, the strand becomes modified and, following cell division, one daughter inherits the modified strand and is therefore competent to switch. The shaded arrows represent DNA strands. Thin arrows represent newly synthesized strands. Modified from Horvitz and Herskowitz (1992).

locus, rather than a consequence of asymmetrical distribution of cyto-plasmic components necessary for switching, as appears to be so in *Sacch. cerevisiae*.

To explain both of these features of the fission-yeast pattern, Klar (1987, 1990) proposed the strand-segregation model (Fig. 10(b)). The *mat1* locus of a cell that has the potential to switch has one DNA strand marked by an unspecified modification. Following semi-conservative replication, two chromatids are produced. The chromatid inheriting the marked strand is cut and subject to switching. The newly switched *mat1* locus must pass through another replication cycle before it can become marked and initiate another switching event. The second chromatid becomes marked after replication is complete, and can initiate a switching event in the next cell cycle.

This model predicts that a strain bearing a tandem inverted duplication of *mat1* should exhibit a different switching pattern. In particular, two of the four granddaughter cells should switch. This expectation has been fulfilled (Klar, 1990) and thus the model has gained substantial support. The challenge for the future is to identify the nature of the mark or imprint.

References

Achstetter, T. (1989). *Molecular and Cellular Biology* **9**, 4507.
Ammerer, G. (1990). *Genes and Development* **4**, 299.
Anderson, N.G., Maller, J.L., Tongs, N.K. and Sturgill, T.W. (1990). *Nature* **343**, 651.
Andrews, B.J. and Herskowitz, I. (1989a). *Cell* **57**, 21.
Andrews, B.J. and Herskowitz, I. (1989b). *Nature* **342**, 830.
Appeltauer, U. and Achstetter, T. (1989). *European Journal of Biochemistry* **181**, 243.
Beach, D.H. (1983). *Nature* **305**, 682.
Bender, A. and Sprague, G.F. (1986). *Cell* **47**, 929.
Bender, A. and Sprague, G.F. (1987). *Cell* **50**, 681.
Blinder, D., Bouvier, S. and Jenness, D.D. (1989). *Cell* **56**, 479.
Blumer, K.J. and Thorner, J. (1990). *Proceedings of the National Academy of Sciences, USA* **87**, 4363.
Blumer, K.J. and Thorner, J. (1991). *Annual Review of Physiology* **53**, 37.
Blumer, K.J., Reneke, J.E. and Thorner, J. (1988). *Journal of Biological Chemistry* **263**, 10836.
Breeden, L. and Nasmyth, K. (1987). *Cell* **48**, 389.
Bruhn, L., Hwang-Shum, J-J. and Sprague, G.F. (1992). *Molecular and Cellular Biology* **12**, 3563.
Burkholder, A.C. and Hartwell, L.H. (1985). *Nucleic Acids Research* **13**, 8463.
Cairns, B.R., Ramer, S.W. and Kornberg, R.D. (1992). *Genes and Development* **6**, 305.
Cartwright, C.P. and Tipper, D.J. (1991). *Molecular and Cellular Biology* **11**, 2620.
Chan, R.K. and Otte, C.A. (1982). *Molecular and Cellular Biology* **2**, 11.
Chang, F. and Herskowitz, I. (1990). *Cell* **63**, 999.
Christ, C. and Tye, B.-K. (1991). *Genes and Development* **5**, 751.
Clark, K.L. and Sprague, G.F. (1989). *Molecular and Cellular Biology* **9**, 2682.
Clark, K.L., David, N.G., Wiest, D.K., Hwang-Shum, J.J. and Sprague, G.F. (1988). *Cold Spring Harbor Symposium on Quantitative Biology* **53**, 611.
Cole, G.M. and Reed, S.I. (1991). *Cell* **64**, 703.
Cole, G.M., Stone, D.E. and Reed, S.I. (1990). *Molecular and Cellular Biology* **10**, 510.
Conde, J. and Fink, G.R. (1976). *Proceedings of the National Academy of Sciences, USA* **73**, 3651.
Courchesne, W.E., Kunisawa, R. and Thorner, J. (1989). *Cell* **58**, 1107.
Covitz, P.A., Herskowitz, I. and Mitchell, A.P. (1991). *Genes and Development* **5**, 1982.
Crews, C.M. and Erickson, R.I. (1992). *Proceedings of the National Academy of Sciences, USA* **89**, 8205.

Cross, R.F. (1988). *Molecular and Cellular Biology* **8**, 4675.
Cross, F., Hartwell, L.H., Jackson, C. and Konopka, J.B. (1988). *Annual Review of Cellular Biology* **4**, 429.
Cross, F.R. and Tinkelenberg, A.H. (1991). *Cell* **65**, 875.
Dalton, S. and Treisman, R. (1992). *Cell* **68**, 597.
Davis, J.L., Kunisawa, R. and Thorner, J. (1992). *Molecular and Cellular Biology* **12**, 1879.
Davis, N.G., Horecka, J.L. and Sprague, G.F. (1993). *Journal of Cell Biology* **122**, 53.
Davey, J. (1991). *Yeast* **7**, 357.
Davey, J. (1992). *EMBO Journal* **11**, 951.
Dietzel, C. and Kurjan, J. (1987a). *Cell* **50**, 1001.
Dietzel, C. and Kurjan, J. (1987b). *Molecular and Cellular Biology* **7**, 4169.
Dirick, L. and Nasmyth, K. (1991). *Nature* **351**, 754.
Dolan, J.W. and Fields, S. (1990). *Genes and Development* **4**, 492.
Dolan, J.W. and Fields, S. (1991). *Biochimica Biophysica Acta* **1088**, 155.
Dolan, J.W., Kirkman, C. and Fields, S. (1989). *Proceedings of the National Academy of Sciences, USA* **86**, 5707.
Dranginis, A.M. (1990). *Nature* **347**, 682.
Egel, R. (1984). *Current Genetics* **8**, 205.
Egel, R. (1989). *In* "Molecular Biology of Fission Yeast" (A. Nasim, B. Johnson and P. Young, eds), pp. 31–73. Academic Press, New York.
Egel, R., Beach, D.H. and Klar, A.J.S. (1984). *Proceedings of the National Academy of Sciences, USA* **81**, 3481.
Egel, R., Nielson, O. and Weilguny, D. (1990a). *Trends in Genetics* **6**, 369.
Egel, R., Nielson, O. and Weilguny, D. (1990b). *Proceedings of the National Academy of Sciences, USA* **88**, 9392.
Elble, R. and Tye, B.-K. (1991). *Proceedings of the National Academy of Sciences, USA* **88**, 10966.
Elion, E.A., Grisafi, P.L. and Fink, G.R. (1990). *Cell* **60**, 649.
Elion, E.A., Brill, J.A. and Fink, G.R. (1991a). *Proceedings of the National Academy of Sciences, USA* **88**, 9392.
Elion, E.A., Brill, J.A. and Fink, G.R. (1991b). *Cold Spring Harbor Symposium on Quantitative Biology* **56**, 41.
Errede, B. and Ammerer, G. (1989). *Genes and Development* **3**, 1349.
Fields, S. (1990). *Trends in Biochemistry* **15**, 270.
Fields, S. and Herskowitz, I. (1987). *Molecular and Cellular Biology* **7**, 3818.
Fields, S. and Herskowitz, I. (1985). *Cell* **42**, 923.
Fields, S., Chaleff, D.T. and Sprague, G.F. (1988). *Molecular and Cellular Biology* **8**, 551.
Finegold, A.A., Schafer, W.R., Rine, J., Whiteway, M. and Tamanoi, F. (1990). *Science* **249**, 165.
Flessel, M.C., Brake, A.J. and Thorner, J. (1989). *Genetics* **121**, 223.
Freyd, G., Kim, S.K. and Horvitz, H.R. (1990). *Nature* **344**, 876.
Fujimura, H.-A. (1989). *Molecular and Cellular Biology* **9**, 152.
Fukui, Y., Kaziro, Y. and Yamamoto, M. (1986a). *EMBO Journal* **5**, 1991.
Fukui, Y., Kozasa, T., Kaziro, Y., Takeda, T. and Yamamoto, M. (1986b). *Cell* **44**, 329.
Gartner, A., Nasmyth, K. and Ammerer, G. (1992). *Genes and Development* **6**, 1280.
Gomez, N. and Cohen, P. (1991). *Nature* **353**, 170.
Goutte, C. and Johnson, A.D. (1988). *Cell* **52**, 875.
Gubbay, J., Colligan, J., Koopman, P., Capel, B., Economun, A., Münsterberg, A., Vivian, N., Goodfellow, P. and Lovell-Badge, R. (1990). *Nature* **346**, 245.
Gutz, H. and Schmidt, H. (1985). *Current Genetics* **9**, 325.
Haber, J.E. (1983). *In* "Mobile Genetic Elements" (J. Shapiro, ed.), pp. 560–619. Academic Press, New York.
Haber, J.E. and Garvik, B. (1977). *Genetics* **87**, 33.
Hadwiger, J.A., Wittenberg, C., Richardson, H.E. and deBarros Lopes, M. (1989). *Proceedings of the National Academy of Sciences, USA* **86**, 6255.
Hagen, D.C. and Sprague, G.F. (1984). *Journal of Molecular Biology* **178**, 835.

Hagen, D.C., McCaffrey, G. and Sprague, G.F. (1986). *Proceedings of the National Academy of Sciences, USA* **83**, 1418.

Hagen, D.C., McCaffrey, G. and Sprague, G.F. (1991). *Molecular and Cellular Biology* **11**, 2952.

Hanks, S.K., Quinn, A.M. and Hunter, T. (1988). *Science* **241**, 42.

Hartig, A., Holly, J., Saari, G. and MacKay, V.L. (1986). *Molecular and Cellular Biology* **6**, 2106.

Hartwell, L.H. (1980). *Journal of Cell Biology* **85**, 811.

Hayes, T.E., Sengupta, P. and Cochran, B.H. (1988). *Genes and Development* **2**, 1713.

Herskowitz, I. (1989). *Nature* **342**, 749.

Herskowitz, I. and Oshima, Y. (1981). *In* "The Molecular Biology of the yeast *Saccharomyces*: Life Cycle and Inheritance" (J.N. Strathern, E.W. Jones and J.R. Broach, eds), pp. 181–209. Cold Spring Harbor Laboratory Press, Cold Spring Harbor.

Hirsch, J.P., Dietzel, C. and Kurjan, J. (1991). *Genes and Development* **5**, 467.

Horvitz, H.R. and Herskowitz, I. (1992). *Cell* **68**, 237.

Hwang-Shum, J-J., Hagen, D.C., Jarvis, E.E., Westby, C.A. and Sprague, G.F. (1991). *Molecular and General Genetics* **227**, 197.

Iida, H., Yagawa, Y. and Anraku, Y. (1990). *Journal of Biological Chemistry* **265**, 13391.

Inokuchi, K., Nakayama, A. and Hishuma, F. (1987). *Molecular and Cellular Biology* **7**, 3185.

Jahng, K.-Y., Ferguson, J. and Reed, S.I. (1988). *Molecular and Cellular Biology* **8**, 2484.

Jarvis, E.E., Hagen, D.C. and Sprague, G.F. (1988). *Molecular and Cellular Biology* **8**, 309.

Jarvis, E.E., Clark, K.L. and Sprague, G.F. (1989). *Genes and Development* **3**, 936.

Jenness, D.D., Burkholder, A.C. and Hartwell, L.H. (1983). *Cell* **35**, 521.

Jenness, D.D., Burkholder, A.C. and Hartwell, L.H. (1986). *Molecular and Cellular Biology* **6**, 318.

Jensen, R. and Herskowitz, I. (1984). *Cold Spring Harbor Symposium on Quantitative Biology* **49**, 97.

Jensen, R., Sprague, G.F. and Herskowitz, I. (1983). *Proceedings of the National Academy of Sciences, USA* **80**, 3035.

Johnson, A.D. and Herskowitz, I. (1985). *Cell* **42**, 237.

Kang, Y.S., Kane, J., Kurgan, J., Stadel, J.M. and Tipper, D.J. (1990). *Molecular and Cellular Biology* **10**, 2582.

Kassir, Y. and Simchen, G. (1976). *Genetics* **82**, 187.

Kassir, Y., Granot, D. and Simchen, G. (1988). *Cell* **52**, 853.

Keleher, C.A., Goutte, C. and Johnson, A.D. (1988). *Cell* **53**, 927.

Keleher, C.A., Passmore, S. and Johnson, A.D. (1989). *Molecular and Cellular Biology* **9**, 5228.

Keleher, C.A., Redd, M.J., Schultz, J., Carlson, M. and Johnson, A.D. (1992). *Cell* **68**, 709.

Kelly, M., Burke, J., Smith, M., Klar, A. and Beach, D. (1988). *EMBO Journal* **7**, 1537.

King, K., Dohlman, H.G., Thorner, J., Caron, M.G. and Lefkowitz, R.J. (1990). *Science* **250**, 121.

Kitamura, K. and Shimoda, C. (1991). *EMBO Journal* **12**, 3743.

Kjelsberg, M.A., Cotecchia, S., Ostrowski, J., Carson, M.G. and Lefkowitz, R.J. (1992). *Journal of Biological Chemistry* **267**, 1430.

Klar, A.J.S. (1987). *Nature* **326**, 466.

Klar, A.J.S. (1989). *In* "Mobile DNA" (D.E. Berg and M.M. Howe, eds), pp. 671–691. American Society for Microbiology. Washington, D.C.

Klar, A.J.S. (1990). *EMBO Journal* **9**, 1407.

Klar, A.J.S. (1992). *Trends in Genetics* **8**, 208.

Klar, A.J.S., Strathern, J.N., Broach, J.R. and Hicks, J.B. (1981). *Nature* **289**, 239.

Klar, A.J.S., Hick, J.B. and Strathern, J.N. (1982). *Cell* **28**, 551.

Konopka, J.B. and Jenness, D.D. (1991). *Cell Regulation* **2**, 439.

Konopka, J.B., Jenness, D.D. and Hartwell, L.H. (1988). *Cell* **54**, 609.

Kosako, H., Gotoh, Y., Matsuda, S., Ishawa, M. and Nichid, E. (1992). *EMBO Journal* **11**, 2903.

Kostriken, R. and Heffron, F. (1984). *Cold Spring Harbor Symposium on Quantitative Biology* 49, 89.

Kostriken, R., Strathern, J.N., Klar, A.J.S., Hicks, B. and Heffron, F. (1983). *Cell* 35, 167.

Kronstad, J.W., Holly, J.S. and MacKay, V.L. (1987). *Cell* 50, 369.

Kruger, W. and Herskowitz, I. (1991). *Molecular and Cellular Biology* 77, 4135.

Kurjan, J. (1992). *Annual Review of Biochemistry* 61, 1097.

Laughon, A. and Scott, M.P. (1984). *Nature* 310, 25.

Leberer, E., Dignard, D., Harcus, D., Thomas, D.Y. and Whiteway, M. (1992). *EMBO Journal* 11, 4815.

Lemontt, J.F., Fugit, D.R. and MacKay, V.L. (1980). *Genetics* 94, 899.

Leupold, U. (1987). *Current Genetics* 12, 543.

Linder, M.E. and Gilman, A.G. (1992). *Scientific American* 267, 56.

Lipke, P.N., Wojciechowicz, D. and Kurjan, J. (1989). *Molecular and Cellular Biology* 9, 3155.

MacKay, V. and Manney, T.R. (1974). *Genetics* 76, 273.

Marsh, L. (1992). *Molecular and Cellular Biology* 12, 3959.

Marsh, L. and Herskowitz, I. (1988). *Proceedings of the National Academy of Sciences, USA* 85, 3855.

Marsh, L., Neiman, A.M. and Herskowitz, I. (1991). *Annual Review of Cell Biology* 7, 699.

Matsuda, S., Kosako, H., Takenaka, K., Moriyama, K., Sakai, H., Akiyama, T., Hotoh, Y. and Nishid, E. (1992). *EMBO Journal* 11, 973.

McCaffrey, G., Clay, F.J., Kelsay, K. and Sprague, G.F. (1987). *Molecular and Cellular Biology* 7, 2680.

McCullough, J. and Herskowitz, I. (1979). *Journal of Bacteriology* 138, 146.

Meluh, P.B. and Rose, M.D. (1990). *Cell* 60, 1029.

Michaelis, S. and Herskowitz, I. (1988). *Molecular and Cellular Biology* 8, 1309.

Miller, A.M., MacKay, V.L. and Nasmyth, K.A. (1985). *Nature* 314, 598.

Mitchell, A.P. and Herskowitz, I. (1986). *Nature* 319, 738.

Miyajima, I., Nakafuku, M., Nakayama, N., Brenner, C., Miyajima, A., Kaibuchi, K., Arai, K.-I., Kaziro, Y. and Matsumoto, K. (1987). *Cell* 50, 1011.

Miyajima, I., Nakayama, N., Nakafuku, M., Kaziro, Y., Arai, K. and Matsumotu, K. (1988). *Genetics* 119, 797.

Miyata, H. and Miyata, M. (1981). *Journal of General and Applied Microbiology* 27, 365.

Moll, T., Tebb, G., Surana, U., Robitsch, H. and Nasmyth, K. (1991). *Cell* 66, 743.

Mumby, S.M., Casey, P.J., Gilman, A.G., Gutowski, S. and Sternweis, P.C. (1990). *Proceedings of the National Academy of Sciences, USA* 87, 5873.

Nadin-Davis, S.A. and Nasim, A. (1988). *EMBO Journal* 7, 985.

Nadin-Davis, S.A. and Nasim, A. (1990). *Molecular and Cellular Biology* 10, 549.

Nadin-Davis, S.A., Nasim, A. and Beach, D. (1986). *EMBO Journal* 5, 2963.

Nagai, K., Nakaseko, Y., Nasmyth, K. and Rhodes, D. (1988). *Nature* 332, 284.

Nakafuku, M., Hoh, H., Nakayama, S. and Kaziro, Y. (1987). *Proceedings of the National Academy of Sciences, USA* 84, 2140.

Nakayama, N., Miyajima, A. and Arai, K. (1985). *EMBO Journal* 4, 2643.

Nakayama, N., Miyajima, A. and Arai, K. (1987). *EMBO Journal* 6, 249.

Nakayama, N., Arai, K. and Matsumoto, K. (1988a). *Molecular and Cellular Biology* 8, 5410.

Nakayama, N., Kaziro, Y., Arai, k. and Matsumoto, K. (1988b). *Molecular and Cellular Biology* 8, 3777.

Nash, R., Tokiwa, G., Anand, S., Erikson, K. and Futcher, A.B. (1988). *EMBO Journal* 7, 4335.

Nasmyth, K. (1983). *Nature* 302, 670.

Nasmyth, K. (1985). *Cell* 42, 225.

Nasmyth, K. (1987). *EMBO Journal* 6, 243.

Nasmyth, K. and Shore, D. (1987). *Science* 237, 1162.

Nasmyth, K., Stillman, D. and Kipling, D. (1987a). *Cell* 48, 579.

Nasmyth, K., Seddon, A. and Ammerer, G. (1987b). *Cell* 49, 549.

458   G.F. Sprague

Nasmyth, K.A., Tatchell, K., Hall, B.D., Astell, C. and Smith, M. (1981). *Nature* **289**, 244.
Neiman, A., Marcus, S., Stevenson, B., Xu, H.-P., Sprague, G.F., Herskowitz, I., Wigler, M. and Marus, S. (1993). *Molecular Biology of the Cell* **4**, 107.
Nielson, A., Davey, J. and Egel, R. (1992). *EMBO Journal* **11**, 1391.
Nielson, O. and Egel, R. (1990). *EMBO Journal* **9**, 1401.
Nomoto, S., Nakayama, N., Arai, K. and Matsumoto, K. (1990). *EMBO Journal* **9**, 691.
Norman, C., Runswich, M., Pollock, R. and Triesman, R. (1988). *Cell* **55**, 989.
Normington, K., Kohno, K., Kozutsumi, Y., Gething, M-J. and Sambrook, J. (1989). *Cell* **57**, 1223.
Obara, T., Nakafuku, M., Yamamoto, M. and Kaziro, Y. (1991). *Proceedings of the National Academy of Sciences, USA* **88**, 5877.
Page, B.D. and Snyder, M. (1992). *Genes and Development* **6**, 1414.
Passmore, S., Elble, R. and Tye, B.-K. (1989). *Genes and Development* **3**, 921.
Payne, D.M., Rossomando, S.J., Martino, P., Erickson, A.K., Her, J.H., Shabonowitz, J., Hunt, D.F., Weber, M.J. and Sturgill, T.W. (1991). *EMBO Journal* **10**, 885.
Pelech, S.L. and Sanghera, J.S. (1992). *Trends in Biochemical Science* **17**, 233.
Peterson, C.L. and Herskowitz, I. (1992). *Cell* **68**, 573.
Polaina, J. and Conde, J. (1982). *Molecular and General Genetics* **186**, 253.
Primig, M., Winkler, H. and Ammerer, G. (1991). *EMBO Journal* **10**, 4209.
Reneke, J.E., Blumer, K.J., Courchesne, W.E. and Thorner, J. (1988). *Cell* **55**, 221.
Rhodes, N., Connell, L. and Errede, B. (1990). *Genes and Development* **4**, 1862.
Richardson, H.E., Wittenberg, C., Cross, F. and Reed, S.I. (1989). *Cell* **59**, 1127.
Rine, J., Jensen, R., Hagen, D., Blair, L. and Herskowitz, I. (1980). *Cold Spring Harbor Symposium on Quantitative Biology* **45**, 951.
Rogers, S., Wells, R. and Rechsteiner, M. (1986). *Science* **234**, 364.
Rose, M.D. (1991). *Annual Review of Microbiology* **45**, 539.
Rose, M.D. and Fink, G.R. (1987). *Cell* **48**, 1047.
Rose, M.D., Misra, L. and Vogel, J.P. (1989). *Cell* **57**, 1211.
Ross, F.M. (1989). *Neuron* **3**, 141.
Russell, D.W., Jensen, R., Zoller, M., Burke, J., Errede, B., Smith, M. and Herskowitz, I. (1986). *Molecular and Cellular Biology* **6**, 4281.
Schmidt, H. (1987). *Molecular and General Genetics* **210**, 486.
Schultz, J. and Carlson, M. (1987). *Molecular and Cellular Biology* **7**, 3637.
Sengupta, P. and Cochran, B.H. (1990). *Molecular and Cellular Biology* **10**, 6809.
Sengupta, P. and Cochran, B.H. (1991). *Genes and Development* **5**, 1924.
Shaw, P.E., Schröter, H. and Nordheim, A. (1989). *Cell* **56**, 563.
Shepherd, J.C.W., McGinnis, W., Carrasco, A.E., DeRobertis, E.M. and Gehring, W.J. (1984). *Nature* **310**, 70.
Sommer, H., Beltran, J.-P., Huigser, P., Pape, H., Lonnig, W.-E., Saedler, H. and Schwarz-Xommer, Z. (1991). *EMBO Journal* **9**, 605.
Song, D., Dolan, J.W., Yuan, Y.L. and Fields, S. (1991). *Genes and Development* **5**, 741.
Sprague, G.F. (1990). *Advances in Genetics* **27**, 33.
Sprague, G.F. (1991). *Trends in Genetics* **7**, 393.
Sprague, G.F. (1992). *Seminars in Developmental Biology* **3**(5).
Sprague, G.F., Jensen, R. and Herskowitz, I. (1983). *Cell* **32**, 409.
Stern, M., Jensen, R.E. and Herskowitz, I. (1984). *Journal of Molecular Biology* **178**, 853.
Sternberg, P.W., Stern, M.J., Clark, I. and Herskowitz, I. (1987). *Cell* **48**, 567.
Stevenson, B.J., Rhodes, N., Errede, B. and Sprague, G.F (1992). *Genes and Development* **6**, 1293.
Stillman, D.J., Bankier, A.T., Seddon, A., Groenhout, E.G. and Nasmyth, K.A. (1988). *EMBO Journal* **7**, 485.
Stone, D.E., Cole, G.M., deBarros Lopes, M., Goebl, M. and Reed, S.I. (1991). *Genes and Development* **5**, 1969.
Strathern, J.N. and Herskowitz, I. (1979). *Cell* **17**, 371.
Strathern, J.N., Hick, J.B. and Herskowitz, I. (1981). *Journal of Molecular Biology* **147**, 357.

Strathern, J.N., Klar, A.J.S., Hicks, J.B., Abraham, J.A., Ivy, J.M., Nasmyth, K.A. and McGill, C. (1982). *Cell* **31**, 183.
Sugimoto, A., Lino, Y., Maeda, T., Watanabe, Y. and Yamamoto, M. (1991). *Genes and Development* **5**, 1990.
Taba, M.R.M., Muroff, I., Lydall, D., Tebb, G. and Nasmyth, K. (1991). *Genes and Development* **5**, 2000.
Tan, S. and Richmond, T.J. (1990). *Cell* **62**, 367.
Tan, S., Ammerer, G. and Richmond, T.J. (1988). *EMBO Journal* **7**, 4255.
Teague, M.A., Chaleff, D.T. and Errede, B. (1986). *Proceedings of the National Academy of Sciences, USA* **83**, 7371.
Thomas, G. (1992). *Cell* **68**, 3.
Toda, T., Shimanuki, M. and Yanagida, M. (1991). *Genes and Development* **5**, 60.
Trueheart, J. and Fink, G.R. (1989). *Proceedings of the National Academy of Sciences, USA* **86**, 9916.
Trueheart, J., Boeke, J.D. and Fink, G.R. (1987). *Molecular and Cellular Biology* **7**, 2316.
Van Arsdell, S.W. and Thorner, J. (1987). *In* "Transcriptional Control Mechanisms" (D. Granner, M.G. Rosenfeld and S. Chang, eds) p. 325. Alan R. Liss, New York.
Wang, H. and Stillman, D. (1990). *Proceedings of the National Academy of Sciences, USA* **87**, 9761.
Wang, Y., Xu, H.-P., Riggs, M., Rodgers, L. and Wigler, M. (1991). *Molecular and Cellular Biology* **11**, 3554.
Watanabe, Y., Iion, Y., Furuhata, K. Shimoda, C. and Yamamoto, M. (1988). *EMBO Journal* **7**, 761.
Whiteway, M., Hougan, L., Dignard, D., Thomas, D.Y., Bell, L., Saari, G.C., Grant, F.J., O'Hara, P. and MacKay, V.L. (1989). *Cell* **56**, 467.
Whiteway, M., Hougan, L. and Thomas, D.Y. (1990). *Molecular and Cellular Biology* **10**, 217.
Williams, F.E., Varanasi, U. and Trumbly, R.J. (1991). *Molecular and Cellular Bioloily* **11**, 3307.
Wittenburg, C., Sugimoto, K. and Reed, S.I. (1990). *Cell* **62**, 225.
Wood, K.W., Sarnecki, C., Roberts, T.M. and Blenis, J. (1992). *Cell* **68**, 1041.
Yamane, H.K., Farnsworth, C.C., Xie, H., Howald, W. and Fung, B.K.-K. (1990). *Proceedings of the National Academy of Sciences, USA* **87**, 5868.
Yanofsky, M.F., Ma, H., Bowman, J.L., Drews, G.M., Feldman, K.A. and Meyerowitz, E.M. (1990). *Nature* **346**, 35.
Yu, L., Blumer, K.J., Davidson, N., Lester, H.A. and Thorner, J. (1989). *Journal of Biological Chemistry* **264**, 20847.
Yuan, Y.L. and Fields, S. (1991). *Molecular and Cellular Biology* **11**, 5910.

Appendix I Genetic Nomenclature for *Saccharomyces cerevisiae*

Fred Sherman

Department of Biochemistry, University of Rochester School of Medicine and Dentistry, Rocherster, NY 14642, USA.

I. THE GENOME OF *SAACHAROMYCES CEREVISIAE*

A genetic nomenclature for *Saccharomyces cerevisiae* has evolved from early recommendations (Sherman and Lawrence, 1974; Sherman, 1981), to a now universally accepted system for at least chromosomal genes (Sherman, 1991). The assignment of gene symbols is dependent on the nature of the associated genetic or physical determinants and if the gene exhibits Mendelian or non-Mendelian inheritance.

S. cerevisiae contains a haploid set of 16 chromosomes that have been well-characterized genetically (Mortimer *et al.*, 1995) and physically (Olson, 1991; Kaback, 1995). A single marker, *KRB1*, has been assigned to chromosome XVII (Wickner *et al.*, 1983), although there is no physical evidence for this chromosome. The 16 chromosomes range in size from 240 to 3500 kb, with a total length of 15 000 kb.

Other nucleic acid entities, listed in Table I, also can be considered part of the yeast genome. Mitochondrial DNA encodes components of the mitochondrial translational machinery and approximately 15% of the mitochondrial proteins (Dujon, 1981; Guérin, 1991). ρ^0 mutants completely lack mitochondrial DNA and are deficient in the respiratory polypeptides synthesized on mitochondrial ribosomes, i.e. cytochrome b and subunits of cytochrome c oxidase and ATPase complexes. Even though ρ^0 mutants are respiratory deficient, they are viable and still retain mitochondria, although morphologically abnormal.

The 2-µm circle plasmids, present in most strains of *S. cerevisiae*, apparently function solely for their own replication (Broach, 1983; Williamson, 1991). Generally *cir*^{0} strains, which lack 2-µm DNA, have no

461

Table I. The genome of a diploid *Sacch. cerevisiae* cell

	← Mendelian →		← Non-Mendelian →					
	Nuclear	← Double-stranded DNA →		← Double-stranded RNA →				
				RNA Molecules				
	Chromosomes	2-μm plasmid	Mitochondrial	L-A	M	L-BC	T	W
Relative amount	85%	5%	10%	80%	10%	9%	0.5%	0.5%
Number of copies	2 sets of 16	60–100	~50 (8–130)	10^3	170	150	10	10
Size (kb)	15 000 (240–3550)	6.318	70–76	4.476	1.8	4.6	2.7	2.25
Deficiencies in mutants	All kinds	None	Cyto. $a \cdot a_3$, etc.	Killer toxin			None	
Wild-type	(Example) ARG2$^+$	cir$^+$	ρ$^+$	KIL-k$_1$				
Mutant or variant	arg2-14	cir^0	ρ$^-$	KIL-O				

Adapted from Carle and Olson, 1985; Olson, 1991; Broach, 1983; Dujon, 1981; Wickner, 1989 observable phenotype; however, a certain chromosomal mutation, *nib1*, causes a reduction in division potential of cir^+ strains (Holm, 1982; Sweeny and Zakian, 1989).

Similarly, almost all *Sacch. cerevisiae* strains contain dsRNA viruses that are not usually extracellularly infective but that are transmitted by mating (Wickner, 1989, 1991, 1995). This dsRNA, constituting approximately 0.1% of total nucleic acid, determines a toxin and components required for the viral transcription and replication. *KIL-o* mutants, lacking M dsRNA and consequently the killer toxin, are readily induced by chemical and physical agents.

Only mutations of chromosomal genes exhibit Mendelian 2:2 segregation in tetrads after sporulation of heterozygous diploids (Cox, 1995), a property dependent on the disjunction of chromosomal centromeres. In contrast, non-Mendelian inheritance is observed for the phenotypes associated with the absence or alteration of other nucleic acids described in Table I.

II. MENDELIAN CHROMOSOMAL GENES

The following genetic nomenclature for chromosomal genes of the yeast *S. cerevisiae* is now universally accepted. Gene symbols are consistent with the proposals of Demerec *et al.* (1966), whenever possible, and are designated by three italicized letters, e.g. *arg*. Contrary to the proposals of Demerec *et al.* (1966) for *Escherichia coli*, yeast genetic loci are identified by numbers, not letters, following the gene symbols, e.g. *arg2*. Dominant alleles are denoted by using uppercase italics for all letters of the gene symbol, e.g. *ARG2*. Lowercase letters denote the recessive allele, e.g. the auxotroph *arg2*. Wild-type genes are designated with a superscript "plus" (*sup6⁺* or *ARG2⁺*). Alleles are designated by a number separated from the locus number by a hyphen, e.g. *arg2-14*. Locus numbers should be consistent with the original assignments, but allele numbers may be specific to a particular laboratory. For example, two different isolates from two different laboratories could be denoted *can1-1*, where both are mutations of the *CAN1* locus. The symbol Δ can denote complete or partial deletions, e.g. *his3-Δ1*. Insertion of genes follow the bacterial nomenclature by using the symbol :: . For example, *cyc1::URA3* denotes the insertion of the *URA3* gene at the *CYC1* locus, in which *URA3* is dominant (and functional), and *cyc1* is recessive (and defective).

Phenotypes are sometimes denoted by cognate symbols in roman type and by the superscripts + and − . For example, the independence and requirement for arginine can be denoted by Arg⁺ and Arg⁻, respectively.

Proteins encoded by *ARG2*, for example, can be denoted Arg2p, or simply Arg2 protein. However, gene symbols are generally used as adjectives for other nouns, for example, *ARG2* mRNA, *ARG2* strains, etc. The conventions used in the genetic nomenclature for *S. cerevisiae* are illustrated in Table II.

Most alleles can be unambiguously assigned as dominant or recessive by examining the phenotype of the heterozygous diploid crosses. However, dominant and recessive traits are defined only with pairs, and a single allele can be both dominant and recessive. For example, because the alleles $CYC1^+$, *cyc1-717* and *cyc1-Δ1* produce, respectively, 100%, 5% and 0% of the gene product, the *cyc1-717* allele can be considered recessive in the *cyc1-717/CYC1⁺* cross and dominant in the *CYC1-717/cyc1-Δ1* cross. Thus, sometimes it is less confusing to denote all mutant alleles in lower case letters, especially when considering a series of mutations having a range of activities.

There are a number of exceptions to these general rules. Gene clusters, complementation groups within a gene, or domains within a gene having different characteristics can be designated by capital letters following the locus number; for example, *HIS4A*, *HIS4B* and *HIS4C* denote three regions of the *HIS4* locus that correspond to three domains of the single polypeptide chain, which in turn determines three different enzymatic steps.

Different members of a gene family should be denoted by the same rules, i.e. a gene symbol with different locus numbers, since the members are non-allelic and are at different loci. For example, overlapping functions and high homology are observed with *SUC1–SUC5*, *SUC7*, six unlinked genes which encode invertase; *CYC1* and *CYC7*, which encode two iso-forms of cytochrome *c*; and with *SSA1*, *SSA2* and *SSA4*, which encode a subgroup of Hsp70 heat shock proteins. We **do not** recommend other designations to denote different members of the same gene families, such as *COX5a* and *COX5b*; *TIF51A* and *TIF51B*; or *TCP1α* and *TCP1β*.

Table II. The conventions used in the genetic nomenclature for *Saccharomyces cerevisiae*

Convention	Meaning
ARG2	A locus or dominant allele
arg2	A locus or recessive allele conferring an arginine requirement
arg2⁻	Any *arg2* allele conferring an arginine requirement
ARG2⁺	The wild-type allele
arg2-9	A specific allele or mutation
Arg⁺	A strain not requiring arginine
Arg⁻	A strain requiring arginine
Arg2p	The protein encoded by *ARG2*
Arg2 protein	The protein encoded by *ARG2*
ARG2 mRNA	The mRNA transcribed from *ARG2*

Although superscript letters should be avoided, it is sometimes expedient to distinguish genes conferring resistance and sensitivity by superscript R and S, respectively. For example, the genes controlling resistance to canavanine sulphate (*can1*) and copper sulphate (*CUP1*) and their sensitive alleles could be denoted, respectively, as *can^{R}1*, *CUP^{R}1*, *CAN^{S}1*, and *cup^{S}1*.

Wild-type and mutant alleles of the mating-type locus and related loci do not follow the standard rules. Most of the genes related to mating-type were defined before acceptance of the current nomenclature. The two wild-type alleles of the mating-type locus are designated *MATa* and *MATα*. The two complementation groups of the *MATα* locus are denoted *MATα1* and *MATα2*. Mutations of the *MAT* genes are denoted, e.g. *mata-1* and *matα1-1*. The wild-type homothallic alleles at the *HMR* and *HML* loci are denoted *HMRa*, *HMRα*, *HMLa* and *HMLα* (Sprague, 1995). Mutations at these loci are denoted, e.g. *hmra-1*, *hmlα-1*. The mating phenotypes of *MATa* and *MATα* cells are denoted simply **a** and α, respectively. The two letters *HO* denote the gene encoding the endonuclease required for homothallic switching.

Dominant and recessive suppressors should be denoted, respectively, by three uppercase or three lowercase letters, followed by a locus designation, e.g. *SUP4*, *SUF1*, *sup35*, *suf11*, etc. In some instances UAA ochre suppressors and UAG amber suppressors are further designated, respectively, **o** and **a** following the locus. For example, *SUP4-o* refers to suppressors of the *SUP4* locus that insert tyrosine residues at UAA sites; *SUP4-a* refers to suppressors of the same *SUP4* locus that insert tyrosine residues at UAG sites. The corresponding wild-type locus coding for the normal tyrosine tRNA and lacking suppressor activity can be referred to as *sup4$^+$*. Intragenic mutations that inactivate suppressor can denoted, for example, *sup4$^-$* or *sup4-o-1*. The nomenclature describing suppressor and wild-type alleles in yeast is unrelated to the bacterial nomenclature. For example, an ochre *E. coli* suppressor that inserts tyrosine residues at both UAA and UAG sites is denoted as *su$_4^+$*, and the wild-type locus coding for the normal tyrosine tRNA and lacking suppressor activity can be referred to as *Su$_4$*, *su$_4^-$*, or *supC*. Frameshift suppressors are denoted as *suf* (or *SUF*), whereas metabolic suppressors are denoted with a variety of specialized symbols, such as *ssn* (suppressor of *snf1*), *srn* (suppressor of *rna1-1*), and *suh* (suppressor os *his2-1*).

Because the sites of mutations are usually used for genetic mapping, published chromosome maps usually contain the mutant allele. For example, chromosome III contains *his4* and *leu2*, whereas chromosome IX contains *SUP22* and *FLD1*. However, multiple dominant wild-type genes that control the same character (for example, *SUC1*, *SUC2*, etc.), are used in genetic mapping, and such chromosomal loci are denoted in capital letters on genetic maps.

Capital letters are also used to designate certain DNA segments whose locations have been determined by a combination of recombinant DNA techniques and classical mapping procedures, e.g. *RDN1*, the segment encoding ribosomal RNA. DNA sequencing has revealed numerous genes encoding open-reading frames or transcripts, whose functions have not been determined. Some such genes have been denoted *UTR* (unidentified transcript) or *FUN* (function unknown). However, the general form YCRXXw is now used to designate genes uncovered by systematically sequencing the yeast genome, where Y designates yeast; C (or A, B, etc.) designates the chromosome III (or I, II, etc.); R (or L) designates the right (or left) arm of the chromosome; XX designates the relative position of the start of the open-reading frame from the centromere; and w (or c) designates the Watson (or Crick) strand. For example, YCR5c denotes *CIT2*, a previously known but unmapped gene situated on the right arm of chromosome III, fifth open reading-frame from the centromere on the Crick strand (Oliver *et al.*, 1992).

E. coli genes inserted into yeast are usually denoted by the prokaryotic nomenclature, e.g. *lacZ*.

In order to prevent duplications, new gene symbols should be approved by the committee for genetic nomenclature, headed by Dr David Botstein (Department of Genetics, School of Medicine, Stanford University, Stanford, CA 94305–5120, USA.; FAX 415 723 7016; E-mail botstein@ genome.stanford.edu).

III. NON-MENDELIAN DETERMINANTS

Where necessary, non-Mendelian genotypes can be distinguished from chromosomal genotypes by enclosure in brackets, e.g. *[KIL-o] MATα trp1-1*. Although it is advisable to employ the above rules for designating non-Mendelian genes and to avoid using Greek letters, the use of well-known and generally accepted Greek symbols should be continued; thus, the original symbols ρ^+, ρ^-, ψ^+ and ψ^- or their transliteration, *rho*$^+$, *rho*$^-$, *psi*$^+$ and *psi*$^-$, respectively, should be retained. Some of the non-Mendelian determinants associated with physically identified and unidentified entities are listed in Tables I and III, respectively. A detailed designation for nuclear and dsRNA genes controlling the killer trait has been described by Wickner (1991, 1995).

IV. MITOCHONDRIAL MUTANTS

Special consideration should be made of the nomenclature describing mutations of mitochondrial components and function that are determined by both nuclear and mitochondrial DNA genes. The growth on media containing non-fermentable substrates (Nfs) as the sole energy and carbon source (such as glycerol or ethanol) is the most convenient operational procedure for testing mitochondrial function. Lack of growth on non-fermentable media (Nfs⁻mutants), as well as other mitochondrial alterations, can be due to either nuclear or mitochondrial mutations as outlined in Table III. Nfs⁻ nuclear mutations are generally denoted by the symbol *pet*; however, more specific designations have been used instead of *pet* when the gene products were known, such as *cox4*, *hem1*, etc.

The complexity of nomenclatures for mitochondrial DNA genes, outlined in Table IV, is due in part to complexity of the system, polymorphic differences of mitochondrial DNA, complementation between exon and intron mutations, the presence of intron-encoded maturases, diverse phenotypes of mutations within the same gene, and the lack of agreement between various workers. Unfortunately, the nomenclature for most mitochondrial mutations do not follow the rules outlined for nuclear mutations. Furthermore, confusion can occur between phenotypic designations, mutant isolation number, allelic designations, loci, and cistrons (complementation groups). Detailed designations for mitochondrial mutants have been presented by Dujon (1981) and Grivell (1984), with recommended modifications by Costanzo and Fox (1990). While further changes may lead to additional confusion, we nevertheless recommend that the following wild-type genes denote the following mitochondrial translated products, presented in Table V.

Table III. Some non-Mendelian determinants of *Sacch. cerevisiae* with unknown genetic elements

Wild-type	Mutant or polymorphic variant	Mutant phenotype
ψ⁺	ψ⁻	Decreased efficiency of certain suppression
ξ⁺	ξ⁻	Decreased efficiency of certain suppression
URE3	ure3⁻	Deficiency in ureidosuccinate utilization

Adapted from Tuite *et al.* (1982); Aigle and Lacroute (1975); Liebman and All-Robyn (1984). Wickner (1994) has suggested that [URE3] and ψ are prion forms of proteins encoded by, respectively, URE2 and PNM2.
Other non-Mendelian determinants have been reported (Sweeny and Zakian, 1989).

Table IV. General classes of mitochondrial mutants with examples (Nfs⁻ denoted lack of growth on non-fermentable substrates)

General mutations with examples	Mutant phenotype or gene product
Nuclear genes mutations	
pet⁻	Nfs⁻
pet1	Unknown function
cox4	Cytochrome c oxidase subunit IV
hem1	δ-Aminolevulinate synthase
cyc3	Cytochrome c haem lyase
Mitochondrial DNA mutations	
Gross aberrations	
ρ^-	Nfs⁻
ρ^0	ρ^- mutants lacking mitochondrial DNA
Single-site mutations	
mit⁻	Nfs⁻, but capable of mitochondrial translation
oxi3 or cox1	Cytochrome c oxidase subunit I
cob or box	Cytochrome b
pho2	ATPase subunit 9
syn⁻	Nfs⁻, deficient in mitochondrial translation
tRNA^{Asp} or M7-37	Mitochondrial tRNA^{Asp} (CUG)
ant^R	Resistant to inhibitors
ery^R or rib1	Resistant to erythromycin, 21S rRNA
cap^R or rib3	Resistant to chloramphenicol, 21S rRNA
par^R or par1	Resistant to paromomycin, 16S rRNA
oli^R or oli1	Resistant to oligomycin, ATPase subunit 9

Table V. Recommended gene symbols for mitochondrial encoded proteins

Gene symbol	Product
[COX1]	cytochrome c oxidase subunit I
[COX2]	cytochrome c oxidase subunit II
[COX3]	cytochrome c oxidase subunit III
[COB1]	cytochrome b
[ATP6]	mitochondrial ATPase subunit 6
[ATP8]	mitochondrial ATPase subunit 8
[ATP9]	mitochondrial ATPase subunit 9
[VAR1]	mitochondrial ribosomal subunit

V. ACKNOWLEDGEMENT

The writing of this chapter was supported by the Public Health Research Grant GM12702 from the National Institutes of Health.

References

Aigle, M. and Lacroute, F. (1975). *Molecular and General Genetics* **136**, 327.
Broach, J.R. (1983). *In* "Methods in Enzymology", Vol. 101, Recombinant DNA, (R. Wu, L. Grossman and K. Moldave, eds.), pp. 307–325. Academic Press, New York.
Carle G.F. and Olson, M. (1985). *Proceedings of the National Academy of Sciences of the United States of America* **82**, 3756.
Costanzo, M.C. and Fox, T.D. (1990). *Annual Review of Genetics* **24**, 91.
Cox, B.S. (1995). *In* "The Yeasts Vol. 6: Yeast Genetics" (A.E. Wheals, A.H. Rose and J.S. Harrison, eds) pp. 7–63. Academic Press, London.
Demerec, M., Adelberg, E.A., Clark, A.J. and Hartman, P.E. (1966). *Genetics* **54**, 61.
Dujon, B. (1981). *In* "Molecular Biology of the Yeast *Saccharomyces*, Life Cycle and Inheritance" (J.N. Strathern, E.W. Jones and J.R. Broach, eds.), pp. 505–651. Cold Spring Harbor Laboratory, Cold Spring Harbor, New York.
Grivell, L.A. (1984). *In* "Genetic Maps" (S.J. O'Brien, ed.), p. 234. Cold Spring Harbor Laboratory, Cold Spring Harbor, New York.
Guérin, B. (1991). *In* "The Yeasts Vol. 4: Yeast Organelles" (A.H. Rose & J.S. Harrison, eds) Academic Press, London.
Holm, C. (1982). *Cell* **29**, 585.
Kaback. D.R., (1995). *In* "The Yeasts Vol. 6: Yeast Genetics" (A.E. Wheals, A.H. Rose & J.S. Harrison, eds), Academic Press, London.
Liebman S.W. and All-Robyn, J.A. (1984). *Current Genetics* **8**, 567.
Mortimer, R.K., Contopoulou, R, and King J.S., (1995). *In* "The Yeasts Vol. 6: Yeast Genetics" (A.E. Wheals, A.H. Rose & J.S. Harrison, eds) Academic Press, London.
Olson, M.V. (1991). *In* "The Molecular and Cellular Biology of the Yeast *Saccharomyces*, Vol. 1. Genome dynamics, protein synthesis, and energetics" (J.R. Broach, E.W. Jones and J.R. Pringle, eds.) pp. 1–39. Cold Spring Harbor Laboratory, Cold Spring Harbor, New York.
Oliver, S.G., *et al.* (1992). *Nature* **357**, 38.
Sherman, F. and Lawrence, C.W. (1974). *In* "Handbook of Genetics" Vol. 1, Bacteria, Bacteriophages, and Fungi (R.C. King, ed.), pp. 359–393. Plenum, New York.
Sherman, F. (1981). *In* "Molecular Biology of the Yeast *Saccharomyces*, Life Cycle and Inheritance" (J.N. Strathem, E.W. Jones and J.R. Broach, eds.), pp. 639–640. Cold Spring Harbor Laboratory, Cold Spring Harbor, New York.
Sherman, F. (1991). *In* "Methods in Enzymology, Guide to Yeast Genetics and Molecular Biology" (C. Guthrie and G.R. Fink, eds.), pp. 3–21. Academic Press, New York.
Sprague, G.F. (1995). *In* "The Yeasts Vol. 6: Yeast Genetics" (A.E. Wheals, A.H. Rose & J.S. Harrison, eds) Academic Press, London.
Sweeny R. and Zakian, V.A. (1989). *Genetics* **122**, 749.
Tuite, M.F., Lund, P.M., Futcher, A.B., Dobson, M.J., Cox, B.S. and McLaughlin, C.S. (1982). *Plasmid* **8**, 103.
Wickner, R.B. (1989). *FASEB* **3**, 2257.
Wickner, R.B. (1991). *In* "The Molecular and Cellular Biology of the Yeast *Saccharomyces*, Vol. 1. Genome dynamics, protein synthesis, and energetics" (J.R. Broach, E.W. Jones and J.R. Pringle, eds.), pp. 263–296. Cold Spring Harbor Laboratory, Cold Spring Harbor, New York.
Wickner, R.B. (1994). *Science* **267**, 566.
Wickner, R.B. (1995). *In* "The Yeasts Vol. 6: Yeast Genetics" (A.E. Wheals, A.H. Rose & J.S. Harrison, eds). Academic Press, London.
Wickner, R.B., Boutelet, F. and Hilger, F., (1983). *Molecular and Cellular Biology* **3**, 415.
Williamson, D.H. (1991). *In* "The Yeasts Vol. 4: Yeast Organelles" (A.E. Wheals, A.H. Rose & J.S. Harrison, eds). Academic Press, London.

Appendix II Genetic and Physical Maps of *Saccharomyces cerevisiae*

Robert K. Mortimer, C. Rebecca Contopoulou and Jeff S. King

Yeast Genetic Stock Center, MCB/Biophysics and Cell Physiology, 102, Donner Laboratory, Berkeley, California 94720 USA

This appendix includes a glossary of gene symbols (Table I), a list of mapped genes (Table II) and a drawing of the genetic and physical map of the nuclear genome. It comprises Edition 11A of the map. Full details can be found in Mortimer, R.K., Contopoulou, C.R. and King, J.S. (1991) *Yeast* **8**, 817.

Table I. Glossary of gene symbols

Symbol	Definition	Symbol	Definition
aaa	Amino terminal, amino acetyl transferase	adr	Alcohol dehydrogenase regulation defective
aac	ATP/ADP carrier, mitochondrial	aep	Required for translation of *ole1*
aar	Amino acid analogue resistance		mRNA and subunit 9 assembly
aar	a1-a2 represion	afi	Alpha factor internalization
aas	Amino acid analogue sensitive (see also gcn)	aga	a cell-specific sexual agglutination
		aga	a-cell specific sexual agglutination
aat	Amino acid toxicity	ags	Aminoglycoside sensitive
abc	Assembly of BC1 complex	agt	Alpha glucoside transporter
abf	Ars binding factor	aky	Adenylate kinase
abp	Actin binding protein	alg	Asparagine-linked glycosylation deficient
ace	Activation of CUP1 expression		
aco	Aconitase, mitochondrial	amc	Artificial minichromosome maintenance
acp	Acidic protein 2		
acs	Acetyl coenzyme A synthetase	amm	Altered minichromosome maintenance
act	Actin		
acu	Aculeacin A resistant	ams	a-Mannosidase
ade	Adenine requiring	amy	Antimycin resistance
adh	Alcohol dehydrogenase defective	amy	Alpha amylase
adk	Adenylate kinase	anb	Anaerobically induced
adp	ATP-dependent permease	ani	Anisomycin resistance

471

The Yeasts Vol. 6, 2nd edition
ISBN 0-12-596416-1

472   Robert K. Mortimer, C. Rebecca Contopoulou and Jeff S. King

Table I. *Continued*

Symbol	Definition	Symbol	Definition
anp	ANP and osmotic sensitive	can	Canavanine resistance
ant	Antibiotic resistance	cap	Capping – addition of actin subunits
apa	Diadenosine 5′, 5‴ p^1, p^4 tetraphosphate phosphorylase	car	Catabolism of arginine defective
ape	Aminopeptidase	cas	Cyclic AMP suppression
apf	Amino acid transport pleiotropic factor	cat	Catabolite repression
aph	3′-Aminoglycoside phosphotransferase	cbp	Cytochrome b pre-mRNA stability; coenzyme QH2-cytochrome c reductase assembly
apn	Apurinic site endonuclease	cbp	Centromere binding protein
aps	Aspartyl-tRNA synthetase	cbs	Cytochrome b pre-mRNA and mRNA translation factor
apt	Aminoglycoside phosphotransferase	cca	ATP (CTP) tRNA specific nucleotidyl transferase
ard	Arrest at start of cell cycle defective		
arf	ADP-ribosylation factor	ccb	Cross-complementation of budding defect (see msb)
arg	Arginine requiring		
aro	Aromatic amino acid requiring	ccp	Cytochrome c peroxidase, mitochondrial
ars	Autonomously replicating sequence		
asp	Aspartic acid requiring	ccr	Carbon catabolite repression
asr	Antisene RNA	ccs	Triethyl-tin resistance, controls oxidative phosphorylation
asu	Antisuppressor		
ata	Sporulation-specific gene characterized by ATA sequences	cdc	Cell division cycle blocked at 36°C
		cdl	Clathrin-deficient lethality
ate	Arginyl-tRNA-protein transferase deficient	cen	Centromere
		cep	Centromere protein
atp	ATPase, mitochondrial	cha	Catabolism of hyroxy amino acids
atr	Aminotriazole resistance	chc	Clathrin heavy chain
axe	Axenomycin resistance	chl	Chromosome loss
bap	Branched chain amino acid permease	cho	Choline requiring
		chs	Chitin synthetase
bar	a cells lack barrier effect on a factor	cid	Cell identity
bas	Basal level control	cid	Constitutive invertase derepression
bck	Bypass of C kinase	cin	Chromosome instability
bcs	Branched chain amino acid sensitivity	cip	Complementer of ipl2
		cit	Citrate synthase, mitochondrial
bcy	Adenylate-cyclase and cAMP-dependent protein kinase deficient	cka	Casein kinase II
		cki	Choline kinase
bem	Bud emergence	cks	Cdc28 kinase subunit
bet	Blocked early in transport	clc	Clathrin light chain
bik	Bilateral defect in karyogamy	cln	Cyclin
blm	Bleomycin sensitive	cls	Calcium sensitive
bls	Blasticidin-5 resistance	cly	Cell lysis at 36°C
bmh	Bovine brain protein homologous	cmd	Calmodulin gene
bor	Borrelidin resistance	cmk	Calmodulin dependent protein kinase
bos	Bet one suppressor		
bot	Biosynthesis of turpenes	cmp	Calmodulin binding protein
btf	Basic transcription factor	cms	Centromere mutation suppressor
bud	Controls sites of bud emergence	cmt	Control of mating type
cad	Cadmium resistance	cna	Calcineurin subunit A
cag	Constitutively agglutinable	cnb	Calcineurin subunit B
cal	Calcium sensitive	cnd	Chromosome non-disjunction
cal	Resistant to calcofluor	coq	Coenzyme Q deficient
cal	Calcium dependent	cor	Coenzyme QH2-cytochrome c

Table I. *Continued*

Symbol	Definition	Symbol	Definition
	reductase	dfr	Dihydrofolate reductase
cox	Cytochrome oxidase, nuclear or mitochondrial	dis	Phosphoserine- and phosphothreonine-specific phosphatase
cpa	Carbamoyl phosphate synthetase, arginine specific	dit	D,L-dityrosine in spore wall
cpe	Constitutive phospholipid expression	dit	Derepressed *INO1* transcription
cpf	Centromere and promoter factor	diu	Diuron resistance
cpr	Cyclophilin, CsA-binding proline rotamase	dka	Downstream of protein kinase A
		dna	DNA synthesis defective
cps	Carboxypeptidase S	doa	Degradation of alpha factor defective
cre	Carbon catabolite repressison effector	dpb	DNA polymerase II, subunit B
crl	Cycloheximide-resistant temperature-sensitive lethal	dpr	Processing of ras protein
		dpm	Dolichol phosphate mannose synthase
cry	Cryptopleurine resistance		
csd	Chitin synthesis defective	dsc	Dominant suppressor of *cdc4*
cse	Chromosome segregation	dsd	Dominant suppressor of *sec14-1*
csr	cyclosporin resistance	dsm	Premeiotic DNA synthesis deficient
cta	Catalase A	dst	DNA strand transfer
ctd	Phosphorylates C-terminal domain of largest RNA polII subunit	dtp	Diadenosine 5′, 5‴ p^1, p^4-tetraphosphate phosphorylase
ctf	Chromosome transmission fidelity	dur	Urea degradation deficient
ctn	Catalase T nutrition	dut	dUTPase
cts	Endochitinase	eam	Endogenous ethanolamine biosynthesis
ctt	Catalase T		
ctt	Cholinephosphate cytidylytransferase	edr	Enhanced delta recombination
cup	Copper resistance	eif	Translation initiation factor
cyb	Cytochrome b_2 deficiency	elm	Elongated bud morphology mutant suppressor
cyc	Cytochrome *c* deficiency		
cyh	Cycloheximide resistance	end	Endocytosis defective
cyp	Modulates expression of *cyc1* and *cyc7*	eno	Enolase
		env	Envelope proteins
cyr	Adenylate cyclase deficient	erd	Endoplasmic reticulum defective
cys	Cysteine requiring	erg	Ergosterol biosynthesis defective; many also nystatin resistant
cyt	Cytochrome c_1 haem lyase; cytochrome *c* (see cyc)		
		esp	Extra spindle pole bodies
dac	Division arrest control for mating pheromones	ess	Essential
		est	Even shorter telomeres
daf	Dominant a factor resistance	eth	Ethionine resistance
dal	Allantoin degradation deficient	exa	Extragenic suppressor subfamily A
dap	Dipeptidyl aminopeptidase B	exg	Exo-1,3-b-glucanase activity
dbf	Dumbell formation; kinase required for late nuclear division	far	Factor arrest
		fas	Fatty acid synthetase deficient
dbl	Alcian blue dye binding deficient	fbp	Fructose-1,6-bisphosphatase
dbp	Dead box protein	fcy	Purine-cytosine permease
dbr	Debranching of excised introns	fdp	Unable to grow on glucose, fructose, sucrose or mannose
dcd	dCMP deaminase		
dea	Deamidation of amino terminal Asn and Gln residues	fen	Fenpropimorph resistant
		fer	Ferric resistance
deg	Depressed growth rate	fes	Ferric sensitivity
ded	Defines essential domain, lethal	fkr	FK-506 resistant
dex	Dextran utilization	flk	Flaky

Table I. Continued

Symbol	Definition	Symbol	Definition
flo	Flocculation		genes; nuclear transcription factor
fol	Folinic acid requiring	hem	Haem synthesis deficient
fox	Fatty acid oxidation	het	Hexose transport
fpr	FKBP proline rotamase (isomerase)	hex	Hexose metabolism regulation
fre	Ferric reductase	him	High induced mutagenesis
fro	Frothing	hip	Histidine specific permease
fum	Fumarase	hir	Histone cell cycle regulation
fun	Function unknown now		defective
fur	Resistant to 5-FU, encodes UPRTase	his	Histidine requiring
fsr	Fluphenazine resistance	hit	High temperature growth
fus	Fusion defective; MAP kinase	hht	Histone
	involved in pheromone signal	hmg	HMG-CoA reductase
	transduction, G1 arrest	hml	Mating type cassette – left
gac	Glycogen accumulation	hmr	Mating type cassette – right
gam	Glucoamylase	ho	Homothallic switching
gal	Galactose non-utilizer	hol	Histidinol uptake proficient
gap	General amino acid permease	hop	Homologue pairing
gap	Glyceraldehyde-3-phosphate	hot	Cis-acting recombination stimulating
	dehydrogenase		sequence for ribosomal DNA
gas	Glycophospholipid-anchored surface	hpl	Haploid lethal
	protein	hpr	Increased intrachromosomal
gcd	General control of amino acids		recombination
	synthesis derepressed	hpr	Hyperresistance to oxidative damage
gcn	General control of amino acid	hps	Hexaprenyl pyrophosphate
	synthesis non-derepressible (see also		synthetase
	aas)	hrr	Regulation of repair DNA damage
gcn	Glucosamine auxotroph	hsf	Heat shock transcription factor
gcr	Glycolysis regulatory protein	hsp	Heat shock protein
gcy	Galactose regulated, homology to	hsr	Heat shock resistance
	aldo/keto-reductases	hta	Histone 2A genes
gda	Guanosine diphosphatase	htb	Histone 2B genes
gdh	Glutamate dehydrogenase, NADP	hts	Histidinyl-tRNA synthetase
	dependent	hxk	Hexokinase deficient
gef	Glycerol ethanol, ferric requiring	hxt	Hexose transport
ger	Spore germination	hyg	Hygromycin resistance
ggp	GPI-glycosylated protein	idh	Isocitrate dehydrogenase
glc	Glycogen storage	ils	Isoleucyl-tRNA synthetase deficient;
glk	Glucokinase deficient		no growth at 36°C
gln	Glutamine synthetase	ilv	Isoleucine-plus-valine requiring
	non-derepressible	ime	Inducer of meiosis
gpa	GTP binding protein	imp	Independent of mitochondrial
gpa	G protein alpha homologous gene		particle
gph	Glycogen phosphorylase	ino	Inositol deficient
grc	Valyl-tRNA-synthetase	ins	Initiation of S phase
grf	General regulatory factor	ipa	Increased accumulation of
grr	Glucose repression resistant		porphyrins
gsh	Glutathione biosynthesis	ipl	Increase in ploidy
gst	G-1 to S transition	ipp	Inorganic pyrophosphatase
gsy	Glycogen synthase	ira	Inhibitory regulator of the RAS-
gut	Glycerol utilization		cAMP pathway
ham	Supermutable	ise	Inhibitor sensitive
hap	Global regulator of respiratory	ivs	Intervening sequence

Table I. *Continued*

Symbol	Definition	Symbol	Definition
kar	Karyogamy defective		chromosome pairing and
kem	kar enhancing mutation		recombination
kex	Killer expression defective	mel	Melibiose fermentation
kin	Protein kinase	mer	Meiotic recombination
kgd	a-Ketoglutarate dehydrogenase, KE1 component; dihydrolipoyl transsuccinylase	mes	Methionyl-tRNA synthetase deficient; no growth at 36°C
kns	Protein kinase	met	Methionine requiring
krb	Suppression of some mak mutations	mfa	Mating factor A
kre	Killer resistance	mfa	a-mating factor
krs	Lysyl tRNA-synthetase	mgl	a-Methylglucoside fermentation
kss	MAP protein kinase homologue involved in pheromone signal transduction	mgm	Mitochondrial genome maintenance
		mgt	Mitochondrial genome transmission
		mif	Mitotic frequency of chromosome transmission
kti	*Kluveromyces lactis* toxin insensitive	mig	Multicopy inhibitor of *Gal* gene expression
ktr	Kre two related		
lap	Leucine aminopeptidase deficient	mih	Protein phosphatase
lcb	Long chain base synthesis defective, an intermediate in sphingolipid biosynthesis	min	Methionine inhibited
		mip	Mitochondrial DNA polymerase
		mis	Mitochondrial c-tetrahydrofolate synthetase
ldd	Lysozyme degradation		
let	Lethal	mkt	Maintenance of K_2 killer factor
leu	Leucine requiring	mnn	Mannan synthesis defective
lip	Lipoic acid deficient	mnt	Mannosyltransferase
lgn	Sporulation-induced transcripts	mod	Modification of tRNA, IPP transferase
lmd	Lanosterol 14 alpha demethylase	mol	Induced on entry into stationary phase on molasses medium
los	Loss of suppression and defective in tRNA processing		
lpd	Lipoamide dehydrogenesis	mom	Matrix organization mutant
lte	Low temperature essential	mos	Modifier of ochre suppressors
lts	Low temperature sensitive	mot	Modifier of transcription
lys	Lysine requiring	mpk	Multicopy suppressor of kinase
lyt	Lysis at 37 (see cly)	mrf	Mitochondrial release factor
mak	Maintenance of killer deficient	mrp	Mitochondrial ribosomal protein
mal	Maltose fermentation	mrs	Mitochondrial RNA splicing
mar	Mating type cassette expression	msb	Multicopy suppressor of a budding defect
mas	Mitochondrial assembly		
mas	Mitochondrial protein import	msd	Mitochondrial aspartyl-tRNA synthetase
mat	Mating type locus		
mbr	Mitochondrial biogenesis of ribosomes	mse	Mitochondrial glutamly-tRNA synthetase
mck	Meiotic and centromere regulatory ser, tyr-kinase	msf	Mitochondrial phenylalanine-tRNA synthetase
mcm	Minichromosome maintenance deficient	msg	Multicopy suppressor of *GPA1*
		msi	Multicopy suppressor of *ira1*
mdh	Malate dehydrogenase	msk	Mitochondrial lysyl-tRNA synthetase
mdm	Mitochondrial distribution and morphology	msl	Mitochondrial leucyl-tRNA synthetase
		msm	Mitochondrial methionyl-tRNA synthetase
mec	Mitosis entry defective		
mef	Mitochondrial elongation factor	mss	Suppression of a mitochondrial RNA splice defect; *COX1* pre-mRNA processing factor
mei	Meiotic gene conversion defective		
mek	Kinase involved in meiotic		

476   Robert K. Mortimer, C. Rebecca Contopoulou and Jeff S. King

Table I. *Continued*

Symbol	Definition	Symbol	Definition
mst	Mitochondrial threonyl-tRNA synthetase	pda	Pyruvate dehydrogenase E1a
		pdb	Pyruvate dehydrogenase E1b
msw	Mitochondrial trytophanyl-tRNA synthetase	pdc	Pyruvate decarboxylase
		pde	Phosphodiesterase (cAMP)
msy	Mitochondrial tyrosyl-tRNA synthetase	pdh	Pyruvate dehydrogenase deficient
		pdi	Protein disulphide isomerase
mtf	Mitochondrial transctiption factor	pdr	Pleiotropic drug resistance
mtf	Mitochondrial translation initiation factor	pdx	Pyridoxin requiring
		pem	Phospholipid methyltransferase
mtp	Melezitose fermentation	pep	Proteinase deficient
mus	Multicopy suppressor of *sis1*	per	Suppressor of *tex1 rad52* lethality
myo	Myosin	pet	Petite; unable to grow on non-fermentable carbon sources
nam	Nuclear accommodation of mitochondria		
		pfk	Phosphofructokinase
nat	N-terminal acetyltransferase	pfy	Profilin of yeast
ndc	Nuclear division cycle	pgi	Phosphoglucose isomerase deficient
net	Negative effect on transcription	pgk	3-Phosphoglycerate kinase deficient
nhp	Non-histone protein	pgm	Phosphoglucomutase
nhs	Hydrogen sulphide production inhibitor	pgm	Phosphoglyceromutase
		pha	Phenylalanine requiring
nib	Nibbled colony phenotype due to 2m DNA	pho	Phosphatase deficient
		phr	Photoreactivation repair deficient
nmt	N-myristoyltransferase	phs	Hydrogen sulphide production deficient
nra	Neutral red accumulation		
nsp	Nuclear membrane SPB protein	pif	Mitochondrial repair and recombination
nsp	Nucleoskeletal protein		
nop	Nucleolar protein	pim	PI and PIP kinase
nov	Novobiocin resistance	pka	Protein kinase catalytic subunit
npi	Nitrogen permease inhibitor	pkc	Protein kinase C homologue
npl	Nuclear protein localization	pma	Plasma membrane ATPase mutations
npr	Nitrogen permease reactivator	pmr	Plasma membrane ATPase related
nub	Nuclear and budding defect	pms	Postmeiotic segregation increased
nuc	DNA double-strand break repair	pol	DNA polymerase
nul	Non-mater	pot	Peroxisomal 3-oxoacyl CoA thiolase
nup	Nuclear pore	pox	Acyl-CoA-oxidase
nur	Nuclease, recombination	ppa	Pyrophosphatase
obf	Origin binding factor	ppc	Phosphoenolpyruvate carboxykinase
odp	Oil drop protein	ppd	Phosphoprotein phosphatase-deficient
ogd	2-Oxoglutarate dehydrogenase		
ole	Oleic acid requiring	pph	Protein phosphatase
oli	Oligomycin resistance	ppr	Defective in pyrimidine biosynthetic pathway regulation
opi	Phopholipid synthesis regulation		
osm	Low osmotic pressure sensitive	ppz	Phosphoserine and phosphothreonine-specific protein phosphatase
oxt	Resistance to oxythiamin		
pab	Poly-A binding protein		
pai	Plasminogen activator inhibitor	pra	Proteinase A deficient
pal	Phenylalanine ammonia-lyase	prb	Proteinase B deficient
pam	PP2A multicopy suppressor	prc	Proteinase C deficient
pan	Poly-A nuclease	pre	Proteinase YSCE deficient
pap	Poly-A polymerase	pri	DNA primase
pas	Peroxisomal assembly	pro	Proline requiring
pbs	Polymyxin B resistance	prp	Pre mRNA processing (homonym for rna)

Table I. *Continued*

Symbol	Definition	Symbol	Definition
prt	Protein synthesis defective at 36°C	rib	Riboflavin biosynthesis
pso	8-Methoxypsoralen sensitive	rip	Iron-sulphur protein of ubiquinol
pss	Phosphatidyl serine synthase		cytochrome c reductase; Rieske
ptp	Phosphotyrosine-specific protein		protein of coenzyme
	phosphatase		QH2-cytochrome c reductase
ptx	Preprotoxin	rgt	Restores glucose transport
pur	Purine excretion	rkt	Restores potassium transport
put	Proline non-utilizer	rme	Repressor of meiosis
pyc	Pyruvate carboxylase	rna	RNA synthesis defective; unable to
pyk	Pyruvate kinase deficient		grow at 36°C (homonym for prp)
qcr	Ubiguinol-cytochrome c	rnh	Ribonuclease H
	oxidoreductase complex subunit	rnr	Ribonucleotide reductase
raa	Resistant to amino acid analogues	roc	Roccal resistance
rad	Radiation (ultraviolet or ionizing	ros	Relaxation of sterility
	sensitive	rox	Regulation of oxygen
ram	RAS protein and a-factor maturation	rpa	Ribosomal protein, acidic
	function	rpa	RNA polymerase
rag	Resistance to antimycin on glucose	rpa	Yeast homologue of human RP-A
rap	Repressor activator protein		needed for *in vitro* SV40 replication
rap	E1a subunit of pyruvate	rpb	RNA polymerase II
	dehydrogenase complex	rpc	RNA polymerase III
rar	Regulation of autonomous	rpd	Reduced potassium dependency
	replication	rpk	Regulatory protein kinase
ras	Homologous to RAS proto-oncogene	rpl	Ribosomal protein, large subunit
rbk	Rinokinase	rpm	Ribonuclease P mitochondrial
rbp	Rapamycin binding protein	rpo	RNA polymerase II, III, IV
rca	Rescue by cAMP of *cyr1* mutants	rpr	RNase P
rcs	Regulation of cell size	rps	Ribosomal protein, small subunit
rdn	Ribosomal RNA structural genes	rrm	rDNA recombination mutation
reb	DNA binding protein	rrp	rRNA processing
reb	RNA polymerase I enhancer binding	r^{s}1	Radiation sensitive
	protein	rsd	Recessive suppressor of *sec14*-1
rec	Recombination deficient	rsf	Regulates septum formation
red	Reduces meiotic interchromosomal	rsr	Ras-related
	crossing over	rvs	Reduced viability upon starvation
ref	Respiration negative on YPE	sac	Suppressor of actin mutations
reg	Regulation of galactose pathway	saf	Suppressor of adr function
	enzymes	sag	Sexual agglutination
reo	Regulation of expression of oxidase	sal	Allosuppression
rep	Replication and equipartion of 2μm	sam	S-Adenosyl methionine synthesis
	DNA to daughter cells	san	Sir antagonist
res	Regulation of sporulation	sap	*sit4* associated proteins
ret	Reduced efficiency of transcription	sar	Suppressor of *sec12*
	termination	sca	Sporulation capable
rev	Revertibily decreased	scc	Suppressor of *cdc65*
rfa	Replication factor A	scc	*Saccharomyces cerevisiae* cyclophilin
rgp	Reduced growth phenotype	scd	Suppressor of clathrin deficiency
rgr	Required for glucose repression	scg	Constitutive arrest of cell division,
rhm	5-Aminolevulinic acid and haem		pheromone-independent mating
	biosynthesis	sch	cAMP-dependent protein kinase
rho	Mitochondrial genome		homologue; suppresses *cdc25*^{ts}
rho	Ras homologous	sck	Suppressor of c kinase

Table I. *Continued*

Symbol	Definition	Symbol	Definition
scl	Dominant suppression of t.s. lethality of *crl3*	spb	Ribosomal biogenesis
scm	Suppressor of chromosome segregation mutation	spd	Sporulation not repressed on rich media
sco	Cytochrome oxidase deficient	spe	Spermidine resistance
sdb	Suppressor of dbf	spg	Suppressor of *GPA1*
sdc	Suppressor of *cdc25*	spk	ser, thr, (tyr)-kinase
sdh	Succinate dehydrogenase deficient	spo	Sporulation deficient
sec	Secretion deficient	spp	Suppressor of prp
sen	Splicing endonuclease	spr	Sporulation regulated genes
sep	Strand exchange protein	sps	Sporulation specific transcript
ser	Serine requiring	spt	Suppressors of Ty transcription
ses	Seryl-tRNA synthetase	spx	Suppressor of X-ray sensitivity of *rad55*
sfa	Sensitive to formaldehyde	sra	Suppressors of the ras mutation
sga	Glucoamylase, intracellular	srb	Suppressor of *rpb1*
sga	Suppression of growth arrest of *cdc25*	srb	Osmotically fragile
		srd	Suppressor of *rrp1*
shi	Spacing between TATA and initiation site	srk	Suppressor of regulatory subunit of protein kinase
sin	Switch independent	srm	Suppressor of rho mutability
sip	*SNF1*-interacting protein	srm	Suppressor of receptor mutations
sir	Silent mating type information regulation	srn	Suppressor of yeast *rna1-1*
		srp	Suppressor of *rp1*, cold-sensitive
sis	*sit4* suppressor	srp	Suppressor of *rpa190*
sit	Sporulation-induced transcripts	srp	Serine rich protein
Sit	Suppression of initiation of transcription	srs	Suppressor of rad six
		srv	Suppressor of RAS^{val19}
skd	Suppressor of *kar2* defect	ssa	Stress-seventy subfamily A
ski	Superkiller	ssb	Stress-seventy subfamily B
sko	Suppressor of kinase overproduction	ssb	Single strand binding protein
slk	Synthetic lethal kinase	ssc	Stress-seventy subfamily C
slp	Small lysine pool (vacuole defective)	ssc	Secretion of heterologous proteins
sls	Sigma like sequence	ssd	Stress-seventy subfamily D
slt	Suppression at low temperature	ssd	Suppression of *sit4* deletion
sly	Suppression of loss of *YPT1*	ssg	Sporulation-specific 1,3-β-glucanase
smc	Stability of minichromosomes	ssl	Super secretion of lysozyme
smd	Suppressor of mating defect of *ste7*	ssn	Suppressor of snf1
sme	Regulation of meiosis	ssp	Suppressor of *rna8*, cold sensitive
smr	Sulphometuron methyl resistance	sst	Supersensitive to a factor
snf	Sucrose non-fermenting	sta	Starch hydrolysis
snq	Sensitivity to 4-nitroquinoline-N-oxide	stc	Suppressor of *tsm*^{409}
snr	Small nuclear RNA	ste	Sterile
snt	Stringency activity; inhibition of leaky mutants on non-ferm substrates	sth	SNF2 homologue
		sti	Stress inducible
soc	Suppressor of cdc8	stp	Ste pseudorevertants
sod	Superoxide dismutase	stp	Species-specific pre-tRNA processing
soe	Suppression of cdc8	str	Sulphur transferasse
sos	Suppressor of *sis1*	suc	Sucrose fermentation
sot	Suppression of deoxythymidine monophosphate uptake	suf	Suppression of frameshift mutation
spa	Spindle pole antigen	suh	Suppression of *his2-1*
		sui	Suppressor of initiator codon

Table I. *Continued*

Symbol	Definition	Symbol	Definition
	mutations	tsv	Temperature-sensitive lethal,
sum	Suppressor of *mar*		UV-induced
sup	Suppression of nonsense mutations	tub	Tubulin; MBC resistance
sur	Suppressor of *radH*	tuf	Mitochondrial elongation factor
sus	Suppression of *ser1*	tup	Deoxythymidine monophosphate
sut	Suppressor of TATA		uptake positive
swi	Homothallic switching deficient	tye	Ty-mediated expression
tal	Transaldolase	tyr	Tyrosine requiring
tar	3-Amino-1,2,4-triazole resistance,	Ty	Transposable element
	gcn4 suppression	uba	Ubiguitin activating enzyme
tcm	Tricodermin resistance	ubc	Ubiquitin conjugating
tcp	t complex polypeptide	ubi	Ubiquitin
tdh	Glyceraldehyde 3-phosphate	uga	Utilization of GABA
	dehydrogenase		(4-aminobutyrate) as a nitrogen
tec	Ty element enhancement control		source
tef	Translational elongation factor	ulc	Zn resistant, cell elongation
tel	Contraction of telomere size	ume	Unscheduled meiotic gene
tel	Telomere		expression
tex	Tn5 excision frequencies increased	umr	Ultraviolet mutability reduced
tflld	Transcription factor gene	ung	Uracil DNA glycosylase
tfp	Trifluoperazine resistance,	ura	Uracil requiring
	Ca sensitive	ure	Ureidosuccinate transport-resistance
tfs	Cdc twenty-five suppressor		to nitrogen repression
thi	Thiamine requiring	urep	Ureidosuccinate permease
thr	Threonine requiring	urr	URS repression
ths	Threonyl tRNA synthetase	utr	Unknoen transcript
tif	Translation initiation factor	uts	Upstream of thymidylate synthetase
til	Thioisoleucine resistance	vac	Vacuolar segregation
tmp	Thymidine monophosphate requiring	van	Vanadate resistant
tof	Target of FK506	vas	Valyl-tRNA-synthetase
top	Topoisomerase	vat	Vacuolar protein ATPase
tor	Target of rapamycin	vma	Vacuolar H-ATPase
tos	Topoisomerase II supprression	vph	Vacuolar pH control
tpd	Phosphoserine and	vpl	Vacuolar protein localization
	phosphothreonine-specific protein	vps	Vacuolar protein sorting
	phosphatase	vpt	Vacuolar protein targeting
tpd	tRNA production deficient	whi	Small cell size
tpi	Triosephosphate isomerase	xrn	Exoribonuclease
tpk	cAMP-dependent protein kinase	xrs	X-ray sensitive
	catalytic subunit	yak	Yet another kinase
tpk	Threonine/serine protein kinase	yal	ORF on chromosome 1L
tra	Triazylalanine resistant	yap	Yeast aspartyl protease
trk	Transport of potassium	yar	ORF on chromosome 1R
trl	tRNA ligase	ybl	ORF on chromosome 2L
trm	tRNA (guanine-N^2,	ybr	ORF on chromosome 2R
	N^2-)-dimethyltransferase	yck	Yeast casein kinase I
trn	Proline-tRNA gene	yel	ORF on chromosome 3L
trp	Tryptophan requiring	yer	ORF on chromosome 3R
trx	Thioredoxin	ydj	Yeast dnaJ
tsl	Temperature sensitive lethal	ydl	ORF on chromosome 4L
tsm	Temperature sensitive lethal	ydr	ORF on chromosome 4R
	mutations	yef	Yeast elongation factor

Table I. *Continued*

Symbol	Definition	Symbol	Definition
yel	ORF on chromosome 5L	ylr	ORF on chromosome 12R
yer	ORF on chromosome 5R	yml	ORF on chromosome 13L
yfl	ORF on chromosome 6L	ymr	ORF on chromosome 13R
yfr	ORF on chromosome 6R	ymr	Yeast mitochondrial ribosomal
ygl	ORF on chromosome 7L		protein
ygr	ORF on chromosome 7R	ynl	ORF on chromosome 14L
yhl	ORF on chromosome 8L	ynr	ORF on chromosome 14R
yhr	ORF on chromosome 8R	yol	ORF on chromosome 15L
yil	ORF on chromosome 9L	yor	ORF on chromosome 15R
yir	ORF on chromosome 9R	ypk	Yeast protein kinase
yjl	ORF on chromosome 10L	ypl	ORF on chromosome 16L
yjr	ORF on chromosome 10R	ypr	ORF on chromosome 16R
ykl	ORF on chromosome 11L	ypt	GTP-binding protein
ykr	Kinase homologue identified with	yuh	Yeast ubiquitin hydrolase
	mammalian cDNA probes	zfh	Zinc finger motif protein
ykr	ORF on chromosome 11R	zwf	Glucose-6-phosphate dehydrogenase
yll	ORF on chromosome 12L		

Table II. List of mapped genes

Gene	Synonyms	Map Position/ Chrom./Arm/(cM)	Gene	Synonyms	Map Position/ Chrom./Arm/(cM)	Gene	Synonyms	Map Position/ Chrom./Arm/(cM)
aaa1		4L18	adk1		4R133	arg1		15L56
aar1	raa1	4R92	adk2		5R146	arg3		10L61
aar2	raa2	6R46	adp1	YCR105	3R1	arg4		8R12
aar2		2	adr1		4R125	arg5,6		5R55
aas1	gcn2	4R172	adr6		16L19	arg8		15L130
aas2	gcn3	11R7	aga1		14R42	arg9	aas3	5L3
aas3	arg9, gcn4	5L3	aga1	sag1	10R2			
			ags1		3L13	arg80	gcn4	13R36
abf1		5	alg1		2R54	arg81		13L63
ABP1		3R80	alg7		7R20	arg82		4R95
abp2		14R14	amc1	ch16	5	arg84		5R55
ace1		7L101	amc3	ch18	4	aro1		4R81
aco1	glu1	12	amc4	ctf12, ch19	7	aro2		7L90
act1		6L59	amm1	tup1, cyc9, flk1, umr1	3R77	aro7	osm2	16R39
ACT2		4	ams1		7L98	asp1		4R190
ade1		1R4	AMY1		7L8	asp5		12R17
ade2		15R64	AMY2		2L	ass1		12
ade3		7R159	ani1		15R46	ATA1		11
ade4		13R213	anp1	pdr1	5L26	ate1		7L12
ade5,7		7L152	ant1	pdr1	7L8	ATP1		7
ade6		7R39	apa1	dtp1	3L33	atr1	snq1	13
ade8		4R256	ape2	lap1	11L144	AXE1		7L4
ade9		15R101	ard1		8R10	bar1	sst1	9L12
ade12		14L51	arf1		4L91	bas1		11R82
ADE15		7R171	arf2		4L64	bas2	pho2	4L59
adh1		15L86				bck1		10L64
adh2		13R216				bcy1		9L28
adh3		13R45				bem1	sra1	2R102
adh4		7R169				bem2	sup9	5R128

Continued

Table II. Continued

Map Positions and Centromere Distances (cM)

Gene	Synonyms	Map Position/Chrom./Arm/(cM)
BIK1		3L21
bls2		11R74
blm1		5
bmh1	pdr1	5R148
BOR1	rsr1	5R47
BOR2		7L8
bud1		7R98
bud3		3L9
bud5		3R30
cad1		4R271
CAD2		2R211
cag1		3R30
CAL1		7
can1		5L50
cap1		11L2
cap2		9L28
car1		16L75
car2	ccr1	12R330
cat1	snf1	4R303
ccb1		15R108
ccr1	cat1, snf1	4R303
ccr4		1L25
cdc2	tex1	4L55
cdc3		12R215
cdc4		6L9
cdc5		13L14
cdc6		10L124
cdc7		4L1
cdc8		10R36
cdc9		4L78
cdc10		3R1
cdc11		10R54
cdc12		8R67
cdc13		4L122
cdc14		6R36
cdc15		1R6
cdc16		11L15
cdc17	pol1	14L74
cdc19	pyk1	1L45
cdc20		7L70
cdc21	tmp1	15R32
cdc24	cls4, tsl1	1L56
cdc25	ctn1	12R213
cdc25		2R108
cdc26		6R52
cdc27		2L58
cdc28	srm5	2R90
cdc29		9L23
cdc31		15R147
cdc34		4R37
cdc35	cyr1, hsr1, SRA4	10R2
cdc36	tsm0185	4L80
cdc37		4R95
cdc39	ros1	3R91
cdc40	xrs2	4R211
cdc42		12R165
cdc43		7L99
cdc53		4L81
cdc55		7L125
cdc60		16L65
cdc62		7R99
cdc63	prt1	15R221
cdc64		15R188
cdc65		13R187
cdc66		15R174
cdc67	myo2	16R58
cdc68	spt16	7L134
cdc77	ndc2	4L5
cep1		10R40
CHA1		3L46
CHC1		7L133
chl1	ctf1	16L6
chl2		12R
chl3	ctf3	12R283
chl4	ctf17	4R
chl5		9L
chl6	AMC1	5
chl8	AMC3, ctf12	4
chl9	AMC4	7
chl10	smc1	6L10
chl15	ctf4	16R103

Gene	Synonym	Location
cho1		5R22
cin1		15R215
cin2		16L143
cin4		13R86
cka1		9L29
cka2		15
cks1		2R73
cln3	WHI1, daf1, FUN10	1L56
cls4	cdc24, tsl1	1L56
cls7		5L15
cly2		2L39
cly3		6R6
cly8		7R76
cmd1		2R53
cmp1	cna1	2
cmp2	cna2	16
cms1		14L
cmt1	mar2, sir3, ste18	12R334
cna1		2
cna2		16
cpa1		15R161
cre2	spt6, ssn20	7R87
crl1		6R5
crl3		7L43
crl4		5R18
crl7		15R71
crl9		2R97
crl10		7R92
crl11		15L41
crl12		7R186
crl13		15R143
crl15		8R30
crl16		4R236
crl17		15R45
crl18		4L1
crl21		10R1
crl22		13R
cry1		3R26
csd2	dit101	2R9
csd3		10
csd4		2R9
ctf1	chl1	16L6
ctf3	chl3	12R283
ctf4	chl15	16R103
ctf12	amc3	4
ctf13	chl8	13
ctf14		7
ctn1	cdc25	12R213
ctn5	ras2	14L79
CUP1		8R54
cup2		7L101
cup3		12R323
cup5		5L18
cup14		4R286
cyb2		13L30
cyc1		10R34
cyc2		15R12
cyc3		1L47
cyc7	cyp3	5L27
cyc8	ssn6	2R56
cyc9	flk1	3R77
cyh1	tup1	2L
cyh2	umr7	7L64
cyh3	amm1	7L8
cyh4		15R34
cyh5	pdr1	11
cyh10		2L
cyr1	cdc35, hsr1, SRA4, tsm0185	10R2
cys1		1L10
cys3		1L16
dac1	gpa1	8R2
dac2		2L5
daf1	FUN10, WHI1, cln3	1L56
dal1		9R32
dal2		9R36
dal3		9R37
dal4		9R34
dal5	urep1	10R97
dal80		11R14
dal81		9R22
DBF1		9
dbi2		7R28
dbi20		16R66
DBF4		4
dbl1		11L84
DBP2		13
ded1		15R114

Continued

Table II. Continued

Map Positions and Centromere Distances (cM)

Gene	Synonyms	Map Position/ Chrom./Arm/(cM)	Gene	Synonyms	Map Position/ Chrom./Arm/(cM)	Gene	Synonyms	Map Position/ Chrom./Arm/(cM)
deg1		6L1	eth2	sam2	4	FUN22		1L
dfr1		15R135	exg1		12R206	fun80		13R36
dit101	csd2	2R9	exg2		4R162	fun81		13L63
dna1		16R3	fas1		11L152	fur1		8R91
dpr1	ram1	4L33	fas2		16L125	fur4		2R8
	scg2		fcy2		5R49	FUS1		3L21
	ste16		fdp1		2R41	fus2		13R160
dsm1		7R14	fen1	cyc9	3R24	fus3		2
dst1		7R24	flk1	tup1	3R77	gac1		15R113
dst2	kem1	7L111		umr7		gal1		2R7
	sep1			amm1		gal2		12R44
	xrn1		FLO1		1R44	gal3		4R1
dtp1	apa1	3L33	FLO5		1	gal4		16L143
dur1		2R112	fol1		14L105	gal5		13R
dur2		2R113	fol2		7R215	gal7		2R7
dur3		8L13	fpr1		14L105	gal10		2R7
dur4		8L12	fro1		7R143	gal11	spt13	15L53
DUR8		2R115	fro2		7R127	gal22	sin4	14L145
DUT1		2	fsr1		2R150	gal80		13L33
eam1		10L108	FUN9	daf1	1L	CAL83		5R11
eam2		4	FUN10	cln3	1L56	gap1		11R27
eam	top3	10L136		WHI1		gcd1	tra3	15R143
edr1		12R163	FUN11		1L	gcd2	gcd12	7R54
erg6		13R2	FUN12		1L	gcd4		3
erg10	tsm0115	16L22	FUN19		1L	gcd5		3
erg11	rar1	8R12	FUN20		1L	gcd11		5R14
erg12		13R76	FUN21		1L	gcd12	gcd2	7R54
esp1		7R60				gcn2	aas1	4R173
ess1		10R17				gcn3	aas2	11R7

Continued

Gene	Synonym	Map
gcn4		5L3
glc1	aas3	2R85
glc3	arg9	5L2
glc4		15L72
glc6		2R64
glk1	ira2	3L47
gln1		16R29
gln3		5R38
gln4		15R88
glu1		12
gpa1	aco1	8R3
gpa2	dac1	5R10
GPH1		13/16
grr1		10R60
GSH1		10
gst1	suf12	4P95
gut1	sup2	8
gut2	sup35	9
hap2	sup36	7L148
hem1		–
hem10		7L35
hex2	reg1	4R0–10
him1		4R275
hip1		7R146
his1		5R49
his2		6R32
his3	his8	15R112
his4		3L22
his5		9L89
his6		9L17

Gene	Synonym	Map
his7		2R146
hit1		10
hit2	pet18	3R17
hmg1		13L50
hmg2		12R343
HML		3L49
HMR		3R94
HO		4L130
hol1		14R48
hom2		4R92
hom3		5R48
hom6		10R87
hop1		9L59
hpr1		4x82
hpr5	radH	10L61
HSC82	srs2	13
HSP90	HSP82	16L137
hsp104		7
hsr1		10R2
HTA2,B2		2R3
HTA1,B1		4R132
hts1	cdc35	16R26
hxk1	cyr1	6R70
hxk2	SRA4	7L167
HYG4	tsm0185	4R201
ils1		2L43
ilv1	tsm4572	5R64
ilv2	SMR1	13R66
ilv3		10R10
ilv5		12R252

Gene	Synonym	Map
ime4		7L130
imp2		9
ino1		10L101
ip11	ppd1	16L116
ira1	glc4	2R78
ira2		15L72
kar1		14L208
kar3		16R94
kem1	dst2	7L111
kem3	sep1	7L83
kex1	sep1	7L131
kex2		14L147
kgd1	ogd1	9L91
KIN3		6
kin28		4
kns1		12L28
KRB1		cen
kre1		14L235
kre2		4R306
kss1		7R
ktr1		15R54
lap1	ape2	11L144
lap3		14L152
lap4		11L131
let1		1R2
let1M		13R79
let3		10L6
let5		10L17
let6		6L7
leu1		7L5
leu2		3L5
leu3		12R346

Table II. Continued

			Map Positions and Centromere Distances (cM)					
Gene	Synonyms	Map Position/ Chrom./Arm/(cM)	Gene	Synonyms	Map Position/ Chrom./Arm/(cM)	Gene	Synonyms	Map Position/ Chrom./Arm/(cM)
leu4		14L76	mak12		12L31		MAL63	
leu5		8C	mak13		9R40+		MAL64	
lgn4	sit4	5L6	mak14		3R61	mar1	sir2	4L16
los1		11	mak15		11R69	mar2	cmt1	12R337
lte1		1L31	mak16		1L31		ste8	
lts1		7L	mak17		10L140+		sir3	
lts3		7L	mak18		8R95	mas2		8R26
lts4		4R145	mak19		8R130	mas5		14
lts10		4R	mak20		8	MAT		3R30
lys1		9R40	mak21		4R	mcm1		13R35
lys2		2R59	mak22		12	mcm2		2L31
lys4		4R145	mak24		7L56	mcm3		5L24
lys5		7L97	mak26		14L114	mei4		5R41
lys7		13R30	mak27		13R156	mek1		15R175
lys9	lys13	14R44	MAL1	MAL11	7R226	MEL1		2L77
lys11		9L69		MAL12		MEL2		7
lys13	lys9	14R44		MAL13		MEL3		16
lys14		4R8	MAL2	MAL21	3R95	MEL4		11
lys15		–		MAL22		MEL5		4
mak1	top1	15R3		MAL23		MEL6		13
mak3		16R34	MAL3	MAL31	2R194	MEL7		6
mak4		2R60		MAL32		MEL8		15
mak5		2R80		MAL33		MEL9		10/14
mak6		16R8		MAL34		MEL10		12
mak7		3L38	MAL4	MAL41	11R85	mer1		14L138
mak8	tcm1	15R22		MAL42		mer2	rec107	10
mak9		11L158		MAL43		mes1		7R204
mak10		5L43	MAL6	MAL61	8	met1		11R54
mak11		11L116		MAL62		met2		14L187

Continued

Gene	Synonym	Position
met3		10R3
met4		14L76
met5		10R84
met6		5R77
met7		15R136
met8		2R119
met10		6R41
met13		7L77
met14		11R1
met20		11R55
mfa1		16L115
mfa2		7L58
MCL2		2R198
mgm1		15R123
mgt1		11L2
mif1		14
mif2		11L43
mig1		7L27
min1		5L37
mis1		2R
mkt1		14L69
mnn1		5C
mnn2		2R4
mnn4		11L136
mnt1		4R307
mos1		2R73
mos3		2R91
mot1		16
mrp1		4
mrp2		16
mrs1		9R
MRS3		10
msb1		15R103
msb2		7R2

Gene	Synonym	Position
msi1		2R89
mss2	nam1	4L
mtf2		4L14
MTP1		7R225
mut1		—
mut2		—
myo2	cdc66	15R175
nam1	mtf2	4L14
nam2		12R325
NAM7		13
NAM8		8
nam9		14
ndc1		13L18
ndc2	cdc77	4L5
NHS1		5R47
nib1		16L18
nmt1		12R142
NOV1		2
NRA2	pdr1	7L8
nul3	ste5	4R63
nup1		15
obi1		4
ogd1	kgd1	9L91
ole1		7L37
oli1	pdr1	7L8
osm1		10R34
osm2	aro7	16R39
oxt1		4R5
pam1		4R
pap1		11L1
pbs2		15R58
pbs2		10L85
pda1		5R15
pdb1		14

Gene	Synonym	Position
pdc1		12L7
pdc2		4R20
pde2		15R220
pdr1	sra5, AMY1, ant1, BOR2, cyh3, NRA2, oli1, smr2, til1	7L8
pdr2		15R89
pdr3	pdr7	2L10
pdr4		2L2
pdr4		13L8
pdr5	pdr4	15R85
pdr6		7L10
pdr7		2L2
pdx2		13R
pep1		2L8
pep3		12R91
pep4	pra1, pho9	16L95
pep5		13R152
pep7		4R192
pep12		15L5
pep16		12R44
pet1		8R7
pet2		14L171
pet3		8R83
pet8		14R3
pet9		2L20
pet11		2R101
pet14		4R112

Table II. Continued

		Map Positions and Centromere Distances (cM)						
Gene	Synonyms	Map Position/ Chrom./Arm/(cM)	Gene	Synonyms	Map Position/ Chrom./Arm/(cM)	Gene	Synonyms	Map Position/ Chrom./Arm/(cM)
pet17		15R23	pho83		3	prc1		13R210
pet18	hit2	3R17	pho84		13L100	PRI1		9
pet54		7R174	pho85		16L34	pro1		4R187
pet56		15R112	phr1		15R205	pro2		15R175
pet111		13R177	phr2		15R224	pro3		5R23
pet122		5R118	PHS1		4R1	prp2	rna2	14R8
pet123		15R84	PKA3	TPK2	16L120	prp3	rna3	4R300
pet494		14R42	PKC1		2L79	prp4	rna4	16
pet-ts1402		5R112	pma1		7L2	prp5	rna5	2R138
pet-ts2858		13R92	pma2		16L	prp6	rna6	2R33
petx		14L138	pmr1	ssc1	7L104	prp11	rna11	4L14
pfk1	pfk2	13R132	pmr2		4R15	prp16		11
pfk2	pfk1	7R184	pms1		14L66	prt1	cdc63	15R221
pfy1		15R64	pol1	cdc17	14L74	prt2		14L171
pgi1		2R89	pol2		14L170	ptp1		15R221
pgk1		3R2	pol30		2L	pur5		4R
pgm1		10	pot1		9	put1		12R73
pha2	bas2	14L227	ppa1		8R27	put2		8R24
pho2		4L59	ppa2		13	put3		11L5
pho3,5		2R47	ppd1	ira1	2R78	pyk1	cdc19	1L45
pho4		6R47	pph1	sit4	4L15	qcr9		7R148
pho8		4R306	pph2		4R96	raa1	aar1	4R92
pho9	pep4, pra1	16L95	pph3		4R53	raa2	aar2	6R46
			pph21		4L66	rad1		16L18
pho11		1R48	pph22		4L102	rad2		7R192
pho12		8	ppr1		12R8	rad3	rem1	5R146
pho80	tup7	15L1	pra1	pep4, pho9	16L95	rad4		5R128
pho81		7				rad5	rev2	12R22
PHO82		6R47	prb1		5L48	rad6		7L44

Gene	Synonym	Location
rad7		10R35
rad9		4R126
rad10		13R48
rad16		2R58
rad17		15R214
rad18	$r^{s}1$	3R67
rad23		5L26
rad24		5R147
rad50		14L170
rad51		5R78
rad52		13L18
rad53		16L109
rad54		7L111
rad55		4R53
rad56		16R85
rad57		4R2
radH	hpr5, srs2	10L61
$r^{s}1$	rad24	5R147
ram1	dpr1, ste16	4L33
rap1	scg2	14L148
rar1	erg12	13R176
RAS1		15R53
ras2	ctn5	14L79
RBK1		3R29
RBP1		14
RDN1		12R105
reb1		2
rec1		7L108
rec3		7L
rec102		12R235
rec104		8R107

Gene	Synonym	Location
rec107	mer2	10
rec114		13R91
reg1	hex2	4RO-10
rem1	rad3	5R146
ret1		15R117
rev1		4
rev2	rad5	12R22
rev3		16L102
rev5		10R32
rev7		9L101
rgr1		12
rgp1		4R82
rgt1		12R70
rib1		2
rib3		4
rib5		2R150
rib7		2
rme1		7R14
rna1		13R159
rna2	prp2	14R8
rna3	prp3	4R300
rna4	prp4	16
rna5	prp5	2R138
rna6	tsm7269	2R33
rna11	prp6	4L14
rna12	prp11	13R216
RNA14		2
RNA15		16
rnh1		13R160
mr2		10C
ROC1		12R23
ros1	cdc39	3R91
ros3		5L11

Gene	Synonym	Location
rpa1		1L3
rpa190		15R192
rpb1	rpo21	4L68
RPB2		15R177
RPB3		9L17
rpb4		10
rpb5		2
rpb6		16
rpb7		12
rpb8		15
rpb9		7
rpb10		15
rpc19		2
rpc31		14L237
rpc34		14R3
RPC40		16R72
rpc53		4L70
RPC80		16
rpc160		15R59
rpd1		15R1
rpd3		14L237
RPK1		4
RPL3		15
rpl16A		7R56
rpl16B		16R51
RPL29		7
RPL44		15
RPL44'		4
RPL45		4
RPL59		3
rpo21	rpb1	4L68
RPO41		6
RPO26		16
rpr1		5L6

Continued

Table II. Continued

Gene	Synonyms	Map Position/Chrom./Arm/(cM)
	Map Positions and Centromere Distances (cM)	
rp28,55,1		15L115
rp28,55,2		14L18
rrm14		13R159
rrp1		4R
rsr1	bud1	7R98
rvs161	spe161	3R1
rvs167		12
sac1		11L115
sac2		4R314
sac3		4R92
sac6		4R84
sac7		4R220
sag1	aga1	10R2
sam1		12
sam2	eth2	4
san1		4R
scc3		3R
scg2	dpr1, ram1, ste16	4L33
SCL1		7L8
SDB21		8
sdc25		12L3
sec1		4R94
sec2		14L176
sec3		5R3
sec4		6L6
sec5		4R95
sec7		4R96
sec11		9R22
sec18		2R48
sec55		3R59
sec59		13R3
sen1	dst2	12R
sep1	kem1, xrn1	7L111
ser1		15R98
ser2		7R168
ses1		4
SFA1		6
SGA		9
shi1		7L
sin1	spt2	5R127
sin4	gal22	14L145
sir1		11R84
sir2	mar1	4L16
sir3	mar2, ste8, cmt1	12R337
sir4	ste9	4R128
sis1		14R3
sis2		11R57
sit3		16L48
sit4	pph1	4L15
	lgr4	5L6
ski1		7L112
ski4		14L
ski8		7L149
sko1		14L142
slk1		10
SLT3		3R
smc1	chl10	6L10
SMR1	ilv2	13R66
smr2	pdr1	7L8
smr3		15R96
snf1	ccr1, cat1	4R304
snf2	swi3, tye3	15
snf3		4L104
SNF5		2
snq1	atr1	13
snr14		5
snr17a		15R136
sod1	sod2	8R12
soe1		7L63
sot1		16L13
spa1		5
spa2		12L28
spd1		15L12
spe2		15L58
spe161	rvs161	3R1
spe1		11L139

Continued

Gene	Synonym	Map	Gene	Synonym	Map	Gene	Synonym	Map
spe10		11L137	spt14		16L113	ssn6	cyc8	2R56
spo1		14L1	spt15	tfIID	5R110	ssn20	spt6	7R97
spo7		1L4	spt16	cdc68	7L134		cre2	
spo11		8L23	spt21		13R		bar1	
spo12		8R83	sra1		9L28	sst1		9L12
spo13		8R11	SRA3	bcy1	10L106	sst2		12R344
spo14		11R18	SRA4	tpk1	10R2	STA1		4
spo15	vps1	11L1		cyr1		STA2	DEX2	2
spo16		8R83		cdc35		STA3	MAL5	14
spo17		7L		hsr1		ste2	DEX1	6L37
spoT1		13C		tsm0185			DEX3	
spoT2		7L142	sra5	pde2	15R220	ste4		15R124
spoT4		4L36	sra6		7R45	ste5		4R63
spoT7		13R161	SRA7		10L100	ste6	nul3	11L117
spoT8		2R31	srb1		15	ste7		4L76
spoT11		15L3	srb2		4L69	ste8		12R342
spoT15		15R57	srd1		3R9	ste9	cmt1	4R128
spoT16		16L140	srk1		4R176	ste11	mar2	12R244
spoT20		16R42	srm5	cdc38	2R90	ste12	sir3	8R65
spoT23		11R19	srn1		4R1	ste13	sir4	15R129
spr41		4R297	srp1		5	ste16		4L33
spr1		15	srp3		16			
SPR2		15	srp5		15R108			
spr6		5R93	srs2		10L61			
spt	sin1	5R127		hpr5		ste18	dpr1	10R59
spt3		4R		radH		ste50	ram1	3L22
spt4		7R40	ssa1	TG100		sth1	scg2	9L91
spt5		13L5	ssb1			stp1		4R298
spt6	ssn20	7R97	srv2		14L101	stp22		3L2
	cre2		ssa1		1L1	stp52		4L2
spt7		2R46	ssb1		4L129	SUC1		7R228
spt8		12R30	ssb20		4	SUC2		9L153
spt10	gal11	10L87	ssb38		6	SUC3		2R196
spt13		15L53	ssc1		10R	SUC5		4L152
			ssc1	pmr1	7L104			
			ssd1		4R176			
			ssg1		15R105			
			ssn2		4R			

Table II. *Continued*

				Map Positions and Centromere Distances (cM)				
Gene	Synonyms	Map Position/ Chrom./Arm/(cM)	Gene	Synonyms	Map Position/ Chrom./Arm/(cM)	Gene	Synonyms	Map Position/ Chrom./Arm/(cM)
SUC7		8	suh2		12R24	sup36	gst1, suf12, sup2, sup35	4R94
SUF1		15L110	sui1		14L171	SUP37		12R86
SUF2		3R2	sui1		10R2	SUP38		7L72
SUF3		4R232	SUP2	gst1	4R142	SUP40		2
SUF4		7R105	sup2	suf12, sup35, sup36	4R94	SUP42		4R20
SUF5		15R76	SUP3		15L16	SUP43		15L14
SUF6		14L192	SUP4		10R36	SUP44		7L72
SUF7		13L32	SUP5		13L34	sup45	sup1	2R81
SUF8		8R82	SUP6		6R38	SUP46	supQ	2R106
SUF9		6L13	SUP7		10L32	sup47	sup47	2R81
SUF10		14L3	SUP8		13R193	SUP50		13R
suf11		15R30	sup9	bem2	5R119	SUP51	sup1	10C
suf12	gst1, sup2, sup35, sup36	4R94	SUP11		6R4	SUP52	sup45	10C
suf13		15R163	SUP15,16		16R48	SUP53	supQ	3L10
suf14		14L125	SUP17		9L53	SUP54		7L24
SUF15		7R171	SUP19	SUP20	5R50	SUP56	SUP52	1R10
SUF16		3R3	SUP20	SUP19	5R50	SUP57	SUP51	6R41
SUF17		15L53	SUP22		9L95	SUP58		11L58
SUF18		6R16	SUP25		11R63	SUP61		3R68
SUF19		5L38	SUP26		12R306			
SUF20		6R6	SUP27		4R9			
SUF21		16R3	SUP28		14R39			
SUF22		13L31	SUP29	SUP30	10C			
SUF23		10R35	SUP30	SUP29	10C			
SUF24		4R160	SUP33		11L147			
SUF25		4L115	sup35	gst1	4R94			

SUP71		5R2	tfIID	spt15	5R110	tsm0115	cdc35	16L23
SUP72		2R41	tfs1		12R124	tsm0119	cyr1	7L
SUP73		10L21	thi1		13R	tsm0120	hrs1	16R32
SUP74		10L32	thr1		8R26	tsm134	SRA4	2R38
SUP75		11L59	thr4		3R51	tsm0139		9L47
SUP76		7R134	tif1		11R22	tsm0151		8R23
SUP77		7R114	tif2		10L92	tsm0185		10R2
SUP78		13R154	til1	pdr1	7L8	tsm0186		8R19
SUP79		13L48	tmp1	cdc21	15R32	tsm0225		4L49
SUP80		4R55	top1	mak1	15R2	tsm437		7L64
SUP85		5R51	top2		14L70	tsm0800		13R12
SUP86		12R304	top3	edr1	12R163	tsm4572	hts1	16R26
SUP87		2R26	tpk1	SRA3	10L106	tsm5162		4R23
SUP88		4R17	tra3	gcd1	15R143	tsm7269		2R33
sup111		8R51	TPK2	PKA3	16L120	tsm8740	rna6	15R55
sup112		7R32	trk1		10L	tsv115	prp6	1L1
sup113		13R52	trk2		11R15	tub1		13L62
SUP138		7L76	trn1		1L2	tub2		6L59
SUP139		5R52	trp1		4R1	tub3		13L100
SUP150		5L1	trp2		5R76	tup1	cyc9	3R77
SUP154		7L42	trp3		11L115	tup4	flk1	15L10
SUP155		7	trp4		4R209	tup7	umr7	15L1
SUP160		15R3	trp5		7L22	tyr1	amm1	2R97
SUS1		5L39	trx1	trl1	12L7	UBA1		11L115
swi1		16L13	trx2	trl	7R161	UBI1		9
tal1		12R	ts26		15L92	UBI2	pho80	11
tcm1	mak8	15R22	tsl1	cdc24	1L56	UBI3		12
tcp1		4R127	tsm1	cls4	3R31			
tdh2		10R2	tsm5		3R60			
tec1		2R	tsm0039		5R4			
tef1		16R46	tsm0070		6L30			
TEF2		2R61	tsm0080		4R7			
tel1		2L60	tsm0111	erg10	13R1			
tex1	cdc2	4L55						

Continued

Table II. *Continued*

Map Positions and Centromere Distances (cM)

Gene	Synonyms	Map Position/ Chrom./Arm/(cM)	Gene	Synonyms	Map Position/ Chrom./Arm/(cM)	Gene	Synonyms	Map Position/ Chrom./Arm/(cM)
UBI4		12	urep1	dal5	10R97	yap3	kem1, sep1	12R67
umr7	cyc9, flk1, tup1, amm1	3R77	van1		13R38	YCR105	adp1	3R1
			vps1	spo15	11L1	yef3		12
ung1		13L16	vps35		10L102	ymr44		13R158
ura1		11L109	vpt3		15R	YG100		1L1
ura2		10L89	vpt15		2R	YP2	ssa1	6L59
ura3		5L7	WHI1	daf1, FUN10, cln3	1L56	YPT1	YPT1	6L59
ura4		12R304	whi2		15R9	383	YP2	3R70
ura6		11R18	xrs2	cdc40	4R211	yuh1		10R57
ure2		14L142	xrn1	dst2	7L111	zwf1		14L153

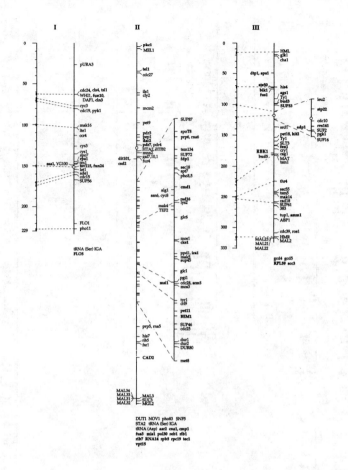

GENETIC AND PHYSICAL MAPS OF
SACCHAROMYCES CEREVISIAE, EDITION 11

compiled by:

ROBERT K. MORTIMER, C. REBECCA CONTOPOULOU
AND JEFF S. KING

*Department of Molecular and Cell Biology,
University of California, Berkeley, CA 94720*

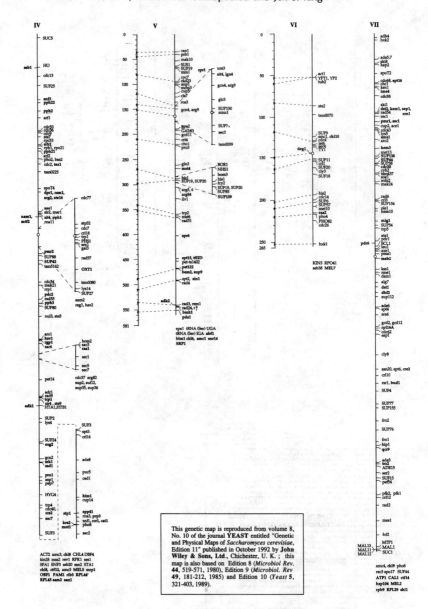

This genetic map is reproduced from volume 8, No. 10 of the journal YEAST entitled "Genetic and Physical Maps of *Saccharomyces cerevisiae*, Edition 11" published in October 1992 by John Wiley & Sons, Ltd., Chichester, U. K. ; this map is also based on Edition 8 (*Microbiol Rev.* 44, 519-571, 1980), Edition 9 (*Microbiol. Rev* 49, 181-212, 1985) and Edition 10 (*Yeast* 5, 321-403, 1989).

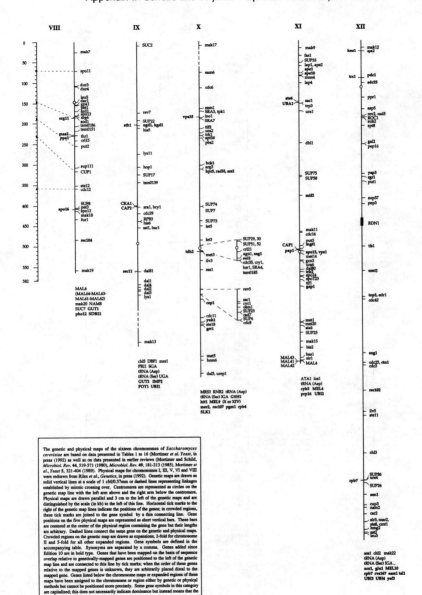

The genetic and physical maps of the sixteen chromosomes of *Saccharomyces cerevisiae* are based on data presented in Tables 1 to 16 (Mortimer *et al. Yeast*, in press (1992) as well as on data presented in earlier reviews (Mortimer and Schild, *Microbiol. Rev.* 44, 519-571 (1980), *Microbiol. Rev.* 49, 181-213 (1985), Mortimer *et al.*, *Yeast* 5, 321-404 (1989). Physical maps for chromosomes I, III, V, VI and VIII were redrawn from Riles *et al.*, *Genetics*, in press (1992). Genetic maps are drawn as solid vertical lines at a scale of 1 cM/0.57mm or dashed lines representing linkages established by mitotic crossing over. Centromeres are represented as circles on the genetic map line with the left arm above and the right arm below the centromere. Physical maps are drawn parallel and 3 cm to the left of the genetic maps and are distinguished by the scale (in kb) to the left of this line. Horizontal tick marks to the right of the genetic map lines indicate the positions of the genes; in crowded regions, these tick marks are joined to the gene symbol by a thin connecting line. Gene positions on the five physical maps are represented as short vertical bars. These bars are centered at the center of the physical region containing the gene but their lengths are arbitrary. Dashed lines connect the same gene on the genetic and physical maps. Crowded regions on the genetic map are drawn as expansions, 2-fold for chromosome II and 5-fold for all other expanded regions. Gene symbols are defined in the accompanying table. Synonyms are separated by a comma. Genes added since Edition 10 are in bold type. Genes that have been mapped on the basis of sequence overlap relative to genetically-mapped genes are positioned to the left of the genetic map line and are connected to this line by tick marks; when the order of these genes relative to the mapped genes is unknown, they are arbitrarily placed distal to the mapped gene. Genes listed below the chromosome maps or expanded regions of these maps have been assigned to the chromosome or region either by genetic or physical methods but cannot be positioned more precisely. Some gene symbols in this category are capitalized; this does not necessarily indicate dominance but instead means that the wild type copy was used to map the gene.

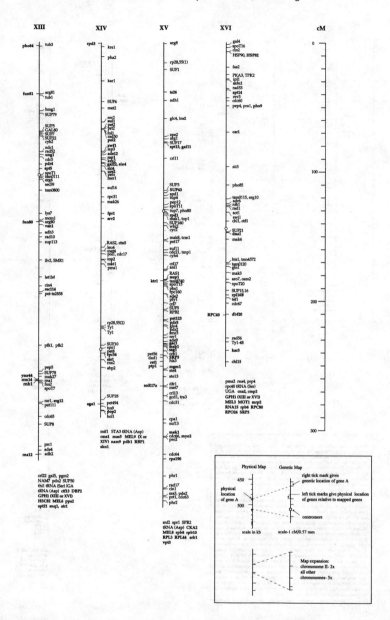

The *Saccharomyces* Genome Database (SGD) project contains genetic maps, physical maps, DNA sequence data and biological information gathered from the published literature a General enquiries should be addressed to Dr Michael Cherry, Department of Genetics, School of Medicine, Stanford University, Stanford, CA 94305–5120, USA. Fax: 415 723 7016, E-mail: yeast-curator@genome.stanford.edu

Appendix III A Comprehensive Compilation of 1400 Nucleotide Sequences Coding for Proteins from the Yeast *Saccharomyces cerevisiae*

Marie-Odile Mossé[a], Reinhard Dölz[b], Jaga Lazowska[a], Piotr P. Slonimski[a] and Patrick Linder[c, d]

[a]*Centre de Génétique Moléculaire, Laboratoire propre du CNRS associé à l'Université Pierre et Marie Curie, F-91190, Gif-sur-Yvette, France;* [b]*Biocomputing, and* [c]*Dept. of Microbiology, Biozentrum, 70, Klingelbergstrasse, CH-4056, Basel, Switzerland.* [d]*Dept. de Biochimiemédicate, CMU, I'rue Michel Servet, CH-1211 Genève 4, Switzerland.*

SUMMARY

The amount of nucleotide sequence data is increasing exponentially. In an effort to make a comprehensive database for the yeast *Saccharomyces*, this database (Table V) includes 1400 protein coding sequences representing 1138 individual genes from this organism. Each sequence has been attributed to a single genetic name and in the case of allelic, duplicated sequences, synonyms are given, if necessary. For the nomenclature a standard principle for naming gene sequences based on priority rules has been introduced. A simple method to distinguish duplicated sequences of one and the same gene from non-allelic sequences of duplicated genes has been applied. Along with the genetic name, the mnemonic from the EMBL data library, the codon bias, reference of the publication of the sequence and the EMBL accession numbers are included in each entry. The database is available by electronic network or on request on diskettes.

In view of the very rapid growth of sequence data an updated compilation is presented of yeast sequences encoding proteins (open reading frames = ORFs) from the yeast *Saccharomyces cerevisiae*. Since the first publication of a comprehensive list (ListA1; Mossé *et al.*, 1988) the database has already increased in size by more than a factor of three. In the present list the com-

The Yeasts Vol. 6, 2nd edition
ISBN 0-12-596416-1

plete set of the new ORFs discovered by the genomic sequencing of chromosome III (EMBL accession number X59720, Oliver *et al.*, 1992) has not been included. As in the previous lists (Mossé *et al.*, 1988, 1993), sequences from the yeasts *Saccharomyces cerevisiae*, *Saccharomyces carlsbergensis* and *Saccharomyces uvarum*, which are believed to constitute the conspecific taxonomic species (Barnett *et al.*, 1983), are included. Sequences from *Saccharomyces douglasii*, believed to be the most recently diverged distinct species from *Sacch. cerevisiae* (Herbert *et al.*, 1988, 1992), are not included. Likewise sequences from *Candida*, *Hansenula* and others are not included in this list. A list of genes of the species *Schizosaccharomyces pombe* is included in Appendix V. Sequences from the 2-micron plasmid (2 µm), mitochondrial sequences, killer sequences and Ty elements are also not included. In some cases wild-type *and* mutant sequences from the same publication are present in the EMBL data library. In these cases only wild-type sequences are listed. Since the goal was to attribute a genetic name to every sequence, unidentified reading frames have not been included, whether complete or partial.

Several problems arose in the preparation of this new list. Many new genes have been isolated by reverse biochemistry or other non-classical genetic methods. In many of these cases either no or incorrect gene designations have been attributed to the sequences. In other cases either the same name is given to different sequences or different names are given to the same sequence. To avoid confusion authors are asked to use the standard yeast genetic nomenclature and to register the names with the "Standing Nomenclature Committee" (Chairman R.K. Mortimer, Dept. of Mol. and Cell. Biology, Division of Genetics, University of California, Berkeley, CA 94720; FAX 510 642-8589).

Sorting out the problem of genetic nomenclature has involved proposing new names in some cases. Each sequence has been given a genetic name according to the entry in the EMBL data library or the publication reporting the sequence. If different names are given to one and the same sequence a *priority rule* has been set up (Table I). According to this rule the name of the first published sequence (date of acceptance of the publication) is used in the list (also see below: sequences of duplicated non-allelic genes), provided it is in accordance with standard genetic nomenclature. Other

Table I. Priority rules as described in the text

Rules for naming gene sequence
• Sequence first published (acceptance date)
• Correct genetic designation: 3 or 4 letters and digit(s)
• Name is not used otherwise

names are included as synonyms. In some cases four letter designations (*ARGR1*, *MRPL20*) or gene names followed by a letter (*RPL4A*, *TIF51A*) have also been used. In the case of historically well established gene designations, such as *HO*, it was self-evident that they should be retained. In a few cases the priority rules would impose a name which is no longer used. Exceptions have been made for those and the currently used name has been adopted. As an example, in a former edition of ListA the name *ADC1* was proposed for a gene encoding an alcohol dehydrogenase. This name is no longer used and this sequence is referred to as *ADH1* with *ADC1* as synonym. Histone genes have been named according to a proposition by Smith (personal communication) and (Smith and Stirling, 1988). If, however, no name is attributed in the first publication or a name which does not follow the rules of the genetic nomenclature, another name is chosen and the original name is given as synonym (see below). Although the proposals are generally in accordance with the gene symbols used in the glossary compiled by Mortimer, Contopoulou and King (Appendix II), some conflicts may still persist. In a few cases the priority rules described in this article preclude this.

DUPLICATED CODING SEQUENCES: DISTINCTION BETWEEN ALLELIC SEQUENCES OF ONE AND THE SAME GENE AND THE SEQUENCES OF DUPLICATED NON-ALLELIC GENES

The present list contains several sequences (identical or almost identical) which are present more than once. The problem of relationships (very often confusing) between names of genes, original mutations used for isolating complementing inserts to be sequenced, mnemonics employed in data libraries corresponding to the sequences and, finally, allelism of genes, can be unambiguously solved (as suggested by Mossé *et al.*, 1988) by comparison of nucleotide sequences upstream of, and downstream from, protein coding ORFs. When two protein coding sequences are completely or almost completely identical, their 5' and 3' flanking protein non-coding sequences are compared. If the degree of identity in flanking sequences is very high and comparable to that of ORFs themselves, the two sequences are considered to represent one and the same gene and are referred to as allelic. Both sequences are included in the database, but retain only one gene name, with indication, when necessary, of a synonym (Table II). It is not possible, at present, to decide whether differences reported are due to natural polymorphic variations, to artefactual sequencing errors or to a combination of both. The presence of differences in sequences are indicated and positions of gaps/inserts are given in individual comments.

Table II. Examples of duplicated sequences of the same gene (allelic sequences). The columns indicate from left to right: the name of the gene; the length of the coding region; the mnemonics; the identity of the coding regions at the DNA level and the presence of gaps; the identity of the protein sequences and the presence of gaps; the identity and the length of the 5′ regions; and the identity and the length of the 3′ regions.

Gene	length cds	seqcds1	seqcds2	coding sequence				flanking sequences			
				ident. bp	gaps	ident. aa	gaps	ident. 5′	length 5′	ident. 3′	length 3′
AAC2	954	scaac2	scaac3	94.9	0	99.1	0	100.0	80	97.1	175
AKY1	666	scaky	scadk	99.7	0	99.6	0	87.0	163	97.5	43
COX5A	459	sccox5	sccox5aa	99.1	0	98.7	0	100.0	26	100.0	99
SPT15	720	scbtfly	sctfiiO1	99.3	0	99.6	0	99.6	275	100.0	147
ADE2	1710	scadeaa	scade2	98.7	3/1	97.5	1/1	98.6	507	98.7	155
LPD1	1497	scdhdha	sclpdh	99.3	0	100.0	0	37.4	534	92.5	80
RPA2	318	scrpa2	scrgap44	100.0	0	100.0	0	95.0	22	91.7	198

If, however, flanking sequences diverge, this is taken as an indication of duplicated genes (non-allelic genes) and thus two different names are used (for further discussion see Mossé *et al.*, 1988). In Table III an extensive list of duplicated non-allelic sequences are given without an analysis of functional relationship. An example of such an analysis is given in Table IV. Four sequences of nuclear genes coding for ADP/ATP carrier proteins located in the mitochondrial membrane are available in data libraries. The sequence *AAC1* (mnemonic scpet9 (Adrian *et al.*, 1986)) was obtained by complementation of the first Oxidative Phosphorylation mutation discovered (mutation *op1⁻* (Kovac *et al.*, 1967)), later rebaptised *pet 9⁻* (Beck *et al.*, 1968). It was subsequently shown that this gene complements the *op1⁻* mutation only on multi-copy vectors, so that it is not allelic to *op1⁻*. By recombination tests it corresponds to a suppressor of *op1⁻* and the cognate *OP1⁺/PET9⁺* gene is *AAC2* (mnemonic scaac2 (Lawson and Douglas, 1988)). Independently, another gene was cloned and sequenced (mnemonic scaac3 (Kolarov *et al.*, 1990)); Table IV shows that this gene is identical and allelic to *AAC2* since the upstream and downstream nucleotide sequences of scaac2 and scaac3 are practically identical. Finally, a third gene *AAC3* was sequenced (mnemonic scaac2a (Kolarov *et al.*, 1990)) which cannot be allelic to the preceding ones since the upstream and downstream sequences are drastically different. Out of 6 pairwise comparisons shown in Table IV only one (scaac2 by scaac3) relates to one and the same gene (*AAC2*); and all the other comparisons deal with non-allelic but structurally closely related genes. It cannot be decided, at present, whether the differences in the two published nucleotide sequences of the gene *AAC2* (5.1% of mismatches in the protein coding sequence) are due to polymorphism or to sequencing errors but the second possibility appears more probable. More recently, in the sequence scbub2q a fragment of sequence scpet9 was reported and found to be identical. From comparisons shown in Table IV it can be deduced that it corresponds to gene *AAC1*. Thus application of this rule is easy and straightforward in most cases. Subtelomeric genes, like *SUC* genes included in this database, are probably an exception (Hohmann and Zimmermann, 1986).

The list presented here contains 1138 individual sequences corresponding to 1138 non-allelic genes with 221 of them present more than once. Of the latter sequences, 185 are present twice, 30 are present three times, four are present four times and two are present five times. The list also contains many duplicated, non-allelic genes (identical or almost identical protein coding sequences and divergent flanking sequences, Table III). The distribution of sequence length clearly shows that a majority of genes has open reading frames of 1000 to 2000 nucleotides (Figure 1). Extremely long genes are also present. The elevated frequency of certain codons in highly

Table III. Examples of sequences of duplicated but different genes (non-allelic sequences). The columns indicate from left to right: the name of the first gene; the mnemonic; the length of the coding region; the name of the second gene; the mnemonic; the length of the coding region; the identity of the coding regions at the DNA level; the identity of the protein sequences; the identity and the length of the 5' regions; and the identity and the length of the 3' regions. In cases where duplicated non-allelic genes are represented more than once in the list only one example is given. In the case of *TIF1/TIF2* it is known from genetic data that the genes are non-allelic

Gene 1	seqcds 1	Length CDS1	Gene 2	seqcds2	Length CDS2	coding sequences Identity bp	coding sequences Identity aa	flanking sequences Identity 5'	flanking sequences Length 5'	flanking sequences Identity 3'	flanking sequences Length 3'
TEF2	sctef1	1374	TEF1	scef1aaa	1374	99.9	100.0	45.5	250	47.0	152
TIF1	sctif1	1185	TIF2	sctif2/scyur1	1185	99.6	100.0	37.8	596	38.5	599
RPL41A	scy141a	75	RPL41B	scy141b	75	98.7	100.0	40.1	356	45.4	150
HHF1	sch3h401-2	309	HHF2	sch4	309	97.7	100.0	43.2	185	47.0	136
SSA1	scssa1	1926	SSA2	scssa2	1917	96.6	98.6	37.1	202	85.2	6
RP51A	scrp51	408	RP51B	scrp51b	408	95.3	99.3	40.8	262	33.1	136
RP28A	scrp28a1	558	RP28B	scrps16a-1	558	95.2	100.0	40.4	301	66.7	15
RP55A	scrp28e1-2	432	RP55B	scrps16a-2	432	95.1	100.0	62.0	975	66.6	65
ENO1	scenoa	1311	ENO2	scenob	1311	95.0	98.4	46.7	357	45.4	386
RPL4B	sc141	768	RPL4A	sc142	768	94.8	97.3	45.3	544	47.1	141
UBI1	scubi1g	384	UBI2	scubi2g	384	94.0	100.0	39.4	736	38.2	490
HTA1	sch2a1	396	HTA2	sch2a2	396	93.9	98.5	44.6	263	46.5	57
HHT3	sch3h401-1	408	HHT2	sch3h402-1	408	93.1	100.0	46.9	210	55.3	684
RPL15A	scrp115a	495	RPL15B	scrp115b	495	91.2	100.0	39.8	553	5.6	30
TIF51A	schyp2	471	ANB1	schyp1	471	91.1	94.9	37.9	441	37.4	508

Table IV. Example of solving the nomenclature problem by comparison of DNA sequences. For details see text. Comparison of different sequences encoding three ADP/ATP carriers. The sequences are indicated by the mnemonics scaac2, scaac2a, scaac3, scpet9. The corresponding genes are indicated in the diagonal. The top right half of the table represents the comparison of the nucleotide sequences. For each pair of sequences, the 5′ ends (top), the coding sequences (middle) and the 3′ ends (bottom) have been compared. The similarity is given in percent identity with the length of the alignment in basepairs. In the bottom (left) half the amino acid sequences have been compared. For each pair the similarity in percent and the length of the alignment is given.

	Scaac2	Scaac2a	Scaac3	Scpet9
Scaac2	AAC2	179 / 957 / 175 — 42.5% / 77.9 / 42.0	78 / 957 / 274 — 100.0% / 94.9 / 98.2	179 / 960 / 293 — 35.8% / 69.1 / 44.2
Scaac2a	318 — 96.4	AAC3	235 / 957 / 175 — 44.9 / 82.8 / 44.9	517 / 930 / 175 — 38.7 / 67.1 / 47.7
Scaac3	318 — 99.1	318 — 97.4	AAC2	517 / 960 / 274 — 44.9 / 68.7 / 40.1
Scpet9	319 — 90.3	309 — 91.2	319 — 89.9	AAC1

Mnemonic

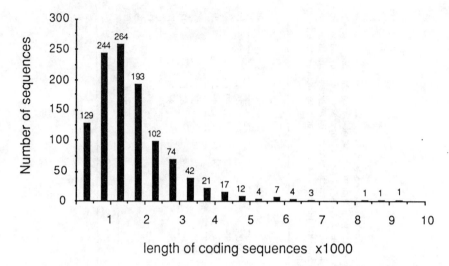

Fig. 1. Distribution of the length of the coding sequences. The length of the protein-coding sequences is indicated in windows of 500 nucleotides. For duplicated sequences only one has been considered. Partial sequences have been excluded.

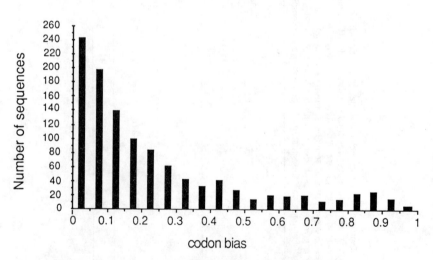

Fig. 2. Distribution of codon bias. The codon bias was calculated according to Bennetzen and Hall (1982). The codon bias is indicated in windows of 0.05. Genes having a bias ≤ 0 were considered as having a value of 0.

expressed genes can be expressed in the so-called codon bias (Bennetzen and Hall, 1982). In highly expressed genes codon bias is close to one, whereas in weakly expressed genes codon bias is close to zero. Figure 2 shows the distribution of codon bias of the genes included in this list. Although the present list is only a subset of total yeast genes this distribution is certainly indicative for the distribution of codon bias.

The present database is easily integrated in a data library environment using the sequence retrieval software (SRS, Etzold and Argos, 1993). In the near future gene functions and chromosomal locations will also be included. We hope that the present list will induce a feed-back response from the community by pointing out inevitable errors, omissions, etc. which will be included in forthcoming editions. Comments, new sequences and suggestions may be sent to Piotr P. Slonimski or Patrick Linder at the above addresses or by electronic mail to "mosse@cgm.vax.cgm.chrs-gif.fr" or "linder@cmu.unige.ch".

PRESENTATION OF THE DATABASE

The database (Table V) is essentially arranged as described before (Mossé *et al.*, 1988). The columns are:

Gene (Primary gene name): The gene designation follows the priority rule described above.

Example: *SUF12*   date of acceptance: August 24, 1987
        GST1   date of acceptance: December 22, 1987
        SUP2   date of acceptance: November 12, 1987

Thus the gene sequence will be named *SUF12*, with *GST1* and *SUP2* as synonyms.

Partial sequences are indicated by an asterisk (*). If more than one sequence of one gene is present the percentage of similarity on the DNA level (n) and the protein level (p) are indicated in the comments.

Synonym (Secondary gene name): If several sequences with different names are present in data libraries and if it has been demonstrated that they concern one and the same gene (see above), the alternative name is given under "synonym". Synonyms are only given if the sequence has two different names. To facilitate searching in the database we include the synonyms also in the gene list and refer to the gene name.

Mnemonic: Under "Mnemonic" the names of the sequences are listed from the EMBL library when available. If more than one open reading frame is present in one sequence, the mnemonics are extended by a hyphen followed by a digit (for example scada2x-1). Sequences not present in the EMBL library obtain a CGMxxx mnemonic. In the electronic form of the database these sequences are included in the comments.

Length: The length is the number of nucleotides of the protein coding sequence without the termination codon. We arbitrarily set the minimal length of sequences to be included in the list to 60 nucleotides (20 amino acids).

Codon bias: It is calculated according to Bennetzen and Hall (1982).

Reference: The publication in which the sequence has been reported. For reasons of typesetting and clarity of the list references to *Proceedings of the National Academy of Sciences of the United States of America*, *Biological Chemistry Hoppe-Seyler* and *Biochemical and Biophysical Research Communications* have been replaced by PNAS, BCHS and BBRC respectively.

Accession Number: It is the number given to each sequence by the data libraries.

Distribution of the database: The database is available by anonymous FTP at bioftp.unibas.ch [131.152.8.1] or on diskettes upon request. It can be obtained also by electronic mail from the EMBL fileserver netserv@embl-heidelberg.de or on the EMBL CD-ROM distribution.

COMMENTS ON INDIVIDUAL SEQUENCES

aaa1   The name is *AAA1* (scnact, accepted 23.1.89), the synonym *NAT1* (scnat, accepted 23.2.89). Both names are present in Appendix II. The sequences are identical.

aac1   The sequences are identical. The reading frame of *AAC1* in scsub2q is partial.

aac2   The sequences are 94.9% (n) and 99.1% (p) identical. The identity in the flanking sequences is 100% and 97.2% for the 5′ and 3′ regions, respectively.

aac3	Is encoded by scaac2a which is in conflict with the header in the EMBL data library. See also Table IV.
acb1	The number "1" has been added to *ACB* to follow genetic nomenclature.
ace1	The sequences are identical.
ach1	The sequences are 99.7% (n) and 99.3% (p) identical.
acp2	The name is *ACP2* (scacp2, accepted 1.12.87), the synonym *RPC31* (RNA polymerase III, scrpc31g, accepted 24.6.90). The sequences are 99.7% (n) and 99.2% (p) identical.
act1	The sequences are 99.7 (scact1/scact2; scact/scact2; scact2/scacti), and 100% (scact1/scacti; scact/scact1; scact/scacti) identical and the resulting proteins are 99.2%, 99.2%, 99.2%, 100%, 100% and 100% identical.
adc1	See adh1.
ade1	The sequences are 99.6% (n) and 99.7% (p) identical. The sequence in scnpk1 is partial.
ade2	The sequences are 98.7% (n) and 97.5% (p) identical. The difference in length is due to the absence of nucleotides 720, 736 and 757 in the sequence scadeaa.
ade5,7	The *ADE5,7* gene encodes two distinct activities (*ADE5* = phosphoribosylglycinamide; *ADE7* = phosphoribosyl-aminoimidazole synthetase).
ade8	The sequences are identical.
adh1	The coding sequences of scadhc1 and scadhi are 99.9% (n) and 100% (p) identical but scadhc1 contains no flanking sequences. Since only one *ADH1* gene exists in yeast (C.L. Denis, personal communication), the name *ADH1* is generally adopted by the yeast community. *ADC1* is retained as synonym.
adh2	The sequences are 99.8% (n) and 100% (p) scadhr2 contains only the coding sequence. The name *ADH2* is generally adopted by the yeast community (C.L. Denis, personal communication) is used here with *ADR2* as synonym.
adr2	see adh2.

agal1   The sequences are 99.8% (n) and 99.7% (p) identical. They code for alpha-agglutinin. In order to avoid conflict with *AGA1*, coding for the **a**-agglutinin core subunit (scaaglcs) and to avoid the Greek symbol alpha, we list the sequence as *AGAL1* with *AGALPHA* as synonym.

aky1   The sequences are named *AKY1* (scaky, accepted 1.6.87, scadk, accepted 3.8.87, scadk1, accepted 26.12.89). They are 99.7% (scaky/scadk; scadk/scadk1) and 100% (scaky/scadk1) identical; the proteins 99.6%, 99.6% and 100% identical. Number "1" is added to *AKY* to follow genetic nomenclature.

ald1   The sequence scalddeh codes for the major aldehyde dehydrogenase and is named *ALD1* (M. Ciriacy, personal communication).

ald2   In analogy to *ALD1*, the gene encoding the major aldehyde dehydrogenase, this sequence, also encoding an aldehyde dehydrogenase, is designated *ALD2*.

amd1   The gene encoding a yeast AMP deaminase is named *AMD1*, whereas the gene for a putative yeast amidase is called *AMDY*.

amdy1   The name of the sequence coding for a putative yeast amidase gene is given as *AMDY1* in order to avoid conflict with the *AMD1* gene coding for AMP deaminase (Meyer *et al.*, 1989).

anb1   The sequences are named *ANB1* (scanbi, accepted 29.11.89) with synonyms *TIF51B* (sctif51b, accepted 23.12.91) and *HYP1* (schyp1, unpublished). The nucleotide sequences are 100% (scanb1/schyp1) and 99.79% (scanb1/sctif51b and schyp1/sctif51b) identical, and the protein sequences 100% identical. The *ANB1* coding sequence is 91.1% identical to *TIF51A*, but the flanking sequences are different (Table III).

apa1   The name of the sequence is *APA1* (scapa1, accepted 28.8.89) with *DTP1* (scdtpa, accepted 17.5.90) as synonym. The genetic name according to Appendix II for diadenosine-5'-5"-p1-p4-tetraphosphate phosphorylase is *DTP*. The sequences are 99.5% (n) and 100% (p) identical.

aps1   The sequences are 99.7% (n) and 100% (p) identical.

Number "1" has been added to *APS* to follow genetic nomenclature.

arf1/arf2   The coding sequences are highly similar, whereas the flanking regions diverge considerably. The coding sequences are 89.1% identical, the proteins 97.2%. We added the number "1" to *ARF* to follow the genetic nomenclature.

arg1   The sequences are identical; scarg1 is a partial sequence.

arg3   The sequences are identical.

arg4   The sequences are 99.8% (n) and 100% (p) identical. scarg4a is a partial sequence.

argr1/argr2/ argr3   The roman numbers have been replaced by the arabic numbers 1, 2 and 3 to avoid confusion between the letters "i" and "l" and the number "1".

aro1   The sequences scaro1a and scaro1b are partial and contain the 5′ and 3′ sequences, respectively. The partial sequences are identical to scaro1.

atp5   The sequences are identical. According to the literature the gene is named *ATP5* with *OSCD* as synonym.

atp13   The name is *ATP13* (scatp13, accepted 29.11.90; scaep2, accepted 4.3.91). The ORF of the scatp13 sequence is, however, considerably shorter than the one in scaep2, due to the absence of a T-residue between coordinates 1421 and 1422 in scatp13, which leads to a stop codon at coordinate 1452. The sequences are 99.6% (n) and 97.3% (p) identical.

baf1   The name is *BAF1* (scbaf1, accepted 30.8.89), with synonyms *ABF1* (scabf1a, accepted 16.10.89) and *OBF1* (scobf1, accepted 30.1.91). The sequences are 99.1% (abf1/baf1), 100% (abf1/obf1) and 99.1% (baf2/obf1) identical; the resulting proteins 99.2%, 100% and 99.2% identical.

bck1   The sequence scorfaa is shorter than scbck1 and is missing the N- and C-terminal extensions. The sequences are 99.7% (n) and 99.4% (p) identical.

bmh1   The sequences are 99.8% (scbmh1/scpda1), 99.7% (scbmh1/scsygp5) and 99.6% (scsygp5/scpda1) identical. The coding sequence in scsygp5 is shorter due to the insertion of a nucleotide between coordinates 864 and 865 of the corresponding scbmh1 sequence and coordinates 497 and 498 of the corresponding scpda1 sequence. The sequence scpda1 is partial.

bud5   The sequences are 99.7% (scbud/scbudiv), 99.9% (scbud/scmatloc) and 99.7% (scbudiv/scmatloc) identical. The sequence scbud has an insertion of three nucleotides (coordinates 18–20) resulting in a longer reading frame.

caf20   Cap associated factor (M. Altmann, personal communication).

cam1   The sequences encoding a calcium- and phospholipid-binding protein homologous to translation elongation factor-1 gamma are named *CAM1* (sccam1g, accepted 22.8.92; scefiga, unpublished). The two sequences are 99.9% (n) and 99.8% (p) identical.

can1   The sequences are 99.9% (n) and 100% (p) identical.

car2   *CARGB* is renamed *CAR2*, according to the proposition of the authors (Degols *et al.*, 1987). The sequences are identical.

cat3   The sequences are named *CAT3* (sccat3, accepted 19.2.88) with *SNF4* as synonym (scsnf4, accepted 31.7.89). The sequences are identical.

cbf1/cp1/   According to priority rules the sequences should be named
cpf1   *CP1*. Since this name does not follow the nomenclature rules *CBF1* has been selected as gene name (sccfb1a, accepted 7.3.90; sccp1a, accepted 11.2.90; sccp1, accepted 12.9.90). The sequences are 99.9 (sccp1a/sccp1), 99.6 (sccp1/scfb1a) and 99.7% (sccp1a/sccfb1a) identical; the proteins 99.7, 99.7 and 100% identical.

ccp1   The sequences are identical. One sequence is partial (scccp, accepted 10.8.82; scccycpx, accepted 25.5.82).

cct1   Number "1" has been added to follow genetic nomenclature.

cdc2   The sequences are 99.5% (n) and 97.8% (p) identical.

cdc6   Published sequences probably belong to one and the same gene. In the sequence sccdc6, the protein coding sequence is considerably shorter than in sccdc601. This is due to the fact that in the sccdc6 sequence an "A" is missing at position 640, which leads to an 87 codon shorter open reading frame in sccdc6. Sccdcaa is a partial sequence. The sequences are 100% (sccdc601/sccdc6g), 98.9% (sccdc6/sccdc601), 99.7% (sccdc601/sccdc6b) and 98.6% (sccdc6/sccdc6b) identical; the proteins are 100%, 98.6%, 99.8% and 98.4% identical.

cdc8	The sequences are 99.2% (sccdc8/sctgysr3), 99.1% (sccdc8/sctkcdc8) and 99.3% (sctgysr3/sctkcdc8) identical; the proteins are 99.5%, 99.5% and 100% identical.
cdc15	The sequences are identical.
cdc23	The sequences are identical.
cdc25	The sequences differ by the absence of nucleotides 2859, 2891 and 2892 in the sequence sccdc25g. The sequences are 99.9% (n) and 99.6% (p) identical.
cdc39	Both sequences give as a gene designation *CDC39* but are clearly different. The sequence sccdc has been mapped to chromosome III, which is in accordance with the genetic map (Appendix II). *CDC39* is adopted for the entry sccdc and *CDC39?* for sccdc39.
cdc68	The sequence sccdc68 overlaps with the sequence of scura2, although the genes were mapped to two different chromosomes. It is possible that this is due to a cloning artefact.
cho1	Since *CHO1* is the correct genetic designation, *CHO1* is adopted with *PSS* as synonym (scppsg, accepted 3.4.87, scchol, accepted 19.5.87). The sequences are 99.8% (n) and 100% (p) identical.
chs3	The sequence encoded by scall coding for chitin synthase 3 is named *CHS3* (A. Durán, personal communication; scall, published 1991, no acceptance date) with *CAL1* and *CSD2* (scpchsy, accepted 18.1.92) as synonyms. Due to absence of a nucleotide in sequence scall the reading frame is shorter at the 5′ end, compared to scpchsy. The sequences are 99.9% (n) and 99.9% (p) identical.
cif1	According to priority rules the gene is designated *CIF1* (sccif1, accepted 20.10.91) with *TPS1* (sctps1, accepted 8.7.92; sctps1a, unpublished) and *GGS1* (scggs1, unpublished) as synonyms. The genes are 99.3% (sccif1/sctps1a),

	98.9% (sccif1/sctps1), 99.3% (sccif1/scggs1), 100% (scggs1/sctps1a), 99.7% (scggs1/sctps1) and 99.7% (sctps1/sctps1a) identical. The proteins are 99.6%, 98.6%, 99.6%, 100%, 99.0% and 99.0% identical.
cit2	The sequences are identical. The sequence sccit2a is partial.
cka1	The sequences are identical.
clb1	The sequences are named *CLB1* (scclb1, accepted 17.1.91) with SCB1 (scscb1g, accepted 6.3.91) as synonym. The sequences are identical.
clb2	The sequences are identical.
clb3	The two sequences are 99.8% (n) and 99.8% (p) identical.
clb4	The sequences are identical.
cmk1	The two sequences are 97.3% (n) and 97.3% (p) identical.
cmk2	The two sequences are 99.7% (n) and 99.8% (p) identical.
cmp1	The sequences are 99.9% (sccmp1/sccmbp1), 99.7% (sccal1/sccmbp1) and 99.7% (sccal1/sccmp1) identical. The proteins are 100%, 99.6% and 99.6% identical.
cmp2	The two sequences are 99.9% (n) and 99.7% (p) identical.
cne1	The sequences are identical.
cof1	The sequences are identical.
cox5a	The sequences are 99.6% (sccox2/sccox5), 99.6% (sccox2/sccox5aa) and 99.1% (sccox5/sccox5aa) identical. The proteins are 98.7%, 100% and 98.7% identical.
cox5b	The sequences are identical; sccox5b is partial.
cox9	The sequences are identical.
cpr1	The sequences are identical. The authors of the sequence sccpha have renamed their sequence to *CPR1* (Heitman *et al.*, 1991). *CPH1* is included as a synonym.
cpr3	According to the priority rules the gene should be named *CYP3*, (sccyp3, accepted 14.11.91; sccpr3, accepted 17.8.92). This gene name has been used for the iso-2-cytochrome *c* structural gene. In order to avoid con-

fusion and to be in accordance with *CPR1*, *CPR3* is adopted as gene designation and *CYP3* used as synonym. The sequences are identical.

cps1   The sequences are 99.9% (n) and 99.8% (p) identical.

ctc1   This gene codes for cytochrome c1 (which should not be confused with iso-1-cytochrome c). The first acronym used was sccyc1, which is confusing with the gene *CYC1* coding for iso-1-cytochrome c. In agreement with the authors (G. Schatz, personal communication) who first established the sequence, this sequence is referred to as *CTC1* for cytochrome c1.

cyc1   The first acronym used for the sequence of this gene coding for iso-1-cytochrome c was sccyt1. This sequence was, however, obtained by the complementation of the mutation cyc1 (originally cy1 (Sherman and Slonimski, 1963)) generally adopted by the yeast community. The historical name *CYC1* is retained. The sequences are identical.

cyp1   The name is *CYP1* (sccyp1, accepted 21.6.88) with *HAP1* (schap1, accepted 18.10.88) as synonym. They are 99.3% (n) and 99.5% (p) identical.

cyp3   The first acronym used for this sequence was sccyt2. As pointed out by the authors who established this sequence (Montgomery *et al.*, 1980) it corresponds to the iso-2-cytochrome c structural gene which was first identified by Verdière and Petrochilo (1975) and named *CYP3*. The same gene was later rebaptised *CYC7* by Downie *et al.* (1977). Since Montgomery *et al.* (1980) did not decide which name should be used for the gene, and according to the priority rule, we adopt *CYP3* as the name and *CYC7* as synonym.

cyr1   The name is *CYR1* with *CDC35* as synonym (sccyr1, accepted 20.8.85; scccdc35, accepted 7.10.85). The sequences are 99.7% (n) and 98.4% (p) identical. The sequence sccyr1 is longer at the 3'-end and has an additional A nucleotide at position 6686, which leads to a large extension at the 3'-end of the coding region and a divergence in the last 18 codons of sccyr1.

daf1   The name is *DAF1* with *WHI1* as synonym (scdaf1a,

accepted 1.8.88; scwhi1, accepted 23.9.88). The sequences are 99.9% identical. The sequence scdaf1a is the *daf1-1* mutant allele which has a C to A change at position 1194.

dal82   The name is *DAL82* (scdal2, accepted 9.10.90; scdurmg, accepted 31.10.90) with *DURM1* as synonym. The sequences are 99.7% (n) and 99.2% (p) identical.

dat1   Number "1" is added to follow genetic nomenclature.

ded1   The name is *DED1* with *SPP81* as synonym (schis3g, accepted 27.2.81; scded1, accepted 5.12.90). *DED1* in schis3g is only a partial sequence. The sequences are 99.4% (n) and 99.1% (p) identical.

dfr1   The sequences are 99.8% (m26668/scdfr1; scdfr1/scdihrg) and 100% (m26668/scdihrg); the proteins are 100% identical.

din1   This is a partial sequence.

dka1   The name *NSP1* of the sequence scnsp1 has been changed in accordance with R. Piñon (personal communication) to avoid conflict with *NSP1* (scns1p).

dmc1   The sequences are identical.

dna43   The name is *DNA43* (scdna43, accepted 3.12.91) with *DBF4* (scdbf4, accepted 20.12.91) as synonym. The two sequences differ in length. The sequence scdna43 has a gap corresponding to coordinates 525–527 of sequence scdbf4, whereas the sequence scdna43 has a gap corresponding to coordinates 1242–1272 of sequence scdbf4. Otherwise the sequences are 99.5% (n) and 99.4% (p) identical.

efc1   The protein of the sequence scelongam is only 65.3% identical to the one of the sequence *CAM1* (scef1ga).

eft1/eft2   The sequences code for exactly the same protein and the coding sequences are 99.8% identical. However, the flanking sequences are clearly different and the two genes are non-allelic (Perentesis *et al.*, 1991).

end1   The sequences are named *END1* (scend1, accepted 6.2.89; scpep5g, accepted 30.4.90). The sequences are 99.8% (n) and 99.8% (p) identical. The sequence scpep5g misses nucleotides at coordinates 1854, 1855 and 1878 in respect to scend1.

eno1/eno2   To follow genetic nomenclature, sequences enoa and enob are renamed *ENO1* and *ENO2* respectively. The coding sequences are 95% identical, but the flanking sequences are different (Table III).

erg3   The sequences are identical.

erg9   The sequences are 99.3% (n) and 99.8% (p) identical.

erg11   The sequences are 99.8% (n) and 98.8% (p) identical. The gene name for lanosterol 14-demethylase is *ERG11*.

fas1   The sequences have not the same length. The sequence scfas1 has a G inserted at position 7002, a T inserted at position 7309 and a T inserted at position 7407. The sequences are 99.5% (n) and 98.9% (p) identical.

fbp1   The sequences are identical.

fkb1   The sequence is named *FKB1* (scfkb1, accepted 29.10.90) with *FPR1* (scfpr1, accepted 9.11.90) and *RBP1* (scrbp1, accepted 7.12.90) as synonyms. The genetic designation *FPR1* reflects the analogy to another isomerase, *CPR1*. The sequences are identical.

fkb2   The sequences are identical.

fus1   The sequences are identical.

fus3   The sequences are 99.6% (n) and 99.7% (p) identical.

gal1   One sequence is partial.

gal2   The sequences are 99.8% (n) and 99.8% (p) identical.

gal83   The sequences are 99.9% (n) and 100% (p) identical.

gas1   The sequences are named *GAS1* (scgas1, accepted 24.9.90) with *GGP1* (scgpp1g, accepted 22.10.90) as synonym. The sequences are 99.9% (n) and 99.8% (p) identical.

gcn4   The sequences are 99.9% (n) and 96% (p) identical. The sequence scgcn4b is missing a nucleotide at coordinate 1675.

gcr1   The sequences are 99.9% (n) and 99.9% (p) identical.

gcy1   Number "1" is added to follow genetic nomenclature.

gdh1   The sequences are 99.9% (n) and 98.5% (p) identical. They differ by an insertion of a C at position 1297 and an AG

	pair at position 1316/1317. The two sequences differ considerably around coordinate 85 in respect to the AUG codon.
glc7	The sequences are 99.7% (n) and 100% (p) identical.
gpa1	The gene is *GPA1* (scgpa1, accepted 20.11.86) with *SCG1* as synonym (ysscg1a, accepted 10.7.87). The sequences are 99.4% (n) and 99.6% (p) identical.
gpm1	The sequences are identical. Number "1" is added to follow genetic nomenclature. The header in the EMBL file of scgpm does not include the ATG!
gsp1/gsp2	The sequences are 83.3% (n) and 99.6% (p) identical, but the flanking sequences are different, therefore the genes *GSP1* and *GSP2* are non-allelic.
gut2	Two different sequences are given for the glycerol-3-phosphate dehydrogenase gene *GUT2* (scgut and scgut2a). The sequence scgut2 is only partial. According to the publication of this sequence, the gene is mapped to Chr XII, whereas on the genetic map gene *GUT2* is mapped to Chr IX. The sequence scgut2a overlaps with the sequence scimp2 coding for *IMP2*, which has been mapped to Chr IX (Appendix II). The name gut2? is provisionally given to scgut2*.
hem15	The sequences are identical.
hta/htb/ hht/hhf	The histone genes are named according the following rules (Smith and Stirling, 1988; Smith, personal communication): H2A1 and H2A2 genes are designated *HTA1* and *HTA2*. H2B1 and H2B2 genes are designated *HTB1* and *HTB2*. H3I and H3II genes are designated *HHT1* and *HHT2* (H̲istone H̲ T̲hree). H4I and H4II genes are designated *HHF1* and *HHF2* (H̲istone H̲ F̲our).
hhf2	The histone sequences are from *Sacch. carlsbergensis* (sch4) and *Sacch. cerevisiae* (sch3h402-2). The *Sacch. carlsbergensis* sequence is 99% identical to *hhf2* from *Sacch. cerevisiae*; the protein is 100% identical.
hsf1	The sequences are 99.8% (n) and 97.7% (p) identical. The sequence schsf1a is missing nucleotide 1666, whereas nucleotide 1738 is missing in schsf. The sequence schsf lacks three codons at the end.

hsp26	The sequences are 99.5% (n) and 100% (p) identical.
hsp60	The sequences are 99.8% (n) and 100% (p) identical.
hsp150	The sequences are 99.5% (n) and 97% (p) identical. The sequence Schsp150a has a gap of two nucleotides between coordinates 527 and 528 and a gap of one nucleotide between coordinates 581 and 582.
hta1	The sequences are identical. One is partial.
hxk1	The sequences are named *HXK1* (schxk1, accepted 12.6.85) with *HKA* as synonym (schka, accepted 13.12.85). They are 98.6% (n) and 98.4% (p) identical. The schka has three nucleotides inserted at positions 1571–1573.
hxk2	The sequences are named *HXK2* (schxk2, accepted 12.6.85) with *HKB* as synonym (schkb, accepted 13.12.85). The sequences are 98.6% (n) and 98.4% (p) identical.
icl1	The sequences are identical.
ime1	The sequences are identical.
ino2	The sequences are identical.
ils1	The sequences are 99.4% (n) and 99.4% (p) identical. The sequence scitrsa has gaps at coordinates 1764, 1765, 1766.
kar2	The sequences are named *KAR2* (sckar2, accepted 24.5.89; sckar2a, accepted 17.3.89) with *GRP78* as synonym (scgrp78, accepted 28.7.89). The sequences are identical.
kem1	All sequences are identical. According to the rules the sequence should be named *XRN1*. However, the paper indicated in the EMBL header gives no sequence data nor an accession number. We therefore chose the name *KEM1* (sckem1g, accepted 20.8.90) with *XRN1* (scxrn1, accepted 13.7.90), *DST2* (scstpb, accepted 14.1.91), *SEP1* (scsep1, accepted 29.1.91), and *RAR5* (unpublished) as synonyms.
kin3	The gene is named *KIN3* (sckin3, accepted 15.2.90) with *NPK1* (scnpk1, accepted 5.3.92) as synonym. The sequences sckin3 and scnpk1 are 99.9% (n) and 100% (p) identical. The 5′ sequences are almost identical over 245 nucleotides but further away (5′) they differ completely suggesting a cloning artefact. The coding sequence of scnpk1 is longer due to a frameshift caused by the absence of a nucleotide between coordinates 4946 and 4947.

kre2   The two sequences are 98.7% (n) and 75.2% (p) identical.
                    The sequence sckre2 has gaps between coordinates 728 and
                    729, between coordinates 949 and 950, and 1163 and 1164,
                    whereas the sequence scmnt1 has gaps between coordinates
                    91 and 92, 117 and 118 (2 nucleotides missing), 299 and 300
                    (2 nucleotides missing), 384 and 385, 778 and 779, 909 and
                    910, leading to a shorter coding sequence in sckre2.

krs1   The sequences are 92.8% (n) and 98.5% (p) identical. The
                    identity in the flanking sequences is astonishingly low
                    (79.2% and 73% for the 5' and the 3' regions).

leu2   The sequences are 99.8% (n) and 100% (p) identical. One
                    sequence is partial.

leu3   The sequences are 99.96% (n) and 100% (p) identical.

lpd1   The sequence sclpdh (*LPD1*), which codes for lipoamide
                    dehydrogenase, is 99.9% identical to the sequence
                    scdhdha, coding also for lipoamide dehydrogenase. The 3'
                    sequences are 92.5% identical (80nt long), whereas the 5'
                    sequences are only 37.5% identical (534nt long) which is
                    due to cDNA cloning artefacts in scdhdha (K.S. Browning,
                    personal communication). Therefore the gene scdhdha is
                    also designated *LPD1*.

lys2   The sequences are identical. Sccmdlys is only partial.

mag1   The two sequences are 99.9% (n) and 100% (p) identical.

mal6s   The sequences are named *MAL6S* (scmal6st, accepted
                    24.10.86; cgm0180, accepted 16.4.88). The sequences are
                    99.6% (n) and 99.8% (p) identical.

mas2   Genetic nomenclature is followed (*MAS2*) despite the order
                    of acceptance dates (scmas2, accepted 6.9.88; miscmpp,
                    accepted 3.8.88) with *MPP* as synonym. The sequences are
                    99.9% (n) and 100% (p) identical.

mata1   The *MATa1* gene is subject to alternative splicing. The
                    MATa1 protein exists in two forms, a spliced version and
                    an incompletely spliced one which differ in their C-terminal
                    sequences. Both sequences are identical up to amino acid
                    residue 117 (Miller, 1984).

MATalpha1 and MATalpha2	The historical names are retained.
mck1	The sequences are identical.
mcm1	Although the gene should be named *FUN80* (fun = function unknown now) (scfun80, accepted 2.4.87) the more meaningful name *MCM1* (scmcm1, accepted 10.6.88; scprtf, accepted 7.11.89) is adopted. The sequences scmcm1 and scprtf are identical. The sequence scfun80 is 99.6% identical and has two gaps between coordinates 465 and 466 and between 508 and 509, leading to a shorter open reading frame.
mek1	The sequences are identical. According to the priority rule, the gene is named *MEK1* with *MRE4* as synonym. (In LISTA2 the sequence was named *MRE4*.)
met4	The sequences are 99.9% (n) and 100% (p) identical. The coding sequence of scmet4 begins at coordinate 1050. The sequence scmet4 has a gap between coordinates 3024 and 3025 leading to a longer open reading frame at the 3' end.
mel1	The sequences are identical.
mes1	The sequences are 99.9% (n) and 100% (p) identical.
met3	The sequences are identical.
mrs3	The sequences are 99.3% (n) and 98.4% (p) identical.
msw1	The two sequences are 99.6% (n) and 97.9% (p) identical. The sequence scmsw has a gap between coordinates 1106 and 1107 leading to a shorter open reading frame.
myo1	The sequences are 98.6% (n) and 96.6% (p) identical. One sequence is partial.
nam1	The gene is named *NAM1* (miscnam1, accepted 30.9.88) with *MTF2* (scmtf2a, accepted 28.6.89) as synonym. The two sequences are 99.9% (n) and 100% (p) identical.
nam2	The gene (scnam2, accepted 7.1.87) is named *NAM2* with *MSL1* as synonym (scmsl1, accepted 8.6.87). The sequences are identical.
nca1	The two sequences are 99.3% (n) and 95.2% (p) identical.

The sequence scaepx has a gap between coordinates 1302 and 1303 leading to a longer open reading frame.

nhp6a/nph6b These names do not follow the genetic nomenclature rules, but, like ribosomal protein genes, names followed by letters are used.

nop1 The sequences are identical.

nop3 The sequences are 99.9% (scnop3/scnp13) and 100% (scnop3/scmts1) (n) and 100% identical (p). The gene is named *NOP3* (scnop3, accepted 10.6.92) with *MTS1* (unpublished) and *NPL3* (unpublished) as synonyms.

nup49 The gene (scnup49, accepted 3.8.92) is named *NUP49* with *NSP49* as synonym (scnsp49, accepted 8.10.92). The two sequences are identical. The two gene designations *NUP* and *NSP* are used for nuclear pore proteins and nucleoskeletal proteins.

nup116 The gene (scnup116, accepted 3.8.92) is named *NUP116* with *NSP116* as synonym (scnsp116, accepted 8.10.92). The genes are 99.6% (n) and 99.6% (p) identical. According to the priority rule the gene is named *NUP116* with *NSP116* as synonym. The two gene designations *NUP* and *NSP* are used for nuclear pore proteins and nucleoskeletal proteins.

ord1 Number "1" is added to follow the genetic nomenclature.

pab1 The sequences are 99.9% (scpabpg/scsygp1; scpabpg/scmrnp) (n), 100% (scmrnp/scsygp1) (n) and 99.8% and 100% (p) identical. The designation *MRNP* is given as synonym.

pak3 The sequences are 99.9% (n) and 100% (p) identical.

pdb1 According to the nomenclature of pyruvate dehydrogenase beta-subunit the gene is named *PDB1*.

pdi1 The sequences are named *PDI1* with *TRG1* and *MFP1* as synonyms. The sequences are 98.9% (sctrg1/scpdiaa, scmfp1/scpdiaa), 99.2% (sctrg1/scpdi1, scmfp1/scpdi1), 99.5% (scpdiaa/scpdi1), 100% (sctrg1/scmfp1) (n) and 95.4%, 98.9%, 96.2% and 100% (p) identical. The sequences scmfp1, sctrg1 and scpdi1 have a nucleotide at coordinate 97 which is missing in scpdiaa and a gap of one nucleotide between coordinates 155 and 156. A gap of three nucleotides between coordinates 425 and 426 is present in the sequences sctrg1 and scmfp1. An insert of 24 nucleotides from nucleotides 1513 to 1536 is present in the sequences sctrg1, scmfp1 and scpdi1.

pdr4
The name of this gene coding for the yeast transcription factor AP1 is in conflict with Appendix II in which *YAP* stands for yeast aspartyl protease. The name *PDR4* suggested by J. Lenard and M. Hussain is used with *YAP1* as synonym. The sequences are 99.6% (n) and 98.3% (p) identical.

pfy1
Sequence encoding yeast profilin. Number "1" is added to follow genetic nomenclature. According to Appendix II this gene is named *PFY*. The two sequences are identical. The sequence scproft has been derived from a cDNA. The sequence M23369 contains an intron. Although the sequence scprofr has been published first, we adopt the genetic name given in Appendix II.

pgi1
The sequences are identical.

pgk1
One sequence is partial. Number "1" is added to follow genetic nomenclature.

pho5
The sequences are 98.2% (scpho5/scpho5a; scpho35-1/scpho5) and 100% (scpho35-1/scpho5a) identical; the proteins are 99.1% and 100% identical. One sequence contains only the open reading frame.

pho80
One sequence is partial.

pho81
The mnemonic scpho81 uses the digit 0 and not the letter O.

phr1
The sequences are 99.5% (n) and 98.9% (p) identical.

pis1
Number "1" is added to follow genetic nomenclature.

plc1
The sequences are 99.9% (n) and 99.6% (p) identical.

pmr2
The sequences are identical. Scpmr2g is partial.

ppc1
The genetic name *PPC1* for phosphoenol-pyruvatecarboxykinase is adopted.

pph1
The sequence named *sit4* (M24395) is renamed *PPH1* (Arndt *et al.*, 1989).

pph22   The sequences are identical.

pre2   The gene is named *PRE2* (scpre2a, accepted 15.7.92) with *PRG1* (scproteo, accepted 22.7.92). The sequence scproteo has a gap between coordinates 842 and 843 leading to a longer reading frame. The proteins are 97.9% identical.

prps1   Since the gene symbol *PRP* is generally used for pre mRNA processing the sequences encoding phosphoribosyl pyrophosphate synthetase are named *PRPS1* with *PRP1* as synonym. The sequences are identical.

ppr2   The sequences are named *PPR2* (scppr2, accepted 29.7.83) with *DST1* (scstpa, accepted 24.1.91) as synonym. The sequence ppr2 misses two nucleotides between coordinates 272 and 273 and is therefore shorter. These differences are due to errors in the sequence (F. Lacroute, personal communication).

prp2   The sequences are identical.

prp4   The sequences are identical.

prs1   The sequence coding for subunit YC1 of the proteasome has been named *PRS1* according to Lee *et al.* (1991).

ptm1   The sequences are 99.9% (n) and 98.3% (p) identical. The sequence scxi12 has a gap between coordinates 1536 and 1537 leading to a longer reading frame.

ptp2   The sequences are 99.9% (n) and 99.7% (p) identical.

pub1   The sequences are 99.5% (n) and 99.5% (p) identical. The sequence scpub1x has a gap between coordinates 1281 and 1282, scrnabind has a gap between coordinates 1345 and 1346, leading to a difference in the length of the reading frames.

pyk1   The sequences are 99.8% (n) and 99.2% (p) identical. The scpko1 sequence has deletions at nucleotides 1143, 1156 and 1157 compared to scpyk1. This also results in a stretch of 5 different amino acids.

qcr2   According to the proposal of L. Grivell the name *QCR2* (subunit 2 ubiquinol:cytochrome *c* oxidoreductase) is adopted.

qcr6   According to the proposal of L. Grivell the name *QCR6* (subunit 6 ubiquinol:cytochrome *c* oxidoreductase) is adopted.

rad1
Deletion of nucleotide 3580 in scrad1 leads to a stop at nucleotide 3884. Scrad1a has deletions at positions 3207 and 3280 in the alignment with the other sequence. The sequences are 99.9% (n) and 97.4% (p) identical.

rad3
The sequences are identical.

rad4
The sequences are 99.9% (scrad4a/scsygp1; scrad4b/scsygp1), 100% (scrad4a/scrad4b) (n), 99.7% and 100% (p) identical. The fourth sequence (scrad4g) is different in length and does not align with the other sequences. The authors do not present genetic evidence that the cloned gene codes for *RAD4*. The sequence scrad4g overlaps with the sequence scret1. A gene *RET1* has been mapped to ChrXV, whereas *RAD4* has been mapped to ChrV. Therefore, the sequence scrad4g most probably does not correspond to the gene *RAD4* and is provisionally named *RAD4?* with ??? as synonym. Since the sequence does not resemble another entry in the EMBL library, no other name could be attributed.

rad4?
see rad4

rad6
The sequences are identical.

rad7
The sequences are 99.5% (n) and 99.7% (p) identical.

rad10
The sequences are identical.

rad16
The sequences are 99.9% (n) and 100% (p) identical. The sequence scrad16a is partial.

rad18
The sequences are identical.

rad51
The sequences are identical.

radh1
Number "1" is added to conform to genetic nomenclature.

rar1
The sequences are named *RAR1* (scrar12, accepted 22.6.87) with *ERG12* (scerg12g, accepted 6.6.91) as synonym. The sequences are identical.

ras2
The sequences are 99.2% (n) and 98.5% (p) identical.

rat1
The sequences are identical.

reg1

The sequences are named *REG1* (screg1, accepted 1.12.86) with *BCY1* (sccapk, accepted 19.12.86) and *SRA1* (scsra1, accepted 1.5.87) as synonyms. The nucleotide sequences are 99.9% and 99.8% (screg1/sccapk) and the proteins 100% identical.

rfa1

The sequences are identical.

rnr2

The sequences are 99.1% (n) and 98% (p) identical. The two sequences are different from coordinates 300 to 324 of the coding sequence. Interestingly, within this region 15 nucleotides are identical in the two sequences, but displaced in respect to the rest of the sequence.

rp28a/rp28b

The sequences are 95.2% identical, but the flanking sequences are different. Therefore the genes *RP28A* and *RP28B* are non-allelic (Table III). The sequences scrp28a1 and scrp28e1 (*RP28A*) are identical.

rp51a/rp51b

The sequences are 95.3% identical, whereas the flanking sequences are different. Therefore the genes *RP51A* and *RP51B* are non-allelic (Table III).

rp55a/rp55b

The sequences are 95.1% identical, whereas the flanking sequences are different. Therefore the genes *RP55A* and *RP55B* are non-allelic (Table III).

rp59

The sequences are named *RP59* (cgm0284, accepted 9.10.84; sccry1, accepted 12.2.87). They are identical.

rpa0

The sequences are 99.9% (scrpa0/scl10e01; scarpa0/scl10e01) and 100% (scrpa0/scarpa0) identical and the proteins are 99.7%, 99.7% and 100% identical.

rpa1

The sequences are 99.7% (scrpa1/scl12e01; sca1/scl10e01) and 100% (scrpa1/sca1) identical and the proteins 99.1% and 100% identical.

rpa2

The sequences coding for the ribosomal protein L44 are identical and named *RPA2* (scrpa2, accepted 18.3.88) with *L12EIB* (scl12ib, accepted 20.10.89) and *RPL44* (scrpgap44 accepted 16.11.87) as synonyms.

rpa135/
rpa190

These genes code not for ribosomal proteins, but for subunits of RNA polymerase A.

rpb

These genes code not for ribosomal proteins, but for subunits of RNA polymerase B (II).

rpb4	The sequences are identical coding for RNA polymerase B (II) subunit 4.
rpl1	The sequences are 99.9% (n) and 100% (p) identical.
rpl4a/rpl4b	The coding sequences are 94.8% identical, whereas the flanking sequences are different. Therefore the genes *RPL4A* and *RPL4B* are non-allelic (Table III).
rpl15a/ rpl15b	The coding sequences are 91.2% identical, whereas flanking regions are different. Therefore the genes *RPL15A* and *RPL15B* are non-allelic (Table III).
rpl16a/ rpl16b	The coding sequences are 96.5% identical, whereas flanking sequences are different. Therefore the genes *RPL16A* and *RPL16B* are non-allelic.
rpl41a/ rpl41b	The coding sequences are 98.7% identical, whereas flanking regions are different. Therefore the genes *RPL41A* and *RPL41B* are non-allelic (Table III).
rpl44	The sequences are identical.
rpl45	The sequences are identical.
rpo41	This gene encodes a mitochondrial RNA polymerase.
rps25	The sequence is named *RPS25* for ribosomal protein S25, and not YS25 as indicated in the header of scrpys25.
sch9	The sequences are 99.9% (n) and 99.8% (p) identical.
sch9uORF	This is an upstream reading frame of 162 nucleotides in the 5′ region of the gene *SCH9*.
scl1	The sequences are named *SCL1* (scscl1a, accepted 16.5.89) with *PRS2* as synonym (ssproty8, accepted 25.10.90 (Lee *et al.*, 1991); scyc7a, accepted 16.4.90). The sequences sccyc7a and ssproty8 are 100% identical, and share 99.9% identity with scscl1a. This coding sequence is longer at the 5′ end due to a frameshift caused by missing nucleotides between coordinates 606 and 607.
sdh1	The sequence coding for succinate dehydrogenase has been named *SDH1* according to Appendix II. This nomenclature is in conflict with the gene name given in the sequence of scsdh1 coding for serine dehydratase (see under sed1).

sed1   The sequence has been named *SED1* to avoid conflict with
                  SDH1 coding for succinate dehydrogenase.

ses1   Although the gene designation given in the header of the
                  EMBL file is serS for seryl-tRNA synthetase, the genetic
                  nomenclature here follows the name given in Appendix II.

sga1   The sequence is named *SGA1* for a sporulation-specific
                  glucoamylase (SGA) gene. SGA stands for intracellular
                  glucoamylase (this sequence) and should not be confused
                  with Suppression of Growth Arrest of *cdc25*.

slp1   The sequences are named *SLP1* (scslp1a, accepted 23.1.90)
                  with *VSP33* (scvps33p, accepted 15.6.90) as synonym. They
                  are 100% identical.

sly2   The sequences are named *SLY2* (scsly2, accepted 20.10.90)
                  standing for suppressor of loss of *YPT1* rather than *TSL26*
                  (cgm0259, accepted 19.6.89) standing for temperature sen-
                  sitive lethal. The sequences are 99.7% (n) and 100% (p)
                  identical.

snf2   The sequences are named *SNF2* (scsnf2a, accepted
                  31.12.90) with *GAM1* (scsnf2, accepted 8.2.91) as synonym.
                  The sequences are identical.

sod1/sod2   The gene encoding the manganese superoxide dismutase has
                  been named *SOD1* and genetically mapped (van Loon *et al.*,
                  1986). Unfortunately the same name is given to the copper,
                  zinc-superoxide dismutase according to the first eukaryotic
                  SOD gene cloned (for review see Gralla and Kosman, 1992)
                  and the gene encoding manganese superoxide dismutase has
                  been renamed *SOD2*. This is the currently used nomen-
                  clature in this list.

spp91   The gene (scspp91a, accepted 28.4.92) is named *SPP91* with
                  PRP21 as synonym (scprp21a, unpublished). The sequences
                  are 99.9% (n) and 99.6% (p) identical. According to our
                  priority rule we adopt *SPP91* as genetic designation and
                  give *PRP21* as synonym.

spt2   The sequences are identical.

spt15   The sequences are named *SPT15* since none of the names
                  follows the nomenclature rules but the identity to *SPT15*
                  has been reported in Hahn *et al.* (1989) (sctfii01, accepted

22.6.89; sctfiida, accepted 24.7.89; sctfiid, accepted 30.8.89; scbtfly, accepted 2.10.89). They are 99.3% (scvtfly/sctfii01; sctfii01/sctfiid; sctfii01/sctfiida) and 100% (p) identical.

srp54   The sequences are named *SRP54* (scsrp4sc, accepted 4.10.89) with *SRH1* as synonym (scsrh1, accepted 23.10.89). The sequences are 99.9% (n) and 99.8% (p) identical.

srv2   The sequences are identical. They encode adenylyl cyclase associated protein. Although the sequence sccap (7.2.90) has been accepted prior to the sequence scsrv2 (9.2.90) we use the name *SRV2* since the name *CAP1* (sccap1g) has already been used for the capping enzyme (see also Appendix II).

ssb1/ssbr1   The name *SSB1* is used for two different genes. One designates heat shock cognate protein. These sequences are 99.95% identical and the proteins 100% identical. The same name is also used for a single stranded binding protein (scssb1). In order to avoid confusion in the present list we *provisionally* name the scssb1 sequence *SSBR1* for single strand binding protein carrying an RNA binding motif (Jong *et al.*, 1987) and added *SSB1* as synonym.

ssc1   The sequences are named *SSC1* (m27229, accepted 5.4.89) with *ENS1* (scens1, accepted 11.12.89) as synonym. The sequences are 99.6% (n) and 100% (p) identical. The coding sequence of *ENS1* is three nucleotides longer due to insertion of AAC at coordinates 2029–2031 which results in an additional asparagine in a run of eight.

ssd1   The sequences are named *SSD1* (scsit4b, accepted 24.1.91) with *SRK1* as synonym (scsrk1a, accepted 13.3.91). The sequences are identical.

ssl2   The gene (scssl2a, accepted 2.4.92) is named *SSL2* with *RAD25* as synonym (scrad25a, accepted 4.9.92). The two sequences are 99.6% (n) and 99.8% (p) identical. According to the priority rule the gene is named *SSL2* with *RAD25* as synonym.

ssn6   The sequences are named *SSN6* (scssn6a, accepted 22.7.87) with *CYC8* as synonym (sccyc8, accepted 5.8.88). The sequences are identical.

ste2	The sequences are identical.
ste6	The sequences are 99.97% (scste6a/scste6g) and 100% (scste6pr/scte6g) identical and the proteins 100% identical. The sequence scste6pr is partial.
ste12	The sequences are identical.
sth1	The gene (scsth1a, accepted 28.1.92) is named *STH1* with *NPS1* as synonym (scnps1, accepted 3.8.92). The two sequences are 99.9% (n) and 99.3% (p) identical. The sequence scsth1a has a gap of 21 nucleotides between co-ordinates 3214 and 3215 leading to a shorter reading frame. According to the priority rule the gene is named *STH1* with *NPS1* as synonym.
suc3/suc5	These genes are more than 99% identical (even in flanking sequences) but according to the authors they correspond to different non-allelic genes located on different chromosomes (Hohmann and Zimmermann, 1986).
suc7	Scsuc7g2 is the 3′ region and scsuc78g1 is the 5′ region.
suf12	The sequences are named *SUF12* (scsuf12, accepted 24.8.87) with *SUP2* (scsup2, accepted 25.2.88) and *GST1* (scgst1, accepted 22.12.87) as synonyms. The sequences are 99.95% (scsuf12/scgst1) and 100% (scsup2/scsuf12) identical and the proteins 100% identical.
sug1	The sequences are 99.9% (n) and 99.8% (p) identical.
swi3	The sequences are identical.
tal1	The sequences are identical.
tef1/tef2	The proteins are identical, whereas the flanking sequences are different. Therefore the genes *TEF1* and *TEF2* are non-allelic (Table III).
tef2	The sequences SCTEF1 and SCEF1ABB are identical. According to the literature they code for *TEF2* (Cottrelle *et al.*, 1985) and not for *TEF1* as indicated in the EMBL data library. The names given by Nagata *et al.* (1984) (EF1αA and EF1αB) were changed to *TEF1* and *TEF2* according to the nomenclature rules.
tdh1/tdh2/ tdh3	The sequences have been named *TDH1*, *TDH2* and *TDH3* in accordance with the currently used genetic nomenclature

(McAlister and Holland (1985); C. Koch, personal communication) to avoid conflict with the name *gap1* (scgap1p) coding for the general amino acid permease. The sequence scg3pda is 99.3% identical to *TDH1* but only 93% identical to *TDH2*. Since no flanking sequences are available for *TDH1* and *TDH2* it is difficult to say with which it is allelic, but according to the high identity to sequence scgap1, it is assumed that it is *TDH1*.

tdh1   The sequences are 99.5% (n) and 100% (p) identical.

tdh2   The sequences are 99.7% (n) and 98.2% (p) identical.

tfp1   The sequences are named *TFP1* (sctfp1, accepted 20.4.88) with *VMA1* as synonym (schatp, accepted 10.10.89). The sequences are 99.9% (n) and 99.9% (p) identical.

thr1   The sequences are identical. The sequence scthr1 has three nucleotides missing between coordinates 254 and 255.

tif1/tif2   The sequences are 99.6% identical and the proteins are 100% identical, but the flanking sequences diverge completely, and the genes map on different chromosomes (Mueller *et al.*, 1989). Thus *TIF1* and *TIF2* are different, non-allelic genes. Scyur1 is a partial sequence of *TIF2*. The sequences sctif2 and scyur1 are 100% identical. Scyur1 is partial.

tif45/cdc33   In accordance with M. Altmann it is proposed to use *TIF45* as a name to replace eIF-4E. *CDC33* is synonym. The sequences are identical.

tif51a   The sequences are named *TIF51A* (sctif51a, accepted 23.2.91) with *HYP2* as synonym (schyp2, unpublished). The sequences are identical.

tpk1   The sequences are identical. The name of the sequence scpk25 (scpk25, accepted 22.8.86) has been changed to *TPK1* (sctpka, accepted 17.4.87) in accordance with H. Küntzel (Lisziewicz *et al.*, 1987; Toda *et al.*, 1987).

tpk2   The sequences are 99.8% (n) and 98.2% (p) identical.

tpk3   The sequences are identical. Scmrp49a is partial. Two conflicts have been noticed in the 3' region of the *TPK3* gene.

trp3	The sequences are 99.1% (n) and 99.9% (p) identical. The sequence scatrp3a is partial.
trp5	The sequences are 99.8% (n) and 100% (p) identical. The sequence sctrp5b is partial.
trx1	The sequences are named *TRX1* in accordance with genetic nomenclature. The name TR-II is given as synonym. The sequences are identical.
tuf1	The gene name for the mitochondrial elongation factor is *TUF* and the sequence is therefore called *TUF1*.
tup1	The sequences have been named *TUP1* according to the indication in the sequence scsfl2 (scsfl2, accepted 4.1.90; sctup1a, accepted 28.8.90; scaer2, accepted 29.8.90). The sequences are 99.6% (sctup1a/scaer2; sctup1a/scsfl2) and 100% (scaer2/scsfl2) identical and the proteins 99.6% and 100% identical. The sequence scsfl2 is missing a nucleotide between coordinates 93 and 94, which leads to a shorter open reading frame.
uga35	The sequences are named *UGA35* (scuga35, accepted 29.9.90) with *DAL81* as synonym (scdal81, accepted 25.10.90). The sequences are 99.7% (n) and 99.7% (p) identical. The sequence scug35 is missing 18 nucleotides between coordinates 631 and 632 in respect to scdal81. This results in six missing amino acids in a row of glutamines.
ura2	The sequences are 98.4% (n) and 97.8% (p) identical. The sequence scura2 is partial.
ura3	The sequences are 97.3% (n) and 97.2% (p) identical. The sequence scura3p is partial.
ura6	The sequences are identical.
vps1	The sequences are named *VPS1* (scvps1a, accepted 2.4.90) with *SPO15* as synonym since in the entry scspo15g the only reference given is the same as the one for *VPS1*. The sequences are 99.8% (n) and 99.7% (p) identical.
yef3	The sequences are identical.
zwf1	The sequences are named *ZWF1* (scg6pd, accepted 18.6.90) with *MET19* as synonym (scmet19, accepted 16.11.90). The sequences are 99.9% (n) and 100% (p) identical.

ACKNOWLEDGEMENTS

The authors would like to thank all the persons who have contributed sequences or helped in a collaborative manner to resolve the nomenclature problems. We are grateful to S. Brouillet, A. Henaut, J.L. Risler from Centre de Génétique Moléculaire for help in the analysis and to the Rechenzentrum of the University of Basel for maintenance of computer facilities. The work was supported by grants from the Ministère de la Recherche et de l'Espace (program GREG) (to P.S.) and grants from the Swiss National Science Foundation and the University of Basel (to P.L. and R.D.).

REFERENCES

Adrian, G.S., McCammon, M.T., Montgomery, D.L. and Douglas, M.G. (1986). *Molecular and Cell Biology* 6, 626.
Arndt, K.T., Styles, C.A. and Fink, G.R. (1989). *Cell* 56, 527.
Barnett, J.A., Payne, R.W. and Yarrow, D. (1983). "Yeasts: Characteristics and Identification". Cambridge University Press, Cambridge
Beck, J.C., Mattoon, J.R., Hawthorne, D.C. and Sherman, F. (1968). *Proceedings of the National Academy of Sciences of the United States of America* 60, 186.
Bennetzen, J.L. and Hall, B.D. (1982). *Journal of Biological Chemistry* 257, 3026.
Cottrelle, P., Cool, M., Thuriaux, P., Price, V.L., Thiele, D., Buhler, J.M. and Fromageot, P. (1985). *Current Genetics* 9, 693.
Degols, G., Jaunniaux, J.C. and Wiame, J.M. (1987). *European Journal of Biochemistry* 165, 289.
Downie, J.A., Stewart, J.W., Brockman, N., Schweingruber, A.M. and Sherman, F. (1977). *Journal of Molecular Biology* 113, 369.
Etzold, T. and Argos, P. (1993). *CABIOS* 9, 49.
Gralla, E.B. and Kosman, D.J. (1992). *Advances in Genetics* 30, 251.
Hahn, S., Buratowski, S., Sharp, P.A. and Guarente, L. (1989). *Cell* 58, 1173.
Heitman, J., Movva, N.R., Hiestand, P.C. and Hall, M.N. (1991). *Proceedings of the National Academy of Sciences of the United States of America* 88, 1948.
Herbert, C.J., Dujardin, G., Labouesse, M. and Slonimski, P.P. (1988). *Molecular and General Genetics* 213, 297.
Herbert, C.J., Macadré, C., Becam, A.M., Lazowska, J. and Slonimski, P.P. (1992). *Gene Expression* 2, 203.
Hohmann, S. and Zimmermann, F.K. (1986). *Current Genetics* 11, 217.
Jong, A.Y., Clark, M.W., Gilbert, M., Oehm, A. and Campbell, J.L. (1987). *Molecular and Cell Biology* 7, 2947.
Kolarov, J., Kolarova, N. and Nelson, N. (1990). *Journal of Biological Chemistry* 256, 12711.
Kovac, L., Lachowicz, T.M. and Slonimski, P.P. (1967). *Science* 158, 1564.
Lawson, J.E. and Douglas, M.G. (1988). *Journal of Biological Chemistry* 263, 14812.
Lee, D.H., Tamura, T., Chung, C.H., Tanaka, K. and Ichihara, A. (1991). *Biochemistry International* 23, 689.
Lisziewicz, J., Godany, A., Forster, H.H. and Kuntzel, H. (1987). *Journal of Biological Chemistry* 262, 2549.
McAlister, L. and Holland, M.J. (1985). *Journal of Biological Chemistry* 260, 15013.
Meyer, S.L., Kvalnes-Krick, K.L. and Schramm, V.L. (1989). *Biochemistry* 28, 8734.

Miller, A.M. (1984). *EMBO Journal* 3, 1061.
Montgomery, D.L., Leung, D.W., Smith, M., Shalit, P., Faye, G. and Hall, B.D. (1980). *Proceedings of the National Academy of Sciences of the United States of America* 77, 541.
Mossé, M.-O., Brouillet, S., Risler, J.L., Lazowska, J. and Slonimski, P.P. (1988). *Current Genetics* 14, 529.
Mossé, M.-O., Linder, P., Lazowska, J.and Slonimski, P.P. (1993). *Current Genetics* 23, 66.
Mueller, P.P., Trachsel, H. and Linder, P. (1989). *Current Genetics* 16, 127.
Nagata, S., Nagashiman, K., Tsunetsugu-Yokota, Y., Fujimura, K., Miyazaki, M. and Kaziro, Y. (1984). *EMBO Journal* 3, 1825.
Oliver, S.G. *et al.* (1992). *Nature* 357, 38.
Perentesis, J.P., Phan, L.D., Gleason, W.B., LaPorte, D.C., Livingston, D.M. and Bodley, J.W. (1991). *Journal of Biological Chemistry* 267, 1190.
Sherman, F. and Slonimski, P.P. (1963). *Biochimica et Biophysica* Acta 90, 1.
Smith, M.M. and Stirling, V.B. (1988). *Journal of Cell Biology* 106, 557.
Toda, T., Cameron, S., Sass, P., Zoller, M. and Wigler, M. (1987). *Cell* 50, 277.
van Loon, A.P.G.M., Pesold-Hurt, B. and Schatz, G. (1986). *Proceedings of the National Academy of Sciences of the United States of America* 83, 3820.
Verdiere, J. and Petrochilo, E. (1975). *Biochemical and Biophysical Research Communications* 67, 1451.

Note added in proof

Since the compilation of the 1400 nucleotide sequences in 1993, a new release list A4 with over 1900 sequences has been made and release list A5 will follow in the near future. In list A4 many nomenclature changes have been introduced to avoid names which are no longer in use.

LIST A3

Gene (Primary gene name)	Synonym	Mnemonic	Length	Codon bias	Reference	Accession
AAA1	NAT1	SCNAT	2562	0,194	EMBO J. 8:2067-2075 (1989).	X15135
AAA1		SCNACT	2562	0,194	J. BIOL. CHEM. 264:12339-12343 (1989).	M23166
AAC1		SCPET9	927	0,173	MOL. CELL. BIOL. 6:626-634 (1986).	M12514
AAC1*		SCBUB2Q-1	135	0,219	CELL 66:507-517 (1991).	M64706
AAC2		SCAA2	954	0,691	J. BIOL. CHEM. 263:14812-14818 (1988).	J04021
AAC2		SCAA3	954	0,622	J. BIOL. CHEM. 265:12711-12716 (1990).	M34075
AAC3		SCAAC2A	921	0,259	J. BIOL. CHEM. 265:12711-12716 (1990).	M34076
AAR2		SCAAR2	1065	0,04	MOL. CELL. BIOL. 10:3262-3267 (1990).	D90455
AAT1		SCAAT1A	1353	0,083	BIOCHIM. BIOPHYS. ACTA 1171:211-214 (1992).	X68052
ABC1		SCABC1G	1503	0,062	EMBO J. 10:2023-2031 (1991).	X59027
ABD1		SCABD1A	1308	0,093	UNPUBLISHED.	L12000
ABF1					see BAF1	
ABP1		SCABP1	1776	0,291	NATURE 343:288-290 (1990).	X51780
ACB1		SCACB	261	0,542	PNAS 89:11287-11291 (1992).	M99489
ACE1		SCACE1A	675	0,132	MOL. CELL. BIOL. 9:421-429 (1989).	M24390
ACE1		SCACEA	675	0,132	CELL 55:705-717 (1988).	M22580
ACE2		SCACE2	2310	0,023	MOL. CELL. BIOL. 11:476-485 (1990).	M55619
ACH1		SC114-3	1578	0,327	YEAST 8:769-776 (1992).	X68577
ACH1		SCACH1A	1578	0,32	J. BIOL. CHEM. 265:7413-7418 (1990).	M31036
ACO1		SCACO1A	2337	0,565	MOL. CELL. BIOL. 10:3551-3561 (1990).	M33131
ACP2	RPC31	SCRPC31G	753	0,056	MOL. CELL. BIOL. 10:4737-4743 (1990).	X51498
ACP2		SCACP2	753	0,053	MOL. CELL. BIOL. 8:1282-1289 (1988).	M20315
ACT		SCACT	1125	0,825	NUCLEIC ACIDS RES. 8:1043-1059 (1980).	L00026
ACT1		SCACT1	1125	0,825	PNAS 77:2546-2550 (1980).	V01288
ACT1		SCACT2	1125	0,778	PNAS 77:3912-3916 (1980).	V01289
ACT1		SCACTI	1125	0,825	J. MOL. APPL. GEN. 1:239-244 (1981).	V01290
ACT2		SCACT2G	1173	0,272	NATURE 355:179-182 (1992).	X61502
ADA2		SCADA2X-1	1302	0,01	CELL 70:251-265 (1992).	M95396
ADC1					see ADH1	

Continued

LIST A3 Continued

Gene (Primary gene name)	Synonym	Mnemonic	Length	Codon bias	Reference	Accession
ADE1		SCADE1B-2	918	0,306	NUCLEIC ACIDS RES. 19:5731–5738 (1991).	M67445
ADE2		SCADE2	1713	0,221	GENE 95:91–98 (1990).	M59824
ADE2		SCADEAA	1710	0,217	YEAST 8:253–259 (1992).	M58324
ADE3		SCADE3	2838	0,423	J. BIOL. CHEM. 261:4629–4637 (1986).	M12878
ADE4		SCADE4GEN	1530	0,285	J. BIOL. CHEM. 266:20453–20456 (1991).	M74309
ADE 5, 7		SCADE57	2046	0,387	J. MOL. BIOL. 190:519–528 (1986).	X04337
ADE8		CGM0163	642	0,104	MOL. CELL. BIOL. 8:1253–1258 (1988).	
ADE8		SCADE8	642	0,104	NATURE 315:350–352 (1985).	M36585
ADH1	ADC1	SCADHC1	1044	0,913	BASIC LIFE SCI. 19:335–361 (1982).	M38456
ADH1		SCADHI	1044	0,913	J. BIOL. CHEM. 257:3018–3025 (1982).	V01292
ADH2	ADR2	SCADHR2	1044	0,711	BASIC LIFE SCI. 19:335–361 (1982).	M38457
ADH2	ADR2	SCADR2-1	1044	0,706	J. BIOL. CHEM. 258:2674–2682 (1983).	V01293
ADH3		SCADH3	1125	0,402	UNPUBLISHED.	K03292
ADH4		SCADH4	1146	0,663	MOL. GEN. GENET. 209:374–381 (1987).	X05992
ADK					see AKY1	
ADK1					see AKY1	
ADK2					see PAK3	
ADR1		SCADR1	3969	0,075	NATURE 320:283–287 (1986).	X03763
ADR2					see ADH2	
ADR6		SCADR6	3942	0,091	NUCLEIC ACIDS RES. 16:10153–10170 (1988).	X12493
AEP1					see NCA1	
AEP2					see ATP13	
AER2					see TUP1	
AFG1		SCAFG	1131	0,083	YEAST 8:787–792 (1992).	M94535
AFG2		SCAFG2A	2340	0,085	UNPUBLISHED.	L14615
AGA1		SCAAGLCS	2175	0,16	MOL. CELL. BIOL. 11:4196–4206 (1991).	M60590
AGA2		SCAGA2MR	261	≤0	EMBO J. 10:4081–4088 (1991).	X62877
AGAL1	AGALPHA1	SCAGA1A	1950	0,071	MOL. CELL. BIOL. 9:3155–3165 (1989).	M28164
AGAL1	AGALPHA1	SCAGALPHA	1950	0,07	FEBS LETT. 255:290–294 (1989).	X16861
AGALPHA1					see AGAL1	
AHT1		SCAHT1	546	0,007	UNPUBLISHED.	X59464

Gene	Synonym	Locus	Length		Reference	Accession
AKY1	ADK	SCADK	666	0,639	NUCLEIC ACIDS RES. 15:7187 (1987).	Y00413
AKY1	ADK1	SCADK1-2	666	0,656	J. BIOL. CHEM. 265:9952-9959 (1990).	M18455
AKY1		SCAKY	666	0,656	CURR. GENET. 12:405-411 (1987).	X06304
ALDH1		SCALDHAA	1599	0,183	J. BACTERIOL. 173:3199-3208 (1991).	M57887
ALDH2		SCALDDEH	1553	0,981	UNPUBLISHED.	Z17314
ALG1		SCALG1	1347	0,063	J. BIOL. CHEM. 265:7042-7049 (1990).	J05416
AMD1		SCAMD	2430	0,178	BIOCHEMISTRY 28:8734-8743 (1989).	M30449
AMDY1		SCAMDY	1647	0,084	NUCLEIC ACIDS RES. 18:7180 (1990).	X56043
AMS1		SCAMS1B	3249	0,126	BBRC 163:908-915 (1989).	M29146
ANB1	HYP1	SCHYP1	471	0,846	UNPUBLISHED.	X56235
ANB1		SCANBI	471	0,846	J. BIOL. CHEM. 265:8802-8807 (1990).	J05455
	TIF51B	SCTIF51B	471	0,833	MOL. CELL. BIOL. 11:3105-3114 (1991).	M63542
APA1	DTP1	SCDTPA	963	0,425	GENE 95:79-84 (1990).	M35204
APA1		SCAPA1	963	0,425	J. BACTERIOL. 171:6437-6445 (1989).	M31791
APA2		SCAPA2	975	0,091	J. BACTERIOL. 172:6892-6899 (1990).	M60265
APE1		SCAPE1	1542	0,245	FEBS LETT. 259:125-129 (1989).	Y07522
APE2	LAP1	SCAPE2G	2583	0,234	EUR. J. BIOCHEM. 202:993-1002 (1991).	X63998
APN1		SCAPN1	1101	0,061	PNAS 87:4193-4197 (1990).	M33667
APS1		SCAPSG	1671	0,495	NUCLEIC ACIDS RES. 14:1657-1666 (1986).	X03606
APS1		SCATS	1671	0,495	NUCLEIC ACIDS RES. 16:1212 (1988).	X06665
ARD1		SCARD1	714	0,155	CELL 43:483-492 (1985).	M11621
ARF1		SCARFA	543	0,679	PNAS 85:4620-4624 (1988).	J03276
ARF2		SCARF2	543	0,554	MOL. CELL. BIOL. 10:6690-6699 (1990).	M35158
ARG1		SCARGGA	1260	0,579	GENE 95:215-221 (1990).	M35237
ARG1*		SCARG1	171	0,548	CURR. GENET. 13:113-124 (1988).	X07070
ARG3		SCARG3	1014	0,181	MOL. CELL. BIOL. 5:3139-3148 (1985).	M11946
ARG3		SCARG30T	1014	0,181	EUR. J. BIOCHEM. 166:371-377 (1987).	M28301
ARG4		SCARG4A	1389	0,315	GENE 29:271-279 (1984).	K01813
ARG4*			529	0,253	NATURE 315:350-352 (1985).	M36586
ARG5, 6		SSCARG56	2589	0,221	MOL. GEN. GENET. 226:154-166 (1991).	X57017
ARG8		SCARG8	1269	0,222	GENE 90:69-78 (1990).	M32795
ARGR1		SCARGI	531	0,026	MOL. GEN. GENET. 207:142-148 (1987).	X05327
ARGR2		SCARGRII	2640	0,066	EUR. J. BIOCHEM. 157:77-81 (1986).	X03940
ARGR3		SCARGIII	1065	0,019	MOL. GEN. GENET. 207:142-148 (1987).	X05328

Continued

LIST A3 Continued

Gene (Primary gene name)	Synonym	Mnemonic	Length	Codon bias	Reference	Accession
ARO1		SCARO1	4764	0,249	BIOCHEM. J. 246:375–386 (1987).	X06077
ARO1*		SCARO1A	132	0,055	FEBS LETT. 241:83–88 (1988).	X13802
ARO1		SCARO1B	96	0,116	FEBS LETT. 241:83–88 (1988).	X13803
ARO2		SCAROA	1128	0,485	MOL. MICROBIOL. 5:2143–2152 (1991).	X60190
ARO3		SCARO3	1110	0,403	MOL. GEN. GENET. 214:165–169 (1988).	X13514
ARO4		SCARO4DAP	1101	0,69	GENE 113:67–74 (1992).	X61107
ARO7		SCARO7A	768	≤0	J. BACTERIOL. 171:1245–1253 (1989).	M24517
ASF1		SCASF	837	0,053	UNPUBLISHED.	L07593
ASF2		SCASF2A	1575	0,058	UNPUBLISHED.	L07649
ASP3		SCASP3A	801	0,18	J. BIOL. CHEM. 263:11948–11953 (1988).	J03926
ATE1		SCATE1	1509	0,046	J. BIOL. CHEM. 265:7464–7471 (1989).	J05404
ATP1		SCATPAMT	1632	0,619	J. BIOL. CHEM. 261:15126–15133 (1986).	J02603
ATP2		SCATP21	1533	0,552	J. BIOL. CHEM. 260:15458–15465 (1985).	M12082
ATP4		SCATP4	732	0,362	EUR. J. BIOCHEM. 170:637–642 (1988).	X06732
ATP5	OSCP	SCOSCP	636	0,303	J. BIOL. CHEM. 265:19047–19052 (1990).	M32487
ATP5		SCATP5	636	0,303	NUCLEIC ACIDS RES. 16:8181 (1988).	X12356
ATP7		SCATP7DA	522	0,246	J. BIOL. CHEM. 266:16541–16549 (1991).	M74048
ATP10		SCATP10-1	792	0,079	J. BIOL. CHEM. 265:9952–9959 (1990).	J05463
ATP11		SCATPMITO-2	954	0,103	J. BIOL. CHEM. 267:7386–7394 (1992).	M87006
ATP12		SCATP12	975	≤0	J. BIOL. CHEM. 266:7517–7523 (1991).	M61773
ATP13	AEP2	SCAEP2	1740	0,004	CURR. GENET. 20:53–61 (1991).	M59860
ATP13		SCATP13	1116	0,012	FEBS LETT. 278:234–238 (1991).	X56215
ATR1		SCATR	1641	0,052	MOL. CELL. BIOL. 8:664–673 (1988).	M20319
AUA1		SCAUA1	282	0,008	MOL. MICROBIOL. 8:167–178 (1993).	X69158
BAF1	ABF1	SCABF1A	2193	0,063	SCIENCE 246:1034–1038 (1989).	M29067
BAF1	OBF1	SCOBF1	2193	0,063	PNAS 88:4089–4093 (1991).	M63578
BAF1		SCBAF1	2193	0,069	EMBO J. 8:4265–4272 (1989).	X16385
BAR1		SCBAR1A	1761	0,038	PNAS 85:55–59 (1988).	J03573
BAS1		SCMYB	2433	0,025	SCIENCE 246:931–935 (1989).	M58057
BCK1		SCBCK1	4434	≤0	MOL. CELL. BIOL. 12:172–182 (1992).	X60227
BCY1					see REG1	
BEL2					see SIN4	

Gene	Alias	Sequence	Length	Value	Reference	Accession
BEM1		SCBEM1G	1653	0,089	NATURE 356:77–79 (1992).	X63826
BEM3*		SCBEMIII	321	0,125	UNPUBLISHED.	L14558
BGL2		SCGLUEXO	939	0,602	J. BACTERIOL. 171:6259–6254 (1989).	M31072
BIK1		SCFUSG-1	1320	0,098	MOL. CELL. BIOL. 7:2316–2328 (1987).	M16717
BMH1		SCBMH1	873	0,387	FEBS LETT. 302:145–150 (1992).	X66206
BMH1		SCPDA1-1	584	0,025	UNPUBLISHED.	X71664
BMH1		SCSYGP5-8	801	0,455	UNPUBLISHED.	L11229
BOS1		SCBOS1	732	0,157	J. CELL BIOL. 113:55–64 (1991).	X57792
BTF1					see SPT15	
BUB2		SCBUB2Q-2	918	0,009	CELL 66:507–517 (1991).	M64706
BUB3		SCBUB3Q	1023	0,039	CELL 66:507–517 (1991).	M64707
BUD5		SCBUD	1614	0,087	CELL 65:1213–1224 (1991).	M63552
BUD5		SCBUDIV	1611	0,09	CELL 65:1225–1231 (1991).	M68938
BUD5	YCR721	SCMATLOC-2	1611	0,089	YEAST 7:881–888 (1991).	M63853
CAD1		SCCAD1AA	927	0,022	YEAST 8: 253–259 (1992).	M58331
CAF20		SCCAP20	531	0,275	NUCLEIC ACIDS RES. 17:7520 (1989).	X15731
CAL1					see CHS3	
CAM1		SCCAM1G	1245	0,596	YEAST 9:151–163 (1993).	X67917
CAM1		SCEF1GA	1245	0,597	UNPUBLISHED.	L01879
CAN1		SCCAN1	1770	0,276	J. BIOL. CHEM. 260:11831–11837 (1985).	M11724
CAN1		SCCAN1G	1770	0,273	CURR. GENET. 10:587–592 (1986).	X03784
CAP					see SRV2	
CAP1		SCCAP1G-2	804	0,164	J. CELL BIOL. 117:1067–1076 (1992).	X61398
CAP2		SCCAP2-1	861	0,154	NATURE 344:352–354 (1990).	X62630
CAR1		SCCAR	999	0,34	J. BACTERIOL. 160:1078–1087 (1984).	M10110
CAR2	CARGB	SCCARGB	1269	0,491	EUR. J. BIOCHEM. 169:193–200 (1987).	X06790
CAR2*		SCCAR2G	162	0,437	EUR. J. BIOCHEM. 165:289–296 (1987).	X05571
CARG					see CAR2	
CAT3		SCCAT3	966	0,164	GENE 67:247–257 (1988).	M21760
CAT3	SNF4	SCSNF4	966	0,164	MOL. CELL. BIOL. 9:5045–5054 (1989).	M30470
CBF1	CP1	SCCP1A	1053	0,101	MOL. CELL. BIOL. 10:2458–2467 (1990).	M34070
CBF1	CP1, CPF1	SCCFB1A	1053	0,09	CELL 61:437–446 (1990).	M33620
CBF1	CP1, CPF1	SCCP1	1053	0,101	EMBO J. 9:4017–4026 (1990).	X52137
CBF2		SCCBF2A	2868	0,02	J. CELL BIOL. 121:513–519 (1993).	Z21627
CBP1		SCCBP1	1962	0,109	J. BIOL. CHEM. 259:4732–4738 (1984).	K02647

Continued

LIST A3 Continued

Gene (Primary gene name)	Synonym	Mnemonic	Length	Codon bias	Reference	Accession
CBP2		SCCBP2	1890	0,094	J. BIOL. CHEM. 258:9459–9468 (1983).	K00138
CBP3		SCCBP3	1005	0,054	J. BIOL. CHEM. 264:11122–11130 (1989).	J04830
CBP6		SCCBP6	486	0,004	J. BIOL. CHEM. 260:1513–1520 (1985).	M10154
CBS1		SCCBS1	699	0,018	MOL. GEN. GENET. 217:162–167 (1989).	X15650
CBS2		SCCBS2-3	1167	0,082	MOL. GEN. GENET. 214:263–270 (1988).	X13523
CCE1		SCCCE1A	1059	0,02	EMBO J. 11:699–704 (1992).	M65275
CCP1		SCCCP	1086	0,275	J. BIOL. CHEM. 257:15054–15058 (1982).	J01468
CCP1*	CYCPX*	SCCYCPX	240	0,317	J. BIOL. CHEM. 257:11186–11190 (1982).	J01321
CCT1		SCCCTA	1272	0,123	EUR. J. BIOCHEM. 169:477–486 (1987).	M36827
CDC2	POL3	SCPOL3	3291	0,14	NUCLEIC ACIDS RES. 20:375 (1992).	X61920
CDC2		SCDC2	3279	0,141	EMBO J. 8:1849–1854 (1989).	X15477
CDC4		SCCDC4G	2337	0,073	J. MOL. BIOL. 195:233–245 (1987).	X05625
CDC5	PKX2	SCPKX2A	2115	0,113	UNPUBLISHED.	M84220
CDC6		SCCDC6	1278	0,095	NUCLEIC ACIDS RES. 16:11507 (1988).	X13118
CDC6		SCCDC601	1539	0,114	MOL. CELL. BIOL. 0:0–0 (0).	M22858
CDC6		SCCDC6B	1539	0,115	J. BIOL. CHEM. 264:9022–9029 (1989).	J04734
CDC6		SCCDC6G	1539	0,114	EMBO J. 11:2167–2176 (1992).	X65299
CDC6*		SCCDCAA	144	0,222	J. BIOL. CHEM. 265:19904–19909 (1990).	M61183
CDC7		SCCDC7	1521	0,026	MOL. CELL. BIOL. 6:1590–1598 (1986).	M12624
CDC8		SCCDC8	648	0,123	MOL. CELL. BIOL. 4:583–590 (1984).	K01783
CDC8		SCTGYSR3	648	0,116	MOL. CELL. BIOL. 7:1198–1207 (1987).	M15468
CDC8		SCTKCDC8	648	0,119	J. BIOL. CHEM. 259:11052–11059 (1984).	K02116
CDC9		SCCDC9-1	2265	0,158	NUCLEIC ACIDS RES. 13:8323–8337 (1985).	X03246
CDC13		SCCDC13	2772	0,001	UNPUBLISHED.	M76550
CDC14		SCCDC14A	1269	0,084	J. BIOL. CHEM. 267:11274–11280 (1992).	M61194
CDC15		SCCDC15	2922	0,084	YEAST 7:265–273 (1991).	X52683
CDC15		SCNPK1-1	2922	0,084	MOL. GEN. GENET. 234:164–167 (1992).	X60549
CDC16		SCCDC16	2520	0,009	NUCLEIC ACIDS RES. 15:8439–8450 (1987).	X06165
CDC20		SCDC20	1557	0,105	MOL. CELL. BIOL. 11:5592–5601 (1991).	X59428
CDC21*		SCCDC21HP	228	0,317	NUCLEIC ACIDS RES. 20:5571–5577 (1992).	Z15032
CD23		SCCDC23	1878	0,006	CELL 60:307–317 (1990).	M31040

Gene	Alt. name	Standard name	Length	Value	Reference	Accession
CDC23		SCCDC23A	1878	0,006	GENE 91:123–126 (1990).	D90081
CDC25		SCCDC25	4767	0,029	CELL 48:789–799 (1987).	M15458
CDC25		SCCDC25G	4764	0,03	EMBO J. 5:375–380 (1986).	X03579
CDC28		SCCDC28	894	0,187	NATURE 307:183–185 (1984).	X00257
CDC31		SCCDC31	483	0,019	PNAS 83:5512–5516 (1986).	M14078
CDC33					see TIF-45	
CDC35					see CYR1	
CDC36		SCCDC36	573	0,09	NUCLEIC ACIDS RES. 14:6681–6697 (1986).	X04287
CDC37		SCCDC37	1347	0,13	NUCLEIC ACIDS RES. 14:6681–6697 (1986).	X04288
CDC39		SCCDC	6324	0,142	EMBO J. 12:177–186 (1993).	X70151
CDC39?		SCCDC39	2502	0,086	NUCLEIC ACIDS RES. 14:6681–6697 (1986).	X04289
CDC42		SCCDC42	573	0,212	J. CELL BIOL. 111:143–152 (1990).	X51906
CDC43		SCCDC43A	639	0,005	GENE 98:149–150 (1991).	M31114
CDC48		SCCDC48-2	2502	0,406	J. CELL BIOL. 114:443–453 (1991).	X56956
CDC55		SCCDC55G	1578	0,033	MOL. CELL. BIOL. 11:5767–5780 (1991).	M72716
CDC60		SCCDC60G	3270	0,361	GENE 120:43–49 (1992).	X62878
CDC68		SCCDC68	3105	0,194	MOL. CELL. BIOL. 11:5718–5726 (1991).	M73533
					see SPM2	
CEG1		SCCEG-1	1377	0,051	J. BIOL. CHEM. 267:9521–9528 (1992).	D10263
CHA1		SCSERTHR	1080	0,131	GENETICS 131:531–539 (1992).	M85194
CHC1		SCCHC1	4959	0,313	J. CELL BIOL. 112:65–80 (1991).	X52900
CHL1		SCCHL1	2583	0,025	EMBO J. 9:4347–4358 (1990).	X56584
CHO1	PSS	SCPSSG	828	0,17	EUR. J. BIOCHEM. 167:7–12 (1987).	X05944
CHO1		SCCHO1	828	0,158	J. BIOCHEM. 102:1089–1100 (1987).	D00001
CHS1		SCCHIS	3393	0,062	CELL 46:213–225 (1986).	M14045
CHS2		SCCHSYNA	2889	0,168	YEAST 5:459–467 (1989).	M23865
CHS3	CAL1	SCAL1	3297	0,16	J. CELL BIOL. 114:101–109 (1991).	X57300
CHS3	CSD2	SCPCHSY	3495	0,142	MOL. CELL. BIOL. 12:1764–1776 (1992).	M73697
CIF1	GGS1	SCGGS1	1485	0,272	MOL. MICROBIOL. 8:927–943 (1993).	X67499
CIF1		SCCIF1	1485	0,248	YEAST 8:183–192 (1992).	X61275
CIF1	TPS1	SCTPS1	1485	0,271	EUR. J. BIOCHEM. 209:951–959 (1992).	X68214
CIF1	TPS1	SCTPS1A	1485	0,272	UNPUBLISHED.	X68496
CIK1		SCCIK1	1782	0,045	GENES DEV. 6:1414–1429 (1992).	M96439
CIN8		SCCIN8A-2	3114	0,053	MOL. CELL. BIOL. 7:4390–4399 (1987).	M90522
CIT1		SCCSO1	1440	0,304	EMBO J. 3:1773–1781 (1984).	X00782

Continued

LIST A3 Continued

Gene (Primary gene name)	Synonym	Mnemonic	Length	Codon bias	Reference	Accession
CIT2		SCCIT2	1380	0,163	MOL. CELL. BIOL. 6:4509–4515 (1986).	M14686
CIT2*		SCCIT2A	72	0,104	MOL. CELL. BIOL. 11:38–46 (1991).	M54982
CKA1		SCCSKAS	1116	0,081	MOL. CELL. BIOL. 8:4981–4990 (1988).	M22473
CKA2		SCCKA2	1017	0,121	MOL. CELL. BIOL. 10:4089–4099 (1990).	M33759
CKI1		SCCK1A	1746	0,154	J. BIOL. CHEM. 264:2053–2059 (1989).	J04454
CKS1		SCPKCKS1	450	0,178	MOL. CELL. BIOL. 9:2034–2041 (1989).	M26033
CLB1	SCB1	SCSCB1G-2	11413	0,071	CELL 65:163–174 (1991).	M62389
CLB1		SCCLB1	1413	0,071	CELL 65:145–161 (1991).	M65069
CLB2		SCCLB2	1473	0,004	CELL 65:145–161 (1991).	M65070
CLB2		SCCLB2G	1473	0,004	GENES DEV. 6:2021–2034 (1992).	X62319
CLB3		SCCLB3	1281	0,066	GENES DEV. 6:2021–2034 (1992).	X69425
CLB3		SCMCYBA	1281	0,071	UNPUBLISHED.	M80302
CLB4		SCCLB4	1380	0,08	GENES DEV. 6:2021–2034 (1992).	X69426
CLB4		SCMCYBB	1380	0,08	UNPUBLISHED.	M80303
CLB5		SCCLB5A	1305	0,05	GENES DEV. 6:1695–1706 (1992).	M91209
CLC1		SCCLC1	699	0,199	J. CELL BIOL. 111:1437–1449 (1990).	X52272
CLN1		SCCLN1A	1638	0,116	PNAS 86:6255–6259 (1989).	M33264
CLN2		SCCLN2A	1635	0,181	PNAS 86:6255–6259 (1989).	M33265
CLS4		SCCLS4A	2208	0,04	GENE 54:125–132 (1987).	M16809
CMD1		SCCMD1	441	0,276	CELL 47:423–431 (1986).	M14760
CMK1		SCCMK1	1335	0,129	EMBO J. 10:1511–1522 (1991).	X57782
CMK1		SCCMK1II	1338	0,121	J. BIOL. CHEM. 266:12784–12794 (1991).	D90375
CMK2		SCCMK2	1341	0,166	EMBO J. 10:1511–1522 (1991).	X56961
CMK2		SCCMK2II	1341	0,166	J. BIOL. CHEM. 266:12784–12794 (1991).	D90376
CMP1	CNA1	SCCALA1	1659	0,118	PNAS 88:7376–7380 (1991).	M64839
CMP1		SCCMBP1	1659	0,114	EUR. J. BIOCHEM. 204:713–723 (1992).	X66490
CMP1		SCCMP1	1659	0,114	MOL. GEN. GENET. 227:52–59 (1991).	X54963
CMP2	CNA2	SCCALA2	1812	0,136	PNAS 88:7376–7380 (1991).	M64840
CMP2		SCCMP2	1812	0,136	MOL. GEN. GENET. 227:52–59 (1991).	X54964
CNA1					see CMP1	
CNA2					see CMP2	

Gene	Synonym	Locus	Length	Fraction	Reference	Accession
CNB1		SCCDPP	525	0,109	BBRC 180:1159-1163 (1991).	D10293
CNE1		SCCALNEX	1506	0,042	UNPUBLISHED.	L11012
CNE1		SCCALNEXH	1506	0,042	YEAST 9:185-188 (1993).	X66470
COF1		SCCOF	429	0,667	GENE 124:115-120 (1993).	D13230
COF1		SCCOF1	429	0,667	J. CELL BIOL. 120:421-435 (1993).	Z14971
COQ1		SCHPS	1419	0,11	J. BIOL. CHEM. 265:13157-13164 (1990).	J05547
COQ2		SCCOQ2A-2	1116	0,142	J. BIOL. CHEM. 267:4128-4136 (1992).	M81698
COQ3		SCDHHBMET	948	0,095	J. BIOL. CHEM. 266:16636-16644 (1991).	M73270
COR1		SCCOR1	1371	0,423	J. BIOL. CHEM. 261:17163-17169 (1986).	J02636
COT1		SCCOT1A	1317	0,108	MOL. CELL. BIOL. 12:3578-3688 (1992).	M88252
COX4		SCCOXIV	465	0,361	EMBO J. 3:2831-2837 (1984).	Y00152
COX5A		SCCOX2-2	459	0,209	CURR. GENET. 9:435-439 (1985).	X02561
COX5A		SCCOX5	459	0,207	J. BIOL. CHEM. 260:9513-9515 (1985).	M11770
COX5A		SCCOX5AA	459	0,185	MOL. CELL. BIOL. 7:3511-3519 (1987).	M17800
COX5B		SCCOX5BA	453	0,033	MOL. CELL. BIOL. 7:3511-3519 (1987).	M17799
COX5B*		SCCOX5B	204	0,111	PNAS 82:2235-2239 (1985).	M11140
COX6		SCCOX6	444	0,311	J. BIOL. CHEM. 259:15401-15407 (1984).	M10138
COX7		SCCYTCVI	180	0,231	J. BIOL. CHEM. 265:16389-16393 (1990).	M31620
COX8		SCCOX8	234	0,248	J. BIOL. CHEM. 261:17192-17197 (1986).	J02634
COX9		SCCB515-1	177	0,503	MOL. GEN. GENET. 218:57-63 (1989).	X16120
COX9		SCCOX9	177	0,503	J. BIOL. CHEM. 261:17183-17191 (1986).	J02633
COX9		SCCOX9A	177	0,503	J. BIOL. CHEM. 265:7273-7283 (1989).	M35260
COX10		SCCOX10	1386	0,053	J. BIOL. CHEM. 265:14220-14226 (1990).	M55566
COX11		SCCOX11-2	831	0,075	EMBO J. 9:2759-2764 (1990).	X55731
COX12		SCCOX12A	249	0,363	J. BIOL. CHEM. 267:22473-22480 (1992).	M98332
COX13		SCCOX13	387	0,126	UNPUBLISHED.	X72970
COXCH2						
CP1					see QCR2	
CPF1					see CBF1	
CPA1		SCCPA1	1233	0,28	EUR. J. BIOCHEM. 146:371-381 (1985).	X01764
CPA2		SCCPA2	3354	0,297	J. BIOL. CHEM. 258:14466-14472 (1983).	K01178
CPH1					see CPR1	
CPR1	CPH1	SCCPHA	486	0,77	GENE 83:39-46 (1989).	M30513
CPR1	CPH1	SCCYCL	486	0,77	NUCLEIC ACIDS RES. 18:373 (1990).	X17505
CPR2	CYP2	SCCYPRE	615	0,104	NUCLEIC ACIDS RES. 18:1643 (1990).	X51497

Continued

LIST A3 Continued

Gene (Primary gene name)	Synonym	Mnemonic	Length	Codon bias	Reference	Accession
CPR3	CYP3	SCCYP3A	546	0,365	GENE 111:85–92 (1992).	M84758
CPR3		SCCPR3	546	0,365	PNAS 89:11169–11173 (1992).	X56962
CPS1		SCCPS1	1728	0,31	EUR. J. BIOCHEM. 197:399 (1991).	X57316
CPS1		SCCPSYSCS	1728	0,311	FEBS LETT. 283:27–32 (1991).	X63068
CPT1		SCCPT1	1221	0,098	J. BIOL. CHEM. 265:1755–1764 (1990).	J05203
CR17					see QCR6	
CRO1		SCCRO1	381	0,266	EUR. J. BIOCHEM. 138:169–177 (1984).	X00256
CRT1		SCCRT1A	813	0,041	UNPUBLISHED.	M86538
CRY1					see RP59	
CSD2					see CHS3	
CTA1		SCCTA1A	1545	0,114	EUR. J. BIOCHEM. 176:159–163 (1988).	M36510
CTC1		SCCYC1	927	0,416	EMBO J. 3:2137–2143 (1984).	X00791
CTK1		SCCTK1A	1581	0,001	GENE EXPRESSION 1:149–167 (1991).	M69024
CTR1		SCCTR	1689	0,282	J. BIOL. CHEM. 265:15996–16003 (1990).	J05603
CTT1		SCCTT1	1686	0,312	EUR. J. BIOCHEM. 160:487–490 (1986).	X04625
CUP1		SCCUP1	183	0,101	PNAS 81:337–341 (1984).	K02204
CYB2		SCCYTB2	1773	0,247	EMBO J. 4:3265–3272 (1985).	X03215
CYC1		SCCORA-1	327	0,467	CELL 87:157–166 (1990).	M37696
CYC1		SCCYT1	327	0,467	CELL 16:753–761 (1979).	X03472
CYC3		SCCYC3G	807	0,111	EMBO J. 6:235–241 (1987).	X04776
CYC7					see CYP3	
CYC8					see SSN6	
CYCPX*					see CCP1*	
CYH2		SCCYH2	447	0,792	NUCLEIC ACIDS RES. 11:3123–3135 (1983).	X01573
CYP1	HAP1	SCHAP1	4449	0,142	CELL 56:291–301 (1989).	J03152
CYP1		SCCYP1	4449	0,135	J. MOL. BIOL. 204:263–272 (1988).	X13793
CYP2					see CPR2	
CYP3	CYC7	SCCYC7A	339	0,164	MOL. GEN. GENET. 199:117–122 (1985).	J01320
CYP3	CYC7	SCCYT2	339	0,164	PNAS 77:541–545 (1980).	V01299
CYP3					see CPR3	

CYR1	CDC35	SCCDC35-1	2967	0,043	CURR. GENET. 10:343-352 (1986).	X03449
CYR1		SCCYR1	6078	0,032	CELL 43:493-505 (1985).	M12057
CYS3		SCCYS3A	1182	0,621	J. BACTERIOL. 174:3339-3347 (1992).	L04459
DAF1		SCDAF1A	1191	0,263	MOL. CELL. BIOL. 8:4675-4684 (1988).	M23359
DAF1	WHI1	SCWHI1	1740	0,264	EMBO J. 7:4335-4346 (1988).	X13964
DAL1		SCDAL1A-1	1416	0,097	YEAST 7:913-923 (1991).	M69294
DAL2		SCDAL2A	1029	0,132	GENE 104:55-62 (1991).	M64720
DAL3		SCDAL3	585	≤0	YEAST 7:693-698 (1991).	M64778
DAL5		SCDAL5A	1629	0,128	J. BACTERIOL. 170:266-271 (1988).	M24098
DAL80		SCDAL80A	807	0,097	MOL. CELL. BIOL. 11:6205-6215 (1991).	M77821
DAL81					see UGA35	
DAL81*	DURM1	SCDAL1A-2	594	0,136	YEAST 7:913-923 (1991).	M69294
DAL82		SCDURMG	765	0,022	NUCLEIC ACIDS RES. 18:7136 (1990).	X54525
DAL82		SCDAL2	765	0,031	J. BACTERIOL. 173:255-261 (1991).	M60414
DAP2		SCDPAPB	2523	0,062	J. CELL. BIOL. 108:1363-1373 (1989).	X15484
DAT1		SCDAT1	744	0,019	EMBO J. 8:1867-1877 (1989).	X15478
DBF2		SCDBF2A	1683	0,082	MOL. CELL. BIOL. 10:1358-1366 (1990).	M34146
DBF4					see DNA43	
DBF20		SCDBFA	1692	0,071	GENE 104:63-70 (1991).	M62506
DBP1		SCDB1G	1851	0,071	MOL. MICROBIOL. 5:805-812 (1991).	X55993
DBP2		SCP68	1638	0,573	MOL. CELL. BIOL. 11:1326-1333 (1991).	X52649
DBR1		SCDBR1	1215	0,046	CELL 65:483-492 (1991).	M62813
DCD1		SCDCD1	936	0,018	MOL. CELL. BIOL. 6:1711-1721 (1986).	X01660
DCG1		SCDCG1A	732	≤0	GENE 104:55-62 (1991).	M64719
DDR48	SPP81	SCDDR48A	1290	0,416	MOL. CELL. BIOL. 10:3174-3184 (1990).	M36110
DED1		SCDED1	1812	0,529	NATURE 349:715-717 (1991).	X57278
DED1*		SCHIS3G-3	336	0,343	J. MOL. BIOL. 152:553-568 (1981).	X03245
DFR1		M26668	633	0,008	GENE 63:175-185 (1988).	M26668
DFR1		SCDFR1	633	0,005	NUCLEIC ACIDS RES. 15:10355 (1987).	Y00887
DFR1		SCDIHRG	633	0,008	GENE 63:165-174 (1988).	M18578
DIN1*		SCDIN1	2267	0,346	MOL. CELL. BIOL. 10:5553-5557 (1990).	M58012
DIT1		SCDIT1	1608	0,088	GENES DEV. 4:1775-1789 (1990).	X55712
DIT2		SCDIT2	1467	0,049	GENES DEV. 4:1775-1789 (1990).	X55713
DKA1	NSP1	SCNSP1	807	0,095	MOL. MICROBIOL. 3:1319-1327 (1989).	X15409

Continued

LIST A3 Continued

Gene (Primary gene name)	Synonym	Mnemonic	Length	Codon bias	Reference	Accession
DLD1		SCDLDG	1728	0,272	MOL. GEN. GENET. 238:315–324 (1993).	X66052
DMC1		SCDMC1A-2	1002	0,145	CELL 69:439–456 (1992)	M87549
DMC1		SCSYGP5-2	1002	0,145	UNPUBLISHED.	L11229
DNA43		SCDNA43A	2112	0,034	YEAST 8:273–289 (1992).	M83539
DNA43	DBF4	SCDBF4	2085	0,031	GENETICS 131:21–29 (1992)	X60279
DNA52		SCDNA52A	1551	≤0	YEAST 8:273–289 (1992).	M83540
DPB2		SSDPB2	2094	0,123	PNAS 88:4601–4605 (1991).	M61710
DPB3		SCDPB3G	603	0,022	NUCLEIC ACIDS RES. 19:4867–4872 (1991).	X58500
DPH2*		SCDPH2A	1527	0,092	UNPUBLISHED.	L01424
DPH5		SCDPHA	900	0,235	MOL. CELL. BIOL. 12:4026–4037 (1992).	M83375
DPM1		SCDPM	801	0,37	J. BIOL. CHEM. 263:17499–17507 (1988).	J04184
DPR1		SCDPR	1293	0,071	YEAST 4:271–281 (1988).	M22753
DRS1		SCHELI	2166	0,259	PNAS 89:11131–11135 (1992).	L00683
DSS4		SCDSS4GNA	429	0,147	NATURE 361:460–463 (1993).	X70495
DST1					see PPR2	
DST2					see KEM1	
DTP1					see APA1	
DURL					see UGA35	
DURM1					see DAL82	
EF1A					see TEF1	
EFB1		SCEFB1DNA	618	0,874	FEBS LETT. 316:165–169 (1993).	D14080
EFC1		SCELONGAM	1233	0,667	UNPUBLISHED.	L01880
EFT1		SCEFT1	2526	0,89	J. BIOL. CHEM. 267:1190–1197 (1992).	M59369
EFT2		SCEFT2	2526	0,888	J. BIOL. CHEM. 267:1190–1197 (1992).	M59370
EGD1		SCEGD	435	0,43	J. BIOL. CHEM. 263:19480–19487 (1988).	L05185
EIF4E					see TIF45	
ELM1		SCELM1A	1689	≤0	UNPUBLISHED.	M81258
EMP24		SCEMP24	609	0,32	UNPUBLISHED.	X67317
EMP70		SCEMP70	2001	0,198	UNPUBLISHED.	X67316
END1	PEP5	SCPEP5G	3087	0,036	GENETICS 125:739–752 (1990).	X54466

Gene	Synonym	Locus	Length	Value	Reference	Accession
END1		SCEND1	3090	0,036	EMBO J. 8:1349–1359 (1989).	X15355
ENO1	ENOA	SCENOA	1311	0,93	J. BIOL. CHEM. 256:1385–1395 (1981).	J01322
ENO2	ENOB	SCENOB	1311	0,965	J. BIOL. CHEM. 256:1385–1395 (1981).	J01323
ENOA					see ENO1	
ENOB					see ENO2	
ENS1					see SSC1	
EPT1		SCEPT1	1173	0,045	J. BIOL. CHEM. 266:5094–5103 (1991).	M59311
ERD1		SCERD1	1086	0,028	EMBO J. 9:623–630 (1990).	X51949
ERD2		SCERD2A	657	0,22	CELL 61:1349–1357 (1990).	M34777
ERG1		SCERG1A-2	1488	0,573	GENE 107:155–160 (1991).	M64994
ERG2		SCERG2A	666	0,47	GENE 107:173–174 (1991).	M74037
ERG3		SCERG3A	1095	0,32	GENE 107:39–44 (1991).	M62623
ERG3		SCERG3AA	1095	0,32	UNPUBLISHED.	M64989
ERG8		SCERG8	1272	0,051	MOL. CELL. BIOL. 11:620–631 (1991).	M63648
ERG9		SCERG9G	1332	0,406	CURR. GENET. 20:365–372 (1991).	X59959
ERG9		SCERG9L	1332	0,413	PNAS 88:6038–6042 (1991).	M63979
ERG10		SUERG10	1194	0,569	CURR. GENET. 13:471–478 (1988).	X07976
ERG11	14DM	SC14DMA	1590	0,634	DNA 6:529–537 (1987).	M18109
ERG11		SCCYLDA	1590	0,638	BBRC 155:317–323	M21483
ERG12					see RAR1	
ERC24		SCC14SRTS	1314	0,166	UNPUBLISHED.	
ERS1		SCERS1	780	0,009	NUCLEIC ACIDS RES. 18:2177 (1990).	M99419
ERV1		SCREGPRG	351	0,137	MOL. GEN. GENET. 232:58–64 (1992).	X52468
ESP1		SCESP1A	4719	≤0	UNPUBLISHED.	X60722
ESP35		SCSUFESP	978	0,546	YEAST 8:699–710 (1992).	L07289
ESS1		CGM0170	516	0,101	YEAST 5:55–72 (1989).	X61669
EST1		SCEST1	2097	0,05	CELL 57:633–643 (1989).	J04849
EUG1*		SCEUG1	1514	0,197	MOL. CELL. BIOL. 12:4601–4611 (1992).	M84796
EXG1		SCEXG1A	1344	0,466	GENE 97:173–182 (1991).	M34341
FAA1		SCFAA1	2100	0,266	J. CELL. BIOL. 117:515–529 (1992).	X66194
FAR1		SCFAR1	2340	0,05	CELL 63:999–1011 (1990).	M60071
FAS1		SCFAS1	5535	0,469	MOL. GEN. GENET. 203:479–486 (1986).	X03977
FAS1		SCFASB	5940	0,476	J. BIOL. CHEM. 262:4231–4240 (1987).	M31034
FAS2		SCFAS2A	5682	0,524	J. BIOL. CHEM. 263:12315–12325 (1988).	J03936
FAS3		SCFAS3A	6735	0,433	PNAS 89:4534–4538 (1992).	M92156

Continued

LIST A3 Continued

Gene (Primary gene name)	Synonym	Mnemonic	Length	Codon bias	Reference	Accession
FBA1		SCFBA1	1077	0,936	EUR. J. BIOCHEM. 180:301–308 (1989).	X15003
FBP1		SCFBP1	1044	0,259	FEBS LETT. 236:195–200 (1988).	Y00754
FBP1		SCFBPA	1044	0,259	J. BIOL. CHEM. 263:6051–6057 (1988).	J03207
FCY2	FPR1	SCFSY2G	1599	0,256	MOL. MICROBIOL. 4:585–596 (1990).	X51751
FKB1	FPR1,RBP1	SSFPR1	342	0,681	PNAS 88:1948–1952 (1991).	M60877
FKB1		SCRBP1	342	0,681	MOL. CELL. BIOL. 11:1718–1723 (1991).	M63892
FKB1		SCFKB1	342	0,681	PNAS 88:1029–1033 (1991).	M57967
FKB2		SCFKB2X	405	0,255	YEAST 8:673–680 (1992).	M90646
FKB2		SCFKBP15A	405	0,255	PNAS 89:7471–7475 (1992).	M90767
FOX2		SCMFBOX	2700	0,103	J. BIOL. CHEM. 267:6646–6653 (1992).	M86456
FOX3		SCFOX3	1251	0,179	EUR. J. BIOCHEM 200:113–122 (1991).	X53946
FPP1		SCFPP	1056	0,516	J. BIOL. CHEM. 264:19176–19184 (1989).	J05091
FPR1					see FKB1	
FPS1		SCFPS1G	2007	0,102	EMBO. J. 10:2014–2095 (1991).	X54157
FRE1		SCFRE1A	2058	0,066	PNAS 89:3869–3873 (1992).	M86908
FRS1		J03964	1785	0,416	J. BIOL. CHEM. 263:15407–15415 (1988).	J03964
FRS2		J03965	1509	0,445	J. BIOL. CHEM. 263:15407–15415 (1988).	J03965
FUM1		SCFUM-1	1464	0,357	J. BIOL. CHEM. 262:12275–12282 (1987).	J02802
FUN80					see MCM1	
FUR1		SCFUR1A	753	0,469	GENE 88:149–157 (1990).	M36485
FUR4		SCFUR4	1899	0,195	EUR. J. BIOCHEM. 171:417–424 (1988).	X06830
FUS1		SCFUS1G	1536	0,005	MOL. CELL. BIOL. 7:2680–2690 (1987).	M17199
FUS1		SCFUSG-2	1536	0,005	MOL. CELL. BIOL. 7:2316–2328 (1987).	M16717
FUS3		SC114-2	1059	0,061	YEAST 8:769–776 (1992).	X68577
FUS3		SCFUS3	1059	0,045	CELL 60:649–664 (1990).	M31132
FZF1		CSFZF1-2	897	0,043	YEAST 9:551–556 (1993).	X67787
GAC1		SCGAC1	2382	0,07	EMBO J. 11:87–96 (1992).	X63941
GAL1		SCGALS2	1584	0,194	J. BACTERIOL. 158:269–278 (1984).	K01609
GAL1*		SCGAL10-2	87	0,262	MOL. CELL. BIOL. 4:1440–1448 (1984).	K02115
GAL2		SCGAL2	1722	0,327	J. BACTERIOL. 171:4484–4493 (1989).	M68547
GAL2		SCGAL2A	1722	0,329	GENE 85:313–319 (1989).	M81879

Gene	Alias	Locus	Length	Value	Reference	Accession
GAL3		SCGAL3	1275	0,028	MOL. CELL. BIOL. 8:3439–3447 (1988).	M21615
GAL4		SCGAL4	2643	0,035	MOL. CELL. BIOL. 4:260–267 (1984).	K01486
GAL7		SCGAL7EN	1095	0,211	YEAST 1:67–77 (1985).	M12348
GAL10*		SCGAL10-1	140	0,181	MOL. CELL. BIOL. 4:1440–1448 (1984).	K02115
GAL11		SCGAL11A	2892	0,153	MOL. CELL. BIOL. 8:4991–4999 (1988).	M22481
GAL80		SCGAL80	1305	0,083	NUCLEIC ACIDS RES. 12:9287–9298 (1984).	X01667
GAL83		SCGAL83	1251	0,042	UNPUBLISHED.	X72893
	SPM1	SCSPM1	1251	0,042	UNPUBLISHED.	Z14127
GAM1					see SNF2	
CAP1		SCGAP1P	1803	0,431	EUR. J. BIOCHEM. 190:39–44 (1990).	X52633
GAR1		SCGAR1	615	0,334	EMBO J. 11:673–682 (1992).	M63617
GAS1	GGP1	SCGGP1C	1677	0,499	J. BIOL. CHEM. 266:12242–12248 (1991).	X56399
GAS1		SCGAS1	1677	0,503	MOL. CELL. BIOL. 11:27–37 (1991).	X53424
GCD1		SCGCD1	1533	0,138	NUCLEIC ACIDS RES. 16:9253–9266 (1988).	X07846
GCD11		SCGCD11NR	1581	0,427	MOL. CELL. BIOL. 13:506–520 (1993).	L04268
GCD2		SCGCD2	1953	0,114	GENETICS 122:551–559 (1989).	X15658
GCD6		SCGCDA-2	2136	0,181	MOL. CELL. BIOL. 13:1920–1932 (1993).	L07115
GCD7		SCGCDB	1143	0,125	MOL. CELL. BIOL. 13:1920–1932 (1993).	L07116
GCN2		SCGCN2	4770	0,063	PNAS 86:4579–4583 (1989).	M27082
GCN3		SCGCN3A	915	0,08	MOL. CELL. BIOL. 8:4808–4820 (1988).	M23356
GCN4		SCGCN4	843	0,3	PNAS 81:6442–6446 (1984).	K02205
GCN4		SCGCN4B	747	0,275	PNAS 81:5096–5100 (1984).	K02649
GCN5		SCGCN5PRA	1317	≤0	EMBO J. 11:4145–4152 (1992).	X68628
GCR1		SCGCR1	2532	≤0	MOL. CELL. BIOL. 7:813–820 (1987).	M15253
GCR1		SCGCR1X	2532	0,001	MOL. CELL. BIOL. 6:3774–3784 (1986).	M14145
GCR2		SCGCR2	1602	≤0	MOL. CELL. BIOL. 12:3834–3842 (1992).	D10104
GCY1		SCGCY	936	0,247	FEBS LETT. 238:123–128 (1988).	X13228
GDH1		SCGDHM	1362	0,744	J. BIOL. CHEM. 260:8502–8508 (1985).	M11297
GDH1		SCGDHN	1359	0,741	GENE 37:247–253 (1985).	M10590
GFA1		SCGFAT	2148	0,323	J. BIOL. CHEM. 264:8753–8758 (1989).	J04719
GGP1					see GAS1	
GS1					see CIF1	
GLC3		SCGBEAA	2112	0,142	J. BIOL. CHEM. 267:15224–15228 (1992).	M76739
GLC7		SCPP1A	936	0,238	CELL 57:997–1007 (1989).	M77175
GLC7		SCSYGP4-1	936	0,238	UNPUBLISHED.	L11120

Continued

LIST A3 Continued

Gene (Primary gene name)	Synonym	Mnemonic	Length	Codon bias	Reference	Accession
GLK1		M24077	1500	0,238	GENE 73:141–152 (1988).	M24077
GLN1		SCGLN1	1038	0,67	J. BACTERIOL. 174:1828–1836 (1992).	M65157
GLN3		SCGLN3	2190	0,008	MOL. CELL. BIOL. 11:6216–6228 (1991).	M35267
GLUCPI					see MAL6S	
GPA1	SCG1	YSSCG1A	1416	0,072	CELL 50:1001–1010 (1987).	M17414
GPA1		SCGPA1	1416	0,075	PNAS 84:2140–2144 (1987).	M15867
GPA2		SCGPA2	1347	0,036	PNAS 85:1374–1378 (1988).	J03609
GPD1					see TDH1	
GPD2					see TDH2	
GPD3					see TDH3	
GPH1		SCPHOSG	2673	0,294	NATURE 324:80–84 (1986).	X04604
GPM1		SCGPM	741	0,91	FEBS LETT. 299:383–387 (1988).	X06408
GPM1		SCGPM1	741	0,91	CURR GENET. 20:167–171 (1991).	X58789
GRP78					see KAR2	
GRR1		SCGRR1	3453	0,015	MOL. CELL. BIOL. 11:5101–5112 (1991).	M59247
GSP1		SCGSP1X	657	0,727	UNPUBLISHED.	L08690
GSP2		SCGSP2X	660	0,276	UNPUBLISHED.	L08691
GST1					see SUF12	
GSY1		SCSYNAA	2124	0,235	J. BIOL. CHEM. 265:20879–20886 (1990).	M60919
GSY2		SCGS	2115	0,134	J. BIOL. CHEM. 266:15602–15607 (1991).	M65206
CTR1		SCGTR1	930	0,039	MOL. CELL. BIOL. 12:2958–2966 (1992).	D10018
GUK1		SCGUKI	561	0,422	J. BIOL. CHEM. 267:25652–25655 (1992).	L04683
GUT2		SCGUT2A	1842	0,275	UNPUBLISHED.	X71660
GUT2*		SCGUT2	323	0,518	GENE 101:89–96 (1991).	M38740
CYP6		SCGYP6	1374	0,144	NATURE 361:736–739 (1993).	X68506
H2A1					see HTA1	
H2A2					see HTA2	
H2B1					see HTB1	
H2B2					see HTB2	
H3I					see HHT1	
H3II					see HHT2	

Gene	Alias	Clone	Length	Value	Reference	Accession
H4					see HHF2	
H4II					see HHF2	
H4I					see HHF1	
HAL1		SCHAL1A	882	≤0	EMBO J. 11:3157–3164 (1992).	X67559
HAL2		SCHAL2	1071	0,219	EMBO J. 7 (1993).	X72847
HAP1					see CYP1	
HAP2		SCHAP2	795	0,055	MOL. CELL. BIOL. 7:578–585 (1987).	M15243
HAP4		SCHAP4	1662	0,169	GENES DEV. 3:1166–1178 (1989).	X16727
HCM1		SCHCM1PA	1692	0,11	UNPUBLISHED.	L08252
HCS26		SCG1CYC	837	0,151	CELL 66:1015–1026 (1991).	M73966
HDF1	NES24	SCNES24	1806	≤0	UNPUBLISHED.	D15052
HDF1		SCHDF1	1806	≤0	UNPUBLISHED.	X70379
HEM1		M26329	1644	0,286	EUR. J. BIOCHEM. 156:511–519 (1989).	M26329
HEM2		SCHEM2	1026	0,235	J. BIOL. CHEM. 262:16822–16829 (1987).	J03493
HEM3		SCHEM3PDG	981	0,216	MOL. GEN. GENET. 234:233–243 (1992).	Z11745
HEM12		SCHEM12	1086	0,182	EUR. J. BIOCHEM. 205:1011–1016 (1992).	X63721
HEM13		SCCPO	984	0,586	J. BIOL. CHEM. 263:9718–9724 (1988).	J03873
HEM15		SCHEM15	1179	0,145	J. BIOL. CHEM. 265:7278–7283 (1990).	J05395
HEM15		SCHEM15T	1179	0,145	NUCLEIC ACIDS RES. 18:6130 (1990).	X54514
HEX2		SCHEX2A	3078	0,032	EUR. J. BIOCHEM. 200:311–319 (1991).	M33703
HEX3		SCHEXMET	1857	0,046	UNPUBLISHED.	L07745
HFA1*		SCHFA1GN	2475	0,074	UNPUBLISHED.	Z22558
HHF1	H4I	SCH3H401-2	309	0,82	J. MOL. BIOL. 169:663–690 (1983).	X00724
HHF2	H4,H4II	SCH4	309	0,771	NUCLEIC ACIDS RES. 11:5347–5360 (1983).	K03154
HHF2	H4,H4II	SCH3H402-2	309	0,788	J. MOL. BIOL. 169:663–690 (1983).	X00725
HHT1	H3I	SCH3H401-1	408	0,772	J. MOL. BIOL. 169:663–690 (1983).	X00724
HHT2	H3II	SCH3H402-1	408	0,709	J. MOL. BIOL. 169:663–690 (1983).	X00725
HIP1		SCHIP1	1596	0,24	GENE 38:205–214 (1985).	M11980
HIR1		SCHIR1X	2364	0,03	MOL. CELL. BIOL. 13:28–38 (1993).	L03838
HIR2		SCHIR2X	2625	0,05	MOL. CELL. BIOL. 13:28–38 (1993).	L03839
HIS1		SCHIO1	891	0,232	J. BIOL. CHEM. 258:5238–5247 (1983).	V01306
HIS3		SCHIS3G-2	657	0,012	J. MOL. BIOL. 152:553–568 (1981).	X03245
HIS4		SCHIS4A	2397	0,37	GENE 18:47–59 (1982).	V01310
HIS5		SCHIS5	1152	0,271	MOL. GEN. GENET. 208:159–167 (1987).	X95650
HKA					see HXK1	
HKB					see HXK2	

Continued

LIST A3 Continued

Gene (Primary gene name)	Synonym	Mnemonic	Length	Codon bias	Reference	Accession
HMG1		SCHMGCR1	3162	0,223	MOL. CELL. BIOL. 8:3797–3808 (1988).	M22002
HMG2		SCHMGCR2	3135	0,16	MOL. CELL. BIOL. 8:3797–3808 (1988).	M22255
HO		SCHORR	1758	0,051	MOL. CELL. BIOL. 6:4281–4294 (1986).	M14678
HOG1		SCHOG1A	1248	0,18	SCIENCE 259:1760–1763 (1993).	L06279
HOM2		SCHOM2A	1095	0,593	MOL. CELL. GENET. 217:149–154 (1989).	X15649
HOM3		SCHOM3	1242	0,315	J. BIOL. CHEM. 263:2146–2151 (1988).	J03526
HOM6		SCHOM6G	1077	0,613	FEBS LETT. 323:289–293 (1993).	X64457
HOP1		SCHOP1	1818	0,015	CELL 61:73–84 (1990).	J04877
HPC2		SCHPC2A	1869	0,073	MOL. CELL. BIOL. 12:5249–5259 (1992).	M94207
HPR1		SCHPR1	2256	0,076	MOL. CELL. BIOL. 10:1439–1451 (1990).	M30484
HRR25		SCHRR25A	1482	0,155	SCIENCE 253:5023 (1991).	M68605
HSF1		SCHSF	2499	0,012	CELL 54:855–864 (1988).	J03139
HSF1		SCHSF1A	2499	0,005	CELL 54:841–853 (1988).	M22040
HSP12		SCHSP	327	0,801	MOL. GEN. GENET. 223:97–106 (1990).	X55785
HSP26		SCHSP26A	642	0,435	MOL. CELL. BIOL. 9:5265–5271 (1989).	M23871
HSP26		SCHSP26B	642	0,432	GENE 78:323–330 (1989).	M26942
HSP30		SCHSP30X	789	0,229	CURR. GENET. 23:435–442 (1993).	M93123
HSP60		CGM0211	1716	0,554	NATURE 337:655–659 (1989).	M33301
HSP60		SCHSP60	1716	0,554	GENE 84:295–302 (1989).	X07886
HSP70		SCILSI-1	153	0,128	NUCLEIC ACIDS RES. 16:2189–2201 (1988).	M26044
HSP82		SCHSP82	2115	0,693	MOL. CELL. BIOL. 9:3919–3930 (1989).	K01387
HSP90		SCHSP90	2127	0,658	J. BIOL. CHEM. 259:5745–5751 (1984).	D13741
HSP150	PIR2	SCPIR2P	1239	0,808	UNPUBLISHED.	
HSP150		SCHSP150A	1236	0,762	PNAS 89:3671–3675 (1992).	M88698
HTA1	H2A1	SCH2A1	396	0,782	PNAS 79:1484–1487 (1982).	V01304
HTA1*		SCADK1-1	354	0,814	J. BIOL. CHEM. 263:19468–19474 (1988).	M18455
HTA2	H2A2	SCH2A2	396	0,679	PNAS 79:1484–1487 (1982).	V01305
HTB1	H2B1	SCH2B1	393	0,774	CELL 22:799–805 (1980).	J01327
HTB2	H2B2	SCHIS2	393	0,705	CELL 22:799–805 (1980).	V01308
HTS1		SCHTS1	1638	0,381	CELL 46:235–243 (1986).	M14048

Gene	Alt	Systematic	Length	Value	Reference	Accession
HXK1		SCHKA	1458	0,598	NUCLEIC ACIDS RES. 14:945–963 (1986).	X03482
HXK1	HKA	SCHXKI	1455	0,605	GENE 35:95–102 (1985).	M11184
HXK2		SCHKB	1458	0,752	NUCLEIC ACIDS RES. 14:945–963 (1986).	X0348383
HXK2	HKB	SCHXK2	1458	0,729	GENE 36:105–111 (1985).	M11181
HXT2		SCHXT2	1623	0,472	MOL. CELL. BIOL. 10:5903–5911 (1990).	M33270
HYP1					see ANB1	
HYP2					see TIF 51A	
ICL1		SCICL1	1671	0,327	EUR. J. BIOCHEM. 204:983–990 (1992).	X61271
ICL1		SCICL1G	1671	0,327	CURR. GENET. 23:375–381 (1993).	X65554
IDH1		SCHISODH	1074	0,33	J. BIOL. CHEM. 267:16417–16423 (1992).	M95203
IDH2		SCIDH2A	1107	0,33	J. BIOL. CHEM. 266:22199–22205 (1991).	M74131
IDP1		SCIDP1	1284	0,432	J. BIOL. CHEM. 266:2339–2345 (1990).	M57229
IFM1		SCIFM1-1	2028	0,069	EUR. J. BIOCHEM. 201:643–652 (1991).	X58379
ILS1		SCILSI-2	3216	0,407	NUCLEIC ACIDS RES. 16:2189–2201 (1988).	X07886
ILS1		SCITRSA	3219	0,406	BCHS 368:971–979 (1987).	M19992
ILV1		SCILV1A	1728	0,428	CARLSBERG RES. COMMUN. 49:567–575 (1984).	M36383
ILV2		SCILV2	2061	0,364	NUCLEIC ACIDS RES. 13:4011–4027 (1985).	X02549
ILV5		SCILV5-1	1185	0,892	NUCLEIC ACIDS RES. 14:9631–9651 (1986).	X04969
IME1		SCIME	1080	0,069	MOL. GEN. GENET. 237:375–384 (1993).	X52152
IME1		SCIME1	1080	0,069	MOL. CELL. BIOL. 10:6103–6113 (1990).	M37188
IMP2		SCIMP2	936	0,139	YEAST 8:83–92 (1992).	X61928
INH1		SCINH1	225	0,138	J. BIOL. CHEM. 265:6274–6278 (1990).	D00443
INO1		SC1INO-2	1659	0,19	J. BIOL. CHEM. 264:1274–1283 (1989).	J04453
INO2		SCINO2	912	0,061	UNPUBLISHED.	X66066
INO2		SCSCSI	912	0,061	UNPUBLISHED.	D90460
INO4		SCINO4X	453	0,019	J. BIOL. CHEM. 265:4736–4745 (1990).	J05267
IRA1		SCIRA1A	8814	0,06	MOL. CELL. BIOL. 9:757–768 (1989).	M24378
IRA2		SCIRA2A	9237	0,073	MOL. CELL. BIOL. 10:4303–4313 (1990).	M33779
IRE1		SCIRE1DNA	3345	0,02	UNPUBLISHED.	Z11701
ISP42		SCISP42	1161	0,375	NATURE 348:605–609 (1990).	X56885
ITR1		SCITR1	1752	0,276	J. BIOL. CHEM. 266:11184–11191 (1991).	D90352
ITR2		SCITR2	1836	0,22	J. BIOL. CHEM. 266:11184–11191 (1991).	D90353
KAR1		SCKAR1	1299	0,127	CELL 48:1047–1060 (1987).	M15683
KAR2	GRP78	SCGRP78	2046	0,585	PNAS 87:1159–1163 (1990).	M31006

Continued

LIST A3 *Continued*

Gene (Primary gene name)	Synonym	Mnemonic	Length	Codon bias	Reference	Accession
KAR2		SCKAR2	2046	0,585	CELL 57:1223–1236 (1989).	M25394
KAR2		SCKAR2A	2046	0,585	CELL 57:1211–1221 (1989).	MM25064
KAR3		SCKAR3AA	2187	0,033	CELL 60:1029–1041 (1990).	M31719
KEM1	DST2	SCSTPB	4584	0,174	MOL. CELL. BIOL. 11:2583–2592 (1991).	M36725
KEM1	RAR2	SCRRAR5	4584	0,174	UNPUBLISHED	X61181
KEM1		SCKEM1G	4584	0,174	GENETICS 126:799–812 (1990).	Z54717
KEM1	SEP1	SCSEP1	4584	0,174	MOL. CELL. BIOL. 11:2593–2608 (1991).	M58367
KEM1	XRN1	SCXRN1	4584	0,174	GENE 95:85–90 (1990).	M62423
KEX1		SCKEX1	2187	0,036	CELL 50:573–584 (1987).	M17231
KEX2		SCKEX2A	2442	0,112	PNAS 86:1434–1438 (1989).	M24201
KGD1		SCKGDA	3042	0,286	MOL. CELL. BIOL. 9:2695–2705 (1989).	M26390
KGD2		SCKGD2	1425	0,294	MOL. CELL. BIOL. 10:4221–4232 (1990).	M34531
KIN1		SCKIN1	3189	0,058	PNAS 84:6035–6039 (1987).	M69017
KIN2		SCKIN2	3459	0,07	PNAS 84:6035–6039 (1987).	M69018
KIN3	NPK1	SCNPK1-2	1305	0,077	MOL. GEN. GENET. 234:164–167 (1992).	X60549
KIN3		SCKIN3	1107	0,071	GENE 90:87–92 (1990).	M55416
KIN28		SCKIN28	918	0,039	EMBO J. 5:2697–2701 (1986).	X04423
KIP1		SCKIP1G	3333	0,064	J. CELL. BIOL. 118:95–108 (1992).	Z11962
KIP2		SCKIP2XVI-2	2118	0,083	J. CELL. BIOL. 118:95–108 (1992).	Z11963
KNS1		SCKNSI	2208	0,054	UNPUBLISHED.	M85200
KRE1		SCKRE1	939	0,207	J. CELL. BIOL. 110:1833–1843 (1990).	X51729
KRE11		SCKTOXR	1680	≤0	GENETICS 133:837–849 (1993).	L10667
KRE2	MNT1	SCMNT1A	1326	0,315	GLYCOBIOLOGY 2:77–84 (1992).	M81110
KRE2		SCKRE2	1299	0,18	GENETICS 130:273:283 (1992).	X62647
KRE5		SCKRE5	4095	0,072	MOL. CELL. BIOL. 10:3013–3019 (1990).	M33556
KRE6		SCKRE6	2160	0,22	PNAS 88:11295–11299 (1991).	M80657
KRS1		SCKRS1	1773	0,537	MOL. CELL. GENET. 227:149–154 (1991).	X56259
KRS1		SCKRS1A	1773	0,54	J. BIOL. CHEM. 263:18443–18451 (1988).	J04186
KSS1		SCPKS	1104	0,024	CELL. 58:1107–1119 (1989).	M26398

Gene	Locus	Length	Value	Reference	Accession
KTR1	SCKTR1G	1179	0,253	GENETICS 130:273–283 (1992).	X62941
L10E				see RPA0	
L12EIA				see RPL45	
L12EIB				see RPA2	
L12EII				see RPL44	
L12EIIA				see RPA1	
L15B				see RPL15B	
L16				see RPL16B	
L2				see RPL2	
LAP1				see APE2	
LCB1	SCLCB1	1674	0,229	J. BACTERIOL. 173:4325–4332 (1991).	M63674
LEU1	SCLEU1	144	0,641	J. BIOL. CHEM. 259:3714–3719 (1984).	K01969
LEU2*	SCTY117X	1092	0,595	NUCLEIC ACIDS RES. 14:3475–3485 (1986).	X03840
LEU2	SCLEU2A	456	0,682	GENE 31:257–261 (1984).	M12909
LEU3	SCLEU	2658	0,023	MOL. CELL. BIOL. 7:2708–2717 (1987).	M17222
LEU3	SCLEU3	2658	0,023	NUCLEIC ACIDS RES. 15:5261–5273 (1987).	Y00360
LEU4	SCLEU4	1857	0,407	J. BIOL. CHEM. 261:5160–5167 (1986).	M12893
LIP5	SCLIPOIC	1242	0,14	UNPUBLISHED.	L11999
LPD1	SCDHDHA	1497	0,298	PNAS 85:1831–1834 (1988).	J03645
LPD1	SCLPDH	1497	0,351	J. GEN. MICROBIOL. 134:1131–1139 (1988).	M20880
LPD1	SCLPD1	924	0,117	YEAST 3:51–57 (1987).	M16076
LTE1	SCLTE1	1833	0,326	UNPUBLISHED.	X67315
LYP1	SCLYP1	4176	0,235	GENE 46:237–245 (1986). GENE 98:141–145 (1991)	M36287
LYS1	SCLYS2A	870	0,111	CELL 66:519–531 (1991).	M73821
MAD2	SCMAD2	888	0,011	EMBO J. 9:4563–4568 (1990).	X56662
MAG1	SCMAG	888	0,004	EMBO J. 9:4569–4575 (1990).	X57781
MAG1	SCMAGG	528	≤ 0	J. MOL. BIOL. 105:427–443 (1976).	M95912
MAK3	SCMAKNHP-1	2199	0,087	GENET. RES. 13:71–83 (1969).	M94533
MAK10	SCMAK10A	1404	0,156	J. BIOL. CHEM. 263:1467–1475 (1988).	J03506
MAK11	SCMAK11	918	0,221	PNAS 85:6007–6111 (1988).	J03852
MAK16	SCMAK16A	1842	0,175	GENETICS 123:477–484 (1989).	X17391
MAL61	SCMAL61	1410	0,043	CURR. GENET. 14:319–323 (1988).	M36537
MAL63	SCMAL63	1419	0,046	MOL. GEN. GENET. 213:56–62 (1988).	X12576
MAL6R	SCMAL6R				
MAL6S (GLUCPI)	CGM0180	1752	0,237	YEAST 5:11–24 (1989).	
MAL6S	SCMAL6ST-2	1752	0,243	GENE 41:75–84 (1986).	M12601

Continued

LIST A3 Continued

Gene (Primary gene name)	Synonym	Mnemonic	Length	Codon bias	Reference	Accession
MAL6T		SCMAL6ST-1	93	0,053	GENE 41:75-84 (1986).	M12601
MAP1		SCMAP1A	1161	0,355	J. BIOL. CHEM. 267:8007-8011 (1992).	M77092
MAS1		SCMAS1	1386	0,109	EMBO J. 7:1439-1447 (1988).	X07649
MAS2	MPP	MISCMPP	1446	0,164	EMBO J. 7:3493-3500 (1988).	X13455
MAS2		SCMAS2	1446	0,161	EMBO J. 7:3863-3871 (1988).	X14105
MAS6		SCMAS6A	666	0,136	UNPUBLISHED.	X71633
MAT1A	MATA	SCMAT1A-2	357	0,042	EMBO J. 3:1061-1065 (1984).	V01313
MAT1A	MATA	SCMAT1A-3	378	0,048	EMBO J. 3:1061-1065 (1984).	V01313
MAT2A	MATALPHA	SCMAT2A-1	630	0,025	CELL 27:15-23 (1981).	V01315
MAT2A	MATALPHA	SCMAT2A-2	522	0,034	CELL 27:15-23 (1981).	V01315
MATA					see MAT1A	
MATALPHA					see MAT2A	
MBR1		SCMBR1	1017	0,056	UNPUBLISHED.	M63309
MTR3		SCMBR3A	1014	0,132	UNPUBLISHED.	X72671
MCK1		SCMCK1A	1125	0,14	MOL. CELL. BIOL. 10:6244-6256 (1990).	M55984
MCK1		SCMCK1G	1125	0,14	GENES DEV. 5:533-548 (1991).	X55054
MCM1	FUN80	SCFUN80	471	0,062	GENE 55:265-275 (1987).	M17511
MCM1		SCMCM1	858	0,096	J. MOL. BIOL. 204:593-606 (1988).	X14187
MCM1		SCPRTF	858	0,096	GENES DEV. 4:299-312 (1990).	X52453
MCM2		SCMCM2	2670	0,184	GENES DEV. 5:944-957 (1991).	X53539
MCM3		SCMCM3G	2913	0,241	MOL. CELL. BIOL. 10:5707-5720 (1990).	X53540
MDH2		SCMDH2	1131	0,279	MOL. CELL. BIOL. 11:370-380 (1991).	M62808
MDH3		SCMDH3A	1029	0,205	J. BIOL. CHEM. 267:24708-24715 (1992).	M98763
MDM1		SCMDM1	1329	0,046	J. CELL. BIOL. 118:385-395 (1992).	X66371
MEF1		SCMEF1	2283	0,22	EUR. J. BIOCHEM. 201:643-652 (1991).	X58378
MEI4		SCMEI4B	1350	≤0	MOL. CELL. BIOL. 12:1340-1351 (1992).	M84765
MEI5		SCMEI5A	669	0,035	UNPUBLISHED.	L03182
MEK1	MRE4	SCMRMNE-1	1491	0,072	NUCLEIC ACIDS RES. 20:449-457 (1992).	X63112
MEK1		SCMSPKH	1491	0,072	GENES DEV. 5:2392-2404 (1991).	X61208
MEL1		SCMEL1	1413	0,23	NUCLEIC ACIDS RES. 13:7257-7268 (1985).	X03102

Gene	Code	Length	Value	Reference	Accession
MEL1	SCMEL1CA	1413	0,23	GENE 36:333–340 (1985).	M10604
MER1	SCMER1	810	0,038	MOL. CELL. BIOL. 10:2379–2389 (1990).	M31304
MER2	SCMER2A	873	0,064	CELL 66:1257–1268 (1991).	M38340
MES1	J01339	2253	0,364	J. BIOL. CHEM. 260:15571–15576 (1985).	J01339
MES1	SCMES1	2253	0,315	PNAS 80:2437–2441 (1983).	V01316
MET2	SCMET2	1314	0,136	GENE 49:283–293 (1986).	M15675
MET3	SCMET3	1563	0,405	MOL. GEN. GENET. 210:307–313 (1987).	X06413
MET3	SCMETTDH-2	1563	0,405	YEAST 7:873–880 (1991).	X60157
MET4	SCMET4-2	2016	≤0	MOL. MICROBIOL. 7:215–228 (1993).	Z12126
MET4	SCMETLZP	1998	≤0	MOL. CELL. BIOL. 12:1719–1727 (1992).	M84455
MET8	SCMET8	822	0,052	NUCLEIC ACIDS RES. 18:659 (1990).	X17271
MET16	SCMET16A	768	0,116	J. BIOL. CHEM. 265:15518 –15524 (1990).	J05591
MET19				see ZWF1	
MET25	SCOASOAS	1332	0,577	NUCLEIC ACIDS RES. 14:7861–7871 (1986).	X04493
MFA1	SCMFA1	495	0,339	NUCLEIC ACIDS RES. 11:4049–4063 (1983).	X01581
MFA2	SCMFA2	360	0,436	NUCLEIC ACIDS RES. 11:4049–4063 (1983).	X01582
MFP1				see PDI1	
MFT1	SCMFT1	765	0,839	MOL. GEN. GENET. 225:483–491 (1991).	X55360
MGM1	SCMGMDNA	2529	0,085	GENES DEV. 6:380–389 (1992).	X62834
MGM101	SCMGM101	797	0,139	UNPUBLISHED.	X68482
MGT1	SCMGT1	630	0,018	EMBO J. 10:2179–2186 (1991).	X60368
MIF2	SCMIF2A	1647	0,07	UNPUBLISHED.	Z18294
MIG1	SCMIG1	1512	0,197	EMBO J. 9:2891–2898 (1990).	X55734
MIH1	J04846	1422	0,009	CELL 57:295–303 (1989).	J04846
MIP1	SCMIP1	3762	0,019	J. BIOL. CHEM. 264:20552–20560 (1989).	J05117
MIR1	SCMIR1	933	0,595	NATURE 347:488–491 (1990).	X57478
MIS1	SCMIS1A	2925	0,256	J. BIOL. CHEM. 263:7717–7725 (1988).	J03724
MKK1	SCD13001	1524	0,001	UNPUBLISHED.	D13001
MKK2	SCMKK2	1518	0,072	UNPUBLISHED.	D13785
MKS1	SCMKS1	1374	0,045	MOL. GEN. GENET. 238:6–16 (1993).	D13715
MLP1	SCMLP	5625	0,109	MOL. GEN. GENET. 237:359–369 (1993).	L01992
MLS1	SCMLSIC	1662	0,297	NUCLEIC ACIDS RES. 20:5677–5686 (1992).	X64407
MNS1	SCMSD1	1647	0,11	J. BIOL. CHEM. 266:15120–15127 (1991).	M63598
MNT1				see KRE2	

Continued

LIST A3 Continued

Gene (Primary gene name)	Synonym	Mnemonic	Length	Codon bias	Reference	Accession
MOD5		SCMOD5	1284	0,057	MOL. CELL. BIOL. 7:185–191 (1987).	M15991
MOT1		SCMOT1	5601	0,08	MOL. CELL. BIOL. 12:1879–1892 (1992).	M83224
MPI1		SCMPI1	1293	0,165	EMBO J. 11:3619–3628 (1992).	X67276
MPP					see MAS2	
MRE11		SCMRE11	1929	0,053	UNPUBLISHED.	D11463
MRE4					see MEK1	
MFR1		SCMRF1-2	1239	0,11	NUCLEIC ACIDS RES. 20:6339–6346 (1992).	X60381
MRNP					see PAB1	
MRP1		SCMRP1	963	0,014	J. BIOL. CHEM. 262:3388–3397 (1987).	M15160
MRP2		SCMRP2A	345	0,171	J. BIOL. CHEM. 262:3388–3397 (1987).	M15161
MRP4		SCMRP4A	1182	0,235	J. BIOL. CHEM. 267:5508–5514 (1992).	M82841
MRP7		SCMRP7	1113	0,077	MOL. CELL. BIOL. 8:3636–3646 (1988).	M22116
MRP13		SCMRP13	972	0,002	MOL. CELL. BIOL. 8:3647–3660 (1988).	M22109
MRP17		SCMRP17	393	0,053	MOL. GEN. GENET. 235:64–73 (1992).	X58362
MRP20		SCMRP20A	789	≤ 0	J. BIOL. CHEM. 267:5162–5170 (1992).	M81696
MRP49		SCMRP49A-2	411	0,051	CELL 50:277–287 (1987).	M81697
MRPL6		SCMTRPL6-2	615	0,08	CURR. GENET. 24:136–140 (1993).	X69480
MRPL8		SCMRPL8	714	0,032	UNPUBLISHED.	X53841
MRPL9		SCMRPL9G	807	0,104	EUR. J. BIOCHEM. 206:373–380 (1992).	X65014
MRPL20		SCMRPL20	585	0,253	NUCLEIC ACIDS RES. 18:1521–1529 (1990).	X53840
MRPL31		SCMRPL31	393	0,004	EUR. J. BIOCHEM. 183:155–160 (1989).	X15099
MRPL33		SCYML33	297	0,041	J. BACTERIOL. 173:4013–4020 (1991).	D90217
MRPS28		SCMRPS28	858	0,049	NUCLEIC ACIDS RES. 18:6895–6901 (1990).	X55977
MRS1		SCMRS1-1	1089	0,169	EMBO J. 6:2123–2129 (1987).	X05509
MRS2		SCMRS2A	1410	0,065	J. BIOL. CHEM. 267:6963–6969 (1992).	M82916
MRS3		SCMRS3	942	0,059	MOL. GEN. GENET. 210:145–152 (1987).	X06239
MRS1		SCMRS3G	942	0,072	J. MOL. BIOL. 217:23–37 (1991).	X56445
MRS4		SCMRS4-2	912	≤ 0	J. MOL. BIOL. 217:23–37 (1991).	X56444
MRS5		SCORF	327	0,113	UNPUBLISHED.	M90689
MRS6*		SCMRS6A	1389		UNPUBLISHED.	M90844

Gene	Synonym	Locus	Length	Value	Reference	Accession
MSB1		SCMSB1R1	3411	0,017	MOL. CELL. BIOL. 11:1295–1305 (1991).	M37767
MSB2		SCMSB2A	3918	0,104	YEAST 8:315–323 (1992).	M77354
MSD1		SCTASPT	1974	0,056	PNAS 86:6023–6027 (1989).	M26020
MSF1		SCMSF1	1422	0,048	J. BIOL. CHEM. 262:3690–3696 (1987).	J02691
MSH1		SCMSH1A	2877	0,003	GENETICS 132:963–973 (1992).	M84169
MSH2		SCMSH2A	2898	0,139	GENETICS 132:963–973 (1992).	M84170
MSH3		SCMSH3A	3141	≤0	UNPUBLISHED.	M96250
MSI1		SCMSI1	1266	0,041	PNAS 86:8778–8782 (1989).	M27300
MSI3		SCMSI3P	2079	0,639	UNPUBLISHED.	D13743
MSL1					see NAM2	
MSM1		SCMSM1GN	1725	0,069	EUR. J. BIOCHEM. 179 365–371 (1989).	X14629
MSN1		SCMSN1	1146	0,013	NUCLEIC ACIDS RES. 18:6959–6964 (1990).	X54324
MSS1		SCMSS1A	1578	0,049	UNPUBLISHED.	X69481
MSS18		SCMSS18	804	0,032	EMBO J. 7:1455–1464 (1988).	X07650
MSS51		MISCCO1A	1308	0,146	CELL 32:77–87 (1983).	J01487
MSS116		CGM0299	1992	0,219	NATURE 337:84–87 (1989).	
MST1		SCMST1A	1386	0,179	J. BIOL. CHEM. 260:15362–15370 (1985).	M12087
MSW1		SCMSW	1122	0,031	J. BIOL. CHEM. 260:15371–15377 (1985).	M12981
MSW1		SCMSW1AA	1137	0,035	CURR. GENET. 21:281–283 (1992).	X66165
MTF1		SCRF1023	1023	0,01	MOL. GEN. GENET. 214:218–223 (1988).	X13513
MTF2					see NAM1	
MTS1					see NOP3	
MYO1		SCMYO1G	5550	0,065	NUCLEIC ACIDS RES. 18:7147 (1990).	X53947
MYO1*		SCMYO1	2280	0,077	EMBO J. †:3499–3505 (1987).	X06187
MYO2		SCMYO2A	4722	0,193	J. CELL BIOL. 113:539–551 (1991).	M35532
MYO4		SCMYO4P	4413	0,108	UNPUBLISHED.	M90057
NAB1		SCNAB1A	756	0,805	UNPUBLISHED.	M88277
NAB2		SCNPRPB	1575	0,175	MOL. CELL BIOL. 13:2730–2741 (1993).	L10288
NAM1		MISCNAM1	1320	0,088	MOL. GEN. GENET. 215:517–528 (1989).	X14719
NAM1	MTF2	SCMTF2A	1320	0,088	MOL. GEN. GENET. 220:186–190 (1990).	X51665
NAM2	MSL1	SCMSL1	2682	0,044	J. BIOL. CHEM. 263:850–856 (1988).	J03495
NAM2		SCNAM2-2	2682	0,044	EMBO J. 6:713–721 (1987).	X05143
NAM7		SCNAM7	2913	0,208	J. MOL. BIOL. 224:575–587 (1992).	X62394
NAM8		SCNAM8-2	1569	0,083	MOL. GEN. GENET. 233:136–144 (1992).	X64763

Continued

LIST A3 *Continued*

Gene (Primary gene name)	Synonym	Mnemonic	Length	Codon bias	Reference	Accession
NAM9		SCNAM9	1452	0,109	MOL. CELL BIOL. 12:402–412 (1992).	M60730
NAP1		SCNAPI	1251	0,18	J. BIOL. CHEM. 266:7025–7029 (1991).	M63555
NAT1					see AAA1	
NCA1	AEP1	SCAEPX	1554	≤0	CURR. GENET. 24:126–135 (1993).	M80615
NCA1		SCMINCA1G	1386	≤0	J. MOL. BIOL. 229:909–916 (1993).	Z12301
NDC1		SCNDC1A	1965	0,074	UNPUBLISHED.	X70281
NDI1		SCNDI1G	1539	0,257	EUR. J. BIOCHEM. 203:587–592 (1992).	X61590
NES24					see HDF1	
NGG1		SCNGG1A	2106	0,03	UNPUBLISHED.	L12137
NGR1		SCNGR1	2016	0,061	UNPUBLISHED.	Z14097
NHP2		SCNHP2GEN	468	0,506	YEAST 7:79–90 (1991).	X57714
NHP6A		SCMAKNHP-2	279	0,537	J. MOL. BIOL. 105:427–443 (1976).	M95912
NHP6A		SCNHP6A	279	0,537	J. BIOL. CHEM. 265:3234–3239 (1990).	X15317
NHP6B		SCNHP6B	297	0,402	J. BIOL. CHEM. 265:3234–3239 (1990).	X15318
NIN1		SCNIN1	822	0,157	UNPUBLISHED.	D10515
NIP1		SCNIP1X	2436	0,368	PNAS 89:10355–10359 (1992).	L02899
NMT1		SCNMT	1365	0,156	J. CELL BIOL. 243:796–800 (1989).	M23726
NOP1		SCBOP1	981	0,664	J. BIOL. CHEM. 265:2209–2215 (1990).	J05230
NOP1		SCNOP1SEQ	981	0,664	EMBO. J 8:4015–4024 (1989).	X51676
NOP3	MTS1	SCMTS1-2	1242	0,546	UNPUBLISHED.	X70951
NOP3	NPL3	SCNPL3A	1242	0,542	UNPUBLISHED.	M86731
NOP3		SCNOP3	1242	0,546	J. CELL BIOL. 119:737–747 (1992).	X66019
NPK1					see KIN3	
NPL1		SCNPL1	1989	0,168	J. CELL BIOL. 109:2665–2675 (1989).	X16388
NPL3					see NOP3	
NPL6		SCNP16X	1305	0,054	UNPUBLISHED.	M98434
NPR1		SCNPR1G	2370	0,07	MOL. GEN. GENET. 222:393–399 (1990).	X56084
NSP1					see STH1	
NSP1		SCNSLP	2469	0,127	EMBO J. 7:4323–4334 (1988).	M37160
NSP1					see DKA1	
NSP116					see NUP116	
NSP49					see NUP49	

Name	SC Name	Alias	Length	Value	Reference	Accession
NSR1	SCNSR1		1242	0,623	J. CELL BIOL. 113:1-12 (1991).	X57185
NTH1	SCNTHC		2079	0,159	J. BIOL. CHEM. 268:4766-4774 (1993).	X65925
NUC1	SCNUC1		987	0,078	NUCLEIC ACIDS RES. 16:3297-3312 (1988).	X06670
NUD1	SCNUD1		2553	0,07	UNPUBLISHED.	X62147
NUF1	SCNUF1G		2832	0,05	J. CELL BIOL. 116:1319-1332 (1992).	Z11582
NUM1	SCNUM1		8244	0,097	MOL. GEN. GENET. 230:277-287 (1991).	X61236
NUP1	SCNUP1		3228	0,023	CELL 61:965-978 (1990).	M33632
NUP2	SCNPTRNA		2160	0,057	MOL. BIOL. CELL. 4:209-222 (1993).	X69964
NUP49	SCNSP49	NSP49	1416	≤0	EMBO J. 11:5051-5061 (1992).	X68109
NUP49	SCNUP49		1416	≤0	J. CELL BIOL. 119:705-723 (1992).	Z15040
NUP100	SCNUP100		2877	0,007	J. CELL BIOL. 119:705-723 (1992).	Z15035
NUP116	SCNSP116	NSP116	3339	0,082	EMBO J. 11:5051-5061 (1992).	X68108
NUP116	SCNUP116		3339	0,087	J. CELL BIOL. 119:705-723 (1992).	Z15036
OBF1					see BAF1	
OCH1	SCOCH1		1440	0,092	EMBO J. 11:2511-2519 (1992).	D11095
OLE1	SCOLE1		1530	0,657	J. BIOL. CHEM. 265:20144-20149 (1990).	J05676
OMP1	SCOMPMI1		1851	0,316	EMBO J. 2:2169-2172 (1983).	X05585
OMP2	SCPORIN		849	0,453	EMBO J. 4:769-774 (1985).	X02324
OPI1	SCOPI1		1212	0,026	J. BIOL CHEM. 266:863-872 (1991).	M57383
ORD1	SCORD1		1398	0,17	J. BIOL CHEM. 262:10127-10133 (1987).	J02777
OSCP					see ATP5	
OSM1	SCCORA-3		903	0,202	GENE 87:157-166 (1990).	M37696
PAB1	SCPABPG	MRNP	1731	0,614	CELL 45:827-835 (1986).	M12780
PAB1	SC		1731	0,616	MOL. CELL. BIOL. 6:2932-2943 (1986).	D00023
PAB1	SCSYGP1-8		1731	0,616	UNPUBLISHED.	L10718
PAI3	SCPAI3		204	≤0	FEBS LETT. 283:78-84 (1991).	X60050
PAK3	SCAKB	ADK2	675	≤0	UNPUBLISHED.	M77757
PAK3	SCPAK3G		675	0,128	MOL. GEN. GENET. 233:363-371 (1992).	X65126
PAN1	SCPAN1A		4407	0,123	CELL 70:961-973 (1992).	M90688
PAP1	SCPAP1G		1704	0,045	NATURE 354:496-498 (1991).	X60307
PAS1	SCPAS1P		3129	≤0	CELL 64:499-510 (1991).	M58676
PAS2*	SCPAS2		957	≤0	NATURE 359:73-76 (1992).	X65470
PAS3	SCPAS3		1323	0,476	J. CELL BIOL. 114:1167-1178 (1991).	X58407
PBI2	SCPBI2		225	0,086	EUR. J. BIOCHEM. 197:1-7 (1991).	X60051
PBS2	SCPBS2		2130	0,356	PNAS 84:5848-5852 (1987).	J02946
PCY1	SCPCB		3534		J. BIOL. CHEM. 263:11493-11497 (1988).	J03889

Continued

LIST A3 Continued

Gene (Primary gene name)	Synonym	Mnemonic	Length	Codon bias	Reference	Accession
PDB1		SCE1B	1098	0,485	PNAS 90:1252-1256 (1993).	M98476
PDC1		SCPDC1	1647	0,928	NUCLEIC ACIDS RES. 14:8963-8977 (1986).	X04675
PDC2		SCPDC2G	2775	0,009	UNPUBLISHED.	X65608
PDC5		SCPDC5	1689	0,842	UNPUBLISHED.	X15668
PDC6		SCPDC6	1689	0,24	J. BACTERIOL. 173:7963-7969 (1991).	X55905
PDE1		SCPPR	1107	0,076	MOL. CELL BIOL. 7:3629-3636 (1987).	M17781
PDE2		SCPDE2	1578	0,072	PNAS 83:9303-9307 (1986).	M14563
PDI1	MFP1	SCMFP1	1587	0,583	UNPUBLISHED.	X52313
PDI1		SCPDI1	1590	0,602	GENE 108:81-89 (1991).	X54535
PDI1		SCPDIAA	1566	0,568	PNAS 88:4453-4457 (1991).	M62815
PDR1	TRG1	SCTRG1	1587	0,583	J. BIOL. CHEM. 266:24557-24563 (1991).	M76982
PDR1		SCPDR1	3189	0,048	J. BIOL. CHEM. 262:16871-16879 (1987).	J03487
PDR4		SCPDR4-2	1950	0,029	UNPUBLISHED.	X53830
PDR4	YAP1	SCYAP1	1950	0,024	GENES DEV. 3:283-292 (1989).	X58693
PEM1		SCPEMA	2607	0,132	J. BIOL. CHEM. 262:15428-15435 (1987).	M16987
PEM2		SCPEMB	618	0,306	J. BIOL. CHEM. 262:15428-15435 (1987).	M16988
PEP1		SC114-1	4737	0,09	YEAST 8:769-776 (1992).	X68577
PEP12P		SCPEP12P	864	≤0	UNPUBLISHED.	M90395
PEP3		SCPEP3	2754	0,053	MOL. CELL BIOL. 11:5801-5812 (1991).	M65144
PEP4		SCPEP4	1215	0,467	MOL. CELL BIOL. 6:2500-2510 (1986).	M13358
PEP5					see END1	
PEPC					see PPC1	
PET3		SCPET3	2154	0,002	MOL. CELL. BIOL. 7:2728-2734 (1987).	M17143
PET54		SCPET54	879	0,052	GENETICS 122:297-305 (1989).	X13427
PET56		SCHIS3G-1	285	0,054	J. MOL. BIOL. 152:553-568 (1981).	X03245
PET117		SCPET117P	321	≤0	CURR. GENET. 13:9-14 (1992).	L06066
PET122		SCPET122-1	726	0,078	NUCLEIC ACIDS RES. 16:10783-10802 (1988).	X07558
PET123		SCPET123	954	0,123	GENETICS 125:495-503 (1990).	X52363
PET127		SCPET127	2130	0,021	MOL. GEN. GENET. 235:64-73 (1992).	X58363
PET191		SCPET191P	324	0,114	CURR. GENET. 13:9-14 (1992).	L06067
PET309		SCPET309A	2895	≤0	UNPUBLISHED.	L06072

PET494	SCPET4941	1467	0,027	MOL. GEN. GENET. 202:294–301 (1986).	K03520
PFK1	SCPFK1AA	2961	0,606	GENE 78:309–321 (1989).	M26943
PFK1	SCPFK2AA	2877	0,66	GENE 78:309–321 (1989).	M26944
PFY1	M23369	378	0,618	MOL. CELL. BIOL. 8:5108–5115 (1988).	M23369
PGI1	SCPG11	1662	0,793	MOL. GEN. GENET. 215:100–106 (1988).	M37267
PGI1	SCPPG	1662	0,793	GENE 73:153–161 (1988).	M21696
PGK1	SCPGK	1248	0,906	NUCLEIC ACIDS RES. 10:7791–7808 (1982).	J01342
PGK1*	SCPGK2	405	0,846	BIOCHEM. J. 211:199–218 (1983).	M14438
PHA2	SCATPMITO-3	278	≤0	J. BIOL. CHEM. 267:7386–7394 (1992).	M87006
PHO2	SCPHO2	1677	0,04	NUCLEIC ACIDS RES. 15:233–246 (1987).	X05062
PHO3	SCPHO35-2	1402	0,474	NUCLEIC ACIDS RES. 12:7721–7739 (1984).	X01080
PHO4	SCPHO4	927	0,015	NUCLEIC ACIDS RES. 14:3059–3073 (1986).	X03719
PHO5	SCPHO5	1401	0,557	NUCLEIC ACIDS RES. 11:1657–1672 (1983).	V01320
PHO5	SCPHO5A	1401	0,579	NUCLEIC ACIDS RES. 12:7721–7739 (1984).	X01079
PHO8	SCPHO8A	1698	0,143	GENE 58:137–148 (1987).	M21134
PHO13	SCPHO13	936	0,258	MOL. GEN. GENET. 220:133–139 (1989).	X51611
PHO80	SCPHO80	879	0,04	NUCLEIC ACIDS RES. 16:2625–2637 (1987).	X07464
PHO80*	SCPHO80R	231	0,05	GENE 96:181–188 (1990).	M60625
PHO81	SCPHO81	3531	0,016	NUCLEIC ACIDS RES. 18:2176 (1990).	X52482
PHO84	SCHO84	1788	0,609	MOL. CELL. BIOL. 11:3229–3238 (1991).	D90346
PHO85	SCPHO85G	906	0,152	NUCLEIC ACIDS RES. 15:10299–10309 (1987).	Y00867
PHR1	SCPHR	1695	0,061	GENE 36:349–355 (1985).	M11578
PHR1	SCPHR1	1695	0,069	NUCLEIC ACIDS RES. 13:8231–8246 (1985).	X03183
PIF1	SCPIF	2571	0,078	EMBO J. 6:1441–1449 (1987).	X05342
PIR1	SCPIR1P	1023	0,688	UNPUBLISHED.	D13740
PIR2				see HSP150	
PIR3	SCPIR3P	1083	0,59	UNPUBLISHED.	D13742
PIS1	SCPIS	660	0,21	J. BIOL. CHEM. 262:4876–4881 (1987).	J02697
PK25				see TPK1	
PKC1	SCPKC1A	3453	0,142	CELL 62:213–224 (1990).	M32491
PKX2				see CDC5	
PLC1	SCPLC-2	2607	0,03	UNPUBLISHED.	L13036
PLC1	SCPLC1	2607	0	PNAS 90:1804–1808 (1993).	D12738
PMA1	SCPMA1	2754	0,836	NATURE 319:689–693 (1986).	X03534

Continued

LIST A3 Continued

Gene (Primary gene name)	Synonym	Mnemonic	Length	Codon bias	Reference	Accession
PMA2		SCPMA2A	2841	0,457	J. BIOL. CHEM. 263:19480-19487 (1988).	J04421
PMI40		SCPHOISOA	1287	0,339	MOL. CELL. BIOL. 12:2924-2930 (1992).	M85238
PMP1		SCPMP1A	120	0,819	J. BIOL. CHEM. 267:6425-6428 (1992).	M77845
PMR1		SCPMR1	2850	0,209	CELL 58:133-145 (1989).	M25488
PMR2		SCPMR2	3273	0,233	CELL 58:133-145 (1989).	M25489
PMR2		SCPMR2G	1674	0,225	MOL. GEN. GENET. 227:149-154 (1991).	X58626
PMS1		SCPMS1A	2712	0,006	J. BACTERIOL. 171:5339-5346 (1989).	M29688
PMT1		SCPMTP	972	0,213	UNPUBLISHED.	L04948
POB1		SCPOB1A	2781	0,074	MOL. CELL. BIOL. 12:5724-5735 (1992).	M94769
POL1		SCPOL1M	4404	0,103	PNAS 85:3772-3776 (1988).	J03268
POL2		SCDNAPOL	6666	0,088	CELL 62:1143-1151 (1990).	M60416
POL3					see CDC2	
POL30		SCPOL30	774	0,297	NUCLEIC ACIDS RES. 18:261-265 (1990).	X16676
POT1		SCPOT1	1251	0,183	YEAST 7:379-389 (1991).	X53395
POX1		SCPOX1	2244	0,121	GENE 88:247-252 (1990).	M27515
PPA1		SCPPA1	639	0,412	BBRC 168:574-579	M35294
PPA2		SCPPA2A	930	0,039	J. BIOL. CHEM. 266:12168-12172 (1991).	M81880
PPC1	PEPC	SCPEPC	1659	0,406	NUCLEIC ACIDS RES. 16:10926 (1988).	X13096
PPG1		SCPPGCS	1104	0,09	J. BIOL. CHEM. 268:1349-1354 (1993).	M94269
PPH1	SIT4	M24395	933	0,139	CELL 56:527-537 (1989).	M24395
PPH3		SCPPH3-2	924	0,036	MOL. CELL. BIOL. 11:4876-4884 (1991).	X58858
PPH21		SCPPH1G	1107	0,103	EMBO J. 9:4339-4346 (1990).	X56261
PPH21		SCPPH21	1107	0,103	MOL. CELL. BIOL. 11:4876-4884 (1991).	X58856
PPH22		SCPPH22-4	1131	0,125	MOL. CELL. BIOL. 11:4876-4884 (1991).	X58857
PPH22		SCPPH22G	1131	0,125	EMBO J. 9:4339-4346 (1990).	X56262
PPH22		SCSIT4A	1131	0,125	MOL. CELL. BIOL. 11:2133-2148 (1991).	M60317
PPR1		SCPPR1	2712	0,009	J. MOL. BIOL. 180:239-250 (1984).	X01739
PPR2	DST1	SCSTPA	927	0,185	MOL. CELL. BIOL. 11:2576-2582 (1991).	M36724
PPR2		SCPPR2	381	0,293	EMBO J. 2:2071-2073 (1983).	X00047
PPZ1		SCPPZ1X	2076	0,098	J. BIOL. CHEM. 267:11734-11740 (1992).	M86242

Gene	Alt.	Sequence	Length	Value	Reference	Accession
PRB1		SCPRB1	1905	0,45	MOL. CELL. BIOL. 7:4390–4399 (1987).	M18097
PRC1		SCPRCCPY	1596	0,371	CELL 48:887–897 (1987).	M15482
PRD1		YSP450R	2073	0,342	J. BIOCHEM. 103:1004–1010 (1989).	D00316
PRE1		SCPRE1	594	0,236	EMBO J. 10:555–562 (1991).	X56812
PRE2		SCPROTEO	864	0,242	GENE 122:203–206 (1992).	M96667
PRE4		SCPRE2A	861	0,262	J. BIOL. CHEM. 268:5115–5120 (1993).	X68662
PRF1		SCPRE4A	798	0,163	J. BIOL. CHEM. 268:3479–3486 (1993).	X68663
PRG1		SCPROFR	351	0,576	NUCLEIC ACIDS RES. 15:9078 (1987).	Y00463
PRG1	PRG1				see PRE2	
PRI1		SCPRI1	1227	0,112	NUCLEIC ACIDS RES. 15:7975–7989 (1987).	Y00458
PRI2		M27209	1584	0,074	MOL. CELL. BIOL. 9:3081–3087 (1989).	M27209
PRO1		SCPRO1A	1284	0,192	UNPUBLISHED.	M85293
PRO3		SCPRO3	858	0,3	J. BACTERIOL. 174:3782–3788 (1992).	M57886
PROFR	PROFR				see PRF1	
PRP1	PRP1				see PRPS1	
PRP2		SCPRP2	2628	0,057	NUCLEIC ACIDS RES. 18:6447 (1990).	X55936
PRP2		SCPRP2G	2628	0,057	UNPUBLISHED.	X55999
PRP4		M28518	1395	0,017	MOL. CELL. BIOL. 9:3710–3719 (1989).	M28518
PRP4		SCPRP4A	1395	0,017	MOL. CELL. BIOL. 9:3698–3709 (1989).	M26597
PRP5		SCPRP5	2547	0,053	PNAS 87:4236–4240 (1990).	M33191
PRP6		SCPRP6	2697	0,026	EMBO J. 9:2775–2781 (1990).	X53465
PRP9		SCPRP9	1590	0,048	EMBO J. 9:2775–2781 (1990).	X53466
PRP16		SCPRP16A	3213	0,052	CELL 60:705–717 (1990).	M31524
PRP19		SCPRP19A	1299	0,007	UNPUBLISHED.	L09721
PRP21					see SPP91	
PRP22		SCPRP22GE	3435	0,028	NATURE 349:487–494 (1991).	X58681
PRP28		SCPRP28C	1764	0,121	GENES DEV. 5:629–641 (1991).	X56934
PRP38		SCPRP38A	726	0,019	MOL. CELL. BIOL. 12:3939–3947 (1992).	M95921
PRPS1		SCPRP1A	1281	0,332	UNPUBLISHED.	L04130
PRPS1		SCPRPSA	1281	0,332	UNPUBLISHED.	X70069
PRS1		SCYC1	864	0,224	J. BIOL. CHEM. 265:16604–16613 (1990).	M55436
PRS2					see SCL1	
PRT1		SCPRT1	2289	0,378	J. BIOL. CHEM. 262:2845–2851 (1987).	J02674
PSE1		SCPSE1G	3267	0,172	J. CELL SCI. 101:709–719 (1992).	Z11538
PSS					see CHO1	

Continued

LIST A3 Continued

Gene (Primary gene name)	Synonym	Mnemonic	Length	Codon bias	Reference	Accession
PTA1		SCPTA1	2355	0,145	MOL. CELL. BIOL. 12:3843–3856 (1992).	M95673
PTM1		SCPTM	1569	0,102	UNPUBLISHED.	L11895
PTM1	YKL252	SCXI12-2	1593	0,103	YEAST 8:977–986 (1992).	X69584
PTP1		SCPTPASE	1005	0,03	J. BIOL. CHEM. 266:12964–12970 (1991).	M64062
PTP2		SCPTP2A	2250	0,023	PNAS 89:2355–2359 (1982).	M82872
PTP2		SCPTPII	2250	0,025	J. BIOL. CHEM. 267:10024–10030 (1992).	M85287
PTR2		SCPTR2	1530	0,327	UNPUBLISHED.	L11994
PUB1		SCPUB1X	1359	0,391	UNPUBLISHED.	L13725
PUB1		SCRNABIND	1287	0,397	UNPUBLISHED.	L01797
PUP1		SCPUP1	783	0,23	NUCLEIC ACIDS RES. 19:5075 (1991).	X61189
PUT1		SCPUT1	1428	0,243	MOL. CELL. BIOL. 7:4431–4440 (1987).	M18107
PUT2		SCPUT2	1725	0,165	MOL. CELL. BIOL. 4:2837–2842 (1984).	M10029
PUT3		SCPUT3	2937	0,092	MOL. CELL. BIOL. 11:2609–2619 (1991).	X55384
PUT4		SCPUT4	1881	0,121	GENE 83:153–159 (1989).	M30583
PWP1		SCPWP1	1728	0,282	UNPUBLISHED.	M37578
PYC2		SCPYC2G	3555	0,425	MOL. GEN. GENET. 229:307–315 (1991).	X59890
PYK1		SCPKO1	1497	0,946	J. BIOL. CHEM. 258:2193–2201 (1983).	V01321
PYK1		SCPYK1	1500	0,965	FEBS LETT. 247:312–316 (1989).	X14400
QCR2	COXCH2	SCCOXCH2	1104	0,326	EUR. J. BIOCHEM. 163:97–103 (1987).	X05120
QCR6	CR17	SCCR17	441	0,222	EMBO J. 3:1039–1043 (1984).	X00551
QCR9		SCQCR9	198	0,057	J. BIOL. CHEM. 265:20813–20821 (1990).	M59797
QRI8		SCQRI8G	495	0,066	BIOCHIM. BIOPHYS. ACTA 1132:211–213 (1992).	X66829
RAD1		SCRAD1	2916	0,013	MOL. CELL. BIOL. 4:2161–2169 (1984).	K02070
RAD1		SCRAD1A	3300	0,018	MOL. CELL. BIOL. 7:1012–1020 (1987).	M15435
RAD2		SCRAD2G	2925	0,038	GENE 36:225–234 (1985).	M10275
RAD3		SCRAD3	2334	0,096	NUCLEIC ACIDS RES. 13:2357–2372 (1985).	X02368
RAD3		SCRAD3G	2334	0,096	MOL. CELL. BIOL. 5:17–26 (1985).	K03293
RAD4		SCRAD4A	2262	0,075	GENE 74:535–541 (1989).	M26050
RAD4		SCRAD4B	2262	0,075	J. BACTERIOL 171:1862–1869 (1989).	M24928

Gene	Alias	Sequence	Size		Reference	Accession
RAD4		SCYGP1-3	2262	0,074	UNPUBLISHED.	L10718
RAD4?	???	SCRAD4G	2190	0,127	NUCLEIC ACIDS RES. 18:7137 (1990).	X55891
RAD5		SCRAD5A	3507	0,053	MOL. CELL. BIOL. 12:3807-3818 (1992).	M96644
RAD6		CGM0286	516	0,22	NATURE 329:131-134 (1987).	K02962
RAD6		SCRAD6	516	0,22	PNAS 82:168-172 (1985).	M37696
RAD7		SCCORA-4	1695	0,109	GENE 87:157-166 (1990).	M13015
RAD7		SCRAD7	1695	0,116	MOL. CELL. BIOL. 6:1497-1507 (1986).	M26049
RAD9		SCRAD9	3927	0,009	MOL. CELL. BIOL. 9:1882-1896 (1989).	X02591
RAD10		SCRAD10-2	630	0,045	EMBO J. 4:1575-1582 (1985).	X05225
RAD10		SCRAD10G	630	0,045	EMBO J. 4:3549-3552 (1985).	X64064
RAD14		SCRAD14	741	0,158	NATURE 355:555-558 (1992).	X66247
RAD16		SCCMDLYS-3	2370	0,127	YEAST 8:397-408 (1992).	M36405
RAD18		SCRAD101-1	1461	0,086	GENE 74:543-547 (1988).	X12588
RAD18		SCRAD18	1461	0,086	NUCLEIC ACIDS RES. 16:7119-7131 (1988).	
RAD25					see SSL2	
RAD50		SCRAD50	3936	0,108	GENETICS 122:47-57 (1989).	X14814
RAD51		SCRAD51-1	1200	0,264	CELL 69:457-470 (1992).	D10023
RAD51		SCRAD51A-3	1200	0,264	MOL. CELL. BIOL. 12:3235-3246 (1992).	M88470
RAD51		SCRADDNA	1200	0,264	MOL. CELL. BIOL. 12:3224-3234 (1992).	X64270
RAD52		SCRAD52	1512	0,093	MOL. CELL. BIOL. 4:2735-2744 (1984).	M10249
RAD54		SCRAD54A	2694	0,149	GENE 104:103-106 (1991).	M63232
RAD57		SCRAD57A	1380	0,004	GENE 105:139-140 (1991).	M65061
RADH1		SCRADH	3525	0,023	NUCLEIC ACIDS RES. 17:7211-7219 (1989).	X15665
RAM2		SCRAM2	948	0,134	PNAS 88:11373-11377 (1991).	M88584
RAP1		YSRAP1A	2481	0,075	CELL 51:721-732 (1987).	M18068
RAR1	ERG12	SCERG12G	1329	0,206	CURR. GENET. 19:9-14 (1991).	X55875
RAR1		SCRAR1	1329	0,206	MOL. GEN. GENET. 210:509-517 (1987).	X06114
RAR5					see KEM1	
RAS1		SCRAS1	927	0,153	CELL 36:607-612 (1984).	K01970
RAS2		SCRAS2	966	0,23	CELL 36:607-612 (1984).	K01971
RAS2	NUCLEASE	SCRASHO2	966	0,218	NUCLEIC ACIDS RES. 12:3611-3618 (1984).	X00528
RAT1		SCEXORIBC	3018	0,16	UNPUBLISHED.	Z11746
RAT1		SCRAT1P	3018	0,16	GENES DEV. 6:1173-1189 (1992).	M95626
RBP1					see FKB1	

Continued

LIST A3 *Continued*

Gene (Primary gene name)	Synonym	Mnemonic	Length	Codon bias	Reference	Accession
RCS1		SCRCS1	963	0,12	YEAST 7:1–14 (1991).	X53046
REB1		SCREB1	2427	0,112	MOL. CELL. BIOL. 10:5226–5234 (1990).	M58728
REC102		SCREC102	600	0,004	MOL. CELL. BIOL. 12:1248–1256 (1992).	M74045
REC114		SCREC114	1305	0,065	UNPUBLISHED.	D11462
RED1		SCRED1	2481	0,067	MOL. GEN. GENET. 218:293–301 (1989).	X16183
REG1	BCY1	SCCAPK	1248	0,257	MOL. CELL. BIOL. 7:1371–1377 (1987).	M15756
REG1		SCREG1	1248	0,262	NUCLEIC ACIDS RES. 15:368–369 (1987).	X05051
REG1	SRA1	SCSRA1	1248	0,257	MOL. CELL. BIOL. 7:2653–2663 (1987).	M17223
RET1		SCRET1-2	3447	0,265	J. BIOL. CHEM. 266:5616–5624 (1991).	M38723
REV1		SCREV1	2955	0,022	J. BACTERIOL. 171:230–237 (1989).	M22222
REV3		SECREV3	4512	0,031	J. BACTERIOL. 171:5659–5667 (1989).	M29683
RFA1	RPA1	SCRPA1A	1863	0,22	EMBO J. 9:2321–2329 (1990).	M60262
RFA1		SCRFA1	1863	0,22	GENES DEV. 5:1589–1600 (1991).	X59748
RFA2		SCRFA2	819	0,197	GENES DEV. 5:1589–1600 (1991).	X59749
RFA3		SCRFA3	366	0,044	GEBES DEV. 5:1589–1600 (1991).	X59750
RGM1		SCRGM1	633	0,15	NUCLEIC ACIDS RES. 19:4873–4877 (1991).	X59861
RGP1		SCRGP1	1983	0,013	NUCLEIC ACIDS RES. 18:1064 (1990).	X52081
RGR1		SCRGR1	3246	0,051	MOL. CELL. BIOL. 10:4103–4138 (1990).	D90051
RHO1		SCRHO1X	627	0,402	PNAS 84:779–783 (1987).	M15189
RHO2		SCRHO2X	576	0,028	PNAS 84:779–783 (1987).	M15190
RHO3		SCRHO3	693	0,163	GENE 114:43–49 (1992).	D10006
RHO4		SCRHO4	690	0,031	GENE 114:43–49 (1992).	D10007
RIF1		SCRIF1	5748	0,008	GENES DEV. 6:801–814 (1992).	X66501
RIP1		SCRIP101	645	0,429	J. BIOL. CHEM. 262:8901–8909 (1987).	M23316
RME1		SCZFP	900	0,005	GENES DEV. 5:1982–1989 (1991).	M76447
RNA1		SCRNA1	1221	0,251	MOL. CELL. BIOL. 9:2989–2999 (1989).	M27142
RNA14		SCRNA14	1908	0,1	MOL. CELL. BIOL. 11:3075–3087 (1991).	M73461
RNA15		SCNA15	888	≤ 0	MOL. CELL. BIOL. 11:3075–3087 (1991).	M73462
RNR2		SCRNR2	1197	0,645	MOL. CELL. BIOL. 7:2783–2793 (1987).	M17221

RNR2		SCRNR2A	1197	0,636	MOL. CELL. BIOL. 7:3673-3677 (1987).	M17789
ROX1		SCROX1R	1104	0,145	UNPUBLISHED.	X60458
ROX3		SCROX3	660	0,011	MOL. CELL. BIOL. 11:5639-5647 (1991).	X58300
RP28A		SCRP28A1	558	0,888	NUCLEIC ACIDS RES. 12:7345-7358 (1984)	X01099
RP28B		SCRPS16A-1	558	0,805	NUCLEIC ACIDS RES. 12:7345-7358 (1984).	X01100
RP29		SCRP29	465	0,827	J. BIOL. CHEM. 259:9218-9224 (1984).	K02650
RP51A		SCRP51	408	0,87	PNAS 80:4403-4407 (1983).	J01349
RP51B		SCRP51B	408	0,832	MOL. CELL. BIOL. 4:1871-1879 (1984).	K02480
RP55A		SCRP28E1-2	432	0,884	NUCLEIC ACIDS RES. 12:7345-7358 (1984).	X02635
RP55B		SCRPS16A-2	432	0,858	NUCLEIC ACIDS RES. 12:7345-7358 (1984).	X01100
RP59		CGM0284	411	0,881	NUCLEIC ACIDS RES. 12:8295-8312 (1984).	X06430
RP59	CRY1	SCCRY1	411	0,881	MOL. CELL. BIOL. 7:1764-1775 (1987).	M37326
RPA0	L10E	SCL10E01	936	0,886	J. BACTERIOL 172:579-588 (1990).	X06959
RPA0		SCRPA0	936	0,886	NUCLEIC ACIDS RES. 16:3573 (1988).	D00529
RPA0	YSCA0	SCARPA0	936	0,886	J. BIOCHEM. 106:223-227 (1989).	M26504
RPA1	L12EIIA	SCL12E01	318	0,835	J. BACTERIOL 172:579-588 (1990).	D90072
RPA1		SCA1	318	0,834	NUCLEIC ACIDS RES. 16:3574 (1988).	M26503
RPA1					see RFA1	
RPA2	L12EIB	SCL12IB	318	0,891	J. BACTERIOL. 172:579-588 (1990).	J03760
RPA2		SCRGAP44	318	0,891	J. BIOL. CHEM. 263:9094-9101 (1988).	X06958
RPA2		SCRPA2	318	0,946	NUCLEIC ACIDS RES. 16:3575 (1988).	L00708
RPA12		SCRPA12II	375	0,229	MOL. CELL. BIOL. 13:114-122 (1993).	M96600
RPA49		SCA49A	1245	0,294	PNAS 89:9302-9305 (1992).	M62804
RPA135		SCRPA135	3609	0,262	MOL. CELL. BIOL. 11:754-764 (1991).	J03530
RPA190		SCPOLAI	4992	0,331	J. BIOL. CHEM. 263:2830-2839 (1988).	M15693
RPB2		SCPOL2R	3672	0,245	PNAS 84:1192-1196 (1987).	M27496
RPB3		SCPOL2RP	954	0,13	MOL. CELL. BIOL. 9:5387-5394 (1989).	M27253
RPB4		SCRPB4A	663	0,078	MOL. CELL. BIOL. 9:2854-2859 (1989).	X58099
RPB4		SCYURI-3	663	0,078	NUCLEIC ACIDS RES. 19:2781 (1991).	X53287
RPB5		SCRPB5	645	0,34	GENES DEV. 4:313-323 (1990).	X53289
RPB8		SCRPB8	438	0,321	GENES DEV. 4:313-323 (1990).	M73060
RPB9		SCRPB9A	366	0,107	J. BIOL. CHEM. 266:19053-19055 (1991).	M60479
RPB10		SCRPB10	138	0,46	J. BIOL. CHEM. 265:17816-17819 (1990).	M64991
RPC10		SCRPC19	426	0,223	J. BIOL. CHEM. 266:15300-15307 (1991).	
RPC31					see ACP2	

Continued

LIST A3 Continued

Gene (Primary gene name)	Synonym	Mnemonic	Length	Codon bias	Reference	Accession
RPC34		SCRPC34-1	951	0,073	J. BIOL. CHEM. 267:21390–21395 (1992).	X63746
RPC40		SCRPC40	1005	0,247	CELL 48:627–637 (1987).	M15499
RPC53		SCRPC53	1272	0,108	MOL. CELL. BIOL. 12:4314–4326 (1992).	X63501
RPC82		SCRPC82	1963	0,115	MOL. CELL. BIOL. 12:4433–4440 (1992).	X63500
RPI1		SCRPI1	1218	0,001	MOL. CELL. BIOL. 11:3894–3904 (1991).	M63178
RPL1		SCRIBPRO	891	0,9	MOL. CELL. BIOL. 13:2835–2845 (1993).	L01796
RPL1	YL3	SCYL3GN-2	891	0,894	EMBO J. 9:2759–2764 (1990).	M94864
RPL2	L2	SCRGL2	1086	0,877	J. BIOL. CHEM. 263:6188–6192 (1988).	J03195
RPL4A		SCL42	768	0,935	MOL. GEN. GENET 227:72–80 (1991).	X56836
RPL4A		SCRPL4A	768	0,928	NUCLEIC ACIDS RES. 18:1447–1450 (1990).	X17204
RPL4B		SCL41	768	0,613	MOL. GEN. GENET. 227:72–80 (1991).	X56835
RPL15A		SCRPL15A	495	0,73	NUCLEIC ACIDS RES. 18:4409–4416 (1990).	C51519
RPL15B	L15B	SCRPL15B	495	0,854	NUCLEIC ACIDS RES. 18:4409–4416 (1990).	X51520
RPL16A		SCRIBL16	522	0,871	FEBS LETT. 175:371–376 (1984).	X01029
RPL16B	L16	CGM0285	522	0,828	NUCLEIC ACIDS RES. 12:8295–8312 (1984).	
RPL17A		SCRIBL17	411	0,786	NUCLEIC ACIDS RES. 12:6685–6700 (1984).	X01694
RPL25		SCRIBL25	411	0,862	NUCLEIC ACIDS RES. 12:6685–6700 (1984).	X01014
RPL32		SCRPL32	315	0,932	J. BIOL. CHEM. 262:16055–16059 (1987).	J03457
RPL34		SCRPL34	339	0,84	CURR. GENET. 9:47–52 (1984).	X01441
RPL37A		SCRPL37A	321	0,868	GENE 105:137–138 (1991).	X57969
RPL41A	YL41A	SCYL41A	75	0,739	CURR. GENET. 17:185–190 (1990).	X16065
RPL41B	YL41B	SCYL41B	75	0,739	CURR. GENET. 17:185–190 (1990).	X16066
RPL44	L122EII	SCL12EII	318	0,789	J. BACTERIOL 172:579–588 (1990).	M26507
RPL44		SCRGAP46	318	0,789	J. BIOL. CHEM. 263:9094–9101 (1988).	M19238
RPL45	L12EIA	SCL12EIA	330	0,88	J. BACTERIOL 172:579–588 (1990).	M26505
RPL45		SCRGAP45	330	0,88	J. BIOL. CHEM. 263:9094–9101 (1988).	J03761
RPL46		SCRPL46	153	0,925	NUCLEIC ACIDS RES. 13:701–709 (1985).	X01963
RPM2		SCRNASE	3606	0,113	MOL. CELL. BIOL. 11:344–353 (1991).	L06209
RPO21		SCRPO21	5178	0,238	CELL 42:599–610 (1985).	X03128
RPO26		SCRPO26A	465	0,294	MOL. CELL. BIOL. 10:6123–6131 (1990).	M33924

Name	Alt	ORF	Length	Fraction	Reference	Accession
RPO31		SCRPO31	4380	0,23	CELL 42:599–610 (1985).	X03219
RPO41		SCRPLA	4053	0,138	CELL 51:89–99 (1987).	M17539
RPS24		SCRPS24	390	0,856	NUCLEIC ACIDS RES. 13:701–709 (1985).	X01962
RPS25		SCRPYS25	261	0,741	NUCLEIC ACIDS RES. 16:6223 (1988).	X07811
RPS28A		SCRPS28A	435	0,821	UNPUBLISHED.	M96570
RPS28B		SCRPS28B	435	0,88	UNPUBLISHED.	M96571
RPS31		SCRPS31	324	0,812	CURR. GENET. 10:1–5 (1985).	X03013
RPS33		SCR33	201	0,632	NUCLEIC ACIDS RES. 11:7759–7768 (1983).	X00128
RPS101		SCRPS10	708	0,94	NUCLEIC ACIDS RES. 10:5869–5878 (1982).	J01350
RPS102		SCRPS102	708	0,94	NUCLEIC ACIDS RES. 13:5027–5039 (1985).	X02746
RRR1		SCRRR1A	1476	0,089	UNPUBLISHED.	X21817
RSD1		SCRSD1	1869	0,234	J. CELL BIOL. 109:2939–2950 (1989).	X51672
RSR1		SCRSR1	816	0,112	PNAS 86:9976–9980 (1989).	M26928
RVS161		SCRVS161	795	0,186	YEAST 6:173–176 (1990).	X63315
SAC6		SCSAC6G	1926	0,298	NATURE 354:404–408 (1991).	X63867
SAC7		SCSAC7P	822	0,017	MOL. CELL. BIOL. 10:2308–2314 (1990).	M32335
SAM1		SCSAM1	1146	0,669	J. BIOL. CHEM. 262:16704–16709 (1987).	J03477
SAM2		M23368	1152	0,642	MOL. CELL. BIOL. 8:5132–5139 (1988).	M23368
SAN1		CGM0216	1830	0,159	GENETICS 122:29–40 (1989).	X51667
SAR1		SCSAR1P	580	0,455	J. CELL BIOL. 109:2677–2691 (1989).	L12028
SBR5		SCSBR5SPR	921	0,042	UNPUBLISHED.	
SCB1					see CLB1	
SCD25	PRS2	SCSCD25-1	3750	0,056	MOL. CELL. BIOL. 11:202–212 (1990).	M26647
SCG1					see GPA1	
SCH9		SCSCH9	2472	0,127	GENES DEV. 2:517–527 (1988).	X12560
SCH9		SCSCH9A-2	2472	0,129	YEAST 9:21–32 (1993).	X57629
SCH9uORF		SCSCH9A-1	162	0,09	YEAST 9:21–32 (1993).	X57629
SCJ1		SCSCJ1	1212	0,165	NATURE 349:627–630 (1991).	X58679
SCL1		SCYC7A	756	0,174	J. BIOL. CHEM. 265:16604–16613 (1990).	M55440
SCL1		SCSCL1A	810	0,081	GENE 83:271–279 (1989).	M31430
SCL1		SSPROTY8	756	0,174	MOL. CELL. BIOL. 11:344–353 (1991).	M63641
SCM4		SCSCM4A	561	0,236	MOL. GEN. GENET. 235:285–291 (1992).	X69566
SCO1		SCSCO1GEN1	885	0,056	MOL. GEN. GENET. 216:37–43 (1989).	X17441
SDC26		SCSDC26A	453	0,078	NUCLEIC ACIDS RES. 17:10491 (1989).	X17118

Continued

LIST A3 Continued

Gene (Primary gene name)	Synonym	Mnemonic	Length	Codon bias	Reference	Accession
SDH1		SCSDH	798	0,312	J. BIOL. CHEM. 265:10419–10423 (1990).	J05487
SEC1		SCSEC1	2172	0,122	YEAST 7:643–650 (1991).	X62451
SEC2		SCSEC2	2277	0,073	J. CELL BIOL. 110:1897–1909 (1990).	X52147
SEC4		SCSEC4A	645	0,214	CELL 49:527–538 (1987).	M16507
SEC6		SCSEC6G	299	0,054	YEAST 8:549–558 (1992).	X64738
SEC7		SCSEC7	6027	0,186	J. BIOL. CHEM. 263:11711–11717 (1988).	J03918
SEC8		SCSEC8	3195	0,053	UNPUBLISHED.	X64693
SEC11		SCSEC11	501	0,173	J. CELL BIOL. 106:1035–1042 (1988).	X07694
SEC12		SCSEC12	1413	0,051	J. CELL BIOL. 107:851–863 (1988).	X13161
SEC13		SCSEC13P	891	0,357	UNPUBLISHED.	L05929
SEC14		SCSEC14G	912	0,374	J. CELL BIOL. 108:1271–1281 (1989).	X15483
SEC15		CGM0280	2730	0,067	J. CELL BIOL. 109:1023–1036 (1989).	
SEC17		SCSEC17P	873	0,056	J. BIOL. CHEM. 267:12106–12115 (1992).	M93104
SEC18		SCSEC18-1	2271	0,206	MOL. CELL. BIOL. 8:4098–4109 (1988).	M20662
SEC20		SCSEC20	1149	0,082	EMBO J. 11:423–432 (1992).	X60215
SEC21		SCERTOCO	2805	0,278	UNPUBLISHED.	M59708
SEC23		SCSEC23P	2304	0,345	EMBO J. 8:1677–1684 (1989).	X15474
SEC53		SCSEC53	762	0,655	J. CELL BIOL. 101:2374–2382 (1985).	X03213
SEC59		SCSEC59A	1557	0,013	MOL. CELL BIOL. 9:1191–1199 (1989).	M25779
SEC61		SCSEC61G	1440	0,394	MOL. BIOL. CELL 3:129–142 (1992).	X62340
SEC62		SCSEC62GN	849	0,199	J. CELL BIOL. 109:2653–2664 (1989).	X51666
SED1		SCSDH1	1014	0,067	NUCLEIC ACIDS RES. 18:3653 (1990).	X52657
SEN1		SCSEN1A	6336	0,091	MOL. CELL. BIOL. 12:2154–2164 (1992).	M74589
SEN2		SCSEN2X	1131	0,031	UNPUBLISHED.	M32336
SEN3		SCSEN3A	2835	0,167	UNPUBLISHED.	L06321
SEP1					see KEM1	
SES1		SCSERS	1386	0,48	NUCLEIC ACIDS RES. 15:1887–1904 (1987).	X04884
SFL1		CGM0266	2298	0,102	GENE 85:321–328 (1989).	
SFL2					see TUP1	
SFP1		SCSFP1AA	2043	0,105	GENE 107:101–110 (1991).	M63577

Gene	Synonym	Locus	Length	Bias	Reference	Accession
SGA1		SCSGA	1530	0,071	J. BACTERIOL 169:2142–2149 (1987).	M16166
SGV1		SCSGV1	1971	0,006	CELL 65:785–795 (1991).	D90317
SHR3		SCSECRE	627	0,266	CELL 71:463–478 (1992).	L01264
SIB1		SCSIBPKNS	2607	0,037	UNPUBLISHED.	M90531
SIN3		SCPAHMA	4614	0,092	MOL. CELL. BIOL. 10:5927–5936 (1990).	M36822
SIN4	BEL2	SCBEL2	2922	0,021	UNPUBLISHED.	D12918
SIR1		SCSIR1	2034	0,079	MOL. CELL. BIOL. 11:2253–2262 (1991).	M38524
SIR2		SCSIR2	1686	0,08	EMBO J. 3:2817–2823 (1984).	X01419
SIR3		SCSIR3	2934	0,012	EMBO J. 3:2817–2823 (1984).	X01420
SIR4		SCSIR4AA	4074	0,032	MOL. CELL. BIOL. 7:4441–4452 (1987).	M37249
SIS1		SCSIS1	1056	0,163	J. CELL BIOL. 114:623–638 (1991).	X58460
SIT4					see PPH1	
SKI3		CGM0219	4296	0,061	YEAST 5:149–158 (1989).	M96058
SKI8		SCSKI8A	1191	0,092	MOL. CELL. BIOL. 4:761–770 (1984).	X67875
SKO1		SCSKO1	1941	0,006	NUCLEIC ACIDS RES. 20:5271–5278 (1992).	M34474
SLP1		SCSLP1A	2073	0,028	MOL. CELL. BIOL. 10:2214–2223 (1990).	M34638
SLP1	VPS33	SCVPS33P	2073	0,028	MOL. CELL. BIOL. 10:4638–4649 (1990).	X59262
SLT2		SCSLT2	1452	0,074	MOL. MICROBIOL. 5:2845–2854 (1991).	X67810
SLU7		SCSLU7A	1146	0,082	GENES DEV. 6:2112–2124 (1992).	—
SLY1		SCSLY1	1998	0,189	MOL. CELL. BIOL. 11:872–885 (1991).	X54323
SLY2	TSL26, TS26	SCSLY2	642	0,231	MOL. CELL. BIOL. 11:872–885 (1991).	X54236
SLY2		CGM0259	642	0,231	YEAST 5:509–524 (1989).	—
SLY12		SCSLY12	426	0,049	MOL. CELL. BIOL. 11:872–885 (1991).	X54237
SLY41		SCSLY41	1356	0,034	MOL. CELL. BIOL. 11:872–885 (1991).	X54238
SMC1		SCSMCI	3675	0,108	UNPUBLISHED.	L00602
SME1		SCSMEIG	1935	0,06	MOL. GEN. GENET. 221:176–186 (1992).	X53262
SMP2		SCSMP2	2586	0,025	MOL. GEN. GENET. 236:283–288 (1993).	D01095
SMP3		SCSMP3	1548	0,051	MOL. GEN. GENET. 225:257–265 (1991).	X58121
SMY1		SCSMY1A	1882	0,02	NATURE 356:358–362 (1992).	M69021
SMY2		SCSMY2P	2370	0,126	UNPUBLISHED.	M90654
SNF1	GAM1	SCSNF1	1899	0,158	SCIENCE 233:1175–1180 (1987).	M13971
SNF2		SCSNF2	5109	0,112	MOL. GEN. GENET. 228:270–280 (1991).	X57837
SNF2		SCSNF2A	5109	0,112	PNAS 88:2687 (1991).	M61703
SNF3		SCSNF3	2652	0,072	PNAS 85:2130–2134 (1988).	J03246
SNF4					see CAT3	

Continued

LIST A3 Continued

Gene (Primary gene name)	Synonym	Mnemonic	Length	Codon bias	Reference	Accession
SNF5		SCSNF5	2715	0,081	MOL. CELL. BIOL. 10:5616–5625 (1990).	M36482
SNF6		SCSNF6A	996	0,188	MOL. CELL. BIOL. 10:2544–2553 (1990).	M37132
SNM1		SCSNM1	1983	≤0	MOL. GEN. GENET. 231:194–200 (1992).	X64004
SNP1		SCSNP1	900	0,189	EMBO J. 10:2627–2634 (1991).	X59986
SNQ2		SCSNQ2	4503	0,182	MOL. GEN. GENET. 236:214–218 (1993).	X66732
SOD1		SCCUZNSD	462	0,555	PNAS 85:4789–4793 (1988).	J03279
SOD2		SCSODMNG2	699	0,344	EUR. J. BIOCHEM. 147:153–161 (1985).	X02156
SON1		SCSON1X	1593	0,043	GENETICS 134:159–173 (1993).	L00928
SPA2		SCSPA2G	4398	0,09	J. CELL BIOL. 111:1451–1464 (1990).	X53731
SPB4		SCSPB4	1818	0,065	SCIENCE 247:1077–1079 (1990).	X16147
SPE2		SCAMDA	1188	0,204	J. BIOL. CHEM. 265:22321–22328 (1990).	M38434
SPK1		SCSPK1	2463	0,024	MOL. CELL. BIOL. 11:987–1001 (1991).	M55623
SPL1		SCNFLP	1491	0,337	UNPUBLISHED.	M98808
SPM1					see GAL83	
SPM2	CDC68	SCSPM2	1245	0,037	MOL. CELL. BIOL. 11:5718–5726 (1991).	Z14128
SPO7		SCSPO7A	777	0,061	GENE 95:65–72 (1990).	M36073
SPO11		SCSPO11	1194	0,128	PNAS 84:8035–8039 (1987).	J02987
SPO12		SCSPOA-2	519	0,011	MOL. CELL. BIOL. 10:2809–2819 (1990).	M32653
SPO13		SCSPO13	873	0,017	PNAS 87:9406–9410 (1990).	M38357
SPO15					see VPS1	
SPO16		SCSPOA-1	594	0,049	MOL. CELL. BIOL. 10:2809–2819 (1990).	M32653
SPP81					see DED1	
SPP91	PRP21	SCPRP21A-1	840	0,048	UNPUBLISHED.	L07744
SPP91		SCSPP91A	840	0,048	EMBO J. 11:3279–3288 (1992).	X67564
SPR6		SCSPR6	573	0,029	CURR. GENET. 18:293–301 (1990).	X57409
SPS1*		SCSPS12-1	291	0,094	MOL. CELL. BIOL. 6:2443–2451 (1986).	M13629
SPS2		SCSPS12-2	1407	0,085	MOL. CELL. BIOL. 6:2443–2451 (1986).	M13629
SPS4		SCSPS4A	1014	0,03	MOL. CELL. BIOL. 6:4478–4485 (1986).	M14684
SPS100		SCSPSWMA	978	0,257	MOL. CELL. BIOL. 8:912–922 (1988).	M20366
SPT2		SCSPT2	999	0,009	MOL. CELL. BIOL. 5:1543–1553 (1985).	M11165

Gene	Factor	Sequence	Length	Value	Reference	Accession
SPT2		SCSYGP1-2	999	0,009	UNPUBLISHED.	L10718
SPT3		SCSPT3	1011	0,064	NUCLEIC ACIDS RES. 14:6885–6900 (1986).	X04383
SPT4		SCSPT4A	306	0,12	MOL. GEN. GENET. 237:449–459 (1993).	M83672
SPT5		SCSPT5	3189	0,227	MOL. CELL. BIOL. 11:3009–3019 (1991).	M62882
SPT6		SCSPT6A	4353	0,216	MOL. CELL. BIOL. 10:4935–4941 (1990).	M34391
SPT14		SCSPT14	1338	0,032	MOL. GEN. GENET. 230:310–320 (1991).	X63290
SPT15	BTF1	SCBTF1Y	720	0,128	PNAS 86:9803–9807 (1989).	M29459
SPT15	TFIID	SCTFII01	720	0,128	PNAS 86:7785–7789 (1989).	M26403
SPT15	TFIID	SCTFIID	720	0,128	NATURE 341:299–303 (1989).	X16860
SPT15	TFIID	SCTFIIDA	720	0,128	CELL 58:1173–1181 (1989).	M27135
SRA1					see REG1	
SRA3		SCSRA3	1212	0,209	MOL. CELL. BIOL. 7:2653–2663 (1987).	M17224
SRB4		SCSRB4SPR	2061	0,06	UNPUBLISHED.	L12026
SRB6		SCSRB6SPR	363	≤0	UNPUBLISHED.	L12027
SRH1					see SRP54	
SRK1					see SSD1	
SRM1		M27013	1446	0,235	MOL. CELL. BIOL. 9:2682–2694 (1989).	M27013
SRP1		SCSRP1	762	0,784	J. MOL. BIOL. 202:455–470 (1988).	X12775
SRP54		SCSRP4SC	1623	0,236	J. CELL BIOL. 109:3223–3230 (1989).	M55517
SRP54	SRH1	SCSRH1	1623	0,235	J. BIOCHEM. 107:457–463 (1990).	X16908
SRP101		SCSRPRA	1863	0,066	UNPUBLISHED.	M77274
SRV2	CAP	SCCAP	1578	0,26	CELL 61:319–327 (1990).	M58284
SRV2		SCSRV2	1578	0,26	CELL 61:329–340 (1990).	M32663
SSA1		SCSSA1	1926	0,833	NUCLEIC ACIDS RES. 12:967–9382 (1984).	X12926
SSA2		SCSSA2	1917	0,893	NUCLEIC ACIDS RES. 17:805–806 (1989).	X12927
SSA3*		SCHSPSSA	105	0,147	MOL. CELL. BIOL. 10:3262–3267 (1990).	M36115
SSA4		SCHSPS01-1	1926	0,238	J. BIOL. CHEM. 265:18912–18921 (1990).	J05637
SSB1		SCSSB1A	1839	0,908	CELL 57:1223–1237 (1989).	M25395
SSB1		SCSSB1G	1839	0,908	NUCLEIC ACIDS RES. 17:4891 (1989).	X13713
SSB1					see SSBR1	
SSBR1	SSB1	SCSSB1	879	0,473	MOL. CELL. BIOL. 7:2947–2955 (1987).	M17244
SSC1	ENS1	SCENS1	1965	0,678	J. BIOL. CHEM. 265:15189–15197 (1990).	M55275
SSC1		M27229	1962	0,684	MOL. CELL. BIOL. 9:3000–3008 (1989).	M27229
SSD1		SCSIT4B	3750	0,205	MOL. CELL. BIOL. 11:2133–2148 (1991).	M60318

Continued

LIST A3 *Continued*

Gene (Primary gene name)	Synonym	Mnemonic	Length	Codon bias	Reference	Accession
SSD1	SRK1	SCSRK1A	3750	0,205	MOL. CELL. BIOL. 11:3369-3373 (1991).	M63004
SSL2	RAD25	SCRAD25A	2529	0,198	PNAS 89:11416-11420 (1992).	L01414
SSL2		SCSSL2A	2529	0,197	CELL 69:1031-1042 (1992).	M94176
SSN6	CYC8	SCCYC8	2898	0,187	GENE 73:97-111 (1988).	M23440
SSN6		SCCMDLYS-2	2898	0,185	YEAST 8:397-408 (1992).	X66247
SSN6		SCSSN6A	2898	0,187	MOL. CELL. BIOL. 7:3637-3645 (1987).	M17826
SST2		SCSST2	2094	0,047	MOL. CELL. BIOL. 7:4169-4177 (1987).	M18105
SSV7		SCSSV7VAC	2781	0,081	UNPUBLISHED.	L08070
STA2		SDSTA2A	2301	0,213	GENE 100:95-103 (1991).	M60650
STE2		M24335	1293	0,085	NUCLEIC ACIDS RES. 13:8463-8475 (1985).	M24335
STE2		SCSTE2	1293	0,085	EMBO J. 4:2643-2648 (1985).	X03010
STE3		SCSTE3	1410	0,037	EMBO J. 4:2643-2648 (1985).	X03011
STE4		SCSTE4	1269	0,021	CELL 56:467-477 (1989).	M23982
STE5		SCSTE5	2751	0,056	MOL. CELL. BIOL. 13:2050-2060 (1993).	D12917
STE6		SCSTE6A	3870	0,055	EMBO J. 8:3973-3984 (1989).	M26376
STE6		SCSTE6G	3870	0,054	NATURE 340:400-404 (1989).	X15428
STE6*		SCSTE6PR	123	0,195	PNAS 83:2536-2540 (1986).	M12842
STE7		SCSTE7	1545	0,042	PNAS 83:7371-7375 (1986).	M14097
STE11		SCSTE11	2151	0,068	GENES DEV. 4:1862-1874 (1990).	X53431
STE12		SCSTE12	2064	0,096	GENES DEV. 3:1349-1361 (1989).	X16112
STE12		SCSTE12A	2064	0,096	PNAS 86:5703-5707 (1989).	M24502
STE14		SCMTSW	717	0,015	UNPUBLISHED.	L07952
STE18		SCSTE18	330	0,031	CELL 56:467-477 (1989).	M23983
STE50		SC5	1038	0,11	MOL. GEN. GENET. 236:145-154 (1992).	Z11116
STF2		SCSTF2	252	0,447	EUR. J. BIOCHEM. 192:49-53 (1990).	D00444
STH1	NPS1	SCNPS1	4077	0,171	EMBO J 11:4012-4026 (1992).	D10595
STH1		SCSTH1A	4056	0,171	MOL. CELL. BIOL. 12:1893-1902 (1992).	M83755
STR4		SCSTR4G	1521	0,444	UNPUBLISHED.	X72922
SUA5		SCSUA5DNA1	1278	0,03	GENETICS 131:791-801 (1992).	X64319
SUA7		SCSUA7A	1035	0,106	J. BIOCHEM. 107:457-463 (1990).	M81380

Gene	Synonym	Sequence	Length	Value	Reference	Accession
SUC1		SCSUC1G	1596	0,399	MOL. GEN. GENET. 211:446–454 (1988).	X07570
SUC2		SCINVE	1596	0,425	MOL. CELL. BIOL. 3:439–447 (1983).	V01311
SUC3		SCSUC3G	222	0,3	MOL. GEN. GENET. 211:446–454 (1988).	X07571
SUC4		SCSUC4C	1596	0,371	MOL. GEN. GENET. 211:446–454 (1988).	X07572
SUC5		SCSUC5G	222	0,3	MOL. GEN. GENET. 211:446–454 (1988).	X07573
SUC7*		SCSUC7G1	213	0,302	NUCLEIC ACIDS RES. 13:6089–6103 (1985).	X02908
SUC7*		SCSUC7C2	66	0,415	NUCLEIC ACIDS RES. 13:6089–6103 (1985).	X02909
SUF12	GST1	SCG1ST1	2055	0,417	EMBO J. 7:1175–1182 (1988).	Y00829
SUF12		SCSUF12	2055	0,417	J. MOL. BIOL. 199:559–573 (1988).	X07163
SUF12	SUP2	SCSUP2	2055	0,349	GENE 66:45–54 (1988).	M21129
SUG1		SCSUG1	1215	0,289	UNPUBLISHED.	X66400
SUG1	TBY1	SCTBY1A	1215	0,288	NATURE 357:698–700 (1992).	L01626
SUI1		SCSUI1A	324	0,265	MOL. CELL. BIOL. 12:248–260 (1992).	M77514
SUI2		M25552	912	0,401	PNAS 86:2784–2788 (1989).	M25552
SUI3		SCELF2B	855	0,361	CELL 54:621–632 (1988).	M21813
SUP1		SCSUP1	1311	0,374	NUCLEIC ACIDS RES. 14:5187–5197 (1986).	X04082
SUP2					see SUF12	
SUP46		CGM0294	585	0,901	UNPUBLISHED.	M91167
SUV3		SCSUV3A	2211	≤0	UNPUBLISHED.	M84390
SWI3		SCSW13A	2475	0,08	CELL 68:573–583 (1992).	X56792
SWI3	TYE2	SCTRFAC-3	2475	0,08	UNPUBLISHED.	X51606
SWI4		SCSW14	3279	0,061	NATURE 342:830–833 (1989).	X06978
SWI5		SCSW15	2127	0,107	EMBO J. 7:485–494 (1988).	X06238
SWI6		SCSW16	2409	0,01	NATURE 329:651 (1987).	X67705
SWP1		SCSWP1	858	0,293	EMBO J. 12:279–284 (1993).	X15953
TAL1		SCTAL1	1005	0,701	EUR. J. BIOCHEM. 188:597–603 (1990).	X04969
TAL1*		SCILV5-2	291	0,706	NUCLEIC ACIDS RES. 14:9631–9651 (1986).	X69394
TBF1		SCTBF1-1	1686	0,076	MOL. CELL. BIOL. 13:1306–1314 (1993).	
TBY1					see SUG1	
TCM1		SCRP13	1161	0,885	J. BACTERIOL. 155:8–14 (1983).	J01351
TCP1		SCTCP1	1677	0,314	GENE 68:267–274 (1988).	M21160
TDH1	GPD1	SCG3PDA	996	0,09	J. BIOL. CHEM. 254:5466–5474 (1979).	J01324
TDH1	GPD1	SCGAP1	996	0,16	J. BIOL. CHEM. 254:9839–9845 (1979).	V01300
TDH2	GPD2	SCGAP2	996	0,379	J. BIOL. CHEM. 255:2596–2605 (1980).	V01301
TDH2	GPD2	SCMETTDH-1	996	0,982	YEAST 7:873–880 (1991).	X60157

Continued

LIST A3 Continued

Gene (Primary gene name)	Synonym	Mnemonic	Length	Codon bias	Reference	Accession
TDH3	GPD3	SCGAP3	996	0,177	J. BIOL. CHEM. 258:5291–5299 (1983).	V01302
TEC1		SCTEC1A	1458	0,056	MOL. CELL. BIOL. 10:3541–3550 (1990).	M32797
TEF1	EF1A	SCEF1AAA	1374	0,929	GENE 45:265–273 (1986).	M15666
TEF1		SCEF1A	1374	0,929	EMBO J. 3:1825–1830 (1984).	X00779
TEF1		SCEF1AB	1374	0,929	J. BIOL. CHEM. 260:3090–3096 (1985).	M10992
TEF2		SCEF1ABB	1374	0,933	GENE 45:265–273 (1986).	M15667
TEF2		SCTEF1	1374	0,933	EMBO J. 3:3311–3315 (1984).	X01638
TFB1		SCRNAPOLY	1926	0,171	SCIENCE 257:1389–1392 (1992).	M95750
TFC1		SCTFC1	1947	0,026	PNAS 88:4887–4891 (1991).	M63385
TFIID					see SPT15	
TFP1		SCTFP1	3093	0,408	MOL. CELL. BIOL. 8:3094–3103 (1988).	M21609
TFP1	VMA1	SCHATP	3213	0,411	J. BIOL. CHEM. 265:6726–6733 (1990).	J05409
TFP1*	VMA1	SCPPH22-1	101	0,478	MOL. CELL. BIOL. 11:4876–4884 (1991).	X58857
TFP3		SCTFP3	306	0,269	MOL. CELL. BIOL. 10:3397–3404 (1990).	M32736
TFS1		SCTFS1G	657	0,147	MOL. GEN. GENET. 230:241–250 (1991).	X62105
THR1		SCTHR1	1068	0,41	EUR. J. BIOCHEM. 191:115–122 (1990).	X52901
THR1		SCTHR101	1071	0,411	GENE 96:177–180 (1990).	M37692
THR4		SCTHR4	1542	0,509	NUCLEIC ACIDS RES. 18:665 (1990).	X17256
THS1		SCTHS1	2199	0,51	NUCLEIC ACIDS RES. 13:6171–6183 (1985).	X02906
TIF1		SCTIF1	1185	0,834	NUCLEIC ACIDS RES. 16:10359 (1988).	X12813
TIF2		SCTIF2	1185	0,834	NUCLEIC ACIDS RES. 16:10359 (1988).	X12814
TIF2*		SCYURI-1	267	0,838	NUCLEIC ACIDS RES. 19:2781 (1991).	X58099
TIF45	CDC33	SCEIF4E	639	0,455	MOL. CELL. BIOL. 7:998–1003 (1987).	M15436
TIF45	EIF4E, CDC33	SCCDC33	639	0,455	MOL. CELL. BIOL. 8:3556–3559 (1988).	M21620
TIF51A	HYP2	SCHYP2	471	0,908	UNPUBLISHED.	X56236
TIF51A		SCTIF51A	471	0,908	MOL. CELL. BIOL. 11:3105–3114 (1991).	M63541
TIF51B					see ANB1	
TIP1		SCTIP1	630	0,663	J. BIOL. CHEM. 266:17537–17544 (1991).	M71216
TMP1		SCTMP1	912	0,192	J. BIOL. CHEM. 262:5298–5307 (1987).	J02706
TNT1		SCTNT	1638	0,054	J. BIOL. CHEM. 265:16216–16220 (1990).	M59780
TOP1		SCTOPI	2307	0,015	PNAS 82:4374–4378 (1985).	K03077
TOP2		SCTOP2	4287	0,115	J. BIOL. CHEM. 261:12448–12454 (1986).	M13814

Gene	Alt.	Locus	Size	Value	Reference	Accession
TOP3		SCTOP3	1968	0,06	CELL 58:409–419 (1989).	M24939
TPD3		SCTPD3X	1905	0,173	MOL. CELL. BIOL. 12:4946–4959 (1992).	M98389
TPI1		SCTPI	744	0,901	J. MOL. APPL. GEN. 1:419–434 (1982).	J01366
TPK1	PK25	SCPK25	1191	0,2	J. BIOL. CHEM. 262:2549–2553 (1987).	J02665
TPK1		SCTPKA	1191	0,2	CELL 50:277–287 (1987).	M17072
TPK2		SCTPK2	1140	0,113	FEBS LETT. 222:279–285 (1987).	Y00694
TPK2		SCTPKB	1140	0,123	CELL 50:277–287 (1987).	M17073
TPK3		SCTPKC	1194	0,148	CELL 50:277–287 (1987).	M17074
TPK3*		SCMRP49A-1	165	0,198	CELL 57:233–242 (1989).	M81697
TPM1		M25501	597	0,489		M25501
TPS1					see CIF1	
TR-II					see TRX1	
TRG1					see PDI1	
TRK1		SCKTR	3705	0,045	MOL. CELL. BIOL. 8:2848–2859 (1988).	M21328
TRK2		SCTRK2Q	2667	0,142	UNPUBLISHED.	M65215
TRL1	TRNL1	SCLIGTR	2481	0,055	J. BIOL. CHEM. 263:3171–3176 (1988).	J03546
TRM1		SCTGM1	1710	0,125	PNAS 84:5172–5176 (1987).	M17193
TRNL1					see TRL1	
TRP1		SCTRP1	672	0,05	GENE 10:157–166 (1980).	V01341
TRP2		SCP2	1584	0,167	J. BIOL. CHEM. 259:3985–3992 (1984).	K01388
TRP3		SCP3	1452	0,218	J. BIOL. CHEM. 259:3985–3992 (1984).	K01386
TRP3*		SCATRP3A	840	0,247	CURR. GENET. 8:165–172 (1984).	M36300
TRP4		SCTRP4	1140	0,146	NUCLEIC ACIDS RES. 14:6357–6373 (1986).	X04273
TRP5		SCTRP5A	2121	0,449	J. BIOL. CHEM. 257:1491–1500 (1982).	V01342
TRP5*		SCTRP5B	518	0,443	GENE 17:223–228 (1982).	V01343
TRX1		SCTRX1	309	0,657	J. BIOL. CHEM. 266:9194–9202 (1991).	M62647
TRX1	TR-II	SCTRIIA	309	0,657	J. BIOL. CHEM. 266:1692–1696 (1991).	M59169
TRX2		SCTRX2	312	0,529	J. BIOL. CHEM. 266:9194–9202 (1991).	M62648
TS26					see SLY2	
TSL26					see SLY2	
TSM1		SCTSM1A	4230	0,048	CURR. GENET. 20:25–31 (1991).	M60486
TUB1		SCTUB1	1341	0,382	MOL. CELL. BIOL. 6:3711–3721 (1986).	M28429
TUB2		SCBTUB	1371	0,407	CELL 33:211–219 (1983).	V01296
TUB3		M28428	1335	0,354	MOL. CELL. BIOL. 6:3711–3721 (1986).	M28428
TUF1		SCTUFM	1311	0,409	PNAS 80:6192–6196 (1983).	K00428

Continued

LIST A3 *Continued*

Gene (Primary gene name)	Synonym	Mnemonic	Length	Codon bias	Reference	Accession
TUP1	AER2	SCAER2	2139	0,238	GENE 97:153–161 (1991).	M35861
TUP1		SCTUP1A	2139	0,231	MOL. CELL. BIOL. 10:6500–6511 (1990).	M31733
TUP1	SFL2	SCSFL2	2007	0,251	GENE 89:93–99 (1990).	X16365
TUR1	UDPNAGT	SCTUNRES	1344	0,074	NUCLEIC ACIDS RES. 15:3627 (1987).	Y00126
TYE2					see SWI3	
TYR1		CGM0251	1323	0,04	GENE 85:303–311 (1989).	
UBA1		SCUBA1G	3072	0,248	EMBO J. 10:227–236 (1991).	X55386
UBC1		SCUBC1G	645	0,117	EMBO J. 9:4535–4541 (1990).	X56402
UBC3		SCUBC3	885	0,142	SCIENCE 241:1331–1335 (1988).	M21877
UBC4		SCUBC4	444	0,415	EMBO J. 9:543–550 (1990).	X17493
UBC5		SCUBC5	444	0,179	EMBO J. 9:543–550 (1990).	X17494
UBC7A		SCUBC7A	495	0,066	NATURE 361:369–371 (1993).	X69100
UBC8		SCUCP	618	≤0	J. BIOL. CHEM. 266:15549–15554 (1991).	M65083
UBI1		SCUBI1G	384	0,789	EMBO J. 6:1429–1439 (1987).	X05728
UBI2		SCUBI2G	384	0,789	EMBO J. 6:1429–1439 (1987).	X05729
UBI3		SCUBI3G	456	0,891	EMBO J. 6:1429–1439 (1987).	X05730
UBI4		SCUBI4G	1143	0,498	EMBO J. 6:1429–1439 (1987).	X05731
UBP1		SCUBP1	2427	0,157	J. BIOL. CHEM. 266:12021–12028 (1991).	M63484
UBP2		SCUBPII	3792	0,085	J. BIOL. CHEM. 267:23364–23375 (1992).	M94916
UBP3		SCUBPIII	2736	0,16	J. BIOL. CHEM. 267:23364–23375 (1992).	M94917
UBR1		SCUBR1G	5850	0,066	EMBO J. 9:3179–3189 (1990).	X53747
UDPNAGT					see TUR1	
UGA1		SCUGA1	1413	0,378	NUCLEIC ACIDS RES. 18:3049 (1990).	X52600
UGA3		SCUGA3G	1584	0,04	MOL. GEN. GENET. 220:269–276 (1990).	X51664
UGA4		SCUGA4	1713	0,293	MOL. GEN. GENET. 237:17–25 (1993).	X66472
UGA35	DAL81	SCDAL81	2910	0,057	MOL. CELL. BIOL. 11:1161–1166 (1991).	M60415
UGA35	DURL, DAL81	SCUGA35	2892	0,054	GENE 97:163–171 (1991).	M63498
UNG1		SCUNG1A	1077	0,041	J. BIOL. CHEM. 264:2593–2598 (1989).	J04470
URA1		SCURA1-1	942	0,238	GENE 118:149–150 (1992).	X59371
URA10		SCURA10	681	0,128	CURR. GENET. 17:105–111 (1990).	X52194

Gene	Alt.	SC locus	bp	CAI	Reference	Accession
URA2		SCURA2B-1	6636	0,359	GENE 79:59–70 (1989).	M27174
URA2*		SCURA2	1527	0,357	MOL. GEN. GENET. 207:314–319 (1987).	X05553
URA3		SCODCD	801	0,207	GENE 29:113–124 (1984).	K02206
URA3*		SCURA3P	218	0,191	MOL. CELL. BIOL. 6:1095–1101 (1986).	M12926
URA4		SCURA4	1092	0,042	MOL. GEN. GENET. 212:134–141 (1988).	X07561
URA5		SCURA5	678	0,36	MOL. GEN. GENET. 215:455–462 (1989).	X14795
URA6		SCUMPK	612	0,23	BBRC 165:464–473	M31455
URA6		SCURA6	612	0,23	J. BIOL. CHEM. 266:18287–18293 (1992).	M69295
URE2		SCURE2	1710	0,389	MOL. GEN. GENET. 231:7–16 (1991).	X53995
URK1		SCURK1-2	1062	0,031	MOL. CELL. BIOL. 11:822–832 (1991).	M35268
USO1		SCUSO1	1503	0,126	NUCLEIC ACIDS RES. 18:5279 (1991).	X53998
UTR3		SCCORA-2	5370	0,014	J. CELL BIOL. 113:245–260 (1991).	X54378
VAC1		SCVAC1A	405	0,034	GENE 87:157–166 (1990).	M37696
VAM7		SCVAM7	1545	0,101	J. BIOL. CHEM. 267:618–623 (1992).	M80596
VAN1		SCVAN1	948	0,085	J. BIOL. CHEM. 267:18671–18675 (1992).	D11379
VAS1		SCVASI	1557	0,468	MOL. CELL. BIOL. 10:898–909 (1990).	M33957
VMA1			3312		J. BIOL. CHEM. 262:7189–7194 (1987).	J02719
					see TFP1	
VMA4		SCVMA4A	699	0,422	J. BIOL. CHEM. 265:18554–18560 (1990).	M60663
VMA6		SCVMA6A	1035	0,311	UNPUBLISHED.	L11584
VPH1		SCVPH1P	2520	0,392	J. BIOL. CHEM. 267:14294–14303 (1992).	M89778
VPH2		SCVPH2A	645	0,056	YEAST 9:175–184 (1993).	M93350
VPS1		SCVPS1A	2112	0,333	CELL 61:1063–1073 (1990).	M33315
VPS1	SPO15	SCSPO15G	2112	0,333	CELL 61:1063–1074 (1990).	X54316
VPS3		SCVPS3	3033	≤0	J. CELL BIOL. 111:877–892 (1990).	X53871
VPS15		SCVPS15	4365	0,036	CELL 64:425–437 (1991).	M59835
VPS16P		SCVPS16P	2391	0,014	J. BIOL. CHEM. 268:4953–4962 (1993).	L07327
VPS17		SCVPS	1653	0,122	J. BIOL. CHEM. 268:559–569 (1993).	L02869
VPS33					see SLP1	
VPS34		SCVPS34	2625	0,038	MOL. CELL. BIOL. 10:6172–6180 (1990).	X53531
WHI1					see DAF1	
WHI2		SCWHI2	1458	0,083	GENE 66:205–213 (1988).	M21089
XRN1					see KEM1	
YAK1		SCYAK1	2421	0,055	GENES DEV. 3:1336–1348 (1989).	X16056
YAP1					see PDR4	

Continued

LIST A3 Continued

Gene (Primary gene name)	Synonym	Mnemonic	Length	Codon bias	Reference	Accession
YAP3		CGM0297	1707	0,171	YEAST 6:127–137 (1990).	M37193
YAP17		SCYAP17	441	0,126	J. BIOL. CHEM. 266:11153–11157 (1991).	M64998
YAP80*		SCYAP80	2100	0,082	MOL. CELL. BIOL. 10:6089–6090 (1990).	M74552
YCK1		SCYCK1	1614	0,187	UNPUBLISHED.	M74453
YCK2		SCYCK2	1638	0,162	UNPUBLISHED.	X59075
YCR591		SCYCR59-1	5499	0,011	YEAST 7:413–424 (1991).	X59075
YCR592		SCYCR59-2	3678	0,006	YEAST 7:413–424 (1991).	X59075
YCR721					see BUD5	
YDJ1		SCYDJ1	1227	0,487	J. CELL BIOL. 114:609–621 (1991).	X56560
YEF3		SCYEF3A	3132	0,858	J. BIOL. CHEM. 265:1903–1912 (1990).	J05197
YEF3		SCYEF3B	3132	0,858	J. BIOL. CHEM. 265:15838–15844 (1990).	J05583
YKL252					see PTM1	
YKR2		SCYKR2A	2031	0,049	GENE 76:177–180 (1989).	M24929
YL3					see RPL1	
YL41A					see RPL41A	
YL41B					see RPL41B	
YMR26		SCYMR26	471	0,117	MOL. GEN. GENET. 225:474–482 (1991).	X56106
YMR31		SCYMR31	369	0,016	MOL. GEN. GENET. 219:119–124 (1989).	X17540
YMR44		SCYMR44	294	0,036	MOL. GEN. GENET. 219:119–124 (1989).	X17552
YP2					see YPT1	
YPK1		SCPKN	2040	0,129	DNA 7:469–474 (1988).	M21307
YPT1	YP2	SCRASX-1	618	0,245	NATURE 306:704–707 (1983).	X00209
YSA1		SCCMDLYS-1	141	0,347	YEAST 8:397–408 (1992).	X66247
YSCA0					see RPA0	
YUR1		SCYURI-2	1284	0,022	NUCLEIC ACIDS RES. 19:2781 (1991).	X58099
ZIP1		SCZIP1A	2625	0,053	CELL 72:365–378 (1993).	L06487
ZRC1		SCZRC1GEN	1326	0,264	MOL. GEN. GENET. 219:161–167 (1989).	X17537
ZWF1	MET19	SCMET19	1515	0,215	EMBO J. 10:547 (1991).	X57336
ZWF1		SCG6PD-2	1515	0,218	GENE 96:161–169 (1990).	M34709
14DM					see ERG11	

Appendix IV *Schizosaccharomyces pombe*: Genetic Nomenclature and Chromosome Maps

Peter Munz and Jürg Kohli

Institute of General Microbiology, University of Bern, Baltzer-Str. 4, CH-3012 Bern, Switzerland

I. INTRODUCTION

The nuclear genome of *Schizosaccharomyces pombe* consists of three chromosomes with sizes of 5.7 Mb (chromosome 1), 4.6 Mb (chromosome II) and 3.5 Mb (chromosome III) (Fan *et al.*, 1989). The mitochondrial genome consists of a circular chromosome of 19 kb length (Wolf, 1987). No further genetical elements have been detected so far.

Information on methods and reviews on research fields has been published in a monograph on the molecular biology of *Schiz. pombe* (edited by Nasim *et al.*, 1989). A large collection of strains useful for genetical research has been assembled by the AFRC Institute of Food Research in Norwich, England. A catalogue of the available strains can be obtained from: National Collection of Yeast Cultures, AFRC Institute of Food Research, Norwich Laboratory, Colney Lane, Norwich, NR4 7UA, UK. A laboratory course manual, *Experiments with Fission Yeast*, is available (Alfa *et al.*, 1993).

II. GENETIC NOMENCLATURE

Here we present a condensed version of the genetic nomenclature rules. They are largely identical with those for *Sacch. cerevisiae*. The few differences are pointed out. A detailed version of genetic nomenclature rules has been published together with a *Schiz. pombe* gene list (Kohli, 1987).

583

The Yeasts Vol. 6, 2nd edition
ISBN 0-12-596416-1

A computerized gene list is included in Appendix V.

Gene symbols are designated by three lower case letters (*arg*). The genetic locus is identified by a number immediately following the gene symbol (*arg1*). Alleles are designated by numbers, or a combination of letters and numbers, separated from the locus number by a hyphen (*arg1-230, ade6-M210*). The whole allele description is italicized. Differently from *Sacch. cerevisiae* only lower case letters are used irrespective of dominance and recessivity of alleles.

Complementation groups of multifunctional genes can be designated by capital letters following the locus number (*trp1A-27*).

Wild-type alleles are identified with a + superscript following the locus number (*arg1⁺*).

Superscripts should be avoided in the basic description of alleles, but when necessary superscripts may be used for identification of alleles confering resistance (r), sensitivity (s), temperature sensitivity (ts) or cold sensitivity (cs), (*can1-1ʳ, cdc2-5ᵗˢ*). — In contrast to *Sacch. cerevisiae* rules, the superscripts are added to the allele number and not directly after the three letter symbol.

Mitochondrial genes (and other non-Mendelian genes) follow the same rules as nuclear genes. When necessary, mitochondrial genes can be distinguished by inclusion in square brackets [*cox1*].

Phenotype and strain designations are not italicized. As an exception the three mating phenotypes of standard strains are italicized: h^{90} is used for homothallic strains, h^+ for heterothallic plus strains and h^- for heterothallic minus strains.

The designations of plasmids are not italicized and it is proposed that they always start with a lower case p (pFL20).

Genes that are characterized by cloning and DNA sequencing are described according to the same rules. The same holds for alleles created by reverse genetics. Integrated sequences are identified with two colons following the designation for the gene that is disrupted by the integration (*ade6::ura4*).

III. CHROMOSOME MAPS

The genetic map of *Schiz. pombe* (Fig. 1) is taken from Munz *et al.* (1989). It is called version 3 since two major versions have been published previously (Kohli *et al.*, 1977; Gygax and Thuriaux, 1984). The following are additions which have not yet been incorporated in version 3 (I, II and

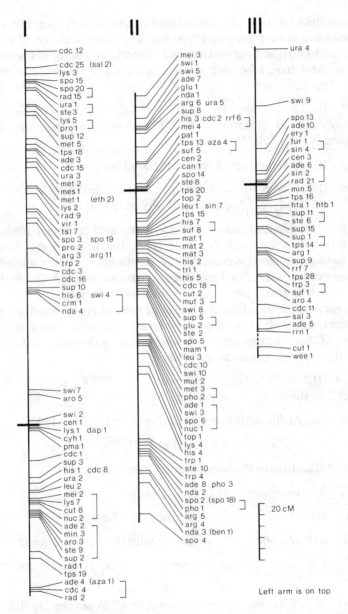

Fig. 1 Genetic map of *Schizosaccharomyces pombe*, taken from Munz *et al.* (1989).

III refer to the respective chromosomes, L and R to the left and the right chromosome arm, respectively). A major revision concerns the segment delimited by *trp1* and *spo4* at the end of the right arm of chromosome II. This segment has to be inverted; the new order is *his4*–29 cM–*spo4*–*trp1* (Egel, 1993).

ste1 IL. The sequence *ura1*–6 cM–*ste1*–18 cM–*lys3* has been established by Leupold and Sipiczki (1991). Thus *ste1* must map in the near neighbourhood of *rad15* and *spo20*.

ste5 IL. *ste5*–1 cM–*pro2* has been found by Lund *et al.* (1987).

ste6 IIIR. The position of *ste6* in version 3 is incorrect. The new allocation is *ste6*–1 cM–*trp3* (Leupold and Sipiczki, 1991).

ste11 IIR. Sipiczki has isolated *ste11* and *aff1* (Sipiczki, 1988a,b) and shown that they are allelic (Leupold *et al.*, 1991). Watanabe *et al.* (1988) have isolated *steX* and Kitamura *et al.* (1990) have a sterile mutant which they also called *ste11*. The latter authors have established allelism between *steX* and their *ste11*. Sipiczki has shown allelism between *steX* and *aff1* (Leupold *et al.*, 1991). Thus the two *ste11* mutant genes as well as *steX* and *aff1* are allelic with each other, and the authors have agreed to name the gene *ste11*. Since linkage *aff1*–1 cM–*arg4* has been demonstrated by Sipiczki (1988b), *ste11* is 1 cM away from *arg4*.

ste12 IIR. *ste12* isolated by Kitamura *et al.* (1990) is separated from *ste2* by 1 cM as shown by the same authors.

byr1 IL. Allelic with *ste1* (Nadin-Davis and Nasim, 1990).

ras1 IL. Allelic with *ste5* (Lund *et al.*, 1987).

gua1 IIL. Inseparable from *aza4* (Oraler *et al.*, 1990).

gua2 IR. Inseparable from *lys1* (Oraler *et al.*, 1990).

tws1 IIL. Allelic with *cdc2* (Grallert and Sipiczki, 1990).

rik1 IIIR. At least 9 cM centromere-distal of *ade6* (Egel *et al.*, 1989).

caf1 IIR. 21 cM centromere-proximal of *leu3* (Benkö and Sipiczki, 1990).

eno1 IIR. This gene has been detected by DNA sequencing. It is located between *cdc18* and *cut2*, the latter genes being 1 cM apart (Uzawa *et al.*, 1990).

rad8 IL. 3 cM centromere-distal of *cdc12* (F.Z. Watts, personal communication).

rad10, rad16, rad20   IIIL. According to Schmidt *et al.* (1989) these genes are allelic with each other as well as with *swi9*.

rpl7   IL. Less than 1 cM from *rad9*, based on DNA sequencing data (Murray and Watts, 1990; F.Z. Watts, personal communication).

cdc18   IIR. The position of *cdc18* with respect to *mat* and *his5* has to be changed. The new order is *mat*–13 cM–*cdc18*–3cM–*his5* (P. Munz, unpublished results).

cwg1   IIIR. 18 cM centromere-proximal of *ade5* (Ribas *et al.*, 1991).

cwg2   IL. Separated from *aro5* by 35 cM (Ribas *et al.*, 1991).

pap1   IIIR. Allelic with or very close to *cwg1* (Ribas *et al.*, 1991).

byr2   IIR. Allelic with *ste8* (Wang *et al.*, 1991; Styrkarsdottir *et al.*, 1992).

ypt1   IIR. Very close to *top1* (Miyake *et al.*, 1991).

ypt2   IR. 2 cM centromere-distal of *ura2* (Miyake *et al.*, 1991).

ypt3   IL. Very close to *arg3* (Miyake *et al.*, 1991).

ryh1   IL. 20 cM centromere-distal of *ade3* (Miyake *et al.*, 1991).

ste3   IL. Allelic with *ste1* (R. Egel and I. Lautrup-Larsen, personal communication).

dis3   IIR. Between *nda3* and *spo4*, separated 1 cM from *nda3* (Kinoshita *et al.*, 1991).

mef1   IL. Separated by 5 cM from *lys5* (Rusu, 1992).

fus1   IR. 14 cM centromere-distal of *ade2* (R. Egel, personal communication).

The conventional approach of mapping by genetic crosses has recently been supplemented by the powerful methods of reverse genetics and physical mapping procedures. This has led to complete maps of YAC, P1 and cosmid clones spanning the *Schiz. pombe* genome (Maier *et al.*, 1992; Hoheisel *et al.*, 1993), including a number of genes not previously mapped genetically.

ACKNOWLEDGEMENTS

Research of the authors has been supported by the Swiss National Science Foundation.

REFERENCES

Alfa, C., Fantes, P., Hyams, J., McLeod, M. and Warbrick, E. (1993). Experiments with Fission Yeast: A Laboratory Course Manual, Cold Spring Harbor Laboratory Press, Cold Spring Harbor.
Benkö, Z. and Sipiczki, M. (1990). *Current Genetics* **18**, 47.
Egel, R. (1993). *Current Genetics* **24**, 179.
Egel, R., Willer, M. and Nielsen, O. (1989). *Current Genetics* **15**, 407.
Fan, J.-B., Chikashige, Y., Smith, C.L., Niwa, O., Yanagida, M. and Cantor, C.R. (1989). *Nucleic Acids Research* **17**, 2801.
Grallert, B. and Sipiczki, M. (1990). *Molecular and General Genetics* **222**, 473.
Gygax, A. and Thuriaux, P. (1984). *Current Genetics* **8**, 85.
Hoheisel, J.D., Maier, E., Mott, R., McCarthy, L., Grigoriev, A.V., Schalkwyk, L.C., Nizetic, D., Francis, F. and Lehrach, H. (1993). *Cell* **73**, 109.
Kinoshita, N., Goebl, M. and Yanagida, M. (1991). *Molecular and Cellular Biology* **11**, 5839.
Kitamura, K., Nakagawa, T. and Shimoda, C. (1990). *Current Genetics* **18**, 315.
Kohli, J. (1987). *Current Genetics* **11**, 575.
Kohli, J., Hottinger, H., Munz, P., Strauss, A. and Thuriaux, P. (1977). *Genetics* **87**, 471.
Leupold, U. and Sipiczki, M. (1991). *Current Genetics* **20**, 67.
Leupold, U., Sipiczki, M. and Egel, R. (1991). *Current Genetics* **20**, 79.
Lund, P.M., Hasegawa, Y., Kitamura, K., Shimoda, C., Fukui, Y. and Yamamoto, M. (1987). *Molecular and General Genetics* **209**, 627.
Maier, E., Hoheisel, J.D., McCarthy, L., Mott, R., Grigoriev, A.V., Monaco, A.P., Larin, Z. and Lehrach, H. (1992). *Nature Genetics* **1**, 273.
Miyake, S., Tanaka, A. and Yamamoto, M. (1991). *Current Genetics* **20**, 277.
Munz, P., Wolf, K., Kohli, J. and Leopold, U. (1989). *In* "Molecular Biology of the Fission Yeast" (A. Nasim, P. Young and B.F. Johnson, eds), pp. 1-30. Academic Press, San Diego.
Murray, M. and Watts, F.Z. (1990). *Nucleic Acids Research* **18**, 4590.
Nadin-Davis, S.A. and Nasim, A. (1990). *Molecular and Cellular Biology* **10**, 549.
Nasim, A., Young, P. and Johnson, B.F. (eds) (1989). "Molecular Biology of the Fission Yeast." Academic Press, San Diego.
Oraler, G., Olgun, A. and Karaer, S. (1990). *Current Genetics* **17**, 543.
Ribas, J.C., Diaz, M., Duran, A. and Perez, P. (1991). *Journal of Bacteriology* **173**, 3456.
Rusu, M. (1992). *Current Genetics* **21**, 17.
Schmidt, H., Kapitza-Fecke, P., Stephen, E.R. and Gutz, H. (1989). *Current Genetics* **16**, 89.
Sipiczki, M. (1988a). *Acta Microbiologica Hungarica* **35**, 200.
Sipiczki, M. (1988b). *Molecular and General Genetics* **213**, 529.
Styrkarsdottir, U., Egel, R. and Nielsen, O. (1992). *Molecular and General Genetics* **235**, 122.
Uzawa, S., Samejima, I., Hirano, T., Tanaka, K. and Yanagida, M. (1990). *Cell* **62**, 913.
Wang, Y., Xu, H.-P., Riggs, M., Rodgers, L. and Wigler, M. (1991). *Molecular and Cellular Biology* **11**, 3554.
Watanabe, Y., Iino, Y., Furuhata, K., Shimoda, C. and Yamamoto, M. (1988). *EMBO Journal* **7**, 761.
Wolf, K. (1987). *In* "Gene Structure in Eukaryotic Microbes" (J.R. Kinghorn, ed.) pp. 69-91. SGM Special Publication **22**, IRL Press, Oxford.

Appendix V Physical and Genetic Mapping Data for the Fission Yeast *Schizosaccharomyces pombe*

Elmar Maier* and Hans Lehrach

Genome Analysis Laboratory, Imperial Cancer Research Fund, P.O. Box 123, 44 Lincoln's Inn Fields, London WC2A 3PX, UK
**Present address:* *Max-Planck-Institut für Molekulare Genetik, Innestrasse 73, D-14195 Berlin-Dahlem*

The 14 million base pairs (Mbp) of the fission yeast *Schizosaccharomyces pombe* genome are organized in three chromsomes of respective size 5.7, 4.6 and 3.5 Mbp. The unicellular fungus shows all the basic characteristics of a eukaryotic genome allowing examination of important biological functions such as the cell cycle division process in a relatively simple model organism. Here we present a summary of both genetic and physical mapping information for the fission yeast which has been collected during our *Schiz. pombe* genome mapping project (Lennon and Lehrach, 1992; Maier *et al.*, 1992; Hoheisel *et al.*, 1993). A figure of a physical clone map of the three chromosomes aligned to the *Not*I restriction map (Fan *et al.*, 1989) and some genetic markers is shown (Fig. 1). Twenty five yeast artificial chromosome (YAC) clones completely span the genome. They are a selected subset of our ordered YAC library consisting of some 47 genome equivalents (Maier *et al.*, 1992).

An included table (Table 1) with an updated gene list of the nuclear and mitochondrial genes of the fission yeast consists of a gene symbol, the physical location, a description of the product or the phenotype, the EMBL accession number (if available), and literature reference to original publications. The gene symbol as well as the gene description is taken from gene lists and their respective rules of nomenclature (Kohli *et al.*, 1977; Kohli, 1987; Lennon and Lehrach, 1992; Appendix IV) or from the submitted gene identification in the EMBL database. The gene list of the identified loci is presented alphabetically by the gene symbol. The gene locations are similar

The Yeasts Vol. 6, 2nd edition
ISBN 0-12-596416-1

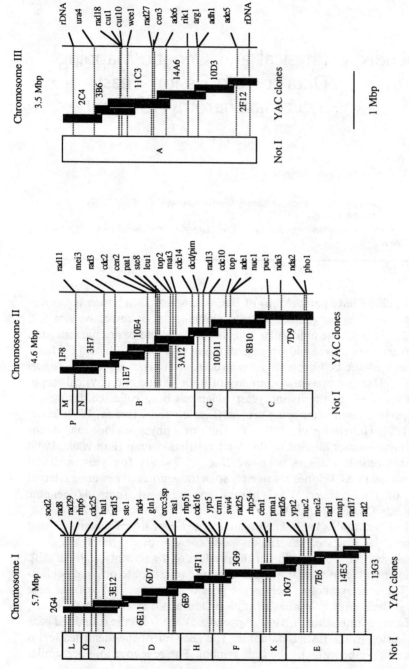

Fig. 1 Physical and genetic map of the *Schizosaccharomyces pombe* genome. A minimal subset of 25 YAC clones spanning the entire genome of the fission yeast *Schiz. pombe* (Maier *et al.*, 1992) is shown aligned to available genetic (Munz *et al.*, 1989, Appendix IV) and physical (Fan *et al.*, 1989) map information.

Table I. Physical and genetic mapping data for *Schizosaccharomyces pombe*.

Gene	Location	Gene description (product or phenotype)	EMBL-no.	Synonyms	Reference
act1		Actin			132
ade01	2.2705	GAR synthetase (EC.6.3.1.3) and AIR synthetase	Y00447		2, 43, 119
ade02	1.4787	SAmP synthetase (EC.6.3.4.4)	X06601	min4	85, 50
ade03	1.0973	FGAR amidotransferase (EC.6.3.5.3)		min10	2, 50
ade04	1.5576	PRPP amidotr. (EC.2.4.2.12)		min11	85, 50
ade05	3.3024	GAR formyltransferase (EC.2.1.2.2)		aza1, min13	85, 50
ade06	3.1566	AIR carboxylase (EC.4.1.4.21)	M37264	min1	85, 50, 115
ade07	2.0359	SAICAR synthetase (EC.6.3.2.6)			85, 50
ade08	2.3905	Adenylosuccinate lyase (EC.4.3.2.2)			85, 50
ade09	2.0000	Formyl THFA synthetase (EC.6.3.4.3)		min6	85, 43
ade10	3.1323	ade10A: AICAR formyltransferase. ade 10B:IMP cyclohydrolase (EC.3.5.4.10)		min14	50, 72
adh1	3.0000	Alcohol dehydrogenase	J01341		150
ala1		Lack of allantoicase			41
all1		Lack of allantoicase			41
all2		Hypoxanthine non-utilization			17
ani1		Anisomycine resistance			154
ani2		Anisomycine resistance		cyh4, tri3	154
arg01	3.2147	Acetylglutamyl aminotransferase (AGTase)			42
arg02	2.0000	Arginine requirement			43
arg03	1.1788	Ornithine carbamoyl transferase (OTCase)	X63577		23, 42
arg04	2.4181	Carbamoylphosphate synthetase (CPSase A)			42
arg05	2.4154	Carbamoylphosphate synthetase (CPSase A)			42
arg06	2.0442	Arginine requirement			43
arg07		Argininosuccinate lyase	X63262		31, 43
arg08		Arginine requirement			43
arg09		Arginine requirement			43
arg10		Arginine requirement			43
arg11	1.0000	Acetylglutamylphosphate reductase			23, 42
aro1		DAHP synthetase (tyr inhibited) (EC4.1.2.15)	X63576		157

Continued

Table I. Continued

Gene	Location	Gene description (product or phenotype)	EMBL-no.	Synonyms	Reference
aro2	1.4826	DAHP synthetase (phe inhibited) (EC4.1.2.15)			157
aro3	3.2565	Aromatic amino acid requirement, complex locus			45, 3,174
aro4	3.5577	Aromatic amino acid requirement			44
aro5	1.3577	Aromatic amino acid requirement			44
aro6		Aromatic amino acid requirement			44
aro7		Aromatic amino acid requirement			44
aro8		Aromatic amino acid requirement			44
ars01	1.0000	High freq. of transformation sequence, pFL20			161, 82
ars02		High freq. of transformation sequence pARS 727			16, 82
ars03		High freq. of transformation sequence pARS 744			16, 82
ars04		High freq. of transformation sequence pARS 745			16, 82
ars05		High freq. of transformation sequence pARS 747			16, 82
ars06		High freq. of transformation sequence pARS 756	X07890		16, 82
ars07		High freq. of transformation sequence pARS 766	X07891		16, 82
ars08		High freq. of transformation sequence pARS 767	X07892		16, 82
ars09		High freq. of transformation sequence pARS 772	X07893		16, 82
ars10	1.0000	High freq. of transformation sequence, next to ura1			16, 82
ars17	1.0000	High freq. of transformation sequence, next to cdc17			149
atb2	2.0000	On 88 kb NotI fragment P			210
atp1	1.0000	Mitochondrial ATPase, alpha subunit	M57955		4, 5, 37
atp2	1.0000	Mitochondrial ATPase, beta subunit	M57956		4, 5, 214
aza1	1.5576	Azaguanine resistance		ade4, min13	85
aza2	3.0000	Azaguanine resistance			43
aza4	2.1035	Azaguanine resistance			43
ben1	2.4278	Thiabendazole resistance		nda3	155
ben2	2.0000	Thiabendazole resistance			155
ben3	2.0000	Thiabendazole resistance			155
ben4	1.0000	Benomyl resistance			48
bip		Heat shock protein	X64416		46
bws1		Bypass of wee suppres,; type 1 protein phosphatase	see ref.		146
byr1		Bypass of ras, protein kinase	X07445		117
byr2		Bypass of ras, protein kinase	M74293		74

Gene	Value	Description	Accession	Synonym	References
cam1	2.1311	Calmodulin	M16475	cal1	123
can1	2.0000	Canavanine resistance, arginine permease			43, 49
can2		Canavanine resistance			43, 49
cat1		Catabolite repression, loss of glycolytic enzymes			32
cdc01	1.4260	Cell division cycle			50
cdc02	2.0745	Cell division cycle	M12912		50, 8
cdc03	1.2249	Cell division cycle			50
cdc04	1.5602	Cell division cycle			50
cdc05	1.0000	Cell division cycle			80
cdc06	2.0000	Cell division cycle			43
cdc07	2.0000	Cell division cycle			43
cdc08	1.4497	Cell division cycle			50
cdc09	3.4970	Cell division cycle		wee1	81
cdc10	2.2167	Cell division cycle; start gene	X02175		50, 11
cdc11	3.2754	Cell division cycle			50
cdc12	1.0026	Cell division cycle			50
cdc13	2.0000	Cell division cycle; cyclin-related	X12557		50, 136
cdc14	2.0000	Cell division cycle			43
cdc15	1.0894	Cell division cycle			50
cdc16	1.2393	Cell division cycle	see ref.		50, 208
cdc17	1.0000	Cell division cycle; DNA ligase	X05107		159, 133
cdc18	2.1814	Cell division cycle			50, 158
cdc19		Cell division cycle			158
cdc20		Cell division cycle			158
cdc21		Cell division cycle	X58824		158, 211
cdc22		Cell division cycle			83
cdc23		Cell division cycle			158
cdc24		Cell division cycle			158
cdc25	1.0184	Cell division cycle	M13158	sal2	50, 33
cdc26		Cell division cycle			158
cdc27		Cell division cycle	M74062		158, 75
cdc28		Cell division cycle			158
cdc2L		Cell division cycle			78
cdc42sp	2.0000	Cell division cycle 42sp gene; GTP-binding protein	M83650		95

Continued

Table I. Continued

Gene	Location	Gene description (product or phenotype)	EMBL-no.	Synonyms	Reference
cdr1		Changed division response	X57549	nim1	86, 205
cdr2		Changed division response			86
cho1	2.0000	Choline requirement			73
cho2	2.0000	Choline requirement			73
chs1		Chitin synthase	M82957		79
cid1		Colcemid resistance			29
cid2		Colcemid resistance			29
cid3		Colcemid resistance			29
cig1		B-type cyclin, G1 phase	M68881		274
coxIII		Mitochondrial coxIII gene	X57664		76
cred		NADPH-cytochrome P450 reductase	X64702		77
crm1	1.2524	Cell division in absence of nuclear division	X15482		114
cta3		Ca^{2+}-ATPase	J05634		40
cut01	3.0000	Cell division in absence of nuclear division	M36179		84, 183
cut02	2.1822	Cell division in absence of nuclear division	M57750		84, 183
cut03		Cell division in absence of nuclear division			84
cut04		Cell division in absence of nuclear division			84
cut05		Cell division in absence of nuclear division			84
cut06		Cell division in absence of nuclear division			84
cut07	1.0000	Cell division in absence of nuclear division			84, 203
cut08	1.4708	Cell division in absence of nuclear division			84
cut09		Cell division in absence of nuclear division			84
cut10	3.0000	Cell division in absence of nuclear division			84
cyc1		Cytochrome c	J01318		160
cyh1	1.3932	Cycloheximide resistance, putative ribosomal protein			152, 153
cyh2		Cycloheximide resistance			152
chy3		Cycloheximide resistance			152
cyh4		Cycloheximide resistance			152, 154
cyr1		Adenylyl cyclase	M24942	ani2, tri3	105,110
dap1	1.3919	Diaminopur. resist., adenine phosphoribosyl transfer.			85
deo1	1.0000	Deoxyglucose resistance			43
dis1	3.0000	Mitotic chromosome disjunction block			112

dis2		Mitotic chromosome disjunction block	see ref.	bws1, pp1	112
dis3		Mitotic control	M74094		99, 112
dna		Actin-related protein gene	M81068		100
end1	1.0000	DNA endonuclease			9
ercc3sp	1.0000	Homologue of human ERCC3 excision repair gene			280
ery1	3.1350	Erythromycin resistance			43
eth1	2.0000	Ethionine resistance			43
eth2	1.1210	Ethionine resistance		metl	50
rth3		Ethionine resistance			50
fbp1		Fructose-1,6-bisphosphatase	J03213		10, 47, 118
fur1	3.1418	Fluorouracil resistance			43
fur2	2.0000	Fluorouracil resistance			43
fur4		Fluorouracil resistance, uracil permease			161
fus1	1.0000	Sexual cell fusion			162
gamt		Gamma-tubulin	M63447		104
gap1		GTPase activating protein	D10457		109
gdh1		Glycerol dehydrogenase			32
glu1	2.0386	Glutamate requirement			43
glu2	2.2029	Glutamate requirement			43
glu3		Glutamate requirement			43
glu4		Glutamate requirement			43
gpa1		G protein alpha-subunit	M64286		127
gpd		Glycerol-3-phosphate deyhdrogenase	M56162		207
gsa		Glutathione synthetase large subunit	M85179		212
gscom		Complements glutamine synthetase def. strain	X07978		213
gtb1		Gamma-tubulin	X62031		128
gtp1		GTP-binding protein, P-loop region	L00671		209
gtpb		GTP-binding, Sar1 protein	M95797		218
gua1	2.1049	IMP dehydrogenase (EC.1.2.1.14)			85, 97
gua2	1.3940	GMP synthetase (EC.6.3.5.2)			85, 97
hat1	1.0000	Gene complements hydroxy. sensitive			279
hcs		Homologue of E. coli RNase III	X53679		184
hfo1		Histone H4, H4.1	X05222		21
hfo2		Histone H4, H4.2	X05223		21

Continued

Table I. Continued

Gene	Location	Gene description (product or phenotype)	EMBL-no.	Synonyms	Reference
hfo3	1.4497	Histone H4, H4.3	X05224		21
his1	2.1684	Histidinol phosphate phosphatase (EC.3.1.3.15)			85, 50
his2	2.0759	Histidinol dehydrogenase (EC.1.1.1.23)			85, 43
his3	2.3091	Imidazole acetol phosphate transaminase (EC.2.6.1.9)			85, 43
his4	2.1766	Histidine requirement			85, 43
his5	1.2485	Imidazole glycerolphosphate dehydrat. (EC.4.2.1.19)			85, 43
his6	2.1546	Histidine requirement			85, 50
his7		Histidine requirement			85, 43
his8		Histidine requirement			85
his9		Histidine requirement			85
hmt1		Heavy metal resistance	Z14055		129
hsf		Heat shock transcription factor	M94683		219
hsp1		Hyperspeckled, suppressor of smt			88
hsp2		Hyperspeckled, suppressor of smt			88
hsp3		Hyperspeckled, suppressor of smt			88
hsp4		Hyperspeckled, suppressor of smt			88
hsp5		Hyperspeckled, suppressor of smt			88
hsp6		Hyperspeckled, suppressor of smt			88
hsp7		Hyperspeckled, suppressor of smt			88
hsp8		Hyperspeckled, suppressor of smt			88
hsp9		Hyperspeckled, suppressor of smt			88
hsp70		Heat shock protein	M60208		135
hta1	3.1701	Histone H2A, H2A.1, H2A-α	M11494		21, 22
hta2		Histone H2A, H2A.1, H2A-β	M11500		21, 22
htb1	3.0000	Histone H2B, H2B.1, H2B-β	X05220		21
hth1		Histone H3, H3.1	X05222		21
hth2		Histone H3, H3.2	X05223		21
hth3		Histone H3, H3.3	X05224		21
hus1	1.0000	Complementing hus1 mutant (checkpoint control)			281
hus2	1.0000	Complementing hus2 mutant (checkpoint control)			281
hyp1		Hypoxanthine non-utilization			41
hyp2		Hypoxanthine non-utilization			41
hyp3		Hypoxanthine non-utilization			41

Gene	Map position	Description	Synonym	Accession	References
hyp4		Hypoxanthine non-utilization			41
hyp5		Hypoxanthine and xanthine non-utilization			41
kin1		Protein kinase		M64999	38
kina		Thymidylate kinase		X65868	139
leu1	2.1490	Beta-isopropylmalate dehydrogenase		M36910	43, 144
leu2	1.4603	Leucine requirement			43
leu3	2.2153	Leucine requirement			43
lys1	1.3919	Lysine requirement			43
lys2	1.1407	Lysine requirement (on 470 kb NotI fragment K)			50
lys3	1.0236	Lysine requirement			50
lys4	2.3008	Lysine requirement			43
lys5	1.0552	Lysine requirement			50
lys6	1.0000	Lysine requirement			43
lys7	1.4681	Lysine requirement			50
lys8		Lysine requirement			43
lys9		Lysine requirement			43
mam1	2.2098	Mating-type auxiliary (minus)		X61672	163, 43
mam2		Pheromone receptor			163, 140
map1	1.0000	Mating-type auxiliary (plus)			163
map2		Mating-type auxiliary (plus)			163
mat1	2.1614	Mating-type expression locus		X07642	52, 141
mat1 smt		smt-o mutation, abolishes mating-type switching		X59127	147, 120
mat2	2.1628	Mating-type storage locus (plus)	mei1	X14001	52, 141
mat3	2.1642	Mating-type storage locus (minus)		X14002	52,141
mcm3hp		S. cerevisiae mcm3 homologue (partial)		Z15034	148
mcs1		Mitotic catastrophe suppressor			137
mcs2		Mitotic catastrophe suppressor			137
mcs3		Mitotic catastrophe suppressor			137
mcs4		Mitotic catastrophe suppressor			137
mcs5		Mitotic catastrophe suppressor			137
mcs6		Mitotic catastrophe suppressor			137
mei1	2.1628	First meiotic division	mat2	X07180	52
mei2	1.4655	First meiotic division		X05142	126, 176
mei3	2.0027	First meiotic division			12, 53, 138

Continued

Table I. *Continued*

Gene	Location	Gene description (product or phenotype)	EMBL-no.	Synonyms	Reference
mei4	2.0773	First meiotic division			12
mes1	1.1157	Second meiotic division			12
mes2		Second meiotic division			69
met1	1.1210	Methionine requirement, allelic with *eth2*			50
met2	1.1118	Methionine requirement			50
met3	2.2650	Methionine requirement			43
met4	2.0000	Methionine requirement			43
met5	1.0671	Methionine requirement			50
mfm1		Mating pheromones	X63627		145
mfm2		Mating pheromones	X63628		145
mik1		Kinase	M60834		177
min01	3.1566	Methionine inhibited		*ade6*	164
min02		Methionine inhibited			164
min03	1.4800	Methionine inhibited			164, 43
min04	2.2705	Methionine inhibited		*ade1*	164
min05	3.1620	Methionine inhibited			164, 43
min06	2.0000	Methionine inhibited		*ade9*	164
min07		Methionine inhibited			164
min08		Methionine inhibited			164
min09		Methionine inhibited			164
min10	1.4787	Methionine inhibited		*ade2*	164
min11	1.0973	Methionine inhibited		*ade3*	164
min12		Methionine inhibited			164
min13	1.5576	Methionine inhibited		*ade4, aza1*	164
min14	3.1323	Methionine inhibited		*ade10*	164
min15		Methionine inhibited			164
min16		Methionine inhibited			164
mut1		Mutator			165
mut2	2.2539	Mutator			43, 165
mut3	2.1849	Mutator			43, 165
nda01	2.0400	Nuclear division arrest			156
nda02	2.0000	Alpha-tubulin 1	K02841		6, 32

Gene		Description	Accession	Alias	Pages
nda03	2.0000	Beta tubulin	M10347	ben1	7, 156
nda04	1.2525	Nuclear division arrest			94
nda05	1.0000	Nuclear division arrest			94
nda06	1.0000	Nuclear division arrest			94
nda07	1.0000	Nuclear division arrest			94
nda08	1.0000	Nuclear division arrest			94
nda09	1.0000	Nuclear division arrest			94
nda10	3.0000	Nuclear division arrest			94
nda11	3.0000	Nuclear division arrest			94
nda12		Nuclear division arrest			94
nim1		Mitotic control element; protein kinase	M16509	cdr1	192, 205
nmt1		No message in thiamine	J05493		100
nuc1	2.2774	Alteration in nucl. struct.; largest RNA pol I subunit	X14783		84, 196
nuc2	1.4734	Alteration in nuclear structure	X07693		94, 178
pab1		p-aminobenzoic acid requirement			44
pabp		Poly (A)-binding protein	M64603		186
pac1		Pac1 ribonuclease	X54998		191
pap1		Ap-1-like transcription factor	X57078		201
pat1	2.0994	Sporulation of haploids		ran1	54
pet1		Respiration (nuclear gene)			175
phe1	1.0000	Phenylalanine requirement			44
phe2	2.0000	Phenylalanine requirement			44
pho1	2.0000	Acid phosphatase	M11857		58, 28
pho2	2.2663	Specific p-nitrophenylphosphatase	X62722		194, 198
pho3	2.3905	Alkaline phosphatase			194
pho4		Acid phosphatase, thiamine repressible	X56939		18, 35
php2		Transcription activation complex subunit	M63639		200
pki1	1.0000	Homologue to 14-3-3 protein			279
pma1	1.3971	Plasma membrane ATPase	J03498		25, 206
pma2		Plasma membrane ATPase	M60471		220
pmd1		ATPase	D10695		221
pol1	2.0000	Catalytic subunit DNA polymerase alpha	X58299		197
pol3		Catalytic subunit DNA polymerase delta	X59278		222
ppa		Inorganic phosphatase (E.C. 3.6.1)	X54301		39

Continued

Table I. *Continued*

Gene	Location	Gene description (product or phenotype)	EMBL-no.	Synonyms	Reference
ppa1		Type 2A protein phosphatase	M58518		215
ppa2		Type 2A protein phosphatase	M58519		215
ppcti		Cyclophilin (peptidyl-prolyl *cis* isomerase)	X53223		185
pro1	1.0579	Proline requirement			43, 50
pro2	1.1736	Proline requirement			43, 50
ptp1		Protein tyrosine phosphatase	L04671		223
pts1		Proteasome subunit pts1+	D13094		224
puc1	2.0000	G1-type cyclin	X59154		1
pur1		Azathioxanthine resistance (EC.2.4.2.8)			85
pyp1		Protein-tyrosine-phosphatase-like protein	M63257		199
rad01	1.5207	Radiation sensitivity	M38132		50, 55, 187
rad02	1.5628	Radiation sensitivity, excision repair			50, 55, 229
rad03	2.0000	Radiation sensitivity, checkpoint repair	X63544		55, 225
rad04	1.0000	Radiation sensitivity, DNA repair	X62676		55, 226
rad05		Radiation sensitivity			55
rad06		Radiation sensitivity			55
rad07		Radiation sensitivity			55
rad08	1.0000	Radiation sensitivity, homol. to S. cerev. *rad5* gene	X58231		43, 55, 278
rad09	1.1420	Radiation sensitivity			50, 55, 227
rad09g	1.1420	Radiation sensitivity	X64648		228
rad10	3.0621	Radiation sensitivity		*rad16*	55, 278
rad11	2.0000	Radiation sensitivity, DNA repair			55, 275
rad12		Radiation sensitivity			55
rad13	2.0000	Radiation sensitivity, homol. to S. cerev. *rad2* gene	X66795		55, 229
rad14		Radiation sensitivity			55
rad15	1.0329	Radiation sensitivity; homol. to S. cerev. *rad3* gene	X60499		50, 55, 230
rad16		Radiation sensitivity			55
rad17	1.0000	Radiation sensitivity, checkpoint repair			55, 282
rad18	3.0000	Radiation sensitivity			55
rad19		Radiation sensitivity			55
rad20		Radiation sensitivity			55
rad21	3.1593	Radiation sensitivity			50,55

Gene		Description	Accession	Alt.	Ref.
rad22		Radiation sensitivity			55
rad23		UV radiation sensitive			96
rad24		Radiation sensitivity			216
rad25	1.0000	Radiation sensitivity, homologue pki1			216, 279
rad26	1.0000	Radiation sensitivity, checkpoint repair			216, 279
rad27	3.0000	Radiation sensitivity, checkpoint repair			216, 279
rad28		Radiation sensitivity, checkpoint repair E2 enzyme			216, 279
rad29	1.0000	Radiation sensitivity, DNA repair			216, 279
rad31	1.0000	Radiation sensitivity			279
ral1		ras-like			113
ral2		ras-like	M30827		113, 103
ral3		ras-like			113
ral4		ras-like			113
ran1	2.0994	Negatove regulator sexual conjugation and meiosis	X04728	pat1	56, 231
ras1	1.1762	ras oncogene related	X03771	ste5	57, 232
rasgtp		ras-related GTP-binding protein	X52732		233
rcc1a		rcc1 gene sequence	M73527		234
rcc1b		rcc1 gene sequence	M73528		234
rec01		Recombination			87
rec02		Recombination			13
rec03		Recombination			13
rec05		Recombination			13
rec06		Meiotic recombination mutants			107
rec07		Meiotic recombination mutants	M85297		107, 235
rec08		Meiotic recombination mutants	M85298		107, 235
rec09		Meiotic recombination mutants			107
rec10		Meiotic recombination mutants			107
rec11		Meiotic recombination mutants			107
rhp3		S. cerevisiae rad3 homologue	X64583		236
rhp6		S. cerevisiae rad6 homologue	see ref.		182, 278
rhp51	1.0000	S. cerevisiae rad51 homologue and recA recomb. repair			237
rhp54	1.0000	S. cerevisiae rad54 homologue, recombination repair			237
ribk5		Ribosomal protein K5	X51659		238
rnl1	3.0000	On NotI-B fragment (function unknown)			279
rpa01		Acidic ribosomal protein gene 1	M33137		188

Continued

Table I. Continued

Gene	Location	Gene description (product or phenotype)	EMBL-no.	Synonyms	Reference
rpa02		Acidic ribosomal protein gene 2	M33138		188, 131
rpa03		Acidic ribosomal protein gene 3	M33139		188
rpa04		Acidic ribosomal protein gene 4	M33142		188
rpa190		RNA polymerase I, large (A190) subunit			142
rpb1		RNA polymerase, largest subunit	X56564		204
rpb2		RNA polymerase, second largest subunit	D13337		239
rpgl29		Ribosomal protein L29	X57207		240
rpgl3		Ribosomal protein L3	X57734		241
rpk37		Ribosomal protein K37	X05036		130, 125
rpkd4		Ribosomal protein KD4	X16392		242
rpl40c		Ribosomal protein L40c	Y00466		131
rpl7	1.1420	Homologue of rat ribosomal protein L7	X53575		186
rpl7g	1.1420	Homologue of rat ribosomal protein L7	X54983		244
rps13g		Ribosomal protein S13	X67030		245
rps6		Ribosomal protein S6	M36382		78, 143
rrf1		5S ribosomal RNA gene, pSPr1, pYM116	see ref.		155, 167
rrf2		5S ribosomal RNA gene, pSPr11	see ref.		166
rrf3		5S ribosomal RNA gene, pSPr36	see ref.		166
rrf4		5S ribosomal RNA gene, pSPr41	see ref.		166
rrf5		5S ribosomal RNA gene, pYM3	see ref.		167
rrf6	2.0759	5S ribosomal RNA gene, close to cdc2			59
rrf7	3.2201	5S ribosomal RNA gene, close to sup9			92
rrk1		Gene for K-RNA (component of RNase P)	X52530		246
rrn1	3.3051	5S ribosomal RNA gene gene repeats (nucleolar organizer)			60, 71, 20
rtptf1		Retrotransposon Tf1-107	M38526		247
ryh1		GTP-binding protein ryh1p	X52475		181
sal1	1.0000	Allosuppressor			50, 51
sal2	1.0184	Allosuppressor		cdc25	50, 51
sal3	3.2889	Allosuppressor			50, 51
sal4		Allosuppressor			51
sal5		Allosuppressor			90
scr1		Suppressor of cycloheximide resistance			151
scr2		Suppressor of cycloheximide resistance			151

Gene	Value	Description	Accession	Ref.
sds21		Suppressor of dis2	see ref. M57495	112, 202
sds22	1.0000	On 600 kb NotI fragment H		112, 202
sds23	1.0000	On 600 kb NotI fragment H		112
sin01	1.0000	Antisuppressor, loss of i6A in tRNA		43, 168
sin02	3.1580	Antisuppressor		43
sin03	2.0000	Antisuppressor, loss of mcm5s2U in tRNA		61
sin04	3.1445	Antisuppressor, loss of mcm5s2U in tRNA		43
sin05	1.0000	Antisuppressor		43
sin06	1.0000	Antisuppressor		43
sin07	2.0000	Antisuppressor		50
sin08	1.0000	Antisuppressor		43
sin09	1.0000	Antisuppressor		43
sin10	2.0000	Antisuppressor		43
sin11	2.0000	Antisuppressor		43
sin12		Antisuppressor		193
sin13		Antisuppressor		193
sin14		Antisuppressor		193
sin15		Antisuppressor, loss of ncm5U in tRNA		62
smt	2.0000	Switching of mating-type sequence		52
snm1		Defective in snRNA maintenance		189
snru3		Fission yeast U3 small nuclear RNA	X13850	248
snu1	2.0000	U1 small nuclear RNA	X55773	106, 249
snu2	1.0000	U2 small nuclear RNA	X55772	174, 250
snu3A	1.0000	U3A small nuclear RNA	X56982	106, 251
snu3B	2.0000	U3B small nuclear RNA	X56189	106, 251
snu4	2.0000	U4 small nuclear RNA	X15491	252, 253
snu5		U5 small nuclear RNA	X16573	102
snu6		U6 small nuclear RNA	X14196	108
sod1		Amiloride resistance	Z14035	254
sod2		Superoxide dismutase	X66722	255
sod2g	1.0000	Putative Na^+/H^+ antiporter	Z11736	256
spk1		Protein kinase	X57334	201
spo01		Sporulation		162
spo02	2.3974	Sporulation		162, 27
spo03	1.1670	Sporulation		162, 27
spo04	2.4623	Sporulation		162, 27

Continued

Table I. *Continued*

Gene	Location	Gene description (product or phenotype)	EMBL-no.	Synonyms	Reference
spo05	2.2084	Sporulation			162, 27
spo06	2.2732	Sporulation			162, 27
spo07		Sporulation			162
spo08		Sporulation			162
spo09		Sporulation			162
spo10		Sporulation			162
spo11		Sporulation			162
spo12		Sporulation			162
spo13	3.1080	Sporulation			162, 27
spo14	2.1380	Sporulation			162, 27
spo15	1.0289	Sporulation			162, 27
spo16		Sporulation			162
spo17		Sporulation			162
spo18	2.0000	Sporulation			162, 27
spo19	1.0000	Sporulation			27
spo20	1.0316	Sporulation			27
spras		ras-gene for spras-protein	X02331		257
srp		Small ribonucleoprot.-assoc. signal rec. particle-like	M20836		258
srp54		Signal recognition particle 54 kDa protein subunit	X51613		259
srp7		7sl RNA component of signal recognition particle	X15220		260
ssp1		Mitochondrial Hsp 70 protein	X59987		261
st11p		Sec 12p-like protein	M95798		262
stb1	3.0000	Plasmid stabilization			24
ste01	1.0000	Sterility, defective meiosis			169
ste02	2.2056	Sterility, defective meiosis			63
ste03	1.0407	Sterility, defective meiosis			63
ste04	1.0000	Sterility, defective meiosis	X61924		63, 263
ste05		Sterility mutant			63
ste06	3.1782	Sterility	X53254		63, 264
ste07	1.0000	Sterility, defective meiosis			89
ste08	2.1408	Sterility, defective meiosis			89
ste09	1.4839	Sterility, defective meiosis			90
ste10	2.3836	Partial sterility			90

Gene		Description			
ste11	2.0000	Sterility locus	Z11156	stex,aff1	116, 265
ste12	2.0000	Sterility locus			101
ste13	2.0000	Sterility locus			101
stf1		Suppressor of cdc25			36
sts01		Staurosporine sensitive	X63549		201, 276
sts02		Staurosporine sensitive			201
sts03		Staurosporine sensitive			201
sts04		Staurosporine sensitive			201
sts05		Staurosporine sensitive			201
sts06		Staurosporine sensitive			201
sts07		Staurosporine sensitive			201
sts08		Staurosporine sensitive			201
sts09		Staurosporine sensitive			201
sts10		Staurosporine sensitive			201
sts11		Staurosporine sensitive			201
suc1		Suppressor of cdc2	M16032		26, 124
suc22		Suppressor of cdc22			83
suf01	3.2457	Frameshift suppressor, group I			64
suf02	2.0000	Frameshift suppressor, group I			64
suf03	1.0000	Frameshift suppressor, group II			64
suf04	3.0000	Frameshift suppressor, group II			64
suf05	2.1049	Frameshift suppressor, group II			64
suf06	3.0000	Frameshift suppressor, group II			64
suf07	3.0000	Frameshift suppressor, group II			64
suf08	2.1573	Frameshift suppressor, group II			50, 64
suf09		Frameshift suppressor, group IV			93
suf10		Frameshift suppressor, group III			93
suf11	2.0000	Frameshift suppressor, group I			64
sup01	3.1917	Nonsense suppressor, omnipotent			45, 193
sup02	1.4852	Nonsense suppressor, omnipotent			50, 193
sup03	1.4445	Nonsense suppress., opal, ochre, tRNAser, tRNAmet	see ref.		50, 14
sup04		Missense suppressor			65
sup05	2.2015	Missense suppressor			50, 65
sup06		Missense suppressor			65
sup07	2.0000	Missense suppressor			43, 65

Continued

Table I. Continued

Gene	Location	Gene description (product or phenotype)	EMBL-no.	Synonyms	Reference
sup08	2.0580	Nonsense suppressor, opal, ochre, tRNAleu	see ref.		43, 66
sup09	3.2187	Nonsense suppressor, opal, tRNAser, tRNAmet	see ref.		43, 14
sup10	1.2472	Nonsense suppressor, opal, tRNAleu			50, 62
sup11	3.1755	Informational suppressor			93
sup12	1.0631	Nonsense suppressor, opal, tRNAser, tRNAmet	see ref.		50, 14
sup13		Nonsense suppressor, amber			15
sup14		Nonsense suppressor, amber			15
sup15		Nonsense suppressor			217
swi01	3.1836	Mating-type switching, class Ia			67
swi02	2.0166	Mating-type switching, class Ib			67
swi03	1.3853	Mating-type switching, class Ia			67
swi04	2.2705	Mating-type switching, class II	X61306		67, 277
swi05	1.2485	Mating-type switching, class Ib			67
swi06	2.0276	Mating-type switching, class Ib			67
swi07	1.3524	Mating-type switching, class Ia			67
swi08	2.1932	Mating-type switching, class II			67
swi09	3.0621	Mating-type switching, class II, allelic with rad10			67
swi10	2.2208	Mating-type switching, class II	X61926		67, 266
sxa1		Put. protease	D10198		267
sxa2		Put. protease	D10199		267
tfIId		TATA box binding-factor gene	X53415		98
tnd1		Aspartate tRNA, anticodon GTc, pYM116	see ref.		167
tne1		Glutamate tRNA, anticodon TTC, pYM104	see ref.		173
tnf1		Phenylalanine tRNA, anticodon AAG, pYM125	see ref.		173
tnh1		Histidine tRNA, anticodon GTG, pYM7.2	see ref.		173
tnk1		Lysine tRNA, anticodon CTT, pYM104	see ref.		173
tnr1		Arginine tRNA, anticodon ACG, pYM104	see ref.		173
tnr2		Arginine tRNA, anticodon ACG, lambda 805	see ref.		173
top1	2.2857	DNA topoisomerase I	X06201		9, 122
top2	2.1477	DNA topoisomerase II	X04326		9, 134
tpi1		Triosephosphate isomerase	M14432		19
tps12		Temperature-sensitive lethal			91
tps13	2.1035	Temperature-sensitive lethal			50

Gene		Description		
tps14	3.1944	Temperature-sensitive lethal		50
tps15	2.1532	Temperature-sensitive lethal		43, 50
tps16	3.1634	Temperature-sensitive lethal		50
tps17	1.0000	Temperature-sensitive lethal		50
tps18	1.0907	Temperature-sensitive lethal		50
tps19	1.5286	Temperature-sensitive lethal		50
tps20	2.1477	Temperature-sensitive lethal		50
tps21		Temperature-sensitive lethal		91
tps22		Temperature-sensitive lethal		91
tps23		Temperature-sensitive lethal		91
tps24	3.0000	Temperature-sensitive lethal		50
tps25	2.0000	Temperature-sensitive lethal		50
tps26	2.0000	Temperature-sensitive lethal		8
tps27		Temperature-sensitive lethal		91
tps28	3.2214	Temperature-sensitive lethal		50
tri1	2.1711	Trichodermine resistance		43
tri2	3.0000	Trichodermine resistance	ani2, cyh4	43
tri3		Trichodermine resistance		154
tri4		Trichodermine resistance		154
tri5		Trichodermine resistance		154
trop		Tropomyosin (cdc8) gene	L04126	268
trp1	2.3795	trp1A:PRA isomerase. trp1B:InGP synthetase. trp1C:anthranilate synthetase		50, 68
trp2	1.2051	Tryptophan synthetase (EC.4.2.1.20)		85, 50
trp3	3.2430	Anthranilate synthetase		85, 50
trp4	2.3864	PR transferase (EC.2.4.2.20)		85, 50
tsl01		Temperature-sensitive lethal		170
tsl02		Temperature-sensitive lethal		170
tsl03		Temperature-sensitive lethal		170
tsl04		Temperature-sensitive lethal		170
tsl05		Temperature-sensitive lethal		170
tsl06		Temperature-sensitive lethal		170
tsl07	1.1670	Temperature-sensitive lethal		50, 170
tsl08		Temperature-sensitive lethal		170

Continued

Table I. Continued

Gene	Location	Gene description (product or phenotype)	EMBL-no.	Synonms	Reference
tsl09		Temperature-sensitive lethal			170
tsl10		Temperature-sensitive lethal			170
tsl11		Temperature-sensitive lethal			170
tuba2		Alpha-tubulin 2	K02842		6
tws1		Two-spored asci			69
tyr1	3.0000	Tyrosine requirement			44
tyr2		Tyrosine requirement			44
ura1	1.0381	Aspartate transcarbamylase			50, 171
ura2	1.4576	Dihydroorotase (EC.3.5.2.3)			85, 50
ura3	1.1026	Dihydroorotate dehydrogenase	X65114		85, 50, 269
ura4	3.0027	Orotidinephosphate decarboxylase	X13976		270
ura5	2.0442	Orotate phosphoribosyl transferase			43
ure1		Lack of urease			41
ure2		Lack of urease			41
ure3		Lack of urease			41
ure4		Lack of urease			41
uro1		Lack of uricase			41
vatp		Vacuolar H$^+$ –ATPase c-6	X59947		271
ver1	3.0000	Verrucarine resistance			43
vir1	1.1486	Vegetative iodine reaction			50, 70
wee1	3.0000	Cell division cycle, small cells	M16508		43, 272
win1		wee interacting			30
wis1		Protein kinase	X62631		273
xan1		Xanthine non-utilization			41
ypt1		Related to S. cerevisiae YPT-1 gene	X15082		180, 111
ypt2	1.0000	Also related to S. cerev. YPT-1 gene; GTP-binding	X52469		190
ypt3		Also related to S. cerev. YPT-1 gene	X52100		180, 179
ypt4		Also related to S. cerev. YPT-1 gene			275
ypt5	1.0000	Also related to S. cerev. YPT-1 gene			275

The gene list consists of six fields: a gene symbol, the physical location, a description of the product or the phenotype, the EMBL accession number (if available) or "see ref." (if the sequence data have not been submitted to the EMBL database), and literature reference to original publications. The gene symbol, the gene location, the gene description, the locus description and the synonyms category are similar to those described earlier (Kohli, J., 1987; Lennon and Lehrach, 1992). Submitted gene identifications from the EMBL database and gene locations from our recent physical mapping effort (Maier *et al.*, 1992; Hoheisel *et al.*, 1993) are included. The exact physical position is shown in the figure of the physical clone map of *Schiz. pombe*. The reference field indicates some of the following key references related to the locus:

1. Forsburg, S.L. and Nurse, P. (1991). *Nature* **351**, 245.
2. Fluri, R., Coddington, A. and Flury, U. (1976). *Molecular and General Genetics* **147**, 271.
3. Nakanishi, N. and Yamamoto, M. (1984). *Molecular and General Genetics* **195**, 164.
4. Vassarotti, A., Boutry, M., Colson, A.M. and Goffeau, A. (1984). *Journal of Biological Chemistry* **259**, 2845.
5. Boutry, M., Vassarotti, A., Ghislain, M., Douglas, M. and Goffeau, A. (1984). *Journal of Biological Chemistry* **259**, 2840.
6. Toda, T., Adachi, Y., Hiraoka, Y. and Yanagida, M. (1984). *Cell* **39**, 233.
7. Hiraoka, Y., Toda, T. and Yanagida, M. (1984). *Cell* **39**, 349.
8. Hindley, J. and Phear, G.A. (1984). *Gene* **31**, 129.
9. Uemura, T. and Yanagida, M. (1984). *EMBO Journal* **3**, 1737.
10. Vassarotti, A. and Friesen, J.D. (1985). *Journal of Biological Chemistry* **260**, 6348.
11. Aves, S.J., Durkacz, B.W., Carr, A. and Nurse, P. (1985). *EMBO Journal* **4**, 457.
12. Shimoda, C., Hirata, A., Kishida, M., Hashida, T. and Tanaka, K. (1985). *Molecular and General Genetics* **200**, 252.
13. Thuriaux, P. (1985). *Molecular and General Genetics* **199**, 365.
14. Amstutz, H., Munz, P., Heyer, W.D., Leupold, U. and Kohli, J. (1985). *Cell* **40**, 879.
15. Krupp, G., Thuriaux, P., Willis, I., Gamulin, V. and Söll, D. (1985). *Molecular and General Genetics* **201**, 82.
16. Maundrell, K., Wright, A.P., Piper, M. and Shall, S. (1985). *Nucleic Acids Research* **13**, 3711.
17. Fluri, R. and Kinghorn, J.R. (1985). *Journal of General Microbiology* **131**, 527.
18. Maundrell, K., Nurse, P., Schonholzer, F. and Schweingruber, M.E. (1985). *Gene* **39**, 223.
19. Russell, P.R. (1985). *Gene* **40**, 125.
20. Balzi, E., Di Pietro, A., Goffeau, A., van Heerikhuizen, H. and Klootwijk, J. (1985). *Gene* **39**, 165.
21. Matsumoto, S. and Yanagida, M. (1985). *EMBO Journal* **4**, 3531.
22. Choe, J., Schuster, T. and Grunstein, M. (1985). *Molecular and Cellular Biology* **5**, 3261.
23. Van Huffel, C., Dubois, E. and Messenguy, F. (1992). *European Journal of Biochemistry* **205**, 33.
24. Heyer, W.D., Sipiczki, M. and Kohli, J. (1986). *Molecular and Cellular Biology* **6**, 80.
25. Ulaszewski, S., Coddington, A. and Goffeau, A. (1986). *Current Genetics* **10**, 359.
26. Hayles, J., Beach, D., Durkacz, B. and Nurse, P. (1986). *Molecular and General Genetics* **202**, 291.
27. Kishida, M. and Shimoda, C. (1986). *Current Genetics* **10**, 443.
28. Elliott, S., Chang, C.W., Schweingruber, M.E., Schaller, J., Rickli, E.E. and Carbon, J. (1986). *Journal of Biological Chemistry* **261**, 2936.
29. Sackett, D.L. and Lederberg, S. (1986). *Experimental Cell Research* **163**, 467.
30. Ogden, J.E. and Fantes, P.A. (1986). *Current Genetics* **10**, 509.
31. Loppes, R., Michels, R., Decroupette, I. and Joris, B. (1991). *Current Genetics* **19**, 255.

Continued

Table 1. *Continued*

32. Gancedo, C., Llobell, A., Ribas, J.C. and Luchi, F. (1986). *European Journal of Biochemistry* **159**, 171.
33. Russell, P. and Nurse, P. (1986). *Cell* **45**, 145.
34. Maundrell, K., Hutchison, A. and Shall, S. (1988). *EMBO Journal* **7**, 2203.
35. Yang, J.W. and Schweingruber, M.E. (1990). *Current Genetics* **18**, 269.
36. Hudson, J.D., Feilotter, H. and Young, P.G. (1990). *Genetics* **126**, 309.
37. Falson, P., Maffey, L., Conrath, K. and Boutry, M. (1991). *Journal of Biological Chemistry* **266**, 287.
38. Levin, D.E. and Bishop, J.M. (1990). *Proceedings of the National Academy of Sciences of the United States of America* **87**, 8272.
39. Kawasaki, L., Adachi, N. and Ikeda, H. (1990). *Nucleic Acids Research* **18**, 5888.
40. Ghislain, M., Goffeau, A., Halachmi, D. and Eilam, Y. (1990). *Journal of Biological Chemistry* **265**, 18400.
41. Kinghorn, J.R. and Fluri, R. (1984). *Current Genetics* **8**, 99.
42. Vissers, S. and Thuriaux, P. (1985). *Current Genetics* **9**, 561.
43. Kohli, J., Hottinger, H., Munz, P., Strauss, A. and Thuriaux, P. (1977). *Genetics* **87**, 471.
44. Strauss, A. (1979). *Journal of General Microbiology* **113**, 172.
45. Strauss, A. (1979). *Molecular and General Genetics* **172**, 233.
46. Pidoux, A.L. and Armstrong, J. (1992). *EMBO Journal* **11**, 1583.
47. Vassarotti, A., Boutry, M. and Colson, A.M. (1982). *Archives of Microbiology* **133**, 131.
48. Roy, D. and Fantes, P.A. (1982). *Current Genetics* **6**, 195.
49. Fantes, P.A. and Creanor, J. (1984). *Journal of General Microbiology* **130**, 3265.
50. Gygax, A. and Thuriaux, P. (1984). *Current Genetics* **8**, 85.
51. Nurse, P. and Thuriaux, P. (1984). *Molecular and General Genetics* **196**, 332.
52. Beach, D.H. (1983). *Nature* **305**, 682.
53. Shimoda, C. and Uehira, M. (1985). *Molecular and General Genetics* **201**, 353.
54. Iino, Y. and Yamamoto, M. (1985). *Molecular and General Genetics* **196**, 416.
55. Nasim, A. and Smith, B.P. (1975). *Genetics* **79**, 573.
56. Nurse, P. (1985). *Molecular and General Genetics* **198**, 497.
57. Fukui, Y. and Kaziro, Y. (1985). *EMBO Journal* **4**, 687.
58. Schweingruber, M.E., Schweingruber, A.M. and Schupbach, M.E. (1982). *Current Genetics* **5**, 109.
59. Durkacz, B., Beach, D., Hayles, J. and Nurse, P. (1985). *Molecular and General Genetics* **201**, 543.
60. Toda, T., Nakaseko, Y., Niwa, O. and Yanagida, M. (1984). *Current Genetics* **8**, 93.
61. Heyer, W.D., Thuriaux, P., Kohli, J., Ebert, P., Kersten, H., Gehrke, C., Kuo, K.C. and Agris, P.F. (1984). *Journal of Biological Chemistry* **259**, 2856.
62. Munz, P., Dorsch-Häsler, K. and Leupold, U. (1983). *Current Genetics* **7**, 101.
63. Girgsdies, O. (1982). *Current Genetics* **6**, 223.
64. Hottinger, H. and Leupold, U. (1981). *Current Genetics* **3**, 133.
65. Barben, H. (1966). *Genetica* **37**, 109.

66. Sumner-Smith, M., Hottinger, H., Willis, I., Koch, T.L., Arentzen, R. and Söll, D. (1984). *Molecular and General Genetics* **197**, 447.
67. Gutz, H. and Schmidt, H. (1985). *Current Genetics* **9**, 325.
68. Thuriaux, P., Heyer, W.D. and Strauss, A. (1982). *Current Genetics* **6**, 13.
69. Nakaseko, Y., Niwa, O. and Yanagida, M. (1983). *Journal of Bacteriology* **157**, 334.
70. Meade, J.H. and Gutz, H. (1978). *Genetics* **88**, 235.
71. Schaack, J., Mao, J. and Söll, D. (1982). *Nucleic Acids Research* **10**, 2851.
72. Richter, R. and Heslot, H. (1982). *Current Genetics* **5**, 233.
73. Hill, J.E., Fernandez, S. and Henry, S. (1986). *Yeast* **2**, s156.
74. Wang, Y., Xu, H.P., Riggs, M., Rodgers, L. and Wigler, M. (1991). *Molecular and Cellular Biology* **11**, 3554.
75. Hughes, D.A., MacNeill, S.A. and Fantes, P.A. (1992). *Molecular and General Genetics* **231**, 401.
76. Schaefer, B., Merlos-Lange, A.M., Anderl, C., Welser, F., Zimmer, M. and Wolf, K. (1991). *Molecular and General Genetics* **225**, 158.
77. Miles, J.S. (1992). *Journal of Biochemistry* **287**, 195.
78. Gross, T., Nischt, R. and Käufer, N.F. (1986). *Molecular and General Genetics* **204**, 543.
79. Bowen, A.R., Chen-Wu, J.L., Momany, M., Young, R., Szaniszlo, P.J. and Robbins, P.W. (1992). *Proceedings of the National Academy of Sciences of the United States of America* **89**, 519.
80. Nurse, P., Thuriaux, P. and Nasmyth, K.A. (1976). *Molecular and General Genetics* **146**, 167.
81. Nurse, P. (1975). *Nature* **256**, 547.
82. Maundrell, K., Hutchinson, A. and Shall, S. (1986). *Yeast* **2**, s233.
83. Gordon, C.B. and Fantes, P.A. (1986). *EMBO Journal* **5**, 2981.
84. Hirano, T., Funahashi, S., Uemura, T. and Yanagida, M. (1986). *EMBO Journal* **5**, 2973.
85. Gutz, H., Heslot, H., Leupold, U. and Loprieno, N. (1974). *In* "Handbook of Genetics", vol. 1, Plenum.
86. Young, P.G. and Fantes, P.A. (1984). *In* "Humana, Growth, Cancer and the Cell Cycle", (P. Skeham & S.J. Friedmann eds) pp 221–228, Human Press, Clifton, New York.
87. Goldman, S.L. and Gutz, H. (1974). *In* "Mechanisms in Recombination", Plenum Press, New York.
88. Fecke, H.C. (1985). *In* "Ph.D. Thesis", Technical University of Braunschweig.
89. Michael, H. (1985). *In* "Ph.D. Thesis", Technical University of Braunschweig.
90. Leupold, U. (1987). Unpublished.
91. Munz, P. (1987). Unpublished.
92. Amstutz, H., Deiss, V. and Kohli, J. (1987). Unpublished.
93. Hottinger, H. (1980). *In* "Ph.D. Thesis", University of Bern.
94. Yanagida, M., Hiraoka, Y., Uemura, T., Miyake, S. and Hirano, T. (1985). *In* "Yeast Cell Biology", Alan Liss Publishers, New York.
95. Fawell, E., Bowden, S. and Armstrong, J. (1992). *Gene* **114**, 153.
96. Lieberman, H.B., Riley, R. and Martel, M. (1989). *Molecular and General Genetics* **218**, 554.
97. Oraler, G., Olgun, A. and Karaer, S. (1990). *Current Genetics* **17**, 543.
98. Hoffmann, A., Horikoshi, M., Wang, C.K., Schroeder, S., Weil, P.A. and Roeder, R.G. (1990). *Genes and Development* **4**, 1141.
99. Kinoshita, N., Goebl, M. and Yanagida, M. (1991). *Molecular and Cellular Biology* **11**, 5839.

Continued

612   E. Maier and H. Lehrach

Table 1. Continued

100. Lees-Miller, J.P., Henry, G.L. and Helfman, D.M. (1992). *Proceedings of the National Academy of Sciences of the United States of America* **89**, 80.
101. Kitamura, K., Nakagawa, T. and Shimoda, C. (1990). *Current Genetics* **18**, 315.
102. Small, K., Brennwald, P., Skinner, H., Schaefer, K. and Wise, J.A. (1989). *Nucleic Acids Research* **17**, 9483.
103. Fukui, Y., Miyake, S., Satoh, M. and Yamamoto, M. (1989). *Molecular and Cellular Biology* **9**, 5617.
104. Stearns, T. and Kirschner, M. (1993). *Cell* **76**, 623.
105. Young, D., Riggs, M., Field, J., Vojtek, A., Broek, D. and Wigler, M. (1989). *Proceedings of the National Academy of Sciences of the United States of America* **86**, 7989.
106. Dandekar, T. and Tollervey, D. (1989). *Gene* **81**, 227.
107. Ponticelli, A.S. and Smith, G.R. (1989). *Genetics* **123**, 45.
108. Potashkin, J. and Frendewey, D. (1989). *Nucleic Acids Research* **17**, 7821.
109. Imai, Y., Miyake, S., Hughes, D. and Yamamoto, M. (1991). *Molecular and Cellular Biology* **11**, 3088.
110. Yamawaki-Kataoka, Y., Tamaoki, T., Choe, H.R., Tanaka, H. and Kataoka, T. (1989). *Proceedings of the National Academy of Sciences of the United States of America* **86**, 5693.
111. Fawell, E., Hook, S. and Armstrong, J. (1989). *Nucleic Acids Research* **17**, 4373.
112. Ohkura, H., Kinoshita, N., Miyatani, S., Toda, T. and Yanagida, M. (1989). *Cell* **57**, 997.
113. Fukui, Y. and Yamamota, M. (1988). *Molecular and General Genetics* **215**, 26.
114. Adachi, Y. and Yanagida, M. (1989). *Journal of Cell Biology* **108**, 1195.
115. Szankasi, P., Heyer, W.D. and Kotyk, A. (1988). *Journal of Molecular Biology* **204**, 917.
116. Sipiczki, M. (1988). *Molecular and General Genetics* **213**, 529.
117. Nadin-Davis, S.A. and Nasim, A. (1988). *EMBO Journal* **7**, 985.
118. Rogers, D.T., Hiller, E., Mitsock, L. and Orr, E. (1988). *Journal of Biological Chemistry* **263**, 6051.
119. McKenzie, R., Schuchert, P. and Kilbey, B. (1987). *Current Genetics* **12**, 591.
120. Bach, M.L. (1987). *Current Genetics* **12**, 527.
121. Lund, P.M., Hasegawa, Y., Kitamura, K., Shimoda, C., Fukui, Y. and Yamamoto, M. (1987). *Molecular and General Genetics* **209**, 627.
122. Uemura, T., Morino, K., Uzawa, S., Shiozaki, K. and Yanagida, M. (1987). *Nucleic Acids Research* **15**, 9727.
123. Takeda, T. and Yamamoto, M. (1987). *Proceedings of the National Academy of Sciences of the United States of America* **84**, 3580.
124. Hindley, J., Phear, G., Stein, M. and Beach, D. (1987). *Molecular and Cellular Biology* **7**, 504.
125. Nischt, R., Gross, T., Gatermann, K., Swida, U. and Kaufer, N. (1987). *Nucleic Acids Research* **15**, 1477.
126. Shimoda, C., Uehira, M., Kishida, M., Fujioka, H., Iino, Y., Watanabe, Y. and Yamamoto, M. (1987). *Journal of Bacteriology* **169**, 93.
127. Obara, T., Nakafuku, M., Yamamoto, M. and Kaziro, Y. (1991). *Proceedings of the National Academy of Sciences of the United States of America* **88**, 5877.
128. Horio, T., Uzawa, S., Jung, M.K., Oakley, B.R., Tanaka, K. and Yanagida, M. (1991). *Journal of Cell Science* **99**, 693.
129. Ortiz, D.F., Kreppel, L., Speiser, D.M., Scheel, G., McDonald G. and Ow, D.W. (1992). *EMBO Journal* **11**, 3491.
130. Nischt, R., Thuroff, E. and Kaeufer, N.F. (1986). *Current Genetics* **10**, 365.

Continued

131. Beltrame, M. and Bianchi, M.E. (1987). *Nucleic Acids Research* **15**, 9089.
132. Mertins, P. and Gallwitz, D. (1987). *Nucleic Acids Research* **15**, 7369.
133. Barker, D.G., White, J.H. and Johnston, L.H. (1987). *European Journal of Biochemistry* **162**, 659.
134. Uemura, T., Morikawa, K. and Yanagida, M. (1986). *EMBO Journal* **5**, 2355.
135. Powell, M.J. and Watts, F.Z. (1990). *Gene* **95**, 105.
136. Booher, R. and Beach D. (1988). *EMBO Journal* **7**, 2321.
137. Molz, L., Booher, R., Young, P. and Beach, D. (1989). *Genetics* **122**, 773.
138. McLeod, M., Stein, M. and Beach, D. (1987). *EMBO Journal* **6**, 729.
139. Fikes, J.D., Becker, D.M., Winston, F. and Guarante, L. (1992). *Nature* **346**, 291.
140. Kitamura, K. and Shimoda, C. (1991). *EMBO Journal* **10**, 3743.
141. Kelly, M., Burke, J., Smith, M., Klar, A. and Beach, D. (1988). *EMBO Journal* **7**, 1537.
142. Yamagishi, M. and Nomura, M. (1988). *Gene* **247**, 242.
143. Gross, T., Nischt, R., Gatermann, K., Swida, U. and Kaeufer, N.F. (1988). *Current Genetics* **13**, 57.
144. Kikuchi, Y., Kitazawa, Y., Shimatake, H. and Yamamoto, M. (1988). *Current Genetics* **14**, 375.
145. Davey, J. (1992). *EMBO Journal* **11**, 951.
146. Booher, R. and Beach, D. (1989). *Cell* **57**, 1009.
147. Nielsen, O. and Egel, R. (1989). *EMBO Journal* **8**, 269.
148. Coxon, A. (1992). *In* "Ph.D. Thesis", University of Oxford.
149. Johnston, L.H. and Barker, D.G. (1987). *Molecular and General Genetics* **207**, 161.
150. Russell, P.R. and Hall, B.D. (1983). *Journal of Biological Chemistry* **258**, 143.
151. Ibrahim, M.A.K. and Coddington, A. (1978). *Molecular and General Genetics* **162**, 213.
152. Ibrahim, M.A.K. and Coddington, A. (1976). *Heredity* **37**, 179.
153. Coddington, A. and Fluri, R. (1977). *Molecular and General Genetics* **158**, 93.
154. Berry, H.J., Ibrahim, M.A.K. and Coddington, A. (1978). *Molecular and General Genetics* **167**, 217.
155. Yamamoto, M. (1980). *Molecular and General Genetics* **180**, 231.
156. Toda, T., Umesono, T., Hayashi, S. and Yanagida, M. (1983). *Journal of Molecular Biology* **168**, 251.
157. Schweingruber, M.E. and Wyssling, H. (1974). *Biochimica et Biophysica Acta* **350**, 319.
158. Nasmyth, K.A. and Nurse, P. (1981). *Molecular and General Genetics* **182**, 119.
159. Nasmyth, K.A. (1977). *Cell* **12**, 1109.
160. Russell, P.R. and Hall, B.D. (1982). *Molecular and Cellular Biology* **2**, 106.
161. Chevallier, M.R. and Lacroute, F. (1982). *EMBO Journal* **1**, 375.
162. Bresch, L., Müller, G. and Egel, R. (1968). *Molecular and General Genetics* **102**, 301.
163. Egel, R. (1973). *Molecular and General Genetics* **122**, 339.
164. Strauss, A. (1979). *Genetical Research* **33**, 261.

Table I. Continued

165. Munz, P. (1975). *Mutation Research* **29**, 155.
166. Tabata, S. (1981). *Nucleic Acids Research* **9**, 6429.
167. Mao, J., Appel, B., Schaack, J., Sharp, S., Yamada, H. and Söll, D. (1982). *Nucleic Acids Research* **10**, 487.
168. Janner, R., Vögeli, G. and Fluri, R. (1980). *Journal of Molecular Biology* **139**, 207.
169. Thuriaux, P., Sipiczki, M. and Fantes, P.A. (1980). *Journal of General Microbiology* **116**, 525.
170. Bonatti, S., Simili, M. and Abbondandolo, A. (1972). *Journal of Bacteriology* **109**, 484.
171. Sakaguchi, J. and Yamamoto, M. (1982). *Proceedings of the National Academy of Sciences of the United States of America* **79**, 7819.
172. Losson, R. and Lacroute, F. (1983). *Cell* **32**, 371.
173. Gamulin, V., Mao, J., Appel, B., Summer-Smith, M., Yamao, F. and Söll, D. (1983). *Nucleic Acids Research* **11**, 8537.
174. Dandekar, T. and Tollervey, D. (1991). *Molecular Microbiology* **5**, 1621.
175. Colson, A.M., Labaille, F. and Goffeau, A. (1976). *Molecular and General Genetics* **149**, 101.
176. Watanabe, Y., Iino, Y., Furuhata, K. Shimoda, C. and Yamamoto, M. (1988). *EMBO Journal* **7**, 761.
177. Lundgren, K., Walworth, N., Booher, R., Dembski, M., Kirschner, M. and Beach, D. (1991). *Cell* **64**, 1111.
178. Hirano, M., Hiraoka, Y. and Yanagida, M. (1988). *Journal of Cell Biology* **106**, 1171.
179. Fawell, E., Hook, S., Sweet, D. and Armstrong, J. (1990). *Nucleic Acids Research* **18**, 4264.
180. Miyake, S. and Yamamoto, M. (1990). *EMBO Journal* **9**, 1417.
181. Hengst, L., Lehmeier, T. and Gallwitz, D. (1990). *EMBO Journal* **9**, 1949.
182. Reynolds, P., Koken, M.H.M., Hoeijmakers, J.H.J., Prakash, S. and Prakash, L. (1990). *EMBO Journal* **9**, 1423.
183. Uzawa, S., Samejima, I., Hirano, T., Tanaka, K. and Yanagida, M. (1990). *Cell* **62**, 913.
184. Xu, H.-P., Riggs, M., Rodgers, L. and Wigler, M. (1990). *Nucleic Acids Research* **18**, 5304.
185. de Martin, R. and Philipson, L. (1990). *Nucleic Acids Research* **18**, 4917.
186. Burd, C.G., Matunis, E.L. and Dreyfuss, G. (1991). *Molecular and Cellular Biology* **11**, 3419.
187. Sunnerhagen, P., Seaton, B.L., Nasim, A. and Subramani, S. (1990). *Molecular and Cellular Biology* **10**, 3750.
188. Beltrame, M. and Bianchi, M.E. (1990). *Molecular and Cellular Biology* **10**, 2341.
189. Potashkin, J. and Frendewey, D. (1990). *EMBO Journal* **9**, 525.
190. Haubruck, H., Engelke, U., Mertins, P. and Gallwitz, D. (1990). *EMBO Journal* **9**, 1957.
191. Iino, Y., Sugimoto, A. and Yamamoto, M. (1991). *EMBO Journal* **10**, 221.
192. Russell, P. and Nurse, P. (1987). *Cell* **49**, 569.
193. Thuriaux, P., Minet, M., Hofer, F. and Leupold, U. (1975). *Molecular and General Genetics* **142**, 251.
194. Dhamija, S., Fluri, R. and Schweingruber, M.E. (1987). *Current Genetics* **11**, 467.
195. Nakaseko, Y., Adachi, Y., Funahashi, S., Niwa, O. and Yanagida, M. (1986). *EMBO Journal* **5**, 1011.
196. Hirano, T., Konoha, G., Toda, T. and Yanagida, M. (1989). *Journal of Cell Biology* **108**, 243.
197. Damagnez, V., Tillit, J., de Recondo, A.M. and Baldacci, G. (1991). *Molecular and General Genetics* **226**, 182.
198. Yang, J., Dhamija, S.S. and Schweingruber, M.E. (1991). *European Journal of Biochemistry* **198**, 493.

199. Ottilie, S., Chernoff, J., Hannig, G., Hoffman, C.S. and Erikson, R.L. (1991). *Proceedings of the National Academy of Sciences of the United States of America* **88**, 3455.
200. Olesen, J.T., Fikes, J.D. and Guarente, L. (1991). *Molecular and Cellular Biology* **11**, 611.
201. Toda, T., Shimanuki, M. and Yanagida, M. (1991). *Genes and Development* **5**, 60.
202. Ohkura, H. and Yanagida, M. (1991). *Cell* **64**, 149.
203. Hagan, I. and Yanagida, M. (1990). *Nature* **347**, 563.
204. Azuma, Y., Yamagishi, M., Ueshima, R. and Ishihama, A. (1991). *Nucleic Acids Research* **19**, 461.
205. Feilotter, H., Nurse, P. and Young, P.G. (1991). *Genetics* **127**, 309.
206. Ghislain, M., Schlesser, A. and Goffeau, A. (1982). *Journal of Biological Chemistry* **262**, 17549.
207. Pidoux, A.L., Fawell, E.H. and Armstrong, J. (1990). *Nucleic Acids Research* **18**, 7145.
208. Simanis, V. (1991). Unpublished.
209. Hudson, J.D. and Young, P.G. (1992). Unpublished.
210. Yanagida, M. (1990). Unpublished.
211. Coxon, A., Maundrell, K. and Kearsey, S.E. (1991). Unpublished.
212. Mutoh, N., Nakagawa, C.W., Ando, S., Tanabe, K. and Hayashi, Y. (1992). Unpublished.
213. Barel, I., Bignell, G., Simpson, A. and MacDonald, D. (1988). *Current Genetics* **13**, 487.
214. Falson, P., Leterme, S. and Boutry, M. (1990). Unpublished.
215. Kinoshita, N., Ohkura, H. and Yanagida, M. (1990). *Cell* **63**, 405.
216. Carr, A. (1993). Unpublished.
217. Mathez, B. and Munz, P. (1989). Unpublished.
218. D'Enfert, C., Gensse, M. and Gaillardin, C. (1992). Unpublished.
219. Gallo, G.J., Prentice, H. and Kingston, R.E. (1992). Unpublished.
220. Styrkasdottir, U., Egel, R. and Nielsen, O. (1992). Unpublished.
220. Ghislain, M. and Goffeau, A. (1991). *Journal of Biological Chemistry* **266**, 18276.
221. Nishi, K., Yoshida, M., Nishikawa, M., Nishiyama, M., Horinouchi, S. and Beppu, T. (1992). *Molecular Microbiology* **6**, 761.
222. Pignede, G., Bouvier, D., Recondo, A.M. and Baldacci, G. (1991). *Journal of Molecular Biology* **222**, 209.
223. Millar, J.B., Russell, P., Dixon, J.E. and Guan, K. (1992). Unpublished.
224. Stone, E.M., Tanaka, K., Ichihara, A. and Yanagida M. (1992). Unpublished.
225. Seaton, B.L., Yucel, J., Sunnerhagen, P. and Subramani, S. (1992). *Gene* **119**, 83.
226. Fenech, M., Carr, A.M., Murray, J., Watts, F.Z. and Lehmann, A.R. (1991). *Nucleic Acids Research* **19**, 6737.
227. Murray, J.M., Carr, A.M., Lehmann, A.R. and Watts, F.Z. (1991). *Nucleic Acids Research* **19**, 3525.
228. Lieberman, H., Hopkins, K.M., Laverty, M. and Chu, H.M. (1992). *Molecular and General Genetics* **232**, 367.
229. Carr, A.M., Sheldrick, K.S., Murray, J.M., Al-Harithy, R., Watts, F.Z. and Lehmann, A.R. (1993). *Nucleic Acids Research* **21**, 1345.
230. Murray, J.M., Doe, C., Schenk, P., Carr, A.M., Lehmann, A.R. and Watts, F.Z. (1992). *Nucleic Acids Research* **20**, 2673.
231. McLeod, M. and Beach, D. (1986). *EMBO Journal* **5**, 3665.

Continued

Table I. Continued

232. Nadin-Davis, S.A., Yang, R.C., Narang, S.A. and Nasim, A. (1986). *Journal of Molecular Evolution* **23**, 41.
233. Fawell, E., Hook, S., Sweet, D. and Armstrong, J. (1990). *Nucleic Acids Research* **18**, 4264.
234. Matsumoto T. and Beach, D.H. (1991). *Cell* **66**, 347.
235. Lin, Y. (1992). *Genetics* **132**, 75.
236. Reynolds, P.R., Biggar, S., Prakash, L. and Prakash, S. (1992). *Nucleic Acids Research* **20**, 2327.
237. Murris, D. (1992). Unpublished.
238. Gatermann, K.B., Teletski, C., Gross, T. and Kaeufer, N.F. (1989). *Current Genetics* **16**, 361.
239. Kawagishi, M., Yamagishi, M. and Ishihama A. (1992). Unpublished.
240. Kwart, M. and Gross, T. (1992). Unpublished.
241. Koehler, G., Liebig, I. and Gross, T. (1992). Unpublished.
242. Teletski, C. and Kaeufer, N.F. (1989). *Nucleic Acids Research* **17**, 10118.
243. Murray, J.M. and Watts, F.Z. (1990). *Nucleic Acids Research* **18**, 4590.
244. Damagnez, V., Recondo, A.M. and Baldacci, G. (1991). *Nucleic Acids Research* **19**, 1099.
245. Marks, J. and Simanis, V. (1992). *Nucleic Acids Research* **20**, 4094.
246. Zimmerly, S., Gamulin, V., Burkard, U. and Soll, D. (1992). Unpublished.
247. Levin, H.L., Weaver, D.C. and Boeke, J.D. (1990). *Molecular and Cellular Biology* **10**, 6791.
248. Porter, G.L., Brennwald, P.J., Holm, K.A. and Wise, J.A. (1988). *Nucleic Acids Research* **16**, 10131.
249. Dandekar, T. and Tollervey, D. (1990). Unpublished.
250. Brennwald, P., Porter, G. and Wise, J.A. (1988). *Molecular and Cellular Biology* **8**, 5575.
251. Selinger, D.A., Porter, G.L., Brennwald, P.J. and Wise, J.A. (1992). *Molecular Biology of Evolution* **9**, 297.
252. Dandekar, T., Ribes, V. and Tollervey, D. (1989). *Journal of Molecular Biology* **208**, 371.
253. Dandekar, T. and Tollervey, D. (1992). *Yeast* **8**, 647.
254. Jia, Z., McCullough, N., Wong, L. and Young, P.G. (1992). Unpublished.
255. O'Dee, K.M. and Snider, M.D. (1992). Unpublished.
256. Jia, Z.P., McCullough, N., Martel, R., Hemmingsen, S. and Young, P.G. (1992). *EMBO Journal* **11**, 1631.
257. Fukui, Y. and Kaziro, Y. (1985). *EMBO Journal* **4**, 687.
258. Poritz, M.A., Siegel, V., Hansen, W. and Walter, P. (1988). *Proceedings of the National Academy of Sciences of the United States of America* **85**, 4315.
259. Hann, B.C., Poritz, M.A. and Walter, P. (1990). *Journal of Cell Biology* **109**, 3223.
260. Ribes, V., Dehoux, P. and Tollervey, D. (1988). *EMBO Journal* **7**, 231.
261. Kasai, H. and Isono, K. (1991). *Nucleic Acids Research* **19**, 5331.
262. D'Enfert, C., Gensse, M. and Gaillardin, C. (1992). Unpublished.
263. Okazaki, N., Okazaki, K., Tanaka, K. and Okayama, H. (1991). *Nucleic Acids Research* **19**, 7043.
264. Hughes, D.A., Fukui, Y. and Yamamoto, M. (1990). *Nature* **344**, 355.

265. Sugimoto, A., Iino, Y., Maeda, T., Watanabe, Y. and Yamamoto, M. (1991). *Genes and Development* **5**, 1990.
266. Roedel, C., Kirchhoff, S. and Schmidt, H. (1991). Unpublished.
267. Imai, Y. and Yamamoto, M. (1992). *Molecular and Cellular Biology* **12**, 1827.
268. Balasubramanian, M.K., Helfman, D.M. and Hemmingsen, S.M. (1992). Unpublished.
269. Nagy, M., Lacroute, F. and Thomas, D. (1992). Unpublished.
270. Grimm, C., Kohli, J., Murray, J. and Maundrell, K. (1988). *Molecular and General Genetics* **215**, 81.
271. Toyama, R., Goldstein, D.J., Schlegel, R. and Dhar, R. (1992). *Yeast* **7**, 989.
272. Russell, P. and Nurse, P. (1987). *Cell* **49**, 559.
273. Warbrick, E. and Fantes, P.A. (1991). *EMBO Journal* **10**, 4291.
274. Bueno, A., Richardson, H.E., Reed, S.I. and Russell, P. (1991). *Cell* **66**, 149.
275. Maier, E., Hoheisel, J., McCarthy, L., Mott, R., Grigoriev, A., Monaco, T., Larin, Z. and Lehrach, H. (1992). *Nature Genetics* **1**, 273.
276. Shimanuki, M., Goebl, M., Yanagida, M. and Toda, T. (1992). *Molecular Biology* **3**, 263.
277. Fleck, O., Michael, H. and Heim, L. (1992). *Nucleic Acids Research* **20**, 2271.
278. Broughton, B.C., Barbet, N., Murray, J., Watts, F.Z., Koken, M.H.M., Lehmann, A.R. and Carr, A.M. (1991). *Molecular and General Genetics* **228**, 470.
279. Carr, A. (1993). Unpublished.
280. Lehmann, A.R. et al. (1992). *Mutation Research* **273**, 1.
281. Enoch, T.R., Carr, A. and Nurse, P. (1992). *Genes and Development* **6**, 2035.
282. Barbet, N.C. and Carr, A.M. (1993). *Nature* **364**, 824.

to those described by Lennon and Lehrach using the genetic map published earlier (Munz *et al.*, 1989). Also included are recent physical mapping data in the gene list indicating the respective chromosome on which a particular gene is located. The exact physical position is shown in Fig. 1. The locus description consists of information about either the gene product or the phenotypes exhibited by mutant alleles of this locus or an indication when a gene has only been physically localized to a *Not*I restriction endonuclease fragment of the genome. The synonyms category indicates other names by which the locus may be known. The reference field indicates some of the key references relating to the locus given in the footnotes of the table. It is not meant to be exhaustive but merely to point the way to a key reference. Of 593 loci identified to date, mapping information is available for 291 and sequence information is available for 199.

Since filters containing all the clones of our ordered genomic *Schiz. pombe* YAC, P1- and cosmid libraries arrayed in a high density format (Lehrach *et al.*, 1990) are freely available on request, it is now very easy to determine where loci map. A large number of different laboratories have already used our genomic *Schiz. pombe* libraries to locate their gene fragments. The mapping data are sent back and included in our relational *Schiz. pombe* database (Mott *et al.*, 1993) in order to collect the locations of all genes and eventually provide that information to different users. Since one particular genome is being used as a reference system by a large number of groups, the data generated are more easily comparable. Also, apart from linking the probes to the physical clone map, and vice versa, other laboratories are being provided with the required genomic clones so that they can work on a more detailed (e.g. sequence) analysis of the genomic region of their interest. As the *Schiz. pombe* genome is thought to comprise around 5000 genes, it is clear that still much remains to be discovered. People working on *Schiz. pombe* are asked to comment and to provide new entries to the gene list for future updates.

ACKNOWLEDGEMENTS

We would like to give all the credit for collecting genetic mapping data of *Schiz. pombe* to Jürg Kohli and Peter Munz. We thank John Armstrong, Antony Carr, Richard Egel, Susan Forsburg, Kathy Gould, Daan Muris, Olaf Nielsen, Tom Patterson, Viesturs Simanis and Mitsuhiro Yanagida for providing genetic probes of *Schiz. pombe*, and Paul Nurse for the *Schiz. pombe* strain 972h⁻. We also thank Greg Lennon and Jane Sandall for the work done on the gene database and finally Richard Mott for helpful computing. Elmar Maier acknowledges the support of the European Community and the Fritz-Thyssen Stiftung, Germany.

REFERENCES

Fan, J.B., Chikashige, Y., Smith, C.L., Niwa, O., Yanagida, M. and Cantor, C.R. (1989). *Nucleic Acids Research* **17**, 2801.
Hoheisel, J.D., Maier, E., Mott, R., McCarthy, L., Grigoriev, A.V., Schalkwyk, L.C., Nizetic, D., Francis, F. and Lehrach, H. (1993). *Cell* **73**, 109.
Kohli, J. (1987). *Current Genetics* **11**, 575.
Kohli, J., Hottinger, H., Munz, P., Strauss, A. and Thuriaux, P. (1977). *Genetics* **87**, 471.
Lehrach, H., Drmanac, R., Hoheisel, J., Larin, Z., Lennon, G., Monaco, A.P., Nizetic, D., Zehetner, G. and Poustka, A. (1990). *In* "Genome Analysis: Genetic and Physical Mapping" (K.E. Davies and S. Tilghman, eds.), pp. 39–81. Cold Spring Harbor Laboratory Press, Cold Spring Harbor, New York.
Lennon, G. and Lehrach, H. (1992). *Current Genetics* **21**, 1.
Maier, E., Hoheisel, J., McCarthy, L., Mott, R., Grigoriev, A., Monaco, T., Larin, Z. and Lehrach, H. (1992). *Nature Genetics* **1**, 273.
Mott, R., Grigoriev, A., Maier, E., Hoheisel, J. and Lehrach, H. (1993). *Nucleic Acids Research* **21**, 1965.
Munz, P., Wolf, K., Kohli, J. and Leupold, U. (1989). *In* "Molecular Biology of the Fission Yeast" (A. Nasim, P. Young and B.F. Johnson, eds.), pp. 1–30. Academic Press, San Diego.

Subject Index

Author Index

Note: Page numbers in **bold** refer to reference sections